MATHEMATICS IN INDUSTRY 9

THE EUROPEAN CONSORTIUM
FOR MATHEMATICS IN INDUSTRY

ECMI

SUBSERIES

A. M. Anile
G. Alì
G. Mascali
Editors

Scientific Computing in Electrical Engineering

With 253 Figures, 86 in Color, and 42 Tables

Springer

Editors

Angelo Marcello Anile
Dept. of Mathematics and Computer Science
University of Catania
Viale Andrea Doria 6
I-95125 Catania, Italy
Email: anile@dmi.unict.it

Giuseppe Alì
Istituto per le Applicazioni del Calcolo "M. Picone", CNR
Via P. Castellino 111
I-80131 Napoli, Italy
Email: g.ali@iac.cnr.it

Giovanni Mascali
Dept. of Mathematics
Università della Calabria
I-87036 Rende (CS), Italy
Email: mascali@dmi.unict.it

Mathematics Subject Classification (2000):
65-06, 65Lxx, 65Mxx, 65Nxx, 65L06, 65L12, 65L15, 65L60, 65L80, 65M06, 65M60, 78-06

ISBN 978-3-642-06941-3 e-ISBN 978-3-540-32862-9

Springer is a part of Springer Science+Business Media

springer.com

Printed in Germany

Cover design: ***design & production*** GmbH, Heidelberg

Preface

The fifth international conference on *Scientific Computing in Electrical Engineering (SCEE)* was held in Capo D'Orlando (ME), Sicily, from 5th to 9th September, 2004. It was sponsored by STMicroelectronics, Italian National Group of Mathematical Physics and National Group of Scientific Computing, Istituto Nazionale di Alta Matematica "Francesco Severi", Philips Research Laboratories Eindhoven, Infineon Technologies A.G. from Munich, Istituto Tecnico Commerciale per Geometri "Francesco Paolo Merendino" at Capo D'Orlando, Synapto from Catania, Fraunhofer Institut für Techno- und Wirtschaftsmathematik at Kaiserslautern, Comune di Capo D'Orlando. The Program committee consisted of:

- Prof. Dr. A. Marcello Anile, Università degli Studi di Catania, Italy.
- Prof. Dr. Flavio Canavero, Politecnico di Torino, Italy.
- Prof. Dr. Ing. Daniel Ioan, "POLITEHNICA" University of Bucharest, Romania.
- Dr. Uwe Feldmann, Infineon Technologies A.G., Munich, Germany.
- Prof. Dr. Michael Günther, Bergische Universität, Wuppertal, Germany.
- Prof. Dr. Ulrich Langer, Johannes Kepler Universität, Linz, Austria.
- Dr. E. Jan W. ter Maten, Philips Research Laboratories Eindhoven, The Netherlands.
- Prof. Dr. Ursula van Rienen, Universität Rostock, Germany.
- Prof. Dr. Wil H.A. Schilders, Technische Universiteit Eindhoven and Philips Research Laboratories Eindhoven, The Netherlands.
- Prof. Dr. Ing. Thomas Weiland, Technische Universität Darmstadt, Germany.

As on all previous occasions, there was a very important support both from industrial and academic sectors, as traditional in this series of conferences. It is precisely the combined effort of industry and academia that assures both the relevance of the work to practical situations and at the same time the presence of long term basic research. For this reason, the interaction between electric or electronic engineers and mathematicians is one of the main aims of the SCEE conferences. This attitude shows up in the areas covered at SCEE-2004, which were: Electromagnetism, Circuit Simulation, Coupled Problems and General mathematical and computational methods.

For each area, two invited speakers were selected by the Organizing Committee, one from industry and one from academia, with the exception of the last area, for which there was only an invited speaker from university. In total, there were 7 Invited Speakers:

- Dr. Augusto Benvenuti, (STMicroelectronics, Agrate Brianza, Italy): "Challenging coupled problems in TCAD".
- Dr. Georg Denk, (Infineon Technologies, Munich, Germany): "Circuit simulation for nanoelectronics".
- Prof. Erion Gjonaj, (Technische Universitaet, Darmstadt, Germany): "Low noise conservative scheme for the solution of Maxwell's equations in PIC simulations".
- Prof. Anne Kværno, (Norwegian Institute of Technology, Trondheim, Norway): "Time integration methods for coupled equations".
- Dr. Ing. Siegbert Martin, (Marconi Communications GmbH, Backnang, Germany): "Microwave issues in EM simulation and design of RF modules, plastic filters and circulators".

- Prof. Giovanni Miano, (Università degli Studi di Napoli Federico II, Italy): "A unified approach for the analysis of networks composed of distributed and lumped circuits".
- Prof. Dave Rodger, (University of Bath, United Kingdom): "Finite element modelling of electrical machines and actuators".

Overall, there were 43 contributed oral presentations, including the talks of the Invited Speakers, and 29 poster presentations. As in the previous edition, there was a session dedicated to short oral introduction of posters where each contributor was given 2 minutes to advertise his/her work.

It has always been the policy of these conferences to encourage participants from all countries, with an emphasis on Europe. Also on this occasion this has been remarkably successful, there were more than one hundred participants from 15 countries. Thus the series of SCEE has confirmed itself as a truly international event.

The papers appearing in this book fall in two categories: the keynote speakers' contributions, and contributions coming both from oral presentations and posters. Each paper was carefully refereed by two suitably chosen referees.
The selected papers have been organized according to the scientific area. Therefore, we have four sections, respectively devoted to Coupled Problems, Circuit Simulation, Electromagnetism and General Mathematical Computational Methods. A fifth section has been added, which comprises all contributions which refer to work in progress, presenting preliminary results on topics of great interest for the Conference.

We would like to thank the organizers of the Conference, the referees of the selected papers, and all the people, both named here and others, whose enthusiasm and hard work ensured the success of this conference SCEE-2004. A special thank goes to Prof. Angelo Santoromita Villa, Headmaster of the Istituto Tecnico Commerciale per Geometri "Francesco Paolo Merendino" at Capo D'Orlando for having offered the facilities of his Institution (our thanks go also to the technical staff of the Istituto) and for his enthusiasm and constant and precious support. Thanks also to the Mayor of the Capo d'Orlando City for having supported the Conference in several ways, particularly with the availability of the premises where the Conference took place.

Arcavacata di Rende,
January 2006

Angelo Marcello Anile
Giuseppe Alì
Giovanni Mascali

Contents

Part I Coupled Problems

Part II Circuit Simulation

List of Contributors

G. Alessi
ST Microelectronics,
Stradale Primosole 50,
I-95121, Catania, Italy.
gesualdo.alessi@st.com

G. Alì
Istituto per le Applicazioni del Calcolo
"M. Picone", CNR, Via P. Castellino 111,
I-80131 Napoli, Italy.
g.ali@iac.cnr.it

A. M. Anile
Università di Catania,
Dipartimento di Matematica e Informatica,
Viale A.Doria 6,
I-95125 Catania, Italy.
anile@dmi.unict.it

Ch. Auer
Graz University of Technology,
Institute of Theoretical and
Computational Physics,
Graz, Austria.
auer@itp.tu.graz.ac.at

A. Bartel
Universität Wuppertal,
Lehrstuhl für Angewandte
Mathematik/Numerische Analysis,
D-42097 Wuppertal, Germany.
bartel@math.uni-wuppertal.de

T. G. J. Beelen
Philips Research Laboratories Eindhoven,
Prof. Holstlaan 4,
5656 AA Eindhoven, the Netherlands.

A. Benvenuti
STMicroelectronics,
Via C. Olivetti 2, Agrate Brianza,
20041 Milano, Italy.
augusto.benvenuti@st.com

E. Bilotta
Università della Calabria,
Dipartimento di Linguistica,
Via P. Bucci, Cubo 17/B,
Arcavacata di Rende (CS), Italy.
bilotta@unical.it

U. Binit Bala
Institute of Electromagnetic Theory
and Microwave Technique,
University of Hannover, Appelstr. 9A,
30167 Hannover, Germany.
bala@tet.uni-hannover.de

O. Bíró
Institute of Fundamentals and Theorie
of Electrical Engineering, IGTE,
Graz University of Technology,
Kopernikusgasse 24, Graz, Austria.
biro@TUGraz.at

L. Bortesi
STMicroelectronics,
Via C. Olivetti 2, Agrate Brianza,
20041 Milano, Italy.

C. Brennan
School of Electronic Engineering,
Dublin City University,
Research Institute for Networks and
Communications Engineering,
Dublin, Ireland.

S. Burger
Konrad-Zuse-Zentrum Berlin,
Takustr. 7, D-14195 Berlin, Germany.
burger@zib.de

R. Campolo
Università della Calabria,
Dipartimento di Matematica,
Via P. Bucci, Cubo 30/B,
Arcavacata di Rende (CS), Italy.
r.campolo@unical.it

G. Carnevale
STMicroelectronics,
Via C. Olivetti 2, Agrate Brianza,
20041 Milano, Italy.

J. A. Carrillo
ICREA - Dpt. Matemàtiques -
UAB, Barcelona, Spain.
carrillo@mat.uab.es

M. C. Carrisi
Università di Cagliari,
Dipartimento di Matematica,
Via Ospedale 72, Cagliari, Italy.

G. Cassano
Politecnico di Milano,
Dipartimento di Matematica "F. Brioschi",
Via Bonardi 9, 20133 Milano, Italy.

S. Castorina
Synapto s.r.l.,
stradale Vincenzo Lancia 57,
95100 Catania, Italy.
scastorina@synapto.com

A. Ciccazzo
STMicroelectronics,
Stradale Primosole 50,
95121 Catania, Italy.
angelo.ciccazzo@st.com

V. Cinnera Martino
ST Microelectronics,
Stradale Primosole 50,
95121 Catania, Italy.
valeria.cinnera-martino@st.com

G. Ciuprina
"Politehnica" University of Bucharest,
CIEAC/LMN, Bucharest, Romania.
lmn@lmn.pub.ro

M. Clemens
Helmut Schimdt Universität,
Holstenhofweg 85,
22043 Hamburg, Germany.
m.clemens@hsu-hh.de

S. Coco
Università di Catania,
DIEES Dipartimento di Ingegneria Elettrica,
Elettronica e dei Sistemi,
Viale A. Doria 6, I-95125 Catania, Italy.
coco@diees.unict.it

P.C. Coles
University of Bath,
Department of Electronic and
Electrical Engineering,
Claverton Down, BA2 7AY Bath, UK.

M. Condon
Research Institute for Networks and
Communications Engineering,
School of Electronic Engineering,
Dublin City University,
Dublin, Ireland.

S. Corsaro
Università di Catania,
Dipartimento di Ingegneria Elettrica
Elettronica e dei Sistemi,
Viale A. Doria 6, I-95125 Catania.

E. Dautbegovic
Research Institute for Networks
and Communications Engineering,
School of Electronic Engineering,
Dublin City University,
Dublin, Ireland.

C. de Falco
Università degli Studi di Milano,
Dipartimento di Matematica "F. Enriques",
via Saldini 50, 20133 Milano, Italy.

M. de Magistris
Università di Napoli "FEDERICO II",
Dipartimento di Ingegneria Elettrica,
Via Claudio 21, I-80125 Napoli, Italy.

L. De Tommasi
Dipartimento di Ingegneria Elettrica,
Università di Napoli "FEDERICO II",
Via Claudio 21, I-80125 Napoli, Italy.

F. Demontis
Università di Cagliari,
Dipartimento di Matematica,
Via Ospedale 72, Cagliari, Italy.

G. Denk
Infineon Technologies, Memory Products,
Balanstr. 73, D-81541 München, Germany.
georg.denk@infineon.com

R. Dionisio
Galileo Avionica,
Via Villagrazia 79, Palermo, Italy.
roberto.dionisio
@galileoavionica.it

C. R. Drago
Università di Catania,
Department of Mathematics
and Computer Sciences,
Viale A. Doria 6, I-95125 Catania, Italy.
drago@dmi.unict.it

A. El Guennouni
Yacht Technology and
Philips Research Laboratories Eindhoven,
Prof. Holstlaan 4,
5656 AA Eindhoven, the Netherlands.

R. A. Ene
Synapto s.r.l.,
stradale Vincenzo Lancia 57,
95100 Catania, Italy.
rene@synapto.com

U. Feldmann
Infineon Technologies AG,
Balanstr. 73, D-81541 München, Germany.
Uwe.Feldmann@infineon.com

G. Frosali
Università di Firenze,
Dipartimento di Matematica Applicata
"G.Sansone",
Via S.Marta 3, I-50139 Firenze, Italy.
giovanni.frosali@unifi.it

M. Galler
Graz University of Technology,
Institute of Theoretical and
Computational Physics,
Graz, Austria.
galler@itp.tu-graz.ac.at

F. Gámiz
Dpt. of Electronics - UGR, Granada, Spain.
fgamiz@ugr.es

D. S. M. Gazzo
Università di Catania,
DIEES Dipartimento di Ingegneria Elettrica,
Elettronica e dei Sistemi,
Viale A. Doria 6, I-95125 Catania, Italy

A. Ghetti
STMicroelectronics,
Via C. Olivetti 2, Agrate Brianza,
20041 Milano, Italy.

C. Giulianetti
Politecnico di Milano,
Dipartimento di Matematica "F. Brioschi",
Via Bonardi 9, 20133 Milano, Italy.

P. González
Dpt. of Applied Mathematics, UGR,
Granada, Spain.
prodelas@ugr.es

G. Greco
STMicroelectronics,
Stradale Primosole 50,
95100 Catania, Italy.
giuseppe-cad.greco@st.com

M. Greiff
University of Hannover,
Institute of Electromagnetic Theory
and Microwave Technique,
Appelstr. 9A, 30167 Hannover, Germany.
mgre@tet.uni-hannover.de

M. Günther
Bergische Universität Wuppertal,
Departement of Mathematics,
Chair of Applied Mathematics/Numerical Analysis,
D-42097 Wuppertal, Germany.
guenther@math.uni-wuppertal.de

T. Halfmann
Fraunhofer Institute for Industrial Mathematics,
67663 Kaiserslautern, Germany.
thomas.halfmann@itwm.fraunhofer.de

A-K. Hamid
University of Sharjah ,
Department of Electrical, Electronics
and Computer Engineering,
P.O. Box 27272, Sharjah, United Arab Emirates.
akhamid@sharjah.ac.ae

M. L. J. Hautus
Technische Universiteit Eindhoven,
Eindhoven, the Netherlands.
averhoev@win.tue.nl

G. Hebermehl
Weierstrass Institute for Applied Analysis
and Stochastics,
Mohrenstr. 39, 10117 Berlin, Germany.
hebermehl@wias-berlin.de

W. Heinrich
Ferdinand-Braun-Institut
für Höchstfrequenztechnik,
Gustav-Kirchhoff-Str. 4, 12489 Berlin, Germany.
w.heinrich@ieee.org

P. J. Heres
Eindhoven University of Technology,
Department of Mathematics and Computer Science,
PO Box 513, 5600 MB Eindhoven, the Netherlands.
p.j.heres@tue.nl

R. J. Hill-Cottingham
University of Bath,
Department of Electronic and
Electrical Engineering,
Claverton Down, BA2 7AY Bath, UK.

D. Ioan
"Politehnica" University of Bucharest,
CIEAC/LMN, Bucharest, Romania.
lmn@lmn.pub.ro

P. Kacor
VSB - Technical University of Ostrava,
Faculty of Electrical Engineering,
Department of Electrical Machines
and Apparatuses,
Ostrava, Czech Republic.
petr.kacor@vsb.cz

R. Klose
Konrad-Zuse-Zentrum Berlin,
Takustr. 7, D-14195 Berlin, Germany.

S. Knorr
Bergische Universität Wuppertal,
Fachbereich C, Gaußstr. 20,
42119 Wuppertal, Germany.
knorr@math.uni-wuppertal.de

J. M. B. Kroot
Eindhoven University of Technology
P.O.Box 513, 5600 MB Eindhoven,
the Netherlands.

A. Kværnø
Norwegian University of Science and Technology,
Department of Mathematical Sciences,
N-7491 Trondheim, Norway.
anne@math.ntnu.no

H. C. Lai
University of Bath,
Department of Electronic and
Electrical Engineering,
Claverton Down, BA2 7AY Bath, UK.

D. Langemann
University of Rostock,
Institute for Mathematics,
Universitätsplatz 1, 18051 Rostock, Germany
dirk.langemann
@mathematik.uni-rostock.de

A. Laudani
Università di Catania,
DIEES Dipartimento di Ingegneria Elettrica,
Elettronica e dei Sistemi,
Viale A. Doria 6, I-95125 Catania, Italy.

Y. Li
National Chiao Tung University,
Microelectronics and Information Systems
Research Center,
Hsinchu 300, Taiwan.
ymli@faculty.nctu.edu.tw

D. Lukáš
University Linz,
SFB F013 "Numerical and Symbolic Scientific
Computing",
Altenberger Strasse 69,
A-4040 Linz, Austria.
dalibor.lukas@vsb.cz,
http://lukas.am.vsb.cz

A. Maffucci
Università di Cassino,
D.A.E.I.M.I.,
Via di Biasio 43,
I-03043 Cassino (FR), Italy.

A. Majorana
Università di Catania,
Dipartimento di Matematica e Informatica,
Viale A.Doria 6,
I-95125 Catania, Italy.
majorana@dmi.unict.it

J. M. Mantas
Software Engineering Department -
UGR, Granada, Spain.
jmmantas@ugr.es

M. Marek
VSB - Technical University of Ostrava,
Faculty of Electrical Engineering,
Department of Electrical Machines
and Apparatuses,
Ostrava, Czech Republic.
martin.marek@vsb.cz

A. Marotta
STMicroelectronics,
Stradale Primosole 50,
95121 Catania, Italy.
angelo.marotta@st.com

A. Marrocco
INRIA, Domaine de Voluceau,
Rocquencourt BP 105,
78153, Le Chesnay, France.
americo.marrocco@inria.fr

R. Martorana
Galileo Avionica,
Via Villagrazia 79, Palermo, Italy.

W. Mathis
University of Hannover,
Institute of Electromagnetic Theory
and Microwave Technique,
Appelstr. 9A, 30167 Hannover, Germany.
mathis@tet.uni-hannover.de

G. Mascali
Università della Calabria,
Dipartimento di Matematica,
Via P. Bucci, Cubo 30/B,
I-87036 Arcavacata di Rende (Cs), Italy.

G. Miano
Università di Napoli "FEDERICO II"
Dipartimento di Ingegneria Elettrica,
Via Claudio 21, I-80125 Napoli, Italy.

S. Micheletti
Politecnico di Milano,
Dipartimento di Matematica "F. Brioschi",
MOX - Modeling and Scientific Computing,
via Bonardi 9, 20133 Milano, Italy.
stefano.micheletti@mate.polimi.it

C. Milazzo
Università di Catania,
Dipartimento di Matematica e Informatica,
viale A. Doria 6, Catania, Italy.

O. Morandi
Università di Firenze,
Dipartimento di Elettronica e Telecomunicazioni,
Via S.Marta 3, I-50139 Firenze, Italy.
omar.morandi@unifi.it

O. Muscato
Università di Catania,
Dipartimento di Matematica e Informatica,
Viale A.Doria 6,
I-95125 Catania, Italy.
muscato@dmi.unict.it

Q. Nasir
University of Sharjah,
Department of Electrical, Electronics
and Computer Engineering,
P.O. Box 27272, Sharjah, United Arab Emirates.
nasir@sharjah.ac.ae

M. E. Oliveri
Università di Catania,
DMFCI Dipartimento di Metodologie Fisiche
e Chimiche per l'Ingegneria,
Viale A. Doria 6,
I-95125 Catania, Italy.
meolive@dmfci.ing.unict.it

P. Pantano
Università della Calabria,
Dipartimento di Matematica,
Via P. Bucci, Cubo 30/B,
I-87036 Arcavacata di Rende (Cs), Italy.
piepa@unical.it

S. Pennisi
Università di Cagliari,
Dipartimento di Matematica,
Via Ospedale 72, Cagliari, Italy.
spennisi@unica.it

S. Perotto
Politecnico di Milano,
Dipartimento di Matematica "F. Brioschi",
MOX - Modeling and Scientific Computing,
via Bonardi 9, 20133 Milano, Italy.
simona.perotto@mate.polimi.it

A. Pirovano
Politecnico di Milano, DEI,
P. zza L. Da Vinci 32,
20133 Milano, Italy.

G. Pollicino
Università di Catania,
DIEES Dipartimento di Ingegneria Elettrica,
Elettronica e dei Sistemi,
Viale A. Doria 6,
I-95125 Catania, Italy.

G. Pöplau
Rostock University,
Institute of General Electrical Engineering,
D-18051 Rostock, Germany.
gisela.poeplau
@etechnik.uni-rostock.de

D. Potts
University of Lübeck,
Institute of Mathematics,
Lübeck, Germany.
potts@math.uni-luebeck.de

R. Pulch
Bergische Universität Wuppertal,
Department of Mathematics,
Chair of Applied Mathematics
and Numerical Analysis,
Gaußstr. 20, D-42119 Wuppertal, Germany.
pulch@math.uni-wuppertal.de

M. Rădulescu
"Politehnica" University of Bucharest,
CIEAC/LMN, Bucharest, Romania.
lmn@lmn.pub.ro

D. Raschka
VSB - Technical University of Ostrava,
Faculty of Electrical Engineering,
Department of Electrical Machines
and Apparatuses,
Ostrava, Czech Republic.
david.raschka.fei@vsb.cz

R. Riaza
Universidad Politécnica de Madrid,
Departamento de Matemática Aplicada
a las Tecnologías de la Información
ETSI Telecomunicación,
Ciudad Universitaria s/n, 28040 Madrid, Spain.
rrr@mat.upm.es

S. Rinaudo
STMicroelectronics,
Stradale Primosole 50,
95100 Catania, Italy.
Salvatore.rinaudo@st.com

D. Rodger
University of Bath,
Department of Electronic
and Electrical Engineering,
Claverton Down, BA2 7AY Bath, UK.
d.rodger@bath.ac.uk

V. Romano
Università di Catania,
Dipartimento di Matematica e Informatica,
Viale A.Doria 6,
I-95125 Catania, Italy.
romano@dmi.unict.it

S. Rinaudo
ST Microelectronics,
Stradale Primosole 50,
95121, Catania, Italy.
salvatore.rinaudo@st.com

R. Sacco
Politecnico di Milano,
Dipartimento di Matematica "F. Brioschi",
Via Bonardi 9, 20133 Milano, Italy.

A. Scanu
Università di Cagliari,
Dipartimento di Matematica,
Via Ospedale 72, Cagliari, Italy.

A. Schädle
Konrad-Zuse-Zentrum Berlin,
Takustr. 7, D-14195 Berlin, Germany.

J. Schefter
Weierstrass Institute for Applied Analysis
and Stochastics,
Mohrenstr. 39, 10117 Berlin, Germany.

W. H. A. Schilders
Philips Research Laboratories Eindhoven,
Prof. Holstlaan 4,
5656 AA Eindhoven, the Netherlands.

R. Schlundt
Weierstrass Institute for Applied Analysis and
Stochastics,
Mohrenstr. 39, 10117 Berlin, Germany.
schlundt@wias-berlin.de

F. Schmidt
Konrad-Zuse-Zentrum Berlin,
Takustr. 7, D-14195 Berlin, Germany.

U. Schreiber
Rostock University,
Institute of General Electrical Engineering,
Rostock, Germany.
`ute.schreiber`
`@etechnik.uni-rostock.de`

F. Schürrer
Graz University of Technology,
Institute of Theoretical and Computational Physics,
Graz, Austria.
`schuerrer@itp.tu-graz.ac.at`

J. M. Sellier
Università di Catania,
Dipartimento di Matematica e Informatica,
Viale A.Doria 6,
I-95125 Catania, Italy.
`sellier@dmi.unict.it`

M. Selva Soto
Humboldt University of Berlin,
Institute of Mathematics,
Berlin, Germany.
`monica@mathematik.hu-berlin.de`

A. Skarlatos
Technische Universität, Schlossgartenstr. 8,
Darmstadt, Germany,
`skarlatos@temf.tu-darmstadt.de`

F. Stranges
Università della Calabria,
Dipartimento di Linguistica,
Via P. Bucci, Cubo 17/B,
87036 Arcavacata di Rende (CS), Italy.
`f.stranges@unical.it`

M. Striebel
Bergische Universität Wuppertal,
Department of Mathematics,
Chair of Applied Mathematics/Numerical Analysis,
D-42097 Wuppertal, Germany.
`striebel@math.uni-wuppertal.de`

S. Spinella
Università della Calabria,
Dipartimento di Linguistica,
Ponte P. Bucci 17B,
87036 Arcavacata di Rende, Italy.

E. J. W. ter Maten
Technische Universiteit Eindhoven
and Philips Research Laboratories Eindhoven,
Prof. Holstlaan 4,
5656 AA Eindhoven, the Netherlands.

T. Tischler
Ferdinand-Braun-Institut
für Höchstfrequenztechnik,
Gustav-Kirchhoff -Str. 4, 12489 Berlin, Germany.

C. Tischendorf
Technische-Universität Berlin,
Institut für Mathematik,
10623 Berlin, Germany.
`tischend@math.tu-berlin.de`

A. G. Tijhuis
Eindhoven University of Technology,
Faculty of Electrical Engineering,
P.O. Box 513, 5600 MB Eindhoven,
the Netherlands.

J. Torres-Ramírez
Universidad Politécnica de Madrid,
Departamento de Matemática Aplicada
a las Tecnologías de la Información
ETSI Telecomunicación,
Ciudad Universitaria s/n, 28040 Madrid, Spain.
`fjtr@mat.upm.es`

M. Torrisi
Università di Catania,
Dipartimento di Matematica e Informatica,
Viale A.Doria 6,
I-95125 Catania, Italy.
`torrisi@dmi.unict.it`

M. C. van Beurden
Eindhoven University of Technology,
Faculty of Electrical Engineering,
P.O. Box 513, 5600 MB Eindhoven,
the Netherlands.

U. van Rienen
Rostock University,
Institute of General Electrical Engineering,
D-18051 Rostock, Germany.
`ursula.van-rienen`
`@etechnik.uni-rostock.de`

L. Vendrame
STMicroelectronics,
Via C. Olivetti 2, Agrate Brianza,
20041 Milano, Italy.

A. Verhoeven
Technische Universiteit Eindhoven,
Eindhoven, the Netherlands.
averhoev@win.tue.nl

S. Viana
University of Bath,
Department of Electronic
and Electrical Engineering,
Claverton Down, BA2 7AY Bath, UK.

S. Voigtmann
Humboldt University Berlin,
Institute of Mathematics,
Unter den Linden 6,
D-10099 Berlin, Germany.
steffen@math.hu-berlin.de

P. K. Vong
University of Bath,
Department of Electronic
and Electrical Engineering,
Claverton Down, BA2 7AY Bath, UK.

C.-S. Wang
Microelectronics and Information Systems
Research Center,
National Chiao Tung University,
Hsinchu 300, Taiwan.

T. Weiland
Technische Universität, Schlossgartenstr. 8,
Darmstadt, Germany.
thomas.weiland
@temf.tu-darmstadt.de

B. Weiß
Institute of Fundamentals and Theories
of Electrical Engineering, IGTE,
Graz University of Technology
Kopernikusgasse 24, Graz, Austria.
bernhard.weiss@TUGraz.at,

T. Wichmann
Fraunhofer Institute for Industrial Mathematics,
67663 Kaiserslautern, Germany.

R. Winkler
Humboldt–Universität zu Berlin,
Institut für Mathematik,
10099 Berlin, Germany.
winkler@mathematik.hu-berlin.de

L. Zschiedrich
Konrad-Zuse-Zentrum Berlin,
Takustr. 7, D-14195 Berlin, Germany.

H. Zscheile
Ferdinand-Braun-Institut
für Höchstfrequenztechnik,
Gustav-Kirchhoff -Str. 4, 12489 Berlin, Germany.

L. Zullino
STMicroelectronics,
Via Tolomeo 1, Cornaredo,
20010 Milano, Italy.

A. P. M. Zwamborn
TNO Physics and Electronics Laboratory
P.O. Box 96864, 2509 JG's-Gravenhage,
the Netherlands.

Part I

Coupled Problems

A Unified Approach for the Analysis of Networks Composed of Transmission Lines and Lumped Circuits*

A. Maffucci[1] and G. Miano[2]

[1] D.A.E.I.M.I, Università di Cassino, Via G. Di Biasio 43, 03043 Cassino, Italy, maffucci@unicas.it
[2] D.I.EL., Università di Napoli Federico II, Via Claudio 21, 80125 Napoli, Italy, miano@unina.it

Abstract The use of transmission line models in high-speed circuit analysis is here reviewed, by means of a unifying approach which allows getting insight on both the numerical simulation and theoretical investigation. Starting from a detailed analysis of the physical meanings of the transmission line models, the paper analyzes the effects of electrical interconnects on signal propagation by using a suitable time-domain equivalent circuit representation of the lines. Qualitative and quantitative analysis are carried out, with particular emphasys to nonlinear dynamics.

1 Introduction

Transmission line (TL) theory is a classic topic of Electromagnetics and several well-assessed analysis techniques are available to study through TL models the effects of propagation in a very wide class of problems, *e.g.*, [1]. Many of such techniques are only suitable for linear problems or for steady-state solutions. However, there are applications such as *high-speed electronic circuits* where the presence of nonlinear devices and the interest on fast transients require time-domain analysis of systems made by distributed and lumped elements. Due to the high operating frequencies and small sizes of such circuits, a reliable design must account for the signal distortion due to the propagation along the *electrical interconnects*, present at various hierarchical levels, *e.g.*: [2]-[6].

Under suitable hypotheses, the interconnects may be described by means of TL models. The TLs of practical interest have losses, parameters depending on the frequency and may be spatially non-uniform. In many cases the physical parameters of the lines are uncertain and a description of statistical type is required, [4]. Lumped circuits may contain dynamic elements (*e.g.*, inductors, capacitors, transformers), resistive elements that may be nonlinear and time-varying (*e.g.*, diodes, transistors, operational amplifiers, logic gates, inverters) and integrated circuits. The interactions between these devices and the TLs, and between the TLs themselves, are described by continuity conditions for voltages and currents at the 'boundaries' between the TLs and the lumped circuit elements, and between the TLs themselves.

To analyze such systems, coupled problems of a profoundly different nature have to be studied: TLs are described by linear and time-invariant partial differential equations, while lumped circuits are modeled by algebraic-ordinary differential equations, eventually time-varying and nonlinear. For such reasons, TL model has recently received renewed attention, focused on important issues concerning both the *qualitative* (well-posedness of the models, convergence of numerical solutions, study of nonlinear dynamics,...) and the *quantitative* point of view (efficient simulation of large systems, model-order reduction,...), *e.g.*: [2]-[6].

Here we present a unifying approach to get an insight on all the above questions. In Sect. 2.1 we first focus on some important *physical properties* of the TL models, in order to highligth the limits of the standard TL model and to suggest a way to *generalize* it. Then in Sect. 2.2 a general method is presented to characterize the terminal behavior of TLs lines, in order to study of networks composed of TLs and

*Invited paper at SCEE-2004

This work is supported in part by Italian Ministry of University under a Program for the development of Research of National Interest (PRIN Grant n.2002093437).

lumped circuits by means of the Circuit Theory approach[7], [8]. To this aim, the most suitable time-domain characterizations of a line is based on an input-state-output representation, where the traveling-wave solutions of TL equations are chosen to represent the 'state'. Such a representation provides a circuit description of the TLs in terms of resistive elements, delayed sources and dynamic elements. Here we refer, for the sake of simplicity, to two-conductor TLs. However, the method is applicable to any kind of line: multiconductor lines, lines with frequency-dependent parameters, and lines with space-varying parameters, [6].

After deriving such a characterization, the analysis of networks composed of TLs and lumped circuits is reduced to the study of networks where TLs are modeled in the same way as the lumped elements: multiports representing the TLs lines differ from multiports representing the lumped elements only in their characteristic relations. In Sect. 3.1 the problem of the well-posedness of both analytical and numerical models describing TLs connected to nonlinear and/or dynamic terminations is addressed. This problem is of a great importance both from a theoretical and from a practical point of view: even if a stable and consistent numerical scheme is adopted, the convergence of the numerical solution is assured only if the analytical and the numerical models are both well-posed: these basic requirements cannot be taken for granted. In Sect. 3.2 some case-studies are presented. The first is intended to highlight the effects of TL modeling on the integrity of propagating signals. The second case-study provides an example of a class of nonlinear circuits, where the role of TLs is crucial to provide a wide richness of nonlinear dynamics, such as multiple steady state solutions, bifurcations and chaotic dynamics.

2 Transmission line models

2.1 Physical interpretation of the TL models

Let us consider the simple interconnect of Fig. 1, made of two perfectly conducting parallel wires of length $2l$ with arbitrary cross-sections, geometrically long, embedded in a homogeneous dielectric. The electromagnetic field can be represented, in the frequency domain, through the potentials φ and $\mathbf{A}$, as

$$\mathbf{E} = -j\omega\mathbf{A} - \nabla\varphi, \mathbf{B} = \nabla \times \mathbf{A}, \tag{1}$$

where ω is the angular frequency and the potentials φ and $\mathbf{A}$ are expressed, assuming Lorentz gauge, in terms of the surface charge σ and current density $\mathbf{J}_s$ by means of the integral relations

$$\mathbf{A}(\mathbf{r}_P) = \mu \int_{\Sigma_1 \cup \Sigma_2} G(r_{PQ})\mathbf{J}_s(\mathbf{r}_Q)ds, \tag{2}$$

$$\varphi(\mathbf{r}_P) = \frac{1}{\epsilon} \int_{\Sigma_1 \cup \Sigma_2} G(r_{PQ})\sigma(\mathbf{r}_Q)ds, \tag{3}$$

where Σ_1, Σ_2 are the conductor surfaces, r_{PQ} is the distance between the field and source points, G is the Green function for the homogeneous space $G(r) = \frac{\exp(-jkr)}{4\pi r}$ and $k = \omega\sqrt{\epsilon\mu}$. Here we assume that the

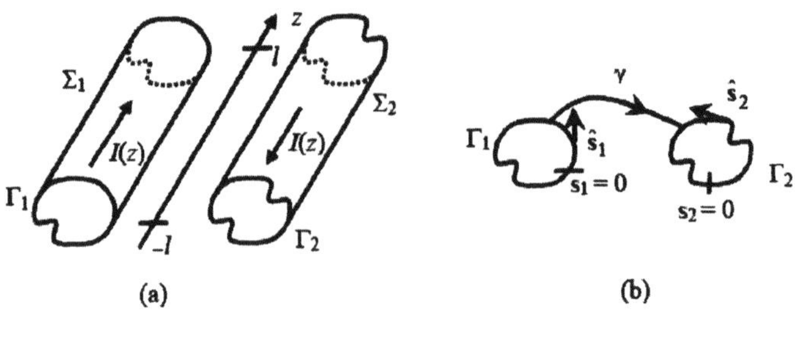

Fig. 1. (a) Schematic representation of the interconnect geometry; (b) cross-section

characteristic dimensions of the devices are small compared to the interconnect length, hence their effects are neglected in (2) and (3).

The unknown distributions σ and $\mathbf{J}_s$ are determined by imposing the *boundary conditions* and the *charge conservation law*:

$$\mathbf{E} \cdot \hat{\mathbf{t}} = \mathbf{0} \quad \textit{on } \Sigma_1 \textit{ and } \Sigma_2, \tag{4}$$

$$\nabla \cdot \mathbf{J}_s = -j\omega\sigma \quad \textit{on } \Sigma_1 \textit{ and } \Sigma_2. \tag{5}$$

The fundamental assumptions to derive the TL models are the following:

1. the current field density has only the longitudinal component;
2. the *common mode* variables are equal to zero;
3. the dependence of σ and $\mathbf{J}_s$ on the transverse and longitudinal coordinates is of a separable type;
4. the interconnect is transversally *electrically short*.

Hypothesis 1 depends on the cylindrical symmetry of the structure and on the way the structure is excited. In such a condition, the magnetic field is of *transverse type* (TM), hence it is possible to define uniquely at each section the voltage between the two conductors $V(z)$, which is related to the per-unit-length (p.u.l.) flux $\Phi(z)$ through:

$$-\frac{dV(z;\omega)}{dz} = j\omega\Phi(z;\omega). \tag{6}$$

Hypothesis 2 is well founded if there are no external sources of electromagnetic field. As a consequence of this assumption and of the conservation equation, the *differential* current at each section $I(z)$ is related to the p.u.l. electric charge $Q(z)$ through:

$$-\frac{dI(z;\omega)}{dz} = j\omega Q(z;\omega). \tag{7}$$

Hypothesis 3 holds if the characteristic dimensions of the conductor sections are *electrically short*, *i.e.* are small compared to the characteristic signal wavelength. The transverse problem may be solved considering the *electrostatic potentials* produced by the same conductor pair, but of *infinite length*. Hypotheses (1)-(3) allow to derive a *transmission line model*, defined by (6), (7) and by the following two integral relations:

$$\Phi(z;\omega) = \mu \int_{l}^{-l} H(z-z';\omega)I(z';\omega)dz', \tag{8}$$

$$V(z;\omega) = \frac{1}{\epsilon} \int_{l}^{-l} H(z-z';\omega)Q(z';\omega)dz', \tag{9}$$

which could be easily derived from (2), (3), as shown in [9]. The kernel of such relations $H(z)$ is expressed in terms of the Green function G and become of impulsive type if hypothesis 4 holds [9]. With such an additional condition we have $\Phi(z) = LI(z)$, $V(z) = Q(z)/C$, where L and C are, respectively, the p.u.l. inductance and capacitance of the interconnect evaluated by solving the transverse 2D problem. By combining the above results we obtain the *standard* TL model described by the *telegrapher's equations*

$$-\frac{dV(z;\omega)}{dz} = j\omega\, LI(z;\omega), \quad -\frac{dI(z;\omega)}{dz} = j\omega\, CV(z;\omega). \tag{10}$$

From a physical point of view, it is well-known that the TL model (10) describe the propagation of a field of transverse electromagnetic type (TEM), *e.g.*, [1]. Instead, the TL model (6), (7), (8), and (9) is a *generalized* model which could describe also the presence of continuum spectrum modes along with the fundamental one. This allows the description of high-frequency effects like *radiation losses* and *dispersion* which are not predicted by the standard TL model [9]. Table 1 summarizes the conditions when the lumped models, the standard TL model (STL) and the above *enhanced* TL model (ETL) have to be used, expressed in terms of operating frequency (through the wavenumber k), characteristic longitudinal ($2l$) and transverse (h) dimensions, and mean radius of conductor section a. A full-wave model is required for the analysis of all those cases not included in Table 1.

Table 1. Interconnect models for different cases

model	$k \cdot 2l$	$k \cdot h$	$k \cdot a$
lumped	$<< 1$	$<< 1$	$<< 1$
STL	≥ 1	$<< 1$	$<< 1$
ETL	≥ 1	≈ 1	$<< 1$

Even in hypothesis 4, when considering non-ideal structures, conductor and dielectric losses have to be taken into account: their effects destroy, in principle, the TEM structure of the field. However, in the quasi-TEM assumption (*e.g.*, [1]) the propagation may be still described by the TL model:

$$-\frac{dV(z;\omega)}{dz} = Z(z,\omega)I(z,\omega), \; -\frac{dI(z;\omega)}{dz} = Y(z,\omega)V(z). \tag{11}$$

where $Z(z,\omega)$ and $Y(z,\omega)$ are the line parameters, *i.e.* the *p.u.l. impedance* and *admittance*. The line parameters depend on the actual physical realization of the line: they can describe the simple ideal case (10) when $Z = j\omega L$ and $Y = j\omega C$. Instead, when $Z = R + j\omega L$ and $Y = G + j\omega C$, they describe the so-called RLGC lines (lossy uniform lines with negligible frequency effects). More generally they could describe non-uniform lines with strong frequency dependence, for instance due to conductor skin-effect and dielectric dispersive behavior: for most cases of practical interest, they could be conveniently described by the following *Laplace domain* model, e.g., [6]:

$$Z(s) = R_\infty + sL_\infty + K\sqrt{s} + Z_r(s) \tag{12}$$

$$Y(s) = G_\infty + sC_\infty + Y_r(s) \tag{13}$$

where $(.)_\infty$ stands for the high-frequency limit, which may be evaluated from the physical model of the line or even from frequency-domain samples of the parameters provided by measurements, *e.g.*, [10]. It is important to stress that $Z_r(s)$ and $Y_r(s)$ tends to zero as $1/s$ for $s \to \infty$.

2.2 Equivalent circuit models

There are many possible two-port equivalent representations of TLs, both in frequency and time domain: the optimal choice strongly depends, of course, on the particular problem to be solved, *e.g.*, [11].

When dealing with high-speed circuits, usually one has to perform time-domain transient analysis of circuits made by linear TLs and non-linear lumped elements. In such cases, among all the possible two-port representations, a very convenient one is provided by the input-state-output representation obtained by assuming forward and backward waves as state-variables of the dynamic system. Such an approach would lead in the ideal-line case to the same result obtained by Branin, [12], by applying the *Method of Characteristics*. In the general case, it provides the following time domain model (*e.g.*, [6]):

$$i_1(t) = y_c(t) * i_1(t) + j_1(t), \quad i_2(t) = y_c(t) * v_2(t) + j_2(t), \tag{14}$$

where $*$ indicates the convolution product, subscripts 1,2 indicates the two line ends and j_1 and j_2 are two controlled current sources given by

$$j_1(t) = p(t) * [-2i_2(t) + j_2(t)], \quad i_2(t) = p(t) * [-2i_1(t) + j_1(t)]. \tag{15}$$

Such a dynamic model is characterized by two *impulse responses*: the *characteristic admittance* $y_c(t)$ and the propagation function $p(t)$, which can be obtained by reverse-transforming their Laplace domain expressions:

$$Y_c(s) = \sqrt{Y(s)/Z(s)}, \quad P(s) = exp\left(-2ls\sqrt{Y(s)Z(s)}\right). \tag{16}$$

The impulse responses may always be split into *irregular* and *regular* parts: the first contains *irregular* functions like Dirac pulses, and may be evaluated analytically from the asymptotic behavior of (16). After such asymptotic behavior is extracted, the regular parts may be easily evaluated numerically by

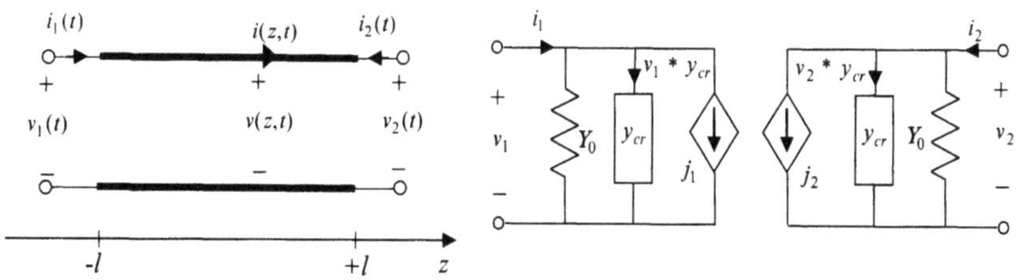

Fig. 2. Norton-type equivalent circuit representation of a two-conductor line

reverse transforming the Laplace domain remainders. We obtain, in the general case, the following decomposition:

$$y_c(t) = Y_0\delta(t) + y_{cr}(t), \quad p(t) = exp(-\mu T)[\delta(t-T) + p_r(t-T)], \tag{17}$$

where $Y_0 = \sqrt{C_\infty / L_\infty}$ is the ideal line characteristic admittance, $T = 2l\sqrt{C_\infty L_\infty}$ is the one-way delay time, μ is a damping factor which is known analytically, $y_{cr}(t)$ and $p_r(t)$ are the regular parts of the impulse responses, often known only numerically. Note that such properties hold for the general case of multiconductor lines with frequency-dependent parameters, with slight differences in the case of pronounced skin-effect, *e.g.* [6]. Such a line representation provides advantages both in the qualitative and numerical analysis, as shown in Sect. 3. Eqs. (14) and (15) describe each line end through the time-domain equivalent circuit of Norton type shown in Fig. 2. Apart for the effect of Y_0, which is always present, the solution at each line end is due to the contribution of the dynamic one-port $y_{cr}(t)$, which describes dispersion effects due to losses and frequency-dependence of line parameters, and of the controlled source $j_k(t)$, $k = 1, 2$, which takes into account the reflection at the other end, the delay and dispersion introduced by the propagation along the line. Note that $j_k(t)$, $k = 1, 2$ vanishes if the line is *matched* at the other end. The most important property of such a model is the fact that, at a given time instant t, $j_1(t)$ only depends on the solution history in the time interval $(0, t-T)$. Therefore, it could be treated as *independent* source, if the problem is solved iteratively.

3 Transmission lines and lumped circuits

3.1 Qualitative properties of the solution

Let us consider a two-conductor lossy line connecting two lumped nonlinear resistors. From the above considerations, the adopted circuit representation is a dynamic system which introduces a *state variable*, namely the current flowing into the dynamic one-port $y_{cr}(t)$. If we solve the problem recursively, at each time instant t we find at the line terminations two *uncoupled networks* which may be simply represented by a resistive circuit, obtained by substituting the dynamic one-port $y_{cr}(t)$ with a constant current source I_y, see Fig. 3a. Note that such a procedure extends to distributed elements the concept of *associated resistive circuit*, introduced in the past for the analysis of lumped circuits, *e.g.*, [7]. Dynamic loads may be easily taken into account in a similar way: their corresponding associated resistive circuits are obtained by substituting each capacitor with a constant voltage source, and each inductor with a constant current source.

The analytical model obtained by combining (14),(15) with the characteristics of the lumped resistors is *well-posed* if it is possible to express all the *non-state circuit variables* as single-valued functions of the *state variables* and of the *source variables*. In fact, in such a case it can be proven that the model may be reduced to a well-posed system of Volterra integral equations in normal form, [6], [13]. For instance, considering two voltage-controlled nonlinear resistors

$$i_k(t) = g_k(v_k(t)), \quad k = 1, 2, \tag{18}$$

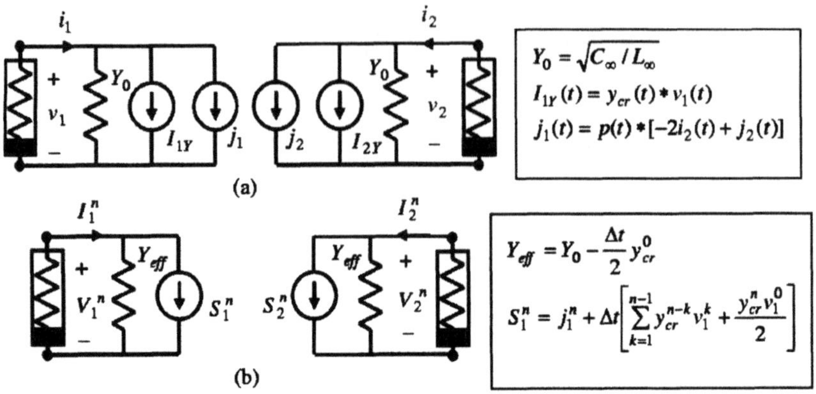

Fig. 3. Transmission line connecting two nonlinear resistors: (a) associated resistive circuit; (b) discrete resistive circuit

the following conditions are sufficient to obtain the well-posedness of the model

1. function g_k is continuous;
2. the resistor is *weakly active*;
3. the following inequality holds:

$$dg_k/dv > -Y_0. \tag{19}$$

Inequality (19) is always satisfied if the characteristics of the resistors are *monotonically increasing*. Instead, it may be not satisfied if the characteristic has tracts with negative slopes. When this occurs, the associated resistive circuit can have more than one solution, so a normal form Volterra integral equation system may not exist and the solution may be not unique. The presence of a capacitor in parallel with the nonlinear resistor (even a parasitic one) ensures uniqueness even when the above condition of the slope of $g(v)$ is not satisfied, [6], [13]. Note that such a result may be easily generalized to multiconductor lines, [14].

Besides the associated resistive circuit, the line representation adopted here allows to introduce also the so-called *discrete resistive circuit*, [8], which describes the problem to be solved at any discrete time-step $t_n = n\Delta t$, where Δt is the time discretization step (let x^n indicate the generic variable $x(t)$ at $t = t_n$). Note that discrete circuits associated with different integration algorithms are different: Fig. 3b shows the circuit obtained by using the trapezoidal rule to integrate the convolutions (the variables at port 2 have analoguos definitions of those of port 1). The transient analysis reduces to the *dc* analysis of the resistive circuit of Fig. 3b: we can solve the associated discrete circuit step by step, by any efficient method, such as modified nodal analysis combined with Newton-Raphson method [8], through recursive updating of S_1^n and S_2^n.

We observe that the associated discrete circuit of Fig. 3b tends to the associated resistive circuit of Fig. 3a for $\Delta t \to 0$. This implies that, if the associated resistive circuit has one and only one solution, and hence the dynamic circuit has one and only one solution, the numerical model has one and only one solution converging to the actual solution, [6], [15]. In fact, it is easy to show that the conditions ensuring the well-posedness of the associated discrete circuit are the same derived above for the associated resistive circuit, provided that Y_0 is replaced with Y_{eff}, see Fig. 3b:

$$dg_k/dv > -Y_{eff}. \tag{20}$$

If (19) is satisfied, there exists a sufficiently small Δt to satisfy (20) also, and vice-versa, and the numerical model admits one and only one solution. If we consider non-linear resistors described through voltage-controlled non-monotone characteristics, (19) could be no longer verified, hence the original equations may admit several solutions, and condition (20) is not satisfied even if Δt is arbitrarily small. As a consequence, the numerical model admits several solutions, and the discrete time sequence approximating the solution is no longer unique. In this case, a well-posed model is again obtained if we take into account the capacitive parasitic effects, neglected during modeling, that have a strong influence on the dynamics of the network, [6], [15].

3.2 Numerical analysis of practical applications

Whatever is the adopted two-port representation, the main drawback of such an approach is the high computational cost of transient analysis, mainly due to time convolution. Therefore, the literature proposes many techniques to obtain convenient reduced-order models, *e.g.*, [3]. It is known in literature that the model adopted here is the most suitable to perform transient analysis of *long* transmission lines, *i.e.* lines for which the propagation delay plays a significant role, *e.g.*, [10]. This is because such a model allows to extract analytically all the unbounded terms contained in the line impulse responses, which are then represented by simple resistive circuits and damped delayed sources. Only the regular remainders are approximated with reduced-order models, and then represented through low-order lumped networks.

Let us now refer to the case-study 1, consisting in a two-conductor microstrip of length $20cm$ analyzed in [4]. The interconnect is modeled as a TL with frequency-dependent parameters by using expressions (12) and (13), with: $C_\infty = 88.25\mu F/m, L_\infty = 0.806\mu H/m, R_\infty = 86.206\Omega/m, G_\infty = 67nS/m, K = 2.4m\Omega s^{-1/2}/m$, while $Z_r(s) = Y_r(s) = 0$. The line presents a characteristic admittance $Y_0 = 10.5mS$ and a delay time $T = 1.69ns$.

The line is synthesized by means of the equivalent circuit model of Fig. 2, by using a rational approximation for the two impulse responses remainders $y_{cr}(t)$ and $p_r(t)$. The near end is terminated on a driver, modeled as a voltage source in series with a resistor $R_1 = 50\Omega$. The voltage source supplies a rectangular pulse of amplitude $1V$, that lasts $2ns$, with rise and fall time $t_r = t_s = 50ps$. The far end is connected to a pn-junction diode, modeled by

$$i = I_s\left(\exp(v/V_T) - 1\right), \tag{21}$$

with $V_T = 1V$ and $I_s = 40\mu A$.

Figure 4 shows the far-end voltage obtained by using three different TL models: the complete one (skin), an approximated one obtained by neglecting the skin-effect (RLGC) and the lossless line limit (ideal). The simulation puts on evidence the strong effect of TL modeling on the signal shape: a correct modeling of TLs is crucial to foresee critical effects for signal integrity, like delay in the receiver switching, and false switching which may be caused by unwanted reflections, [16].

The second case-study analyzed here is intended to highlight the richness of behavior which could be observed when TLs connect nonlinear devices: multiple steady state solutions, bifurcations and chaotic dynamics. Let us consider an ideal TL connecting two nonlinear resistors: such a line is described by (17) with $y_{cr}(t) = p_r(t) = 0,\ \mu = 0$, hence the circuit state equations, obtained by combining (14) and (15) with the resistor characteristics, are nonlinear difference equations with one delay. The dynamics of the problem may be studied by analyzing the behavior of a *nonlinear one-dimensional map*, in which the time is no longer a continuous variable but a sequence of discrete values:

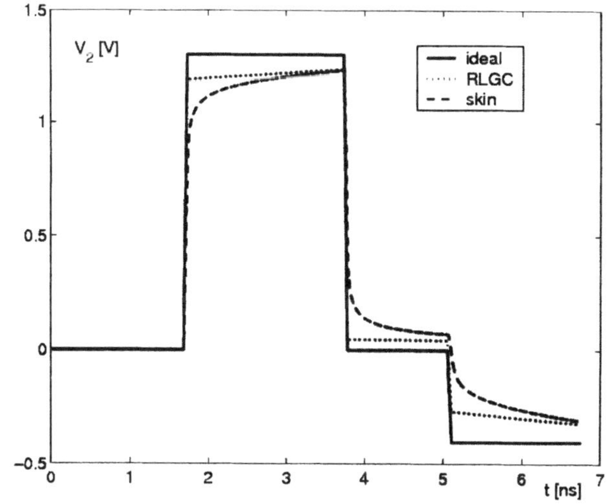

Fig. 4. Far-end voltage for case-study 1 predicted by different line models

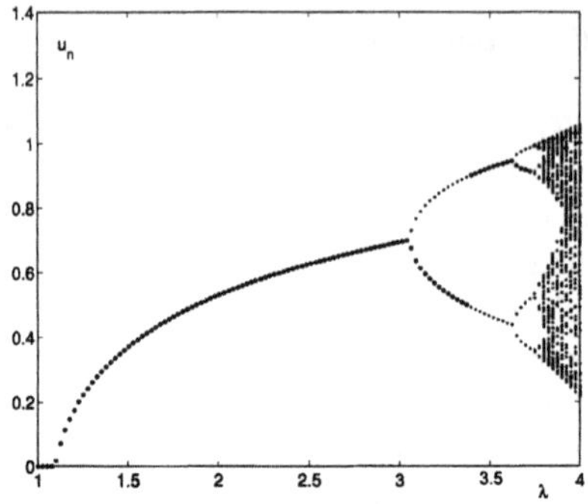

Fig. 5. Bifurcation diagram for case-study 2

$$u_{n+1} = f(u_n), \tag{22}$$

with a proper definition of the state variable u_n, [6]. By studying the main properties of these maps, bifurcations, periodic oscillations and chaos may observed when at least one terminal resistor is active and the other is nonlinear, *e.g.*, [6], [17].

Let us consider an ideal line of length $1m$, with the following parameters: $C = 3.00pF/m, L = 3.70\mu H/m$, leading to $Y_0 = 0.90mS$ and $T = 3.33ns$. Let us assume that the far-end is connected to an active linear resistor of conductance G_1, and the near-end is terminated on the diode of Eq.(21). A non-zero initial condition is imposed, by applying a unit voltage pulse at the far-end. In such a case, the map f in (22) reduces to the *logistic map* $f = \lambda u_n(1 - u_n)$, where the parameter λ is given by:

$$\lambda = \frac{Y_0 - G_1}{Y_0 + G_1}. \tag{23}$$

and the state variable is expressed in terms of the backward voltage wave at the far-end:

$$u_n = \lambda \frac{v_1 - i_1/Y_0}{2} \frac{1}{(1 - \beta - \ln(\beta))V_T}, \quad \beta = \frac{I_s}{V_T Y_0}. \tag{24}$$

Figure 5 shows the *bifurcation diagram* of such a map. For $0 \leq \lambda \leq 1$ the only fixed point is $u = 0$, while for increasing λ we enter a region where a non-zero asymptotically stable fixed point may be observed. Then stable periodic orbits of period 2, 4, and so on may be observed, until λ reaches a value such to excite chaotic dynamics. Note that the chaotic regime is interrupted by some windows where the asymptotic behavior of the orbits is again periodic.

References

[1] Collin R.E., Foundation of Microwave Engineering. Mc Graw-Hill, New York (1992)

[2] Tripathi, V.K., Sturdivant, R., *Guest Editors*, Special Issue on: Interconnects and Packaging. IEEE Trans. on Microwave Theory and Techniques, **45** (1997)

[3] Schutt-Ainé, J. S., Kang, S. S., *Guest Editors*, Special Issues on: Interconnections - Addressing the Next Challenge of IC Technology. IEEE Proceedings, **89** (2001)

[4] Paul, C.: Analysis of Multiconductor Transmission Lines. Wiley, New-York (1994)

[5] Matick R.E.: Transmission Lines for Digital and Communication Networks. IEEE press, Piscataway, N.J. (1995)

[6] Maffucci, A., Miano, G.: Transmission Lines and Lumped Circuits. Academic Press, San Diego (2001)

[7] Hasler, M., Neirynck, J.: Nonlinear Circuits. Norwood, Artech House (1986)

[8] Chua, L.O., Lin, P.M.: Computer Aided Analysis of Electronic Circuits. Prentice-Hall, Englewood Cliffs. (1975)

[9] Maffucci, A., Miano, G., Villone, F.: An enhanced transmission line model for conductors with arbitrary cross-sections. IEEE Transactions on Advanced Packaging, **28**, 174-188 (2005)

[10] Grivet-Talocia S., *et al.*: Transient analysis of lossy transmission lines: an efficient approach based on the method of characteristics. IEEE Trans. on Advanced Packaging, **27**, 45-56 (2004)

[11] Kuznetsov D.B., Schutt-Ainé J.E.: Optimal transient simulation of transmission lines, IEEE Trans. Circuits Systems-I, **43**, 110-121 (1996)
[12] Branin F.H., Jr: Transient analysis of lossless transmission lines. Proceedings of IEEE, **55**, 2012-2013 (1967)
[13] Miano, G.: Uniqueness of solution for linear transmission lines with nonlinear terminal resistors. IEEE Trans. on Circuit and Systems-I, **44**, 569-582 (1997)
[14] Maffucci, A., Miano, G.: On the dynamic equations of linear multiconductor transmission lines with terminal nonlinear multiport resistors, IEEE Transactions on Circuit and Systems-I, **45**, 812-829 (1998)
[15] Maffucci, A., Miano, G.: On the uniqueness of the numerical solutions of non linearly loaded lossy transmission lines, Intern. Journal of Circuit Theory and Applications, **27**, 455-472 (1999)
[16] Deutsch, A., *et al.*: High-speed signal propagation on lossy transmission lines. IBM Journal Research Development. **34**, 601-616 (1990)
[17] Corti, L., De Menna, L., Miano, G., Verolino, L.: Chaotic dynamics in an infinite-dimensional electromagnetic system. IEEE Trans. Circuits Systems-I, **41**, 730-736 (1994)

Circuit Simulation for Nanoelectronics*

G. Denk

Infineon Technologies, Memory Products, Balanstr. 73, D-81541 München, georg.denk@infineon.com

Abstract Circuit simulation is a well-established and important tool for the design of integrated circuits. However, the current challenges of today's technology give rise to new requirements for analog simulators. This paper tries to show some mathematical research topics necessary for future survival of circuit simulation: One of the main issues is the coupling of the circuit equations (differential-algebraic equations) with equations originating from other domains like thermal models, semiconductor models, wire models (partial differential equations), and noise models (stochastic differential equations). Other topics are multi-scale problems and separation of time constants, model order reduction, diagnosis, and finally efficiency and robustness.

Key words: circuit simulation, analog simulation, coupled domains, hierarchical modeling, multi-scale problems, model order reduction, differential-algebraic equations (DAEs), partial differential-algebraic equations (PDAEs), stochastic differential-algebraic equations (SDAEs)

1 Introduction

Since more than 100 years semiconductor devices are used in industrial applications, it started with the discovery of the rectifier effect of crystalline sulfides by Ferdinand Braun in 1874. But it took till 1940, when the first semiconductor transistor was developed at Bell Labs. In 1951 Shockley presented the first junction transistor, and ten years later the first integrated circuit (IC) was presented by Fairchild. It contained only a few elements, but with this IC an evolution was initiated. More and more devices were integrated on a single chip, today more than a billion MOSFETs are on one memory chip. The increasing number of elements was already predicted in 1965 and is now well-known as Moore's law. It states that the number of MOSFETs per chip doubles every 2 years.

The development of more and more complex chips was accomplished with circuit simulation. In 1967 one of the first simulation programs was written, namely BIAS by Howard. So circuit simulation is nearly as old as the design of integrated circuits. Some years later, another simulation program, CANCER, was developed by Nagel in the research group of Rohrer. Later the development of both programs was combined under the guidance of Pederson, and in 1972 the first version of SPICE was released. Due to the free availability of SPICE, it became widely used and some kind of an industry standard (see [14] for further information about the history of SPICE). Meanwhile SPICE-like simulators are known to yield realistic and reliable simulation results. They are universally applicable and often used for "golden" simulations.

Though SPICE is still available from Berkeley University, a variety of commercial and in-house industrial circuit simulators have been released. One of these is TITAN used at Infineon. A lot of research was done to keep up with the increasing design sizes and technology demands. Similar to Moore's law, we need a corresponding development in the field of analog circuit simulators and their underlying algorithms. This paper tries to present some of the challenges of current chip design and their implications on analog simulators. It does not claim to be exhaustive and sometimes it is biased towards Infineon's TITAN. We will not go into details of the mathematical problems but try to give an idea what is needed for further application of analog simulation. Nevertheless, we hope that these research topics will be picked up by the

*Invited paper at SCEE-2004

scientific community, so that the results will enable the simulators to continue to provide a reliable tool for circuit designers.

2 Functionality challenge

Moore's law states that the number of MOSFETs on an integrated chip doubles every two years. This means that more and more functions are integrated onto complex chips. In most cases we have analog functions (like sensors, actuators, converters) on a digital chip which provide the interface between the digital part and the analog environment. As these two worlds interact, a simultaneous simulation of the analog and digital parts is necessary. One way to achieve this is to use an analog simulator for both areas, but complexity does not allow this as an analog simulation is usually some orders of magnitude slower than a digital simulation due to the higher level of accuracy. Therefore, we need a mixed-signal simulation and a hierarchical modeling of the chip.

2.1 Mixed-signal simulation

In a mixed-signal simulation the circuit is decoupled into an analog part and a digital part, and the analog part is simulated by an analog simulator and the digital one by a digital simulator. Normally – due to the decoupling of both parts – the simulation works quite well, but this is not guaranteed. The handling of events plays an important role: a change of a digital signal may cause a discontinuity of an analog signal, which may then trigger a digital event and so on. This iteration loop between analog and digital may prevent convergence.

In analog/digital converters (ADC) or phase looked loops (PLL) the feedback loop implies a strong coupling between the analog and the digital parts of the circuit. The high speed digital clock drives a slowly settling analog part, and therefore a transient simulation has to follow the digital part with very small step sizes for a long period of time. As the main interest is the locking behavior (lock-in stability, lock-in time, lock-in solution) it would be more efficient, if envelope schemes would be available in the analog simulator.

Another issue is finding the correct analog time point which corresponds to a digital event. Due to the continuous time scale in the analog part, we need some kind of iterations to get the switching time accurately enough. As a digital event may occur quite often, an efficient implementation is essential.

Research topics:

- envelope solver
- switching-point computation

2.2 Hierarchical modeling

The analog part of a mixed-signal simulation usually determines the simulation speed, as it is several orders of magnitude slower than the digital simulation. To improve performance behavioral models are developed (e. g. using VHDL-AMS, Matlab/Simulink) which describe possibly large analog building blocks. The same approach can be used for parts which are not yet fully designed, while a simulation of the interaction is already needed. Similar arguments hold for large building blocks, for which a reduced model has been constructed by model-order reduction [9, 13, 18]. In any case we get an hierarchical modeling, where the different models of the same building block should describe the same functionality. But this has to be verified.

The formal verification of behavioral models is quite difficult, despite the well-established application of formal verification in the digital design. Due to the analog nature of these building blocks continuous input/output signal have to be checked. As the dimension of the behavioral model is in general different from the analog model, the relevant parts of the differential-algebraic equations (DAEs) have to be matched. First promising results can be found in [11].

The hierarchical structure of a chip – independent of whether it comes from a mixed-signal approach or just from a bottom-up design approach – can be exploited for simulation. If the hierarchy of the circuit is reflected by the data structure, identical building blocks need to be saved only once. Also the evaluation has to be done only once for identical bias conditions. This helps to speedup simulation but of course requires new concepts in circuit evaluation. However, due to parasitics present in real circuits, the same building blocks are not really identical, usually they differ a little bit. An even better and more general approach is to map the hierarchy of the circuit to the numerical algorithms, namely the solver and the integration scheme.

Hierarchical structures allow for speeding-up and parallelization of the linear solver which is used for solving the circuit equations for every time point of a transient simulation. But this requires fill-in strategies which make a compromise between minimal fill-in and parallelization, so that large portions of the operations can be performed in parallel [19, 20].

Also the integration scheme may benefit from the hierarchical structures: Different building blocks usually have different activity, and this would allow different step sizes. For accuracy reason, the most active block with the smallest step size determines the overall step size in conventional simulation. Using a multi-rate scheme allows larger step sizes in latent blocks, while only a (hopefully) small part of the circuit has to be computed with a small step size. For an efficient multi-rate integration we need specialized schemes like the mixed multi-rate schemes [10, 22].

Research topics:

- automatic partitioning of circuits into hierarchical systems
- hierarchical linear-algebra solver
- multi-rate integration scheme
- formal verification of analog systems

2.3 Diagnosis

A circuit simulator should always give reasonable results for the circuit to be analyzed. But sometimes there is an error in the design, and due to the many functions integrated on one chip, it can be quite difficult for the designer to detect the fault. It does not help if the simulator gives "no convergence", it should provide hints where to look for. As the programming languages used for behavioral modeling allow much more freedom than a SPICE-like language, it is much easier to introduce errors. On the other side, it is more complicated to detect them. Therefore, the diagnostic part of an analog simulator gets more and more important. In TITAN a combination of numerical methods and graph methods has been implemented [8] which provides significant support to the designer in detecting flaws in the chip design.

Research topic:

- detailed diagnosis of numerical problems, coupled to circuit design

3 Frequency challenge

Increasing the frequency of a chip helps to speedup operations like data transfer rates, and therefore the frequency is an often used marketing argument, especially for CPUs. On average the frequency increases with a factor of 5 each 3 years. Voltage-controlled oscillators (VCOs) are driven up to 50 GHz, typical rise/fall times and gate delays are ≤ 50 ps. But the increase of frequency induces problems not only in the design but also in the simulator.

3.1 Modeling

Due to the high frequency the wave character of signal propagation becomes more and more important, we do no longer have "ideal" connections between the circuit elements. This also means that simple approximations of wires by resistance/capacitance elements do not correctly reflect the physical situation on high-frequency chips. To get a realistic model of the interconnect, either a more complex extraction with inductances or even the solution of the telegrapher's equation is necessary. While the former approach significantly increases the dimension of the circuit's equation due to the lumped elements, the latter one requires the coupling to a solver for partial differential equations.

Not only the connection is affected by the high frequency but also the models for the elements. For instance, the MOSFET model needs extensions to include non-quasi static (NQS) effects making the model more complex.

Though these necessary extensions do not cause any problems in principle, they result in additional modeling and simulation effort.

Research topics:
- modeling of signal propagation
- modeling of non-quasi static effects

3.2 Frequency- and time-domain simulation

In many cases a mixture of high-frequency and moderate-frequency signals is present in a circuit, which makes both the frequency-domain and the time-domain simulation quite expensive and requires special approaches. For a simulation in the time domain such a mixture of frequencies means that widely separated time constants are present in the circuit, and – similar to Sect. 2.2 – the fastest one determines the rather small step size, while the slowest one determines the rather large simulation interval. By a separation of the time constants with multivariate functions [5, 17] it is possible to transform the DAEs of the circuit's system into partial differential-algebraic equations (PDAEs). Specialized integration schemes [12, 15, 16] for this type of equation are currently being developed which will allow a very efficient simulation of mixed-frequency circuits.

Research topics:
- efficient integration of multi-tone circuits in frequency and time domain

4 Shrinking challenge

Following Moore's law is that the devices on a chip will get smaller and smaller while the number of devices increases. This shrinking allows a reduction of the production costs per transistor, and due to the smaller devices a faster switching. But shrinking has drawbacks, too: The compact modeling used for devices like MOSFETs is no longer accurate enough, and the power density on the chip increases. Both issues have their impact on circuit simulation. While it is possible to shrink the active devices, this is not true for interconnect. This means that parasitics elements get more dominant with increased shrinking and have to be considered, see Sect. 5.1.

4.1 Modeling

With decreasing device geometry more and more effects have to be considered by the compact models used for MOSFETs and other devices. The modeling of these effects are packed onto the existing model which gets more and more complicated. A sign of this complexity is the number of model parameters – the BSIM4 model from Berkeley University [6], which is used as a quasi standard, has more than 800 parameters! For simulation the parameters have to be chosen in such a way that the model reflects the current technology. But due to the high fitting dimension this task is quite difficult and may lead to reasonable matching characteristics with unphysical parameter settings, which cause numerical problems during simulation. In addition, the device characteristics are heavily affected by transistor geometry and even by neighborhood relations which can not be reflected by compact models.

A remedy for this situation is to switch back to the full set of semiconductor equations for critical MOSFETs (like high frequency settings, MOSFETs used for electrostatic discharge (ESD) protection). This approach allows full accuracy but it has a severe drawback: a large computational effort is needed for the device evaluation even compared to complex compact models, there are some orders of magnitude between compact modeling and solving the semiconductor equations. Therefore, some criteria are needed for the decision which MOSFETs can be simulated using the compact model and which devices require the semiconductor equations for appropriate modeling. Currently the decision is made by the designer.

Assuming that it is known which model fits to which device, there is still an open issue: how should the coupling be done between the circuit simulation (DAEs) and the device simulation (PDEs)? Using both simulators as black box may work, but there is a need to analyze the interaction of them. This will allow also a more efficient coupling, especially when transient simulations are performed. The analysis requires the extension of the index definition of the DAE and of the computation of consistent initial values. A theoretical foundation for this are abstract differential-algebraic systems (ADASs) [3, 23]. First results [4, 21, 24] indicate that the integration of device simulation into circuit simulation will be successful.

Research topics:

- analysis of the coupled systems
- efficient solution of coupled semiconductor (PDEs) and circuit equations (DAEs)

4.2 Power density challenge

A consequence of shrinking is that the power density on the chip increases as the currents, which are necessary for charging or discharging the capacitances of a device, flow within smaller areas. Though this is partially compensated by a decrease of the power supply voltage, the power density has exceeded $100\,\mathrm{Watt/cm^2}$. The designer has to take care to bring the heat off the chip, and he must avoid hot spots, i. e. chip areas which are too hot while the average temperature is still okay. In order to do so, the designer needs simulation which regards for both the electrical and the thermal properties.

Compared with electrical changes within a circuit, the thermal interaction is 3–6 orders of magnitude slower. A naive approach would therefore require very long transient simulation intervals leading to unacceptable run times. The situation is even worse, as the power density issue is especially important for large chips which increases the effort again. Therefore specialized methods are needed which couple the thermal simulation (PDEs) and circuit simulation (DAEs) in an intelligent way and use multi-rate techniques to perform the co-simulation efficiently [1].

Research topics:

- efficient simulation of thermal effects
- efficient solution of coupled thermal (PDEs) and circuit equations (DAEs)

5 Power supply challenge

One mean to accomplish smaller devices and higher frequency is the reduction of the power-supply voltage. While supplies of 5 V have been used in the past, the supply voltage has been reduced down to 1 V or even below. The advantages of this approach is the avoidance of breakthrough and punch in the small devices, and – as the voltage swing is reduced – a higher switching frequency. The lower supply voltages also help in the field of mobile applications. But reducing the power supply has also a drawback: The signal-to-noise ratio decreases which means that parasitic effects and noise become more and more significant and can no longer be omitted from circuit simulation.

5.1 Parasitics

During the design phase of a chip the connections between the elements are treated as ideal, i. e. it is assumed that the elements do not influence each other and there are no delays between devices. But this is not true on a real chip, due to the length and the neighborhood of the wires there is interference like crosstalk, and the elements themselves suffer from the actual placement on silicon. Therefore the circuit is extracted from of the layout of a chip, and this post-layout circuit containing all the parasitic elements has to be simulated and cross-checked against the original design. Currently it is still sufficient in many cases that only resistances and capacitances are used to model the parasitic effects. However, to get all relevant effects due to the small structures and currents present on a chip, the accuracy of the extraction has to be improved by extracting more elements and using additional element types like inductances.

Of course, the quantity of parasitic elements has an impact on circuit simulation: Due to the parasitics the number of nodes present in a circuit increases significantly, which increases the dimension of the system to be solved. This effect is amplified by the fact that due to fill-in the sparsity of the system decreases, therefore the effort necessary for solving the underlying linear equation system becomes dominant and the simulation's run times are no longer acceptable. To speedup the simulation it is necessary to perform a parasitics reduction. Several approaches are possible: One possibility is to rearrange the parasitic elements in such a way that the influence on the circuit is unchanged but the number of nodes is reduced, possibly at the cost of additional elements. A typical example for this is the so-called star-delta conversion. Another way is to remove and rearrange parasitic elements which do not significantly contribute to the circuit's response, for example long RC trees are replaced by shorter ones. As some errors are introduced by this reduction, a reliable error control is necessary for this approach. A third possibility which tackles the fill-in problem is to discard some of the fill-ins. From a mathematical point of view, this resembles some kind of incomplete LU-factorization (ILU) of the matrix. Here care must be taken so that convergence of the non-linear equation solver is still guaranteed.

Research topics:

- error control and new approaches for parasitic reductions
- fill-in minimization strategies for post-layout circuits

5.2 Noise analysis

Reduced signal-to-noise ratio means that the difference between the wanted signal and noise is getting smaller. A consequence of this is that the circuit simulation has to take noise into account. Usually noise simulation is performed in the frequency domain, either as small-signal noise analysis in conjunction with an AC analysis or as large-signal noise analysis as part of an harmonic balance or shooting method. These noise analyses are well-established in the meantime. But noise analysis is also possible in the context of transient noise analysis for non-oscillatory circuits. For an implementation of an efficient transient noise analysis in an analog simulator, both an appropriate modeling and integration scheme is necessary.

Modeling of transient noise

A noisy element in transient simulation is usually modeled as an ideal, non-noisy element and a stochastic current source which is shunt in parallel to the ideal element. As a consequence of this approach the underlying circuit equations are extended by an additional stochastic part, which extends the DAE to a stochastic differential-algebraic equation (SDAE) (for details refer to [26]). The current supplied by the current source is modeled as a stochastic process.

Depending on the cause of noise there are mainly three different noise models in use: thermal noise, shot noise and flicker noise. While the stochastic current source for thermal and shot noise can be simulated using Gaussian white noise, this is not possible for flicker noise. The memory of this process – corresponding to the $1/f^\beta$ dependency for low frequencies f and $\beta \approx 1$ – does not allow a white noise modeling where the increments are stochastically independent. One possibility to represent flicker noise for transient analysis is to use fractional Brownian motion (fBm) for $0 < \beta < 1$. Fractional Brownian motion is a Gaussian stochastic process, and the increments of the fBm required for a transient simulation can be realized with normal-distributed random numbers, for details see [7].

Research topic:

- transient model for flicker noise

Integration of stochastic differential-algebraic equations

The modeling of transient noise is only one part of a transient noise simulation, the other one is the integration of the SDAEs. Though there are some numerical schemes available for stochastic differential equations (SDEs), they do not fit properly to the context of circuit simulation. Besides the fact that the standard schemes are for SDEs and not for SDAEs, they mostly require high-order Itô integrals and/or higher order derivatives of the noise densities. Both is not possible in circuit simulation, this is either too expensive or even not available. For efficiency reasons we need specialized integration schemes which exploit the special structure of the equations. Fortunately, even in current chip designs, the noise level is still smaller in magnitude compared to the wanted signal, which can be used for the construction of efficient numerical schemes [25].

Most currently available numerical schemes aim at the solution of Gaussian white noise processes. But flicker noise is not such a process, so there is the need of a calculus for fBm, which would allow the development of an appropriate integration method. Though there are first results [2], they are not yet applicable in circuit simulation.

Normally it is not sufficient to compute a single path of the transient noise but several paths are necessary to get reliable stochastic conclusions. It would help to improve the efficiency of a transient noise analysis, if these paths could be computed simultaneously, which requires a unique step-size control for all paths. This is currently investigated.

Research topics:

- understanding of stochastic differential-algebraic equations
- efficient integration schemes for transient noise analysis
- step-size control for simultaneous computation of several paths
- integration schemes for flicker noise

6 Conclusion

Circuit simulation is one of the most important tools used in the design process of integrated circuits. As the chips are getting more and more complex and advanced, the requirements for the simulation are getting harder. We have stated some challenges of chip design and tried to conclude the corresponding challenges for analog simulation and their mathematical foundations. Though circuit simulation is done for many years, there still remain a lot of research topics. These can be summarized as:

Coupling of different domains gets more and more important as models from different domains have to be used in order to get accurate enough results. These models are formulated as PDEs (thermal model, semiconductor model, wire model) or SDEs (transient noise model). It is no longer possible to treat the domains separately, so an analysis of how to set up and discretize the equations is necessary in order to get favorable mathematical properties (index, stability, uniqueness, efficiency).

Hierarchical modeling is a possibility to cope with the large design complexity. To exploit this property also in simulation it must be reflected by the algorithms like solver and integration scheme.

Multi-scale problems arise at several places. Here we have time constants which differ by some orders of magnitude. Multi-scale problems may be tackled by multi-rate integration schemes which exploit the latency in the slower parts, by envelope methods, or by separation of the time constants by reformulating the problem.

Model order reduction helps to speedup the simulation, either by reducing parasitic elements or by creating behavioral models. We need a thorough error analysis for parasitic reduction in order to maintain the accuracy. Formal verification of analog models ensures that two models of different level match sufficiently.

Diagnosis helps to find errors in the design which show up as numerical problems in the simulator. The simulator must map them back to the circuit elements.

Efficiency and robustness of an analog simulation is a never-ending challenge. In order to keep simulation times acceptable it is necessary to constantly think about possibilities to speedup the simulation.

References

1. Bartel, A.: First order thermal-electric interaction in chip-design. Presentation at SCEE 2004. See also contribution in this issue
2. Bender, C.: An Itô formula for generalized functionals of a fractional Brownian motion with arbitrary Hurst parameter. Stochastic Process. Appl. **104**, 81–106 (2003)
3. Bodestedt, M., Tischendorf, C.: PDAE models of integrated circuits and perturbation analysis. Preprint 2004-8, Inst. für Math., Humboldt-Univ. zu Berlin, 2004. To appear in Math. Comput. Model. Dyn. Syst
4. Bodestedt, M.: The index of an integrated circuit, analysis and simulations. Presentation at SCEE 2004. See also contribution in this issue
5. Brachtendorf, H.G., Welsch, G., Laur, R., Bunse-Gerstner, A.: Numerical steady state analysis of electronic circuits driven by multi-tone signals. Electrical engineering **79**, 103–112 (1996)
6. BSIM4 manual. Department of Electrical Engineering and Computer Science, University of California, Berkeley (2000). http://www-device.EECS.Berkeley.EDU/~bsim/
7. Denk, G., Meintrup, D., Schäffler, S.: Transient noise simulation: Modeling and simulation of $1/f$-noise. In: Antreich, K., Bulirsch, R., Gilg, A., Rentrop, P. (eds) Modeling, Simulation and Optimization of Integrated Circuits. ISNM Vol. 146, 251–267, Birkhäuser, Basel, (2003)
8. Estévez Schwarz, D., Feldmann, U.: Actual problems in circuit simulation. In: Antreich, K., Bulirsch, R., Gilg, A., Rentrop, P. (eds) Modeling, Simulation and Optimization of Integrated Circuits. ISNM Vol. 146, 83–99, Birkhäuser, Basel, (2003)
9. Feldmann, P., Freund, R.W.: Efficient linear circuit analysis by Padé approximation via the Lanczos process. IEEE Trans. Computer-Aided Design. **14**, 639–649 (1995)
10. Günther, M., Striebel, M.: A charge oriented mixed multirate method for a special class of index-1 network equations in chip design. Appl. Numer. Math. **53**, 489–507 (2005)
11. Hartong, W., Klausen, R., Hedrich, L.: Formal verification for nonlinear analog systems: Approaches to model and equivalence checking. In: Drechsel, R. (ed) Advanced Formal Verification. Kluwer Academic Publishers, Boston (2004)
12. Knorr, S., Günther, M.: Analysis of multirate partial differential-algebraic systems for RF-circuit design. Presentation at SCEE 2004. See also contribution in this issue
13. Odabasioglu, A., Celik, M., Pileggi, L.: PRIMA: Passive reduced-order interconnect macromodeling algorithm. IEEE Trans. Computer-Aided Design. **17**, 645–654 (1998)
14. Perry, T.: Donald O. Pederson. IEEE Spectrum, June 1998, 22–27
15. Pulch, R.: PDAE models for simulating frequency modulated signals. Presentation at SCEE 2004. See also contribution in this issue
16. Pulch, R., Günther, M.: A method of characteristics for solving multirate partial differential equations in radio frequency application. Appl. Numer. Math. **42**, 397–409 (2002)
17. Roychowdhury, J.: Analyzing circuits with widely separated time scales using numerical PDE methods. IEEE Trans. Circuits Syst. I **48** 578–594 (2001)
18. Roychowdhury, J.: Reduced-order modeling of time-varying systems. IEEE Trans. Circuits Syst. II **46**, 1273–1288 (1999)
19. Schenk, O., Röllin, S., Gupta, A.: The effects of unsymmetric matrix permutations and scalings in semiconductor and circuit simulation. IEEE Trans. Computer-Aided Design. **23**, 400–411 (2004)
20. Schenk, O., Gärtner, K.: Solving unsymmetric sparse systems of linear equations with PARDISO. Journal of Future Generation Computer Systems **20**, 475–487 (2004)
21. Selva Soto, M.: Numerical simulation of electrical circuits containing semiconductor devices. Presentation at SCEE 2004. See also contribution in this issue
22. Striebel, M.: Towards one-step multirate-methods in full chip design. Presentation at SCEE 2004. See also contribution in this issue
23. Tischendorf, C.: Coupled systems of differential algebraic and partial differential equations in circuit and device simulation. Modeling and numerical analysis. Habil. thesis, Humboldt-Universität, Berlin, 2003
24. Tischendorf, C.: Numerical analysis of coupled circuit and device models. Presentation at SCEE 2004. See also contribution in this issue
25. Winkler, R.: Stochastic differential algebraic equations in transient noise. Presentation at SCEE 2004. See also contribution in this issue
26. Winkler, R.: Stochastic differential algebraic equations of index 1 and applications in circuit simulation. J. Comp. Appl. Math., **157**, 477–505 (2003)

Hot-Phonon Effects on the Transport Properties of an Indium Phosphide $n^+ - n - n^+$ Diode

Ch. Auer and F. Schürrer

Institute of Theoretical and Computational Physics, Graz University of Technology, Austria, {auer,schuerrer}@itp.tu.graz.ac.at

Abstract This paper presents studies of hot-phonon effects in an indium phosphide $n^+ - n - n^+$ diode. A direct solver for the system of the Boltzmann equations for electrons and polar optical phonons coupled with the Poisson equation is applied. Remarkable differences between calculations with a hot-phonon gas and with equilibrium phonons are discussed.

Key words: Boltzmann-Poisson system, hot-phonons, semiconductor devices, indium phosphide

1 Introduction

Direct numerical solutions of the Boltzmann-Poisson system are very popular approaches for simulating semiconductor devices [MP01, CGMS03]. These methods allow for the investigation of far-from-equilibrium electron systems from a mesoscopic point of view [ES03, GS04]. Most of the kinetic models consider a phonon background gas with a fixed temperature. However, several investigations [AS04, GS04] prove that the phonon distributions can drastically deviate from the Bose-Einstein distribution in coupled electron-phonon systems. Especially when electrons interact with polar optical (pop) phonons is the phonon system strongly driven out of equilibrium.

In this paper, we present simulations of an indium phosphide (InP) $n^+ - n - n^+$ diode taking into account the hot-phonon effects. We numerically solve the coupled system consisting of the Boltzmann equations for electrons and pop phonons and the Poisson equation. Our scheme is a suitable extension of the multigroup model [GS04], which is able to cope with spatial inhomogeneous problems. To reconstruct the spatial derivatives, the shock-capturing WENO method [LOC94] is applied. The obtained results show that the non-equilibrium phonons re affect the electron distribution and significantly changes the macroscopic quantities of the electrons inside the channel of the diode.

2 Kinetic Equations

We consider a two-valley approximation of the conduction band of InP. The dispersion relations $e_\nu(k)$ for the central Γ-valley ($\nu = 1$) and the four equivalent L-valleys ($\nu = 2$) satisfy

$$e_\nu(k)[1 + \alpha_\nu e_\nu(k)] = \frac{\hbar^2 k^2}{2m_\nu}. \tag{1}$$

Here, $\hbar$ denotes the Planck constant, m_ν is the effective mass, and the positive parameters $\alpha_\nu > 0$ are the nonparabolicity factors.

In our kinetic approach, the distribution function $f^\nu = f^\nu(\boldsymbol{k}, \boldsymbol{r}, t)$ depending on the quasi-momentum $\boldsymbol{k}$, the position vector $\boldsymbol{r}$ and time t characterizes the electrons in the ν-th energy valley of the conduction band. For the pop phonons we introduce the phonon distribution function $g = g(\boldsymbol{k}, \boldsymbol{r}, t)$. The temporal evolution of f^ν and g is governed by the coupled system of Boltzmann equations

$$\partial_t f^\nu + \boldsymbol{v}_\nu \cdot \nabla_r f^\nu - \frac{e_0}{\hbar} \boldsymbol{E} \cdot \nabla_k f^\nu = C^\nu(\{f^\nu\}, g), \tag{2}$$

$$\partial_t g = C^p(\{f^\nu\}, g) \tag{3}$$

with the electron velocities $\boldsymbol{v}_\nu(\boldsymbol{k}) = 1/\hbar \nabla_k e_\nu(k)$ and the elementary charge e_0. In equation (2), the electric field vector $\boldsymbol{E} = -\nabla_r V(\boldsymbol{r}, t)$ is coupled with the electron density

$$n(\boldsymbol{r}, t) = \sum_{\nu=1}^{2} \int_{\mathbb{R}} f^\nu(\boldsymbol{k}, \boldsymbol{r}, t) d\boldsymbol{k} \tag{4}$$

via the Poisson equation

$$\Delta_r V(\boldsymbol{r}, t) = \frac{e_0}{\epsilon_0} [n(\boldsymbol{r}, t) - n_0(\boldsymbol{r})], \tag{5}$$

where $e_0 n_0(\boldsymbol{r})$ represents the fixed charge density of the donors and ϵ_0 the dielectric constant. Temporal changes of the electron distribution functions due to the scattering processes are determined by the collision operators $C^\nu(\{f^\nu\}, g)$. In addition to the interaction of electrons with the pop phonons, we consider acoustic phonon scattering, deformation potential inter- and intravalley scattering with optical phonons and impurity scattering. It should be pointed out that the hot-phonon effects are taken into account only for the pop mode, because the pop interaction is the most efficient one. A detailed description of all the scattering processes and a table summarizing the material parameters used in our calculations can be found in [GS04]. The operator $C^p(\{f^\nu\}, g)$ allows for the changes of the phonon distribution due to absorption and emission by electrons as well as phonon-phonon interactions. For a discussion of the structure and the properties of the phonon collision operators we refer to [AS04, GS04].

3 Numerical Method

We treat the simulation of the InP diode as a one-dimensional problem in the physical space. Hence, we assume that the distribution functions f^ν and g, the electric field vector $\boldsymbol{E}$ as well as the densities $n(\boldsymbol{r}, t)$ and $n_0(\boldsymbol{r})$ only depend on the space variable $z = r_3$. Next, we consider the representation

$$\boldsymbol{k}_\nu(\varepsilon, \mu, \varphi) = \frac{\sqrt{2m_\nu}}{\hbar} \sqrt{\varepsilon(1 + \alpha_\nu \varepsilon)} \left(\sqrt{1 - \mu^2} \cos\varphi, \sqrt{1 - \mu^2} \sin\varphi, \mu \right) \tag{6}$$

of the momentum vector $\boldsymbol{k}$ in terms of the electron energy $\varepsilon = e_\nu(k)$, the cosine $\mu = k_3/|\boldsymbol{k}|$ of the angle between $\boldsymbol{k}$ and the z-axis and the polar angle φ. Following the derivations in [CGMS03], we introduce the distribution functions

$$\phi^\nu(\varepsilon, \mu, z, t) = \sigma_\nu(\varepsilon) f^\nu[\boldsymbol{k}_\nu(\varepsilon, \mu, \varphi), z, t], \tag{7}$$

where the dependence of ϕ^ν on the polar angle φ is omitted due to the cylindrical symmetry in the momentum space. The quantity $\sigma_\nu(\varepsilon)$ obeys the relation $d\boldsymbol{k}_\nu = \sigma_\nu(\varepsilon) d\varepsilon d\mu d\varphi$.

For the phonon system, we write the distribution g as

$$g_1(k, \mu, z, t) = g[\boldsymbol{k}(k, \mu, \varphi), z, t] - g_0, \tag{8}$$

with the Bose-Einstein distribution g_0 and the function $\boldsymbol{k}(k, \mu, \varphi)$ determining the vector $\boldsymbol{k}$ in terms of its modulus k and the angular variables μ and φ. Finally, the kinetic equations for the new distribution functions ϕ^ν and g_1 result in

$$\partial_t \phi^\nu + \partial_z(a_1^\nu \phi^\nu) + \partial_\varepsilon(a_2^\nu \phi^\nu) + \partial_\mu(a_3^\nu \phi^\nu) = \sigma_\nu C^\nu(\{\phi^\nu/\sigma_\nu\}, g_1 + g_0), \tag{9}$$

$$\partial_t g_1 = C^p(\{\phi^\nu/\sigma_\nu\}, g_1 + g_0) \tag{10}$$

with the coefficients

$$a_1^\nu = \frac{\hbar^3\sigma_\nu\mu}{m_\nu^2(1+2\alpha_\nu\varepsilon)^2}, \quad a_2^\nu = -e_0E_za_1^\nu, \quad a_3^\nu = \frac{-e_0E_z(1-\mu^2)}{\sqrt{2m_\nu\varepsilon(1+\alpha_\nu\varepsilon)}}. \tag{11}$$

We perform the phase-space discretization for the electrons by introducing the intervals $I_i^z = [(i-1)\Delta z, i\Delta z]$ for $i = 1, \ldots, N_z$, $I_j^{\varepsilon,\nu} = [(j-1)\Delta\varepsilon, j\Delta\varepsilon]$ for $j = 1, \ldots, N_\varepsilon^\nu$ and $I_k^\mu = [(k-1)\Delta\mu - 1, k\Delta\mu - 1]$ for $k = 1, \ldots, N_\mu$. Further, we define the cells $\mathcal{I}_{ijk}^\nu = I_i^z \times I_j^{\varepsilon,\nu} \times I_k^\mu$. The length of the energy interval is defined by $\Delta\varepsilon = \hbar\omega_0/\beta$ with $\beta \in \mathbb{N}$ and the pop phonon energy $\hbar\omega_0$. This choice of the energy discretization mainly simplifies the treatment of the collision operators. The electron cut-off energy $\varepsilon_{\max} = N_\varepsilon^1\Delta\varepsilon$ is chosen high enough to ensure that the number of electrons with $\varepsilon > \varepsilon_{max}$ can be neglected. Concerning the polar optical phonons, we additionally introduce the intervals $I_j^q = [(j-1)\Delta q + q_0, j\Delta q + q_0]$ for $j = 1, \ldots, N_q$ and the cells $\mathcal{I}_{ijk}^p = I_i^z \times I_j^q \times I_k^\mu$. The minimal and maximal moduli of the phonon momentum q_0 and $q_{max} = q_0 + N_q\Delta q$ are determined so that $g_1(q, \mu, z, t) \approx 0$ for $q < q_0$ and for $q > q_{max}$.

Our numerical scheme is based on the set of equations

$$\partial_t\phi_{ijk}^\nu + \frac{h^1_{i+\frac{1}{2}jk} - h^1_{i-\frac{1}{2}jk}}{\Delta z} + \frac{h^2_{ij+\frac{1}{2}k} - h^2_{ij-\frac{1}{2}k}}{\Delta\varepsilon} + \frac{h^3_{ijk+\frac{1}{2}} - h^3_{ijk-\frac{1}{2}}}{\Delta\mu} = C_{ijk}^\nu, \tag{12}$$

$$\partial_t g_{1ilk} = C_{ilk}^p, \tag{13}$$

governing the time evolution of the cell averages ϕ_{ijk}^ν and g_{1ilk}. In (12) the functions $h^1_{i+\frac{1}{2}jk}$, $h^2_{ij+\frac{1}{2}k}$ and $h^3_{ijk+\frac{1}{2}}$ represent numerical fluxes. The quantities C_{ijk}^ν and C_{ilk}^p are approximations of the cell-averaged collision operators. To determine the numerical fluxes, we apply an upwind scheme combined with high-order ENO and WENO reconstruction techniques. The third-order WENO method [LOC94] is used to obtain $h^1_{i+\frac{1}{2}jk}$, while the numerical fluxes in the ε- and μ-direction, $h^2_{ij+\frac{1}{2}k}$ and $h^3_{ijk+\frac{1}{2}}$, are calculated according to the second-order ENO procedure [SO88, LOC94]. For the treatment of the collision operators we use the ansatz

$$\phi^\nu(\varepsilon, \mu, z, t) \approx \sum_{i,j,k} \phi_{ijk}^\nu(t)\chi_{ijk}^\nu, \quad g_1(q, \mu, z, t) \approx \sum_{i,l,k} g_{1ilk}(t)\chi_{ilk}^p \tag{14}$$

with the characteristic functions $\chi_{ijk}^\nu = \chi_{ijk}^\nu(z, \varepsilon, \mu)$ and $\chi_{ilk}^p = \chi_{ilk}^p(z, q, \mu)$ of the domains $\mathcal{I}_{ijk}^\nu$ and $\mathcal{I}_{ilk}^p$. In (14) the summation is performed over all cells. The right-hand sides of (12) and (13) are then calculated by inserting (14) into the cell-averaged collision operators and carrying out the integrations of the collision kernels. In our treatment the midpoint rule is used to evaluate these integrals numerically. Concerning a detailed description of the approximation procedure for the collision operators we refer to [GS04]. The Poisson equation (5) is solved by inserting (14) into the integral representation [MP01] of the exact solution for given boundary values of the electric potential. The time integration of the coupled equations (12) and (13) is performed by applying the second-order Runge-Kutta type TVD scheme [SO88].

4 Results

The considered InP diode is 1 μm long and has a channel length of 0.4 μm. In the n^+ region, i.e., in the intervals $[0, 0.3]$ μm and $[0.7, 1.0]$ μm, the doping concentration is 5×10^{17} cm^{-3}. Inside the channel, ranging from 0.3 μm to 0.7 μm, we have a donor concentration of 2×10^{15} cm^{-3}. The grid used for our calculations has the dimensions $N_z = 150$, $N_\varepsilon^1 = 90$, $N_\varepsilon^2 = 34$, $N_\mu = 16$ and $N_q = 25$. For the energy discretization length we choose $\beta = 4$, which corresponds to $\Delta\varepsilon = 10.8$ meV. The simulation is performed at room temperature, $T = 300$ K, for an applied bias of 1 V.

In Fig. 1 we present the electron density and the electric field strength at $t = 10.0$ ps. Since at this instant of time the temporal change of these quantities almost vanish, we interpret them as stationary state results. It should be mentioned that the initial electron density coincides with the discontinuous donor density $n_0(z)$. The right-hand plot shows a comparison of the electric field obtained under the assumption of equilibrium and non-equilibrium pop phonons. Noticeable differences of the electric field are localized

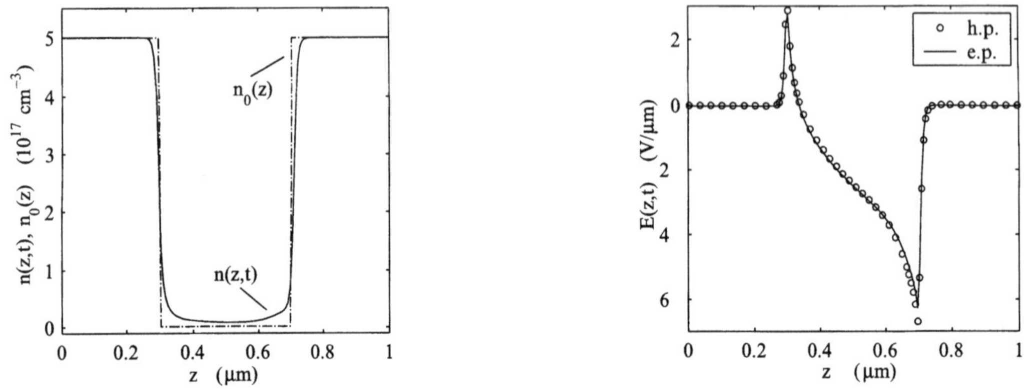

Fig. 1. Electron density $n(z, t)$ and doping concentration $n_0(z)$ as functions of z at $t = 10.0$ ps; z-component of the electric field $E(z, t)$ for equilibrium pop phonons (solid line) and for hot phonons (circles)

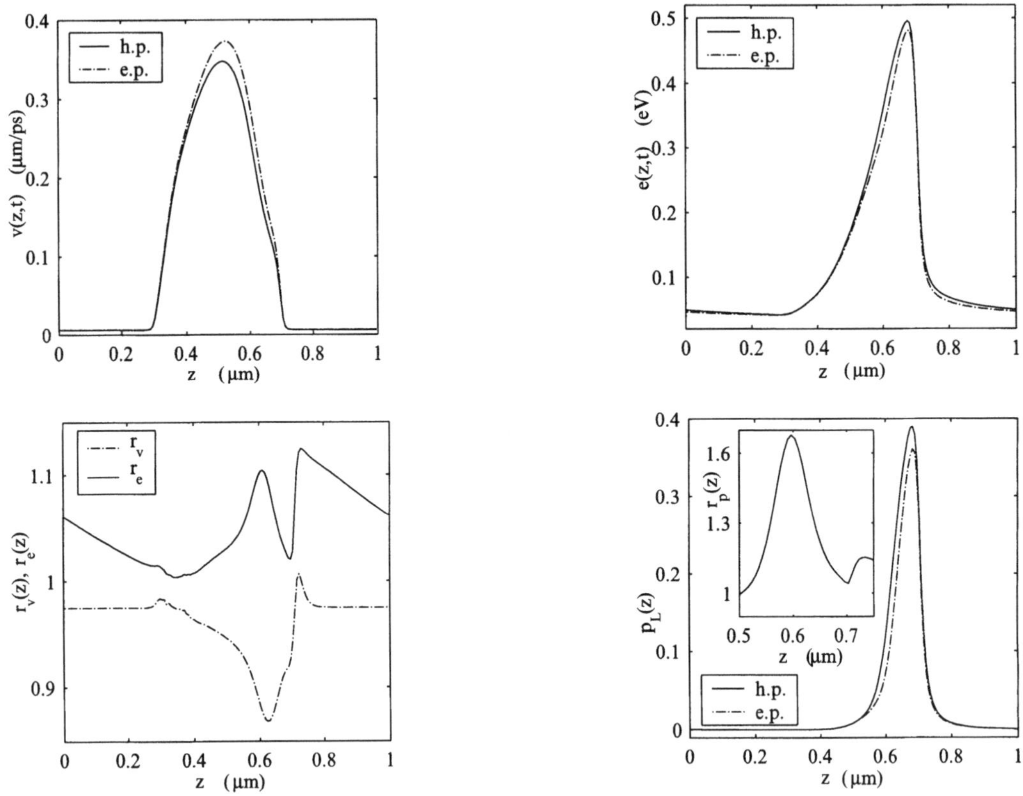

Fig. 2. Mean electron velocity (top-left) and mean electron energy (top-right) obtained for hot phonons (h.p.) and equilibrium phonons (e.p.); ratios r_v and r_e (bottom-left) and the L-valley population p_L as well as ration r_p (bottom-right) as functions of z at $t = 10.0$ ps

near the junctions of the diode at $z \approx 0.3\ \mu\text{m}$ and $z \approx 0.7\ \mu\text{m}$. In these regions, we obtain higher absolute values of the electric field for the hot-phonon calculations.

More pronounced effects of the hot pop phonons on the electrons are found for the mean electron velocity and the mean electron energy. The two plots on the top of Fig. 2 again show results for $t = 10.0$ ps. It should be noted that the non-equilibrium behavior of the phonon gas lowers the average electron velocity and increases the average electron energy inside the channel. This effect can also be observed in calculations for the bulk-case [GS04].

To quantify the deviations due to the non-equilibrium phonons we plot the ratios $r_v(z,t) = v(z,t)/v^*(z,t)$ and $r_e(z,t) = e(z,t)/e^*(z,t)$ bottom left in Fig. 2. The star refers to calculations based on equilibrium phonons. Distinct hot-phonon effects are found in the right half of the channel. In this region, strong electric fields and the spatial diffusion at the junction lead to hot electron distributions, which drive the phonon system out of equilibrium. The ratios r_v and r_e indicate deviations of more than 10 percent for the average electron velocity and energy. Outside of the channel, the differences between the mean velocities resulting from calculations considering hot-phonons and equilibrium phonons are smaller but still noticeable. The mean electron energy is also strongly influenced by non-equilibrium phonons in the n^+-regions of the diode.

The right-hand plot on the bottom of Fig. 2 depicts the population of the L-valleys $p_L = n_2(z,t)/n(z,t)$. In the region where the electron energy (see top-right plot) is high, the L-valley population reaches nearly 40 percent. Since the energy difference between the minima of the L- and the Γ-valley is 0.61 eV, p_L tends to zero if the average electron energy is small. The sub-plot displays the quantity $r_p = p_L/p_L^*$ representing the ratio between the results for p_L in the case of hot phonons and equilibrium phonons. It should be noted that the non-equilibrium phonon gas increases the L-valley population up to 60 percent.

Figure 3 shows the electron energy density $e_e(z,t)$ (left) and the change of the phonon energy $\Delta e_p(z,t) = e_p(z,t) - e_p(z,0)$ (right) as functions of z at different times. The electron energy strongly increases at the right junction of the diode. Consequently, a sharp phonon energy peak appears at the same local position. However, we realize differences in the temporal evolution of the electron and phonon energies at the hot spot. The peak of the electron energy builds up much faster than that of the phonon energy.

Our kinetic approach allows us to directly investigate the distribution functions (7) and (8). Figure 4 presents a comparison of the Γ-valley distribution functions of electrons interacting with hot phonons

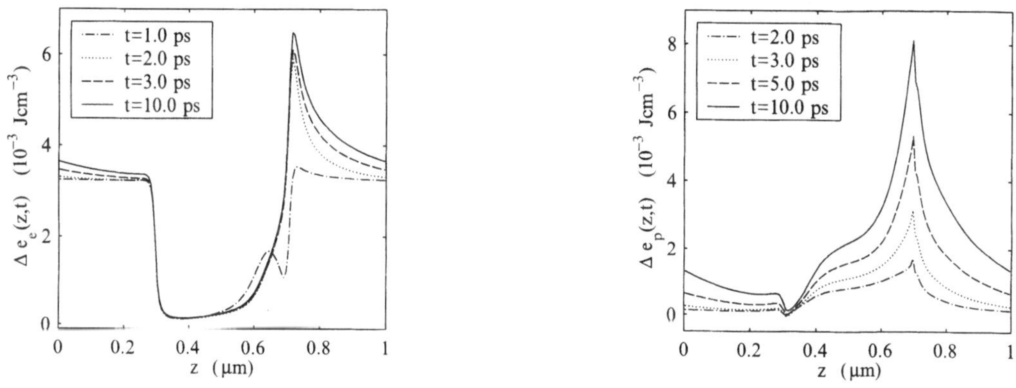

Fig. 3. Time evolution of the electron energy (left) and the change of the pop phonon energy Δe_p (right)

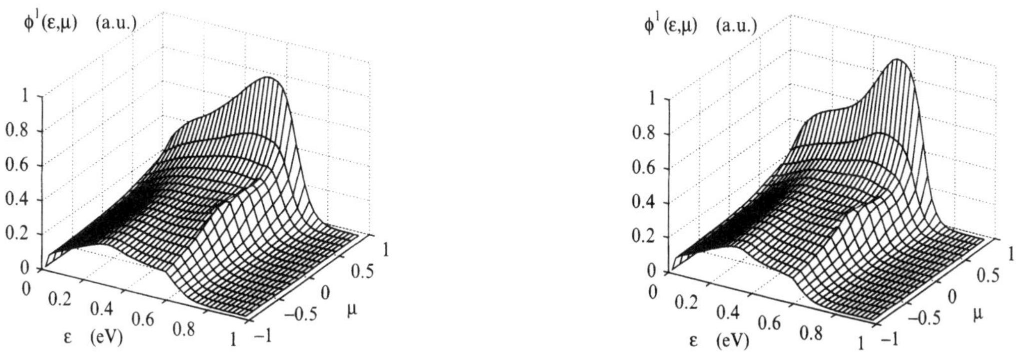

Fig. 4. Γ-electron distributions at $t = 10.0$ ps and $z = 0.65$ μm for hot pop phonons (left-hand plot) and for equilibrium pop phonons (right-hand plot) as function of ε and μ

(left-hand side) and equilibrium phonons (right-hand side). Both distribution functions show a far-from-equilibrium behavior. This proves the necessity of applying a kinetic description to investigate the transport properties of the considered InP diode. The steep decent of the Γ-valley distribution functions at $\varepsilon \approx 0.6$ eV results from the strong transfer of electrons to the L-valleys. The higher maximum value of the distribution function of electrons interacting with equilibrium phonons represents the most significant difference between the plots in Fig. 4.

5 Conclusion

The obtained results show that hot phonons lower the average electron velocity and increase the mean electron energy inside the channel of the diode. The maximal deviations reach 10 percent. The most significant changes are found for the population of the L-valleys. Non-equilibrium phonons increase the L-valley population up to 60 percent. The total electron density and the electric field are only slightly influenced by the hot phonons. Finally, our investigations prove that non-equilibrium phonon effects must be taken into account to simulate InP devices accurately.

6 Acknowledgement

This work has been supported by the Fonds zur Förderung der wissenschaftlichen Forschung, Vienna, under contract number P14669-TPH.

References

[MP01] Majorana A., Pidatella R.M.: A finite difference scheme solving the Boltzmann-Poisson system for semiconductor devices. J. Comp. Phys., **174**, 649–668 (2001)

[CGMS03] Carrillo J.A., Gamba I.M., Majorana A., Shu C.: A WENO-solver for the transients of Boltzmann-Poisson system for semiconductor devices: performance and comparisons with Monte Carlo Methods. J. Comp. Phys., **184**, 498–525 (2003)

[ES03] Ertler C., Schürrer F.: A multicell matrix solution to the Boltzmann equation applied to the anisotropic electron transport in silicon. J. Phys. A: Math. Gen., **36**, 8759–8774 (2003)

[GS04] Galler M., Schürrer F.: A deterministic solution method for the coupled system of transport equations for the electrons and phonons in polar semiconductors. J. Phys. A: Math. Gen., **37**, 1479–1497 (2004)

[AS04] Auer Ch., Schürrer F., Koller W.: A semi-continuous formulation of the Bloch-Boltzmann-Peierls equations. SIAM J. Appl. Math., **64**, 1457–1475 (2004)

[LOC94] Liu X., Osher, S., Chan, T.: Weighted essentially non-oscillatory schemes. J. Comp. Phys., **115**, 200–212 (1994)

[SO88] Shu C., Osher, S.: Efficient implementation of essentially non-oscillatory shock-capturing schemes. J. Comp. Phys., **77**, 439–471 (1988)

Modeling and Simulation for Thermal-Electric Coupling in an SOI-Circuit

A. Bartel[1] and U. Feldmann[2]

[1] Lehrstuhl für Angewandte Mathematik/Numerische Analysis, Universität Wuppertal, D-42097 Wuppertal, Germany, bartel@math.uni-wuppertal.de
[2] Infineon Technologies AG, Balanstr. 73, D-81541 München, Germany, Uwe.Feldmann@infineon.com

Abstract In Silicon on Insulator (SOI) circuits thermal effects are of particular relevance due to restricted cooling via the substrate. The accompanied thermal network enables to include one dimensional heat conduction effects to the lumped electric network equations. In this framework for thermal-electric coupling, we model an industrial-like benchmark based on a simple ring oscillator circuit. This is to picture abstractly the on-chip behavior by simple means. After a rough description of an applied simulation technique multirate results for this example are given. These underline the huge saving potential according to the widely separated timescales of electric networks and heat conduction.

Key words: Electric circuit simulation, heat conduction, temperature dependence, parabolic partial differential algebraic equations.

1 Introduction

Due to miniaturization of devices and the increasing package densities, the dissipated power per chip area increases. Since semiconducting devices and interconnects in chip technology are temperature sensitive and even may be destroyed in hot spots, it is important to include the heat evolution into circuit analysis. Usually, this is done by thermal networks, which include the local temperature and its cooling towards environment. Since this cooling is limited especially in SOI (Silicon on Insulator) technologies, the heat conduction phenomenon becomes more pronounced. And this is aggravated by decreasing spacing between devices in higher package densities.

Therefore, an approach to include one dimensional thermal effects is introduced in the next section. It is the so-called accompanying thermal network (AN), which completes the network equations. As a benchmark example we then discuss a ring oscillator circuit and its thermal description using the AN. The fourth section presents numerical results, which are computed by exploiting the multirate setting of this system. Finally we draw some conclusions.

2 Mathematical Model

To extend the more or less standard lumped thermal approach for modeling heat conduction, a spatial model needs to be provided. As a first step towards full 3D-modeling, the accompanied thermal network (AN) [BaGü03, Ba03] includes heat conduction in one spatial dimension. These can be macro-structures on chip into any preferred direction of conduction: for instance, cell arrays, interconnects, or others. Also one can picture this model as order reduced real world, where a designer has specified macro-structures. In power circuits with a relatively small number of power dissipators, such an order reduction can be performed on the set of equations, see [WCSW97].

To have both lumped and spatial 1D-thermal element models, the AN needs to couple both descriptions. The interface is established by a flux condition, which enables the use of standard schemes for setting up equations within this formulation. For the overall thermal-electric problem, we obtain the system of equations in Box 1. The concurring systems are the following: first we have the common electric network equations [Ti99] in terms of the node voltages $\mathbf{u}$ and branch currents $\jmath_L, \jmath_V$ (with topology $\mathbf{A}$); the second part, the AN, is basically a coupled system of energy balance equations for both types of elements; these use 1D and 0D temperatures $\mathbf{T}$ and $\widehat{\mathbf{T}}$, respectively. The temperatures enter the network equations via

Box 1: COUPLED THERMAL-ELECTRIC PROBLEM.

electric network: (DAE-IVP) $\mathbf{A} = (\mathbf{A}_C, \mathbf{A}_G, \mathbf{A}_L, \mathbf{A}_S, \mathbf{A}_I, \mathbf{A}_V)$

$$\begin{aligned} \mathbf{0} &= \mathbf{A}_C \dot{\mathbf{q}}(\mathbf{A}_C^\top \mathbf{u}(t)) + \mathbf{A}_G \mathbf{r}(\mathbf{A}_G^\top \mathbf{u}(t), \boxed{\mathbf{T}^{\text{br}}, \mathbf{F}}) + \mathbf{A}_L \jmath_L(t) \\ &\quad + \mathbf{A}_S \jmath(\mathbf{A}_C^\top \mathbf{u}, \boxed{\mathbf{T}^{\text{br}}, \mathbf{F}}) + \mathbf{A}_I \imath(t) + \mathbf{A}_V \jmath_V(t) \\ \mathbf{0} &= \dot{\boldsymbol{\phi}}(\jmath_L(t)) - \mathbf{A}_L^\top \mathbf{u}(t) \\ \mathbf{0} &= \mathbf{A}_V^\top \mathbf{u}(t) - \boldsymbol{v}(t) \end{aligned} \tag{1a}$$

$$\text{(IV)} \qquad \mathbf{x}(t_0) = (\mathbf{u}_0, \jmath_{L,0}, \jmath_{V,0})^\top \tag{1b}$$

coupling interface: $(\boldsymbol{\lambda}_P = \boldsymbol{\lambda}_P(\mathbf{u}, \jmath_L, \jmath_V))$

$$(\mathbf{P}_{\text{tr}}, \mathbf{P}_{\text{lp}})^\top = \mathbf{P} = \text{diag}(\mathbf{K}\boldsymbol{\lambda}_P)\mathbf{A}_{\text{tp}}^\top \mathbf{u}, \quad \mathbf{F} = \mathbf{F}(\mathbf{T}), \quad \mathbf{T}^{\text{br}} = \mathbf{Q}^\top \widehat{\mathbf{T}} \tag{1c}$$

thermal network: (PDAE-BIVP) $i = 1, \ldots, m$

$$\text{(1D)} \quad M_i \dot{T}_i(x,t) = \partial_x(\Lambda_i \partial_x T_i(x,t)) - \gamma S_i \cdot (T_i(x,t) - T_{\text{env}}) + \widetilde{P}_i(x,t) \tag{1d}$$

$$\widetilde{P}_i(x,t) = \sum_{k=k_i}^{l_i} \boxed{P_{\text{tr},k}(t)} \cdot \frac{\widetilde{\rho}_k(x, T_i)}{R_k(t, T_i)}, \quad R_k = \int_0^1 \widetilde{\rho}_k(x, T_i(x,.))\,\mathrm{d}x \tag{1e}$$

$$\text{(0D)} \quad \widehat{\mathbf{M}}\dot{\widehat{\mathbf{T}}}(t) = \mathbf{A}_{\text{AN}} \begin{pmatrix} \Lambda(0)\partial_x \mathbf{T}(0,t) \\ -\Lambda(1)\partial_x \mathbf{T}(1,t) \end{pmatrix} - \gamma \widehat{\mathbf{S}}(\widehat{\mathbf{T}} - T_{\text{env}} \mathbb{1}_k) + \mathbf{Q} \boxed{\mathbf{P}_{\text{lp}}(t)} \tag{1f}$$

$$\text{(BC)} \quad \begin{pmatrix} \mathbf{T}(0,t) \\ \mathbf{T}(1,t) \end{pmatrix} = \mathbf{A}_{\text{AN}}^\top \widehat{\mathbf{T}}(t) \tag{1g}$$

$$\text{(IC)} \quad \mathbf{T}(x,0) = \mathbf{T}_0(x) \geq T_{\text{env}} \mathbb{1}_m \qquad \widehat{\mathbf{T}}(0) = \widehat{\mathbf{T}}_0 \geq T_{\text{env}} \mathbb{1}_m \tag{1h}$$

parameters of the static part (using branch temperatures $\mathbf{T}^{\text{br}}$ and functionals $\mathbf{F}$ – $\mathbf{Q}$ identifies the according thermal 0D-unit for the thermally lumped electric elements); vice versa, the dissipated powers P of passive electric elements ($\mathbf{A}_{\text{tp}}$) yield source terms for the AN. – The existence of solutions of this system and its well-posedness will be the topic of [BaGJ04].

3 AN for SOI Circuits

In the following, we construct a benchmark to reflect the complex on-chip behavior of SOI circuits in a simplified way. One main component is a standard ring oscillator, which is composed CMOS-inverters in SOI technology as depicted in Fig. 1. In the left-hand part, several inverters are connected in feed back configuration to form an autonomous oscillator. This unit drives a cascade of inverter stages in the right-hand part. In total, this configuration may model an inner chip signal-flow (e.g. a critical path): part one provides the logic or analog functionality, while part two serves for signal amplification or for driving output signals. For simplicity, the involved CMOS-inverters are here described using the standard level-1 transistor model

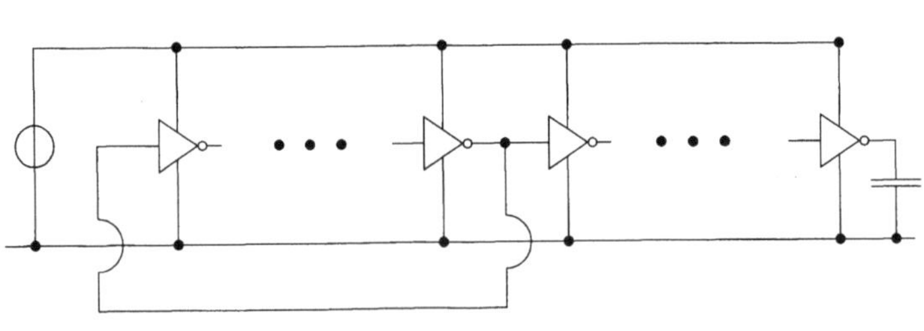

Fig. 1. Industrial benchmark – electric network

of Shichman-Hodges [ShHo68]. Consequently, the benchmark circuit in Fig. 1 represents a network of basic elements, where semiconducting devices are replaced by capacitances, diodes and controlled current sources.

Here the major temperature T dependence is given by the mobility of charge carriers. It enters the system by the scaling factor β of the controlled current source modeling the transistor channel between drain and source

$$\imath_{ds}(u_{gate}, u_{source}, u_{drain}, u_{bulk}, T) = \beta(T) \cdot \hat{\imath}_{ds}(u_{gate}, u_{source}, u_{drain}, u_{bulk}),$$

(with $u_{gate}, u_{source}, u_{drain}, u_{bulk}$: potential of gate, source, drain, bulk node)

$$\beta = \mu(T) \cdot C'_{ox} W/L,$$

where mobility μ strongly depends on device temperature T; W and L refer to the transistor's width and length, and C'_{ox} is the capacitance per unit area for the oxide layer between gate and channel (see Box 2). Now, mobility decreases with temperature nonlinearly, which can be approximated as [MaAn93]

$$\mu(T) = \mu(300\,\mathrm{K}) \left(\tfrac{T}{300\,\mathrm{K}}\right)^{-1.5}.$$

In this example, the concurring p- and n-type devices (in CMOS technology) are electrically separated in one-dimensional arrays. Since the thermal and electric insulation comes along with each other, we have in first order only a thermal link for transistors of the same type. Heat transfer to the bottom of the chip is small here due to high thermal resistance of the insulating oxide layer in SOI technology. Therefore the AN for this example consists of two decoupled 1D-lines following the arrays of n- and p-type transistors. The situation is sketched in Fig. 2; the dotted lines signify the electrical connection to recognize the layout of this benchmark circuit. The devices' main currents pass just below the gates, through the channel area. There the main power is dissipated. Since the driver-units are scaled to amplify electric signals, we expect a large heat production there. This will further heat the remaining circuit, and cause a signal delay in the oscillatory part.

For simplicity, we consider a reasonably sized test circuit with only three inverter stages in the oscillator and a cascade of five bootstrap inverters. The latter are scaled to drive the load capacitance (see Fig. 1). The scaling is applied to the width for both n- and p-type transistors and takes the following values from left-to-right: 1, 2, 5, 10, 25.

To form a thermal-electric coupling the geometric data of the 1D-line are necessary. Here we assume that successive devices are spaced by a distance of $4\,W_n (= 2\mu\mathrm{m})$, where W_n is the width of n-channel device in the oscillator (see Box 2). Oscillator and driver unit are separated by an additional spacing of distance $8\,W_n$.

In this setup, the driver stages exhibit a thermal 1D-extension, which excites the ring oscillator stages. Therefore the formation of a special 0D-unit is inappropriate and we embed the driver stages to the 1D-lines. Consequently, the two 1D-lines in the AN have attached artificial 0D-units [Ba03], which form a zero flux condition (von-Neumann BC).

For the electric-to-thermal coupling, we equally distribute the (lumped) dissipated power on the respective transistor's location (width): for type $i \in \{\mathrm{n, p}\}$ and the kth device in line, we have

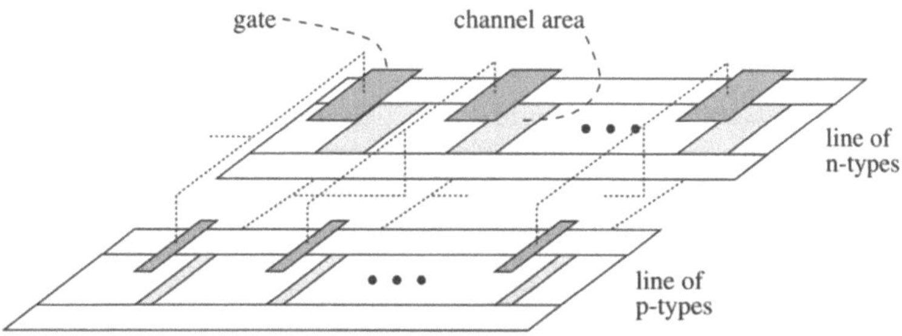

Fig. 2. Thermal 1D-lines

Box 2:		TYPICAL TRANSISTOR PARAMETERS.
Geometric parameters:		
width:	$W_n = 0.5\,\mu\text{m}$	$W_p = 1.5\,\mu\text{m}$
length:	$L = 0.2\,\mu\text{m}$	
thickness channel-gate oxide:	$t_{ox} = 3.0\,\text{nm}$	
thickness buried oxide:	$t_{box} = 34.5\,\text{nm}$	
Model parameters:		
mobility:	$\mu_n(300) = 400\,\text{cm}^2/(\text{Vs})$	$\mu_p(300) = 130\,\text{cm}^2/(\text{Vs})$
threshold voltage:	$V_{th}^n = +250\,\text{mV}$	$V_{th}^p = -250\,\text{mV}$
overlap capacitance per width:	$C'_{ov} = 0.4\,\text{nF/m}$	
junction capacitance per width:	$C'_j = 2.0\,\text{nF/m}$	
saturation current:	$I_S = 1.0 \cdot 10^{-15}\,\text{A}$	

$$P_{i,k} = i_{ds}^{i,k} \cdot u_{ds}^{i,k}$$

with the corresponding 1D-indicator function $\widetilde{\rho} = \chi_{i,k}$ (this simply locates the device in 1D-line segment). In this way, we obtain the continuous thermal model, equation (1d), where the local power source term is given as

$$\widetilde{P}_i(x,t) = \sum_{k=1}^{s} P_{i,k}(t) \cdot \frac{\chi_{i,k}(x)}{R_{i,k}}, \qquad R_{i,k}(= W_{i,k}) = \int_0^1 \chi_{i,k}(x)\,\mathrm{d}x$$

for the two lines ($i \in \{\text{n, p}\}$) and a total number of $s = 8$ inverter stages. In turn, the lumped temperatures can be obtained by averaging the temperature with the indicator function as weight (thermal-to-electric coupling). Thus mobilities are obtained by evaluation at the derived temperature.

The second source term in (1d) is cooling. It is proportional to the local surface (perimeter) S. Here several sides of the 1D-line are covered by oxide limiting the heat flow to environment. But there are additional electric contacts at the devices. These metal interfaces can be treated as additional surfaces whose transmission coefficient γ is several orders of magnitude larger.

The next step is discretization. A simple and applied choice are finite volumes; these fit to the AN-setting: each device and each interspacing gets an own cell, cf. Fig. 2, giving a rough scale of thermal resolution. Here the device's total length is condensed to a point in the 1D-line, the width is represented by the according line segment. Parameters for this benchmark circuit are summarized in Box 3.

Box 3:	RING OSCILLATOR WITH THERMAL FEEDBACK.
Geometry	1D-quantities
load capacitance: $C_L = 200\,\text{fF}$	heat mass (Si): $M = 3.5\,10^{-8}\,\text{J/m K}$
surface: $\gamma S = 2.4\,\mu\text{W/mK}\ (+4.8\,10^{-2})$	conductivity: $\Lambda = 3.18\,10^{-12}\,\text{Wm/K}$

4 Simulation Results

To enable a simple inclusion and coupling to a circuit simulator, we consider for the semi-discretized system a co-simulation approach. The spatially discrete system does not suffer from contraction conditions [ArGü00] due to DAE effects [Ba03], and enables a multirate procedure, where iteration is not necessary. To this end, an energy coupling is formed [DeTü99, Ba03]: additionally the dissipated powers are integrated together with the network equations

$$\dot{E} = i_{ds} \cdot u_{ds}, \quad E(0) = 0$$

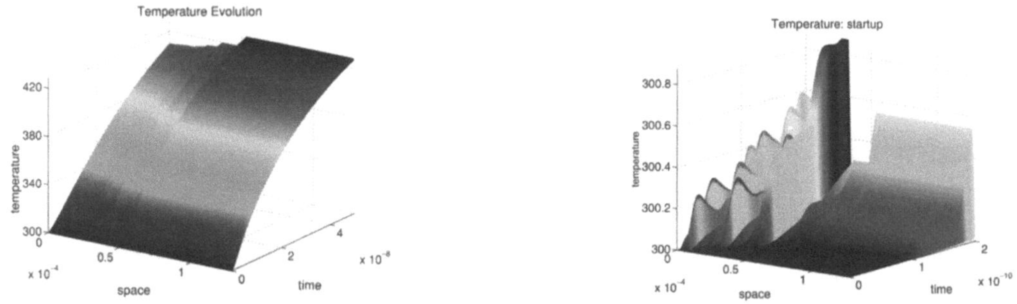

Fig. 3. Temperature distribution $[0.\,\mathrm{s}, 50.\mathrm{ns}]$ (left), startup $[0.\,\mathrm{s}, 0.2\mathrm{ns}]$ (right)

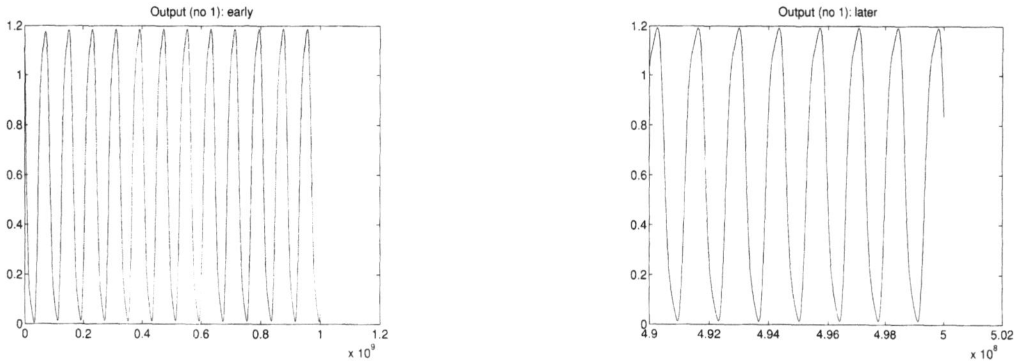

Fig. 4. Output inverter (no. 1): $[0.\mathrm{s},\ 1.\mathrm{ns}]$ (left), $[49.\mathrm{ns},\ 50.\mathrm{ns}]$ (right)

over a communication step H ($[0, H]$, for simplicity). Here temperature is kept fix (or can be extrapolated). In a second step, this energy is transmitted to the AN and is equally distributed in time during the computation of the temperature. Due to on-chip dimensions, capacitances in the network equations are tiny and yield small time constants. Thus the network-power equations form a multiscaled subsystem, and rescaling is necessary.

Next, we discuss simulation. The ring oscillator begins its autonomous oscillation fast (its shape depends on the parameters). With varying temperature, signals will traverse the circuit with different speed due to the temperature dependence of mobility. – This can be seen in the output signals in Fig. 4: a lower temperature at an early time and a higher temperature at a later time. Clearly, the two signals diverge.

For MATLAB simulations, we scaled both thermal mass (M) and heat conduction (Λ) by a factor of $1/50$ and 10, respectively. Of course, this gives unphysical parameters, but the thermal-electric system gets tighter coupled. Therefore it is harder for a multirate scheme to work properly. On the other hand, we need smaller simulation times to recognize thermal effects. Thus the scaling is good for demonstration purposes. Here the time window $[0\mathrm{s}, 50\mathrm{ns}]$ is considered. After semi-discretization, MATLAB routine `ode15s` was employed for time-integration. First, the system was integrated in a singlerate mode, i.e. electric and thermal part were solved simultaneously using the same timestep. Figure 3 depicts the overall temperature evolution and a startup-phase. Furthermore, Fig. 4 gives the output signal at the first inverter, showing the temperature dependence of the electric signal. Inclusion of temperature effects is indeed necessary in simulation, since their impact on signal delays is significant and may even cause malfunctions of the chip.

In Fig. 3 (right), we see the development of the temperature in our benchmark at a very early time. We can precisely identify the devices in our line, and we recognize the scaling and spacing of the devices. Actually, the larger p-channel devices are depicted, here.

In a second step electric and thermal subsystem were solved in a multirate co-simulation, using different timesteps for each subsystem, as described above. Here we have chosen a communication step of $0.2\,\mathrm{ns}$.

Table 1. Results: Singlerate vs. Multirate

		steps	comm.-steps
single-rate	(total)	70 924	–
multi-rate	(network)	71 630	125
co-simulation	(heat)	129	

At the final time, we obtained a very good agreement with the singlerate solution: the error is less than $0.16\,\mathrm{K}$ (thus the relative error is less than 10^{-3}). Since model evaluation is the most costly part in circuit simulation, we contrast the number of time steps for both algorithms in Table 1. Indeed multirating is achieved. Notice, per communication step there is only about one step of the AN solver necessary; actually, the remaining four steps are all spent in the startup phase. Therefore due to averaging an order two method based on the mid-point rule can be constructed [Ba03]. However, recall that this multirating has its price. Additionally to the electric network, the energy equations have to be integrated. Fortunately, there was no iteration necessary for getting these accurate results.

For a validation of our 1D-thermal approach, numerical simulations of the thermal-electric system have also been executed in a 3D-setting and compared to corresponding 1D-results. These tests were performed with Infineon's circuit simulator TITAN, using a coarse spatial discretization with an equivalent thermal network and running TITAN in an electrothermal interaction mode. The temperature difference between 3D- and 1D-solutions was at most $10K$ (about 8% error in centigrade). Thus, regarding first order thermal effects, a 1D-coupling is valuable (for this accuracy), and, of course, better than a pure 0D-thermal set-up. Furthermore, it is also much more efficient than the full 3D simulation.

5 Conclusions

We have addressed the multirate behavior of the thermal-electric problem in our benchmark circuit. Since the discretized coupled system with energy coupling does not suffer from additional contractivity conditions in co-simulation, an adapted multirate strategy is applicable. Numerical tests for this benchmark example verify that indeed multirate is achieved.

Acknowledgement. This work is part of the project "Numerische Simulation von elektrischen Netzwerken mit Wärmeleitungseffekten" (03GUM3W1) supported by the German Federal Ministry of Education and Research.

References

[ArGü00] Arnold, M., Günther, M.: Preconditioned dynamic iteration for coupled differential algebraic systems. BIT, **41**:1, 1–25 (2000)

[Ba03] Bartel, A.: Partial Differential-Algebraic Models in Chip Design — Thermal and Semiconductor Problems. PhD Thesis, TU München (2003)

[BaGü03] Bartel, A., Günther, M.: From SOI to abstract electric-thermal-1D multiscale modeling for first order thermal effects. Math. Comput. Modell. Dynam. Syst., **9**:1, 25–44 (2003)

[BaGJ04] Bartel, A., Günther, M., Jüngel, A.: Existence analysis for a thermal-electric problem. In preparation

[DeTü99] Deml, Ch., Türkes, P.: Fast Simulation Technique for Power Electronic Circuits with Widely Different Time Constants. IEEE Transactions on Industry Applications **35**:3, 657–662 (1999)

[MaAn93] Massobrio, G., Antognetti, P.: Semiconductor Device Modeling with SPICE, 2nd ed., McGraw-Hill, New York (1993)

[ShHo68] Shichman, H. and Hodges, D. A.: Insulated-gate field-effect transistor switching circuits. IEEE J. Solid State Circuits **SC-3**, 285–289 (1968)

[Ti99] Tischendorf, C.: Topological index calculation of differential-algebraic equations in circuit simulation. Surv. Math. Ind. **8**, 187–199 (1999)

[WCSW97] Wünsche, W., Clauß Ch., Schwarz, P., Winkler, F.: Electro-Thermal Circuit Simulation Using Simulator Coupling. IEEE Trans. VLSI Syst. **5**:3, 277–282 (1997)

A Staggered ALE Approach for Coupled Electromechanical Systems

M. Greiff , U. Binit Bala and W. Mathis

Institute of Electromagnetic Theory and Microwave Technique, University of Hannover, Appelstr. 9A, 30167 Hannover, Germany, {mgre,bala,mathis}@tet.uni-hannover.de

Abstract In this paper several numerical methods for the simulation of a EFM (Electrostatic Force Microscope) are presented. An approach to couple these methods is proposed in order to improve the modeling.

1 Introduction

For modeling and simulating Micro-Electro-Mechnical Systems (MEMS), multi physics aspects must be taken into consideration. From the numerical point of view additional problems arise since frequently we are confronted with multi-scale problems. Therefore advanced numerical methods have to be applied. An interesting example for an MEMS is the so-called atomic force microscope (AFM) which can be used for scanning samples with nearly atomic resolution. For a complete investigation of the AFM, quantum mechanical and classical effects have to be considered, but in some cases the quantum mechanical effects can be neglected. For instance the Kelvin force microscope (EFM) is used at a relatively large distance from the sample [MWM02]. Therefore the interaction between the probe and sample is mainly determined by the Coulomb force. Several approaches to calculate the electric field in order to model this interaction have been made, such as in [JLHS] the author carries out a multipole expansion by using the program MMP. In this paper we will present a concept for physical and numerical modeling of EFMs which can be extended to other types of AFMs. This method takes into consideration the classical interaction of the cantilever tip with the sample surface. The goal is to develop a simulation tool which can be used for the design of AFM probes and for the interpretation of measurement results.

2 A Modeling Concept for the EFM

The EFM in non-contact mode is mainly used for scanning surfaces holding an electric potential or charge distribution. For our investigations the distance between the probe and sample is assumed to be relatively large. Therefore all other forces can be neglected due to the significantly larger influence of the electrostatic force. Nevertheless, many aspects must be taken into account to develop an accurate model for the EFM. Firstly one has to deal with the coupled nature of the problem. Any variation of the electric field will change the forces acting on the cantilever, thereby causing the cantilever to move, and effecting a variation in the electric field and vice versa. Secondly since the tip is very small compared to the cantilever, this multi-scale aspect has to be considered in the model. The goal of this project is to create an algorithm that is able to simulate the EFM and takes these aspects into account.

For modeling the EFM it is convenient to partition it to its different physical aspects and calculate them separately. Therefore they have to be coupled to each other during the calculation.

For calculating the electrostatic field $\mathbf{E}$ in an uncharged region G bounded by $\partial\mathbf{G} = \partial\mathbf{G_D} \cup \partial\mathbf{G_N}$ (Fig. 1), Laplace's equation $\Delta\varphi = 0$ has to be solved using the Dirichlet boundary conditions on $\partial\mathbf{G_D}$ and the Neumann boundary conditions on $\partial\mathbf{G_N}$ [Zhou93]. The electrostatic field $\mathbf{E}$ is determined by $\mathbf{E} = -\text{grad}\,\varphi$. In this paper some numerical methods for the calculation of the electric field will be presented.

In the example given in this paper the cantilever is assumed to hold the potential 1V while the sample's potential is 0V. On ∂G_N the normal component of the electric field is assumed to be zero. From the definition of the uniform load using the Maxwell stress tensor the following equation can be found.

Using Eq. 1 and the Maxwell stress tensor $\mathbf{T}_e$ [HYT00] the uniform load $\mathbf{f}$ which acts on the cantilever (width $= z_0$) can be found.

$$\mathbf{f} = \int_{z_0} \mathbf{T}_e \, \mathbf{dz} = \int_{z_0} \begin{pmatrix} \varepsilon(E_x^2 - \frac{1}{2}\|\mathbf{E}\|^2) & \varepsilon E_x E_y & \varepsilon E_x E_z \\ \varepsilon E_x E_y & \varepsilon(E_y^2 - \frac{1}{2}\|\mathbf{E}\|^2) & \varepsilon E_y E_z \\ \varepsilon E_x E_z & \varepsilon E_y E_z & \varepsilon(E_z^2 - \frac{1}{2}\|\mathbf{E}\|^2) \end{pmatrix} \mathbf{dz} \tag{1}$$

For the computation of the cantilever deflection $u(t)$ the y component of the uniform load f_y can be used in a beam model.

$$\frac{\partial^2}{\partial x^2}\left(EI\frac{\partial^2 u(t)}{\partial x^2}\right) + \rho A \frac{\partial^2 u(t)}{\partial t^2} = f_y(x), \qquad \begin{aligned} u(0,t) &= u_x(0,t) = 0 \\ u_{xx}(L,t) &= u_{xxx}(L,t) = 0 \end{aligned} \tag{2}$$

Here E is the elastic modulus, I is the moment of inertia, ρ is the mass density and A is the cross sectional area of the cantilever.

During the treatment of the equations given in the last section one has to deal with different kinds of problems. Therefore it is convenient to split the calculation domain into several parts each of which is calculated using a different numerical method (Fig. 3). In region 2 a versatile method such as FEM has

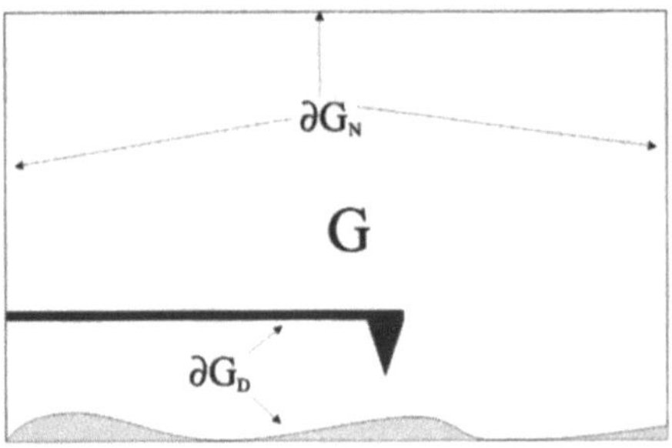

Fig. 1. Electrostatic calculation domain

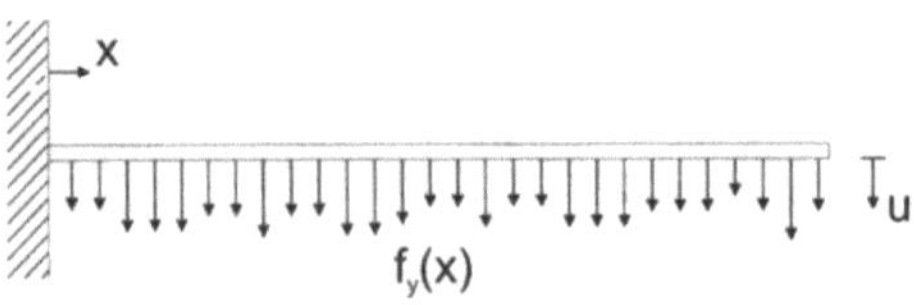

Fig. 2. Beam model of the cantilever

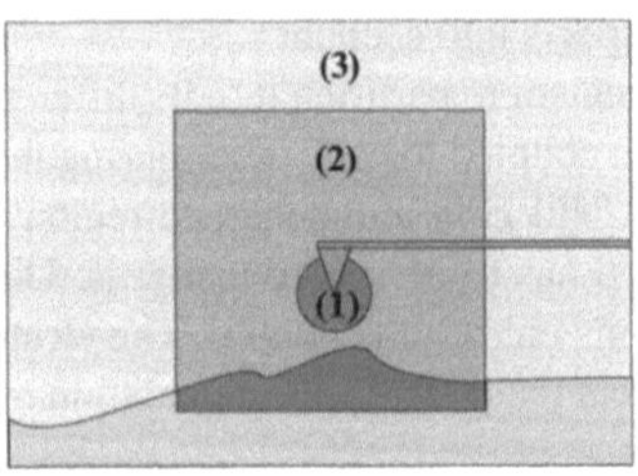

Fig. 3. 2D model including various numerical methods

to be applied because one has to deal with different materials and charge distributions in the sample. For modeling the high values of the electric field near the tip (region 1) more accurately the Ritz-Galerkin method (RGM) is used while the long distance interaction (region 3) will be treated by the boundary element method (BEM). During the calculation of the electric potential the methods mentioned above have to be coupled to each other.

Not yet taking into account possible charge distributions in region 2 (Fig. 3) the FEM formulation of the setup leads to the following system of equations.

$$\sum_j K_{ij}\varphi_j + \sum_n K_{in}\varphi_n = 0 \tag{3}$$

$$K_{ik} = \int_\Omega \left[\frac{\partial \psi_i}{\partial x}\left(\varepsilon \sum_{k=1}^{n} \frac{\partial \psi_k}{\partial x}\right) + \frac{\partial \psi_i}{\partial y}\left(\varepsilon \sum_{k=1}^{n} \frac{\partial \psi_k}{\partial y}\right)\right] d\Omega \tag{4}$$

Where φ_j are the values of the electric potential to be calculated and φ_n are the potential values on the nodes with Dirichlet boundary condition.

Starting from the two dimensional Laplace equation in cylindrical coordinates and using the separation approach one can find the following term for the electric potential φ in region 1 [Jack75].

$$\varphi(\rho,\phi) = \varphi_0 + \sum_{m=1}^{\infty} a_m \rho^{\frac{m\pi}{\beta}} \sin(\frac{m\pi\phi}{\beta}) \tag{5}$$

For the derivation of Eq. 5 the whole tip was assumed to hold a constant electric potential φ_0.

To compute Eq. 5 numerically the number of coefficients a_m is limited by the number and arrangement of the points at which the potential is known. In order to find the unknown coefficients a_m Eq. 5 is applied to the points where the potential is known.

$$\sum_{m=1}^{m_{max}} \left(\rho_j^{\frac{m\pi}{\beta}} \sin(\frac{m\pi\phi_j}{\beta})\right) a_m = \varphi_j - \varphi_0, \qquad j = 1,2,..,N \tag{6}$$

This over determined linear system of equations can be solved by standard methods.

To couple the RGM to the FEM Eq. 6 is applied to the coupling points and used in Eq. 3. This leads to

$$\sum_j K_{ij}\varphi_j + \sum_l K_{il}\left(\varphi_0 + \sum_{m=1}^{m_{max}} a_m \rho_j^{\frac{m\pi}{\beta}} \sin(\frac{m\pi\phi_j}{\beta})\right) = \sum_n K_{in}\varphi_n \tag{7}$$

where l are the coupling nodes and n are the nodes with Dirichlet boundary condition. The sum j includes the nodes that are neither coupling nodes nor nodes with Dirichlet boundary conditions. Figure 4 shows the resulting electric potential and field ($m_{max} = 9$).

The BEM formulation of the setup leads to the following set of equations

$$\sum_{i=1}^{M}\sum_{j=1}^{N} H^{ij}\varphi^j = \sum_{i=1}^{M}\sum_{j=1}^{N} G^{ij} q^j \tag{8}$$

$$H^{ij} = -\frac{1}{2\pi}\frac{l_e}{2}\frac{(\underline{x}_j - \underline{\xi}_i)\cdot \underline{n}}{|\underline{x}_j - \underline{\xi}_i|^2} \tag{9}$$

$$G^{ij} = -\frac{1}{2\pi}\frac{l_e}{2} \ln|\underline{x}_j - \underline{\xi}_i| \tag{10}$$

where N and M are the total number of elements and nodes, respectively, φ is the potential, q is the potential derivative, l_e is the length of the element, i and j are node indices. Each boundary node must have either a Dirichlet or Neumann boundary condition. Applying these boundary conditions in Eq. 8 the potential and potential derivative at the boundary nodes can be obtained. By using these values the internal

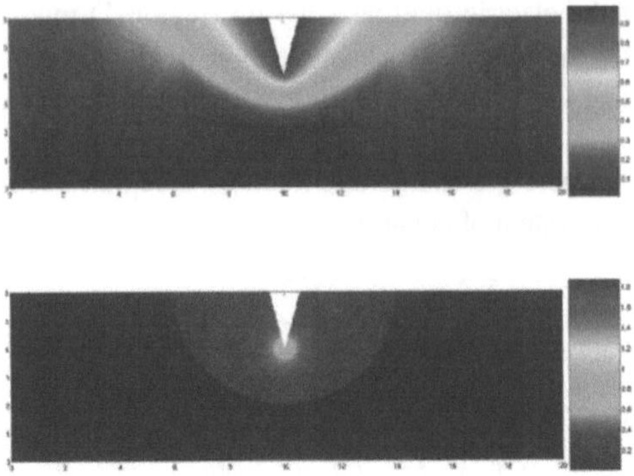

Fig. 4. Simulated potential and electric field by coupled FEM-RGM

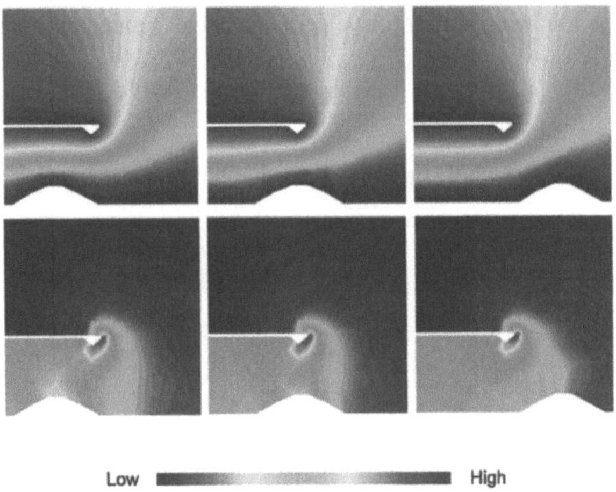

Fig. 5. Simulated potential and electric field by BEM

potential at all internal points i can be obtained from the following equation and the resulting simulation of EFM is shown in Fig. 5.

$$\sum_{i=1}^{E} \varphi^i = \sum_{i=1}^{E}\sum_{j=1}^{N} G^{ij} q^j - \sum_{i=1}^{E}\sum_{j=1}^{N} \hat{H}^{ij} \varphi^j \tag{11}$$

As BEM uses the fundamental solutions, it possesses improved accuracy in the calculation of electric field and exterior problems. But when the observation point comes very near to the boundary, some errors occur as the boundary integrals tend closer to singularity. For this reason a high potential is observed near the tip. Such error doesn't occur when the simulation of EFM is done by FEM. So for more accurate simulation results the use of another numerical method near the tip would be preferable. In this way an improved accuracy can be obtained in both regions, near the tip and far away. For this purpose the necessary coupling equation after implying coupling conditions and matrix M is

$$\begin{bmatrix} [\mathbf{H}]^{II} & [\mathbf{H}]^{IB} & -[\mathbf{G}]^{II} & -[\mathbf{G}]^{IB} & 0 \\ [\mathbf{H}]^{BI} & [\mathbf{H}]^{BB} & -[\mathbf{G}]^{BI} & -[\mathbf{G}]^{BB} & 0 \\ [\mathbf{K}]^{II} & 0 & \mathbf{M} & 0 & [\mathbf{K}]^{IF} \\ [\mathbf{K}]^{FI} & 0 & 0 & 0 & [\mathbf{K}]^{FF} \end{bmatrix} \begin{Bmatrix} \{\varphi\}^I \\ \{\varphi\}^B \\ \{q\}^I \\ \{q\}^B \\ \{\varphi\}^F \end{Bmatrix} = \begin{Bmatrix} 0 \\ 0 \\ 0 \\ 0 \\ 0 \end{Bmatrix} \tag{12}$$

where I indicates the coupling nodes, F and B indicate the nodes which are in the FEM and BEM region respectively.

3 Numerical Calculation of the Cantilever Deflection

The application of Eq. 1 to calculate the force acting on the i-th mesh element of the cantilever leads to

$$\mathbf{F}_i = \begin{pmatrix} \varepsilon(E_x^2 - \frac{1}{2}\|\mathbf{E}\|^2) & \varepsilon E_x\, E_y & 0 \\ \varepsilon E_x\, E_y & \varepsilon(E_y^2 - \frac{1}{2}\|\mathbf{E}\|^2) & 0 \\ 0 & 0 & 0 \end{pmatrix} \begin{pmatrix} y_{i+1} - y_i \\ -(x_{i+1} - x_i) \\ 0 \end{pmatrix} z_0 \tag{13}$$

where $(x_i, y_i, 0)$ is the position of the i-th node on the cantilever. The FEM approach using beam elements results in

$$\sum_j K_{ij} u_j + M_{ij} \ddot{u}_j = F_{i,y} \tag{14}$$

Here K_{ij} is the stiffness matrix, M_{ij} is the mass matrix and F_i is the load vector [Red93]. The deflections u_j are calculated using the Galerkin method.

4 ALE Implementations

During the FEM calculation described in the last section the whole domain and therefore the FEM mesh was assumed to be time independent. Since the scanning process of a EFM is dynamic, one has to deal with a moving geometry and therefore with moving boundaries. This brings up the problem that the mesh has to be changed during the calculation to fit the geometry. As a brute force method one could choose to call the mesh generator in each time step. In the approach presented here the mesh update is achieved by using the arbitrary Lagrangian Eulerian (ALE) method which means that the mesh is neither fixed in space (Eulerian) nor are all its nodes attached to the material (Lagrangian). In this work the mesh is modelled as a massless elastic which is deformed by the changing boundaries on $\partial\mathbf{G}$ (Fig. 1). Therefore in each time step the new positions of the mesh nodes are calculated by solving a vector Laplace equation for the mesh deflection. The solution is obtained by FEM [Red93]. In Fig. 6 the movement of the mesh can be observed. Since the governing equation for the electrostatic potential does not include any time derivatives, no modification of the FEM is necessary and ALE is reduced to only a mesh update method [BKM04].

The cantilever deflection obtained by using this approach shows the same behavior as in [Witt00] (Fig. 7).

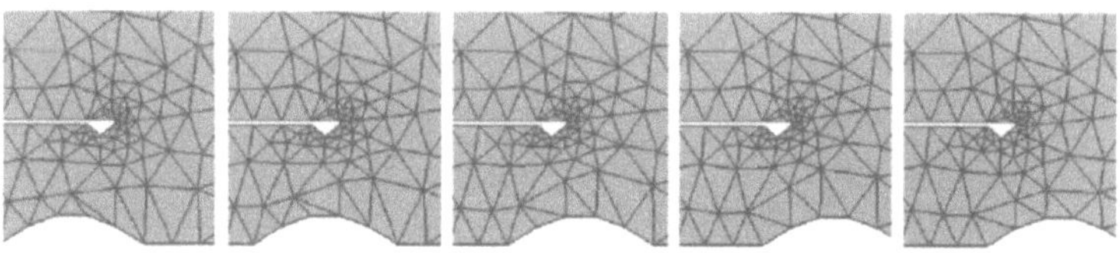

Fig. 6. ALE mesh update

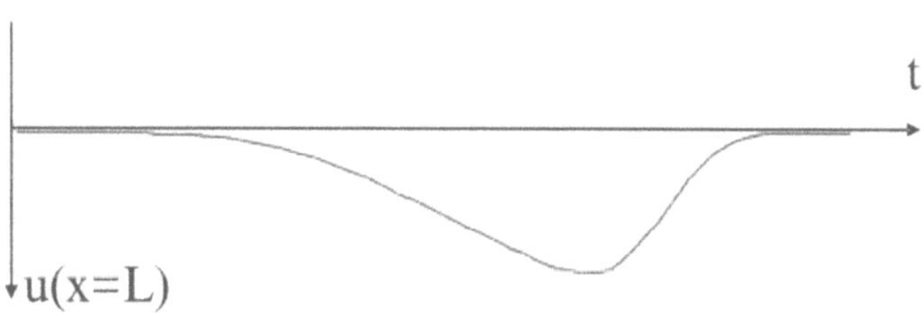

Fig. 7. Simulated cantilever deflection

5 Conclusion

In this paper a concept of physical and numerical modeling for a 2D EFM is presented. It is assumed that its interaction with the sample is determined by the Coulomb force. Some results obtained by using FEM and BEM are presented. An ALE approach is used for updating the mesh. In order to improve the results a coupling scheme for RGM and the other numerical methods mentioned above is proposed. A typical simulation result obtained by a coupled FEM-RGM is presented.

Acknowledgements The authors thank the DFG GRK 615 for their financial support. The authors also thank Ilker Basol and Dominik Mente for their input.

References

[BD89] Brebbia, C.A., Dominguez, J.: Boundary Elements An Introductory Course. McGraw-Hill Book Company (1989)

[BE04] Bhushan Editor: Handbook of Nano-technology. Springer (2004)

[BKM04] Belytschko, T., Kam Liu, W., Moran, B.: Nonlinear Finite Elements for Continua and Structures. Wiley, Sussex (2004)

[BW92] Beer, G., Watson, J.O.: Introduction to Finite and Boundary Element Methods for Engineers. Wiley (1992)

[GKW03] Gaul, L., Kögl, M., Wagner, M.: Boundary Element Methods for Engineers and Scientists. Springer Verlag (2003)

[HYT00] Hui, C.Y., Yeh, J.L., Tien, N.C. : Calculation of electrostatic forces and torques in MEMS using path-independent integrals. J. Micromech. Microeng. 10, 477-482 (2000)

[Jack75] Jackson, J.D.: Classical Electrodynamics. Wiley (1975)

[JLHS] Jacobs, H.O., Leuchtmann P., Homan O.J., Stemmer A.: Resolution and contrast in Kelvin probe force microscopy. Journal of Applied Physics Vol. 84, num. 3 (1998)

[MWM02] Morita, S., Wiesendanger, R., Meyer, E.: Noncontract Atomic Force Microscopy. Springer (2002)

[Red93] Reddy, J.N.: Finite Element Method. McGraw-Hill International Editions, Singapore (1993)

[Witt00] Wittpahl, V.: Entwicklung eines neuartigen elektrischen Rasterkraftmikroskopie-Tests zur quantitativen Bestimmung von Gleich- und Wechselspannungen bis über 100 GHz in integrierten mikroelektronischen Schaltungen. Dissertation, University of Duisburg, VDI Verlag (2000)

[ZCL87] Zi-CAi Li: A Nonconforming Combined Method for Solving Laplace's Boundary Value Problems with Singularities. Numerische Mathematik 49, 475-497, Springer-Verlag (1986)

[ZCL90] Zi Cai Li: Numerical Methods for Elliptic Problems with Singularities. World Scientific (1990)

[Zhou93] Zhou, P. : Numerical Analysis of Electromagnetic Fields. Springer Verlag (1993)

Orthogonalisation in Krylov Subspace Methods for Model Order Reduction

P. J. Heres[1] and W. H. A. Schilders[2]

[1] Eindhoven University of Technology, Department of Mathematics and Computer Science, PO Box 513, 5600 MB Eindhoven, The Netherlands, p.j.heres@tue.nl
[2] Philips Research Laboratories Eindhoven, Prof. Holstlaan 4, 5656 AA Eindhoven, The Netherlands

1 Introduction

The modelling of the EM behaviour of electronic structures nowadays involves a broad frequency range and coupling of analog and digital behaviour. Much research and increasing computational resources enabled the designers in the past decades to simulate complicated and large structures. One of the approaches to make this modelling feasible is Model Order Reduction. In this approach one tries to capture the essential features of a large model, into a smaller, a more easy to handle model. A wide range of different techniques has been proposed and investigated in the last few decades. Especially Krylov-subspace methods have proved themselves to be very suitable for this area of application (eg. [1], [3], [5] and [6]). Many of these methods guarantee preservation of passivity, which makes them even more interesting.

However, implementing the methods straightforwardly is not enough to make them applicable for real-life applications. In order to make the methods accurate, efficient and suitable for large systems, extra attention and mathematical knowledge is needed. In this paper we will focus on the orthogonalisation of the Krylov space, which is seen to be of importance. Special attention is paid to the orthogonalisation of a Block Krylov space. Also some directions to cheaply avoid parts of the redundancy in the Krylov space methods are pointed out in this paper.

2 Krylov subspace methods

Modelling of an electronic structure can lead to a Differential Algebraic Equation (DAE), which form now on will be considered in this form:

$$\begin{aligned}(\mathbf{C}\frac{d}{dt} + \mathbf{G})\mathbf{x}(t) &= \mathbf{B}_i\boldsymbol{u}(t)\\ \boldsymbol{y}(t) &= \mathbf{B}_o^T\mathbf{x}(t),\end{aligned} \tag{1}$$

where $\mathbf{C} \in \mathbf{R}^{n\times n}$,$\mathbf{G} \in \mathbf{R}^{n\times n}$, $\mathbf{B}_i \in \mathbf{R}^{n\times p}$ and $\mathbf{B}_o \in \mathbf{R}^{n\times p}$ In the very common case that $\mathbf{C}$ is singular this model is not an ODE, but a DAE. The models we consider here can be derived in several ways. It can for instance be a transmission line model, a PEEC model or an FDTD model with spatial discretizations. In general the matrices $\mathbf{G}$ and $\mathbf{C}$ are real and constant in time.

This system of equations can be transformed to the frequency domain with a Laplace transform:

$$\begin{aligned}(s\mathbf{C} + \mathbf{G})\boldsymbol{X}(s) &= \mathbf{B}_i\boldsymbol{U}(s)\\ \boldsymbol{Y}(s) &= \mathbf{B}_o^T\boldsymbol{X}(s)\end{aligned} \tag{2}$$

When the state space vector in frequency domain $\boldsymbol{X}(s)$ is eliminated, a transfer function is obtained:

$$\mathbf{H}(s) = \mathbf{B}_o^T(\mathbf{G} + s\mathbf{C})^{-1}\mathbf{B}_i, \tag{3}$$

$\mathbf{H}(s) \in \mathbb{C}^{p \times p}$. This transfer function gives a direct relation between input and output of the system and is therefore a compact description of the system behaviour in the frequency domain.

Model Order Reduction methods attempt to approximate the behaviour of the system with a smaller model. A Krylov-subspace method generates a Krylov subspace based on some input matrix $\mathbf{B}$ and some generating matrix $\mathbf{A}$:

$$\mathcal{K}_q(\mathbf{B}, \mathbf{A}) = [\mathbf{B}, \mathbf{AB}, \ldots, \mathbf{A}^q\mathbf{B}] \tag{4}$$

The actual definition of $\mathbf{B}$ and $\mathbf{A}$ depends on the method of choice. For instance, in the method Laguerre-SVD [3] for some choice of $\alpha \in \mathbf{R}$, the input matrix is defined as:

$$(\mathbf{G} + \alpha\mathbf{C})^{-1}\mathbf{B}_i \tag{5}$$

and the generating matrix is:

$$(\mathbf{G} + \alpha\mathbf{C})^{-1}(\mathbf{G} - \alpha\mathbf{C}) \tag{6}$$

In general, for the basis of the Krylov space, say $\mathbf{V}$, the following basic property holds:

$$\mathbf{AV}_m = \mathbf{V}_{m+1}\mathbf{H} \text{ for all } m, \tag{7}$$

for some matrix $\mathbf{H}$. Here the notation $\mathbf{A}_m$ means the first m columns of the matrix $\mathbf{A}$.

In a next step the system matrices are projected onto an orthonormal basis of the Krylov space. This can be done explicitly; the matrices of the reduced system are then defined as:

$$\mathbf{G}_q = \mathbf{V}^T\mathbf{G}\mathbf{V}\mathbf{C}_q = \mathbf{V}^T\mathbf{C}\mathbf{V}$$

$$\mathbf{B}_{iq} = \mathbf{V}^T\mathbf{B}_i\mathbf{B}_{oq} = \mathbf{V}^T\mathbf{B}_o$$

If the dimensions of the space are smaller than the dimensions of the original system, an order reduction is achieved. Some methods, like [7] make use of the matrix $\mathbf{H}$ as defined in (7). The projection is then implicit. Others define two Krylov spaces [1], which are orthogonal with respect to each other. Other details about Krylov subspace methods can be found in [3], [5] and [6] and many other papers.

3 Orthogonalisation

The columns in the Krylov space

$$\mathcal{K}_q(\boldsymbol{b}, \mathbf{A}) = [\boldsymbol{b}, \mathbf{A}\boldsymbol{b}, \ldots, \mathbf{A}^q\boldsymbol{b}] \tag{8}$$

gradually converge to the dominant eigenvector of the matrix $\mathbf{A}$, i.e. the eigenvector of $\mathbf{A}$ associated to the largest eigenvalue. This causes the Krylov space to be very ill-conditioned. Next to that, it becomes hard to calculate an accurate orthogonal basis of this space, because the columns become similar to each other. If the orthogonalisation is done after the generation of the space, as proposed in the Laguerre-SVD method [3], the convergence of the method stagnates. We advocate here to orthogonalizes during the generation of the columns. In that case more directions than only the dominant eigenvector can be calculated accurately and severe numerical artefacts are avoided. We therefore propose to orthogonalize the newly generated vectors immediately after generation. We have been using Modified Gram-Schmidt for this and in there we orthogonalize against all previously generated vectors. After the newly generated columns are made orthogonal with respect to all previously generated columns, they are normalized. This procedure costs some computation time, but the accuracy of the method is drastically increased in all directions. Also numerical artefacts are avoided.

Next to this, we propose to apply a second refinement on the orthogonalisation, in order to ensure orthogonality up to the machine precision. This is needed in some critical problems, to ensure the preservation of stability during time domain simulations of the reduced model.

4 Block Arnoldi Orthogonalisation

When a system has more than one, say p ports, $\mathbf{B}_i$ has more has than one column:

$$\mathbf{B}_i = [\boldsymbol{b}_1, \mathbf{b}_2, \ldots, \mathbf{b}_p] \tag{9}$$

For this system a *Block* Krylov space is built:

$$\mathcal{K}_q(\boldsymbol{B}_i, \mathbf{A}) = [\boldsymbol{b}_1, \mathbf{b}_2, \ldots, \mathbf{b}_p, \mathbf{A}\boldsymbol{b}_1, \ldots, \mathbf{A}\mathbf{b}_p, \ldots, \mathbf{A}^q\boldsymbol{b}_1, \ldots, \mathbf{A}^q\mathbf{b}_p] \tag{10}$$

One can imagine that the size of the Krylov space grows with p and so the approximation will be larger if the number of ports grows. Orthogonalisation and normalization in a Block Krylov space can be done in several orders. For instance, one can add columns to the space one column at the time, or one can add them in blocks. We state that in this case it is important to preserve the basic property of a Krylov space given in (7). If this property is violated, the generated approximation can be totally wrong. In experiments we saw that for a corrupted Krylov space, already for very small Krylov spaces of 8 columns, the transfer function of the approximation differed dramatically from the original function. The order of orthogonalisation in the Block Arnoldi Algorithm, as proposed in PRIMA [5] is seen as a right order to orthogonalize a Block Krylov subspace. Here, we also applied a second orthogonalisation step, to ensure exact orthogonality up to machine precision. The function $qr()$ represents a QR-decompostion.

The Block Arnoldi algorithm, to generate a Block Krylov space for Laguerre-SVD, looks like this:

Solve $\mathbf{V}_1$ from $(\mathbf{G} + \alpha\mathbf{C})\mathbf{V}_1 = \mathbf{B}$
$\mathbf{V}_1\mathbf{R} = qr(\mathbf{V}_1)$
for $j = 1...q-1$
  Solve $\mathbf{W}$ from $(\mathbf{G} + \alpha\mathbf{C})\mathbf{W} = (\mathbf{G} - \alpha\mathbf{C})\mathbf{B}$
  for $i = 1...j$
    $\mathbf{H}_{ij} = \mathbf{V}_i^T\mathbf{W}$
    $\mathbf{W} = \mathbf{W} - \mathbf{V}_i\mathbf{H}_{ij}$
  end
  for $j = 1...j$
    $\boldsymbol{\Theta} = \mathbf{V}_i^T\mathbf{W}$
    $\mathbf{W} = \mathbf{W} - \mathbf{V}_i\boldsymbol{\Theta}$
    $\mathbf{H}_{ij} = \mathbf{H}_{ij} + \boldsymbol{\Theta}$
  end
  $\mathbf{V}_{j+1}\mathbf{H}_{i+1,j} = qr(\mathbf{W})$
end
$\mathbf{V}_{tot} = [\mathbf{V}_1, ..., \mathbf{V}_q]$

5 Redundancy

Krylov-subspace methods are known for their redundancy. The method is relatively cheap, but it can contain a lot of information which is not really needed for an accurate approximation. This is even worse if one realizes that there is no known error bound for Arnoldi methods: Easily too large approximations are generated. But even if we were able to stop in time, the Block structure of the Krylov space leads to redundant approximation. Many authors proposed therefore a combination of a Krylov-subspace method with another method, to form a two-step method. In that approach, first a course approximation is calculated with a cheap Krylov-subspace method. In a second step the order of this approximation is decreased by a more expensive but more controllable method like a Truncated Balanced Realization method [4] or by Proper Orthogonal Decomposition [8]. An interesting approach for a two-step Krylov method is given in [9]. In our research we discovered that a lot can already be done, very cheaply, during the first run of the Krylov-subspace method.

If a Block Krylov-space method is to be generated, it can occur that one of the columns in a new block is almost zero or almost completely spanned by the other columns in the block. In that case we want to stop iterating with this columns, while proceeding with the others. Simply removing information from the space we project on, can lead to the same problems we saw with careless orthogonalisation. With a modified way to calculate a QR-decomposition in the Block Arnoldi Algorithm we are now able to stop iterating with any wanted column, at any wanted time, because still the basic property of Krylov spaces holds for this algorithm. Details can be found in [2].

6 Results

For example, we tested the proposed algorithm on a MNA formulation of an RLC-circuit. The formulation consisted of matrices with size 695. The system has 11 input/output ports. We generated a reduced model with 7 iteration of the Block Arnoldi algorithm. In the standard algorithm this leads to a 77-sized system. Columns with norm smaller than the tolerance 10^{-12} were removed. Then in the 2-nd and 3-th iteration a column is removed and the total system size is eventually 66. The approximation, however, is identical to the approximation of size 77, generated by the ordinary PRIMA algorithm. In Fig. 1, the magnitude of the (1,2) entry of the transfer function of this system (dashed line) is compared with the transfer function of the system of the same size, but generated by ordinary PRIMA (dotted line) and with the transfer function of the full system (solid line). The transfer functions are plotted for values of the frequency ranging from 0 to 2 GHz. We see that the approximation of the system where the redundant columns are removed, forms a better approximation of the original transfer function than an approximation of the same size, but without removal of redundant columns.

Apart from the removal of columns, we also propose a way to remove unwanted poles from the system, without distroying the Krylov space property. This can be done by an eigendecomposition. The reduced system is reasonably small to make the calculation of a full eigendecomposition feasible. This decomposition gives us direct access to the poles of the reduced system and the associated residues.

The most important reason to implement Krylov subspace methods was their preservation of stability and passivity. This makes stable time domain analysis of very large models of real-life electronic structures

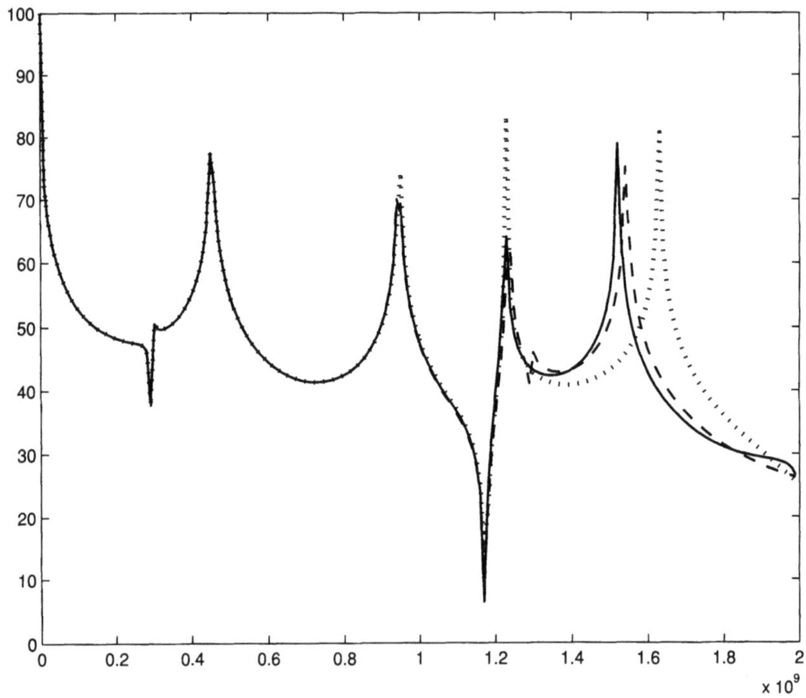

Fig. 1. Comparison of three transfer functions

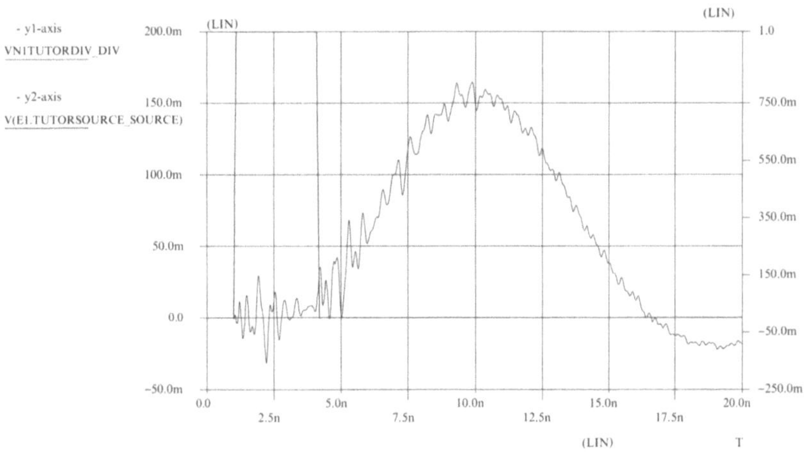

Fig. 2. Output response of a very steep input pulse

possible. The preservation of stability is shown by an example of a model of the printed circuit board, in Fig. 2. The input is a very steep input pulse with a rise-time of 100 ps.

7 Conclusions

We have shown that, to be able to apply Krylov subspace methods for Model Order Reduction to large real-life problems, extra effort is needed. Firstly, the accuracy of the method can be improved by orthogonalisation during the generation of the Krylov space. The Block Arnoldi algorithm is one way to do the orthogonlisation in a correct way. This orthogonalisation is sometimes needed twice. Further, converged columns can be removed during the orthogonalisation step. This can be done without violating the basic Krylov subspace properties. The proposed removal makes the reduced models smaller and therefore less redundant.

All these improvements can be implemented easily in existing methods. This all makes the application of existing methods to large real-life problems feasible.

References

1. P. Feldmann and R. W. Freund. Efficient Linear Circuit Analysis by Padé Approximation via the Lanczos Process. *IEEE Trans. Computer-Aided Design*, 14:137–158, 1993
2. P.J. Heres and W.H.A. Schilders. Deflation of Converged Columns in Krylov Subspace Methods. *To be published*, 2004
3. L. Knockaert and D. De Zutter. Passive Reduced Order Multiport Modeling: The Padé-Arnoldi-SVD Connection. *Int. J. Electronics and Communications*, 53:254–260, 1999
4. B. C. Moore. Principal component analysis in linear systems: Controllability, observability, and model reduction. *IEEE Trans. Automatic Control*, AC-26(1):17–31, 1981
5. A. Odabasioglu and M. Celik. PRIMA: Passive Reduced-order Interconnect Macromodeling Algorithm. *IEEE. Trans. Computer-Aided Design*, 17(8):645–654, August 1998
6. A. Ruhe. Rational krylov methods for eigenvalue computation. *Linear Algebr. Appl.*, 58:391–405, 1984
7. L. M. Silveira, M. Kamon, and White. J. Efficient Reduced-Order Modeling of Frequency-Dependent coupling inductances associated with 3-D interconnect structures. In *Design Automation Conference*, pages 376–380, 1995
8. L. Sirovich. Turbulence and the Dynamics of Coherent Structures. part i: Coherent structures. *Quarterly of Applied Mathematics*, 45(3):561–571, Oct. 1987
9. T. Wittig, I. Munteanu, R. Schuhmann, and T. Weiland. Model order reduction and equivalent circuit extraction for FIT discretized electromagnetic systems. *International Journal of Numerical Modelling: Electronic Networks, Devices and Fields*, 15(5–6):517–533, September–December 2002

Algebraic Sparsefied Partial Equivalent Electric Circuit (ASPEEC)

D. Ioan, G. Ciuprina, and M. Rădulescu

"Politehnica" University of Bucharest -CIEAC/LMN, lmn@lmn.pub.ro

1 Introduction

Due to the increase of the operation frequency and the down-scaling of the on-chip size, the parasitic effects of the electromagnetic field cannot be neglected any longer in the design of ICs. The high frequency field modeling of on-chip passive components and interconnects was one of the topics addressed by the FP5/IST/Codestar project [1].

The reference method for the modeling of passive structures is considered to be PEEC, based on Green function [2, 3]. In this method, the conductors are discretized in filaments, in which constant current densities flow, and their surfaces are discretized in panels having constant charge density. An equivalent RLC circuit containing a resistance for each filament, coupling inductances between whatever two filaments and capacitors between whatever two nodes can be conceived. The inductances and capacitances of such a circuit are described by full matrices. One of the main disadvantages of PEEC is that the accurate modeling of the skin effect needs detailed discretization of conductors. Thus, the method is relatively expensive from the memory requirement point of view. For instance, a 64 b bus with 10 segments per line and 6 filaments per segment conduces to $n = 6 \times 10 \times 64 = 3840$ RL branches, $n(n-1)/2 = 7,370,880$ couplings and $(11 \times 64) \times 12 = 495,616$ C branches, yielding a total number of 7,874,176 elements. Several acceleration techniques, such as fast multipole [4], SVD [5], hierarchical approach [6], FFT [7], etc., are proposed to manage this difficulty.

An alternative approach for the electromagnetic field modeling is proposed in this paper. It is based on the Finite Integration Technique (FIT), which does not use Green functions and which generates a model having a number of degrees of freedom at least as small as PEEC.

FIT is a numerical method able to solve field problems [8], based on spatial discretization "without shape functions". FIT starts from the global form of electromagnetic field equations. Its degrees of freedom (dofs) are not local field components, but the global variables i.e. voltages and fluxes assigned to grid elements (edges and faces, respectively). Two Yee type staggered grids are used as discretization mesh. They are usually orthogonal, but they can be non-orthogonal Delaunay/Veronoi meshes as well.

The Maxwell Grid Equations (MGE) obtained by FIT are

$$\mathbf{D} \cdot \boldsymbol{d} = \mathbf{q}, \quad \mathbf{D}' \cdot \boldsymbol{b} = 0, \quad \mathbf{C} \cdot \boldsymbol{e} = -\frac{\mathrm{d}\mathbf{b}}{dt}, \quad \mathbf{C}' \cdot \boldsymbol{h} = \mathbf{j} + \frac{\mathrm{d}\mathbf{d}}{dt}, \tag{1}$$

where $\mathbf{e}$ is the vector of emfs along the edges of the primary grid, $\mathbf{d}$ is the vector of electric fluxes through the faces of the secondary grid, $\mathbf{h}$ is the vector of mmfs along the edges of the secondary grid, $\mathbf{b}$ is the vector of magnetic fluxes through the faces of the primary grid, $\mathbf{j}$ is the vector of currents through the faces of the secondary grid, $\mathbf{q}$ is the vector of charges in the secondary grid cells. The $\mathbf{D}$ operator is the discrete divergence and the $\mathbf{C}$ operator is the discrete curl. The $'$ notation refers to the secondary grid.

One important feature of FIT is that there are no discretization errors in the fundamental (metric-free) MGE. The equations are sparse, mimetic and conservative. Due to this, no spurious modes arise in the numerical solution.

The material behavior is described by means of the Hodge's operators

$$\mathbf{d} = \mathbf{M}_\varepsilon \mathbf{e}, \quad \mathbf{b} = \mathbf{M}_\mu \mathbf{h}, \quad \mathbf{j} = \mathbf{M}_\sigma \mathbf{e}. \tag{2}$$

These constitutive equations are metric-dependent and they hold the discretization error. An effective a-priori approximation of the modelling error is given in [9].

Due to the huge size of the unknown vectors, classical FIT (MGE+Hodge) must be improved and adapted in order to be effective for the compact model extraction in real life configurations. In this respect, we conceived a new strategy called "All Levels Reduced Order Model" (ALLROM). The ASPEEC technique detailed in this paper is part of the ALLROM strategy, developed within the CODESTAR project.

2 Magneto-Electric Equivalent Circuits (MEEC)

From (1) and (2), an equivalent circuit can be derived (actually two mutual coupled circuits, as in Fig. 1). Thus, the discrete form of charge conservation law is similar to Kirchhoff Current Law (KCL) for the electrical circuit (having as graph the primary grid): $\mathbf{Ai} = \mathbf{0}, \mathbf{i} = \mathbf{i}' + \mathbf{i}''$; the discrete form of magnetic flux law is similar to KCL for the magnetic circuit (having as graph the secondary grid): $\mathbf{A}'\varphi = \mathbf{0}$; the discrete form of Faraday's law is similar to Kirchhoff Voltage Law (KVL) for the electric circuit: $\mathbf{Bu} = \mathbf{0}, \mathbf{u} = \mathbf{u}' + \mathbf{F}\mathrm{d}\,\varphi/\mathrm{d}\,t$; and the discrete form of Ampere's law is similar to KVL for the magnetic circuit: $\mathbf{B}'\mathbf{u}_m = \mathbf{0}, \mathbf{u}_m = \mathbf{u}'_m + \mathbf{Si}$.

Relations (2) are conducing to the following constitutive relationships expressed in terms of circuits' quantities: $\mathbf{i}' = \mathbf{C}\mathrm{d}\,\mathbf{u}'/\mathrm{d}\,t, \mathbf{u}'_m = \mathbf{R}_m\varphi, \mathbf{i}'' = \mathbf{Gu}'$.

Standard SPICE does not accept voltage sources controlled in the time derivative of currents (actually magnetic fluxes in the case of MEEC). That is why we modeled such sources by means of a "derivative circuit" (Fig. 2) which will provide the emfs induced by the magnetic flux. The total currents which control the sources of the magnetic circuit are obtained by means of a "summation circuit" having a ladder topology, similar to the "derivative circuit".

Thus, the SPICE equivalent circuit for the full-wave distributed model consists of four mutual coupled sub-circuits: electric, magnetic, summation and derivative circuits. The SPICE equivalent circuit thus derived has linear complexity (nodes and branches number versus the number of FIT grid cells), while the PEEC model has a quadratic complexity due to their full RL matrices. However, the number of dofs is still large, as comparing to PEEC based on Electro-Magneto-Quasi-Static field.

In order to reduce the number of dofs associated to the MEEC model, the conductive domains (metal and poly-silicon) are modeled with magneto-quasi-static field (MQS) with frequency dependent Hodge operators [10]. In this way, the grid on the cross section does not need to be refined in order to take into account the frequency effect. In this case, the equivalent electric circuit has no parallel capacitances, but three series RL cells with non-coupled inductances replacing R and R_m [10].

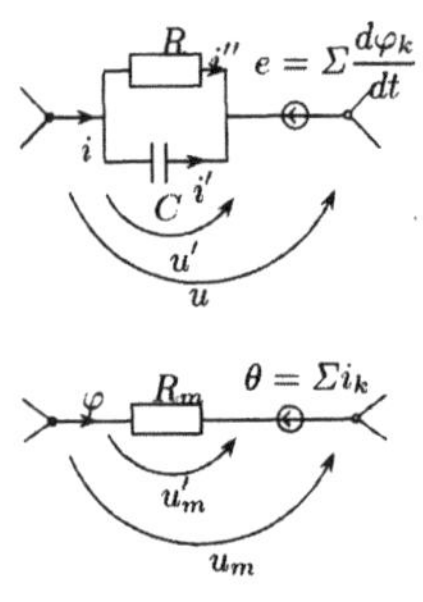

Fig. 1. Typical branches of electric (top) and magnetic (bottom) circuits

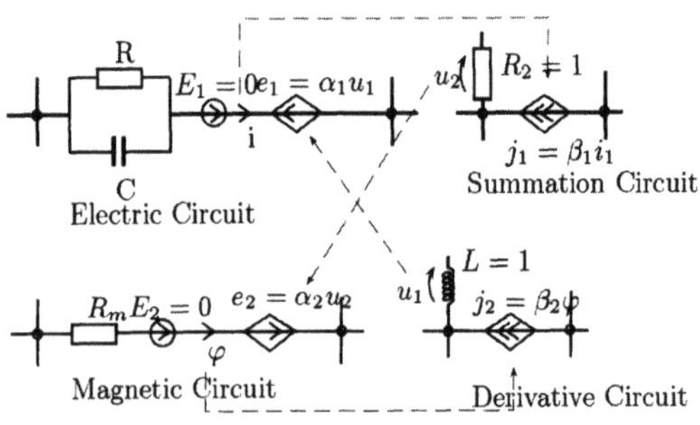

Fig. 2. Typical branches of the four SPICE-like subcircuits: electric, magnetic, summation, derivative

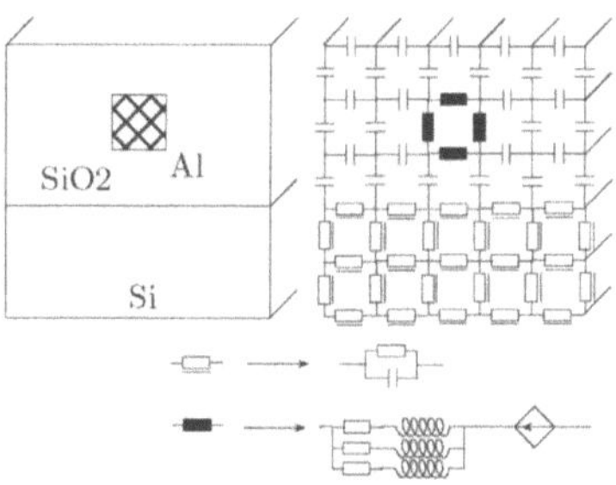

Fig. 3. Example of distributed circuit equivalent to EMQS model

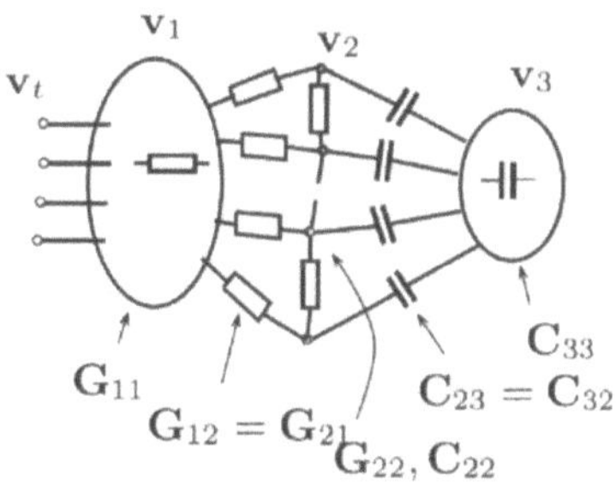

Fig. 4. The distributed circuit, with RC separated parts

The sub-domains with low conductivity (e.g. low doped Si) can be modeled with electro-quasi-static field (EQS) superposed with magneto-static field. Both induced emf and total current are vanished. Resistance in the electric circuit can be also disregarded in order to model insulating domains (e.g. SiO2 and low k).

The obtained EMQS model (Fig. 3) is smaller than the Full Wave model, but still larger than PEEC due to the nodes in the insulator sub-domain.

3 Algebraic reduction of Partial Electro-magnetic Equivalent Circuit (APEEC)

To reduce the model size to that of PEEC's, the generalized delta-star transforms of capacitors and magnetic reluctances in EMQS-MEEC can be carry out. In this way, all internal electric nodes in insulators and internal magnetic nodes in non-conductors are removed.This static condensation procedure eliminates nodes that are non-essential, i.e. nodes having no state variables associated to them. The equivalent reduced circuits obtained (we call them APEEC) are similar to those obtained by the VPEC technique based on integral equations of EMQS field [11].

Each node elimination in APEEC is equivalent to one step of algebraic Gauss-elimination. After the elimination of a node, a fill-in appears in the matrix involved. The fill-in depends very much of the elimination order. In order to preserve the matrix symmetry, only diagonal permutations (equivalent to node re-ordering) are allowed. To find optimal re-ordering (minimal fill-in) a problem with NP complexity should be solved. Therefore, only heuristic techniques to find pseudo-optimal ordering can be used (e.g. the Marcowitz technique). After algebraic reduction, the capacitors and magnetic reluctances in APEEC are described by full $\mathbf{C}$ and $\mathbf{G}_m$ matrices, which are the Schur complements of the initial sparse nodal matrices.

Let us take for instance the simple electro-quasi-static case, and assume that the distributed RC circuit obtained by discretization has the resistive part separated from the capacitive one (Fig. 4). This case, often encountered in practical devices incorporating metals and dielectrics, is described by a system of differential algebraic equations: $\mathbf{C}\mathrm{d}\,\mathbf{v}/\mathrm{d}\,t = -\mathbf{G}\mathbf{v}+\mathbf{S}\mathbf{i}_t$, where both nodal capacitances $\mathbf{C}$ and nodal conductances $\mathbf{G}$ are singular matrices.

Partitioning the semi-state space vector in $\mathbf{v} = [\mathbf{v}_1, \mathbf{v}_2, \mathbf{v}_3]^T$ as in Fig. 4, the following sub-matrices will be null: $\mathbf{C_{11}}$, $\mathbf{C_{12}} = \mathbf{C_{21}}$, $\mathbf{C_{13}} = \mathbf{C_{31}}$, $\mathbf{G_{33}}$, $\mathbf{G_{13}} = \mathbf{G_{31}}$, $\mathbf{G_{23}} = \mathbf{G_{32}}$. Consequently, the semi-state-space equations will conduce to

$$\mathbf{0} = -\mathbf{G}_{11}\mathbf{v}_1 - \mathbf{G}_{12}\mathbf{v}_2 + \mathbf{S}_1\mathbf{i}_t; \tag{3}$$

$$\mathbf{C}_{22}\frac{\mathrm{d}\mathbf{v}_2}{\mathrm{d}t} + \mathbf{C}_{23}\frac{\mathrm{d}\mathbf{v}_3}{\mathrm{d}t} = -\mathbf{G}_{21}\mathbf{v}_1 - \mathbf{G}_{22}\mathbf{v}_2 + \mathbf{S}_2\mathbf{i}_t; \tag{4}$$

$$\mathbf{C}_{32}\frac{\mathrm{d}\mathbf{v}_2}{\mathrm{d}t} + \mathbf{C}_{33}\frac{\mathrm{d}\mathbf{v}_3}{\mathrm{d}t} = \mathbf{0}; \tag{5}$$

and the terminal voltages are $\mathbf{v}_t = \mathbf{S}_1^T\mathbf{v}_1 + \mathbf{S}_2^T\mathbf{v}_2$.

Assuming that $\mathbf{G}_{11}$ and $\mathbf{C}_{33}$ are non-singular, from (3) and (5) it follows that $\mathbf{v}_1 = \mathbf{G}_{11}^{-1}[-\mathbf{G}_{12}\mathbf{v}_2 + \mathbf{S}_1\mathbf{i}_t]$, and $\mathbf{v}_3 = -\mathbf{C}_{33}^{-1}\mathbf{C}_{32}\mathbf{v}_2$. Therefore, from (4), the following state-space equations can be derived

$$(\mathbf{C}_{22} - \mathbf{C}_{23}\mathbf{C}_{33}^{-1}\mathbf{C}_{32})\frac{d\mathbf{v}_2}{dt} = -(\mathbf{G}_{22} - \mathbf{G}_{21}\mathbf{G}_{11}^{-1}\mathbf{G}_{12})\mathbf{v}_2 - \mathbf{G}_{21}\mathbf{G}_{11}^{-1}\mathbf{S}_1\mathbf{i}_t + \mathbf{S}_2\mathbf{i}_t,$$

$$\mathbf{v}_t = (\mathbf{S}_2^T - \mathbf{S}_1^T\mathbf{G}_{11}^{-1}\mathbf{G}_{12})\mathbf{v}_2 + \mathbf{S}_1 i^T\mathbf{G}_{11}^{-1}\mathbf{S}_1\mathbf{i}_t. \tag{6}$$

This is the proof that, in the case of EQS field in conductor and dielectric structures (each cell is either a perfect insulator or a conductor), the state variables are the potentials of the nodes placed on the conductor-dielectric interfaces ($\mathbf{v}_2$).

The state equations of this minimal model are obtained by computing the Schur complements of the matrices $\mathbf{C}_{11}$ (nodal capacitances of the dielectric part) and $\mathbf{G}_{11}$ (nodal conductances of the conductive part).

In order to compute the Schur complement, the LU factorization algorithm (e.g. MUMPS [12] sparse implementation) is applied to the $\mathbf{C}$ and $\mathbf{G}$ matrices. If this algorithm is interrupted after the internal node elimination, then the not-yet factorized block is exactly the desired Schur complement.

4 Sparsefication of Algebraic PEEC

The nodal capacitance matrix $\mathbf{C}$ which describes the capacitive part of APEEC is a full, symmetric, positive definite, diagonal dominant, M-matrix (the diagonal has positive elements and off diagonal elements are negative). The nodal magnetic susceptance matrix $\mathbf{G}_m$ ($G_{mij} = -1/R_{mij}$) which describes the inductive part of MEEC has similar properties as $\mathbf{C}$, and therefore it can be sparsified using similar techniques. It is similar to K - element method used to describe coupled inductors [13], having their advantages.

The problem of *sparsefication* is to find sparse approximations of the matrices $\mathbf{C}$ or $\mathbf{G}_m$ (or a representation by a sparse matrix, such as SVD truncation), which keep their proprieties (e.g. if passivity is preserved, it is called *passivity-guaranteed sparsefication*). It would be ideal if circuit representations will be kept after sparsefication (if capacitive/resistive equivalent circuit having lower number of elements can be synthesized, it is called *realizable sparsefication*).

Two kinds of sparsefication are known. The *geometric sparsefication* is based on the observation that close interactions are stronger than far interactions, and therefore the former should be accurately described. In this type of sparsefication, the "distance" between nodes plays an important role. In *numeric sparsefication*, the neglectable elements of the matrices are dropped-off. In both matrices, any neglectable non-diagonal element and its symmetric can be vanished without loosing the desired properties. The preferable criterion to detect if an element is neglectable or not, is to compare its value with the corresponding diagonal element.

Any acceleration method encountered in the numerical solving of the electromagnetic field integral equations can be considered as a sparsefication technique. However, we prefer a simpler but effective technique called *hierarchical geometric sparsefication (HGS)*, followed by a numeric sparsefication. The idea behind HGS is to use fine grids for close interactions and coarse grids for far interactions, as in the hierarchical matrices (Hlib) approach [14]. For n nodes, the number of non-zero elements after sparsefication is of the order $O(n \log n)$.

The ASPEEC (Algebraic Sparsefied Partial Equivalent Element Circuit) model generated by the sparsefication of the APEEC model can be further reduced, using Krylov ROM techniques [15] or by circuit transform such as TICER [16], as aposteriori ROM.

5 Numerical results

This section holds numerical results related to the application of ASPEEC technique to one of Codestar benchmark, the meander resistor (Fig. 5).

The computational domain has the dimensions (in μm) $48 \times 43.5 \times 2.937$, discretized with an initial mesh having 368,200 nodes, which corresponds to 2,209,680 dofs. A macromodel with 5,940 nodes

Fig. 5. Codestar meander resistor benchmark - RPOLY2_ME

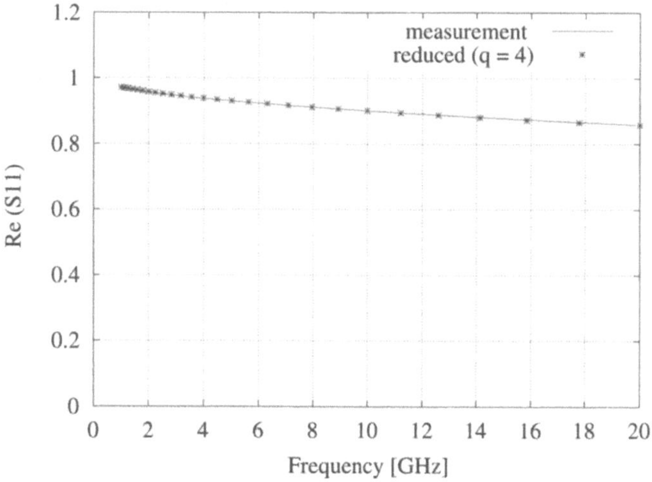

Fig. 6. Real part of S_{11} versus frequency

(19, 510 dofs) was extracted by the ALLROM strategy. After applying the ASPEEC technique, the number of dofs decreased to 1,882. The evaluation of the frequency characteristics was carried out in accordance with an adaptive frequency sampling technique in 11 points. The final model order, obtained at the end of the aposteriori ROM was $q = 4$. The whole ALLROM computing time on a standard PC is 145 s, and the relative error $\varepsilon = \text{rms}\|\mathbf{S}_{ref} - \mathbf{S}\|_F / \max_f \|\mathbf{S}_{ref}\|_F$ between the measurement and the simulation being 1.4 % (Fig. 6). In the error computation, the Frobenius norm is used, $\mathbf{S}_{ref}$ are the reference scattering parameters, and the maximum is computed with respect to the frequency range of interest (e.g. $0 < f < 20$ GHz in our case).

6 Conclusions

The paper presents a powerful technique to extract reduced order models of on-chip passive structures, included in a new compact modeling technology. The *distributed* equivalent circuit we propose has a linear complexity, it is similar to VPEC, but is based on FIT, not on the integral approach (PEEC). Using algebraic techniques (Schur complement), APEEC method reduces the FIT equations (and the associated equivalent circuit) to ones similar to PEEC (having the same number of dofs). To be effective in simulation, the APEEC matrices are approximated by sparse ones, conducing to the ASPEEC model.

The proposed approach combines advantages of FIT with those of PEEC, providing

- more flexibility in the modeling of conductor/insulator/substrate non-homogeneous structures;
- Green functions are not required;
- accurate models for skin effects, without significant increase of computational effort;
- fast and accurate direct SPICE equivalent circuits with low complexity for any full-wave, EMQS, MQS or EQS model;
- when applying the proposed method, the explicit build of equivalent circuits is not a compulsory step; they can be used as software objects in order to represent the model of the device or for checking purposes (however, its theoretical importance is without any doubt);
- structural passivity preservation;
- same (realizable and passivity guaranteed) sparsefication technique is applied for both capacitance and inductance components of the extracted model.

The proposed approach proved to be suitable for the Codestar benchmarks, most of them being simulated with an accuracy better than 5 %.

References

1. Codestar Website, "`http://www.imec.be/codestar`,"
2. P.J. Restle, A.E. Ruehli, S.G. Walker, and G. Papadopoulos, "Full-wave PEEC time-domain method for the modeling of on-chip interconnects," *IEEE Trans. on CAD of Integrated Circuits and Systems*, vol. 20, no. 7, pp. 877–86, 2001
3. A.M. Niknejad, R. Gharpurey, and R.G. Meyer, "Numerically Stable Green Function for Modeling and Analysis of Substrate Coupling in Integrated Circuits," *IEEE Trans. on CAD of Integrated Circuits and Systems*, vol. 17, no. 4, pp. 305–315, 1998
4. K. Nabors and J. White, "FastCap: A multipole-accelerated 3-d capacitance extraction program," *IEEE Trans. CAD*, vol. 10, pp. 1447–1459, 1991
5. S. Kapur and D.E. Long, "IES3: a fast integral equation solver for efficient 3-dimensional extraction," in *Int. Conf. on CAD*, 1997, pp. 448–455
6. Weiping Shi, Jianguo Liu, Naveen Kakani, and Tiejun Yu, "A fast hierarchical algorithm for three-dimensinal capacitance extraction," *IEEE Transactions on CAD of Integrated Circuits and Systems*, vol. 21, no. 3, 2002
7. Zhenhai Zhu, Ben Song, and Jacob White, "Algorithms in fastimp: A fast and wideband impedance extraction program for complicated 3d geometries," in *IEEE/ACM DAC*, Anaheim, CA, USA, 2003
8. M. Clemens and T. Weiland, "Discrete Electromagnetism with the Finite Integration Technique ," *Progress In Electromagnetics Research, PIER*, vol. 32, pp. 65–87, 2001
9. D. Ioan, M. Rădulescu, and G. Ciuprina, "Fast extraction of static electric parameters with accuracy control," *Scientific Computing in Electrical Engineering*, pp. 248–256, 2004
10. D. Ioan and M. Piper, "Fit models with frequency dependent Hodge operators for HF effects in metallic conductors," in *Progress In Electromagnetics Research, PIERS*, Pisa, Italy, 2004
11. Hao Yu and Lei He, "Vector potential equivalent circuit based on peec inversion," in *40th ACM/IEEE DAC*, Anaheim, CA, USA, 2003, pp. 718–723
12. A. Guermouche, J.Y. L'Excellent, and G.Utard, "Impact of reordering on the memory of a multifrontal solver," *Parallel Computing, Report RR2003-08/INRIA/RR-4729*, vol. 29, pp. 1191–1218, 2003
13. A. Devgan, H. Ji, and W. Dai, "How to efficiently capture on-chip inductance effects: Introducing a new circuit element k," in *IEEE/ACM Int. Conf. on CAD*, 2000, pp. 150–155
14. Wolfgang Hackbusch, "A sparse matrix arithmetic based on h-matrices," *Computing*, vol. 62, pp. 89–108, 1999
15. Mustafa Celik, Lawrence Pileggi, and Altan Odabasioglu, *IC Interconnect Analysis*, Kluwer Academic Publishers, 2002
16. Bernard N. Sheehan, "Ticer: realizable reduction of extracted rc circuits," in *IEEE/ACM Int. Conf. on CAD*, San Jose, California, 1999, pp. 200–203

Analytical and Numerical Techniques for Simulating a 3D Rainwater Droplet in a Strong Electric Field

D. Langemann

University of Rostock, Institute for Mathematics, Universitätsplatz 1, 18051 Rostock, Germany,
dirk.langemann@mathematik.uni-rostock.de

Abstract Outdoor high-voltage equipment is exposed to moisture, rain and pollution. Water droplets on insulators influence negatively the material-aging process. A numerical procedure is presented which simulates the droplet behavior in a strong electric field. It consists of an iteration over an electric and a mechanical sub-problem to solve a coupled system of boundary value problems on the free domain of the droplet. Finally we give the resulting droplet shapes for 2D and 3D models, and we mention the behavior of a droplet in an inhomogeneous electric field.

1 Introduction to the Coupled Problem

The experimentally observed droplets [5, 12] become lengthened and flattened, they oscillate with double the frequency of an applied alternating voltage and their changes are visible by naked eyes. A model of the experimental set-up [5] is shown in Fig. 1. A conductive water droplet of $V = 50\,\mu l$ lies on a solid support made of resin which contains two electrodes with the applied voltage of $2U$ between them.

The electric field and the ponderomotoric force density are dealt in [11, 12] for the case of an undeformable droplet and an alternating voltage U. The present paper concentrates on the simulation of deformable droplets in an electric field which is generated by a time-constant voltage U.

The question of the behavior of deformable droplet in a stationary electric field is a feed-back problem. The droplet shape determines the ponderomotoric force density $\boldsymbol{p}_{\mathrm{e}}$ caused by the electric field, and thus the electric field changes the equilibrium of forces at the droplet surface and influences the droplet shape. The mechanical sub-problem of the droplet shape and the electric sub-problem of finding $\boldsymbol{p}_{\mathrm{e}}$ are decoupled via an iteration in Sect. 2. After evolving both sub-problems stationary 3D and 2D results are given in Sect. 5.

The paper finishes with a short remark on non-stationary droplets in inhomogeneous electric field. In the conclusion, we give an outlook to the simulation of time-dependent deformable droplets.

2 Decoupling Strategy

The two basic sub-problems of the coupled problem are the electric sub-problem and mechanical sub-problem. The electric sub-problem consists of finding the outer force density $\boldsymbol{p}_{\mathrm{e}}$ for a given upper droplet surface Γ_{u} which is parameterized by spherical co-ordinates $r(\varphi, \zeta)$, cf. Fig. 1 (a). The mechanical sub-problem is the search for Γ_{u} depending on $\boldsymbol{p}_{\mathrm{e}}$. In Sect. 4.1 the force density $\boldsymbol{p}_{\mathrm{e}}$ is assigned to the unknown Γ_{u}. The sub-problems are expressed by the operators $\mathcal{P} : \Gamma_{\mathrm{u}} \to \boldsymbol{p}_{\mathrm{e}}$ and $\mathcal{R} : \boldsymbol{p}_{\mathrm{e}} \to \Gamma_{\mathrm{u}}$. In this formalism, we search for a fixed point $\Gamma_{\mathrm{u}}^{\mathrm{fix}} = \mathcal{RP}\Gamma_{\mathrm{u}}^{\mathrm{fix}}$ described by r^{fix} by a Banach-like iteration [1, 13] with a relaxation $\omega \in (0, 1]$, i.e.

$$r^{(k+1)} = \omega \mathcal{RP} r^{(k)} + (1-\omega) r^{(k)} \quad \text{with} \quad \lim_{k\to\infty} r^{(k)} = r^{\mathrm{fix}} . \tag{1}$$

If Eq. (1) converges, the iteration $\boldsymbol{p}_{\mathrm{e}}^{(k+1)} = \omega \mathcal{PR} \boldsymbol{p}_{\mathrm{e}}^{(k)} + (1-\omega) \boldsymbol{p}_{\mathrm{e}}^{(k)}$ converges similarly, and it holds $\lim_{k\to\infty} \boldsymbol{p}_{\mathrm{e}}^{(k)} = \boldsymbol{p}_{\mathrm{e}}^{\mathrm{fix}}$ with the ponderomotoric force density $\boldsymbol{p}_{\mathrm{e}}^{\mathrm{fix}}$ which belongs to Γ_{u} described by r^{fix}.

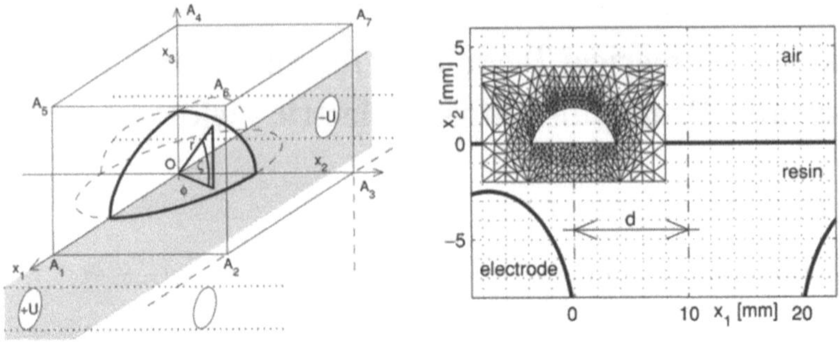

Fig. 1. **(a)** Experimental set-up. The bold lines mark the quarter droplet lying on the surface $OA_1A_2A_3A_4$ which coincides with the (x_1, x_2)-plane. **(b)** 2D intersection through a set-up with non-centered droplet. FE/FD discretization scheme

3 The Electric Sub-Problem

This section deals with the boundary value problem for the stationary potential Φ, its numerics and the determination of the force density $\boldsymbol{p}_\mathrm{e}$.

3.1 The Stationary Electric Field Around the Droplet

The support Ω of the electric field is the resin and the air. For formal simplification we write $\varepsilon(\boldsymbol{x}) = \varepsilon_\mathrm{air} = 1.00058$ on Γ_u and $\varepsilon(\boldsymbol{x}) = \varepsilon_\mathrm{res} = 4$ on the ground patch Γ_s of the droplet on the support.

Rainwater is conductive, and the droplet is free of charge. We get the linear elliptic boundary value problem with Dirichlet-conditions at the electrodes Γ_e

$$\begin{aligned} \nabla \cdot [\varepsilon(\boldsymbol{x}) \nabla \Phi(\boldsymbol{x})] &= 0 \quad \text{in } \Omega\,, \\ \Phi(\boldsymbol{x}) &= \pm U \text{ on } \Gamma_\mathrm{e}\,, \\ \Phi(\boldsymbol{x}) &= c \quad\ \text{on } \Gamma_\mathrm{u} \cup \Gamma_\mathrm{s}\,, \\ \int_{\Gamma_\mathrm{u} \cup \Gamma_\mathrm{s}} \varepsilon(\boldsymbol{x}) \frac{\partial}{\partial \boldsymbol{n}} \Phi(\boldsymbol{x})\, \mathrm{d}\boldsymbol{x} &= 0\,. \end{aligned} \tag{2}$$

The integral over the density of free charge in (2) determines the constant c. The potential vanishes at infinity and has a finite energy. The boundary problem (2) is linear for fixed Ω, but Ω depends on the searched droplet shape.

The electric field is $\boldsymbol{E}(\boldsymbol{x}) = -\nabla\Phi(\boldsymbol{x})$ and the dielectric displacement is $\boldsymbol{D}(\boldsymbol{x}) = \varepsilon_0\varepsilon(\boldsymbol{x})\boldsymbol{E}(\boldsymbol{x})$. The Maxwell stress tensor and the ponderomotoric surface force density are [4]

$$\mathsf{T} = \boldsymbol{E}\boldsymbol{D}^\mathrm{T} - \frac{1}{2}(\boldsymbol{E}^\mathrm{T}\boldsymbol{D})I \text{ and } \boldsymbol{p}_\mathrm{e}(\boldsymbol{x}) = \mathsf{T}^+(\boldsymbol{x})\boldsymbol{n} = \frac{1}{2}\varepsilon_0\varepsilon(\boldsymbol{x}) \left(\frac{\partial\Phi(\boldsymbol{x})}{\partial\boldsymbol{n}}\right)^2 \boldsymbol{n}$$

with the unilateral limit $\mathsf{T}^+(\boldsymbol{x})$ of the stress tensor at the droplet surface.

3.2 Numerics of the Electric Sub-Problem

The numerical solution of (2) requires some care. Finite differences or finite integration techniques [11] on a rectangular grid like in [12] approximate well the potential Φ, but the outer force density $\boldsymbol{p}_\mathrm{e}$ depending on $\nabla\Phi$ cannot be evaluated on the curved surface outside the meshes in satisfactory accuracy. The components of the electric field $\boldsymbol{E} = -\nabla\Phi$ oscillate numerically in the neighborhood of Γ_u, i. e. in the only part of Ω, where $\boldsymbol{E}$ is really searched. Sophisticated interpolation and averaging methods would be necessary.

Boundary element methods need a domain decomposition with an unbounded skeleton due to the non-constant $\varepsilon(\boldsymbol{x})$. Finite elements in whole the 3D domain do not justify the costly effort to find $\boldsymbol{p}_\mathrm{e}$ on Γ_u only.

Thus, we use a combination of finite elements on an adapted tetrahedral grid refined near $\Gamma_u \cup \Gamma_s$ in a parallelepiped around the droplet and finite differences remote from it. A sketch of the hybrid discretization scheme is given in Fig. 1 (b).

The numerical errors in the finite element approximation of Φ and $\nabla\Phi$ are small, and they do so for $\boldsymbol{p}_e$. Any local disturbances caused by the finite differences outside the parallelepiped – e.g. on the curved Γ_e – are levelled out in its neighborhood, and they do not perturb $\boldsymbol{p}_e$. The computational costs are restricted. In the examples, a tetrahedral grid with 150,000 elements was used which could be handled on a standard desktop computer.

3.3 Scattered $\boldsymbol{p}_e$-Data on the Droplet Surface

The extrapolation of the $\boldsymbol{p}_e$ data found by the finite element computation onto whole the droplet surface occurs in particular in three-dimensional models. A relatively small number of tetrahedral elements borders on the two-dimensional Γ_u. Their indices are collected in J and the outer force density $\boldsymbol{p}_e(\boldsymbol{x}_\Gamma^{(j)})$ is known in their centres $\boldsymbol{x}_\Gamma^{(j)}, j \in J$.

For solving the $\mathcal{R}$-problem, the outer force density $\boldsymbol{p}_e$ is required at mesh-points $\boldsymbol{y}_\Gamma^{(i)}, i = 1, \ldots, N$ of a fixed (φ, ζ)-grid. The tetrahedral grid is adapted to the changing surface Γ_u in each step of the iteration (1) and thus the points $\boldsymbol{x}_\Gamma^{(j)}$ do not have fixed (φ, ζ)-co-ordinates. All points $\boldsymbol{x}_\Gamma^{(j)}$ are inside Γ_u. An extrapolation should continue $\boldsymbol{p}_e$ reasonably to whole the surface Γ_u. Therefore we use the weighted average [3]

$$\boldsymbol{p}_e(\boldsymbol{y}_\Gamma^{(i)}) = \boldsymbol{p}_e^{\text{out}}(\boldsymbol{y}_\Gamma^{(i)}) + \left[\sum_{j\in J} \gamma(\boldsymbol{x}_\Gamma^{(j)}, \boldsymbol{y}_\Gamma^{(i)})\right]^{-1} \sum_{j\in J} \gamma(\boldsymbol{x}_\Gamma^{(j)}, \boldsymbol{y}_\Gamma^{(i)})\, \boldsymbol{p}_e(\boldsymbol{x}_\Gamma^{(j)})$$

with a decreasing function γ of the distance between $\boldsymbol{x}_\Gamma, \boldsymbol{y}_\Gamma \in \Gamma_u$ and a corrector term $\boldsymbol{p}_e^{\text{out}}(\boldsymbol{y}_\Gamma)$ containing the known asymptotic behavior of the outer force density near the triple line $\partial\Gamma_u$ with $\zeta = 0$. A homogenization [10] near the triple line and a series expansion yields

$$\boldsymbol{p}_e(\boldsymbol{y}(\varphi, \zeta)) \sim \zeta^{2(a-1)} \quad \text{for } \zeta \to 0 \text{ and all } \varphi \tag{3}$$

with the smallest positive $a \approx 0.54$ in $\varepsilon_{\text{res}} \tan(a(\pi - \vartheta)) = -\varepsilon_{\text{air}} \tan(a\pi)$ [6] and the contact angle $\vartheta = 1.1$. The relation (3) assures the non-existence of an essential concentration of free charge and thus of forces on $\partial\Gamma_u$. The stationary balance between the surface tensions inside the interfaces air/water, water/resin and resin/air is not be disturbed by the ponderomotoric force density.

4 The Mechanical Sub-Problem

4.1 The Non-Linear Boundary Value Problem on the Droplet

The force densities acting on Γ_u are the capillary pressure $\boldsymbol{p}_k(\boldsymbol{x})$, the hydrostatic pressure $p_h(\boldsymbol{x}) + p_0$ and the outer force density $\boldsymbol{p}_e(\boldsymbol{x})$ caused by the electric field. The capillary pressure is given by the Young-Laplace equation

$$\boldsymbol{p}_k(\boldsymbol{x}) = -2\sigma\kappa(\boldsymbol{x})\boldsymbol{n}$$

with the mean curvature $\kappa(\boldsymbol{x}) \geq 0$ of the droplet surface. The hydrostatic pressure p_h depends on the height of the droplet. With the mass density ϱ and the gravitational acceleration g we get

$$p_h(\boldsymbol{x}) = g\varrho\left(\max_{\boldsymbol{x}'\in\Gamma_u} x_3' - x_3\right).$$

The incompressibility of the water yields the constraint condition of a constant volume V and thus the Lagrangian multiplier p_0. The equilibrium of forces leads to a boundary value problem

$$\boldsymbol{p}_e(\boldsymbol{x}) + \boldsymbol{p}_k(\boldsymbol{x}) + (p_h(\boldsymbol{x}) + p_0)\,\boldsymbol{n} = 0 \tag{4}$$

with the boundary conditions of a constant contact angle ϑ on $\partial \Gamma_\mathrm{u}$. The problem (4) is formulated on the free Γ_u but the parameterization maps it to the fixed (φ, ζ)-domain $\varphi \in [0, 2\pi)$ and $\zeta \in [0, \pi/2]$.

4.2 Numerics of the Mechanical Sub-Problem

The parametrization $\boldsymbol{x} = \boldsymbol{x}(\varphi, \zeta)$ generates a bijective map $\Gamma_\mathrm{u} \leftrightarrow \Gamma'_\mathrm{u}$ between variable surfaces Γ_u and Γ'_u. The given outer force $\boldsymbol{p}_\mathrm{e}$ is assigned to (φ, ζ). In opposite the force densities $\boldsymbol{p}_\mathrm{k}$ and $p_\mathrm{h}\boldsymbol{n}$ depend on Γ_u, and they can be expressed directly for every occurring surface.

Let τ be an auxiliary time. We simulate a transient process of a damped droplet deformation by the artificial evolution problem

$$\frac{\partial}{\partial \tau} r(\varphi, \zeta, \tau) = \left(\boldsymbol{p}_\mathrm{e}(\varphi, \zeta) + \boldsymbol{p}_\mathrm{k}(\varphi, \zeta, \tau)\right)^T \boldsymbol{n} + p_\mathrm{h}(\varphi, \zeta, \tau) + p_0(\tau) \tag{5}$$

with the boundary condition of a fixed contact angle ϑ and initial conditions $r(\varphi, \zeta, 0) = r_0(\varphi, \zeta)$ with $\|\nabla_{(\varphi,\zeta)} r_0\|_{C(\Gamma_\mathrm{u})} \leq C$ with a constant C. The limit solution $r_\mathrm{lim}(\varphi, \zeta) = \lim_{\tau \to \infty} r(\varphi, \zeta, \tau)$ of the parabolic system (5) is the solution of the non-linear elliptic boundary value problem (4). Cause of $p_0(\tau)$ we solve a differential-algebraic system. The limit solution is found with small numerical costs within $\tau < 1\,\mathrm{ms}$.

If $p_0(\tau)$ is replaced by a penalty force assuring an incompressible droplet volume V [8], the discretized problem (5) becomes a system of stiff ordinary differential equations [2].

5 Results in 3D Compared with the 2D Case

For small voltages U, the iteration (1) with $\omega = 1$ reaches a stationary droplet shape after five steps where $r^{(k+1)}$ and $r^{(k)}$ do not differ numerically. Larger voltages require up to 20 iteration steps with the relaxation parameter $\omega = 1$.

Let be U_max the voltages which tears up a conductive droplet into two smaller droplets. Very large voltages $U/U_\mathrm{max} \in [0.8, 1]$ require $\omega < 1$, but iteration (1) converges even for $U > U_\mathrm{max}$ to unphysical shapes with $p_0 < 0$ and $\kappa(\boldsymbol{x}) < 0$ for some $\boldsymbol{x} \in \Gamma_\mathrm{u}$ in less than 50 steps with a fixed $\omega > 1/2$.

The given numbers of iteration steps are independent of the number N of discretization points $\boldsymbol{y}_\Gamma^{(i)}$, $i = 1, \ldots, N$ on Γ_u. This independency is proven by the existence of a constant L with $\|\boldsymbol{p}_\mathrm{e}^{(k)} - \boldsymbol{p}_\mathrm{e}^\mathrm{fix}\|_{C^{0,\alpha}(\Gamma_\mathrm{u})} \leq L\|r^{(k)} - r^\mathrm{fix}\|_{C^{2,\alpha}(\Gamma_\mathrm{u})}$ in a neighborhood of the stationary droplet shape.

Thus, the iteration (1) is much more effective than a Newton-type method to solve the coupled problem. Only the evaluation of the Jacobian of a non-linear system for $r(\varphi_i, \zeta_i)$, $i = 1, \ldots, N$ describing the points $\boldsymbol{y}_\Gamma^{(i)}$ needs N solutions of the electric sub-problem (2), and $N \approx 1000$ in the examples.

Figure 2 (a) shows a quarter droplet in the absence of an electric field. It is axial-symmetric. After the application of a strong electric field, it becomes lengthened and flattened, Fig. 2 (b). The width diminishes.

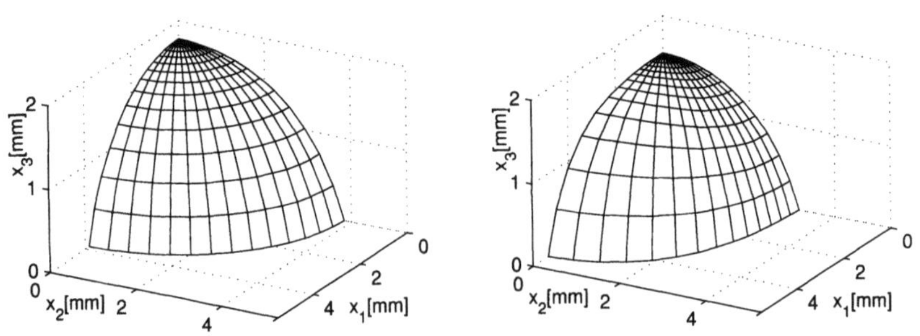

Fig. 2. Quarter droplets (**a**) in the absence of an electric field and (**b**) in a strong electric field, $U = 24\,\mathrm{kV}$. Plots are not true in scale, height is exaggerated

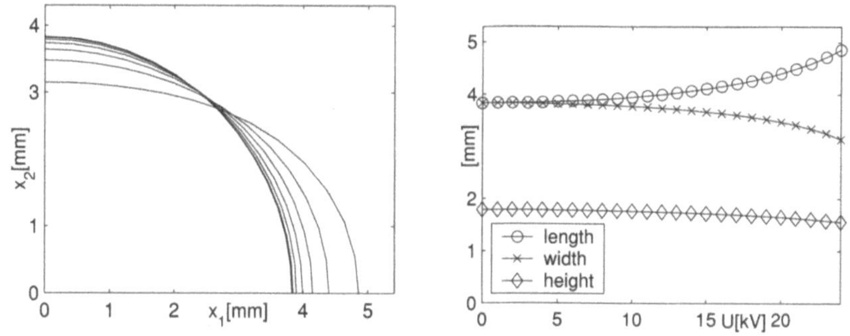

Fig. 3. (**a**) Quarter ground patches of 3D droplets for $U = 0, 4, \ldots, 20, 24\,\mathrm{kV}$ from above. (**b**) Heights, widths and lengths of 50 µl-droplets depending on U

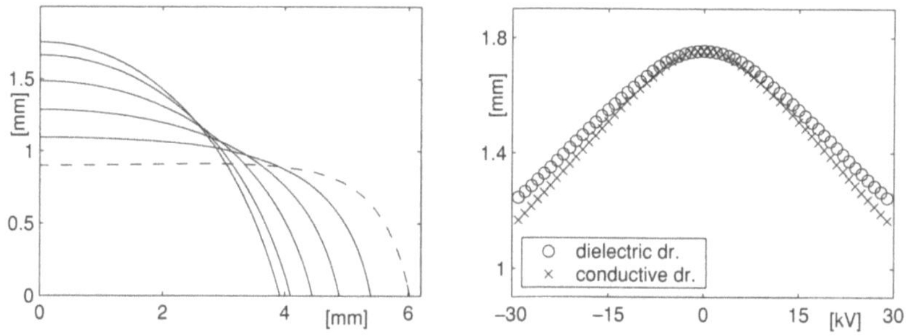

Fig. 4. (**a**) 2D droplet shapes for $U = 0, 8, 16, 32\,\mathrm{kV}$ and a droplet being torn up by $40\,\mathrm{kV}$ in dashed line. (**b**) Heights of conductive and dielectric 2D droplets depending on U

This effect is illustrated by the ground patches in Fig. 3 together with the dimensions of the droplets. The droplet height is independent of sign U. Hence a droplet in a low frequent alternating electric field oscillates with twice the frequency [6, 12].

Figure 4 gives the respective 2D results for comparison. In addition to the calculations for deformable rainwater droplets, analogous simulations are done for the theoretical case of 2D pure-water droplets. In this case the domain Ω in the respective electric sub-problem (2) includes the droplet and the H_2O-dipols are oriented at the droplet surface. We get a line concentration of polarization charge, and a respective expression for T and $\boldsymbol{p}_\mathrm{e}$ at Γ_u.

The difference between conductive and dielectric droplets is discussed in [7]. 2D droplet models react more sensitive to an applied field. The 2D model lacks the second curvature term, and the incompressibility of the fluid couples length and height of a 2D droplet model in an enforced manner [8].

The integral over the ponderomotoric force density $\boldsymbol{p}_\mathrm{e}$ is not necessarily vanishing. In general the droplet suffers a total force $\boldsymbol{F}$ and moves leaving a water film. Without real charge and for $\varepsilon(\boldsymbol{x}) = 1$, we show analytically that

$$\lim_{V \to 0} \frac{\boldsymbol{F}(\bar{\boldsymbol{x}})}{V} = \varepsilon_0 \nabla \left| \nabla \tilde{\Phi}(\bar{\boldsymbol{x}}) \right|^2 = 2\varepsilon_0 \nabla \tilde{\boldsymbol{E}}(\bar{\boldsymbol{x}}) \cdot \tilde{\boldsymbol{E}}(\bar{\boldsymbol{x}}) \tag{6}$$

holds with the undisturbed electric potential $\tilde{\Phi}$ in the absence of the droplet and with the droplet's centre of gravity $\bar{\boldsymbol{x}}$ [9]. The case $\boldsymbol{F} = 0$ is the rather extra-ordinary one, e.g. if $\tilde{\boldsymbol{E}}$ is perfectly homogeneous or if the particle is dimensionless $V = 0$. Eq. (6) can be used to approximate the motion of whole droplets on realistically shaped insulators with only one computation of an undisturbed electric field. Neglecting weather conditions, we find that droplets move into the thin part of insulators and form larger droplets there.

6 Conclusion

An algorithm was presented which enables us to simulate the behavior of a deformable droplet in a strong stationary electric fields. It is based on an iteration decoupling the sub-problems which can be applied to a class of free boundary problems. Particular features of the 3D simulations are the restriction of the finite element approximation to a domain close to the droplet and an extrapolation of the discretized ponderomotoric force density on the droplet surface by the use of the analytically known growth behavior.

The observation of moving droplets in non-homogeneous electric fields motivates a forthcoming hydrodynamical investigation of the droplet fluid including adhesion to the material of the support. The simulation of time-dependent deformable droplets in an alternating electric field involves new difficulties like flux inside the droplet fluid, inertial effects, induced currents in the fluid and so on. Such a combined solution of the time-dependent Maxwell's equations and the Navier-Stokes equation with boundary conditions from the Young-Laplace equation would require enormous numerical costs.

A further challenge in the simulation of the droplet behavior is the consideration of specific surface properties of aging insulating material.

References

1. Frischmuth, K., Hänler, M.: Numerical analysis of the closed osmometer problem. ZAMM **79** (2), 107–116 (1999)
2. Hairer, E., Wanner, G.: Solving Ordinary Differential Equations. Part 2. Stiff and Differential-Algebraic Problems. Springer, Berlin (1991)
3. Iske, A., Quak, E., Floater, M.S.: Tutorials on Multiresolution in Geometric Modelling. Springer, Berlin (2002)
4. Jackson, J.D.: Classical Electrodynamics. Wiley, New York (1999)
5. Keim, S., König, D.: Study of the behavior of droplets on polymeric surfaces under the influence of an applied electrical field. In: Proceedings of the IEEE Conference on Electrical Insulation and Dielectric Phenomena, Austin. 707–710 (1999)
6. Langemann, D.: A droplet in a stationary field. Mathematics and Computers in Simulation **63** (6) 529–539 (2003)
7. Langemann, D.: The free transmission problem for a water droplet. In: Proceedings of the 4th International Conference on Large-Scale Scientific Computing, Sozopol, Bulgaria. Springer, Berlin 387–395 (2004)
8. Langemann, D., Krüger, M.: 3D model of a droplet in an electric field. Mathematics and Computers in Simulation **66** (6) 539–549 (2004)
9. Langemann, D.: Modelling a droplet moving in an electric field. Mathematics and Computers in Simulation **68** (2) 157–169 (2004)
10. Mazja, V.G., Nazarov, S.A., Plamenevskij, B.A.: Asymptotic Theory of Elliptic Boundary Value Problems in Singularly Pertubated Domains. Birkhäuser, Basel (2000)
11. van Rienen, U., Clemens, M., Wendland, T.: Simulation of Low-Frequency Fields on Insulators with Light Contaminations. IEEE Trans, Magn. **32** (3) 816–819 (1996)
12. Schreiber, U., van Rienen, U.: Simulation of the behavior of droplets on polymeric surfaces under the influence of an applied electrical field. In: Proceedings of the 9th Biennial IEEE Conference CEFC, Milwaukee (2000)
13. Sethian, J.A.: Level Set Methods: Evolving Interfaces In Geometry, Fluid Mechanics, Computer Vision and Material Science. Cambridge Univ. Press, Cambridge (1998)

3-D FE Particle Based Model of Ion Transport Across Ionic Channels

M. E. Oliveri[1], S. Coco[2], D. S. M. Gazzo[2], A. Laudani[2], and G. Pollicino[2]

[1] DMFCI Dipartimento di Metodologie Fisiche e Chimiche per l'Ingegneria, Viale A. Doria 6, Catania, Italy I-95125, meolive@dmfci.ing.unict.it
[2] DIEES Dipartimento di Ingegneria Elettrica, Elettronica e dei Sistemi, Viale A. Doria 6, Catania, Italy I-95125, coco@diees.unict.it

Abstract In this paper a novel 3-D Finite Element (FE) particle based approach is presented to investigate the ion flow across ionic channels. This consistent model foresees direct integration of the dynamical equations of ions subject to electromagnetic forces inside membrane channels, considering ion-ion interactions and taking into account explicitly the effects of molecular friction and thermal noise. The simulation results presented show that the mechanism of opening and closing of the membrane channels (Ca^{++}) as a function of the membrane voltage can be correctly reproduced by a particle model.

1 Introduction

The exchange of signals between living cells takes place mainly through the cellular membrane, which represents a selective permeable barrier between the cell and the extracellular environment. The communication network of chemical signals between cells rules and coordinates various critical cellular functions including differentiation, apoptosis, etc. The flow of substances across cell membranes takes place through membrane channels, which are typical hydrophobic regions having a size of the order of few Å, where the membrane lipid bilayer exhibits 'openings'. Among interesting substances, ions are of paramount importance since activation of several critical signalling pathways depends on ionic concentrations (especially Ca^{++} and K^{+}) and therefore a number of cellular functions are activated by specific ion concentrations. For these reasons, the ion transport across cell membranes and the electrolytic equilibrium between the cells and their environment have a fundamental role in biological systems.

The simulation of the mechanism of ion flow across ionic channels is a very complicated task, for a number of reasons including the lack of accurate descriptions of channel structure, the difficulty of modeling the behavior of the proteinaceous chains constituting the channel walls, the very high number of atoms, the very short time scale of the involved dynamical phenomena, etc. Nevertheless several attempts have been made to build coherent representations of ion flow across ionic channels, in accordance with experimental measurements. In literature several approaches have been followed for this purpose. The most used technique for the analysis of the interactions between ions in a biological environment is Molecular Dynamic (MD), which is based on the atomic model of macromolecular systems, where the microscopic forces between atoms are represented by potential functions. The motion of all the atoms and particles in the system is obtained by the integration of Newton classical equations. In this way the macroscopic properties are deduced from microscopic observations. The link between microscopic and macroscopic properties is supplied by statistical mechanics. The drawback of such an approach is the huge computational effort, required as soon as more than few particles are considered and extremely short time steps are needed. This has prevented MD application to complete simulation of ion transport across ionic channels.

The Brownian Dynamics (BD) is a computational method well suited for the analysis of the ion permeation process in the long time scale (ns). The BD considers integration of stochastic equations of ion motion, where the ion-ion and ion-channel interactions are represented by potential functions. The main hypothesis of BD is that the solvent molecules are not dealt with explicitly and are represented as a continuous dielectric.

The Poisson-Nernst-Planck (PNP) model is based on the electrodiffusion theory that describes the average ionic flux due to gradients of ion concentrations and electric fields. This method is different from MD and BD because ion motions are not explicitly considered.

In this paper a 3-D Finite Element (FE) particle based model is presented to investigate the ion flow mechanism by direct integration of the dynamical equations of ions subject to electromagnetic forces inside the ionic channels. In our approach attention is entirely focused on the moving species and their mutual interaction (ion-ion interaction) which is believed to be dominant in the description of transport phenomena. In our model the channel environment is represented synthetically using few quantities, summarizing all the influencing factors by means of a continuum equation (Poisson equation). This approach has been especially set up in order to take into account the ion-ion interaction inside ionic channels in a simple way. This model is able to describe the behavior of ionic channels (K^+, Ca^{++}) in terms of total current carried by ions at various membrane voltages.

The paper is structured as follows: in Section II the mathematical model and finite element discretization is presented; in Section III simulation of the Ca^{++} ion flow across a simple calcium channel is illustrated; the authors' conclusions follow in Section IV.

2 Mathematical Model and Finite Element Discretization

The behavior of ions inside membrane ionic channels is governed by the following system of coupled equations (Langevin-Lorentz-Poisson), in which the effects of the spatial charge are modelled by assuming stationary conditions and considering a Poisson problem for the scalar potential φ

$$m \cdot \frac{d\mathbf{v}}{dt} = -h \cdot \mathbf{v} + q \cdot (\mathbf{E}_T + \mathbf{v} \times \mathbf{B}_T) + N(t) \tag{1}$$

$$\nabla^2 \varphi + \iiint_{\Omega} f dV = 0 \tag{2}$$

$$\mathbf{E}_T = -\nabla\varphi + \mathbf{E}_{eso} \tag{3}$$

where m is the generic ion mass, **v** is its velocity, h is a viscous friction coefficient, modeling ion interactions with water molecules, q is ion charge, $\mathbf{E}_T$ and $\mathbf{B}_T$, if any, are the total electric and magnetic fields respectively, N(t) is a random force which takes into account the thermal effects, f is an unknown function describing the space charge distribution. The total electric field $\mathbf{E}_T$ consist of two terms: the first is the contribution due to the scalar potential, the second, E_{eso}, takes into account the exogenous electric forces, if any, to which the ion is subject in the cell environment.

In our approach the numerical solution of the above coupled electromagnetic-motional problem is performed according to a self-consistent scheme in which the time-domain integration of the ion motion equations alternates with the FE solution of the 3-D Poisson problem. The resulting discretized problem consists of two systems of equations: the first is an FE linear algebraic system regarding the spatial distribution of unknown potential values at a certain time instant t, the other regards the displacement, occurring at a certain time step Δt, of all the moving particles (ions) used in the modelization of the ionic channel. The FE linear algebraic system is obtained from the minimization of the energy functional associated with the Poisson equation, in which it is supposed that the space charge distribution at the generic time instant t is known.

In this way the procedure allows us to determine all the ion trajectories during a certain time interval, within which the following two steps alternate:

1. In the first step the discretized Poisson equation is solved (initially guessing an ion distribution at time t=0).

2. In the other step an estimate for the electrical field distribution is derived by suitably post-processing the obtained scalar potential values; this electrical field is used to determine the ion displacements by time integrating the Langevin-Lorentz equations, subject to thermal noise. From this a new configuration for the ion distribution is computed and then used to perform the successive Poisson-solver step.

It is worth noticing that the region where the FE analysis is performed is dielectrically homogeneous and encloses all the moving ions involved in the flow mechanism.

The boundary conditions for the FE problem are imposed by evaluating all the contributions due to sources external to the selected surfaces constituting the boundary. In particular Dirichlet conditions are used mainly to take into account membrane voltages, whereas nonhomogeneous Neumann conditions model the effects of surface charge distributions on channel walls.

3 Simulation of ion flow across a Ca++ channel

Some simulations regarding the description of ionic flow across a typical Ca^{++} channel are shown in order to illustrate the application and the advantageous features of the above 3-D FE particle model.

In the simulations a commonly used description of the channel geometry and schematization of its environment are assumed. In particular a cylindrical geometry is considered as shown in Fig. 1 (height 10 Å and radius 5 Å respectively), the cylinder axis is coincident with the z axis.

The channel walls are made of polypeptide chains, which are assumed fixed and are represented by nonhomogeneous Neumann conditions in the FE analysis.

Two charge reservoirs are considered, one at the beginning and the other at the end of the channel [CHA98] . These reservoirs represent the ion populations in the close proximities of channel extremities, ready to access or leave the channel region. The behaviour of reservoirs is approximated by resorting to fixed point charge configurations placed in the neighborhood of the channel extremities (Fig. 1).

In the following simulations the contributions of these charges is accounted for by means of the external field $\mathbf{E}_{eso}$. The aim of the analysis is to investigate the influence of membrane voltage on the opening or closing of the channel. Assuming that the typical transit time of a Ca^{++} ion is about $10^{-9}s$, the simulation time interval was chosen one order of magnitude greater that this typical transit time. In order to make the analysis compatible with the hypothized stationary conditions the integration time step was chosen equal to $10^{-15}s$. During analysis it is also assumed that the ionic channel is always fully occupied by a number of ions that saturates its geometrical capacity; for the presented geometry this capacity has been estimated to be four Ca^{++} ions. For this reason at the beginning ($t = 0$) four Ca^{++} are placed inside the cylindrical region in a non-interfering configuration as shown in Fig. 1. In addition, when one ion exits the cylindrical region, another ion is added in the successive integration time step on the opposite side, in such a way to maintain channel full occupancy during analysis. The various computations have been performed for several values of the membrane voltage. The results are expressed in terms of number of Ca^{++} ions crossing the channel during the aforementioned simulation time interval as a function of membrane voltage, and are shown in Fig. 2. It is worth noticing the influence of the membrane voltage on the channel gating: for membrane voltages in the range below 30 mV no ion flow is observed, and the ion trajectories are all confined inside the cylindrical region, as shown in Fig. 3; when membrane voltages greater than 30 mV are applied, a net charge flow is observed which increases for increasing values of membrane voltages, and the ion trajectories are no longer confined inside the ionic channel (see Fig. 4).

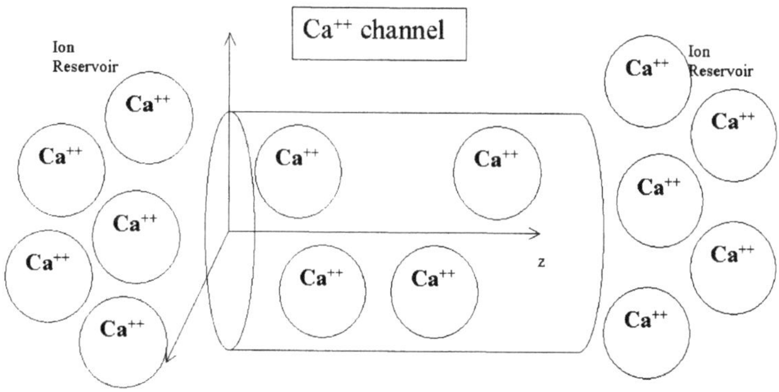

Fig. 1. Schematization of the adopted channel model

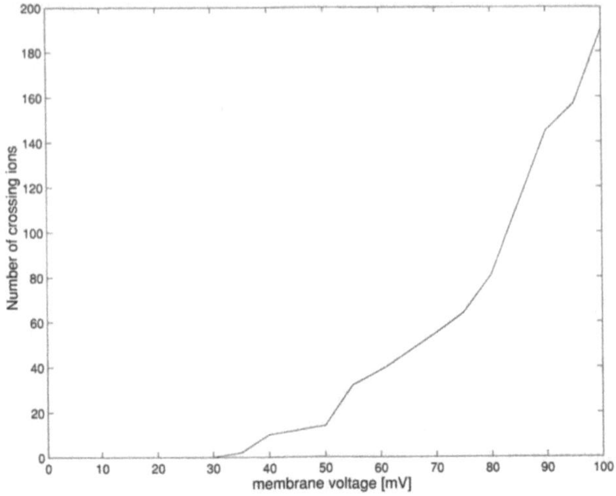

Fig. 2. Number of ions the channel versus membrane voltage

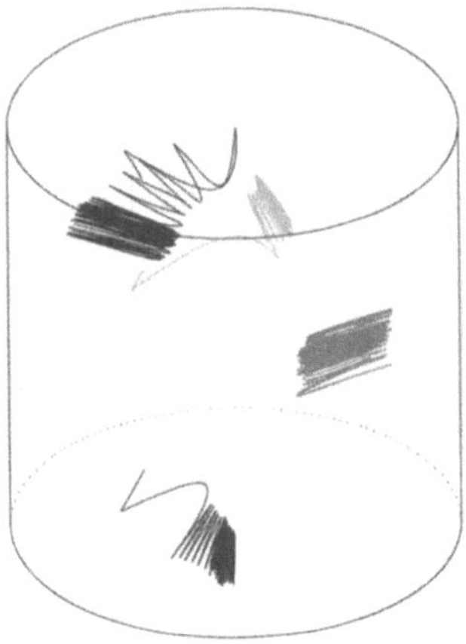

Fig. 3. Ion trajectories all confined inside the channel for a membrane voltage of 0 mV (simulation interval 10ps)

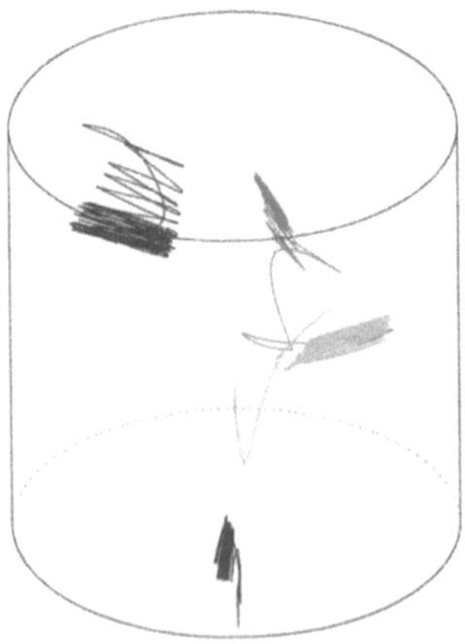

Fig. 4. Ion trajectories inside the channel for a membrane voltage of 100 mV in the event of an ion exiting the channel (simulation interval 10ps)

4 Conclusions

The presented 3-D FE particle model has proven effective for the simulations of the behaviour of ionic channels. In particular gating of a Ca^{++} channel due to membrane voltage has been successfully described in accordance to expected results. The results achieved are in good agreement with analogous simulations available in literature obtained by using other techniques (PNP) [GNE02]. The main advantage of this 3-D FE particle approach is the simplicity of treatment of moving ion interactions.

References

[OCG04] Oliveri, M. E., Coco, S., Gazzo, D. S. M., Giuffrida, C., and Laudani, A.: A 3D Stationary Langevin-Lorentz-Poisson Model for the Analysis of Ion Transport across Cell Membranes, International Journal Of Applied Electromagnetics And Mechanics,**19**, 165–168 (2004)

[SWS02] Saraniti, M. , Wigger, S. J., Schuss, Z. and Eisenberg, R.S.: Towards a reliable model of ion channels: three-dimensional simulation of ionic solutions, Proceedings of the Second International Conference on Computational Nanoscience and Nanotechnology, Puerto Rico, U.S.A. April 2002

[GNE02] Gillespie, D., Nonner, W. and Eisenberg, R. S.: Coupling Poisson-Nerst-Planck and Density functional theory to calculate ion flux, Journal of Physics: Condensed Matter, **14**, 12129–12145, (2002)

[Jak98] Jakobsson, E.: Using Theory and Simulation to Understand Permeation and Selectivity in Ion Channels, Methods: A Companion to Methods in Enzymology, **14**, 342–351, (1998)

[KCG99] Kurnikova, M. G., Coalson, R. D., Graf, P. and Nitzan, A.: A Lattice Relaxation Algorithm for Three-Dimensional Poisson-Nerst-Planck Theory with Application to Ion Transport through the Gramicidin A Channel, Biophysical Journal, **76**, 642–656, (1999)

[BMC97] Bianco, B., Moggia, E. and Chiabrera, A.: Fokker-Planck analysis of the Langevin-Lorentz equation: Application to ligand-receptor binding under electromagnetics exposure,J. Appl. Phys. **9**, 4669–4677, (1997)

[CHA98] Chung, S., Hoyles, M., Allen, T. and Kuyucak, S.: Study of Ionic Current across a Model Membrane Channel using Brownian Dinamics, Biophisical Journal, **75**, 793–809, 1998

[GBM01] Gomulkiewicz, J., Bartoskiewicz, M. and Miekisz, S.: Some remarks on ion transport across excitable membranes. The stationary state. Current Topics in Biophisic, **25**, 3–9, (2001)

[Ris89] Risken, H.: The Fokker-Planck Equation, Berlin (1989)

4 Conclusions

References

Coupled Calculation of Electromagnetic Fields and Mechanical Deformation

U. Schreiber and U. van Rienen

Institute of General Electrical Engineering, Rostock University, Germany,
{ute.schreiber,ursula.van-rienen}@etechnik.uni-rostock.de

Abstract Interest in multi-disciplinary simulations as a means of solving coupled electromagnetic-mechanical problems is increasing. The development of adapted simulation codes is an answer but these codes are often very specialized and are not always applicable to physically similar problems. For this reason it is advisable to create coupling software for the existing codes. The MpCCI-library is such a coupling software. In this paper we will present the results for a coupled simulation of electromagnetic fields in an accelerating cavity and its structural deformation by Lorentz force via the MpCCI-library.

1 Introduction

Today, numerical simulation plays a key role both in industry and research. Simulation tools and the results they are providing are an integral part in the design and development of new and better products. Many aspects of a system/product behavior are affected by the interaction of different physical phenomena. Two ways are possible to realize such complex simulations. The development of *one* software package adapted to the multi-physic problem is one way, a so called strong coupled calculation. It is also possible to use existing software codes coupled via a coupling software which is called weak coupling. Such a weak coupling is utilized for the following simulation.

The electromagnetic fields cause forces which may lead to a significant mechanical deformation of a structure. Thus a feedback effect exists between the field distribution and the geometrical shape of the studied device. In different cases, where highly accurate fields are needed, this deformation has to be taken into account and a dynamic "self-consistent" calculation is required.

This paper first describes the general weak coupling procedure with the coupling code MpCCI. Then one example calculation is explained. In this example we determine the frequency shift caused by the deformation in a superconducting cavity. We show that the coupled calculation is applicable to our own specific problem.

2 Coupling

Many codes for the computation of different physical problems may result in different kinds of meshes adapted to the type of problem that they are being applied for (see Fig. 1). Using weak coupling, a primary task of the coupling software is the transfer of data between these different meshes and hence the interpolation of the quantities to be exchanged. A second important task of the coupling software is in synchronizing the calculations produced at different stages of the coupled computational process or in synchronizing calculations performed by separate processing modules (see Fig. 2).

To couple two or more mesh-based numerical codes, a library called MpCCI (Mesh-based parallel Code Coupling Interface) [1] can be used. The MpCCI library realizes the main tasks mentioned, the interpolation and the synchronization of the coupling.

MpCCI's main purpose lies in sending and receiving messages to synchronize the computation process. It performs neighborhood search and interpolation between the meshes to achieve the reliable exchange of

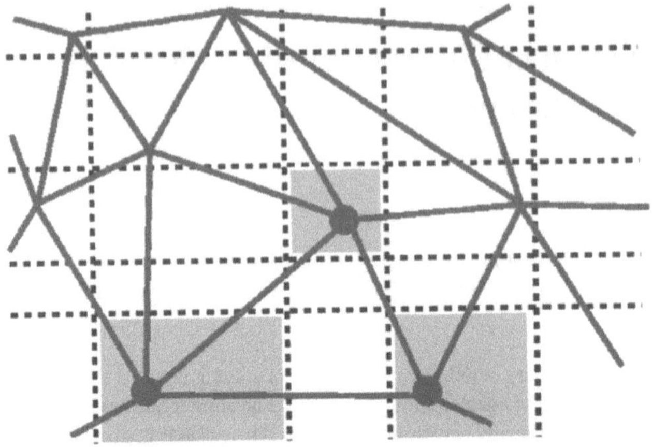

Fig. 1. Coupled Simulation Meshes

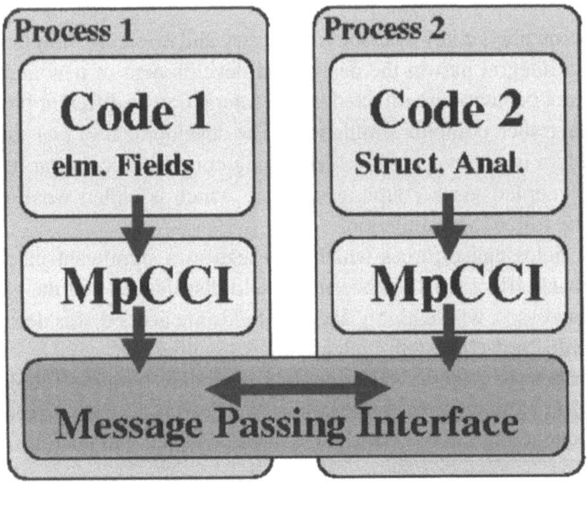

Fig. 2. MpCCI Software-Layers

coupling data. The neighborhood computation determined in a setup phase is used to establish the inter-process communication, and to assist the codes in the interpolation of coupling values between different grids. MpCCI realizes the exchange of mesh-based data (any type of quantities) in one or more specified coupling areas where the interactions between the specified physical properties take place.

The software MpCCI facilitates the coupling of any of the mesh based simulation software on different softwares systems (linux, unix, windows,...) simultaneously. The different coupled codes can be run on different computers and are coupled via the net. The coupled codes generally need an interface to MpCCI.

Here, the electromagnetic simulation is carried out with the software package MAFIA [2] based on the Finite Integration Technique (FIT, [3, 4]). An MpCCI-interface in MAFIA has been newly implemented.

The software package ParaFep [5] is an object oriented Finite Element program to calculate stability problems in 2D/3D structures. ParaFep is applicable to different problem types, e.g. linear and non-linear static. The MpCCI interface is readily available in ParaFep.

3 Example

Present fundamental research in particle physics and in nuclear physics needs high energy experimental setups. The use of superconducting cavities with high gradients constitutes an important technological advance for such facilities. TESLA (**Te**V **E**nergy **S**uperconducting **L**inear **A**ccelerator) is a proposal for a superconducting linear accelerator [6]. The TESLA collaboration operates a test facility at DESY (Deutsches Elektronen-Synchrotron) which will be utilised as a second generation test facility.

The resonant, time harmonic electromagnetic fields inside of superconducting cavities (see Figs. 3 and 4) can be calculated using the so called "curl-curl-equation" (wave equation)

$$\mathrm{curl}\frac{1}{\mu}\mathrm{curl}\underline{\boldsymbol{E}} - \omega^2\varepsilon\underline{\boldsymbol{E}} = 0, \tag{1}$$

$$\mathrm{curl}\frac{1}{\mu}\mathrm{curl}\underline{\boldsymbol{H}} - \omega^2\varepsilon\underline{\boldsymbol{H}} = 0 \tag{2}$$

where $\underline{\boldsymbol{E}} = \boldsymbol{E}\cdot e^{i\omega t}$ is the complex electric field amplitude and $\underline{\boldsymbol{H}} = \boldsymbol{H}\cdot e^{i\omega t}$ denotes the complex magnetic field amplitude. The quantities μ and ε are the permeability and the permittivity (material parameters), ω is the resonant circular frequency ($\omega = 2\pi f$) searched for.

The electromagnetic field excerts a Lorentz force on the currents induced in a thin surface layer. The resulting pressure acting on the cavity wall (see Fig. 5)

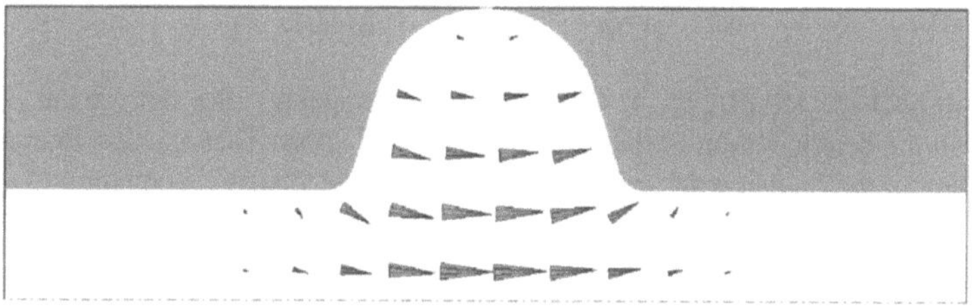

Fig. 3. Electric field distribution of the fundamental mode at f = 1.3 GHz in a one-cell-cavitiy type TESLA. The maximal field strength on axis is 25 MV/m

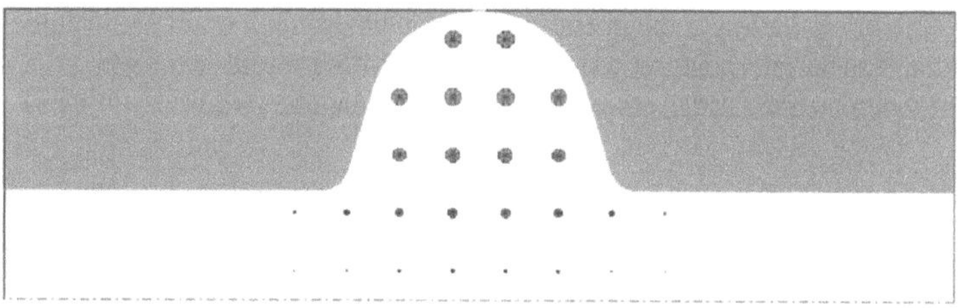

Fig. 4. Magnetic field distribution of the fundamental mode at f = 1.3 GHz in a one-cell-cavitiy type TESLA. The maximal field strength on axis is 25 MV/m

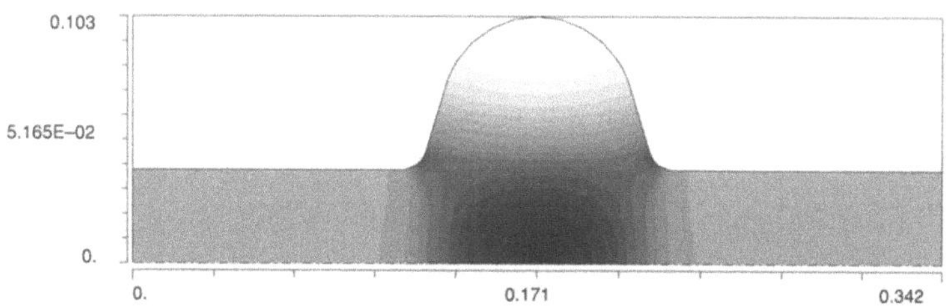

Fig. 5. Pressure distribution p obtained from the electromagnetic field of the fundamental mode at 1.3 GHz in a one-cell-cavitiy type TESLA. The pressure p ranges between $-2.2\cdot10^{-12}$ N/m^2 (black) and $1.02\cdot10^{-12}$ N/m^2 (white)

$$p = \frac{1}{4}\left(\mu_0|H|^2 - \epsilon_0|E|^2\right) \tag{3}$$

leads to a deformation of the cells in the μm range and a change ΔV of their volume. The result is a frequency shift Δf according to

$$\frac{f - f_0}{f_0} = \frac{\int_{\Delta V}(\mu_0 H^2 - \varepsilon_0 E^2)dV}{\int_v(\varepsilon_0 E^2 + \mu_0 H^2)dV}.$$

Here, f_0 is the resonant frequency of the unperturbed cavity. The parameters ε_0 and μ_0 are the absolute permittivity and the absolut permeability. The cavities could be driven out of resonance by this infinitesimal mechanical deformation, since such deformations increase in size with the square of the accelerating gradient [7]:

$$\Delta f = -K \cdot \underline{\boldsymbol{E}}^2$$

were K is a constant called the detuning factor.

To implement reliable measures that will prevent frequency shift, knowledge of the force distribution is necessary as well as the field distribution after the deformation. A coupled computation of the electromagnetic field and mechanical deformation is advisable to get this informations. Calculation of the frequency shift in a coupled simulation for one cell of a TESLA cavity were performed in the following way.

Firstly, we transformed the curl-curl-equation (1) to an analogous discrete eigenvalue problem

$$(\widetilde{\mathbf{C}}\mathbf{M}_{\mu^{-1}}\mathbf{C} - \mathbf{M}_{\epsilon}\,\omega^2)\,\widehat{\mathbf{e}} = 0$$

with the software package MAFIA [2, 4] (equation (2) analogously). The discretized curl-curl-equation obtained served in the determination of the eigenmodes of our cavity. Then the distribution of the pressure inside of the cavity in the fundamental mode of 1.3 GHz was calculated via equation (3) (see Fig. 5). The pressure vector was converted to the force vector

$$\boldsymbol{f} = \boldsymbol{p} \cdot d\boldsymbol{A}$$

by multiplying the grid cell area with the normal component of the pressure.

Secondly, we opened the MpCCI interface and sent the force data at the boundary of the cavity to the coupling software. MpCCI interpolated the force values of the FIT-grid to ParaFep-grid and sent the new data to ParaFep. The interpolation of the quantities was established in a setup phase at the beginning of the coupled calculation. Linear interpolation was used within MpCCI for this example.

Then ParaFep calculated the displcement u of the cavity boundary via

$$K \cdot u = R$$

from the outer forces R and a stiffness matrix K. The displacements were given back to MAFIA via MpCCI.

The Finite Integration Technique allocate material parameters to each grid cell. Now new material parameters were determined for grid cells with a displaced contour inside. Therefore the material parameter utilized were averaged proportionally to their modified volume rate (see Fig. 6). This procedure was reflected in the change of the material matrices $\mathbf{M}_\epsilon$ and $\mathbf{M}_{\mu^{-1}}$ of equation (1) and it's done in an additional routine outside of the participated coupled softwares. This method is also called Conformal FIT (CFIT) which is useful for a better approximation of boundaries in Cartesian grids.

Finally, equation (1) was solved again with modified material matrices $\mathbf{M}_\epsilon$ and $\mathbf{M}_{\mu^{-1}}$. The result for the resonant frequency was shifted relative to the first calculations.

4 Results

The cavity has been discretized in MAFIA with 21×51 =1071 mesh points in rz-geometry. The wave equation (1) was solved. We got for the primary calculation a resonant frequency of

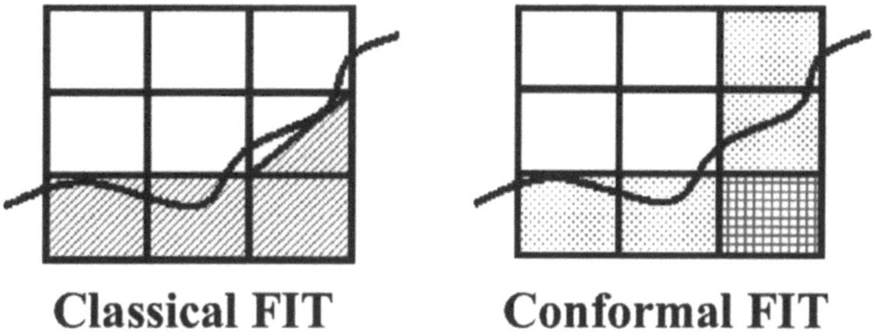

Fig. 6. Differently used material parameters for grid cells in Mafia

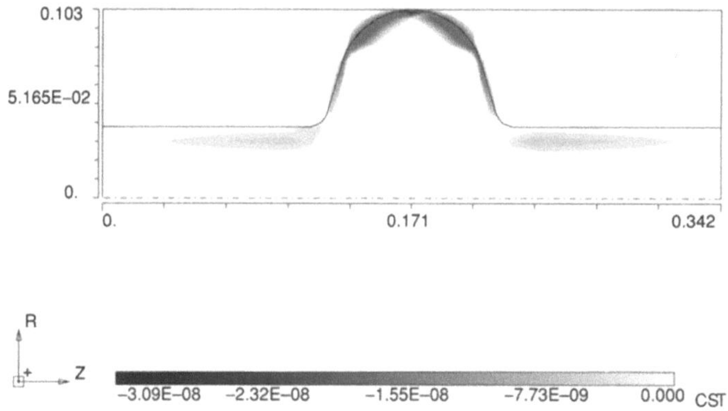

Fig. 7. Calculated deformation in r-direction of a cavity with 3 mm niobium wall thickness for the fundamental mode of 1.3 GHz

$$f_1 = 1\ 303\ 393\ 294.144 Hz.$$

Since the geometry is axially symmetric we extrapolated the determined force values to a 3D geometry with $120 \times 21 \times 51 = 128\ 520$ mesh points. Only the $2 \times 51 \times 120 = 12240$ force values of the grid points at the boundary was sent to ParaFep.

The ParaFep geometry (only the cavity structure of niobium with 3 mm thickness, Youngs Modulus = 105 GPa, Poisson's ratio = 0.38) with 14 958 meshpoints received the force data. The displacements of the geometry was calculated with fixed edges and were sent back to MAFIA. The results are shown in the Figs. 7 and 8.

The resonant frequency calculated after changing the geometry (meshfill) was

$$f_2 = 1\ 303\ 393\ 140.324 Hz.$$

This results in a frequency shift of

$$\Delta f \approx 145 Hz.$$

Reference values are the frequency shift of [8] with a value of $\Delta f = 150$ Hz given for the same dimensions of cavity with a thickness of 4 mm and fixed edges. Another reference value is given in [6] which gives a frequency shift of $\Delta f = 900$ Hz for a cavity thickness of 2.5 mm and free edges. In spite of very few mesh points we approximated the frequency shift with a relative good degree of accuracy.

5 Summary

In this paper we described a weak coupled calculation of electromagnetic fields and mechanical deformation. A simple example was used as a proof of principle. Expanded simulations which means more steps of

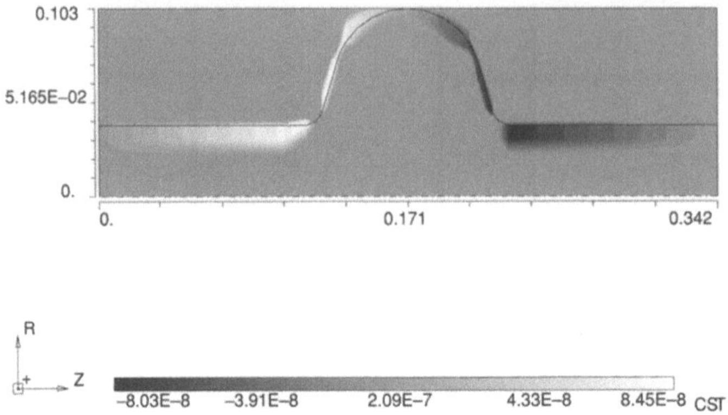

Fig. 8. Calculated deformation in z-direction of a cavity with 3 mm niobium wall thickness for the fundamental mode of 1.3 GHz

one coupled calculations will be carried out. Hereby the results of different coupling algorithms like Jacobi or Gauß-Seidel has to be compared. Additionally a mesh size convergence study will be performed.

References

1. MpCCI - Mesh-based parallel Code Coupling Interface. Version 1.3,PALLAS GmbH, Hermülheimer Str. 10, D-50321 Brühl, Version 1.3 (2002)
2. MAFIA V4.106. CST GmbH, Lautenschlägerstraße 38, D-64289 Darmstadt, Büdinger Straße 2a, D-64289 Darmstadt, Germany
3. Weiland, T.: A discretization method for the solution of Maxwell's equation for six–component fields. Electron. Commun. AEÜ, **31**(3), 116–120 (1977)
4. Schuhmann, R., Clemens, M., Thoma, P., Weiland, T.: Frequency and Time Domain Computations of S-Parameters Using the Finite Integration Technique. Proc. of the 12th Annual Review of Progress in Applied Computational Electromagnetics (ACES Conference), Monterey, 1295-1302 (1996)
5. Niekamp, R., Stein, E.: An object oriented approach for parallel 2– and 3–dimensional adaptive nonlinear Finite-Element-computations. Int. Journal of Computer and Structures, **80**, 317–328 (2002)
6. TESLA - Technical Design Report (II), http://tesla.desy.de/new_pages/TDR_CD/
7. Bousson, S. et. al.: SRF Cavity Stiffening by Thermal Spraying. Proceedings of the EPAC (2002) http://accelconf.web.cern.ch/AccelConf/
8. Gassot, H.: Mechanical Stability of the RF Superconducting Cavities. Proceedings of the EPAC (2002) http://accelconf.web.cern.ch/AccelConf/

Part II

Circuit Simulation

Challenging Coupled Problems in TCAD*

A. Benvenuti[1], L. Bortesi[1], G. Carnevale[1], A. Ghetti[1], A. Pirovano[1,2], L. Vendrame[1], and L. Zullino[3]

[1] STMicroelectronics, Via C. Olivetti 2, Agrate Brianza, 20041 Milano, Italy, augusto.benvenuti@st.com
[2] DEI, Politecnico di Milano, P. zza L. Da Vinci 32, 20133 Milano, Italy
[3] STMicroelectronics, Via Tolomeo 1, Cornaredo, 20010 Milano, Italy

1 Abstract

Many challenging coupled modelling problems arise in microelectronics; this paper illustrates some examples in the so-called Technology CAD (TCAD) area, encompassing process and device modeling, and will mention additional issues in the closely related fields of equipment and circuit modeling.

2 Introduction

On the eve of the "nano-electronics" era, the integrated circuit technology scenario is continuously evolving along the Moore scaling rule, affecting the complexity of the physical and mathematical problems to be addressed for numerical modelling.

The scaling of geometrical dimensions is emphasizing the importance of quantum effects, such as charge carriers confinement in the channel of MOS transistors, and tunneling across the gate dielectric.

As some critical dimensions are approaching the nanometer range statistical fluctuations of discrete particles can no longer be accurately described in terms of average concentrations only. As in device simulation a statistical description of the behaviour of individual electrons and holes can be achieved solving the Boltzmann Transport Equation by means of MonteCarlo techniques, a comprehension of the atomic scale mechanisms underlying macroscopic dopants diffusion, activation and clustering phenomena requires an insight not achievable by continuum models only.

Reduction of the elementary devices (i.e. MOS transistors) footprint allows to drastically increase their packing density, therefore undesired interactions between adjacent transistors or memory cells, such as proximity effects during fabrication and electrical or thermal disturbs during device operation are becoming more important.

Furthermore as vertical dimensions (e.g. gate oxide thickness and junction depth) shrink, effects related to interfaces between different materials keep increasing, such as dopant segregation, diffusion along interfaces and grain boundaries, and carriers scattering due to surface roughness.

On the other hand the die size is not scaling, as the increased density is typically exploited to add new functionality for the final product, that is increasing circuit complexity and the number of transistors; the consequence is that circuit speed is becoming limited by the delay due to global interconnect wires with respect to the intrinsic active device delay contribution (gate and junction capacitances).

Last but not least, as pure geometrical scaling is becoming more and more difficult (since state-of-the-art transistors are approaching fundamental scaling limits), in order to keep the same performance improvement rate the microelectronics industry is making use of aggressive technology and elementary devices engineering, introducing at each new generation new dopants, new materials, new architectures. In particular non-equilibrium phenomena are often exploited, with overshoot and ballistic effects in carrier transport in device operation, as well as the evolution towards rapid, high temperature processing in dopants annealing technology.

*Invited paper at SCEE 2004

These trends can – at least for what concerns modelling – be largely interpreted as a drive towards an increase in problems *coupling*:

- **between different physical effects**: electrical / thermal / mechanical / optical / electromagnetic interactions are common, and in some cases (e.g. sensors or equipment modelling) also chemistry and fluido-dynamics play an essential role;
- **between different elementary devices**, due to their direct interaction because of their spatial proximity, or through the parasitic effects introduced by their non-ideal interconnection (e.g. cross-talk, inductive effects and in general noise injection and propagation either in the substrate or through the metal interconnect lines);
- **between materials**, due to cross-contamination or integration issues, through the increasing effects related to material interfaces, or even through the influence between the bulk of two different regions like in the case of the remote Coulomb scattering contribution to mobility degradation;
- **between problems occurring on different space or time scales**, like in the interaction between gas dynamics and chemistry occurring on a reactor scale with the deposition or growth rate at the feature-size scale, or between atoms jump attempt frequency, microscopic dopants migration and macroscopic diffusion.

In the following we will describe shortly perceived needs, status and/or some recent advances for a few examples of such coupled problems.

3 Examples of coupled problems

3.1 Oxidation and mechanical stress

Fully accounting for the 3D mechanical stress distribution during the whole process flow is a very tough numerical problem still largely unsolved. In fact even 3D oxidation by itself requires the solution of a diffusion/reaction problem with moving boundaries, which for the complex layer system corresponding to realistic microelectronic devices is still not feasible with the required level of accuracy and stability. As a consequence, only simplified problems can - at present - be realistically tackled in an industrial environment. A non-trivial example is the description of the planar stress distribution dependence on the active area layout (Fig. 1). Alternatively the robustness (from a mechanical point of view) of several different processing options can be assessed by simulating a 2D vertical cross section in a critical region of the device and monitoring extremal quantities like the maximum or average resolved shear stress in silicon (Fig. 2). By comparing their time evolution during the process flow the most critical technological modules and/or recipes can be identified, and useful guidelines for reducing reliability concerns related to extended defect formation can be obtained.

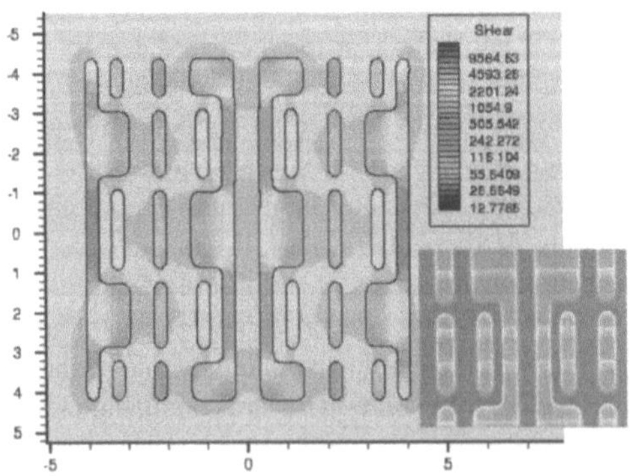

Fig. 1. Simulated final shear stress distribution on SRAM layout (planar stress approximation). Inset: SEM planar view of active area layout after delayering

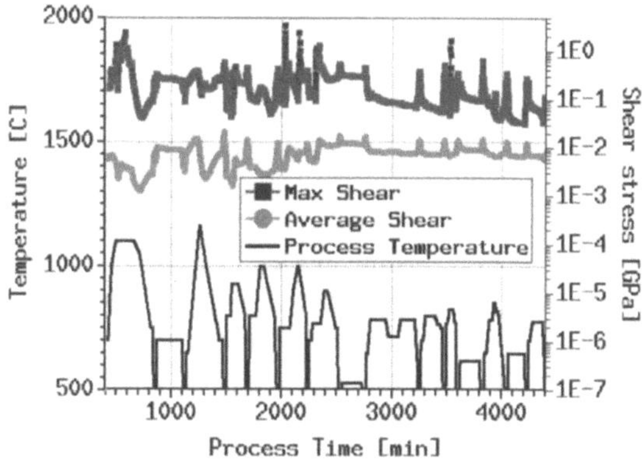

Fig. 2. Simulated maximum and average shear stress (right Y axis) during full process flow (left Y axis: thermal budget profile)

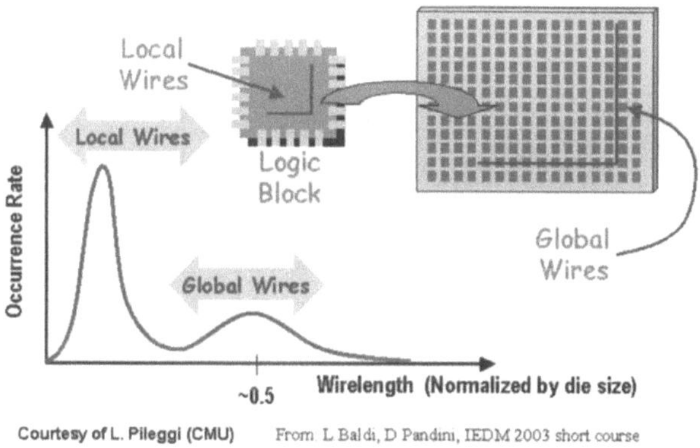

Fig. 3. Typical distribution of wire lengths (normalized to chip size) for a block-based design

3.2 Interconnect parasitics extraction

With continuous scaling of transistor dimensions and chip size increase, interconnect lines play a key role in the design of modern chips. In deep submicron technologies with multi-million transistors, it is not trivial to manage very large circuit layouts in order to extract interconnect parasitics which affects signal timing and integrity. In a flat extraction approach, which considers the layout spread over a single level, it is difficult to handle at the same time the whole layout description and the extracted RC values. This approach generates large netlists composed of millions nodes impractical for an electrical Spice-like simulation. To cope with these problems, different methods have been implemented. A hierarchical approach, which considers the layout divided in different levels and blocks, is useful to avoid an impractical pure flat extraction and to distinguish between global and local interconnects (Fig. 3).

Depending on the sub-problem size and according to the required level of accuracy different simulation methodologies are applied, as detailed below.

Global interconnects link different functional blocks inside the chip, therefore details at transistor level can be neglected, while small blocks of elementary circuits where accuracy plays an important role can be extracted and simulated at transistor level. In this case the focus is on local interconnects, where coupling capacitances and resistances in general constitute the dominant elements of interconnect parasitic lines.

On the other hand between different blocks signals are generally distributed with busses (global routing). According to the required data rate the model used to describe such global interconnect behavior can range from a simple RC ladder to a distributed RLC transmission line.

The most sensitive case is represented by so-called *critical nets* (for example clock distribution across a microprocessor), for which a high accuracy is needed also for long interconnects, since their performance deeply affect the global performance of the circuit.

A smart approach to extract the parasitic elements of critical paths with the desired precision is the Floating Random Walk (FRW) algorithm. It is based on a Monte Carlo approach which looks for the neighboring conductors along the critical net and evaluates the coupling capacitance by estimating the electric field with a recursive formula applied for each "hop" of the random walk between the conductor of interest and its neighbors [Bra03]. In this case only the RC parasitic components related to the critical net of interest are extracted, so it is not necessary to mesh the whole layout.

In Fig. 4 a schematic of RC parasitics extraction methodology is shown. Numerical calculation engines, i.e. field solvers and/or FRW tools, can be used directly on critical nets or to build libraries of analytical formulas (or look-up tables) for typical structures.

A special case for extraction is represented by repetitive structures such as an array of SRAM, DRAM or Non-Volatile (e.g. Flash EEPROMs, Fig. 5) memory cells. Due to the symmetry of the cell array and using reflecting boundary conditions, a solution of the Laplace equation with a standard finite-element based solver becomes feasible and attractive. Such a small structure also allows a fully 3D parasitics extraction including vias and contacts. FRW due to its general working principle can also be used for the parasitic extractions of this particular layout.

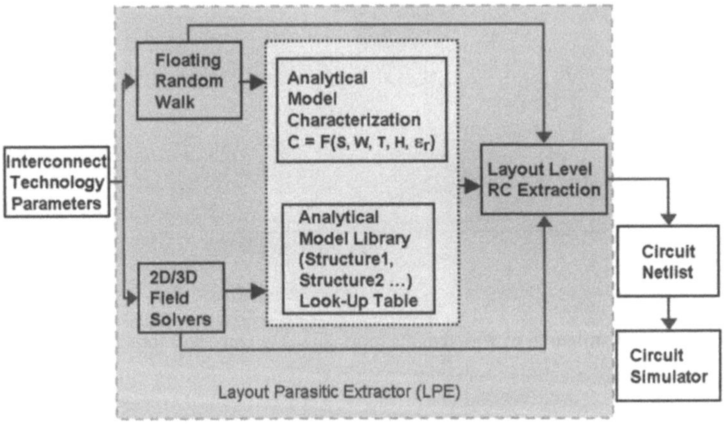

Fig. 4. Schematic of RC parasitics extraction methodology

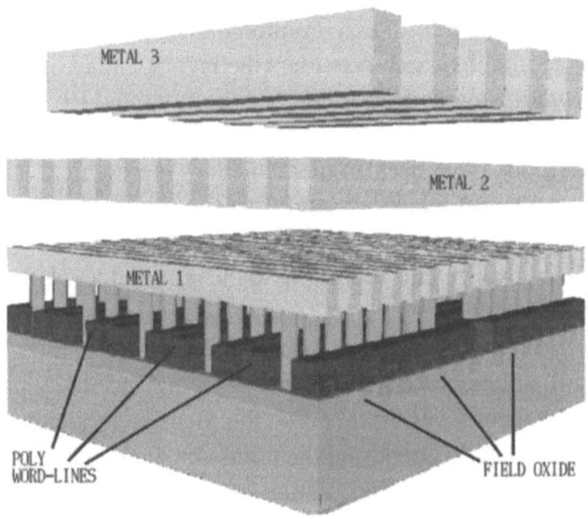

Fig. 5. 3D view of interconnect lines on a portion of Flash array (10 x 16 cells). The parasitic capacitances have been extracted both with a conventional field solver and with an efficient Floating Random Walk code [Bra03]

Structure	Target Cap	Measure (fF)	3D Field Solver (fF)	3D FRW (fF)
M1 bus over Si plate	Coupling M1	4.54	4.3	4.23
M3 bus	Inner M3 line	24.47	24.49	23.2
Orthogonal buses from Poly up to M3	Inner M2 line	8.21	8.11	7.9
Orthogonal buses from Poly up to M3	Total M3 bus	8.82	9.03	9.1
		CPU:	> 1hr	60-130"

Fig. 6. Comparison between parasitic capacitances simulated with a 3D Laplace equation solver, the Floating Random Walk algorithm and on-chip measurements for geometrically regular structures

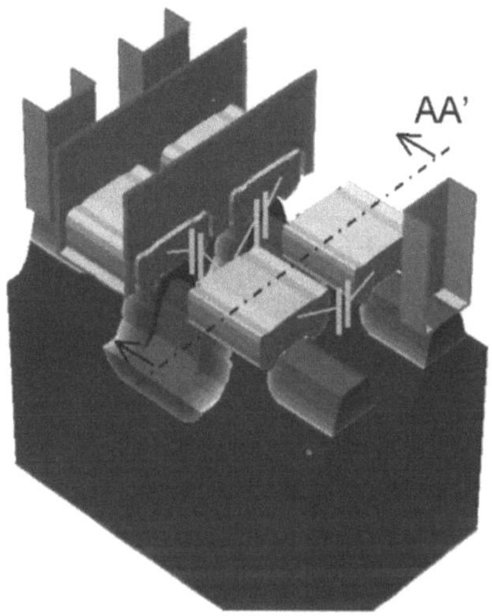

Fig. 7. Pictorial view of four adjacent Flash cells illustrating parasitic capacitive coupling between floating gates. Dielectrics, Word Line along cutplane AA' and one drain contact are not shown to allow better visibility of the floating gates

In Fig. 6 capacitances extracted on regular patterns of intermediate complexity with different techniques are compared against accurate on-chip measurements. The FRW technique compares favorably in terms of CPU/accuracy trade-off with respect to conventional Laplace equation solvers – which become too time consuming already for such medium-size problems – on one hand (Fig. 6), and with respect to layout pattern-based extraction tools – which are not suitable for complex, geometrically irregular layer structures – on the other hand.

However problems where the coupling with transport in silicon is of primary concern can be more thoroughly investigated by means of device simulation; an example is depicted in Fig. 7–8, where a 3D finite element-based solution of the Drift-Diffusion model is applied to evaluate the capacitive disturbs induced on a Flash cell by the charge stored on the floating gates of neighboring cells [Ghe05]. This approach, although more CPU intensive, leads to a more straight-forward and realistic coupling with technology options via process simulation, and allows to take into account localized perturbation effects on the current flowing in active devices.

3.3 Phase Change Memory modelling

The chalcogenide-based Phase-Change Memory (PCM), also called Ovonic Unified Memory (OUM), is a promising non-volatile semiconductor memory technology for stand alone and embedded applications. Such devices rely on reversible thermally-induced phase changes of thin-film chalcogenide materials, like Ge-Sb-Te (GST) alloys. Design and optimization of PCM cells require numerical models accounting for

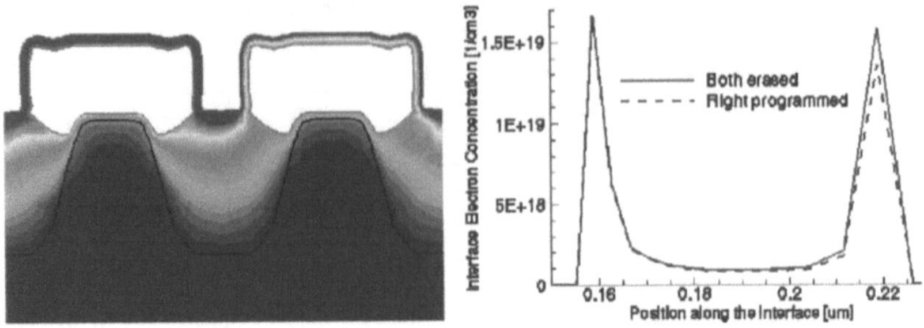

Fig. 8. Left: electrostatic potential distribution along the cutplane AA' of Fig. ref4celle, when the disturbed cell (left) is erased and the disturbing cell (right) is programmed. An asymmetry in the channel potential of the disturbed cell can be qualitatively seen. Right: corresponding electron concentration along the channel of the disturbed cell as a function of the charge stored in the floating gate of the disturbing cell

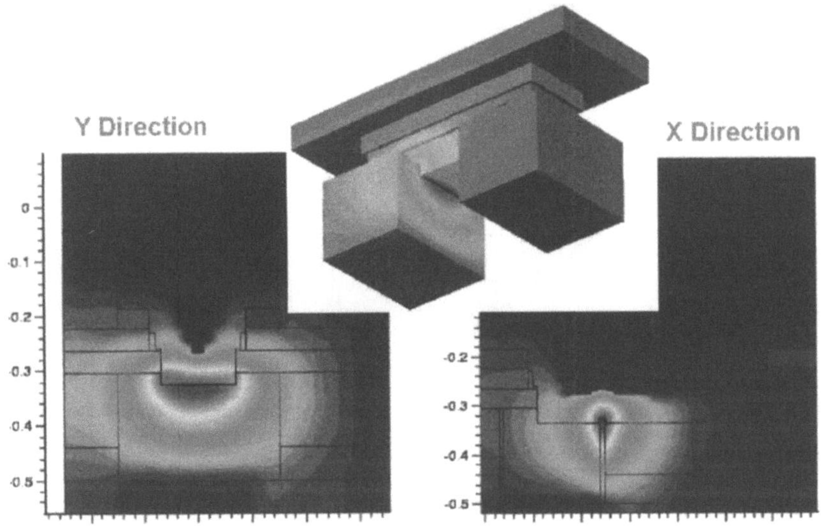

Fig. 9. 3D temperature distribution (top) and 2D cross sections on two adjacent Phase Change Memory (PCM) cells during "reset" operation on the left cell

carrier transport and phase change dynamics in the GST chalcogenide material. Recently, a model for GST band structure and carrier transport for both crystalline and amorphous material was developed [Pir02], and successfully implemented in a standard semiconductor device simulator. The electrical model has been self-consistently coupled to heat conduction equations; thermal conductivities for the GST, SiO_2 and other materials used in such devices were taken from literature, while a Monte Carlo nucleation and growth model [Pen97] was implemented to describe the phase transition from amorphous to crystalline GST.

Such a comprehensive electro-thermal model can describe complex phenomena such as electronic switching to the on-state, non-equilibrium, localized self-heating, and the crystallization kinetics.

Figure 9 reports the simulated thermal profile for two adjacent "μ-trench" PCM cells (please refer to [Pel04] for a detailed description of the cell structure). Relying on the available experimental data, it was possible to accurately tailor the simulation parameters, to perform predictive simulations on the PCM technology scaling capabilities [Pir03] and to investigate the impact on thermal cross-talk between neightboor bits down to the 45 nm technological node. In Fig. 10 the programming curve of the PCM cell is reported, showing the phase change distribution inside the cell. The amorphized (high resistivity) regions

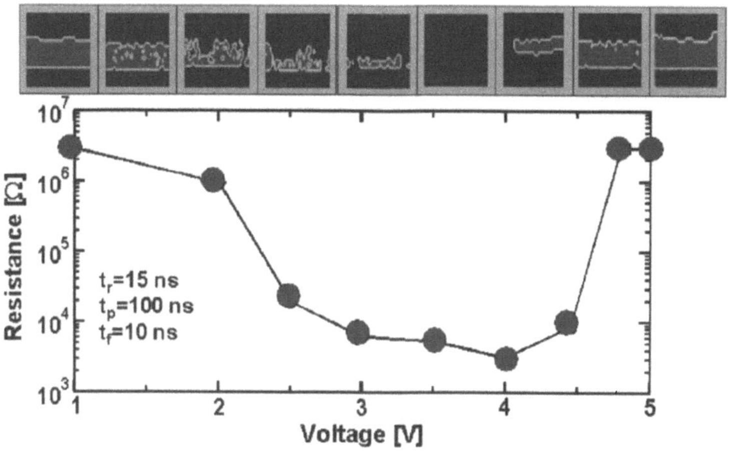

Fig. 10. Simulated "programming curve" for a PCM cell, showing the programmed resistance as a function of the programming pulse voltage. On the top the corresponding self-consistently simulated amorphous region is shown in red for each of the nine bias points

are shown in red, highlighting the existence of parallel conductive paths that cause a low resistance state even with a not-fully crystalline cell.

3.4 Electro-Static Discharge modelling

The modelling of Electro-Static Discharge (ESD) events is a particularly challenging task, since it involves the interaction of the protection device (e.g. a BJT transistor) with the surrounding circuit on the chip, and with the external environment through the package.

In the protection device the high intensity transient current peaks cause strong current crowding which in turn leads to the formation of hot spots due to localized self-heating, while external parasitic capacitances and inductances play an important role in determining the transient current overshoot.

In particular in the case of so-called *Charged Device Model* (CDM) discharge events the tester-package electro-magnetic interaction has to be taken into account. The problem can be modeled extracting for each package type an equivalent compact circuit describing the pin-to-pin interactions in terms of their coupling capacitances and mutual inductances, modelling the tester-package interaction with a lumped equivalent circuit (Fig. 11), and simulating the integrated circuit by means of a mixed-mode approach where the snapback of the pnp BJT ESD protection device is described with a thermo-electric Drift-Diffusion or Hydrodynamic transport model.

With such an approach it is possible to investigate internal discharge current paths during CDM events, which could not be easily achievable experimentally, and the corresponding current density and self-heating distributions inside the protection device can be compared with those occurring during the more established *Human Body Model* (HBM) discharge events (Fig. 12).

3.5 Full 2D quantum-mechanical charge confinement

Starting from the works of Stern [Ste70], many attempts have been made in the last two decades to correctly include quantum mechanical (QM) effects in MOSFET device simulation. The investigation of the quantized inversion layer has been initially addressed by one-dimensional (1D) models that take into account QM effects in the direction normal to the silicon-oxide interface. These approaches are reliable enough as far as a quantitative description of the threshold voltage shift and of the effective oxide thickness is concerned, which are the primary effects in long channel devices. However, QM effects along the transport

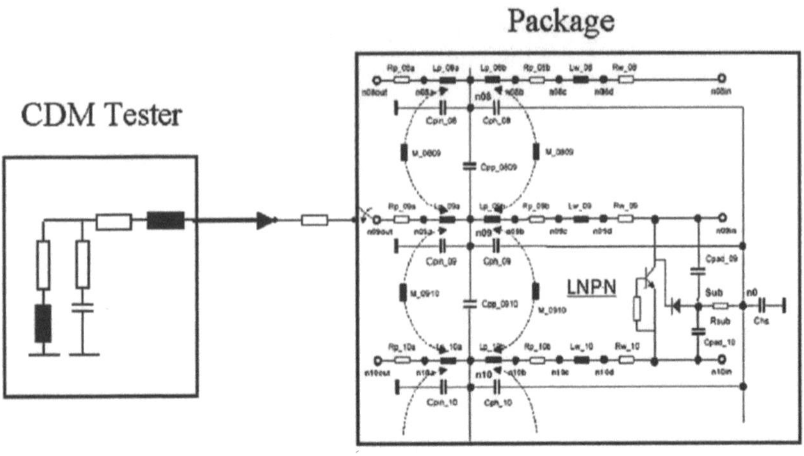

Fig. 11. Simplified schematic of equivalent circuit used in Charged Device Model ESD event simulation. The lateral npn BJT (in red) is described numerically with a Drift-Diffusion model

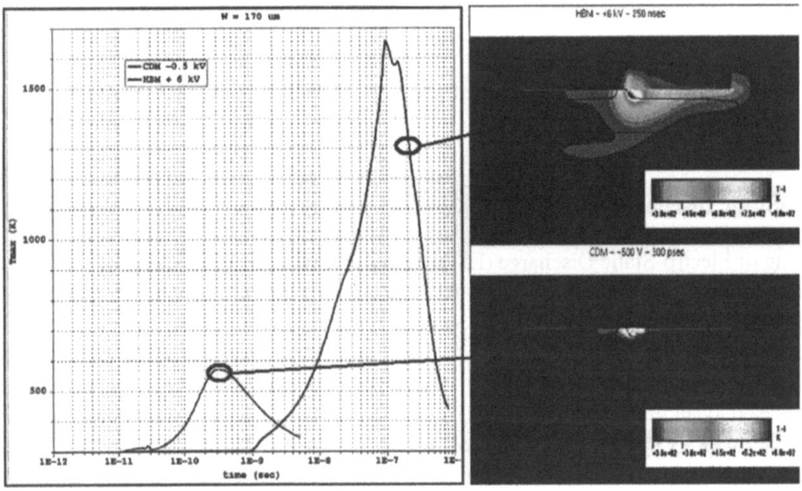

Fig. 12. Maximum device temperature during ESD events for Human-Body Model (black, top inset) and Charged Device Model (red, bottom inset) discharge event simulation

direction cannot be neglected in sub-50-nm transistors, thus requiring a fully 2D approach to solve the Schrödinger equation in order to take into account the confinement effects near the source/drain potential barriers.

A full-2D treatment of charge quantization in the channel of MOS transistors [Pir02b, Gus03] is achieved by solving a classical (Drift-Diffusion) transport model self-consistently with the 2D Poisson-Schrödinger equation. Figure 13 shows the classical and the QM charge distributions obtained in a deca-nanometer MOS device. It is worth noting that the peak charge concentration is located at the interface for a classical (CL) solution, but is shifted about 1 nm toward the bulk in the QM model. In the latter case the strong confinement leads to a much lower value for the QM charge in the channel and in the gate overlap regions; on the other hand, in the source/drain regions the QM confinement is removed, and the classical distribution is exactly recovered at a slightly larger distance from the silicon-oxide interface.

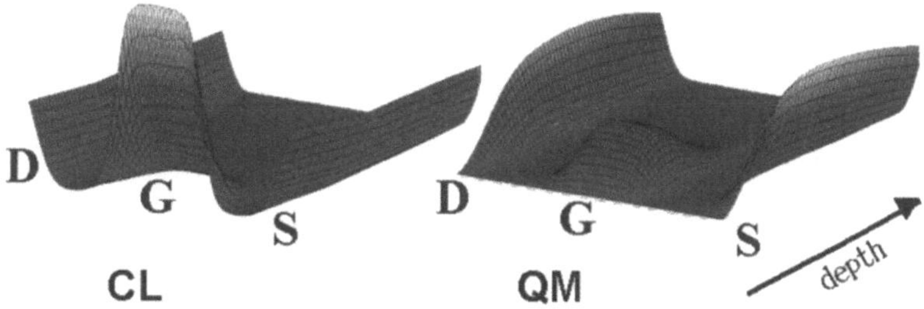

Fig. 13. Comparison between electron distribution in a MOS vertical cross section under low V_{ds} bias condition. Left: classic result; right: full-2D Quantum-Mechanical solution. The position of gate (G), source (S) and drain (D) electrodes is schematically marked on both figures; the depth direction has been stretched to highlight the different distance from the gate oxide of the peak channel concentration

4 Conclusions

We have presented a few cases of challenging problems in TCAD, emphasizing the increasing importance of coupling between physical effects, between different devices, between different materials, and between sub-problems occurring on different space or time scales.

In most cases further advances in modelling of such problems would require improvements in the identification of the most suitable physical model and of their mathematical formulation, along with the exploitation of more efficient numerical algorithms, but would also largely benefit from the availability of more flexible and modular modelling tools.

For the future it is expected that the trends related to the fast evolving scenario of the microelectronic industry will further enhance the complexity and lead to even stronger sub-problems coupling; to address the resulting modelling needs a comprehensive, flexible computational modelling platform able to cope with different physical phenomena and spatial dimensionality, and exploiting a variety of deterministic and statistical numerical techniques will be needed.

Acknowledgements

The efficient FRW code used for interconnect parasitics extraction has been developed by A. Brambilla, P. Maffezzoni and L. Codecasa (Politecnico di Milano). The full-2D Quantum-Mechanical MOS model has been developed in cooperation with A. Spinelli (Univ. Insubria, Como), R. Gusmeroli and A. Lacaita (Politecnico di Milano). The CDM ESD modelling methodology has been developed in the framework of the MEDEA+ ASDESE european project.

References

[Bra03] Brambilla, A., *et al.*, Measurements and Extractions of Parasitic Capacitances in ULSI Layouts, IEEE Trans. on Electron Devices, **ED-50**, 2236–2247 (2003)

[Ghe05] Ghetti, A., *et al.*, 3D simulation study of gate and noise coupling in advanced floating gate non volatile memories, Proc. ICMTD 2005, 157–160 (2005)

[Pir02] Pirovano, A., *et al.*, Electronic Switching Effect in Phase-Change Memory Cells, IEDM Tech. Dig. 2002, 923–926 (2002)

[Pen97] Peng, C., *et al.*, Experimental and theoretical investigations of laser-induced crystallization and amorphization in phase-change optical recording media, J. Appl. Phys., **82**, 4183–4191 (1997)

[Pel04] Pellizzer, F., *et al.*, Novel μ-trench phase-change memory cell for embedded and stand-alone non-volatile memory applications, Symp. on VLSI Tech. 2004 (2004)

[Pir03] Pirovano, A., *et al.*, Scaling Analysis of Phase-Change Memory Technology, IEDM Tech. Dig. 2003, 699–702 (2003)

[Ste70] Stern, F., Iteration methods for calculating self-consistent fields in semiconductor inversion layers, J. Comput. Phys., **6**, 56–67 (1970)

[Pir02b] Pirovano, A., *et al.*, Two-Dimensional Quantum Effects in Nanoscale MOSFETs, IEEE Trans. on Electron Devices, **ED-49**, 25–31 (2002)

[Gus03] Gusmeroli, R., *et al.*, 2D QM simulation and optimization of decanano non-overlapped MOS devices, IEDM Tech. Dig. 2003, 225–228 (2003)

On the Formulation and Lumped Equivalents Extraction Techniques for the Efficient Modeling of Long Interconnects

M. de Magistris[1], L. De Tommasi[1], A. Maffucci[2], and G. Miano[1]

[1] Dipartimento di Ingegneria Elettrica, Università di Napoli "FEDERICO II", Via Claudio 21, I-80125 Napoli, Italy
[2] D.A.E.I.M.I., Università di Cassino, Via di Biasio 43, I-03043 Cassino (FR), Italy

Abstract The paper investigates some crucial aspects in the derivation of efficient time-domain equivalent circuits of lossy multiconductor transmission lines. We firstly highlight the possibility to achieve an exact extraction of the delays from the propagation operators in the Multi Transmission Line model. This provides best properties for defining regular remainders of describing operators to be identified. Secondly, we address the problem of representation of such remainders, showing how a proper "rank" condition on the residue matrices leads directly to a minimal order circuit, so improving the accuracy of the procedure. The proposed formulation and identification procedure are then applied to a reference test case and results are compared to a more traditional approach.

1 Introduction

The modern design and verification of complex high-speed circuits is crucially based on simulation tools which are required to be at the same time accurate and efficient. Usually the overall electromagnetic system can be modeled as a network composed of distributed and lumped elements: the problem is then reduced to the derivation of a suitable model from the electromagnetic characterization of each element, often based on time and/or frequency-domain samples of the port variables (e.g., [1]–[5]). As long as complexity, number of components and clock frequencies increases, the simulation of distributed elements affects the computational burden, posing severe limitations to the complete system analysis. Therefore, accurate and efficient tools should be characterized by: (*i*) a satisfactory electromagnetic description of the different subsystems and of the coupling between them; (*ii*) an efficient model-order reduction procedure to allow simulation of large systems.

There are two possible approaches to the problem (see [6] for a comprehensive review). In the first one all the linear subsystems (lumped and distributed) are modeled together, and a "blind" model order reduction is applied to the complete linear subsystem before simulation of the entire system. The second approach requires the separate modeling and the model-order reduction of each distributed and lumped element. Although this latter approach produces in the first step a more complex model, it allows exploiting usefully valuable qualitative information on the physical properties of each element in the subsequent process of model-order reduction. On this line there are some challenging tasks, mainly due to the coupling between elements for which the propagation (and the related delays) plays a significant role (the electrically long interconnects) and elements for which the propagation may be neglected (short 3D interconnects, lumped terminal devices, ...).

It is known in literature that the generalized method of characteristics (MoC) provides the most suitable model to perform transient analysis of electrically long transmission lines, i.e. lines for which the propagation delays play a significant role (e.g., [2]-[5]). In the frequency domain, which is the natural domain to take into account the frequency-dependence of the line p.u.l. parameters, the model obtained by using MoC is described by operators characterized by a rather complicated behavior. These operators, in particular, show a singular behavior at infinity due to the presence of irregular terms mainly arising from the delays associated to propagation. Furthermore, the corresponding time-domain model suffers from the drawback of costly time convolutions.

For such reasons, an accurate and efficient model can be derived only if the delay terms are properly extracted, and the regular remainders are approximated with reduced-order models. The most commonly adopted approaches to extract these delays are based on an asymptotic evaluation of the behavior of the frequency domain operators (e.g., [5]-[8]).

In this paper we will firstly show how an "exact" delay extraction procedure, based on the theory of the perturbation of the spectrum of symmetric operators [5], can be performed in the general case of lossy multiconductor lines with frequency dependent parameters. After the exact delays extraction, the regular remainders, are identified with reduced order equivalent circuits by exploiting valuable qualitative information on their smoothness properties. Moreover, a discussion on the problem of minimal order circuit synthesis is considered, leading to a "rank" condition which is shown to improve the accuracy of the identification. The proposed method is then applied to a test case of multiconductor transmission line and compared to standard techniques.

2 The Mathematical Model

Let us consider a line of length *d* consisting of *m* signal conductors and a reference one. The frequency-domain currents distributions $\boldsymbol{I}$ and voltages $\boldsymbol{V}$ along the line are solutions of the telegrapher's equations:

$$-\frac{d\boldsymbol{V}}{dz} = \mathsf{Z}(s)\boldsymbol{I}, -\frac{d\boldsymbol{I}}{dz} = \mathsf{Y}(s)\boldsymbol{V}. \tag{1}$$

With a suitable definition of the per-unit-length impedance matrix $\mathsf{Z}(s)$ and admittance matrix $\mathsf{Y}(s)$ these equations describe the most general case of lossy multiconductor lines with frequency dependent parameters. Having defined the terminal variables as $\boldsymbol{V}_k,\ \boldsymbol{I}_k,\ k = 1, 2$ the following equivalent multiport representation may be derived (e.g. [5]):

$$\boldsymbol{I}_1(s) = \mathsf{Y}_c(s)\boldsymbol{V}_1(s) + \boldsymbol{J}_1(s), \quad \boldsymbol{I}_2(s) = \mathsf{Y}_c(s)\boldsymbol{V}_2(s) + \boldsymbol{J}_2(s), \tag{2}$$

$$\boldsymbol{J}_1(s) = \mathsf{P}(s)[-2\boldsymbol{I}_2(s) + \boldsymbol{J}_2(s)], \quad \boldsymbol{J}_2(s) = \mathsf{P}(s)[-2\boldsymbol{I}_1(s) + \boldsymbol{J}_1(s)]. \tag{3}$$

Equations (2) are the network equations at the two line ends, while equations (3) describe the control laws of two controlled current sources. The *characteristic admittance matrix* $\mathsf{Y}_c(s)$ and the *propagation operator* $\mathsf{P}(s)$ are given by

$$\mathsf{Y}_c(s) = (\sqrt{\mathsf{Z}(s)^{-1}\mathsf{Y}(s)^{-1}})\mathsf{Y}(s),\ \mathsf{P}(s) = e^{-d\sqrt{\mathsf{Y}(s)\mathsf{Z}(s)}}. \tag{4}$$

The time-domain model is obtained by reverse transforming (2) and (3), therefore involving the time convolution product between the voltage and current waveforms and the inverse transforms of (4), i.e. the line impulse responses: $\mathsf{y}_c(t) = L^{-1}[\mathsf{Y}_c(s)],\ \mathsf{p}(t) = L^{-1}[\mathsf{P}(s)]$.

This model is extremely accurate since it fits naturally the propagation: for instance, when port 2 is matched $\boldsymbol{J}_1 = 0$ and the model exactly provides the characteristic admittance as the input admittance at port 1. Weak points of this model are the difficult evaluation of the impulse responses and the high computational cost of the time convolutions, which lower the efficiency.

The impulse responses $\mathsf{y}_c(t),\ \mathsf{p}(t)$ cannot be evaluated analytically even when the p.u.l. parameters are given in analytic form. On the other hand, they cannot be computed numerically because of the presence of delayed Dirac pulses and/or highly irregular terms that are "unbounded". Therefore the literature has proposed a semi-analytical approach with the following steps: (i) evaluate analytically the irregular terms of these functions; (ii) extract them; (iii) perform a numerical evaluation of the remainders, possibly associated to a model-order reduction [5]-[8]. The key point in this approach is given by the asymptotic evaluation of the behavior of $\mathsf{Y}_c(s)$ and $\mathsf{P}(s)$ for $s \to \infty$, which has been already proposed, for instance, in [7]. Here we underline the possibility to extract exactly these irregular terms by applying the perturbation theory of symmetric matrices as described in [5]. First the operators are decomposed in a way such to highlight their principal parts $\mathsf{Y}_{\mathsf{cp}}(s)$ and $\mathsf{P}_{\mathsf{p}}(s)$, namely the leading parts as $s \to \infty$:

$$\mathsf{Y}_{\mathsf{c}}(s) = \mathsf{Y}_{\mathsf{cp}}(s) + \mathsf{Y}_{\mathsf{cr}}(s),\ \mathsf{P}(s) = \mathsf{P}_{\mathsf{p}}(s) + \mathsf{P}_{\mathsf{r}}(s). \tag{5}$$

If the principal parts are computed exactly, the remainders are low-pass functions, with the following asymptotic behavior

$$\mathsf{Y}_{\mathsf{cr}}(s) = \mathsf{O}(1/s),\ \mathsf{P}_{\mathsf{r}}(s) = \mathsf{O}(1/s), \quad for\ s \to \infty. \tag{6}$$

To briefly review the main steps of the analytical evaluation of the principal parts, let us define the matrix:

$$\Lambda(s) = \mathsf{Y}(s)\mathsf{Z}(s)/s^2. \tag{7}$$

The functions $\mathsf{Y}_{\mathsf{c}}(s)$ and $\mathsf{P}(s)$ may be diagonalized as

$$\mathsf{P}(s) = \mathsf{U}(s) diag\left(e^{-ds\sqrt{\lambda_1(s)}}, \ldots, e^{-ds\sqrt{\lambda_n(s)}}\right)\mathsf{U}^{-1}(s), \tag{8}$$

$$\mathsf{Y}_{\mathsf{c}}(s) = \mathsf{U}(s) diag\left(\frac{1}{s\sqrt{\lambda_1(s)}}, \ldots, \frac{1}{s\sqrt{\lambda_n(s)}}\right)\mathsf{U}^{-1}(s), \tag{9}$$

where $\lambda_i(s)$ and the columns of $\mathsf{U}(s)$ are, respectively, the eigenvalues and the eigenvectors of the matrix (7). For a large class of lines of practical interest, which include the RLGC lines and the lines with dispersive dielectric, the starting point is the expansion of eigenvalues and eigenvectors for $s \to \infty$ as follows: $\lambda_i(s) = \lambda_i^{(0)} + \frac{\lambda_i^{(1)}}{s} + \ldots \quad , u_i(s) = u_i^{(0)} + \frac{u_i^{(1)}}{s} + \ldots$

The zero-order terms $\lambda_i^{(0)}$ and $u_i^{(0)}$ are, respectively, the eigenvalues and eigenvectors of

$$\Lambda_\infty = \lim_{s \to \infty} \Lambda(s), \tag{10}$$

which can be easily obtained from the high-frequency behavior of the p.u.l. parameters $\mathsf{Z}(s)$ and $\mathsf{Y}(s)$. This information allows a straightforward extraction of the principal part of $\mathsf{Y}_{\mathsf{c}}(s)$:

$$\mathsf{Y}_{\mathsf{cp}}(s) = \mathsf{Y}_\infty. \tag{11}$$

The principal part of $\mathsf{P}(s)$ is analytically computed as:

$$\mathsf{P}_{\mathsf{p}}(s) = \sum_{i=1}^{n} e^{-\mu_i T_i} \mathsf{A}_{\mathsf{i}} e^{-sT_i}, \tag{12}$$

where T_i is the delay associated to the ith mode, A_{i} is a matrix given by the product between the ith the right column and left row eigenvectors of matrix (10) and $\mu_i(s)$ is the damping coefficient of the ith mode that could be evaluated from the knowledge of the first order term $\lambda_i^{(1)}$ in the expansion of $\lambda_i(s)$ [5]. Note that the procedure is based on the possibility to know the high-frequency behavior or the p.u.l. parameters. This information may be easily obtained in a very consistent way also in cases when the p.u.l. parameters are not given analytically but only known in terms of frequency samples, as shown in [8]. The difference between the delay extraction procedure presented here and those at present proposed in literature (e.g., [8]), is the analytical determination of the amplitude of the damping factors in equation (12), besides the delays T_i. For instance, by applying the method proposed in [8], only the delays are analytical evaluated and extracted, so not all the informations achievable from the knowledge of p.u.l. parameters are exploited at the best. Finally, it is important to stress that a similar approach may be adopted also for another class of dispersive lines, to which belong the lines with pronounced skin effect. In such a case the starting point is the expansion in Laurent series in the neighborhood of $s = \infty$ by powers of $1/\sqrt{s}$ [5].

3 Efficient Identification of the Reduced Lumped Equivalent

By comparing (2)-(3) with (5), (11) and (12) it is evident that the irregular terms may be exactly implemented by a resistive multiport at each termination and by damped delayed sources. Therefore the complete equivalent circuit could be sinthezised once the regular remainders are approximated by means of a reduced-order model. Efficient identification of the regular parts of the responses demands some requirements to be fulfilled [10]:

1. extraction of irregular terms has to be performed in a way leaving the "simplest" remainder to be identified;
2. valuable qualitative information on the properties of these reminders have to be "a-priori" exploited in the closed form expansion;
3. the most appropriate minimization procedure have to be applied;
4. the identified expression must lead in a straight-forward way to the synthesis of passive lumped equivalent circuits.

With respect to the first point, we just mention that the "exact" delay extraction as described by eq. (11) guarantees the remainders to be regular functions, with a low-pass asymptotic behaviour. Several expressions have been proposed as expansions for the identification of the regular remainders. We consider a rational approximation in the classical pole-residue form as in equation (13), where for example $\mathsf{Y}_{\mathsf{cr}}(s)$ is given as:

$$\mathsf{Y}_{\mathsf{cr}}(j\omega) = \sum_{i=1}^{N} \frac{\mathsf{R}_{\mathsf{i}}}{1 + j\omega\tau_i}, \tag{13}$$

where τ_i are positive time costants, R_{i} residue matrices, N is the order of the expansion. An important issue to be clarified here is the order of the dynamic circuit to be synthesized after the identification procedure has been successfully applied. Using any expansion in the form of equation (13), the number of dynamic elements in the equivalent circuit is given by:

$$N^* = \sum_{i=1}^{N} rank(\mathsf{R}_{\mathsf{i}}). \tag{14}$$

It is important to understand that, since normally N^* is fixed as degree of complexity of the equivalent dynamic network, the case of $rank(\mathsf{R}_{\mathsf{i}}) = r > 1$ have to be avoided since it corresponds to constrain r time constants τ_i to be equal, so limiting the performance of the identification at a fixed order. Consequently, the best way to derive an N^* order-approximated circuit model from given responses is to add, in the identification procedure, the rank-1 constraint on residue matrices R_{i}. This condition cannot be explicitly enforced if identification procedures based on some linearization are used, such as the Model Based Parameter Estimation method [10], the Subspace System Identification method [12], or the Vector Fitting method [9],[13]. In fact, in the general case, using any of these methods the residues R_{i} come out with full rank. Some nonlinear optimization is therefore needed if the afore mentioned condition has to be fulfilled. In this work we combine the Vector Fitting algorithm to a Nonlinear Least Square for the identification of the regular part of a $m \times m$ MTL characteristic admittance. Best results are in fact obtained when the non linear procedure is launched after a good starting point has been estimated by means of Vector Fitting. The procedure is then as follows: firstly a vector fitting algorithm with $N = N^*/m$ poles is launched; then, singular value decomposition of each full rank residue R_{i} is performed, and each residue R_{i} is written as the sum of m rank-1 residues, R_{ij} where j=1,..m. Each of them is obtained by including in the SVD decomposition only a singular value at one time, and replacing the others with zero. At this point the nonlinear optimization procedure gets, as inputs, N poles determined by Vector Fitting repeated m times, and as corresponding rank-1 residues, those determined by the decomposition described above. In this way, the equality constrains between poles determined by Vector Fitting are removed and fitting error decrease, but the corresponding circuit order N^* does not increase.

Synthesis of the equivalent circuit can be performed easily [10] once a state space representation has been obtained from the Laplace domain representation.

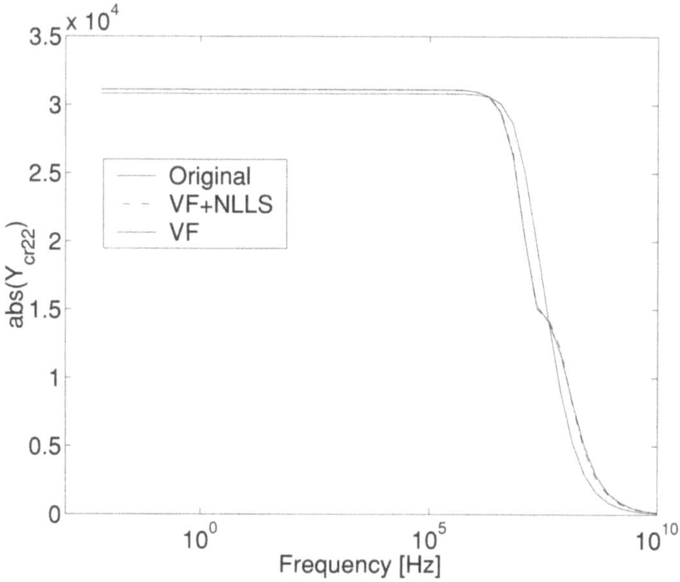

Fig. 1. regular part of RLGC trasmission line characteristic admittance Y_{cr22}

Table 1. rms errors on components of Y_{cr}

rms error	VF	VF + NLLS
σ_{11}	7.60e-6	6.47e-6
σ_{12}	3.13e-6	2.32e-6
σ_{13}	1.46e-5	6.21e-6
σ_{22}	1.78e-5	1.46e-6
σ_{23}	3.13e-6	2.32e-6
σ_{33}	7.60e-6	6.47e-6

4 A Test Case

The formulation and identification procedure proposed and discussed above have been tested with reference to a multi-conductor RLGC transmission line ($m = 3$) considered in reference [14]. After the exact extraction has been applied to the describing functions, as described in Sect. 2, we simply compare the identification results at a fixed order of the equivalent circuit (6 dynamic components), as obtained by applying the standard Vector Fitting technique and our method. The results of the identification are shown in Fig. 1 for the Y_{cr22} component, and summarized in Table 1 as rms error for each component, where the advantage in terms of accuracy ranges from 0.1 to 1 order of magnitude. Note that NLLS refinement step leads to a more uniform distribution of rms errors, since after applying this procedure, all the errors are in the same order of magnitude. Furthermore we underline that, in this example, both VF and NLLS steps have been performed with all matrix components unitarily weighted.

References

1. V.K. Tripathi, R. Sturdivant, Guest Editors, Special Issue on "Interconnects and Packaging", IEEE Trans. on Microwave Theory and Techniques, Vol. 45, Issue 10, Oct. 1997, pp. 1817 - 1818
2. J.S. Schutt-Ainè, S.S. Kang, Guest Editors, Scanning the Issue, "Interconnections - Addressing the Next Challenge of IC Technology (part II: design, characterization, and modeling)", IEEE Proceedings, Vol. 89, Issue 5, May 2001, pp. 583 - 585

3. F.G. Canavero, Foreword: special issue on "Recent Advances in EMC of Printed Circuit Boards", IEEE Trans. on Electromagnetic Compatibility, Vol. 43, Issue 4, Nov. 2001, pp. 414 - 415
4. M. Celik, L. Pileggi, A. Odabasioglu, IC Interconnects Analysis, Kluwer, 2002
5. G. Miano, A. Maffucci, Transmission lines and lumped circuits, Academic Press, U.S.A., 2001
6. Achar R., Nakhla M.S., "Simulation of High Speed Interconnects", IEEE Proceedings, Vol. 89, Issue 5, May 2001, pp. 693 - 728
7. C. Gordon, T. Blazeck, R. Mittra, "Time-Domain Simulation of Multiconductor Transmission Lines with Frequency-Dependent Losses", IEEE Trans. on Computer-Aided Design of Integrated Circuits and Systems, Vol.11, Issue 11, Nov. 2002, pp. 1372 - 1387
8. S. Grivet-Talocia et al., "Transient Analysis of Lossy Transmission Lines: An Efficient Approach Based on the Method of Characteristics", IEEE Trans. on Advanced Packaging, Vol. 27, Issue 1, Feb. 2004, pp. 45 - 56
9. B. Gustavsen, A. Semlyen, "Rational Approximation of Frequency Domain Responses by Vector Fitting" IEEE Trans. on Power Delivery, Vol. 14, Issue 3, Jul. 1999, pp. 1052 - 1061
10. R. Neumayer, A. Stelzer, F.Haslinger, R. Weigel, "On the Syntesis of Equivalent-Circuit Models for Multiports Characterized by Frequency Dependent Parameters", IEEE Trans. on Microwave Theory and Techniques, Vol. 50, Issue 12, Dec. 2002, pp. 2789 - 2796
11. M. de Magistris, A. Maffucci, "Identification of a Spice Reduced-Order Model for Lossy Interconnects from Terminal Behavior", Proc. of EMC-Zurich 2003, Zurich, CH, Feb. 2003, pp. 443 - 448
12. S. Grivet Talocia, F. Canavero, I. Maio, I.S. Stievano, "Reduced Order Macromodeling of Complex Multiport Interconnects", URSI General Assembly, Maastricht, Belgium, Aug. 2002
13. B. Gustavsen, "Computer Code for Rational Approximation of Frequency Dependant Admittance Matrices", IEEE Trans. on Power Delivery, Vol. 17, Issue 4, Oct. 2002, pp. 1093 - 1098
14. E.C. Chang, S.M. Kang, "Transient Simulation of Lossy Coupled Transmission Lines Using Iterative Linear Least Square Fitting and Piecewise Recursive Convolution", IEEE Trans. on Circuits and Systems I, Vol. 43, Issue 11, Nov. 1996, pp. 923 - 932

Symbolic Methods in Industrial Analog Circuit Design

T. Halfmann and T. Wichmann

Fraunhofer Institute for Industrial Mathematics, 67663 Kaiserslautern, Germany,
thomas.halfmann@itwm.fraunhofer.de

Abstract Industrial analog circuits are usually designed using numerical simulation tools. To obtain a deeper circuit understanding, symbolic analysis techniques can additionally be applied. Approximation methods which reduce the complexity of symbolic expressions are needed in order to handle industrial-sized problems. This paper describes aspects of the field of symbolic analog circuit analysis. Some state-of-the-art simplification algorithms for linear and nonlinear circuits are presented. The basic ideas behind the different techniques are described and two application examples for the linear and nonlinear case will be demonstrated.

1 Introduction to Symbolic Circuit Analysis

The motivation for applying symbolic techniques to the field of analog circuit design has been to gain insight into circuit behavior by interpreting analytic formulas instead of using traditional numerical design and simulation tools which lack in providing deeper design understanding. However, it becomes apparent quite quickly that exact symbolic analysis yields expressions which are too complex to be of any use. Obviously, for industrial circuits with more than just one transistor it is impossible to obtain useful results due to the extreme computational complexity of symbolic calculations. This contradicts the initial intention of symbolic analysis, namely to gain insight into unknown circuit behavior. This motivated the development of simplification algorithms which lead to a breakthrough in the field of symbolic circuit analysis.

The equation system describing the behavior of an analog circuit consists of equations originating from Kirchhoff's current and voltage laws as well as of the circuit element characteristics. It can be set up automatically using standard formulation methods such as the Modified Nodal Analysis or the Sparse Tableau Analysis. In general, the circuit equations are given by a differential-algebraic equation system (DAE system)

$$F = \begin{pmatrix} f(x(t), x'(t), y(t), u(t)\,; p) \\ g(x(t), y(t), u(t)\,; p) \end{pmatrix} = 0 \quad \text{for all } t \in I \ . \tag{1}$$

Here, $u : \mathbb{R} \to \mathbb{R}^r$ denotes the inputs, $x = (v, i) : \mathbb{R} \to \mathbb{R}^k$ denotes the vector of dependent variables, $y : \mathbb{R} \to \mathbb{R}^s$ denotes the outputs, and $I \subset \mathbb{R}$ denotes a time interval. Since we are working with symbolic equations, F is parameterized by symbolic element parameters $p = (p_1, \ldots, p_N)$ (like a resistor value R_1, a voltage source value V_0, or a transistor parameter β_F).

1.1 Symbolic Simplification Algorithms

As mentioned above, symbolic analysis of large analog circuits seems to be senseless as long as the complexity problem has not been solved. Thus, in order to reduce the complexity of the symbolic expression, one needs to simplify it.

In general, the term *symbolic simplification* or *symbolic approximation* refers to a whole family of hybrid symbolic/numeric algorithms for expression simplification. These techniques require more numerical knowledge about the investigated circuit than manual simplifications but yield compact expressions with predictable error in a fully automated way. In manual circuit analysis the decisions on which expressions to keep and which ones to discard are based on vague and only qualitative assumptions (e.g. $R_1 \ll R_2$) that do not allow for assigning precise error figures to simplified expressions. For automating the designer's behavior within a computer program one needs exact figures to simplify an expression because qualitative relations between elements are not sufficient for determining the importance of a term especially if the expression to be simplified consists of non-trivial combinations of symbols.

The basic idea behind the simplification algorithms can be outlined as follows: starting with a symbolic equation system F describing the circuit's behavior, the user chooses one or more numerical reference solutions f_i as well as an error bound ε. The algorithms then apply symbolic simplifications to the system (e.g. the deletion of an entire expression in a sum) and solve this simplified system numerically. The hereby obtained solutions $\tilde{f}_i$ are compared to the reference solutions using an appropriate error norm: $\delta_i = \|f_i - \tilde{f}_i\|$. If the error bound is exceeded, i.e. $\max \delta_i > \varepsilon$, the simplification is undone. This is repeated until no more simplifications are possible without a violation of the error bound and the simplified symbolic system $\widetilde{F}$ is returned.

The simplification algorithms assure that the numerical behavior (with respect to the chosen references f_i) of the simplified system coincides with that of the original system within the user-given error bound. Depending on the analysis task, the reference solutions f_i can for example be a numerical transfer function, its poles and zeros, or a time-dependent solution.

1.2 Ranking Methods

The order in which to simplify terms from the equation system is one of the crucial points: It is quite clear that those terms should be simplified first which have only a minor influence on the output behavior. Terms with a large influence should not be removed at all. To achieve a maximum number of simplifications and to avoid unnecessary modifications an optimized order, the so called *ranking*, should be used. For this, a ranking algorithm is needed which predicts the influence on the output a modification would cause. As the number of possible simplifications is very large it is inconvenient to exactly compute the influence and therefore estimation methods have to be used. The design of a good ranking algorithm is a trade off between accurate error prediction and computational effort.

2 Linear Symbolic Analysis

The transfer function is the main object of interest in linear symbolic analysis. It allows for obtaining insights into the circuit's behavior and parameter dependencies. By post-processing the transfer function one can for example symbolically compute its poles and zeros to investigate the circuit stability. The research on this topic started in the early 1990's (e.g. [GS91], [Som94]).

2.1 Linear Simplification Techniques

Basically, one distinguishes three types of linear simplification methods: Simplification Before Generation methods (SBG) simplify the matrix equations before computing the transfer function. Simplification During Generation methods (SDG) apply simplifications during the process of transfer function calculation. Simplification After Generation methods (SAG) simplify the transfer function directly. Now, we will describe SAG and SBG methods only.

Simplification After Generation. This technique [GS91] is based on the manipulation of the symbolic transfer function given as a rational expression

$$H(s\,;p) = \sum a_i(p)\, s^i \;/\; \sum b_i(p)\, s^i \quad , \tag{2}$$

where the coefficients $a_i = \sum a_{ij}$ and $b_i = \sum b_{ij}$ are symbolic functions of the parameter vector p given in canonical sum-of-products form. For a given error bound, those terms a_{ij} and b_{ij} are removed from the transfer function which cause a negligible deviation on a_i and b_i, respectively. By this, one can drastically reduce the symbolic complexity of the transfer function.

Simplification Before Generation. Even for circuits of small size it is not possible to calculate the full symbolic transfer function (2). For example, the $\mu A741$ operational amplifier yields a transfer function whose expanded denominator consists of more than 10^{34} terms [Hen00]. Thus, the linear equation system itself has to be simplified before computing the symbolic transfer function. This can be done by rewriting each entry of the system matrix in sum-of-products form and sequentially removing terms from the matrix. The error is checked by computing the magnitude and phase of the (numerical) transfer function at certain frequency points. SBG methods reduce both the complexity of the transfer function as well as its polynomial order. For SBG techniques a dedicated ranking method has been developed [Hen00] which makes use of the Sherman-Morrison formula.

Poles and Zeros. The extraction of symbolic expressions for poles and zeros is rarely possible without simplifications. In [Hen00], a matrix-based SBG method for direct approximation of a linear system with respect to a selected eigenvalue of a generalized eigenvalue problem was presented. By means of eigenvalue sensitivity the symbolic parameters with negligible influence on the eigenvalue are discarded from the linear system resulting in a simplified generalized eigenvalue problem whose determinant yields a reduced-order approximation of the characteristic polynomial. To detect potentially false eigenvalue pairings during approximation, the modal assurance criterion (MAC) [FM95] is applied, which constitutes a measure for the correlation of two eigenvectors u_1 and u_2 and which is defined as

$$\mathrm{MAC}(u_1, u_2) = \frac{\left|u_1^H u_2\right|^2}{(u_1^H u_1)(u_2^H u_2)} \,. \tag{3}$$

The value of the MAC ranges from 0 (orthogonal vectors) to 1 (parallel vectors). Hence, the MAC must be very close to 1 for considering a valid approximation. Since in this context one is interested in a single eigenvalue only, it is appropriate to use an iterative generalized eigenvalue problem solver like the Jacobi orthogonal correction method [SBFV96] instead of the QZ algorithm. As an additional benefit, the MAC can be integrated within the Jacobi correction iteration. This results in a very efficient and reliable approximation method for the extraction of approximated symbolic poles and zeros.

2.2 Industrial Application

The application of the pole/zero extraction algorithm will be demonstrated on the CMOS folded-cascode operational amplifier shown in Fig. 1. The frequency response of the operational amplifier's open-loop differential-mode voltage gain (solid curve) shows a peak near 10 MHz, caused by a parasitic complex pole pair close to the imaginary axis. The analysis task is to extract a symbolic expression for the parasitic pole pair which allows to determine those circuit parameters which have a dominant influence on the peak.

Using a SPICE Level 3 AC model for the MOS devices [GM93] yields a system of 29 equations. The differential-mode voltage transfer function has 19 poles and 19 zeros and contains more than 5×10^{19} product terms. The symbolic approximation routines are applied to extract the parasitic pole pair at $s_p = (-2.1 \pm 8.3j) \times 10^7$ using a relative error bound $\varepsilon = 0.1$. The resulting simplified equation system can be algebraically reduced to a system of dimension 4 from which the wanted pole pair $s_p^{1,2}$ can be easily computed to the expression shown in Fig. 2. The overall computation time (including netlist import and

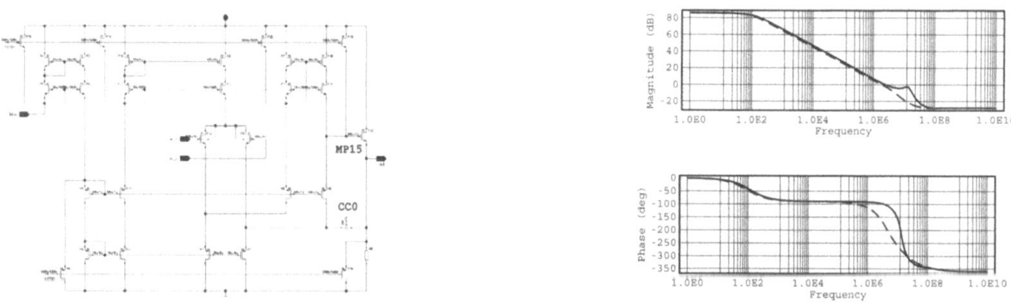

Fig. 1. CMOS folded-cascode operational amplifier

$$\mathtt{S_p}^{1,2} == -\frac{\mathtt{(CC0+CL)\ gm\$MN6}}{\mathtt{2\ CC0\ CL}} \pm \frac{\sqrt{\mathtt{Cgs\$MP15gm\$MN6\,(Cgs\$MP15\ (CC0+CL)^{2}\ gm\$MN6 - 4\ CC0^{2}CLgm\$MP15)}}}{\mathtt{2\ CC0\ Cgs\$MP15CL}}$$

Fig. 2. Computed formula for the complex pole pair

equation setup) to approximate the equation system and to extract this formula is about 8 seconds running the routines under Mathematica 4.0 on an AMD Athlon 1200 with 512 MB memory. By interpretation of the computed formula for the complex pole pair it turns out that an increased value for the gate-source capacitance `Cgs$MP15` of the transistor MP15 allows for decreasing the imaginary parts of the pole pair. As a consequence one could add an additional capacitor between the gate and the source terminals of the corresponding transistor and by that remove the peak in the transfer function of the voltage gain of the operational amplifier (dashed curve).

3 Nonlinear Symbolic Analysis

Simplification methods for linear analog circuits have been successfully applied to industrial applications for several years. In the following, we want to describe how the presented methods can be extended to the analysis of nonlinear circuits. Research on this topic started a few years ago and is still in progress (e.g. [Bor98], [PHHB98], [WPHH99], [Wic01]).

3.1 Nonlinear Simplification Techniques

As opposed to the linear case, for nonlinear circuits in general we can not expect to obtain explicit formulas for the solution of the output variables. In contrast to linear symbolic analysis, nonlinear simplification techniques are therefore mainly used for automated behavioral model generation. However, for small circuits it can be possible to obtain an interpretable result.

Behavioral model generation is a technique for speeding up numerical simulation of large circuits. The idea is to replace each frequently used subblock in the circuit by a single simplified model description and by that reducing the complexity of the whole circuit and decreasing the computation time. Nonlinear simplifications can be used to automatically generate behavioral models from netlist descriptions of the subblocks [NHB02].

The general outline of the nonlinear simplification algorithms is described in Sect. 1.1. The following simplification techniques have been developed to reduce the complexity of nonlinear DAE systems:

- *Elimination*: Solving equations explicitly for one variable and substituting this variable in the remaining system, thus reducing the number of equations.
- *Simplification of piecewise-defined functions*: Detecting and removing branches of piecewise-defined functions which are unused for the given input-value set.
- *Deletion of terms*: Omitting terms of sums which do not participate a significant part to the sum.
- *Substitution of terms by constant values*: Substituting a constant value for terms which do not participate a significant part to the sum.

3.2 Operational Amplifier Example

Figure 3 shows the schematic of an operational amplifier consisting of eight bipolar transistors. The input signal is given by the pulse wave voltage source VSig, the node voltage v_9 at node 9 is the output signal (dashed curve). Using the Gummel-Poon equations for modeling the bipolar transistors, the transient behavior of the operational amplifier can be described by an equation system consisting of 73 equations and variables. Nonlinear symbolic simplfications techniques are now used for generating a behavioral model of the circuit. For this, a maximum transient error of 1 V for the output variable is chosen (gray shaded area).

To automatically reduce the complexity of the equation system, symbolic simplification techniques are applied in the following order: Elimination, simplification of piecewise-defined function, cancellation of

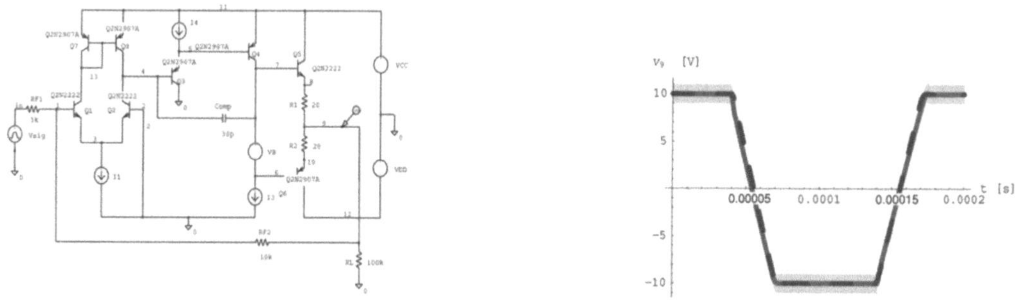

Fig. 3. Bipolar operational amplifier

terms, and again elimination. This results in a system of 6 equations only with an overall computation time of 792 s. The numerical solution of the simplifed system (solid curve) lies within the specified error margin. Solving the original DAE system numerically on the time interval $t \in [0\,\mathrm{s}, 0.002\,\mathrm{s}]$ takes about 166 s, whereas solving the reduced equation system takes about 2.3 s. Thus, the generated behavioral model yields a simulation speed-up of more than a factor of 70.

3.3 Nonlinear Ranking Methods

In principle, nonlinear ranking methods have to be designed for each simplification method and each analysis method separately. In the following we will briefly describe the *one-step solver ranking*, a ranking method which measures the influence of term cancellations on the transient behaviour. Additional ranking methods can for example be found in [PHHB98, WPHH99].

Let F denote the original DAE system with transient solution x_F, let G denote the simplified system, and let G_S denote the static system which results from G by replacing differentials by finite difference expressions according to the chosen integration scheme. An estimation $\tilde{x}_G$ of the (unknown) solution x_G of G is computed as follows: At each time instance t_i a Newton iteration to solve G_S is started for the initial value $x_F(t_i)$. The first Newston step

$$\tilde{x}_G^{[1]}(t_i) = x_F(t_i) - J_{G_\mathrm{S}}(x_F(t_i))^{-1} G_\mathrm{S}(x_F(t_i)) \ , \tag{4}$$

is then used as an estimation for the true solution $x_G(t_i)$. Finally, the obtained values $\tilde{x}_G^{[1]}(t_i)$ are interpolated yielding the estimation $\tilde{x}_G$ for x_G. The ranking value is then given by $\delta_G = \|\tilde{x}_G - x_F\|$. In our applications this ranking method yields very accurate error estimates with moderate computational effort.

3.4 Index Calculation

The index plays an important role in the theory of DAE systems [BCP89]. It is well known that numerical solving of systems with an index higher than 1 is an ill-posed problem. Since during nonlinear symbolic simplifications the index may increase, we want to compute the index in order to avoid index changes. For this, from the wide variety of different index concepts we have chosen to control the *tractability index* [GM86] and the *strangeness index* [KM98] during simplification. They are both defined for general nonlinear DAE systems and can be computed numerically. In our applications it turned out that the singular value decomposition yields the best numerical results for computing both the tractability index and the strangeness index. The Gram-Schmidt orthonormalization can also be used to calculate the tractability index symbolically, but the resulting expressions are too complex even for small circuits. For the operational amplifier example in Sect. 3.2 a number of 11 simplifications out of 254 had to be rejected due to an invalid increase of the index of the simplified system.

4 Conclusions

We have provided an insight into the area of symbolic techniques for the analysis and design of analog circuits. It was motivated that due to the complexity problem simplification methods are indispensable for handling industrial-sized problems. The basic ideas behind these simplification methods – a combination of symbolic and numeric algorithms – have been shown. The described techniques are integrated in the software *Analog Insydes* ([HHTW01], `www.analog-insydes.de`) which is an add-on package to the computer-algebra system Mathematica for the analysis, modeling, sizing, and optimization of linear and nonlinear circuits of industrial size.

During several years of application the symbolic simplification algorithms have proven to be applicable to industrial-sized problems and by that making symbolic analysis a powerful technique in industrial analog circuit design.

References

[BCP89] Brenan, K.E., Campbell, S.L., Petzold, L.R.: The Numerical Solution of Initial Value Problems in Ordinary Differential-Algebraic Equations. North Holland Publishing Co. (1989)

[Bor98] Borchers, C.: The symbolic behavioral model generation of nonlinear analog circuits. IEEE Transactions on Circuits and Systems II **45:10**, 1362–1371 (1998)

[FM95] Friswell, M.I., Mottershead, I.E.: Finite Element Model Updating in Structural Mechanics. Kluwer Academic Publishers, Dordrecht (1995)

[GM86] Griepentrog, E., März, R.: Differential-Algebraic Equations and their Numerical Treatment. BSB Teubner (1986)

[GM93] Gray, P.R., Meyer, R.G.: Analysis and Design of Analog Integrated Circuits, 3rd edition. John Wiley & Sons, Inc. (1993)

[GS91] Gielen, G., Sansen, W.: Symbolic Analysis for Automated Design of Analog Integrated Circuits. Kluwer Academic Publishers, Boston (1991)

[Hen00] Hennig, E.: Symbolic Approximation and Modeling Techniques for Analysis and Design of Analog Circuits. Shaker Verlag, Aachen (2000)

[HHTW01] Halfmann, T., Hennig, E., Thole, M., Wichmann, T.: Analog Insydes 2 Manual. Fraunhofer ITWM, Kaiserslautern, Germany (2001)

[KM98] Kunkel, P., Mehrmann, V.: Regular solutions of nonlinear differential-algebraic equations and their numerical treatment. Numerische Mathematik **79:4**, 581–600 (1998)

[NHB02] Näthke, L., Hedrich, L., Barke, E.: Betrachtungen zur Simulationsgeschwindigkeit von Verhaltensmodellen nichtlinearer integrierter Analogschaltungen. In Analog 2002, Bremen, Germany, 107–112 (2002)

[PHHB98] Popp, R., Hartong, W., Hedrich, L., Barke, E.: Error estimation on symbolic behavioral models of nonlinear analog circuits. In SMACD 1998, Kaiserslautern, Germany, 223–226 (1998)

[SBFV96] Sleijpen, G.L., Booten, A.G.L., Fokkema, D.R., Van der Vorst, H.A.: Jacobi-Davidson type methods for generalized eigenproblems and polynomial eigenproblems: Part I. BIT **36:3**, 595–633 (1996)

[Som94] Sommer, R.: Konzepte und Verfahren für den rechnergestützten Entwurf von Analogschaltungen. VDI-Verlag, Düsseldorf, Germany (1994)

[Wic01] Wichmann, T.: Simplification of nonlinear DAE systems with index tracking. In ECCTD'01. volume II, Espoo, Finland, 173–176 (2001)

[WPHH99] Wichmann, T., Popp, R., Hartong, W., Hedrich, L.: On the Simplification of Nonlinear DAE Systems in Analog Circuit Design. In: CASC 99. Springer Verlag, Berlin, Germany (1999)

Index Analysis of Multirate Partial Differential-Algebraic Systems in RF-Circuits

S. Knorr and M. Günther

Bergische Universität Wuppertal, Fachbereich C, Gaußstr. 20, 42119 Wuppertal,
{knorr,guenther}@math.uni-wuppertal.de

Abstract In radio frequency (RF) design, signals with widely separated time scales arise. To describe those circuits efficiently, a multidimensional signal model was developed. This approach transfers the circuit's differential-algebraic equations (DAE) to a multirate system of partial differential-algebraic equations (MPDAE). A structural analysis, based on the concept of underlying PDE systems and the index characterization of DAE systems, emphasises the entitlement of MPDAE-modeling.

1 Introduction - multidimensional signal model

In electronic circuit design the classical modified nodal analysis (MNA) leads to a system of differential-algebraic equations (DAE). Excluding controlled sources, the charge-flux oriented formulation of the network equations in terms of charges q, fluxes Φ, node potentials u, currents $\jmath_L$ and $\jmath_V$ through inductances and voltage sources, respectively, yields [ET98]

$$\begin{aligned}
A_C\dot{q} + A_R r(A_R^\top u(t), t) + A_L \jmath_L(t) + A_V \jmath_V(t) + A_I \imath(t) &= 0, && \text{(1a)}\\
\dot{\Phi} - A_L^\top u(t) &= 0, && \text{(1b)}\\
A_V^\top u(t) - v(t) &= 0, && \text{(1c)}\\
q - q_C(A_C^\top u(t), t) &= 0, && \text{(1d)}\\
\Phi - \Phi_L(\jmath_L(t), t) &= 0. && \text{(1e)}
\end{aligned}$$

In the following we will consider quasiperiodic input signals. To face widely separated time scales, that occur frequently in RF application, the quasiperiodic functions are transferred to multivariate functions (MVF), where for each time scale a corresponding variable is introduced [BWLB96]. A signal with m fundamental frequencies $\omega_i = 2\pi/T_i$, $i = 1, \ldots, m$ and $X(k_1, \ldots, k_m) \in \mathbb{C}$

$$x(t) = \sum_{k_1=-\infty}^{\infty} \cdots \sum_{k_m=-\infty}^{\infty} X(k_1, \ldots, k_m) \exp((jk_1\omega_1 + \cdots + jk_m\omega_m)t)$$

is generalized to its MVF

$$\hat{x}(t_1, \ldots, t_m) = \sum_{k_1=-\infty}^{\infty} \cdots \sum_{k_m=-\infty}^{\infty} X(k_1, \ldots, k_m) \exp(jk_1\omega_1 t_1 + \cdots + jk_m\omega_m t_m).$$

Now, the time scales are decoupled and the MVF is periodic in each coordinate direction. The original signal is contained on the diagonal of the MVF and can be reconstructed by $x(t) = \hat{x}(t, \ldots, t)$.

We apply the multidimensional signal model to the network equations and introduce MVFs of charges, fluxes, sources and of the state variables. Looking at the MVF of the charge function

$$\hat{q}_C(w, t_1, \ldots, t_m) \quad \text{with} \quad \frac{\partial \hat{q}_C}{\partial w} =: \hat{C}(w, t_1, \ldots, t_m),$$

we define $\tau_m := (t, \ldots, t)^\top \in \mathbb{R}^m$ and get for the time derivative

$$\begin{aligned}
&\frac{d}{dt} q_C(A_C^\top u(t), t) \\
&\quad = \frac{d}{dt} \hat{q}_C(A_C^\top \hat{u}(\tau_m), \tau_m) \\
&\quad = \hat{C}(A_C^\top \hat{u}(\tau_m), \tau_m) A_C^\top \cdot \sum_{i=1}^{m} \frac{\partial \hat{u}(\tau_m)}{\partial t_i} + \sum_{i=1}^{m} \frac{\partial}{\partial t_i} \hat{q}_C(A_C^\top \hat{u}(\tau_m), \tau_m).
\end{aligned}$$

Therefore, we define $\hat{\tau}_m := (t_1, \ldots, t_m)^\top$ and introduce the differential operator D_m with

$$D_m f(x(\hat{\tau}_m), \hat{\tau}_m) := \frac{df}{d\hat{\tau}_m} \cdot \mathbb{1} = \sum_{i=1}^{m} \left(\frac{\partial f}{\partial x} \cdot \frac{\partial x}{\partial t_i} + \frac{\partial f}{\partial t_i} \right).$$

Now, we are able to generalize the original DAE to the multirate system of partial differential-algebraic equations (MPDAE)

$$\begin{aligned}
A_C D_m \hat{q} + A_R \hat{r}(A_R^\top \hat{u}(\hat{\tau}_m), \hat{\tau}_m) + A_L \hat{\jmath}_L(\hat{\tau}_m) + A_V \hat{\jmath}_V(\hat{\tau}_m) & \\
+ A_I \hat{\imath}(\hat{\tau}_m) &= 0, \qquad (2a)\\
D_m \hat{\Phi} - A_L^\top \hat{u}(\hat{\tau}_m) &= 0, \qquad (2b)\\
A_V^\top \hat{u}(\hat{\tau}_m) - \hat{v}(\hat{\tau}_m) &= 0, \qquad (2c)\\
\hat{q} - \hat{q}_C(A_C^\top \hat{u}(\hat{\tau}_m), \hat{\tau}_m) &= 0, \qquad (2d)\\
\hat{\Phi} - \hat{\Phi}_L(\hat{\jmath}_L(\hat{\tau}_m), \hat{\tau}_m) &= 0. \qquad (2e)
\end{aligned}$$

As the MVF $\hat{x}$ contains the original signal on its diagonal, the DAE-solution x with $x = (u, \jmath_L, \jmath_V)^\top$ can be reconstructed by $x(t) = \hat{x}(t_m)$ via the MPDAE-solution $\hat{x} = (\hat{u}, \hat{\jmath}_L, \hat{\jmath}_V)^\top$. For more details we refer to [BWLB96].

In order to resolve structural properties for this transferred system , we apply the index concept to extract the algebraic and differential parts of the MPDAE as it was done for the original DAE in [ET98].

2 Index-1 networks

The differential-algebraic network equations (1) have differential index 1, if the following two topological conditions are fulfilled (see [Ti99]):

T1: There are no cutsets consisting of inductances and/or current sources only: $\ker(A_C, A_R, A_V)^\top = \{0\}$.
T2: There are no loops consisting of only capacitances and at least one voltage source: $\ker Q_C^\top A_V = \{0\}$.

To transfer this context to our partial differential-algebraic system, we rewrite (2) in a semi-explicit form. We assume passivity for the network elements; in contrast to the DAE-case, we need the sharper condition, that the capacitance, inductance and conductance matrices

$$C(w, \hat{\tau}_m) := \frac{\partial \hat{q}_C(w, \hat{\tau}_m)}{\partial w}, \; L(w, \hat{\tau}_m) := \frac{\partial \hat{\Phi}_L(w, \hat{\tau}_m)}{\partial w}, \; G(w, \hat{\tau}_m) := \frac{\partial \hat{r}(w, \hat{\tau}_m)}{\partial w}$$

are positive definite (but not necessarily symmetric) with a globally bounded inverse on the domain $[0, T_1] \times \cdots \times [0, T_m]$ defined by the time scales.

Let Q_C be an orthogonal projector onto the kernel of $A_C^\top$ and its complement P_C such that $P_C = I - Q_C$, with the identity matrix I. Hence, equation (2a) only contains information about $P_C\hat{u}$ as

$$A_C^\top \hat{u} = A_C^\top (P_C + Q_C)\hat{u} = A_C^\top P_C \hat{u}.$$

Subsequently, we define two sets of network variables

$$\hat{y} = \begin{pmatrix} \hat{y}_1 \\ \hat{y}_2 \end{pmatrix} = \begin{pmatrix} P_C\hat{u} \\ \hat{\jmath}_L \end{pmatrix} \quad \text{and} \quad \hat{z} = \begin{pmatrix} \hat{z}_1 \\ \hat{z}_2 \end{pmatrix} = \begin{pmatrix} Q_C\hat{u} \\ \hat{\jmath}_V \end{pmatrix}$$

(to shorten notations, we skip the arguments of the multivariate functions).

We insert the charges (2d) in (2a) and multiply the equation by $P_C^\top$ from the left. With adding the orthogonal complement $Q_C^\top Q_C \hat{y}_1 = 0$ (for regularity)

$$A_C \hat{q}_C(A_C^\top \hat{u}, \hat{\tau}_m) = A_C \hat{q}_C(A_C^\top \hat{y}_1, \hat{\tau}_m) + Q_C^\top Q_C \hat{y}_1 =: H(\hat{y}_1, \hat{\tau}_m)$$

we obtain a PDE for $\hat{y}_1$:

$$D_m H(\hat{y}_1, \hat{\tau}_m) = -P_C^\top \left(A_R \hat{r}(A_R^\top [\hat{y}_1 + \hat{z}_1], \hat{\tau}_m) + A_L \hat{y}_2 + A_V \hat{z}_2 + A_I \hat{\imath} \right). \tag{3}$$

The Jacobian $H_1 := \frac{\partial H}{\partial \hat{y}_1} = A_C C(A_C^\top [\hat{y}_1 + \hat{z}_1], \hat{\tau}_m) A_C^\top + Q_C^\top Q_C$ is positive definite by construction. Inserting the fluxes (2e) in (2b), we directly obtain a PDE for $\hat{y}_2$:

$$D_m \hat{\Phi}_L(\hat{y}_2, \hat{\tau}_m) = A_L^\top [\hat{y}_1 + \hat{z}_1]. \tag{4}$$

Besides the differential equations (3) for $\hat{y}_1$ and (4) for $\hat{y}_2$ we are left with equation (2a) multiplied by $Q_C^\top$ from the left and (2c). Using $Q_C \hat{z}_1 = \hat{z}_1$ and $P_C^\top P_C \hat{z}_1 = 0$, we have

$$\begin{pmatrix} Q_C^\top \left(A_R \hat{r}(A_R^\top [\hat{y}_1 + Q_C \hat{z}_1], \hat{\tau}_m) + A_L \hat{y}_2 + A_V \hat{z}_2 + A_I \hat{\imath} \right) + P_C^\top P_C \hat{z}_1 \\ A_V^\top [\hat{y}_1 + Q_C \hat{z}_1] - \hat{v} \end{pmatrix} = 0. \tag{5}$$

The Jacobian with respect to $\hat{z}$

$$B := \begin{pmatrix} Q_C^\top A_R G(A_R^\top [\hat{y}_1 + \hat{z}_1], \hat{\tau}_m) A_R^\top Q_C + P_C^\top P_C & Q_C^\top A_V \\ A_V^\top Q_C & 0 \end{pmatrix}$$

is regular, iff T1 and T2 hold, see [Ti99]. Thus, demanding the topological conditions, we are able to rewrite (2) in a semi-explicit form:

$$\begin{aligned} D_m \hat{y} &= F(\hat{y}, \hat{z}, \hat{\tau}_m), \\ 0 &= h(\hat{y}, \hat{z}, \hat{\tau}_m), \end{aligned}$$

where the algebraic equation is resolvable for $\hat{z} = \varphi(\hat{y}, \hat{\tau}_m)$. Hence, we are able to derive (in analogy to the underlying ODE introduced in [HW96]) the underlying PDE

$$D_m \hat{y} = F(\hat{y}, \varphi(\hat{y}, \hat{\tau}_m), \hat{\tau}_m).$$

3 Index-2 networks

To investigate the differences in the index-2 case, we split system (2) until it is possible to resolve it for all the different sets of network variables.

After the first splitting $\hat{u} = P_C\hat{u} + Q_C\hat{u}$ in the index-1 case, we determined the algebraic equations (5)

$$Q_C^\top \left(A_R \hat{r}(A_R^\top [P_C\hat{u} + Q_C\hat{u}], \hat{\tau}_m) + A_L \hat{\jmath}_L + A_V \hat{\jmath}_V + A_I \hat{\imath} \right) = 0, \tag{5a}$$

$$A_V^\top [P_C\hat{u} + Q_C\hat{u}] - \hat{v} = 0. \tag{5b}$$

In the index-2 case T1 and/or T2 are violated and the Jacobian relating to $Q_C\hat{u}$ and $\hat{\jmath}_V$ is not regular anymore. Therefore, (5a) and (5b) contain hidden constraints and further splittings of the network variables are necessary.

Lemma 1. *If T1 and/or T2 are violated, the MPDAE (2) is equivalent to the semi-explicit system*

$$A_C D_m \hat{q}_C(A_C^\top \hat{u}, \hat{\tau}_m) + P_C^\top \left(A_R \hat{r}(A_R^\top \hat{u}, \hat{\tau}_m) + A_L \hat{\jmath}_L + A_V \hat{\jmath}_V + A_I \hat{\imath}\right) = 0,$$

$$D_m \hat{\Phi}_L(\hat{\jmath}_L, \hat{\tau}_m) - A_L^\top \hat{u} = 0.$$

$$\text{Index 1} \begin{cases} \bar{P}_{V-C}^\top \left(A_V^\top \hat{u} - \hat{v}\right) = 0, \\ P_{R-CV}^\top Q_{V-C}^\top Q_C^\top \left(A_R \hat{r}(A_R^\top \hat{u}, \hat{\tau}_m) + A_L \hat{\jmath}_L + A_I \hat{\imath}\right) = 0, \\ P_{V-C}^\top Q_C^\top \left(A_R \hat{r}(A_R^\top \hat{u}, \hat{\tau}_m) + A_L \hat{\jmath}_L + A_V \hat{\jmath}_V + A_I \hat{\imath}\right) = 0. \end{cases}$$

$$\text{Index 2} \begin{cases} Q_{CRV}^\top \left(A_L \hat{\jmath}_L + A_I \hat{\imath}\right) = 0, \\ \bar{Q}_{V-C}^\top \left(A_V^\top P_C \hat{u} - \hat{v}\right) = 0. \end{cases}$$

Proof. The orthogonal projectors used to obtain this semi-explicit description are defined as follows (see [ET98]):

projector	Q_{V-C}	$\bar{Q}_{V-C}$	Q_{R-CV}	Q_{CRV}
onto	$\ker A_V^\top Q_C$	$\ker Q_C^\top A_V$	$\ker A_R^\top Q_C Q_{V-C}$	$\ker(A_C, A_R, A_V)^\top$

with complements denoted by P and the corresponding subindex.

In the following, we will use the just defined projectors to filter out nontrivial information from the algebraic equations, as the variables of interest lie in the kernel of the antecedent matrices. To make the successive steps more comprehensible, equations extracted from (5a) and (5b) are denoted using a subindex: (5a$_i$) and (5b$_i$). If the differential operator is applied to an equation (x), it is denoted by (x′).

Regarding equation (5b), we only get information about $Q_C P_{V-C} \hat{u}$ as $A_V^\top Q_C Q_{V-C} = 0$. Furthermore, multiplying (5b) by $\bar{Q}_{V-C}^\top$ from the left reveals the linear combination

$$\bar{Q}_{V-C}^\top \left(A_V^\top P_C \hat{u} - \hat{v}\right) = 0, \tag{5b$_1$}$$

which does not appear in the index-1 case, as T2 implies $\bar{Q}_{V-C} = 0$. We will refer to this equation later.

To determine $Q_C P_{V-C} \hat{u}$ from the part $\bar{P}_{V-C}^\top \cdot (5b)$ of the equation, we have to multiply by $Q_C^\top A_V$ from the left and add $Q_{V-C}^\top Q_{V-C} P_{V-C} \hat{u} = 0$:

$$\left(Q_C^\top A_V A_V^\top Q_C + Q_{V-C}^\top Q_{V-C}\right) P_{V-C} \hat{u} = Q_C^\top A_V \bar{P}_{V-C}^\top \left(\hat{v} - A_V^\top P_C \hat{u}\right). \tag{5b$_2$}$$

With $H_2 := Q_C^\top A_V A_V^\top Q_C + Q_{V-C}^\top Q_{V-C}$ positive definite, we can resolve for $P_{V-C} \hat{u}$, which leads to

$$Q_C P_{V-C} \hat{u} = Q_C H_2^{-1} Q_C^\top A_V \bar{P}_{V-C}^\top \left(\hat{v} - A_V^\top P_C \hat{u}\right).$$

At the moment we have the splitting

$$\hat{u} = \left[P_C + Q_C(P_{V-C} + Q_{V-C})\right] \hat{u}$$

and still need equations for $Q_C Q_{V-C} \hat{u}$ and $\hat{\jmath}_V$.

To split equation (5a) in the right manner, we have a look at its derivative, as $\hat{u}$ is the argument of the nonlinear function $\hat{r}(\cdot)$. In our case, we apply the differential operator D_m, which yields

$$D_m \hat{u}(\hat{\tau}_m) = \frac{\partial \hat{u}}{\partial t_1} + \cdots + \frac{\partial \hat{u}}{\partial t_m}.$$

With the abbreviation $G := G(A_R^\top \hat{u}, \hat{\tau}_m)$, we obtain

$$Q_C^\top A_R G A_R^\top [P_C + Q_C P_{V-C} + Q_C Q_{V-C}] D_m \hat{u} + Q_C^\top (A_L D_m \hat{\jmath}_L + A_I D_m \hat{\imath}) + Q_C^\top A_V D_m \hat{\jmath}_V = 0. \tag{5a′}$$

Multiplying by $Q_{V-C}^\top$ from the left strikes off $D_m\hat{\jmath}_V$ and we obtain an equation for $P_{R-CV}D_m\hat{u}$ as $Q_CQ_{V-C}Q_{R-CV} =: Q_{CRV}$ and $A_R^\top Q_{CRV} = 0$. Thus, we also multiply by $P_{R-CV}^\top$ from the left and get

$$\begin{aligned} &Q_{V-C}^\top Q_C^\top A_R G A_R^\top Q_C Q_{V-C} P_{R-CV} D_m\hat{u} \\ &= -P_{R-CV}^\top Q_{V-C}^\top Q_C^\top \left(A_R G A_R^\top [P_C + Q_C P_{V-C}] D_m\hat{u} + A_L D_m \hat{\jmath}_L + A_I D_m \hat{\imath}\right). \end{aligned} \tag{5a$'_1$}$$

To resolve for $P_{R-CV}D_m\hat{u}$, we add $Q_{R-CV}^\top Q_{R-CV}P_{R-CV}D_m\hat{u} = 0$, which leads to

$$H_4 := H_4(A_R^\top\hat{u}, \hat{\tau}_m) := Q_{V-C}^\top Q_C^\top A_R G A_R^\top Q_C Q_{V-C} + Q_{R-CV}^\top Q_{R-CV}$$

positive definite.

Now, we have to regard the splitting

$$\hat{u} = [P_C + Q_C(P_{V-C} + Q_{V-C}(P_{R-CV} + Q_{R-CV}))]\,\hat{u}$$

and have left the two equations $Q_{R-CV}^\top Q_{V-C}^\top\cdot$(5a′) as well as $P_{V-C}^\top\cdot$(5a′).

The first one is a hidden constraint, which the index-1 equations are lacking as T1 implies $Q_{CRV} = 0$. Using the PDE (4) for $\hat{\jmath}_L$ we obtain with the abbreviation $L := L(\hat{\jmath}_L, \hat{\tau}_m)$

$$\begin{aligned} &Q_{CRV}^\top \left(A_L L^{-1} A_L^\top [P_C + Q_C P_{V-C} + Q_C Q_{V-C} P_{R-CV} + Q_{CRV}]\hat{u} + A_I D_m \hat{\imath}\right) \\ &\quad = 0. \end{aligned} \tag{5a$'_2$}$$

Replacing $Q_{CRV}\hat{u}$ by $Q_{CRV}Q_{CRV}\hat{u}$ and adding $P_{CRV}^\top P_{CRV}Q_{CRV}\hat{u} = 0$ yields

$$\begin{aligned} Q_{CRV}\hat{u} \;&= \\ -\,H_5^{-1}Q_{CRV}^\top &\left(A_L L^{-1} A_L^\top [P_C + Q_C P_{V-C} + Q_C Q_{V-C} P_{R-CV}]\hat{u} + A_I D_m \hat{\imath}\right) \end{aligned}$$

with $H_5 := H_5(\hat{\jmath}_L, \hat{\tau}_m) := Q_{CRV}^\top A_L L^{-1} A_L^\top Q_{CRV} + P_{CRV}^\top P_{CRV}$ positive definite. Here, we have to apply the differential operator D_m one more time to obtain an equation for $Q_{CRV}D_m\hat{u}$.

As we now have determined all parts of $D_m\hat{u}$, the second equation $P_{V-C}^\top\cdot$(5a′) yields $\bar{P}_{V-C}D_m\hat{\jmath}_V$:

$$\begin{aligned} P_{V-C}^\top Q_C^\top &\left(A_R G A_R^\top D_m\hat{u} + A_L D_m\hat{\jmath}_L + A_I D_m\hat{\imath}\right) \\ &+ Q_C^\top A_V[\bar{P}_{V-C} + \bar{Q}_{V-C}]D_m\hat{\jmath}_V \;=\; 0. \end{aligned} \tag{5a$'_3$}$$

We multiply by $A_V^\top Q_C$ from the left and add $\bar{Q}_{V-C}^\top \bar{Q}_{V-C}\bar{P}_{V-C}D_m\hat{\jmath}_V = 0$ to obtain the positive dcfinite matrix $H_3 := A_V^\top Q_C Q_C^\top A_V + \bar{Q}_{V-C}^\top\bar{Q}_{V-C}$ and

$$\bar{P}_{V-C}D_m\hat{\jmath}_V = -H_3^{-1}A_V^\top Q_C P_{V-C}^\top Q_C^\top \left(A_R G A_R^\top D_m\hat{u} + A_L D_m\hat{\jmath}_L + A_I D_m\hat{\imath}\right).$$

Finally, we have a look at the derivative of equation (5b$_1$):

$$\bar{Q}_{V-C}^\top A_V^\top P_C D_m\hat{u} - \bar{Q}_{V-C}^\top D_m\hat{v} = 0. \tag{5b$'_1$}$$

With $P_C D_m\hat{u} = -H_1^{-1}P_C^\top \left(A_R\hat{r}(A_R^\top\hat{u}) + A_L\hat{\jmath}_L + A_V\hat{\jmath}_V + A_I\hat{\imath}\right)$ from (3), we get

$$\begin{aligned} &\bar{Q}_{V-C}^\top A_V^\top H_1^{-1}P_C^\top A_V[\bar{P}_{V-C} + \bar{Q}_{V-C}]\hat{\jmath}_V \\ &\qquad = -\bar{Q}_{V-C}^\top \left(D_m\hat{v} + A_V^\top H_1^{-1}P_C^\top \left(A_R\hat{r}(A_R^\top\hat{u}) + A_L\hat{\jmath}_L + A_I\hat{\imath}\right)\right). \end{aligned}$$

Replacing $\bar{Q}_{V-C}\hat{\jmath}_V$ by $\bar{Q}_{V-C}\bar{Q}_{V-C}\hat{\jmath}_V$ and adding $\bar{P}_{V-C}^\top\bar{P}_{V-C}\bar{Q}_{V-C}\hat{\jmath}_V = 0$ yields

$$\begin{aligned} \bar{Q}_{V-C}\hat{\jmath}_V \;&= \\ -\,H_6^{-1}\bar{Q}_{V-C}^\top &\left[D_m\hat{v} + A_V^\top H_1^{-1}P_C^\top \left(A_R\hat{r}(A_R^\top\hat{u}) + A_L\hat{\jmath}_L + A_V\bar{P}_{V-C}\hat{\jmath}_V + A_I\hat{\imath}\right)\right] \end{aligned}$$

with $H_6 := H_6(A_C^\top\hat{u}, \hat{\tau}_m) := \bar{Q}_{V-C}^\top A_V^\top H_1^{-1}A_V\bar{Q}_{V-C} + \bar{P}_{V-C}^\top\bar{P}_{V-C}$ positive definite. Again, another differentiation is needed to obtain an expression for $\bar{Q}_{V-C}D_m\hat{\jmath}_V$. □

Remark 1. The system defined in Lemma 1 can also be obtained by starting first from the semi-explicit formulation of the original DAE network equations following [ET98] and then introducing the multidimensional signal model.

Corollary 1. *The system defined in Lemma 1 is equivalent to the index-2 semi-explicit (but not Hessenberg) system*

$$\begin{aligned} D_m\hat{y} &= f(\hat{y},\hat{v},\hat{w},\hat{\tau}_m), &(9a)\\ 0 &= g_1(\hat{y},\hat{v},\hat{w},\hat{\tau}_m), &(9b)\\ 0 &= g_2(\hat{y},\hat{\tau}_m), &(9c)\end{aligned}$$

with three sets of network variables

$$\hat{y} = \begin{pmatrix} P_C\hat{u} \\ \hat{\jmath}_L \end{pmatrix}, \quad \hat{v} = \begin{pmatrix} Q_C P_{V-C}\hat{u} \\ Q_C Q_{V-C} P_{R-CV}\hat{u} \\ \bar{P}_{V-C}\hat{\jmath}_V \end{pmatrix} \quad \textit{and} \quad \hat{w} = \begin{pmatrix} Q_{CRV}\hat{u} \\ \bar{Q}_{V-C}\hat{\jmath}_V \end{pmatrix}.$$

Now applying the differential operator D_m to (9c), we are able to resolve $g := (g_1,g_2)^\top$ for $\hat{z} := (\hat{v},\hat{w})^\top = \Psi(\hat{y},\hat{\tau}_m)$ and to derive the underlying PDE

$$D_m\hat{y} = f(\hat{y},\Psi(\hat{y},\hat{\tau}_m),\hat{\tau}_m).$$

Of course, when thinking of solving the MPDAE, we do not use the concept of the underlying PDE, but it is a helpful tool to use the analogy of the MPDAE network equations to DAE-systems when transferring the index concept.

A special characteristics index was proposed in [Wa00] for linear hyperbolic PDAEs. As our network MPDAE is of hyperbolic type, we can proceed similarly. Defining a characteristic system leads to a continuous set of DAEs. In our special case, the characteristic curves are straight lines in the direction of the diagonal and the DAEs have the same structure as the original system (1). Thus, it is natural to use the index for this DAE system to characterize the MPDAE. Perturbation estimates and other suitable PDAE index concepts proposed in [GW00] are reserved to future work.

4 Conclusions

In this paper we have analysed a system of multirate partial differential-algebraic equations, which arises when a multidimensional signal model is applied to the MNA network equations. We showed, that the MPDAE inherits all the characteristics of the original network DAE. In both index-1 and index-2 cases, an underlying PDE can be found, i.e. the MPDAE can be reduced to a PDE on a manifold. Index concepts can be transferred and therefore no additional stability problems have to be expected when solving the network equations via the multidimensional approach. And, exploiting its special structure, the MPDAE can be solved very efficiently, e.g. with a method of characteristics proposed in [PG02].

Acknowledgement. This work has been supported within the federal BMBF project with the grant number 03GU-NAVN.

References

[BWLB96] Brachtendorf, H.G., Welsch, G., Laur, R., Bunse-Gerstner, A.: Numerical steady state analysis of electronic circuits driven by multi-tone signals. Electrical Engineering, **79**, 103–112 (1996)

[ET98] Estèvez Schwarz, D., Tischendorf, C.: Structural analysis for electric circuits and consequences for modified nodal analysis. Int. J. Circ. Theor. Appl., **28**, 131–162 (2000)

[GW00] Günther, M., Wagner, Y.: Index concepts for linear mixed systems of differential-algebraic and hyperbolic-type equations. SIAM J. Sci. Comp., **22**:5, 1610–1629 (2000)

[HW96] Hairer, E., Wanner, G.: Solving Ordinary Differential Equations II: Stiff and Differential-Algebraic Problems. Springer, Berlin, 2nd rev. ed. (1996)

[PG02] Pulch, R., Günther, M.: A method of characteristics for solving multirate partial differential equations in radio frequency application. Appl. Numer. Math., **42**, 397–409 (2002)

[Ti99] Tischendorf, C.: Topological index calculation of differential-algebraic equations in circuit simulation. Surv. Math. Ind., **8**, 187–199 (1999)

[Wa00] Wagner, Y.: A further index concept for linear PDAEs of hyperbolic type. Mathematics and Computers in Simulation, **53**, 287–291 (2000)

Semidiscretisation Methods for Warped MPDAEs

R. Pulch

Bergische Universität Wuppertal, Department of Mathematics, Chair of Applied Mathematics and Numerical Analysis, Gaußstr. 20, D-42119 Wuppertal, Germany, pulch@math.uni-wuppertal.de

Abstract The numerical simulation of electric circuits including signals with largely differing time scales demands specific strategies. A multivariate model for signals, which exhibit amplitude as well as frequency modulation, yields a warped multirate partial differential algebraic equation (MPDAE). Corresponding initial boundary value problems lead to particular solution types. Two strategies for numerical simulation are discussed, which use contrary semidiscretisation techniques.

1 Introduction

Signals acting at widely separated time scales arise in radio frequency applications. The mathematical model of corresponding electric circuits consists in differential algebraic equations (DAEs). Integrating these systems demands a huge computational effort, since the fastest time scale restricts the step sizes. Consequently, numerical methods have to incorporate the specific structure of arising solutions in order to be efficient.

A multidimensional model yields a strategy for the simulation of amplitude and/or frequency modulated signals. Narayan and Roychowdhury [5] introduced an according warped multirate partial differential algebraic equation (MPDAE). The MPDAE solution of an initial boundary value problem reproduces a multitone DAE solution. Solving the MPDAE demands less effort than handling the DAE directly, since the model omits the computation of all fast oscillations. However, the warped MPDAE system includes a local frequency function, which is a priori unspecified. The determination of appropriate local frequencies is crucial for the efficiency of the model. Continuous phase conditions can be applied as additional boundary constraints to obtain suitable solutions.

We present two approaches for the numerical simulation of the initial boundary value problem, which both apply a semidiscretisation of the warped MPDAE system. On the one hand, we consider a method of Rothe type, which performs a discretisation in the slow time scale. On the other hand, we arrange a method of lines, which discretises the fast time scale. The properties of these two antipodal techniques are analysed. In particular, we discuss the inclusion of a continuous phase condition in view of an unknown local frequency. Finally, numerical results illustrate the performance of the two methods.

2 Warped MPDAE Model

To explicate the multidimensional model, we consider a multitone signal, which includes amplitude as well as frequency modulation, namely

$$x(t) = \left[1 + \alpha \sin\left(\tfrac{2\pi}{T_1} t\right)\right] \sin\left(\tfrac{2\pi}{T_2} t + \beta \sin\left(\tfrac{2\pi}{T_1} t\right)\right) \tag{1}$$

with parameters $0 < \alpha < 1$, $\beta > 0$. Fig. 1 illustrates this signal qualitatively. Assuming $T_1 \gg T_2$, many fast oscillations proceed during one slow oscillation of the modulation. Thus the number of time points

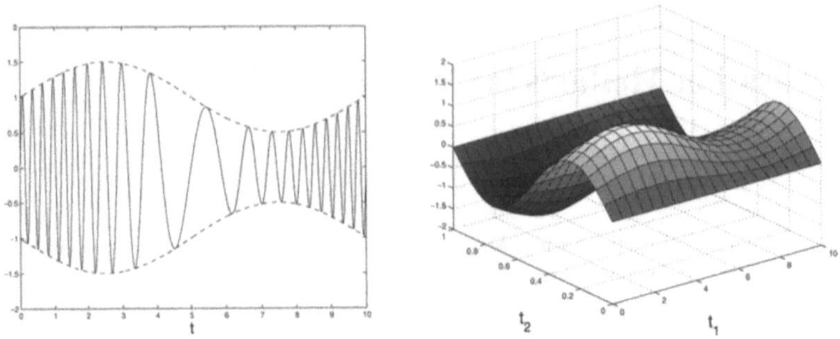

Fig. 1. Frequency modulated signal (left) and corresponding MVF (right)

to represent this signal increases drastically. Alternatively, we introduce an own variable for each separate time scale to model the amplitude modulation part

$$\hat{x}(t_1, t_2) = \left[1 + \alpha \sin\left(\tfrac{2\pi}{T_1} t_1\right)\right] \sin\left(2\pi t_2\right). \tag{2}$$

This representation is called the *multivariate function* (*MVF*) of the signal (1). Now the MVF is biperiodic and exhibits a simple structure in the rectangle $[0, T_1] \times [0, 1]$, which is also shown in Fig. 1. Hence we can resolve the MVF using relatively few grid points. The frequency modulation part is modelled by an additional time-dependent function

$$\Psi(t) = \frac{t}{T_2} + \frac{\beta}{2\pi} \sin\left(\tfrac{2\pi}{T_1} t\right). \tag{3}$$

The derivative $\nu := \Psi'$ plays the role of a *local frequency* belonging to the multitone signal (1). The function ν is T_1-periodic and features a simple behaviour, too. Nevertheless, we completely reconstruct the original signal via

$$x(t) = \hat{x}(t, \Psi(t)). \tag{4}$$

Thereby, Ψ is called a *warping function*, since it stretches the second time scale. Consequently, we obtain an efficient representation of the multitone signal by means of MVF and warping function/local frequency.

However, the multidimensional model is not unique. A family of MVFs and respective warping functions can describe the same signal. An inappropriate choice of the local frequency may yield a MVF, which exhibits many oscillations. Hence the identification of a local frequency with simple MVF determines the benefit of this representation.

Remark: The MVF concept is also convenient, if the first time scale is aperiodic and slowly varying. Consequently, the local frequency becomes aperiodic, too. In this case, we arrange the MVF in the domain $\mathbb{R}^+ \times [0, 1]$. Thus performing a step in t_1-direction already reproduces many fast oscillations.

Now we apply the multidimensional model in electric circuit simulation. A network approach yields *differential algebraic equations* (*DAEs*), which describe the transient behaviour of all node voltages and some branch currents, see [2]. In the following, we consider a semiexplicit DAE of index 1

$$\begin{aligned} \frac{\mathrm{d}\boldsymbol{y}}{\mathrm{d}t} &= \boldsymbol{f}(\boldsymbol{y}, \boldsymbol{z}) + \boldsymbol{b}(t) \\ \boldsymbol{0} &= \boldsymbol{g}(\boldsymbol{y}, \boldsymbol{z}) + \boldsymbol{c}(t) \end{aligned} \tag{5}$$

with solutions $\boldsymbol{y}(t) \in \mathbb{R}^d$, $\boldsymbol{z}(t) \in \mathbb{R}^a$. The functions $\boldsymbol{b}(t) \in \mathbb{R}^d$, $\boldsymbol{c}(t) \in \mathbb{R}^a$ represent independent input signals. We assume that the input varies slowly the amplitude and frequency of fast oscillations in the solution. Thus the above multivariate representation becomes feasible. A transformation with respect to the reconstruction (4) changes the DAE model (5) into a *warped multirate partial differential algebraic equation* (*MPDAE*)

$$\begin{aligned}\frac{\partial \hat{\boldsymbol{y}}}{\partial t_1} + \nu(t_1)\frac{\partial \hat{\boldsymbol{y}}}{\partial t_2} &= \boldsymbol{f}(\hat{\boldsymbol{y}}, \hat{\boldsymbol{z}}) + \boldsymbol{b}(t_1)\\ \boldsymbol{0} &= \boldsymbol{g}(\hat{\boldsymbol{y}}, \hat{\boldsymbol{z}}) + \boldsymbol{c}(t_1).\end{aligned} \tag{6}$$

Now we solve the MPDAE system in a domain $[0, T_f] \times [0, 1]$ with arbitrary final time $T_f > 0$. Therefore we consider the initial boundary value problem (6) together with

$$\begin{aligned}&\hat{\boldsymbol{y}}(0, t_2) = \boldsymbol{v}(t_2), \qquad \hat{\boldsymbol{z}}(0, t_2) = \boldsymbol{w}(t_2) \qquad \text{for all } t_2 \in \mathbb{R},\\ &\hat{\boldsymbol{y}}(t_1, t_2) = \hat{\boldsymbol{y}}(t_1, t_2 + 1),\ \hat{\boldsymbol{z}}(t_1, t_2) = \hat{\boldsymbol{z}}(t_1, t_2 + 1) \text{ for all } t_1 \geq 0,\ t_2 \in \mathbb{R}.\end{aligned} \tag{7}$$

Thereby, the choice of the periodic initial values $\boldsymbol{v}$,$\boldsymbol{w}$ has to be consistent with respect to the DAE (5). An according MPDAE solution yields a complete DAE solution applying (4) with $\Psi(t) = \int_0^t \nu(\tau)\mathrm{d}\tau$. The reconstructed signal is uniquely defined by the initial values $\boldsymbol{y}(0) = \boldsymbol{v}(0)$, $\boldsymbol{z}(0) = \boldsymbol{w}(0)$. Hence the choice of the other values in $\boldsymbol{v}, \boldsymbol{w}$ just influence the efficiency of the model, since the resulting MVF depends on these initial functions. Using constant input $\boldsymbol{b} \equiv \boldsymbol{b}(0)$, $\boldsymbol{c} \equiv \boldsymbol{c}(0)$ in the DAE, a corresponding periodic solution represents a suitable initial state in general.

Assuming T_1-periodic input signals, biperiodic MPDAE solutions may exist. We can apply the problem (6),(7) to compute a biperiodic solution, too. We solve the MPDAE proceeding in t_1-direction until the solution enters a biperiodic steady state response. This strategy represents an advancement of transient analysis by using more information about the problem structure.

Since the local frequency ν stands for an a priori unknown function, the system (6),(7) is underdetermined. Hence we require additional conditions to isolate special solutions. In [5], *continuous phase conditions* are proposed to achieve this purpose. Thereby, the idea is to control the phase in each cross section $t_1 = \text{const}$ of a MVF. In the following, we apply a specific phase condition to the (without loss of generality) first component of the solution $\hat{\boldsymbol{y}} = (\hat{y}^1, \ldots, \hat{y}^d)^T$, namely

$$\left.\frac{\partial \hat{y}^1}{\partial t_2}\right|_{t_2=0} = 0 \quad \text{for all } t_1. \tag{8}$$

If the involved functions are sufficiently smooth, then differentiating (8) with respect to t_1 and (6) with respect to t_2 implies

$$\left.\frac{\partial^2 \hat{y}^1}{\partial t_1 \partial t_2}\right|_{t_2=0} = 0 \quad \Rightarrow \quad \nu(t_1)\left.\frac{\partial^2 \hat{y}^1}{\partial {t_2}^2}\right|_{t_2=0} = \left.\frac{\partial f^1(\hat{\boldsymbol{y}}, \hat{\boldsymbol{z}})}{\partial t_2}\right|_{t_2=0} \qquad \text{for all } t_1. \tag{9}$$

Thus to ensure that the phase condition determines the local frequency uniquely, we assume the existence of a solution satisfying (8) and

$$\left|\left.\frac{\partial^2 \hat{y}^1}{\partial {t_2}^2}\right|_{t_2=0}\right| \geq \delta \quad \text{for all } t_1 \tag{10}$$

with a constant $\delta > 0$ in the following.

Alternatively, Houben [3] introduces minimum demands, which shall reduce oscillations in MVFs. Using these criteria, the determination of a relatively simple MVF representation is guaranteed. However, minimum demands cause more computation work in comparison to the elementary condition (8), which we add directly to the boundary conditions in the underlying domain.

3 Semidiscretisation Techniques

Now we examine two numerical techniques for solving the MPDAE initial boundary value problem (6),(7), which both apply semidiscretisation.

Firstly, we perform a *Rothe method* (*RM*). For parabolic PDEs including a time and a space variable, this means that the time derivative is discretised and thus a sequence of ODE boundary value problems in space arises. Accordingly, a difference scheme replaces the derivative with respect to t_1 in (6). Assuming a positive local frequency, the implicit Euler scheme, for example, yields the subsequent DAE systems

$$\begin{aligned} \frac{\mathrm{d}\tilde{\boldsymbol{y}}_j}{\mathrm{d}t_2}(t_2) &= \frac{1}{\nu_j}\left\{\boldsymbol{f}(\tilde{\boldsymbol{y}}_j(t_2),\tilde{\boldsymbol{z}}_j(t_2)) + \boldsymbol{b}(jh_1) - \tfrac{1}{h_1}\left[\tilde{\boldsymbol{y}}_j(t_2) - \tilde{\boldsymbol{y}}_{j-1}(t_2)\right]\right\} \\ \mathbf{0} &= \boldsymbol{g}(\tilde{\boldsymbol{y}}_j(t_2),\tilde{\boldsymbol{z}}_j(t_2)) + \mathbf{c}(jh_1) \end{aligned} \tag{11}$$

for $j = 1, 2, \ldots$ with step size h_1, where the jth part is an approximation of the MPDAE solution in the layer $t_1 = jh_1$. The initial values correspond to $j = 0$. The periodicity and the phase condition (8) generate the boundary constraints

$$\tilde{\boldsymbol{y}}_j(0) = \tilde{\boldsymbol{y}}_j(1), \quad \tilde{\boldsymbol{z}}_j(0) = \tilde{\boldsymbol{z}}_j(1), \quad \tfrac{\mathrm{d}\tilde{y}_j^1}{\mathrm{d}t_2}(0) = 0. \tag{12}$$

The local frequency ν_j represents an unknown parameter in each system. Hence the RM consists in the successive handling of boundary value problems corresponding to parameter-dependent DAEs with $d + a$ unknown functions. The DAEs (11) inherit the index 1 from the DAE (5). Moreover, specific techniques can be used to determine the periodic solution $\tilde{\boldsymbol{y}}_j, \tilde{\boldsymbol{z}}_j$ and the parameter ν_j in view of phase conditions, see [4].

Secondly, we apply a *method of lines (ML)*. Now the derivative with respect to t_2 is substituted by a difference formula in the MPDAE. We employ symmetric differences and obtain a large DAE system including the subunits

$$\begin{aligned} \frac{\mathrm{d}\bar{\boldsymbol{y}}_i}{\mathrm{d}t_1}(t_1) &= \boldsymbol{f}(\bar{\boldsymbol{y}}_i(t_1),\bar{\boldsymbol{z}}_i(t_1)) + \boldsymbol{b}(t_1) - \nu(t_1)\tfrac{1}{2h_2}\left[\bar{\boldsymbol{y}}_{i+1}(t_1) - \bar{\boldsymbol{y}}_{i-1}(t_1)\right] \\ \mathbf{0} &= \boldsymbol{g}(\bar{\boldsymbol{y}}_i(t_1),\bar{\boldsymbol{z}}_i(t_1)) + \mathbf{c}(t_1) \end{aligned} \tag{13}$$

for $i = 1, \ldots, n_2$ with step size $h_2 = 1/n_2$. The ith component represents an approximation of the MPDAE solution in the layer $t_2 = (i-1)h_2$. The periodicity allows to identify $\bar{\boldsymbol{y}}_{n_2+1} = \bar{\boldsymbol{y}}_1$, $\bar{\boldsymbol{y}}_0 = \bar{\boldsymbol{y}}_{n_2}$ and thus to eliminate these unknown. Since the local frequency ν is unidentified, too, we have to incorporate the phase condition (8) via a difference formula. For example,

$$0 = \tfrac{\partial \hat{y}^1}{\partial t_2}(t_1, 0) \doteq \tfrac{1}{2h_2}\left[\bar{y}_2^1(t_1) - \bar{y}_{n_2}^1(t_1)\right] \tag{14}$$

gives an additional algebraic relation. Consequently, the ML yields an initial value problem of DAEs with dimension $n_2(d + a) + 1$. However, if we see ν as a part of the solution, then the index of the system (13),(14) is at least 2 even for an original DAE (5) of index 1. Furthermore, a suitable consistent choice of a starting value $\nu(0)$ is necessary.

As mentioned in the previous section, the initial boundary value problem can be used to determine a biperiodic solution by transient analysis. If we want to compute this steady state response directly, then a method of characteristics becomes favourable, see [6]. Moreover, the employed information transfer generates an inherent potential for parallelism. In contrast, the solution of the initial boundary value problem implies a sequential structure.

4 Numerical Results

We apply both semidiscretisation methods for the numerical simulation of a voltage controlled Van der Pol oscillator. The corresponding system reads

$$\begin{aligned} \dot{u} &= v \\ \dot{v} &= -10(u^2 - 1)v - (2\pi w)^2 u \\ 0 &= w - b(t), \end{aligned} \tag{15}$$

which represents a semiexplicit DAE of index 1. If the input signal b is constant, a periodic steady state response arises. Otherwise, a time-dependent input signal produces frequency modulation. We choose the function

$$b(t) = 1 + \tfrac{1}{2} \sin\left(\tfrac{2\pi}{T_1} t\right) \quad \text{with} \quad T_1 = 1000. \tag{16}$$

Since the involved time scales are widely separated, we use the corresponding MPDAE model and treat problem (6),(7). As initial values, the periodic response of (15) corresponding to $b \equiv 1$ is employed. In the RM (11), we solve the periodic boundary value problems via a finite difference method including trapezoidal rule. In the ML (13), the initial value problems are integrated by the implicit Euler scheme. The used step sizes are equidistant, namely $h_1 = 20$ and $h_2 = 0.01$ in both techniques.

Figure 2 illustrates the computed local frequencies. Since both functions respond to the input signal, the local frequencies are physically reasonable. Fig. 3 and Fig. 4 show the MPDAE solutions for u and v, respectively. The MVF of u features a constant amplitude, whereas the MVF of v includes amplitude modulation. The component w just reproduces the input signal. Investigating these MVFs, we recognise that assumption (10) is satisfied with $\delta \approx 80$.

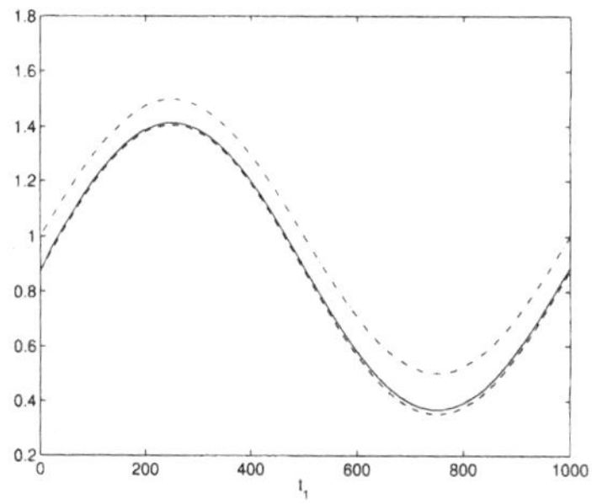

Fig. 2. Local frequency computed by RM (—) and ML (- -), respectively, together with input signal (- · · -)

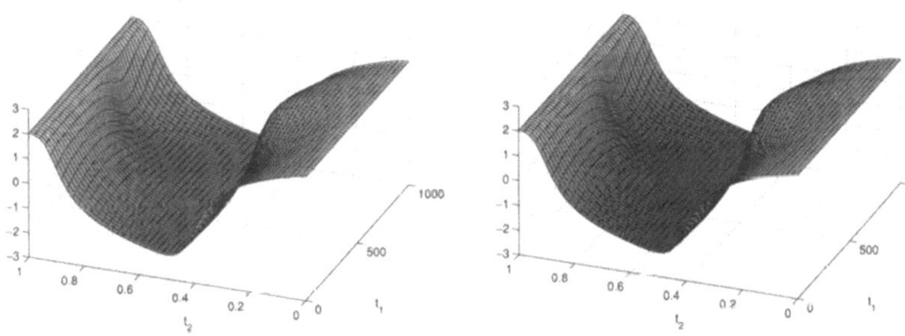

Fig. 3. MPDAE solution for u computed by RM (left) and ML (right)

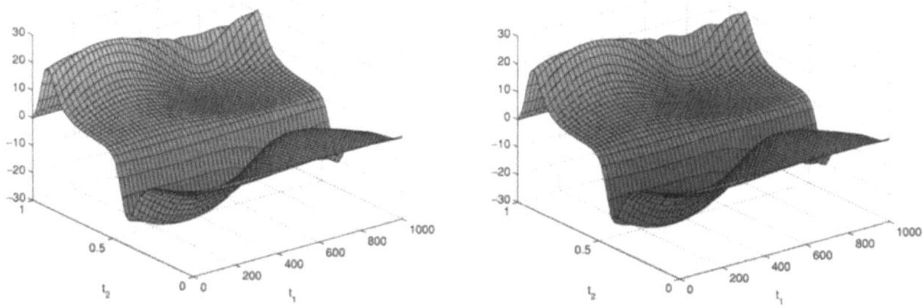

Fig. 4. MPDAE solution for v computed by RM (left) and ML (right)

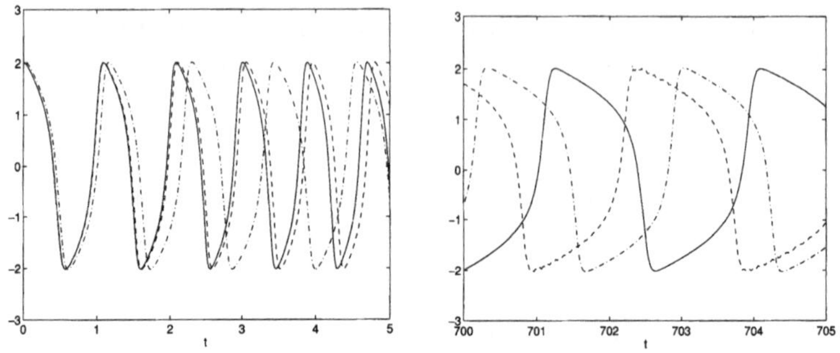

Fig. 5. DAE solution for u in time intervals $[0, 5]$ (left) and $[700, 705]$ (right) from RM (—), ML (- -) and transient integration (- · -)

Finally, we observe the corresponding DAE solutions. The results of the RM and the ML are used in the reconstruction (4). The outcome for u is shown in Fig. 5. Thereby, a reference solution was computed via an initial value problem of (15) using trapezoidal rule. In the first few cycles, both semidiscretisation methods exhibit a frequency, which is too high in comparison to the reference signal. In the RM, the local frequency even increases incorrectly for smaller step sizes h_1, whereas the frequency remains the same in the ML. In later cycles, all signals exhibit a significant phase shift to each other, which reflects a certain sensitivity, see [6]. Nevertheless, amplitude and shape agree in all three signals.

Other simulations, for example using a smaller value T_1, indicate an even more problematical behaviour of the semidiscretisation methods, where also too high amplitudes may arise. Moreover, the use of a BDF2 scheme, see [1], to proceed in t_1-direction leads to less accurate results in both RM and ML. Applying trapezoidal rule in the ML causes significant inaccuracies, which reflect the higher index of the semidiscretised system. Thus the application of semidiscretisation techniques seems to be critical, at least if the boundary constraint (8) is involved.

5 Conclusions

The MPDAE model provides an alternative approach for the numerical simulation of multitone signals. Two techniques based on semidiscretisation for solving initial boundary value problems of MPDAEs have been presented, namely a Rothe method and a method of lines. Thereby, a specific boundary constraint is applied to identify the local frequency function. Numerical results demonstrate that both techniques exhibit problems in computing an accurate solution. Hence further theoretical examinations with respect to feasibility and stability of semidiscretisation methods are necessary in this context.

References

1. C.W. Gear. Simultaneous numerical solution of differential-algebraic equations. *IEEE Trans. on Circuit Theory*, 18:89–95, 1971
2. M. Günther and U. Feldmann. CAD based electric circuit modeling in industry I: mathematical structure and index of network equations. *Surv. Math. Ind.*, 8:97–129, 1999
3. S.H.M.J. Houben. Simulating multi-tone free-running oscillators with optimal sweep following. In W.H.A. Schilders, E.J.W. terMaten, and S.H.M.J. Houben, editors, *Scientific Computing in Electrical Engineering*, Mathematics in Industry, pages 240–247. Springer, 2004
4. K.S. Kundert, A. Sangiovanni-Vincentelli, and T. Sugawara. Techniques for finding the periodic steady-state response of circuits. In T. Ozawa, editor, *Analog methods for computer-aided circuit analysis and diagnosis*, pages 169–203. Marcel Dekker Inc., 1988
5. O. Narayan and J. Roychowdhury. Analyzing oscillators using multitime PDEs. *IEEE Trans. CAS I*, 50(7):894–903, 2003
6. R. Pulch. Multi time scale differential equations for simulating frequency modulated signals. *Appl. Numer. Math.*, 53(2-4):421–436, 2005

Qualitative Properties of Equilibria in MNA Models of Electrical Circuits

R. Riaza[1] and C. Tischendorf[2]

[1] Departamento de Matemática Aplicada a las Tecnologías de la Información
ETSI Telecomunicación, Universidad Politécnica de Madrid
Ciudad Universitaria s/n - 28040 Madrid, Spain, `rrr@mat.upm.es`
[2] Institut für Mathematik, Technische-Universität Berlin
10623 Berlin, Germany, `tischend@math.tu-berlin.de`

Abstract We present in this communication some tools for the qualitative analysis of lumped circuits directed to differential-algebraic MNA models. The attention is focused on equilibria, which describe operating points of the circuit. Specifically, hyperbolicity and asymptotic stability of linearized models are analyzed in terms of the circuit topology and device characteristics. The topological conditions arising in this qualitative study are proved independent of those supporting the index of the differential-algebraic circuit model. An example containing a Josephson junction circuit illustrates the discussion.

1 Introduction

Qualitative properties of nonlinear circuits have been often discussed assuming that a state-space model describing network dynamics is available [Chu80, GW92]. However, such a state model does not always exist or is difficult to obtain in practice; this has led to semistate formalisms based on differential-algebraic equations (DAEs), which currently frame approaches such as Modified Nodal Analysis (MNA) or Tableau Analysis [ET00, GF99, Tis99]. In this differential-algebraic context, we address in the present communication several qualitative properties of equilibria in MNA-modeled nonlinear circuits, using and extending previous results from [Ria04, Tis99].

Qualitative features of circuits have been also addressed in the last decades within a geometric framework. This stems from the work [BM64]; see also [DW72, HB84, HB86, Mat87, Sma72, WM97, WMT98]. This approach provides a coordinate-free point of view for the analysis of several intrinsic properties of circuit dynamics. Our approach, in contrast, uses the natural coordinates arising in the widely-used MNA models of electrical circuits.

We work with nonlinear RLC circuits assuming that capacitors, resistors and inductors are respectively controlled through C^1 relations of the form $q = \psi(v_c)$, $i_r = \gamma(v_r)$, $\phi = \varphi(i_l)$. Denote the capacitance, inductance, and conductance matrices as $C(v_c) = \psi'(v_c)$, $L(i_l) = \varphi'(i_l)$, $G(v_r) = \gamma'(v_r)$. In circuit-theoretic terms, symmetric capacitance or inductance matrices will be said to describe *reciprocal* devices, whereas positive definite capacitance, inductance or conductance matrices will be said to yield *strictly locally passive* elements [Chu80]; positive definiteness of an $n \times n$ matrix B means in this work that $x^{\mathrm{T}} Bx > 0$ for any $x \in \mathbb{R}^n - \{0\}$, not implying that B is symmetric.

Conventional MNA equations for circuits without controlled sources read

$$A_C C(A_C^{\mathrm{T}} e) A_C^{\mathrm{T}} e' + A_R \gamma(A_R^{\mathrm{T}} e) + A_L i_l + A_V i_v = -A_I i_s(t) \tag{1a}$$

$$L(i_l) i_l' - A_L^{\mathrm{T}} e = 0 \tag{1b}$$

$$-A_V^{\mathrm{T}} e = -v_s(t). \tag{1c}$$

Here, e stands for node voltages; i_l, i_v represent currents in inductors and voltage sources, respectively, and $i_s(t)$, $v_s(t)$ denote currents and voltages in the (independent) sources. A_R (resp. A_L, A_C, A_V, A_I)

describes the *incidence* between resistive (resp. inductive, capacitive, voltage source, current source) branches and nodes in the circuit, once a reference node has been chosen. Specifically, the incidence matrix $(a_{ij}) \in \mathbb{R}^{(n-1)\times b}$ (n and b being the number of nodes and branches in the circuit, respectively) is given by

$$a_{ij} = \begin{cases} 1 & \text{if branch } j \text{ leaves node } i \\ -1 & \text{if branch } j \text{ enters node } i \\ 0 & \text{if branch } j \text{ is not incident with node } i. \end{cases}$$

Note that (1) is a quasilinear DAE of the form

$$A(x)x' + f(x) = s(t), \tag{2}$$

where $x = (e,\ i_l,\ i_v)^{\mathrm{T}}$, s is the excitation term $(-A_I i_s,\ 0,\ -v_s)^{\mathrm{T}}$, and

$$A = \begin{pmatrix} A_C C(A_C^{\mathrm{T}} e) A_C^{\mathrm{T}} & 0 & 0 \\ 0 & L(i_l) & 0 \\ 0 & 0 & 0 \end{pmatrix},\ f = \begin{pmatrix} A_R \gamma(A_R^{\mathrm{T}} e) + A_L i_l + A_V i_v \\ -A_L^{\mathrm{T}} e \\ -A_V^{\mathrm{T}} e \end{pmatrix}.$$

Many analytical and numerical features of a semistate circuit model rely upon its *index* (see [ET00, GF99, Ria04, Tis99] and references therein). We compile in Proposition 1 below Theorems 4 and 5 of [Tis99], replacing in the first claim positive definiteness by just non-singularity on L:

Proposition 1. *Assume that the capacitance and conductance matrices are positive definite, and that the inductance matrix is non-singular.*

1. *If the network contains neither I-L cutsets nor V-C loops (except for C-loops), then the MNA system (1) has index ≤ 1.*
2. *Assume additionally that the inductance matrix is positive definite. If the network contains an I-L cutset or a V-C loop (with at least one voltage source), then the MNA system (1) has index 2.*

Assume that a given circuit has only independent DC sources, so that s in (2) is a constant vector. We may hence rewrite this equations as the quasilinear autonomous DAE

$$A(x)x' + g(x) = 0, \tag{3}$$

with $g(x) = f(x) - s$.

Equilibrium points of (3) are defined by the condition $g(x^*) = 0$, and the linearization of the DAE at equilibrium leads to the *matrix pencil* $\lambda A(x^*) + g'(x^*)$, i.e.,

$$\lambda \begin{pmatrix} A_C C(A_C^{\mathrm{T}} e^*) A_C^{\mathrm{T}} & 0 & 0 \\ 0 & L(i_l^*) & 0 \\ 0 & 0 & 0 \end{pmatrix} + \begin{pmatrix} A_R G(A_R^{\mathrm{T}} e^*) A_R^{\mathrm{T}} & A_L & A_V \\ -A_L^{\mathrm{T}} & 0 & 0 \\ -A_V^{\mathrm{T}} & 0 & 0 \end{pmatrix}. \tag{4}$$

Several qualitative properties of equilibria can be characterized in terms of the spectrum $\sigma\{A(x^*), g'(x^*)\} = \{\lambda \in \mathbb{C}\ /\ \det(\lambda A(x^*) + g'(x^*)) = 0\}$ of the matrix pencil depicted in (4). The reader is referred to [Ria04] and references therein for background in this regard. The purpose of the present work is to characterize the spectrum of (4) in terms of the circuit topology.

We compile below some results coming from graph theory which will be useful in this regard. $\mathcal{K}$ will represent a subset of the set of branches of a connected graph $\mathcal{G}$. We denote as $A_{\mathcal{K}}$ (resp. $A_{\mathcal{G}-\mathcal{K}}$) the submatrix of A formed by the columns corresponding to the branches in $\mathcal{K}$ (resp. not in $\mathcal{K}$).

Lemma 1. *$\mathcal{K}$ does not contain loops if and only if $A_{\mathcal{K}} y = 0 \Rightarrow y = 0$, that is,* $\mathrm{Ker}\ A_{\mathcal{K}} = \{0\}$.

The subset $\mathcal{K}$ is a *cutset* if the deletion of $\mathcal{K}$ results in a disconnected graph, and it is minimal with respect to this property (i.e., removing any proper subset of $\mathcal{K}$ does not disconnect the graph).

Lemma 2. *$\mathcal{K}$ does not contain cutsets if and only if $x^{\mathrm{T}} A_{\mathcal{G}-\mathcal{K}} = 0 \Rightarrow x = 0$ or, equivalently, $A^{\mathrm{T}}_{\mathcal{G}-\mathcal{K}} x = 0 \Rightarrow x = 0$, that is,* Ker $A^{\mathrm{T}}_{\mathcal{G}-\mathcal{K}} = \{0\}$.

The following two properties will be useful later on.

Lemma 3. *Let $\mathcal{J}_1$, $\mathcal{J}_2$ be two sets of branches of a connected graph $\mathcal{G}$, $\mathcal{J}_1 \subseteq \mathcal{J}_2$. If all loops within $\mathcal{J}_2$ are contained in $\mathcal{J}_1$, then $A_{\mathcal{J}_1} w_1 + A_{\mathcal{J}_2 - \mathcal{J}_1} w_2 = 0 \Rightarrow w_2 = 0$. Equivalently, letting the first columns of $A_{\mathcal{J}_2}$ be those of $A_{\mathcal{J}_1}$,* Ker $A_{\mathcal{J}_2}$ = Ker $A_{\mathcal{J}_1} \times \{0\}$.

Lemma 4. *Let $\mathcal{K}_1$, $\mathcal{K}_2$ be two sets of branches of a connected graph $\mathcal{G}$, $\mathcal{K}_1 \subseteq \mathcal{K}_2$. If all cutsets within $\mathcal{K}_2$ are contained in $\mathcal{K}_1$, then $w^{\mathrm{T}} A_{\mathcal{G}-\mathcal{K}_2} = 0 \Rightarrow w^{\mathrm{T}} A_{\mathcal{K}_2 - \mathcal{K}_1} = 0$. Equivalently,* Ker $A^{\mathrm{T}}_{\mathcal{G}-\mathcal{K}_2}$ = Ker $A^{\mathrm{T}}_{\mathcal{G}-\mathcal{K}_1}$.

2 Hyperbolicity

Equilibrium points of (3) are defined by the vanishing of $g(x)$. An equilibrium x^* is said to be *hyperbolic* if the spectrum of the linearization has no purely imaginary eigenvalues. Null eigenvalues are depicted if and only if $g'(x^*)$ is singular; non-singularity of $g'(x^*)$ guarantees the isolation of this equilibrium and follows, in circuits with definite conductance, from the topological conditions of Theorem 1 below. We skip the proof of this result for the sake of brevity; note that it is a restatement, in a matrix pencil setting, of a known result [HB86, MCM79]. Non-vanishing, purely imaginary eigenvalues will be ruled out by the conditions in Theorem 2.

Theorem 1. *Let $x^* = (e^*, i_l^*, i_v^*)$ be an equilibrium point of (3). Denote $G = G(A_R^{\mathrm{T}} e^*)$, and assume that G is (positive or negative) definite. Then x^* is non-singular (equivalently, $0 \notin \sigma\{A(x^*), g'(x^*)\}$) if and only if there are neither V-L loops nor I-C cutsets in the circuit.*

Theorem 2. *If G is (positive or negative) definite, both $C = C(A_C^{\mathrm{T}} e^*)$ and $L = L(i_l^*)$ are symmetric and non-singular, and any one of the conditions*

a) there are no I-C-L cutsets; or
b) there are no V-C-L loops;

holds, then there are no purely imaginary eigenvalues $\lambda = \alpha j$ with $\alpha \in \mathbb{R} - \{0\}$.

Proof: $\lambda \in \mathbb{C}$ is an eigenvalue if and only if there exists a nonvanishing vector $w = (w_e, w_l, w_v)$ such that $(\lambda A(x^*) + g'(x^*))w = 0$, what yields

$$\lambda A_C C A_C^{\mathrm{T}} w_e + A_R G A_R^{\mathrm{T}} w_e + A_L w_l + A_V w_v = 0 \tag{5a}$$

$$-A_L^{\mathrm{T}} w_e + \lambda L w_l = 0 \tag{5b}$$

$$-A_V^{\mathrm{T}} w_e = 0. \tag{5c}$$

Multiplying (5a) by the conjugate transpose $w_e^\star$, we get

$$\lambda w_e^\star A_C C A_C^{\mathrm{T}} w_e + w_e^\star A_R G A_R^{\mathrm{T}} w_e + w_e^\star A_L w_l + w_e^\star A_V w_v = 0. \tag{6}$$

Note that (5b) yields $w_e^\star A_L = \overline{\lambda} w_l^\star L$, where we have used the symmetry of L. On the other hand, from (5c), it follows that $w_e^\star A_V = 0$. Some simple computations lead to

$$(\mathrm{Re}\lambda) w_e^\star A_C C A_C^{\mathrm{T}} w_e + w_e^\star A_R \frac{G + G^{\mathrm{T}}}{2} A_R^{\mathrm{T}} w_e + (\mathrm{Re}\lambda) w_l^\star L w_l = 0. \tag{7}$$

Let λ be a non-vanishing eigenvalue with $\mathrm{Re}\lambda = 0$. Equation (7) then leads to $A_R^{\mathrm{T}} w_e = 0$, due to the definiteness of G. Now, assume first that condition a) is satisfied. The exclusion of I-C-L cutsets, together with $A_R^{\mathrm{T}} w_e = 0$ and $A_V^{\mathrm{T}} w_e = 0$ (from (5c)), implies that $w_e = 0$. From (5b), the assumption $\lambda \neq 0$, and the non-singularity of L, we get $w_l = 0$. Then, from (5a), we get $A_V w_v = 0$, and the exclusion of V-loops in well-posed circuits would yield $w_v = 0$.

Assume now that condition b) is satisfied, and write (5a) as

$$A_C(\lambda C A_C^{\mathrm{T}} w_e) + A_L w_l + A_V w_v = 0,$$

since $A_R^{\mathrm{T}} w_e = 0$. From the V-C-L loop exclusion property, it follows that $\lambda C A_C^{\mathrm{T}} w_e = 0$, $w_l = 0$, $w_v = 0$. From the first identity, the non-vanishing of λ, and the non-singularity of C, we get $A_C^{\mathrm{T}} w_e = 0$. On the other hand, $w_l = 0$ yields, in the light of (5b), $A_L^{\mathrm{T}} w_e = 0$. Together with the conditions $A_C^{\mathrm{T}} w_e = 0$, $A_R^{\mathrm{T}} w_e = 0$, $A_V^{\mathrm{T}} w_e = 0$, and the exclusion of I cutsets in well-posed circuits, we would get $w_e = 0$. □

Theorems 1 and 2 together provide a sufficient condition for the hyperbolicity of the matrix pencil. Merging the topological conditions and using Lemmas 3 and 4, we may assert hyperbolicity allowing for the existence of V-C loops and I-L cutsets, so that the resulting topological conditions be entirely independent of the index conditions appearing in Proposition 1. Therefore, Theorem 3 will naturally apply to both index-1 and index-2 problems.

Theorem 3. *If G is (positive or negative) definite, both C and L are symmetric and non-singular, and any one of the two pairs of conditions*

a) there are neither V-L loops nor I-C-L cutsets (except maybe I-L cutsets); or
b) there are neither I-C cutsets nor V-C-L loops (except maybe V-C loops);

is satisfied, then $\mathrm{Re}\lambda \neq 0,\ \forall \lambda \in \sigma\{A(x^*), g'(x^*)\}$.

Proof: Since I-C-L cutsets include in particular I-C cutsets, and so do V-C-L loops with regard to V-L loops, the only cases which do not follow automatically from Theorem 1 and Theorem 2 are those in which either I-L cutsets or V-C loops are present. We have to show that purely imaginary non-vanishing eigenvalues may not exist in this situation.

Let us first consider case a). Proceeding as in the proof of Theorem 2, we get $A_R^{\mathrm{T}} w_e = 0$ and $A_V^{\mathrm{T}} w_e = 0$. Denote as $\mathcal{K}_1$ the set of branches corresponding to inductors and current sources, and as $\mathcal{K}_2$ the ones corresponding to capacitors, inductors and current sources. If $\mathcal{G}$ stands for the graph of the circuit, the branches in $\mathcal{G} - \mathcal{K}_2$ correspond to resistors and voltage sources, whereas those in $\mathcal{K}_2 - \mathcal{K}_1$ are the capacitive ones. With this notation, and in the light of Lemma 4, we get that $w_e^{\mathrm{T}}(A_R\ A_V) = 0 \Rightarrow w_e^{\mathrm{T}} A_C = 0$, that is, $A_C^{\mathrm{T}} w_e = 0$. From this property, (5a) reads $A_L w_l + A_V w_v = 0$, and the exclusion of V-L loops in a) yields $w_l = 0$, $w_v = 0$. Additionally, (5b) implies $A_L^{\mathrm{T}} w_e = 0$, and the absence of I cutsets in well-posed circuits implies $w_e = 0$.

Now consider case b). Again, $A_R^{\mathrm{T}} w_e = 0$ and $A_V^{\mathrm{T}} w_e = 0$ hold. Using $A_R^{\mathrm{T}} w_e = 0$, equation (5a) reads $\lambda A_C C A_C^{\mathrm{T}} w_e + A_L w_l + A_V w_v = 0$. Let $\mathcal{J}_1$ stand for the capacitor and voltage source branches, and assume that $\mathcal{J}_2$ includes these and, additionally, the inductive branches. Based upon the absence of V-C-L loops except for V-C loops, application of Lemma 3 yields $w_l = 0$. In virtue of (5b), it is $A_L^{\mathrm{T}} w_e = 0$, and the properties $A_R^{\mathrm{T}} w_e = 0$, $A_V^{\mathrm{T}} w_e = 0$, together with the exclusion of I-C cutsets, lead to $w_e = 0$. Finally, $w_v = 0$ from (5a) and the absence of V-loops in well-posed circuits. □

3 Asymptotic stability

Proposition 2. *If G is positive definite, and both C and L are symmetric positive definite, then* $\mathrm{Re}\lambda \leq 0,\ \forall \lambda \in \sigma\{A(x^*), g'(x^*)\}$.

Proof: The derivation of (7) in Theorem 2 is still valid under the current working assumptions. Let λ be an eigenvalue with $\mathrm{Re}\lambda > 0$. From the assumption of symmetry and positive definiteness on C and L, it follows that

$$w_e^{\star} A_C C A_C^{\mathrm{T}} w_e = w_e^{\star} A_R \frac{G + G^{\mathrm{T}}}{2} A_R^{\mathrm{T}} w_e = w_l^{\star} L w_l = 0, \tag{8}$$

so that $A_C^{\mathrm{T}} w_e = 0$, $A_R^{\mathrm{T}} w_e = 0$, $w_l = 0$ and (using (5b)) $A_L^{\mathrm{T}} w_e = 0$. Additionally, $A_V^{\mathrm{T}} w_e = 0$ as displayed in (5c). Since current source cutsets are forbidden in well-posed circuits, it follows that $w_e = 0$. From (5a), we get $A_V w_v = 0$ and, since voltage source loops are also excluded in well-posed circuits, it follows that $w_v = 0$. This would yield the contradiction $w = 0$, meaning that it must be $\text{Re}\lambda \leq 0$. □

Adding to Proposition 2 the topological conditions of Theorem 3, we get the following asymptotic stability criterion, where the topological conditions are again independent of those characterizing the index in Proposition 1.

Theorem 4. *Assume that:*

1) G is positive definite, and both C and L are symmetric positive definite.
2) At least one of the two pairs of topological conditions holds:
2a) There are neither V-L loops nor I-C-L cutsets (except maybe I-L cutsets); or
2b) There are neither I-C cutsets nor V-C-L loops (except maybe V-C loops).

Then, all eigenvalues in the spectrum $\sigma\{A(x^*), g'(x^*)\}$ *verify* $\text{Re}\lambda < 0$. □

4 Example

Consider the nonlinear circuit depicted in Fig. 1. The device labeled as L_2 is a *Josephson junction*, which can be treated as a nonlinear inductor with a current-flux characteristic $i_2 = I_0 \sin k\phi_2$, where $I_0 > 0$ is a device parameter, and k is a positive physical constant. The incremental inductance of this device is $L_2 = (I_0 k \cos k\phi_2)^{-1}$.

The two resistors are linear with conductances $G_1 > 0$, $G_2 \geq 0$, and the inductor is linear with inductance $L_1 > 0$. MNA equations read

$$L_1 i_1' = e_1 \tag{9a}$$
$$L_2 i_2' = e_2 \tag{9b}$$
$$0 = i_1 + G_1(e_1 - e_2) - I \tag{9c}$$
$$0 = i_2 - G_1(e_1 - e_2) + G_2 e_2. \tag{9d}$$

Equilibrium points are given by $e_1 = e_2 = 0$, $i_1 = I$, $i_2 = 0$. The latter yields $\sin k\phi_2 = 0$, i.e., $\phi_2 = n\pi/k$, $n \in \mathbb{Z}$, so that the incremental inductance L_2 at equilibrium is $\pm(I_0 k)^{-1}$, the sign depending on the parity of n.

Stability properties have been analyzed in [Ria04] via a DAE model of the circuit. Our present goal is to illustrate that this qualitative analysis can be performed checking only device characteristics and circuit topology, without making explicit use of any model. We will distinguish the two cases $G_2 > 0$ and $G_2 = 0$. Note that, in both cases, the (symmetric) inductance matrix $L = \text{diag}(L_1, L_2)$ is positive definite (resp. indefinite) at equilibria for which $L_2 = (I_0 k)^{-1}$ (resp. $L_2 = -(I_0 k)^{-1}$).

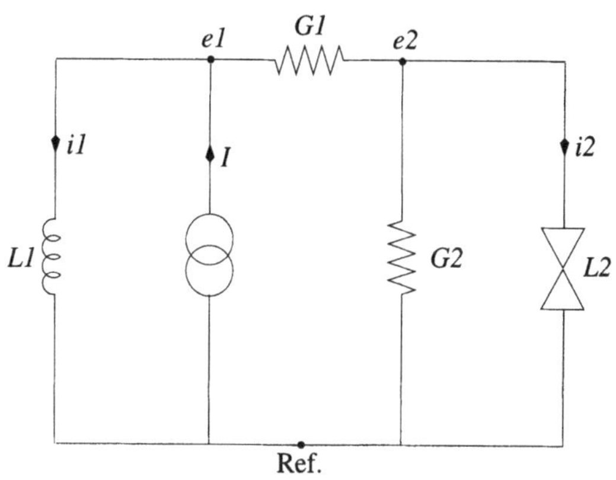

Fig. 1. A Josephson junction circuit

Index. In the absence of capacitors and voltage sources, according to Proposition 1 it suffices to check for I-L cutsets in order to compute the index of (9). This yields index-1 regardless of the sign of L_2 when $G_2 > 0$. In contrast, the case $G_2 = 0$ yields an I-L cutset defined by the linear inductor, the current source and the Josephson junction. In this situation, Proposition 1 only allows one to conclude that the index is 2 if L is positive definite, that is, around equilibria in which $L_2 = (I_0 k)^{-1} > 0$. Using (9), it is not difficult to check that, at the remaining equilibria (for which $L_2 = -(I_0 k)^{-1} < 0$), the index is 2 if and only if the additional condition $L_1 \neq -L_2$ is satisfied.

Hyperbolicity. The absence of capacitors and voltage sources make the topological conditions in Theorem 3 amount to the absence of L-loops, which is verified for all equilibria independently of the value of G_2, making all of them hyperbolic regardless of the sign of L_2.

Asymptotic stability. Theorem 4 guarantees that equilibria with $L_2 = (I_0 k)^{-1} > 0$ are asymptotically stable, since for them the inductance matrix is symmetric positive definite. The case $L_2 = -(I_0 k)^{-1} < 0$ cannot be assessed in these terms. It can be checked that, actually, when $G_2 > 0$, these equilibria are unstable; in contrast, if $G_2 = 0$, these equilibria are asymptotically stable if $-L_2 = (I_0 k)^{-1} < L_1$, and unstable if $-L_2 = (I_0 k)^{-1} > L_1$.

Acknowledgements. Work supported by Project 14583 of Universidad Politécnica de Madrid, and by the DFG-Research Center MATHEON in Berlin.

References

[BM64] Brayton, R.K., Moser, J.K.: A theory of nonlinear networks, I, Quarterly of Applied Mathematics **22**, 1–33 (1964); *ibid* II **22**, 81–104 (1964)

[Chu80] Chua, L.O.: Dynamic nonlinear networks: state-of-the-art, IEEE Trans. Cir. Sys. **27**, 1059–1087 (1980)

[DW72] Desoer, Ch.A., Wu, F.F.: Trajectories of nonlinear RLC networks: A geometric approach, IEEE Trans. Circuit Theory **19**, 562–571 (1972)

[ET00] Estévez-Schwarz, D., Tischendorf, C.: Structural analysis of electric circuits and consequences for MNA, Internat. J. Circuit Theory Appl. **28**, 131–162 (2000)

[GW92] Green, M.M., Willson Jr, A.N.: How to identify unstable dc operating points, IEEE Trans. Cir. Sys. I **39**, 820–832 (1992)

[GF99] Günther, M., Feldmann, U.: CAD-based electric-circuit modeling in industry. I: Mathematical structure and index of network equations, Surv. Math. Ind. **8**, 97–129 (1999); *ibid* II: Impact of circuit configurations and parameters, Surv. Math. Ind. **8**, 131–157 (1999)

[HB84] Haggman, B.C., Bryant, P.R.: Solutions of singular constrained differential equations: A generalization of circuits containing capacitor-only loops and inductor-only cutsets, IEEE Trans. Cir. Sys. **31**, 1015–1029 (1984)

[HB86] Haggman, B.C., Bryant, P.R.: Geometric properties of nonlinear networks containing capacitor-only cutsets and/or inductor-only loops. Part I: Conservation laws, Cir. Sys. Signal Process. **5**, 279–319 (1986)

[Mat87] Mathis, W.: Theorie Nichtlinearer Netzwerke. Springer-Verlag, 1987

[MCM79] Matsumoto, T., Chua, L.O., Makino, A.: On the implications of capacitor-only cutsets and inductor-only loops in nonlinear networks, IEEE Trans. Cir. Sys. **26**, 828–845 (1979)

[Ria04] Riaza, R.: A matrix pencil approach to the local stability analysis of nonlinear circuits, Internat. J. Circuit Theory Appl. **32**, 23–46 (2004)

[Sma72] Smale, S.: On the mathematical foundations of electrical circuit theory, J. Diff. Geometry **7**, 193–210 (1972)

[Tis99] Tischendorf, C.: Topological index calculation of DAEs in circuit simulation, Surv. Math. Ind. **8**, 187–199 (1999)

[WM97] Weiss, L., Mahtis, W.: A Hamiltonian formulation for complete nonlinear RLC-networks, IEEE Trans. Cir. Sys. I **44**, 843–846 (1997)

[WMT98] Weiss, L., Mahtis, W., Trajkovic, L.: A generalization of Brayton-Moser's mixed potential function, IEEE Trans. Cir. Sys. I **45**, 423–427 (1998)

State and Semistate Models of Lumped Circuits

R. Riaza and J. Torres-Ramírez

Departamento de Matemática Aplicada a las Tecnologías de la Información
ETSI Telecomunicación, Universidad Politécnica de Madrid
Ciudad Universitaria s/n - 28040 Madrid, Spain,
rrr@mat.upm.es, fjtr@mat.upm.es

Abstract The formulation of state equations for nonlinear circuits is tackled in this work as a reduction problem for semistate (differential-algebraic) models. We show how the differential-algebraic approach to state-space modeling makes it possible to give precise assumptions under which certain state reductions are feasible. Different semistate approaches are surveyed, addressing several relations among them and letting the above-mentioned state model be an end-point of a hierarchy of nodal analysis methods for lumped circuits. Special attention is paid to so-called augmented node analysis (ANA) models.

1 Introduction

The derivation of state-space models for lumped circuits in terms of ordinary differential equations (ODEs) has attracted considerable recent attention, in both linear and non-linear contexts: see [CDK87, LW02, Nat91, Som01] and references therein. Nevertheless, state formulations have several known drawbacks, which have driven much interest to semistate models defined by differential-algebraic equations (DAEs) [ET00, GF99, Rei96, Ria04, Tis99].

In this semistate context, so-called *augmented node analysis* (ANA) models (see [LW02]) have received much less attention in the nonlinear setting than tableau or MNA systems [ET00, GF99, Tis99]. However, from the authors' point of view, ANA models seem to somehow link the tableau/MNA families: on the one hand, ANA models are obtained as a reduction of tableau systems, having index one if and only if the corresponding tableau model does. On the other hand, MNA models can be easily obtained from ANA, but the semiexplicit structure of augmented systems simplifies the formulation of index-1 conditions and allows for just non-singular reactances, in contrast to MNA, where positive definiteness is required in the index analysis [Tis99].

In this work we address several properties of ANA systems and, in particular, the state reduction problem within the ANA context. For brevity we focus on conventional models, which use capacitor voltages and inductor currents as dynamic variables, and restrict the discussion to index-1 cases.

2 Node tableau analysis

Broadly speaking, node analysis of lumped circuits is based on the formulation of Kirchhoff current law in the form $Ai = 0$, A describing the *incidence* between branches and nodes in the circuit. Branch currents of voltage controlled elements are expressed as far as possible in terms of branch voltages, and these are in turn written in terms of node voltages using Kirchhoff voltage law $v = A^T e$.

If we assume that capacitors and inductors are locally voltage/current controlled through certain C^1 relations $q = \psi(v_c)$, $\phi = \varphi(i_l)$, respectively, we may define the incremental capacitance and inductance matrices as

$$C(v_c) = \psi'(v_c),\ L(i_l) = \varphi'(i_l). \tag{1}$$

Assume additionally that the resistors are voltage-controlled by $i_r = \gamma(v_r)$, and split the incidence matrix A as $(A_R\ A_L\ A_C\ A_V\ A_I)$, where A_R (resp. $A_L,\ A_C,\ A_V,\ A_I$) describes the incidence between resistive (resp. inductive, capacitive, voltage source, current source) branches and nodes. The conventional node tableau analysis (NTA) model can be then written as the following quasilinear DAE:

$$C(v_c)v_c' = i_c \tag{2a}$$
$$L(i_l)i_l' = v_l \tag{2b}$$
$$0 = i_j - j(t) \tag{2c}$$
$$0 = v_u - u(t) \tag{2d}$$
$$0 = i_r - \gamma(v_r) \tag{2e}$$
$$0 = A_R i_r + A_L i_l + A_C i_c + A_V i_v + A_I i_j \tag{2f}$$
$$0 = v_r - A_R^T e \tag{2g}$$
$$0 = v_l - A_L^T e \tag{2h}$$
$$0 = v_c - A_C^T e \tag{2i}$$
$$0 = v_u - A_V^T e \tag{2j}$$
$$0 = v_i - A_I^T e. \tag{2k}$$

3 Augmented node analysis (ANA)

For simplicity and comparative purposes we assume below that resistors are locally voltage-controlled by $i_r = \gamma(v_r)$. In Section 5 we show how to extend the results to problems without this restriction. In the sequel we also assume that $C(v_c)$ and $L(i_l)$ are non-singular, so that the conventional NTA model (2) admits a semiexplicit form.

Let us eliminate resistive currents and voltages using (2e) and (2g). Inductive voltages will be substituted by means of (2h), and current and voltage variables in the corresponding sources can be trivially eliminated using (2c) and (2d). Finally, (2k) will be considered as an output equation giving voltages in current source branches and will therefore be removed from the model. This way we get

$$C(v_c)v_c' = i_c \tag{3a}$$
$$L(i_l)i_l' = A_L^T e \tag{3b}$$
$$0 = A_R\gamma(A_R^T e) + A_L i_l + A_C i_c + A_V i_v + A_I j(t) \tag{3c}$$
$$0 = v_c - A_C^T e \tag{3d}$$
$$0 = u(t) - A_V^T e. \tag{3e}$$

Equations (3c)-(3e) can be understood as a time-domain analogue of [LW02, eq. (2.2)], and the method yielding this system will be therefore called augmented node analysis (ANA), here formulated without the need for branch replacements. The differential relations in the form (3a)-(3b) are those used in [LW02, §2, step 4].

The additional interest of system (3) stems from the fact that it provides an intermediate formulation between NTA and several different methods, having index-1 if and only if so it has NTA, as shown in Theorem 1 below. It can be understood as the result of eliminating "superfluous" variables from NTA. Using (3), the key step in the state-space formulation of Li and Woo [LW02, §2, step 3] may be seen as an index-1 assumption on this differential-algebraic system. Such an index-1 condition can be rephrased in circuit-theoretic terms: this will be performed in Section 4. Furthermore, MNA can be seen as a reduction of (3), as shown at the end of this section.

Theorem 1. *If the capacitance and inductance matrices $C(v_c)$, $L(i_l)$ are nonsingular, the conventional node tableau analysis (NTA) system (2) has index 1 if and only if the augmented node analysis (ANA) system (3) has index 1.*

Proof: Let G stand for the incremental conductance γ'. The derivative of the algebraic restrictions of NTA (2) with respect to algebraic variables $(i_j,\ v_u,\ e,\ i_v,\ i_c,\ i_r,\ v_r,\ v_l,\ v_i)$ reads

$$\begin{pmatrix} I & 0 & 0 & 0 & 0 & 0 & 0 & 0 & 0 \\ 0 & I & 0 & 0 & 0 & 0 & 0 & 0 & 0 \\ 0 & 0 & 0 & 0 & 0 & I & -G & 0 & 0 \\ A_I & 0 & 0 & A_V & A_C & A_R & 0 & 0 & 0 \\ 0 & I & -A_V^T & 0 & 0 & 0 & 0 & 0 & 0 \\ 0 & 0 & -A_C^T & 0 & 0 & 0 & 0 & 0 & 0 \\ 0 & 0 & -A_R^T & 0 & 0 & 0 & I & 0 & 0 \\ 0 & 0 & -A_L^T & 0 & 0 & 0 & 0 & I & 0 \\ 0 & 0 & -A_I^T & 0 & 0 & 0 & 0 & 0 & I \end{pmatrix}. \tag{4}$$

This matrix is non-singular if and only if

$$D = \begin{pmatrix} 0 & 0 & 0 & I & -G \\ 0 & A_V & A_C & A_R & 0 \\ -A_V^T & 0 & 0 & 0 & 0 \\ -A_C^T & 0 & 0 & 0 & 0 \\ -A_R^T & 0 & 0 & 0 & I \end{pmatrix} \text{ and } E = \begin{pmatrix} 0 & A_V & A_C & A_R & 0 \\ -A_V^T & 0 & 0 & 0 & 0 \\ -A_C^T & 0 & 0 & 0 & 0 \\ 0 & 0 & 0 & I & -G \\ -A_R^T & 0 & 0 & 0 & I \end{pmatrix} \tag{5}$$

are also non-singular. Note that E results from a row reordering in D. Writing

$$E_{11} = \begin{pmatrix} 0 & A_V & A_C \\ -A_V^T & 0 & 0 \\ -A_C^T & 0 & 0 \end{pmatrix},\ E_{12} = \begin{pmatrix} A_R & 0 \\ 0 & 0 \\ 0 & 0 \end{pmatrix},$$

$$E_{21} = \begin{pmatrix} 0 & 0 & 0 \\ -A_R^T & 0 & 0 \end{pmatrix},\quad E_{22} = \begin{pmatrix} I & -G \\ 0 & I \end{pmatrix},$$

the Schur complement [HJ85] of E_{22} may be easily checked to read

$$\begin{pmatrix} A_R G A_R^T & A_V & A_C \\ -A_V^T & 0 & 0 \\ -A_C^T & 0 & 0 \end{pmatrix}, \tag{6}$$

which is the derivative of the algebraic relations in ANA (3) with respect to the algebraic variables (e, i_v, i_c). Since the Schur complement of E_{22} is non-singular if and only if so it is E, this completes the proof. □

3.1 MNA

Modified node analysis (MNA) [ET00, GF99, Tis99] is easily obtained from ANA by eliminating capacitive currents and voltages. This is done via (3a) and (3d), and yields

$$L(i_l)i_l' = A_L^T e \tag{7a}$$

$$A_C C(A_C^T e) A_C^T e' = -A_R \gamma(A_R^T e) - A_L i_l - A_V i_v - A_I j(t) \tag{7b}$$

$$0 = u(t) - A_V^T e. \tag{7c}$$

This is a quasilinear system, for which the matrix $A_C C(A_C^T e) A_C^T$ will typically be singular. Its analysis requires more sophisticated techniques; specifically, index-1 and index-2 conditions have been obtained for these systems using projector methods [ET00, Tis99].

For later comparison, we restate here [Tis99, Th. 4], providing index-1 conditions under positive definiteness assumptions on G, C and L. A square matrix F is positive definite if $x^T F x > 0$ for all $x \neq 0$; we do not assume it to be symmetric. Additionally, a V-C loop (resp. an I-L cutset) is a loop (resp. cutset) which consists only of voltage sources and/or capacitors (resp. current sources and/or inductors). Note that index-0 cases are only displayed if there are no voltage sources and there exists a capacitive spanning tree [Tis03]. In other cases the "index $\leq$ 1" condition below amounts to "index-1."

Theorem 2. *Assume that the capacitance, inductance, and conductance matrices are positive definite. If the network contains neither V-C loops (except for C-loops) nor I-L cutsets, then the MNA system (7) has index ≤ 1.*

4 Index-1 conditions for ANA/NTA and state reduction

A result analogous to Theorem 2 can be stated for ANA systems. Note that, below, we do not need to restrict the analysis to problems with positive definite reactances. We also emphasize that the reasoning in this case is easier due to the semiexplicit form of the ANA system, *versus* the quasilinear one of MNA.

Theorem 3. *Assume that the conductance matrix G is (positive or negative) definite, and that the local capacitance and inductance matrices C, L are non-singular. Then, the ANA system (3) has index-1 if and only if there are neither V-C loops nor I-L cutsets in the circuit.*

Proof: Note that the derivative of the algebraic relations (3c)-(3e) with respect to the algebraic variables e, i_c, i_v reads

$$J = \begin{pmatrix} A_R G A_R^T & A_C & A_V \\ -A_C^T & 0 & 0 \\ -A_V^T & 0 & 0 \end{pmatrix}.$$

Non-singularity of this matrix is equivalent to index-1 in ANA. Such non-singularity condition holds if and only if the homogeneous linear system

$$A_R G A_R^T x + A_C y + A_V z = 0 \tag{8a}$$
$$-A_C^T x = 0 \tag{8b}$$
$$-A_V^T x = 0 \tag{8c}$$

has only the zero solution. If we premultiply (8a) by x^T and use (8b) and (8c), we get

$$x^T A_R G A_R^T x = 0, \tag{9}$$

which implies

$$A_R^T x = 0, \tag{10}$$

because of the definiteness of G, whereas (8a) amounts to

$$A_C y + A_V z = 0. \tag{11}$$

The existence of a non-vanishing solution holds simultaneously for (8) and for (8b), (8c), (10), (11) altogether. It can be shown that, in turn, (8b), (8c), (10) having only the trivial solution is equivalent to the absence of I-L cutsets, and the same holds for (11) with respect to V-C loops. This means that index-1 in the ANA model is equivalent to the absence of V-C loops and I-L cutsets.

□

Corollary 1. *If the local conductance matrix G is (positive or negative) definite, and the local capacitance and inductance matrices C, L are non-singular, then the NTA system (2) has index-1 if and only if there are neither V-C loops nor I-L cutsets in the circuit.* □

Corollary 2. *If the local conductance matrix G is (positive or negative) definite, and there are neither V-C loops nor I-L cutsets in the circuit, then (3c)-(3e) yield* $i_c = \psi_1(i_l, v_c, j(t), u(t))$, $e = \psi_2(i_l, v_c, j(t), u(t))$, *and an output equation* $i_v = \psi_3(i_l, v_c, j(t), u(t))$, *for locally well-defined functions* ψ_1, ψ_2, ψ_3. *Inserting these into (3a)-(3b), we get the state-space equation*

$$C(v_c)v_c' = \psi_1(i_l, v_c, j(t), u(t)) \tag{12a}$$

$$L(i_l)i_l' = A_L^T\psi_2(i_l, v_c, j(t), u(t)), \tag{12b}$$

This system trivially amounts to an explicit ODE if, additionally, the local capacitance and inductance matrices C, L are non-singular. □

Corollary 2 follows immediately from the Implicit Function Theorem, and provides precise assumptions under which the state-space formulation of [LW02] is feasible. More precisely, the above-stated index-1 condition guarantees that the matrix $\mathbf{Y}_n'$ in [LW02, §2, step 3] is invertible.

5 Current-controlled and semidefinite resistors

It is of interest to extend the previous approach in order to accommodate current-controlled resistors, and also non-definite problems. Let us assume in this regard that, instead of the voltage-controlled representation (2e), resistors are governed by a relation of the form $g_r(i_r, v_r) = 0$ which splits into four different subequations, describing four uncoupled groups (some of which might be empty) with characteristics $v_{r1} = \rho_1(i_{r1})$, $v_{r2} = \rho_2(i_{r2})$, $i_{r3} = \gamma_1(v_{r3})$, $i_{r4} = \gamma_2(v_{r4})$. The first two are current-controlled resistors, and both groups are distinguished by the fact that, at the operating point, we will assume that $R_1 = \rho_1'(i_{r1}^*)$ is definite, and $R_2 = \rho_2'(i_{r2}^*)$ is symmetric and semidefinite, whereas for the last two (which are voltage-controlled), we assume that $G_1 = \gamma_1'(v_{r3}^*)$ is definite, and $G_2 = \gamma_2'(v_{r4}^*)$ is symmetric and semidefinite. All matrices are assumed simultaneously either positive or negative (semi)definite.

In this context, the conventional ANA system can be written as (3a)-(3b)-(3d)-(3e), together with

$$\begin{aligned} 0 = {} & A_{R_1}i_{r1} + A_{R_2}i_{r2} + A_{G_1}\gamma_1(A_{G_1}^T e) + A_{G_2}\gamma_2(A_{G_2}^T e) + \\ & + A_L i_l + A_C i_c + A_V i_v + A_I j(t) \end{aligned} \tag{13a}$$

$$0 = \rho_1(i_{r1}) - A_{R_1}^T e \tag{13b}$$

$$0 = \rho_2(i_{r2}) - A_{R_2}^T e \tag{13c}$$

instead of (3c). We have split the previous A_R in an obvious manner. Note that this model includes as new variables the currents i_{r1}, i_{r2} of current-controlled resistors. The following result can be proved along the lines defined by Theorem 3. We omit the proof for the sake of brevity.

Theorem 4. *Assume that the local capacitance and inductance matrices C, L are non-singular, and that the above-indicated assumptions on R_1, R_2, G_1, G_2 hold. Then, the ANA system (13) has index-1 if there are neither V-C-R_2 loops nor I-L-G_2 cutsets in the circuit.* □

6 Example

Consider the series RLC circuit displayed in Figure 1(a). The capacitor and the inductor are linear. We assume that the nonlinear resistor is voltage-controlled through a characteristic $i = \gamma(v_r)$ which may display critical points $\gamma'(v_r) = 0$. Since there are neither V-C loops nor I-L cutsets in the circuit, Theorem 3 predicts that this DAE has index-1 in the regions where the conductance matrix is definite, what amounts in this case to the condition $\gamma' \neq 0$. It can be easily checked that this is indeed the case, and that the condition $\gamma' \neq 0$ is actually necessary for the system to have index-1.

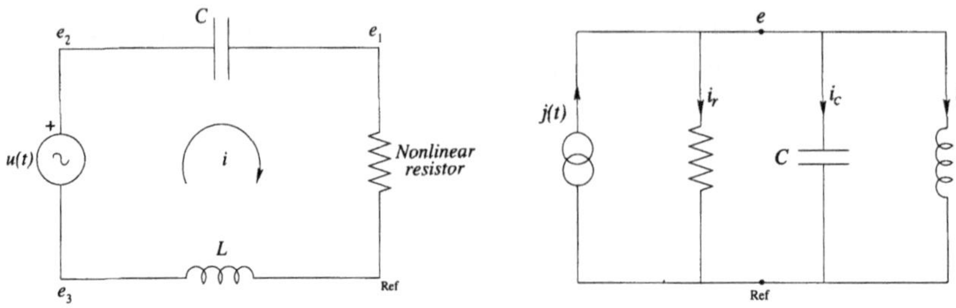

Fig. 1. (a) Series and (b) parallel nonlinear RLC circuits

A reader might conjecture that such critical points in the characteristic should always prevent the model from being index-1, regardless of the topology. This is not the case, as illustrated by the circuit in Figure 1(b), where again the resistor is voltage-controlled. To consider the local behavior at critical points, we may use Theorem 4 framing the resistor in the group G_2. In Figure 1(a), there appears a pathological L-G_2 cutset, and this explains why the index-1 condition is not met at critical points. Note that, away from critical points, the resistor can be included in the G_1 group and therefore the cutset causes no problem. On the contrary, in Figure 1(b) there are no pathological configurations and hence the model has index 1 even at critical points. In particular, the C-G_2 loop does not cause any difficulty.

Dual examples illustrating the role of current-controlled resistors can be easily constructed along the same lines.

7 Concluding remarks

State and semistate formulations for lumped circuits have been discussed in this work. Among the latter, augmented node analysis (ANA) models capture explicitly the circuit topology while keeping capacitive voltages and inductor currents as variables; this allows for a direct discussion of state-space reductions in terms of certain conditions in the ANA system, in contrast to modified node analysis (MNA) in which capacitor voltages are expressed in terms of node voltages. A drawback of ANA w.r.t. MNA is the additional computational cost due to the capacitive branch variables in the model. Also, ANA inherits from tableau analysis the property that C-loops yield an index greater than one; from a different perspective, ANA displays a symmetry in the topologies precluded for index-1, in contrast to MNA, what might be of theoretical interest in the discussion of duality aspects and related issues.

Acknowledgements. Work supported by Research Project 14583 of Universidad Politécnica de Madrid.

References

[CDK87] Chua, L.O., Desoer, C.A., Kuh, E.S.: Linear and Nonlinear Circuits. McGraw-Hill, 1987

[ET00] Estévez-Schwarz, D., Tischendorf, C.: Structural analysis of electric circuits and consequences for MNA, Internat. J. Circuit Theory Appl. **28**, 131–162 (2000)

[GF99] Günther, M., Feldmann, U.: CAD-based electric-circuit modeling in industry. I: Mathematical structure and index of network equations, Surv. Math. Ind. **8**, 97–129 (1999); II: Impact of circuit configurations and parameters, *ibid* 131–157

[HJ85] Horn, R.A., Johnson, Ch.R.: Matrix Analysis. Cambridge U. P., 1985

[LW02] Li, F., Woo, P.Y.: A new method for establishing state equations: the branch replacement and augmented node-voltage equation approach, Cir. Sys. Signal Process. **21**, 149–161 (2002)

[Nat91] Natarajan, S.: A systematic method for obtaining state equations using MNA, IEE Proceedings-G **138**, 341–346 (1991)

[Rei96] Reiszig, G.: Differential-algebraic equations and impasse points, IEEE Trans. Circuits and Systems, Part I **43**, 122–133 (1996)

[Ria04] Riaza, R.: A matrix pencil approach to the local stability analysis of nonlinear circuits, Internat. J. Circuit Theory Appl. **32**, 23–46 (2004)

[Som01] Sommariva, A.: State-space equations of regular and strictly topologically degenerate linear lumped time-invariant networks: the multiport method, Internat. J. Circuit Theory Appl. **29**, 435–453 (2001)

[Tis99] Tischendorf, C.: Topological index calculation of DAEs in circuit simulation, Surv. Math. Ind. **8**, 187–199 (1999)

[Tis03] Tischendorf, C.: Coupled systems of differential algebraic and partial differential equations in circuit and device simulation. Modeling and numerical analysis. Habilitationsschrift, Humboldt-Univ. Berlin, 2003

An Index Analysis from Coupled Circuit and Device Simulation

M. Selva Soto *

Institute of Mathematics, Humboldt University of Berlin, Germany, monica@mathematik.hu-berlin.de

1 Introduction

Nowadays the semiconductor devices in an electrical circuit are modelled by equivalent circuits containing basic network elements described by algebraic and ordinary differential equations. But the correct adjustment of these circuits has become a very difficult task for the network design. In [2] a new model for electrical circuits containing semiconductor devices is proposed and in [1] its well-posedness is studied. In both articles the differential algebraic equations (DAEs) for the basic circuit's elements are coupled to partial differential equations (PDEs), more specifically to one-dimensional Drift-Diffusion (DD) equations, modelling the semiconductor devices in it. Systems of this type are called Abstract Differential Algebraic Systems (ADAS). In [9] the tractability index [5, 9] of this model is analysed and in [8] it is proved that the DAE obtained after discretization in space of the DD equations in it has the same index as the abstract system. In this work we study the tractability index of an abstract system where higher dimensional PDEs describe the behavior of the semiconductor devices in the circuit. The index of the DAE obtained after discretization in space of the PDEs in the system is also analysed.

In the next section the model is briefly described. The Sect. 3 is devoted to the study of the index of the system, as ADAS. Finally, in Sect. 4 it is shown that the DAE that is obtained after discretization in space of the DD equations has the same index as the abstract system.

In what follows we consider electrical circuits with only one semiconductor device, the results can easily be generalized to circuits containing more semiconductor devices.

2 Abstract Differential Algebraic System for the simulation of electrical circuits

Suppose Ω is a bounded domain in $\mathbb{R}^d$, $d \in \{1,2,3\}$, $x \in \Omega$ represents the space variable and t is the time variable, $t \in [t_0, t_F]$. The system proposed in [9] for the simulation of electrical circuits containing semiconductor devices couples the Modified Nodal Analysis (MNA) equations for electrical circuits to the DD equations for semiconductor devices.

The MNA equations for an electrical circuit have the form

$$A_C \frac{\mathrm{d}\, q_C(A_C^T e, t)}{\mathrm{d}t} + A_R g(A_R^T e, t) + A_L j_L + A_V j_V + A_S j_S + A_I i_S = 0, \tag{1a}$$

$$\frac{\mathrm{d}\,\phi(j_L, t)}{\mathrm{d}t} - A_L^T e = 0, \tag{1b}$$

$$A_V^T e - v_S = 0, \tag{1c}$$

where A_C, A_R, A_L, A_V, A_S and A_I are the element related reduced incidence matrices, $v_S(t)$, $i_S(t)$, $q_C(u,t)$, $g(u,t)$ and $\phi(j,t)$ are given functions and the unknowns are the node potentials, excepting the mass node $e(t) : \mathbb{R} \to \mathbb{R}^{n_N}$ and the currents through inductors, voltage sources and semiconductor devices

*This work is supported by the DFG Research Center MATHEON "Mathematics for key technologies", Berlin.

$j_L(t):\mathbb{R}\to\mathbb{R}^{n_L}$, $j_V(t):\mathbb{R}\to\mathbb{R}^{n_V}$ and $j_S:\mathbb{R}\to\mathbb{R}^{n_S}$ respectively. The DD equations are given by the following set of PDEs for the electrostatic potential $\psi(x,t)$ and the electrons and holes densities, $n(x,t)$ and $p(x,t)$ respectively

$$\nabla(-\varepsilon\nabla\psi)-q(N-n+p)=0, \tag{1d}$$

$$-\frac{\partial n}{\partial t}+\frac{1}{q}\mathrm{div}J_n-R=0,\quad J_n-q\mu_n(U_T\nabla n-n\nabla\psi)=0, \tag{1e}$$

$$\frac{\partial p}{\partial t}+\frac{1}{q}\mathrm{div}J_p+R=0,\quad J_p+q\mu_p(U_T\nabla p+p\nabla\psi)=0. \tag{1f}$$

We consider $R=R(n,p),\mu_n=\mu_n(x),\mu_p=\mu_p(x)$ and ε,q and U_T as constants. Unlike [9], in (1d)-(1f) we replace the Poisson equation by the energy conservation equation

$$\nabla\cdot(J_n+J_p-\varepsilon\partial_t\nabla\psi)=0. \tag{1g}$$

This replacement not only facilitates the theoretical analysis of the DD equations, but has also proved advantageous in numerical simulations [4]. Nevertheless the results in the next sections remain the same if we consider (1d)-(1f) instead of (1e)-(1g). We assume the boundary of the semiconductor device can be divided in two disjoint parts $\Gamma=\Gamma_O\cup\Gamma_A$. The boundary conditions are

$$n=n_D(x),\ p=p_D(x),\ \psi=\psi_{bi}(x)+\psi_D(x,e)\ \text{ on }\Gamma_O \tag{1h}$$

$$\text{and }\frac{\partial\psi}{\partial\nu}=\frac{\partial n}{\partial\nu}=\frac{\partial p}{\partial\nu}=0\ \text{ on }\Gamma_A, \tag{1i}$$

where ψ_D denotes the externally applied bias, it depends on the node potentials of the circuit. The functions $\psi_{bi}(x),n_D(x)$ and $p_D(x)$ are given functions of x that depend on the doping concentration of the semiconductor.

Suppose $\Gamma_O=\cup_{j=1}^{n_S+1}\Gamma_j$. The current flowing through the contact $\Gamma_i\subset\Gamma_O$ of the semiconductor is

$$j_i=\int_{\Gamma_i}J_{tot}\cdot\nu\,\mathrm{d}\sigma,\quad\text{with}\quad J_{tot}=J_n+J_p-\varepsilon\frac{\partial}{\partial t}\nabla\psi \tag{1j}$$

Unlike [9], where the currents thorugh the semiconductor are calculated as in (1j), here we transform the integrals over the boundary into integrals over the domain. Let us introduce the auxiliary functions $f_i(x),\ i=1,2,\dots,n_S$ that satisfy [4]

$$\Delta f_i=0\ \text{ in }\Omega,\ \ f_i|_{\Gamma_j}=\begin{cases}1 & \text{if } i=j\\ 0 & \text{else}\end{cases},\ j=1,\dots,n_S+1,\ \ (\nabla f_i\cdot\nu)|_{\Gamma_A}=0. \tag{1k}$$

The current through $\Gamma_i,\ i=1,2,\dots,n_S$ can be calculated as

$$j_i=\int_{\Gamma_i}J_{tot}\cdot\nu\,\mathrm{d}s=\int_{\Gamma}J_{tot}\cdot\nu f_i\,\mathrm{d}s=\int_{\Omega}J_{tot}\cdot\nabla f_i\mathrm{d}x$$

$$j_i=-\varepsilon\frac{\mathrm{d}}{\mathrm{d}t}\int_{\Omega}\nabla\psi\cdot\nabla f_i\mathrm{d}x+\int_{\Omega}(J_n+J_p)\cdot\nabla f_i\mathrm{d}x.$$

The current at Γ_{n_S+1} is the negative sum of the currents through the other contacts†. Suppose the contact Γ_i of the semiconductor device is joined to the k_i-th node of the circuit for $i=1,2,\dots,n_S+1$. We set $\psi_D(x,e)=e_{k_i}-e_{k_{n_S+1}}\ \forall x\in\Gamma_i,\ i=1,2,\dots,n_S$, and $\psi_D(x,e)=0,\ \forall x\in\Gamma_{n_S+1}$. Following the convention in [3] the vector of the branch currents of the semiconductor device $j_S=\left(j_{S_1},\dots,j_{S_{n_S}}\right)^T$ must be such that

†The sum of the currents at the contacts of the semiconductor is zero, $\sum_{i=1}^{n_S+1}j_i=\sum_{i=1}^{n_S+1}\int_{\Gamma_i}J_{tot}\cdot\nu\mathrm{d}s=\int_{\Gamma}J_{tot}\cdot\nu\mathrm{d}s=\int_{\Omega}\nabla\cdot J_{tot}\mathrm{d}x=0$.

$$j_{S_i} = -j_i = -\int_\Omega (J_n + J_p) \cdot \nabla f_i \mathrm{d}x + \frac{\mathrm{d}}{\mathrm{d}t}\int_\Omega \varepsilon \nabla\psi \cdot \nabla f_i \,\mathrm{d}x,\ i = 1, 2, \dots, n_S$$

$$j_{S_i} = -\int_\Omega (J_n + J_p) \cdot \nabla f_i \mathrm{d}x - \frac{\mathrm{d}}{\mathrm{d}t} j^d_{S_i},\quad j^d_{S_i} = -\int_\Omega \varepsilon \nabla\psi \cdot \nabla f_i \,\mathrm{d}x. \tag{11}$$

With a matrix $A_S \in \mathbb{R}^{n_N \times n_S}$ such that

$$A_S(k,i) = \begin{cases} 1, & \text{if } \Gamma_i \text{ is joined to the node } k \\ -1, & \text{if } \Gamma_{n_S+1} \text{ is joined to the node } k \\ 0, & \text{else} \end{cases},$$

the product $A_S j_S$ describes the incidence of the current at the semiconductor's contacts in the circuit and the potential applied to the semiconductor's boundaries can be written as $\psi_D(x,e) = f(x) \cdot A_S^T e$ where $f(x) = (f_1(x) \ \dots f_{n_S}(x))$.

Let the following assumptions on the circuit equations be satisfied in the forthcoming sections:

1. the input functions $v_S(t)$ and $i_S(t)$, associated to the independent voltage and current sources respectively, are continuous,
2. the functions $q_C(u,t)$, $\phi(j,t)$ and $g(u,t)$ are continuously differentiable and have positive definite partial Jacobians

$$C(u,t) = \frac{\partial q_C(u,t)}{\partial u},\quad L(j,t) = \frac{\partial \phi(j,t)}{\partial j},\quad G(u,t) = \frac{\partial g(u,t)}{\partial u},$$

3. the circuit contains neither loops of voltage sources only nor cut sets of current sources only. These two conditions hold if and only if the matrices A_V and $(A_C\ A_R\ A_L\ A_V\ A_S)^T$ have full column rank, respectively,
4. the function $R(n,p)$ is continuously differentiable,
5. the functions $\mu_n(x)$ and $\mu_p(x)$ are bounded.

3 Tractability index of the Abstract Differential Algebraic System

The tractability index concept for ADAS was introduced in [5] and [9]. This is a kind of time index as the uniform differential time index in [6], but it is not restricted to time-invariant linear Partial Differential Algebraic Equations.

We decided to study the tractability index of (1) because we are mostly interested in the transient behaviour of the electrical circuit. Besides the tractability index concept for DAEs has proved very useful in the analysis of electrical circuits described by algebraic and differential equations only.

As mentioned in the introduction, in [9] the index of an ADAS where one-dimensional DD equations describe the behaviour of the semiconductor devices in the circuit was studied. There it was proved that the index of the coupled system is always less or equal to two and it is two only if the circuit contains CVS-loops[‡] or LI-cut sets[§]. In this section the same results are proved, but related to an ADAS where higher dimensional DD equations model the semiconductor devices in the circuit.

After homogenization of the electrostatic potential and the densities of electrons and holes,

$$\tilde{\psi} = \psi - f(x) \cdot A_S^T e - g_1(x),\ \tilde{n} = n - g_2,\ \tilde{p} = p - g_3$$

with functions $g_i,\ i = 1,2,3$ such that

$$(\nabla g_i \cdot \nu)|_{\Gamma_A} = 0, i = 1,2,3,\ g_1|_{\Gamma_O} = \psi_{bi}(x),\ g_2|_{\Gamma_O} = n_D,\ g_3|_{\Gamma_O} = p_D$$

[‡] Loops of capacitors, voltage sources and semiconductor devices with at least one voltage source or one semiconductor device.

[§] Cut sets of inductors and current sources.

and if we rename the homogenized variables $\tilde{\psi}, \tilde{n}$ and $\tilde{p}$ as originally, the above described model can be written as $\mathcal{A}\frac{\mathrm{d}}{\mathrm{d}t}\mathcal{D}(u,t)+\mathcal{B}(u,t)=0$ with $u=(e,\, j_L,\, j_V,\, j_S,\, j_S^d,\, \psi(\cdot,t),\, n(\cdot,t),\, p(\cdot,t))$,

$$\mathcal{A}=\begin{pmatrix} A_C & 0\,0\,0\,0\,0 \\ 0 & I\,0\,0\,0\,0 \\ 0 & 0\,0\,0\,0\,0 \\ 0 & 0\,0\,0\,0\,0 \\ 0 & 0\,I\,0\,0\,0 \\ 0 & 0\,0\,I\,0\,0 \\ 0 & 0\,0\,0\,I\,0 \\ 0 & 0\,0\,0\,0\,I \end{pmatrix},\quad \mathcal{D}(u,t)=\begin{pmatrix} A_C^+A_Cq_C(A_C^Te,t) \\ \phi(j_L,t) \\ j_S^d \\ \nabla\cdot(-\varepsilon\nabla\psi) \\ -n \\ p \end{pmatrix} \tag{2a}$$

and

$$\mathcal{B}(u,t)=\begin{pmatrix} A_Rg(A_R^Te,t)+A_Lj_L+A_Vj_V+A_Sj_S+A_Ii_s(t) \\ -A_L^Te \\ A_V^Te-v_S(t) \\ j_S^d+\varepsilon\begin{pmatrix} \int_\Omega\nabla\left(\psi+f\cdot A_S^Te\right)\cdot\nabla f_1\mathrm{d}x \\ \vdots \\ \int_\Omega\nabla\left(\psi+f\cdot A_S^Te\right)\cdot\nabla f_{n_S}\mathrm{d}x \end{pmatrix} \\ j_S+\begin{pmatrix} \int_\Omega(J_n+J_p)\cdot\nabla f_1\mathrm{d}x \\ \vdots \\ \int_\Omega(J_n+J_p)\cdot\nabla f_{n_S}\mathrm{d}x \end{pmatrix} \\ \nabla\cdot(J_n+J_p) \\ \frac{1}{q}\nabla\cdot J_n-R \\ \frac{1}{q}\nabla\cdot J_p+R \end{pmatrix}. \tag{2b}$$

In (2) $J_n=q\mu_n\left(U_T\nabla(n+g_2)-(n+g_2)\nabla\left(\psi+f\cdot A_S^Te+g_1\right)\right)$, J_p has a similar structure and A_C^+ denotes the Moore-Penrose pseudo-inverse of A_C. The operators $\mathcal{A},\mathcal{D}$ and $\mathcal{B}$ are acting on Hilbert spaces $\mathcal{A}:\mathcal{Z}\to\mathcal{Y}$, $\mathcal{D}:\mathcal{X}\to\mathcal{Z}$ and $\mathcal{B}:\mathcal{X}\to\mathcal{Y}$ with

$$\begin{aligned} \mathcal{X}&=\mathbb{R}^{n_N}\times\mathbb{R}^{n_L}\times\mathbb{R}^{n_V}\times\mathbb{R}^{n_S}\times\mathbb{R}^{n_S}\times V\times L^2(\Omega)\times L^2(\Omega),\\ \mathcal{Y}&=\mathbb{R}^{n_N}\times\mathbb{R}^{n_L}\times\mathbb{R}^{n_V}\times\mathbb{R}^{n_S}\times\mathbb{R}^{n_S}\times L^2(\Omega)\times L^2(\Omega)\times L^2(\Omega),\\ \mathcal{Z}&=\mathbb{R}^{n_C}\times\mathbb{R}^{n_L}\times\mathbb{R}^{n_S}\times L^2(\Omega)\times L^2(\Omega)\times L^2(\Omega), \end{aligned}$$

where $V=\left\{v\in H^2(\Omega)\,|\,v|_{\Gamma_O}=0,(\nabla v\cdot\nu)|_{\Gamma_A}=0\right\}$. Note that the definition domain $\mathcal{D}_\mathcal{B}$ of $\mathcal{B}(u,t)$,

$$\mathcal{D}_\mathcal{B}=\mathbb{R}^{n_N}\times\mathbb{R}^{n_L}\times\mathbb{R}^{n_V}\times\mathbb{R}^{n_S}\times\mathbb{R}^{n_S}\times V\times V\times V,$$

is dense in $\mathcal{X}$. The Fréchet derivative of $\mathcal{D}(u,t)$ is

$$\mathcal{D}_0(u,t)=\begin{pmatrix} A_C^+A_CC(A_C^Te,t)A_C^T & 0 & 0\,0\,0 & 0 & 0 & 0 \\ 0 & L(j_L,t) & 0\,0\,0 & 0 & 0 & 0 \\ 0 & 0 & 0\,0\,I & 0 & 0 & 0 \\ 0 & 0 & 0\,0\,0 & -\varepsilon\Delta & 0 & 0 \\ 0 & 0 & 0\,0\,0 & 0 & -I & 0 \\ 0 & 0 & 0\,0\,0 & 0 & 0 & I \end{pmatrix}$$

and because the equation $-\varepsilon\Delta u=f$, completed with homogeneous Dirichlet and Neumann conditions, has a unique solution for all $f\in L^2(\Omega)$,

$$\begin{aligned} \operatorname{im}\mathcal{D}_0(u,t)&=\operatorname{im}A_C^T\times\mathbb{R}^{n_L}\times\mathbb{R}^{n_S}\times L^2(\Omega)\times L^2(\Omega)\times L^2(\Omega),\\ \ker\mathcal{D}_0(u,t)&=\ker A_C^T\times\{0\}\times\mathbb{R}^{n_V}\times\mathbb{R}^{n_S}\times\{0\}\times\{0\}\times\{0\}\times\{0\} \end{aligned}$$

On the other hand, the operator $\mathcal{A}$ satisfies

$$\ker \mathcal{A} = \ker A_C \times \{0\} \times \{0\} \times \{0\} \times \{0\} \times \{0\},$$
$$\operatorname{im} \mathcal{A} = \operatorname{im} A_C \times \mathbb{R}^{n_L} \times \{0\} \times \{0\} \times \mathbb{R}^{n_S} \times L^2(\Omega) \times L^2(\Omega) \times L^2(\Omega).$$

The operators $\mathcal{A}$ and $\mathcal{D}_0$ are well matched [5, 9], i.e., they satisfy

$$\ker \mathcal{A} \oplus \operatorname{im} \mathcal{D}_0(u,t) = \mathcal{Z}, \quad \forall u \in \mathcal{X},\ \forall t \in [t_0, t_F]$$

and there is a projector $\mathcal{R} \in \mathcal{L}(\mathcal{Z})$ such that $\operatorname{im} \mathcal{R} = \operatorname{im} \mathcal{D}_0(u,t)$ and $\ker \mathcal{R} = \ker \mathcal{A}$, where $\mathcal{L}(\mathcal{Z})$ denotes the space of linear operators $L : \mathcal{Z} \to \mathcal{Z}$.

It was the introduction of the variables $j^d_{S_i}$ in (11) what allowed us to write the ADAS in such a way that $\mathcal{A}$ and $\mathcal{D}_0$ are well matched. In [9] the Poisson equation, instead of the energy conservation equation, is considered. There it is not necessary to introduce $j^d_{S_i}$ in order to have well matched operators $\mathcal{A}$ and $\mathcal{D}_0$.

Remark *The functions $f_1, \dots, f_{n_S}$ defined above are a basis of the linear space*

$$\mathcal{F} = \left\{ v \in H^2(\Omega) \mid \Delta v = 0 \text{ in } \Omega,\ (\nabla v \cdot \nu)|_{\Gamma_A} = 0,\ v|_{\Gamma_j} = a_j,\ v|_{\Gamma_{n_S+1}} = 0 \right\},$$

where $j = 1, 2, \dots, n_S$ and $a_j \in \mathbb{R}\ \forall j$. Because $(u,v)_{\mathcal{F}} = \int_\Omega \nabla u \cdot \nabla v\, dx$ is a scalar product in $\mathcal{F}$, the matrix

$$J = \begin{pmatrix} \int_\Omega \nabla f_1 \cdot \nabla f_1\, dx & \dots & \int_\Omega \nabla f_1 \cdot \nabla f_{n_S}\, dx \\ \vdots & \vdots & \vdots \\ \int_\Omega \nabla f_{n_S} \cdot \nabla f_1\, dx & \dots & \int_\Omega \nabla f_{n_S} \cdot \nabla f_{n_S}\, dx \end{pmatrix}, \tag{3}$$

is positive definite.

Note that the fourth equation of the abstract system is

$$j^d_S + \varepsilon J A_S^T e + \varepsilon \begin{pmatrix} \int_\Omega \nabla \psi \cdot \nabla f_1\, \mathrm{d}x \\ \vdots \\ \int_\Omega \nabla \psi \cdot \nabla f_{n_S}\, \mathrm{d}x \end{pmatrix} = 0, \tag{4}$$

with the matrix J in (3). The positive definitness of J is very important in the proofs below.

Theorem 1. *If the conditions on the circuit mentioned in Sect. 2 are satisfied and the circuit contains neither LI-cut sets nor CVS-loops, the abstract system has tractability index one.*

Proof: Let $\mathcal{G}_0(u,t) = \mathcal{A}\mathcal{D}_0(u,t)$ and $\mathcal{B}_0(u,t)$ denote the Fréchet-derivative of $\mathcal{B}$. Under the conditions in Sect. 2 the operator $\mathcal{B}_0(u,t)$ exists. The system has tractability index one if there is a projection operator $\mathcal{Q}_0 \in \mathcal{L}(\mathcal{X}')$ onto $\ker \mathcal{G}_0(u,t)$ such that $\mathcal{G}_1(u,t) = \mathcal{G}_0(u,t) + \mathcal{B}_0(u,t)\mathcal{Q}_0$ is injective and $\operatorname{im} \mathcal{G}_1(u,t) = \mathcal{Y}$ for all $u \in \mathcal{X}$ and $t \in [t_0, t_F]$.

Because the system has a properly stated leading term, $\ker \mathcal{G}_0(u,t) = \ker \mathcal{D}_0(u,t)$ and $\mathcal{Q}_0 = \begin{pmatrix} Q_C & 0 & 0 & 0 & 0 & 0 & 0 & 0 \\ 0 & 0 & 0 & 0 & 0 & 0 & 0 & 0 \\ 0 & 0 & I & 0 & 0 & 0 & 0 & 0 \\ 0 & 0 & 0 & I & 0 & 0 & 0 & 0 \\ 0 & 0 & 0 & 0 & 0 & 0 & 0 & 0 \\ 0 & 0 & 0 & 0 & 0 & 0 & 0 & 0 \\ 0 & 0 & 0 & 0 & 0 & 0 & 0 & 0 \\ 0 & 0 & 0 & 0 & 0 & 0 & 0 & 0 \end{pmatrix}$ is a projection operator onto $\ker \mathcal{G}_0(u,t)$ if Q_C is a projector onto $\ker A_C^T$. The operator $\mathcal{G}_1$ can easily be calculated. Let $w = (w_e, w_L, w_V, w_S, w_S^d, w_\psi, w_n, w_p) \in \ker \mathcal{G}_1(u,t)$. The fourth equation of $\mathcal{G}_1(u,t)w = 0$ is $\varepsilon J A_S^T Q_C w_e = 0$ where J is the matrix in (3), then $\varepsilon J A_S^T Q_C w_e = 0$ iff $A_S^T Q_C w_e = 0$. The sixth equation of $\mathcal{G}_1(u,t)w = 0$ is $-\varepsilon \Delta w_\psi = 0$, it implies that $w_\psi = 0$. The rest of the proof is very similar to the one in [9]. We arrive to

$$\begin{aligned} \ker \mathcal{G}_1(u,t) = \Big\{ w \mid w_\psi = 0, w_n = 0, w_p = 0, Q_C w_e \in \ker (A_C\ A_R\ A_V\ A_S)^T, \\ P_C w_e = -H_C(\cdot)^{-1} (A_V \quad A_S) \left(\begin{smallmatrix} w_V \\ w_S \end{smallmatrix} \right), w_L = L(\cdot)^{-1} A_L^T Q_C w_e, \\ \left(\begin{smallmatrix} w_V \\ w_S \end{smallmatrix} \right) \in \ker \left(Q_C^T A_V\ Q_C^T A_S \right), w_S^d = -(0 \quad I) \left(\begin{smallmatrix} w_V \\ w_S \end{smallmatrix} \right) \Big\}, \end{aligned}$$

where $H_C(A_C^T e, t) = A_C C(A_C^T e, t) A_C^T + Q_C^T Q_C$ is positive definite. If the circuit contains neither LI-cut sets ($\left(A_C\ A_R\ A_V\ A_S\right)^T$ has full column rank) nor CVS-loops with at least one voltage source or one semiconductor device ($\left(Q_C^T A_V\ Q_C^T A_S\right)$ has full column rank), then $\ker \mathcal{G}_1(u,t) = \{0\}$, i.e. $\mathcal{G}_1(u,t)$ is injective. The dense solvability of $\mathcal{G}_1(u,t)$ ($\overline{\operatorname{im}\mathcal{G}_1(u,t)} = \mathcal{Y}$) can be shown using similar arguments as those in [9] and taking into account that J is nonsingular □.

Suppose the circuit contains LI-cut sets or CVS-loops with at least one voltage source or one semiconductor device. Let Q_{CRVS} be a projector onto $\ker\left(A_C\ A_R\ A_V\ A_S\right)^T$ and Q_{C-VS}, a projector onto $\ker\left(Q_C^T A_V\ Q_C^T A_S\right)$. Because $\operatorname{im} Q_{CRVS} \subset \operatorname{im} Q_C$, Q_{CRVS} can be constructed so that $\ker Q_C \subset \ker Q_{CRVS}$. A projector $\mathcal{Q}_1(u,t)$ onto $\ker \mathcal{G}_1(u,t)$ is then

$$\mathcal{Q}_1(u,t) = \begin{pmatrix} Q_{CRVS} & 0 & -H_C(\cdot)^{-1}(A_V\ A_S)Q_{C-VS} & 0\ 0\ 0\ 0 \\ L(\cdot)^{-1}A_L^T Q_{CRVS} & 0 & 0 & 0\ 0\ 0\ 0 \\ 0 & 0 & Q_{C-VS} & 0\ 0\ 0\ 0 \\ 0 & 0 & -(0\ I)Q_{C-VS} & 0\ 0\ 0\ 0 \\ 0 & 0 & 0 & 0\ 0\ 0\ 0 \\ 0 & 0 & 0 & 0\ 0\ 0\ 0 \\ 0 & 0 & 0 & 0\ 0\ 0\ 0 \end{pmatrix}.$$

Theorem 2. *Under the conditions mentioned in Sect. 2 and if the circuit contains LI-cut sets or CVS-loops, the coupled system has tractability index two.*

Proof: The ADAS has index two if the operator $\mathcal{G}_2(u,t) = \mathcal{G}_1(u,t) + \mathcal{B}_0(u,t)(I - \mathcal{Q}_0)\mathcal{Q}_1(u,t)$ is injective and densely solvable for all $u \in \mathcal{X}$ and $t \in [t_0, t_F]$.

The operator $\mathcal{G}_2(u,t)$ can easily be calculated. Let w be an element in $\ker \mathcal{G}_2(u,t)$. The third and fourth equations of $\mathcal{G}_2(u,t)w = 0$, pre-multiplied by Q_{C-VS}^T, can be written as

$$-Q_{C-VS}^T \left\{ \begin{pmatrix} A_V^T \\ mathA_S^T \end{pmatrix} H_C(\cdot)^{-1} \left(A_V\ A_S\right) + \begin{pmatrix} 0 & 0 \\ 0 & \frac{1}{\varepsilon}J^{-1} \end{pmatrix} \right\} Q_{C-VS} \begin{pmatrix} w_V \\ w_S \end{pmatrix} = 0. \tag{5}$$

Because $\begin{pmatrix} 0 & 0 \\ 0 & \frac{1}{\varepsilon}J^{-1} \end{pmatrix}$ is positive semidefinite and $H_C^{-1}(\cdot)$ is positive definite, equation (5) is satisfied iff $Q_{C-VS}\begin{pmatrix} w_V \\ w_S \end{pmatrix} = 0$. The rest of the proof is very similar to the one in [9]. We arrive to $\ker \mathcal{G}_2(u,t) = \{0\}$. The dense solvability of $\mathcal{G}_2(u,t)$ can be proved following the lines in [9] □.

4 Index of the Discrete System

Suppose that the coupled system, after discretization in space of the Drift-Diffusion equations has the following form

$$A_C \frac{\mathrm{d}\, q_C(A_C^T e, t)}{\mathrm{d}t} + A_R g(A_R^T e, t) + A_L j_L + A_V j_V + A_S j_S + A_I i_S = 0, \tag{6a}$$

$$\frac{\mathrm{d}\,\phi(j_L, t)}{\mathrm{d}t} - A_L^T e = 0, \tag{6b}$$

$$A_V^T e - v_S = 0, \tag{6c}$$

$$j_S^d + J_h A_S^T e + g(y) = 0, \tag{6d}$$

$$j_S + j_S^c(A_S^T e, y) + \frac{\mathrm{d}\, j_S^d}{\mathrm{d}t} = 0, \tag{6e}$$

$$A\frac{\mathrm{d}y}{\mathrm{d}t} + b(A_S^T e, y) = 0, \tag{6f}$$

where A is a nonsingular matrix, J_h is positive definite and $b(u,y)$, $j_S^c(u,y)$ and $g(y)$ are continuously differentiable functions. The vector y is $y = (\Psi,\ N,\ P)^T$ and Ψ, N and P define the approximations to $\psi(x,t)$, $n(x,t)$ and $p(x,t)$ by the discretization method. Then, in a similar way as in the previous section it can be shown that its index is always less or equal to two and it is two only if the circuit contains LI-cut sets or CVS-loops.

4.1 The Scharfetter-Gummel discretization of the Drift-Diffusion equations

If the so-called Scharfetter-Gummel Discretization is applied to the DD equations in (2) the resulting DAE has the same structure as (6). The Scharfetter-Gummel scheme can be described as a Finite Element Method (FEM) for the discretization of the Drift-Diffusion equations that is based on the assumption that the current densities J_n and J_p are constant on each element (triangles, tetrahedrons, etc) of the spatial mesh. For a detailed description of this method we refer to [7].

Suppose $\mathcal{T} = \{T_1, T_2, \ldots, T_K\}$ is a conforming triangulation of Ω and $\mathcal{P} = \{P_1, P_2, \ldots, P_M, \ldots, P_N\}$ denotes the set of vertices of elements in $\mathcal{T}$, where $P_i \in \Omega \cup \Gamma_A$ for $i = 1, 2, \ldots, M$. Let $\{\varphi_1, \varphi_2, \ldots, \varphi_N\}$ be continuous functions that are linear on each $T_i \in \mathcal{T}$ and satisfy $\varphi_i(P_j) = \begin{cases} 1, \text{ if } i = j \\ 0, \text{ else} \end{cases}$. The coefficients that define the approximation $\psi_h(x,t) = \sum_{j=1}^{N} \Psi_j(t)\varphi_j(x)$ are given by

$$\varepsilon \frac{\mathrm{d}}{\mathrm{d}t} \sum_{j=1}^{N} \Psi_j \int_{\Omega} \nabla\varphi_j \cdot \nabla\varphi_i \, \mathrm{d}x - \int_{\Omega} (J_n + J_p) \cdot \nabla\varphi_i \, \mathrm{d}x = 0, \tag{7}$$

where $i = 1, 2, \ldots, M$. The last $N - M$ values of Ψ_j are $\Psi_j = \psi_{bi}(P_j) + f_h(P_j) \cdot A_S^T e$ where $f_h = (f_{1,h}, f_{2,h}, \ldots, f_{n_S,h})$ are approximations to the functions f_i defined in (1k). Suppose the functions $f_{i,h}$ are calculated as $\sum_{j=1}^{N} f_{i,h}(P_j)\varphi_j(x)$. If we substitute Ψ_j, $j = M+1, M+2, \ldots, N$ in (7) by their values and introduce the change of variables $\tilde{\Psi}_j = \Psi_j - f_h(P_j) \cdot A_S^T e$, $j = 1, 2, \ldots, M$ (the tractability index of a DAE is invariant under regular variable transformations) the following equations are obtained

$$\varepsilon \frac{\mathrm{d}}{\mathrm{d}t} \sum_{j=1}^{M} \tilde{\Psi}_j \int_{\Omega} \nabla\varphi_j \cdot \nabla\varphi_i \, \mathrm{d}x - \int_{\Omega} (J_n + J_p) \cdot \nabla\varphi_i \, \mathrm{d}x = 0. \tag{8a}$$

The discretized continuity equations are

$$-\frac{\mathrm{d}}{\mathrm{d}t} \sum_{j=1}^{M} N_j \int_{\Omega} \varphi_j\varphi_i \, \mathrm{d}x - \frac{1}{q} \int_{\Omega} J_n \cdot \nabla\varphi_i \, \mathrm{d}x - \int_{\Omega} R\varphi_i \, \mathrm{d}x = 0, \tag{8b}$$

$$\frac{\mathrm{d}}{\mathrm{d}t} \sum_{j=1}^{M} P_j \int_{\Omega} \varphi_j\varphi_i \, \mathrm{d}x - \frac{1}{q} \int_{\Omega} J_p \cdot \nabla\varphi_i \, \mathrm{d}x + \int_{\Omega} R\varphi_i \, \mathrm{d}x = 0, \tag{8c}$$

where $i = 1, 2, \ldots, M$ and N_j and P_j define the approximations $n_h(x,t) = \sum_{j=1}^{N} N_j(t)\varphi_j(x)$, $p_h(x,t) = \sum_{j=1}^{N} P_j(t)\varphi_j(x)$ to $n(x,t)$ and $p(x,t)$ respectively. Equations (8) were derived applying a FEM to the DD equations. The difference between this discretization and the Scharfetter-Gummel discretization is in the way the integrals involving J_n and J_p are approximated. In both cases the resulting system has the form $A\frac{\mathrm{d}y}{\mathrm{d}t} + b(A_S^T e, y) = 0$ where A is nonsingular and $b(u, y)$ is continuous and differentiable. The equations for the current are as in (6d)-(6e). The matrix J_h has the same form as (3) but with the functions $f_{i,h}(x)$ instead of $f_i(x)$.

5 Conclusions

In this work we have studied the tractability index of a coupled system for the simulation of electrical circuits. The results in this paper generalize those presented in [9] where one dimensional Drift-Diffusion (DD) equations model the behaviour of the semiconductor devices in the circuit, here DD equations in higher space dimensions were considered. It was proved that the ADAS has always index smaller or equal to two and it can be determined by topological conditions on the circuit only. The DAE that is obtained after spatial discretization of the DD equations in the system has also index smaller or equal to two and under the same topological conditions on the circuit as the ADAS assuming that the semi-discretized DD equations have a certain structure. After the Scharfetter-Gummel discretization of the DD-equations a DAE with the required structure is obtained.

References

1. Ali G., Bartel A., Günther M.: Parabolic Differential-Algebraic Models in Electrical Network Design, submitted to SIAM MMS, (2004)
2. Ali G., Bartel A., Günther M., Tischendorf C.: Elliptic Partial Differential Algebraic Multiphysics Models in Electrical Network Design, Math. Models Meth. Appl. Sci., **13(9)**, pp. 1261–1278, (2003)
3. Chua L. O., Desoer C. A., Kuh E. S.: Linear and Nonlinear Circuits. McGraw-Hill Book Company, (1987)
4. Gajewski H.: Analysis und Numerik von Ladungstransport in Halbleitern, Technical Report, Weierstrass Institut für Andgewandte Analysis und Stochastik, **6**, (1993)
5. Lamour R., März R., Tischendorf C.: PDAEs and Further Mixed Systems as Abstract Differential Algebraic Systems, Technical Report, Institute of Mathematics, Humboldt University of Berlin, **11**, (2001)
6. Lutch W., Strehmel K., Eichler-Liebenow C.: Indexed and special discretization methods for linear partial differential algebraic equation, BIT, **39(3)**, pp. 484–512, (1999)
7. Markowich P.: The stationary Semiconductor Device Equations, Springer Verlag, (1986)
8. Selva Soto M., Tischendorf C.: Numerical analysis of DAEs from coupled circuit and semiconductor simulation, APPNUM, to appear
9. Tischendorf C.: Coupled Systems of Differential Algebraic and Partial Differential Equations in Circuit and Device Simulation, Habilitation Thesis, Humboldt University of Berlin, Berlin (2003)

Multirate Methods in Chip Design: Interface Treatment and Multi Domain Extension *

M. Striebel and M. Günther

Bergische Universität Wuppertal, Departement of Mathematics, Chair of Applied Mathematics/Numerical Analysis, D-42097 Wuppertal, Germany,{striebel,guenther}@math.uni-wuppertal.de

Abstract Multirate methods make use of latency that occurs in electrical circuits to simulate more efficiently the transient behaviour of networks: different stepsizes are used for subcircuits according to the different levels of activity. As modelling is usually done by applying modified nodal analysis (MNA), the network equations are given by coupled systems of stiff differential-algebraic equations. Following the idea of 2-level mixed multirate for ordinary differential equations, a hierarchical ROW-based multirate method that can deal with an arbitrary amount subsystems is developed.

1 Introduction

Large integrated electrical networks are usually build up by numerous coupled subcircuits of different functionality. These subcircuits are modelled independently and composed to one macro system by connecting them at the respective terminals, i. e. each pair of connected terminal nodes merge to one node (see Fig. 1, left).

From a modelling point of view, this procedure can be described by introducing virtual voltage sources at the boundary nodes (see Fig. 1, right). This approach preserves the macro circuits block structure and produces additional variables: *branch currents w through the coupling voltage sources*. These currents are determined by the property, that the node potentials of each pair of connected boundary nodes have to coincide.

Regarding r subcircuits, r systems of differential-algebraic equations (DAE), coupled by algebraic equations arise:

$$\mathcal{F}_\lambda(x_\lambda, \frac{d}{dt}q_\lambda(x_\lambda), w, t) = 0, \quad (\lambda = 1, \ldots, r) \tag{1a}$$

$$\mathcal{G}(x_1, \ldots, x_r) = 0, \tag{1b}$$

where x_λ describes the node potentials and currents and q_λ the charges and fluxes of the $\lambda-$th subcircuit and w the coupling currents.

As the subcircuits constitute different functional units, the macro system often shows *multirate behaviour*, i. e. the subcircuits behave on different timescales. Thus *multirate methods* can be applied, that integrate subsystems showing different transient behaviour with different stepsizes adjusted to each subcircuits activity level.

*This work is part of the project "Partielle Differential-Algebraische Multiskalensysteme für die Numerische Simulation von Hochfrequenz-Schaltungen" (No. 03GUNAVN), which is founded by the BMBF program "Multiskalensysteme in Mikro- und Optoelektronik ".

*The author is indebted to Infineon Technologies München, and especially to Drs. Feldmann and Schultz, for supporting his PhD project.

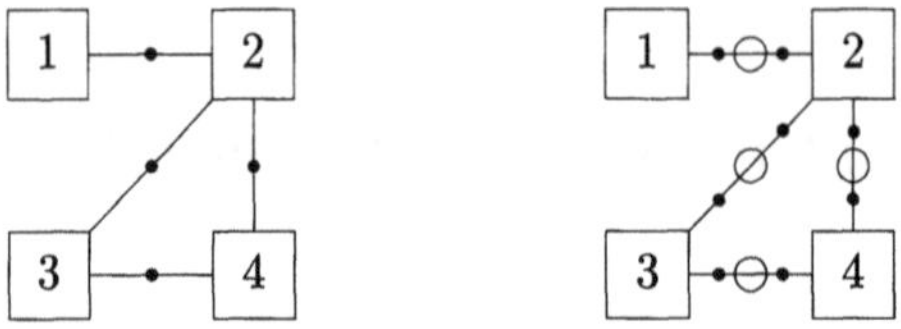

Fig. 1. Coupling: technical and modelling point of view

2 Partitioned Network

Coupled problems that can be described by the abstract model (1a, 1b) also occur in other applications (e. g. multi-body physics). To set up numerical methods that are adapted to simulating electrical networks, a closer look at their special properties is required.

2.1 Network Equations

For circuits that are designed in the described manner, charge oriented modified nodal analysis (MNA) yields network equations of the following form (see also [1]):

$$0 = A_{C_\lambda}\dot{q}_\lambda + A_{R_\lambda} r_\lambda(A_{R_\lambda}^t e_\lambda, t) + A_{L_\lambda}\jmath_{L_\lambda} + A_{V_\lambda}\jmath_{V_\lambda} + \\ + A_{I_\lambda}\imath_\lambda(t) + \boxed{A_{w_\lambda} w}, \qquad \text{(2a)}$$

$$0 = \dot{\phi}_\lambda - A_{L_\lambda}^t e_\lambda, \qquad \text{(2b)}$$

$$0 = A_{V_\lambda}^t e_\lambda - v_\lambda(t), \qquad \text{(2c)}$$

$$0 = q_\lambda - q_{C_\lambda}(A_{C_\lambda}^t e_\lambda, t), \qquad \text{(2d)}$$

$$0 = \phi_\lambda - \varphi_{L_\lambda}(\jmath_{L_\lambda}, t) \qquad \text{(2e)}$$

for the λ–th subcircuit ($\lambda = 1, \ldots, r$) and the overall coupling equation

$$0 = \sum_{\lambda=1}^{r} A_{w_\lambda}^t \cdot e_\lambda. \qquad \text{(3)}$$

The unknowns this DAE-system has to be solved for are the node potentials e_λ, the currents $\jmath_{L_\lambda}$ and $\jmath_{V_\lambda}$ through inductances and voltage sources respectively, the charges q_λ and magnetic fluxes Φ_λ for each subcircuit and the overall coupling currents w.
(2a) constitutes the current balance for each node that belongs to the λ-th subcircuit. The incidence matrices $A_{C_\lambda}, A_{R_\lambda}, A_{L_\lambda}, A_{V_\lambda}, A_{I_\lambda}$ assemble the element related currents $q_\lambda, r_\lambda(\cdot,\cdot), \jmath_{L_\lambda}, \jmath_{V_\lambda}, \imath_\lambda(\cdot)$ through capacitances, resistances, inductances, voltage and current sources respectively. The additional (boxed) term $A_{w_\lambda} w$ reflects the coupling currents to adjoined subcircuits, i. e. through the virtual voltage sources. The appropriate incidence matrix A_{w_λ} filters out the adequate boundary nodes.

The flux–node potential correlation (2b),the node potential – voltage source dependency (2c) and the charge and flux defining equations (2d,2e) are not affected by coupling to other subcircuits, as the information exchange is done solely via coupling currents.

Also controlled current and voltage sources can be included in (2a-2e) by replacing $\imath_\lambda(t)$ and $v_\lambda(t)$ by $\imath_\lambda(A_\lambda^t e_\lambda, \dot{q}_\lambda, \jmath_{L_\lambda}, \jmath_{V_\lambda}, t)$ and $v_\lambda(A_\lambda^t e_\lambda, \dot{q}_\lambda, \jmath_{L_\lambda}, \jmath_{V_\lambda}, t)$ with $A_\lambda^t e_\lambda$ describing the controlling branch voltages.

The linear coupling equation (3) states, that the potentials at the boundary nodes of connected subcircuits have to coincide.

Remark 1. The decomposition of large electrical circuits into subcircuits in the above described manner introduces artifical voltage sources (shorts). Hence additional unknowns w (the terminal currents) emerge. This may not be wanted in general but it yields several benefits:

- The terminal currents are explicitly available and have not to be collected by running through big parts of the hierarchie. This allows to decouple the time domain analysis ([2]). In multirate methods the latent part has to be bypassed when performing integration of the active part ([3]). This can easily be done by interpolation of the terminal currents.
- The coupling equation (3) that is needed to determine the additional unknowns can be helpful for error control.

2.2 Index properties

The overall system (1a,1b) is made up of r subsystems – each with inner variables x_λ ($\lambda = 1, \ldots, r$) – that are coupled by one equation and one variable w respectively. Hence, several index-1 conditions are assumed to be fulfilled, according to the subsystems and the overall system:

(C1) The overall system (1a,1b) has index 1 (with respect to $x_1, \ldots, x_r, w$).
(C2) All systems (1a) define index-1 systems with respect to x_λ (and w given as input).
(C3) For all $\lambda \in \{1, \ldots, r\}$, the system $[\mathcal{F}_\lambda = 0, \mathcal{G} = 0]$ has index-1 with respect to x_λ and w (and $x_i, \forall i \neq \lambda$ given as input).

Topological Conditions

In analogy to the procedure described in [8] and [9], topological conditions to guarantee the index conditions (C1)-(C3) can be derived. Therefore (1a,1b) is transformed into the semi-explicit systems for $\lambda = 1, \ldots, r$

$$\begin{aligned} \dot{y}_\lambda(t) &= f_\lambda(z_\lambda, w, t), \\ 0 &= h_\lambda(y_\lambda, z_\lambda, w, t) \end{aligned} \tag{4a}$$

coupled by the algebraic equation

$$0 = g(z_1, \ldots, z_r). \tag{4b}$$

With fixed projectors $Q_{C_\lambda}, Q_{V-C_\lambda}, Q_{R-CV_\lambda}$ onto $\ker A^t_{C_\lambda}$, $\ker A^t_{V_\lambda} Q_{C_\lambda}$, $\ker A^t_{R_\lambda} Q_{C_\lambda} Q_{V-C_\lambda}$ and their complementary projectors $P_\star = I - Q_\star$, z_λ identifies the node potentials and inner currents and y_λ defines the charges and fluxes (see also [4]):

$$z_\lambda = \hat{\vartheta}_\lambda((e^t_\lambda, \jmath^t_{L_\lambda}, \jmath^t_{V_\lambda})^t), \qquad y_\lambda = \begin{pmatrix} A_{C_\lambda} q_\lambda \\ \Phi_\lambda \end{pmatrix}$$

$$\text{with} \quad \hat{\vartheta}_\lambda((e^t_\lambda, \jmath^t_{L_\lambda}, \jmath^t_{V_\lambda})^t) = \begin{pmatrix} P_{C_\lambda} & 0 & 0 \\ 0 & I_{\jmath L_\lambda} & 0 \\ P_{V-C_\lambda} & 0 & 0 \\ P_{R-CV_\lambda} & 0 & 0 \\ 0 & 0 & I_{\jmath V_\lambda} \end{pmatrix} \cdot \begin{pmatrix} e_\lambda \\ \jmath_{L_\lambda} \\ \jmath_{V_\lambda} \end{pmatrix} \tag{5}$$

The topological conditions read: (C2) holds, if the λ-th subcircuit neither contains CV-loops nor LI-cutsets ([8, 9]). In addition (C1) and (C3) hold, if the according composition of subcircuits does not contain loops of only capacitors, voltage sources and at least one virtual voltage source.

3 Multirate Methods

The basic idea of *multirate methods* is to prevent parts to be integrated with smaller stepsizes than necessary to guarantee given error tolerances. This is done by using different stepsizes that are suitable for the

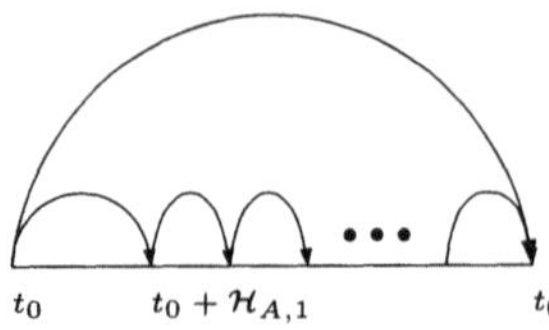

Fig. 2. Macro- and microsteps

different levels of activity at each time. In the case of problems that are already given in the form of coupled subsystems like (1a,1b) it is convenient to assume, that these subsystems have no intrinsic *multirate potential* (but they may show multitone behaviour)

Remark 2. Multirate methods have to interweave approximations working on different time grids. This causes an overhead that has to be outbalanced by the reduction of computational costs for the discretisation of the less active (latent) parts. Hence systems showing multirate behaviour are said to have *multirate potential* if the different timescales are widley seperated, the latent parts are larger than the active ones and the coupling amongst subsystems representing different activity levels is weak.

3.1 Multirate schemes for ODE systems

The concept of onestep multirate methods can be described with a system of two coupled ODEs:

$$\dot{y}_L = f_L(y_L, y_A), \quad y_L(t_0) = y_{L,0}, \tag{6a}$$
$$\dot{y}_A = f_A(y_L, y_A), \quad y_A(t_0) = y_{A,0}. \tag{6b}$$

The idea is to compute one *macrostep* of the latent† part (subscript L) with the stepsize $\mathcal{H}_L$, i. e. get an approximation $y_{L,1} \approx y_L(t_0 + \mathcal{H}_l)$ and to perform q *microsteps* with stepsizes $\mathcal{H}_{A,\mu}$ (with $\mathcal{H}_l = \mathcal{H}_{a,1} + \ldots + \mathcal{H}_{a,q}$) for the active* part (see Fig. 2). In its most general way this procedure can be defined as follows:

$$y_{L,1} = y_{L,0} + \sum_{i=1}^{s_L} b_i^L \cdot k_i^L,$$
$$y_{A,\mu} = y_{A,\mu-1} + \sum_{i=1}^{s_A} b_i^A \cdot k_i^{A,\mu} \quad (\mu = 1, \ldots, q),$$
$$k_i^L = \Phi_L(\mathcal{H}_L; y_{L,0}, Y_i^A, k_1^L, \ldots, k_{s_L}^L) \quad (i = 1, \ldots, s_L),$$
$$k_i^{A,\mu} = \Phi_A(\mathcal{H}_{A,\mu}; y_{A,\mu-1}, Y_i^{L,\mu}, k_1^{A,\mu}, \ldots, k_{s_A}^{A,\mu}) \quad (i = 1, \ldots, s_A),$$

where Φ_* denotes an s_* stage IRK or ROW scheme with coefficients $\alpha^*, \beta^*, \gamma^*, \nu^*$ ($* \in \{$L,A$\}$).

As the subsystems are coupled, the computation of the weights for each part depends on information on the other one at some supporting timepoints:

$$Y_i^A \approx y_A(t_0 + \alpha_i^L \mathcal{H}_L) \quad (i = 1, \ldots, s_L), \tag{7a}$$
$$Y_i^{L,\mu} \approx y_L(t_0 + \sum_{\nu=1}^{\mu-1} \mathcal{H}_{A,\nu} + \alpha_i^A \mathcal{H}_{A,\mu}) \quad (i = 1, \ldots, s_A; \mu = 1, \ldots, q). \tag{7b}$$

There are different strategies to compute these values. Explicitly done *extra-/interpolation* [6] destroyes the onestep character of the method. *Generalised multirate* [7], a RK-based method, calculates $Y_i^{L,\mu}$ and

†Subsystems for which the actual step is computed with a large optimal stepsize a called *latent*, subsystems that need small stepsizes are denoted *active*.

Y_i^A in RK-like manner using the stage increments $k_i^L, k_i^{A,\mu}$. *Mixed multirate* [3], ROW-based, builds up on generalised multirate. It decomposes the computation of one macrostep with its inner microstep to a so-called "compound step" and "later microsteps" and therefore is a slowest first approach ([6]). For the former the incremental formulation of generalise multirate is used. In the latter dense-output is used for the coupling.

3.2 Mixed multirate scheme for coupled index-1 DAE systems

First consider a system of two ($r = 2$) coupled index-1 DAEs of semi-explicit form (4a,4b):

$$\begin{aligned} \dot{y}_L &= f_L(z_L, w) & \dot{y}_A &= f_A(z_A, w) \\ 0 &= h_L(y_L, z_L, w) & 0 &= h_A(y_A, z_A, w) \\ & & \hspace{-6em} 0 = g(z_L, z_A). \end{aligned} \tag{8}$$

As the coupling current affects both subsytems it is natural to assume that the according variable w behaves like the latent part with y_L, z_l. If there are more than two subsystems w may also contain couplings amongst active parts and may be decomposed to latent and active parts itself.
Due to the index assumptions (C3) and (C2) $[\dot{y}_L = f_L, 0 = h_L, 0 = g]$ and $[\dot{y}_A = f_A, 0 = h_A]$ are index-1 problems with respect to z_L, w and z_A respectively.

The mixed multirate ansatz for ODEs can be brought forward to the coupled semi-explicit problem (8). The *compound step regulations* read :

$$\begin{pmatrix} y_{L,1} \\ z_{L,1} \\ w_1 \end{pmatrix} = \begin{pmatrix} y_{L,0} \\ z_{L,0} \\ w_0 \end{pmatrix} + b_L^t \begin{pmatrix} l_L \\ k_L \\ p \end{pmatrix}, \quad \begin{pmatrix} y_{A,1} \\ z_{A,1} \end{pmatrix} = \begin{pmatrix} y_{A,0} \\ z_{A,0} \end{pmatrix} + b_A^t \begin{pmatrix} l_A \\ k_A \end{pmatrix}. \tag{9a}$$

with $b_\lambda := (b_{\lambda,1}, \ldots, b_{\lambda,s})^t$ denoting the weights and $l_\lambda := (l_{\lambda,1}^t, \ldots, l_{\lambda,s}^t)^t$, $k_\lambda := (k_{\lambda,1}^t, \ldots, k_{\lambda,s}^t)^t$ and $p_\lambda := (p_{\lambda,1}^t, \ldots, p_{\lambda,s}^t)^t$ denoting the increments to y_λ, z_λ and w.

The incremtents for the i-th stage are defined by the linear equation

$$M^\star \cdot \begin{pmatrix} l_{L,i} \\ k_{L,i} \\ \hline l_{A,i} \\ k_{A,i} \\ \hline p_i \end{pmatrix} = \mathrm{RHS}_i, \tag{9b}$$

with $M^\star =$

$$\left(\begin{array}{cc|cc|c} \mathrm{I}_{y_L} & -\mathcal{H}_L \gamma^{(L)} \frac{\partial f_L}{\partial z_L} & & & -\mathcal{H}_L \gamma^{(L)} \frac{\partial f_L}{\partial w} \\ -\gamma^{(L)} \frac{\partial h_L}{\partial y_L} & -\gamma^{(L)} \frac{\partial h_L}{\partial z_L} & & & -\gamma^{(L)} \frac{\partial h_L}{\partial w} \\ \hline & & \mathrm{I}_{y_A} & -\mathcal{H}_A \gamma^{(A)} \frac{\partial f_A}{\partial z_A} & -\frac{1}{m} \cdot \mathcal{H}_A \nu^{(A,L)} \frac{\partial f_A}{\partial w} \\ & & -\gamma^{(A)} \frac{\partial h_A}{\partial y_a} & -\gamma^{(A)} \frac{\partial h_A}{\partial z_A} & -\frac{1}{m} \cdot \nu^{(A,L)} \frac{\partial h_A}{\partial w} \\ \hline & -\gamma^{(L)} \frac{\partial g}{\partial z_L} & & -m \cdot \nu^{(L,A)} \frac{\partial g}{\partial z_A} & \end{array}\right)$$

and a right-hand side RHS_i depending on both stepsizes $\mathcal{H}_L, \mathcal{H}_{A,1}$, the stepsizeratio $m := \frac{\mathcal{H}_L}{\mathcal{H}_A}$, the incremtents $l_{L,j}, k_{L,j}, l_{A,j}, k_{A,j}, p_j$ of the former steps $j = 1, \ldots, i-1$ and a set of coefficients that includes $\gamma^{(L)}, \nu^{(L,A)}, \ldots$.

Remark 3. In the *later microsteps* it remains, to solve the system $[\dot{y}_A = f_a, 0 = h_a]$ with respect to y_A, z_A and $w(t)$ entering the right-hand-side via dense output: $w(t_0 + \xi \cdot \mathcal{H}_L) \approx w_0 + \sum_{i=1}^{s} b_{L,i}(\xi) \cdot p_i$ with $\xi \in (0, 1)$

In circuit simulation methods are wanted that directly operate on the network equations. The semi-explicit formulation (4a,4b) is just introduced to do analytical studies.

3.3 Mixed multirate for coupled network (2a-e,3)

The coupled network equations (2a-e,3) are transfered to the semi-explicit formulation (4a,4b).

According to (5) the network equations' variables e_λ, $\jmath_{L_\lambda}$, $\jmath_{V_\lambda}$ identify the semi-explicit formulation's one z_λ via the linear operator $\hat{\vartheta}_\lambda$.

With $\vartheta_\lambda(z_\lambda) = \begin{pmatrix} I_{e_\lambda} & 0 & Q_{C_\lambda} & Q_{C_\lambda}Q_{V-C_\lambda} & 0 \\ 0 & I_{\jmath_{L_\lambda}} & 0 & 0 & 0 \\ 0 & 0 & 0 & 0 & I_{\jmath_{V_\lambda}} \end{pmatrix} \cdot z_\lambda$

also the network equation's variables can be reconstructed from the "semi-explicit" ones. Hence

$$\begin{aligned} z_\lambda &= \hat{\vartheta}_\lambda((e_\lambda^t, \jmath_{L_\lambda}^t, \jmath_{V_\lambda}^t)^t) \\ &= \hat{\vartheta}_\lambda \circ \vartheta_\lambda(z_\lambda) \end{aligned} \quad \Bigg| \quad \begin{aligned} (e_\lambda^t, \jmath_{L_\lambda}^t, \jmath_{V_\lambda}^t)^t &= \vartheta_\lambda(z_\lambda) \\ &= \vartheta_\lambda \circ \hat{\vartheta}_\lambda((e_\lambda^t, \jmath_{L_\lambda}^t, \jmath_{V_\lambda}^t)^t). \end{aligned} \tag{10}$$

The semi-explicit problem (4a,4b) and its associated method (9a,9b) is suitable to derive order conditions to get adequate coefficients, such that $\|\text{err}_\lambda^{(\text{se})}\| = \mathcal{O}(\mathcal{H}_\lambda^{p+1})$§

As the transformation between the two formulations is not invertible, it is not possible to carry forward the attained method to a method that draws directly on the coupled network. To obtain such "network-regulations" with demanded accuracy ($\|\text{err}_\lambda^{(\text{mna})}\| = \mathcal{O}(\mathcal{H}_\lambda^{p+1})$)§ in terms of node voltages and currents $(e, \jmath_L, \jmath_V)$ however, there is another way:

- Based upon the idea of (9a,9b) regulations with an (undefined) coefficient set can be deduced from the network formulation (2a-e,3).
- The same transformation that carries over the network formulation to the semi-explicit one, applied to the above regulations yields instructions that coincide with (9a,9b).
- Regarding (10), it holds that $\|\text{err}_\lambda^{(\text{mna})}\| \le \|\vartheta\| \cdot \|\text{err}_\lambda^{(\text{se})}\|$ for the same coefficientset. Hence, if a coefficientset is chosen properly for the semi-explicit formulation, it is also suitable for the network formulation.
- Finally a Block-Gaussian elimination and some linear transformation allow to eliminate the charges and fluxes q_λ, ϕ_λ. This guarantees charge- and flux-conservation and enables error-check and stepsize control based directly on the node potentials and currents $(e_\lambda, \jmath_{L_\lambda}, \jmath_{V_\lambda})$ (see also [5]).

4 Hierarchical mixed multirate

The already known multirate schemes deal with two different levels of activity. However coupled problems like (1a,1b) need *n-level-multirate schemes*¶ with stepsizes $\mathcal{H}_1 > \ldots > \mathcal{H}_n$. Transfering the 2-level-mixed multirate to n-level-schemes in a straight forward way produces a bunch of coupling coefficients. *Hierarchical mixed multirate* is a new approach in dealing with an arbitrary amount of activity levels and still limits the amount of coupling-factors. The main idea is to nest compound steps and later micro-steps in a way, that at each time merely a two-level multirate scheme is engaged: (see Fig. 3)

(i) Group remaining subsystems in terms of activity level. This yields k_i virtual blocks consisting of subsystems showing similar behaviour. If $k_i = 1$ employ later micro steps – the coupling to other subsystems/blocks is given by dense output – until endpoint is reached, skip to (iv).
(ii) Build up a sorted stack (top down, decreasing stepsizes). Apply a compound step with the stacks top as latent block and its associated stepsize and all the other blocks combined to one as active block with the stepsize associated to the stacks bottom. The coupling to already integrated subsystems is given by dense output.
(iii) Remove the stacks top. The new endpoint is the one reached by the macrostep. Skip to (i).
(iv) Enlarge the set of remainig subsystems by the ones that produced the last endpoint. If the endpoint is the endpoint of integration as demanded it is finished. Else forget the endpoint and skip to (i).

§ $\text{err}_\lambda^\star$ denotes the local error after one step for the λ-th subsystem, with "se" short form for semi-explicit formulation, "mna" short form for network's equation formulation

¶ $n \le r$ as some subsystems may show the same activity level.

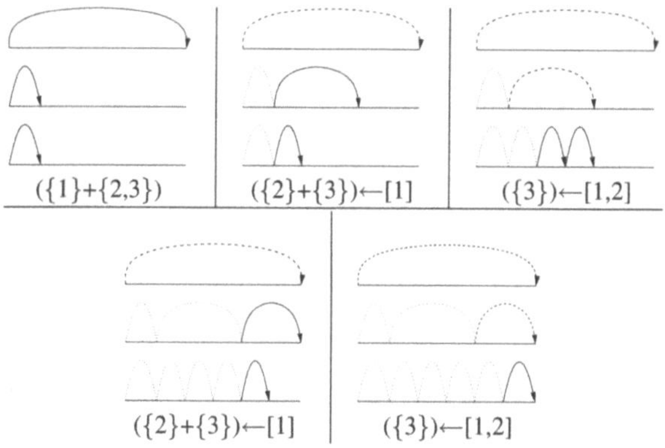

Fig. 3. Hierarchical mixed multirate for three blocks

Table 1. accepted (refused) steps, time intervall $[0, 2\pi]$

	Multirate	Singlerate
Block I ($\approx \sin(t)$)	$255(9)$	
Block II ($\approx \sin(10t)$)	$1391(22)$	$7801(150)$
Block III ($\approx \sin(100t)$)	$8322(150)$	

Numerical Tests

A first hierarchical multirate-method of order 2 has recently been implemented in MATLAB. It can deal with an arbitrary amount of subcircuits with grouping them in terms of activity levels.

First testruns were done with a three-block circuit (with 3/5/3 nodes) "behaving like" $\sin(\omega t)$ with $\omega = 1, 10, 100$ respectively. This yields promising results (see Table 1) as the mid-latent and latent block are calculated about 30 and 6 times less than in a corresponding singlerate.

5 Conclusions

A multirate scheme for circuit simulation that can deal with an arbitrary amount of subsystems has been derived. Domain decomposition of large electrical circuits has been reached by introducing extra variables. Now numerical tests have to be done with industry related examples to demonstrate the qualities of the method.

Stepsize controll for multirate schemes has to be improved using the additional unknowns and the order of the method has to be enlarged to at least order three.

References

1. M. Günther, U. Feldmann and J. ter Maten: Modelling and Discretization of Circuit Problems. In: P. G. Ciarlet, W. H. A. Schilders, E. J. W. ter Maten(Ed.): Numerical Methods in Electromagnetics, Special Volume of HANDBOOK OF NUMERICAL ANALYSIS, VOL. XIII, Elsevier B. V. 2005
2. U. Wever and Q. Zheng: Parallel Circuit Simulation on Workstation Clusters In Progress in Industrial Mathematics at ECMI 94, John Wiley & Sons Ltd and B. G. Teubner: 274-284
3. A. Bartel, M. Günther and A. Kværnø: Multirate Methods in Electrical Circuit Simulation. In: A.M. Anile et al. (Ed.): Progress in Industrial Mathematics at ECMI 2000, Springer 2002, 258-265
4. Günther, M., Arnold, M.: Coupled simulation of partitioned differential-algebraic network models. In preparation

5. M. Günther: Simulating digital circuits numerically – a charge-oriented ROW approach. Numer. Math. (1998) **79**: 203-212
6. C. W. Gear and R. R. Wells: Multirate linear multistep methods, BIT **24**, 484-502 (1984)
7. A. Kværnø and P. Rentrop: Low Order Multirate Runge-Kutta Methods in Electric Circuit Simulation. Preprint Nr. 99/1, IWRMM Universität Karlsruhe
8. C. Tischendorf: Topological index calculation of differential-algebraic equations in circuit simulation. Surv. Math. Ind., **8**, 187–199 (1999)
9. D. Estévez Schwarz and C. Tischendorf: Structural analysis for electrical circuits and consequences for MNA. Int. J. Circ. Theor. Appl. 28 (2000), 131-162

Digital Linear Control Theory for Automatic Stepsize Control

A. Verhoeven[1], T. G. J. Beelen[2], M. L. J. Hautus[1], and E. J. W. ter Maten[3]

[1] Technische Universiteit Eindhoven, averhoev@win.tue.nl
[2] Philips Research Laboratories
[3] Technische Universiteit Eindhoven and Philips Research Laboratories

Abstract In transient analysis of electrical circuits the solution is computed by means of numerical integration methods. Adaptive stepsize control is used to control the local errors of the numerical solution. For optimization purposes smoother stepsize controllers can ensure that the errors and stepsizes also behave smoothly. For onestep methods, the stepsize control process can be viewed as a digital (i.e. discrete) linear control system for the logarithms of the errors and steps. For the multistep BDF-method this control process can be approximated by such a linear control system.

1 Introduction

Electrical circuits can be modelled by the following Differential-Algebraic Equation

$$\frac{d}{dt}\left[\mathbf{q}(t,\mathbf{x})\right]+\mathbf{j}(t,\mathbf{x})=\mathbf{0}, \tag{1}$$

where $\mathbf{q},\mathbf{j}:\mathbb{R}\times\mathbb{R}^n\to\mathbb{R}^n$ represent the charges on capacitors and currents through resistors and sources in the circuit and $\mathbf{x}$ is the state vector. In transient analysis an Initial Value Problem has to be solved for this DAE, which is done by implicit integration methods (usually BDF methods).

The accuracy of integration methods depends on the magnitude of the stepsizes. Adaptive stepsize control is used to handle the trade-off between the computational work load and the accuracy. Therefore, each step the magnitude of the local error must be estimated. If this estimate $\hat{r}_n$ is larger than a given tolerance level TOL, the current step is rejected. Otherwise, the numerical solution can be computed at a next timepoint $t_{n+1}=t_n+h_n$.

The following stepsize controller is very commonly used for integration methods of order p:

$$h_n=\left(\frac{\epsilon}{\hat{r}_{n-1}}\right)^{\frac{1}{p+1}}h_{n-1}. \tag{2}$$

This controller tries to keep the error $\hat{r}_n$ close to a reference level ϵ by means of the stepsize h_n. The reference level ϵ is equal to θ TOL, where $0<\theta<1$ is a safety factor, which reduces the number of rejections.

The stepsize controller is based on the assumption that the error estimate satisfies the model

$$\hat{r}_n=\hat{\varphi}_n h_n^{p+1}, \tag{3}$$

where φ_n is an unknown variable which is independent of h_n. This model is a good description for onestep methods and also a first order approximation for the multistep BDF-methods. In practice some bounds and limiters are always added to this controller in order to avoid numerical problems.

Important properties of a good simulator are speed, accuracy and robustness. It appears that the controller (2) produces rather irregular error and stepsize sequences, which will decrease the robustness.

2 Application of control theory

It seems attractive to use control-theoretic techniques for error control. In [1, 5] this idea has been applied to onestep methods where we have the simple model (3). Figure 1 shows the block diagram of this feedback control system. The process model $G(q)$ and the controller model $C(q)$ are described in the next subsections.

2.1 Process model $G(q)$

The logarithmic version of the onestep error model (3) is

$$\log \hat{r}_n = (p+1)\log h_n + \log \hat{\varphi}_n. \tag{4}$$

Writing $\log \hat{r} = \{\log \hat{r}_n\}_{n\in\mathbb{N}}, \log h = \{\log h_n\}_{n\in\mathbb{N}}$ and $\log \hat{\varphi} = \{\log \hat{\varphi}_n\}_{n\in\mathbb{N}}$, this implies that the sequence $\log \hat{r}$ can be viewed as the output of a digital (i.e. discrete) linear control system, where $\log h$ is the input signal and $\log \hat{\varphi}$ is an unknown output disturbance. In general, we can denote all linear models with finite recursions for $\log \hat{r}$ by

$$\log \hat{r} = G(q)\log h + \log \hat{\varphi}, \tag{5}$$

where q is the shift-operator, with $q(\log h_n) = \log h_{n+1}$ and where $G(q)$ is a rational function of q:

$$G(q) = \frac{L(q)}{K(q)} = \frac{\lambda_0 q^M + \cdots + \lambda_M}{q^M + \kappa_1 q^{M-1} + \cdots + \kappa_M}. \tag{6}$$

For the onestep model, we just have that $G(q) = p+1$. However, it is not possible to derive a linear model of this form for the multistep BDF methods. In this case for a p-step method, we have the following nonlinear model for $\log \hat{r}$ [6]

$$\log \hat{r}_n = 2\log h_n + \log(h_{n-1}+h_n) + \cdots + \log(h_{n-p+1}+\cdots+h_n) + \log \hat{\varphi}_n - \log p!. \tag{7}$$

Note that $\log \hat{r}_n$ also depends on the previous stepsizes, because it is a multistep method. In [8] it is tried to approximate this model by the previous model for onestep methods. Another possibility is to use an adaptive process model which is based on parameter identification [3].

If the stepsizes only have small variations, also linearization can be used [4]. In [6] it is proved that the linearized model is equal to

$$\log \hat{r}_n = (1+\gamma_p)\log h_n + (\gamma_p-\gamma_1)\log h_{n-1} + \cdots + (\gamma_p-\gamma_{p-1})\log h_{n-p+1} + \log \hat{\varphi}_n, \tag{8}$$

where $\gamma_m = \sum_{n=1}^{m} \frac{1}{n}$ for $m \in \mathbb{N}$.

This model can also be cast in (5), where

$$G(q) = \frac{(1+\gamma_p)q^{p-1} + (\gamma_p-\gamma_1)q^{p-2} + \cdots + (\gamma_p-\gamma_{p-1})}{q^{p-1}}. \tag{9}$$

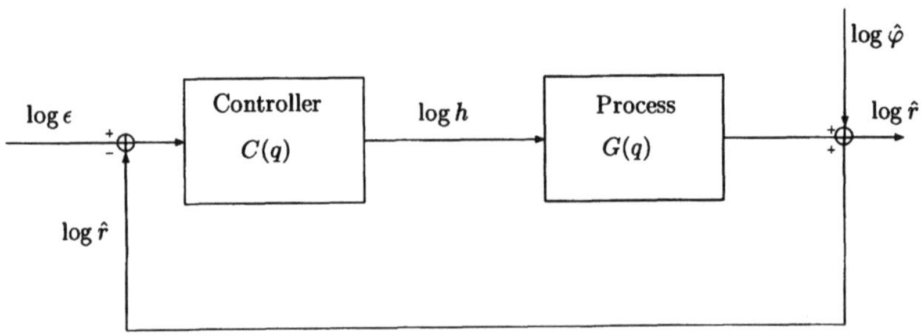

Fig. 1. Diagram of adaptive stepsize control viewed as a feedback control system

2.2 Controller model $C(q)$

The logarithmic version of the controller in eqn. (2) is

$$\log h_n - \log h_{n-1} = \frac{1}{p+1}(\log \epsilon - \log \hat{r}_{n-1}). \tag{10}$$

So, also the control action can be viewed as a linear feedback controller for the same linear system. The input $\log h$ is computed based on the previous values of the output $\log \hat{r}$ and the reference $\log \epsilon$. All linear controllers can be denoted by

$$\log h = C(q)(\log \epsilon - \log \hat{r}), \tag{11}$$

where $C(q)$ is a rational function of q:

$$C(q) = \frac{B(q)}{A(q)} = \frac{\beta_0 q^{N-1} + \cdots + \beta_{N-1}}{q^N + \alpha_1 q^{N-1} + \cdots + \alpha_N}. \tag{12}$$

For the controller of eqn. (2) we just have that $C(q) = \frac{1}{p+1}\frac{1}{q-1}$.

3 Design of finite order digital linear stepsize controller

Consider the error model (5), which is controlled by the linear controller (11). It is assumed that the error model is already available, while the controller still must be designed. This means that K, L are known, while A, B are unknown. Now, the closed loop dynamics are described by the following equations:

$$\begin{cases} \log h = U_r(q)\log \epsilon + U_w(q)\log \hat{\varphi}, \\ \log \hat{r} = Y_r(q)\log \epsilon + Y_w(q)\log \hat{\varphi}. \end{cases} \tag{13}$$

The transfer functions satisfy

$$\begin{aligned} U_r(q) &= \tfrac{B(q)K(q)}{R(q)}, \; U_w(q) = \tfrac{-B(q)K(q)}{R(q)}, \\ Y_r(q) &= \tfrac{B(q)L(q)}{R(q)}, \;\; Y_w(q) = \tfrac{A(q)K(q)}{R(q)}, \end{aligned} \tag{14}$$

where $R(q) = A(q)K(q) + B(q)L(q)$. In this section we will derive conditions for A, B such that the closed loop dynamics have some preferred properties.

3.1 Adaptivity and filter properties

The output $\log \hat{r}$ depends on the reference signal $\log \epsilon$ and the disturbance $\log \hat{\varphi}$. This means that in general the control error $\log \epsilon - \log \hat{r}$ deviates from zero. However, there is no control error if $Y_w(q)\log \hat{\varphi} = 0$ and $Y_r(1) = 1$ [6]. If $\log \hat{\varphi}$ is a polynomial of degree $p_A - 1$ and $Y_w(q)\log \hat{\varphi} = 0$, we call the order of adaptivity p_A. It is always required that $p_A \geq 1$ in order to have no control error for a constant disturbance. For higher order adaptivity the controller is capable to follow linear or other polynomial trends of the disturbance $\log \hat{\varphi}$. It can be proved that the controller is adaptive with adaptivity order p_A if $(q-1)^{p_A}$ is a divisor of $A(q)$.

$$A(q) = (q-1)^{p_A}\hat{A}(q)$$

Because of numerical errors, the disturbance $\log \hat{\varphi}$ can contain alternating noise with frequency near π. The controller acts like a filter for the stepsizes or the errors if

$$|U_w(e^{i\omega})| = O(|\omega - \pi|^{p_F}), \quad \omega \to \pi$$

or

$$|Y_w(e^{i\omega})| = O(|\omega - \pi|^{p_R}), \quad \omega \to \pi.$$

Here p_F and p_R are the orders of the stepsize filter and the error filter. It is not possible to combine an error filter with a stepsize filter. The controller is a stepsize filter of order p_F if $(q+1)^{p_F}$ is a divisor of $B(q)$.

$$B(q) = (q+1)^{p_F}\hat{B}(q)$$

The controller is an error filter of order p_R if $(q+1)^{p_R}$ is a divisor of $A(q)$.

$$A(q) = (q+1)^{p_R}\check{A}(q)$$

3.2 Position of the poles

The poles of the system are determined by the $N+M$ roots of the characteristic equation

$$A(q)K(q) + B(q)L(q) = 0.$$

If the poles lie inside the complex unity circle, the closed loop system is stable. The absolute values determine the reaction speed of the controllers, while the angles determine the eigenfrequencies. This means that real positive poles will produce smoother behaviour.

If the controller is adaptive, we know that the error always will be equal to the reference level if the disturbance is a low degree polynomial. However, this will never be the case in practice. Thus it is still possible that the next error will be larger than the tolerance level TOL.

Let R, S be polynomials of degree $N+M$, such that

$$\begin{aligned} S(q) &= A(q)K(q) &&= q^{N+M} + \sigma_1 q^{N+M-1} + \cdots + \sigma_{N+M} \\ R(q) &= A(q)K(q) + B(q)L(q) &&= q^{N+M} + \rho_1 q^{N+M-1} + \cdots + \rho_{N+M} \end{aligned}$$

In [6] it is proved that there are no rejections, such that $\hat{r}_n \geq$TOL if

- The disturbance $\hat{\varphi}$ satisfies the inequality:

$$\theta^{R(1)} \hat{\varphi}_n \hat{\varphi}_{n-1}^{\sigma_1} \cdots \hat{\varphi}_{n-N-M}^{\sigma_{N+M}} \leq 1. \tag{15}$$

- The coefficients of $R(q)$ satisfy: $\rho_i \leq 0, \quad i \in \{1, \ldots, N+M\}$, e.g. $R(q) = q^{N+M} - r^{N+M}$.
- The previous $N+M$ stepsizes have been accepted.

The first condition for the disturbance also depends on θ. Note that a small θ will indeed decrease the number of future rejections. The second condition is not true if all poles are real positive.

3.3 Computation of the control parameters

In order to get the optimal control parameters, in [1, 5] a systemetic investigation is done for a large range of possible control parameters. Below we propose a theoretical approach which is only based on the closed loop dynamics. Assume that A, B can be factorized like $A(q) = (q-1)^{p_A}(q+1)^{p_R}\tilde{A}(q)$ and $B(q) = (q+1)^{p_F}\tilde{B}(q)$. Then the order of adaptivity is equal to p_A, while the filter orders are p_R and p_F. Because $q = 1$ and $q = -1$ are not stable poles, it is not allowed that $A(1) = B(1) = 0$ or $A(-1) = B(-1) = 0$. Thus it follows that $p_R = 0 \vee p_F = 0$. Let $R(q)$ be the polynomial whose roots are equal to the wanted poles, then the polynomials A, B are determined by

$$(q-1)^{p_A}(q+1)^{p_R}\tilde{A}(q)K(q) + (q+1)^{p_F}\tilde{B}(q)L(q) = R(q). \tag{16}$$

The coefficients of A, B are the control parameters, which can be computed from (16). Instead of this theoretical approach, in [1, 4] a systemetic investigation is done for a large range of possible control parameters.

$$\begin{aligned} &\tfrac{dV_1}{dt} + i_L + 30V_1\left(\tfrac{V_1^2}{3} - 1\right) = 0 \quad && V_1(0) = 0 \\ &\tfrac{di_L}{dt} - V_1 = 0 && i_L(0) = 1 \end{aligned}$$

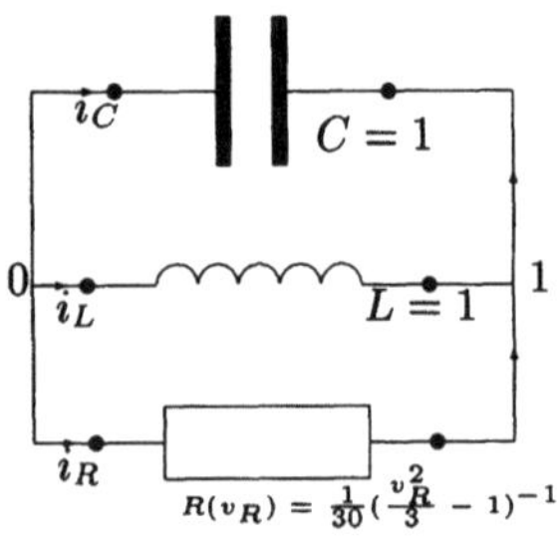

Fig. 2. Diagram of adaptive stepsize control viewed as a feedback control system

4 Numerical experiments

Consider the initial value problem (van der Pol equation) for the following electrical circuit:

This IVP is solved on $[0, 100]$ by means of the BDF2 method (order $p = 2$) with tolerance level TOL = 1e-4 and reference level $\epsilon = 0.3$TOL. A frequently used controller is (2) with $p_A = 1$ and having a pole equal to zero.

$$\text{I:} \quad h_n = \left(\frac{\epsilon}{\hat{r}_{n-1}}\right)^{\frac{1}{p+1}} h_{n-1} \quad (p_A = 1)$$

Often, this controller is used in combination with a buffer, e.g.

$$\frac{h_n}{h_{n-1}} \in [0.8, 2] \Rightarrow h_n = h_{n-1}.$$

Consider the next second order adaptive stepsize controller, whose closed loop poles are equal to 0.2. This means that it is able to predict linear trends of the disturbance $\log \hat{\phi}$.

$$\text{II:} \quad \frac{h_n}{h_{n-1}} = \left(\frac{\epsilon}{\hat{r}_{n-1}}\right)^{\frac{16}{25}\frac{1}{p+1}} \left(\frac{\hat{r}_{n-2}}{\hat{r}_{n-1}}\right)^{\frac{24}{25}\frac{1}{p+1}} \frac{h_{n-1}}{h_{n-2}} \quad (p_A = 2)$$

The IVP has been solved by controller I with buffer (case 1) and Controller II (case 2). These cases require 1000, 1080 stepsizes and 1686, 2054 Newton iterations, respectively. Figure 3 shows the resulting stepsize and error sequences. The best results are obtained in case 2, because of the better adaptivity at the cost of an increase of Newton iterations. Because of the higher smoothness of case 2, the safety factor could be increased for case 2. Indeed, for $\epsilon = 0.6$TOL, the cases need 1847 and 1667 Newton iterations, respectively.

An important question is whether the new designed controllers also have a better performance for a real circuit simulator. Therefore, in the next three cases a real circuit is simulated, while a variable integration order is used [6]. In case 1 the default stepsize controller of the simulator is used. In the other cases, the stepsize controllers are based on digital linear control theory applied to the onestep model and the multistep model (9) respectively. The closed loop poles are equal to 0.5 while $p_A = 1$ and $p_F = p_R = 0$. For all three cases, the safety factor is variable. The smoothness of the stepsize and error sequences is quantified by means of the number $s(x) = \sqrt{\sum_{m=1}^{N}(x_m - x_{m-1})^2/\|x\|_2}$. Table 1 shows the results of these three cases. Note that for the cases 2 and 3 the smoothness of the error sequence is improved, while the computational work is about the same. Furthermore the performance is even better (8%) than for case 1. Case 3 leads to the least smooth stepsize sequence which can be improved by a more expensive stepsize filter.

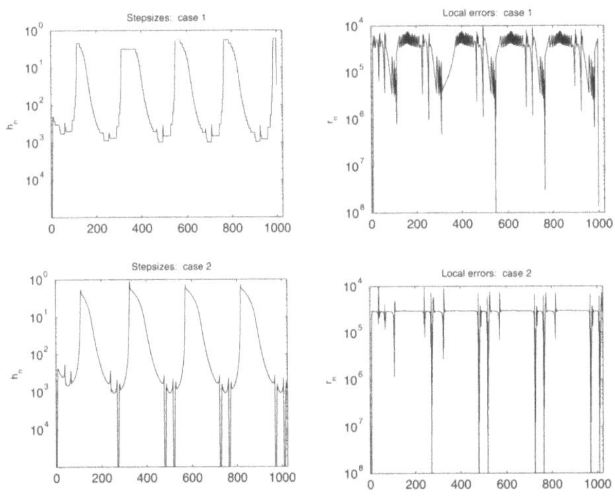

Fig. 3. Stepsize and error sequences for the two tested controllers

Table 1. Numerical results for perf_mos7_qubic_6953 ($p_A = 1, p_F = p_R = 0$)

Case	# stepsizes	# rejections	# Newton iterations	$s(\hat{r})$	$s(h)$
1	6465	947	43232	0.85	0.58
2	6934	777	40234	0.79	0.48
3	6423	714	39619	0.74	0.85

5 Conclusions

It has been tried to derive a linear model for the behaviour of the local error. For onestep methods this is less complex, because then the local error only depends on the last stepsize. But because circuit simulators use the multistep BDF-methods, also the application for BDF-methods has been studied. In that case, a linearized linear model can be derived, which is only correct for small variations of the stepsizes.

From the experiments it seems not always attractive to use higher order adaptive controllers. However, filtering appears to be attractive because it reduces the high-frequent noise, which makes the behaviour of the stepsizes and the errors much smoother.

Because the described method is only developed for a fixed order of integration, the theoretical results for a variable integration order are not known yet. Clearly the local error also depends on the integration order and this affects the process model. It seems not possible to describe this behaviour by means of a linear model. Despite this application in the variable integration order case works satisfactorily.

To deal with the trade-off between the smoothness and the speed, optimal control could be applied. In this case, a cost function should be defined which is dependent on the stepsize sequence and the error sequence.

References

1. Gustafsson, K.: Control Theoretic Techniques for Stepsize Selection in Implicit Runge-Kutta Methods. ACM TOMS **20** 496–517, (1994)
2. Kuo, B.C.: Digital Control Systems. Holt-Saunders International Editions, New York (1980)
3. Mauritz, H., Mathis, W.: Integration System as Adaptive Control System. Circuits and Systems (ISCAS94) 101–104, (1994)
4. Sjö, A.: Analysis of computational algorithms for linear multistep methods. PhD Thesis, Lund University, Lund (1999)
5. Söderlind, G.: Digital filters in adaptive time-stepping. ACM Tr. on Math.Softw., **Vol.V, No.N** 1–24 (2002)
6. Verhoeven, A.: Automatic control for adaptive time stepping in electrical circuit simulation. MSc Thesis, Technische Universiteit Eindhoven, Eindhoven (2004), Technical Note TN-2004/00033, Philips Research Laboratories, Eindhoven (2004)
7. Verhoeven, A., Beelen, T.G.J., M.L.J. Hautus, Maten, E.J.W. ter: Digital linear control theory applied to automatic stepsize control in electrical circuit simulation. ECMI Proceedings, (2004)
8. Zhuang, M., Mathis, W.: Research on Stepsize Control in the BDF Method for Solving Differential-Algebraic Equations. IEEE Proceedings (ISCAS94) 229–232, (1994)

A General Compound Multirate Method for Circuit Simulation Problems

A. Verhoeven[1], A. El Guennouni[2], E. J. W. ter Maten[3] and R. M. M. Mattheij[1]

[1] Technische Universiteit Eindhoven, averhoev@win.tue.nl
[2] Yacht Technology and Philips Research Laboratories
[3] Technische Universiteit Eindhoven and Philips Research Laboratories

Abstract The "General Compound" multirate methods are attractive integration methods for the transient analysis of mixed analog-digital circuits. From a stability analysis, it follows that they have good stability properties.

1 Introduction

Electrical circuits consist of analog and digital sub-circuits. In analog circuits, the exact values of the voltages and currents are important, but in digital circuits only the logical state is important.

If the mixed analog-digital circuits have to be simulated in high accuracy, it is necessary to simulate the complete circuit on electrical level. In this case, the complete electrical circuit is modeled by the following differential-algebraic equation

$$\frac{d}{dt}[\mathbf{q}(t,\mathbf{x})] + \mathbf{j}(t,\mathbf{x}) = \mathbf{0}, \quad \mathbf{j}(0,\mathbf{x}(0)) = \mathbf{0}, \tag{1}$$

where $\mathbf{x}$ consists of nodal voltages and some currents in the circuit.

Commonly, this IVP is solved by means of implicit integration methods, like BDF-methods. In each iteration all equations are discretized with the same step h_n. Often, parts of electrical circuits have latency or multirate behaviour. Latency means that parts of the circuit are constant during a certain time interval. Multirate behaviour means that some variables are slowly-varying, compared to other variables. In both cases, it would be attractive to integrate the latent or slowly-varying sub-circuit with a larger step.

In Sect. 2 we will show an attractive class of multirate methods for electrical circuits. Next we will study the stability for a two-dimensional linear test equation.

2 Multirate methods for circuits

2.1 Partitioning of variables and equations

For a multirate method it is necessary to partition the variables and equations into an active (A) and a latent (L) part. This can be done by the user or automatically. Then the DAE (1) is equivalent to the coupled system

$$\frac{d}{dt}[\mathbf{q}_A(t,\mathbf{x}_A,\mathbf{x}_L)] + \mathbf{j}_A(t,\mathbf{x}_A,\mathbf{x}_L) = \mathbf{0}, \tag{2}$$

$$\frac{d}{dt}[\mathbf{q}_L(t,\mathbf{x}_A,\mathbf{x}_L)] + \mathbf{j}_L(t,\mathbf{x}_A,\mathbf{x}_L) = \mathbf{0}. \tag{3}$$

It is necessary that the equations (2) and (3) are uniquely solvable. The partitioning is very important, because it affects the stability and the accuracy of the multirate method. Decomposing the DAE (1) into two nearly decoupled parts requires too much effort and hence approximation methods should be used.

2.2 Different multirate algorithms

There are many multirate methods for the system of equations (2),(3) [2, 6]. We will restrict our attention to multirate versions of the Euler Backward method. The time interval $[0,T]$ is discretized into the multirate time-grid $\{t_n = nh = n\frac{H}{q} : n = 0,\dots,N\}$ where the number q is called the multirate factor. The latent equations are integrated with one large step H, but the active equations are integrated with a much smaller step $h = \frac{H}{q}$ on a refinement of $[t_n, t_{n+q}]$.

The "Slowest First" (SF) method (algorithm 1) first integrates (3) with one large step H, while $\mathbf{x}_A$ is approximated by means of extrapolation. Then equation (2) is integrated with the small step h, while $\mathbf{x}_L$ is approximated by linear interpolation.

ALGORITHM 1 *The Slowest First (SF) method*

Solve for $\boldsymbol{x}_L^{n+q}$:

$$\boldsymbol{q}_L(\hat{\boldsymbol{x}}_A^{n+q},\boldsymbol{x}_L^{n+q}) - \boldsymbol{q}_L(\boldsymbol{x}_A^n,\boldsymbol{x}_L^n) + H\boldsymbol{j}_L(\hat{\boldsymbol{x}}_A^{n+q},\boldsymbol{x}_L^{n+q}) = \boldsymbol{0} \tag{4}$$

$$\hat{\boldsymbol{x}}_A^{n+q} - \boldsymbol{x}_A^n = \boldsymbol{0} \tag{5}$$

Solve for $\boldsymbol{x}_A^{n+j+1}$ $(j = 0,\dots,q-1)$:

$$\boldsymbol{q}_A(\boldsymbol{x}_A^{n+j+1},\hat{\boldsymbol{x}}_L^{n+j+1}) - \boldsymbol{q}_A(\boldsymbol{x}_A^{n+j},\hat{\boldsymbol{x}}_L^{n+j}) + h\boldsymbol{j}_A(\boldsymbol{x}_A^{n+j+1},\hat{\boldsymbol{x}}_L^{n+j+1}) = \boldsymbol{0} \tag{6}$$

$$\hat{\boldsymbol{x}}_L^{n+j} - \boldsymbol{x}_L^n - \frac{j}{q}(\boldsymbol{x}_L^{n+q} - \boldsymbol{x}_L^n) = \boldsymbol{0} \tag{7}$$

To improve the stability, the latent part can be integrated by an implicit compound step [4]. This "Compound Step" (CS) method first integrates (2) and (3) together with one large step H, which results in $\mathbf{x}_A^{n+q}$ and $\mathbf{x}_L^{n+q}$. Then only equation (2) is integrated with the small step h, while $\mathbf{x}_L^{n+j}$ is found by linear interpolation. Note that $\mathbf{x}_A^{n+q}$ is twice computed by the "Compound Step" method, which could be used to estimate the error. Another possibility is the "Mixed Compound Step" (MCS) method, which computes $\mathbf{x}_A^{n+1}$ and $\mathbf{x}_L^{n+q}$ simultaneously. This method corresponds to the multirate method for the Rosenbrock-Wanner methods described in [1]. The "Compound Step" has the advantage that it is easier to implement, while the "Mixed Compound Step" method is better scaled.

A generalized version is the "General Compound" (GC) method (algorithm 2) with $\alpha \in \mathbb{R}$. This GC method contains the CS method ($\alpha = 1$) and the MCS method ($\alpha = \frac{1}{q}$).

ALGORITHM 2 *The General Compound (GC) method*

Solve for $\boldsymbol{x}_L^{n+q}$ *and* $\boldsymbol{x}_A^{n+\alpha q}$:

$$\boldsymbol{q}_A(\boldsymbol{x}_A^{n+\alpha q},\hat{\boldsymbol{x}}_L^{n+\alpha q}) - \boldsymbol{q}_A(\boldsymbol{x}_A^n,\boldsymbol{x}_L^n) + \alpha H\boldsymbol{j}_A(\boldsymbol{x}_A^{n+\alpha q},\hat{\boldsymbol{x}}_L^{n+\alpha q}) = \boldsymbol{0} \tag{8}$$

$$\hat{\boldsymbol{x}}_L^{n+\alpha q} - \boldsymbol{x}_L^n - \alpha(\boldsymbol{x}_L^{n+q} - \boldsymbol{x}_L^n) = \boldsymbol{0} \tag{9}$$

$$\boldsymbol{q}_L(\hat{\boldsymbol{x}}_A^{n+q},\boldsymbol{x}_L^{n+q}) - \boldsymbol{q}_L(\boldsymbol{x}_A^n,\boldsymbol{x}_L^n) + H\boldsymbol{j}_L(\hat{\boldsymbol{x}}_A^{n+q},\boldsymbol{x}_L^{n+q}) = \boldsymbol{0} \tag{10}$$

$$\hat{\boldsymbol{x}}_A^{n+q} - \boldsymbol{x}_A^n - \frac{1}{\alpha}(\boldsymbol{x}_A^{n+\alpha q} - \boldsymbol{x}_A^n) = \boldsymbol{0} \tag{11}$$

Solve for $\boldsymbol{x}_A^{n+j+1}$ $(j = 0,\dots,q-1)$:

$$\boldsymbol{q}_A(\boldsymbol{x}_A^{n+j+1},\hat{\boldsymbol{x}}_L^{n+j+1}) - \boldsymbol{q}_A(\boldsymbol{x}_A^{n+j},\hat{\boldsymbol{x}}_L^{n+j}) + h\boldsymbol{j}_A(\boldsymbol{x}_A^{n+j+1},\hat{\boldsymbol{x}}_L^{n+j+1}) = \boldsymbol{0} \tag{12}$$

$$\hat{\boldsymbol{x}}_L^{n+j} - \boldsymbol{x}_L^n - \frac{j}{q}(\boldsymbol{x}_L^{n+q} - \boldsymbol{x}_L^n) = \boldsymbol{0} \tag{13}$$

---

3 Stability analysis of the SF and GC methods

Multirate methods have less good stability properties than ordinary integration methods. Therefore this section contains a stability analysis of the SF and GC methods.

3.1 A test equation

For ordinary integration methods absolute stability can be studied by looking at the scalar test equation $\dot{x} = \lambda x$ with $\lambda \in \mathbb{C}$. For multirate methods with two time-steps h and H, the following (real) linear test equation is studied [5, 6], where x_A and x_L are the active and latent variable respectively.

$$\begin{pmatrix} \dot{x}_A \\ \dot{x}_L \end{pmatrix} = \underbrace{\begin{pmatrix} a_{11} & a_{12} \\ a_{21} & a_{22} \end{pmatrix}}_{\mathbf{A}} \begin{pmatrix} x_A \\ x_L \end{pmatrix} \tag{14}$$

Let x_A^n and x_L^n be the numerical approximations at the time-point $t_n = nh = \frac{n}{q}H$. The multirate method is absolutely stable when x_A^n and x_L^n tend to zero for $n \to \infty$ if A is a stable matrix.

For $q = 1$, the stability behaviour of the multirate methods is independent of the used coordinate system. However, for $q > 1$ the stability does not only depend on the eigenvalues but also on the eigenvectors of the matrix $\mathbf{A}$.

3.2 Analysis of the compound step

In both the SF and the GC methods the latent variable is first integrated. Using constant extrapolation of x_A^n for the SF method we obtain the system

$$\frac{x_L^{n+q} - x_L^n}{H} = a_{21}x_A^n + a_{22}x_L^{n+q}. \tag{15}$$

From the equations (15), it follows that

$$x_L^{n+q} = \rho x_A^n + \sigma x_L^n, \tag{16}$$

where

$$\rho = \frac{a_{21}H}{1-a_{22}H}, \ \sigma = \frac{1}{1-a_{22}H}. \tag{17}$$

For the GC method, we get another complete system of equations for $\bar{x}_A^{n+\alpha q}$ and x_L^{n+q}:

$$\begin{cases} \frac{\bar{x}_A^{n+\alpha q} - x_A^n}{\alpha H} = a_{11}\bar{x}_A^{n+\alpha q} + a_{12}(x_L^n + \alpha(x_L^{n+q} - x_L^n)), \\ \frac{x_L^{n+q} - x_L^n}{H} = a_{21}(x_A^n + \frac{1}{\alpha}(\bar{x}_A^{n+\alpha q} - x_A^n)) + a_{22}x_L^{n+q}. \end{cases} \tag{18}$$

The solution satisfies again equation (16) with different values for ρ and σ:

$$\rho = \frac{a_{21}H + a_{11}a_{21}(1-\alpha)H^2}{1-(\alpha a_{11}+a_{22})H+(a_{11}a_{22}-a_{12}a_{21})\alpha H^2}, \ \sigma = \frac{1-\alpha a_{11}H + a_{12}a_{21}(1-\alpha)H^2}{1-(\alpha a_{11}+a_{22})H+(a_{11}a_{22}-a_{12}a_{21})\alpha H^2}. \tag{19}$$

3.3 Stability conditions

For both methods x_L^{n+j} is estimated for $j \in \{1, \ldots, q-1\}$ employing x_L^n and x_L^{n+q}.

$$\hat{x}_L^{n+j} = x_L^n + \frac{j}{q}(x_L^{n+q} - x_L^n) = \frac{q-j}{q}x_L^n + \frac{j}{q}x_L^{n+q}. \tag{20}$$

Finally, the active part is integrated along the time window $[t_n, t_n + H]$ with q steps h.

$$\frac{x_A^{n+j+1} - x_A^{n+j}}{h} = a_{11} x_A^{n+j+1} + a_{12} \hat{x}_L^{n+j+1}. \tag{21}$$

Equation (21) is equivalent to

$$x_A^{n+j+1} = \frac{\frac{1}{h}}{\frac{1}{h} - a_{11}} x_A^{n+j} + \frac{a_{12}}{\frac{1}{h} - a_{11}} \hat{x}_L^{n+j+1} = \gamma x_A^{n+j} + \delta \hat{x}_L^{n+j+1}, \tag{22}$$

where

$$\gamma = \tfrac{1}{1-a_{11}h}, \delta = \tfrac{a_{12}h}{1-a_{11}h}. \tag{23}$$

For $j \in \{0, \ldots, q-1\}$ we have

$$\begin{aligned} x_A^{n+j+1} &= \gamma x_A^{n+j} + \delta(1 - \tfrac{j+1}{q}) x_L^n + \delta \tfrac{j+1}{q} x_L^{n+q} \\ &= \gamma^{j+1} x_A^n + \textstyle\sum_{k=0}^{j} \gamma^{j-k} \left(\delta(1 - \tfrac{k+1}{q}) x_L^n + \delta \tfrac{k+1}{q} x_L^{n+q} \right). \end{aligned} \tag{24}$$

Inserting (16) into (24) for $j = q - 1$ results in

$$\begin{aligned} x_A^{n+q} &= \gamma^q x_A^n + \left(\textstyle\sum_{k=0}^{q-1} \gamma^{q-1-k} \delta (1 - \tfrac{k+1}{q}) \right) x_L^n + \\ &\quad \left(\textstyle\sum_{k=0}^{q-1} \gamma^{q-1-k} \delta \tfrac{k+1}{q} \right) (\rho x_A^n + \sigma x_L^n) \\ &= \nu x_A^n + \tau x_L^n, \end{aligned} \tag{25}$$

where

$$\nu = \gamma^q + \textstyle\sum_{l=0}^{q-1} \gamma^l \rho \delta (1 - \tfrac{l}{q}), \tau = \textstyle\sum_{l=0}^{q-1} \gamma^l \delta (\tfrac{l}{q}(1-\sigma) + \sigma). \tag{26}$$

From equations (16) and (25) it follows that

$$\begin{pmatrix} x_L^{n+q} \\ x_A^{n+q} \end{pmatrix} = \underbrace{\begin{pmatrix} \sigma & \rho \\ \tau & \nu \end{pmatrix}}_{\mathbf{M}} \begin{pmatrix} x_L^n \\ x_A^n \end{pmatrix}. \tag{27}$$

The methods are A-stable if $\rho(\mathbf{M}) < 1$ for all $H, q > 0$ and stable matrices $\mathbf{A}$ [6]. Let $\phi(\lambda) = \det(\mathbf{M} - \lambda \mathbf{I}) = \lambda^2 - \mathrm{tr}(\mathbf{M})\lambda + \det(\mathbf{M})$, where $\mathbf{M} \in \mathbb{R}^{2\times 2}$. Using the Routh-Hurwitz criterion one can easily show that [3]

$$\rho(\mathbf{M}) < 1 \Leftrightarrow \begin{cases} \phi(-1) = 1 + \mathrm{tr}(\mathbf{M}) + \det(\mathbf{M}) > 0, \\ \phi(0) = \det(\mathbf{M}) < 1, \\ \phi(1) = 1 - \mathrm{tr}(\mathbf{M}) + \det(\mathbf{M}) > 0. \end{cases} \tag{28}$$

Because

$$\begin{aligned} \mathrm{tr}(\mathbf{M}) &= \sigma + \gamma^q + \textstyle\sum_{l=0}^{q-1} \gamma^l \rho \delta (1 - \tfrac{l}{q}), \\ \det(\mathbf{M}) &= \sigma \gamma^q + \sigma \textstyle\sum_{l=0}^{q-1} \gamma^l \rho \delta (1 - \tfrac{l}{q}) - \rho \textstyle\sum_{l=0}^{q-1} \gamma^l \delta (\tfrac{l}{q}(1-\sigma) + \sigma) \\ &= \sigma \gamma^q - \tfrac{\rho\delta}{q} \textstyle\sum_{l=0}^{q-1} \gamma^l l, \end{aligned} \tag{29}$$

we obtain the following three constraints which ensure absolutely stability

$$\begin{aligned} 1 + \mathrm{tr}(\mathbf{M}) + \det(\mathbf{M}) &= 1 + (1+\sigma)\gamma^q + \sigma - \rho\delta \textstyle\sum_{l=0}^{q-1} \gamma^l (\tfrac{2l}{q} - 1) > 0, \\ \det(\mathbf{M}) &= \sigma\gamma^q - \tfrac{\rho\delta}{q} \textstyle\sum_{l=0}^{q-1} \gamma^l l < 1, \\ 1 - \mathrm{tr}(\mathbf{M}) + \det(\mathbf{M}) &= 1 + (\sigma - 1)\gamma^q - \sigma - \rho\delta \textstyle\sum_{l=0}^{q-1} \gamma^l > 0. \end{aligned} \tag{30}$$

Thus we get the following stability conditions for the investigated multirate methods

$$\begin{aligned} (1+\sigma)(1+\gamma^q) - \rho\delta \textstyle\sum_{l=0}^{q-1} \gamma^l (\tfrac{2l}{q} - 1) &> 0, \\ \tfrac{\rho\delta}{q} \textstyle\sum_{l=0}^{q-1} \gamma^l l - \sigma\gamma^q + 1 &> 0, \\ (1-\sigma)(1-\gamma^q) - \rho\delta \textstyle\sum_{l=0}^{q-1} \gamma^l &> 0. \end{aligned} \tag{31}$$

3.4 Asymptotic stability conditions

Because the stability conditions (31) are rather complex, we will derive more compact stability conditions by means of asymptotical analysis.

Stability for $H \to 0$ (fixed q)

The multirate methods are conditionally stable if the stability conditions are valid for $H \to 0$. Therefore we will derive asymptotic approximations of these conditions. It easily follows that $\rho\delta = \frac{a_{12}a_{21}}{q}H^2 + O(H^3)$, $\gamma = 1 + \frac{a_{11}}{q}H + O(H^2)$, $\sigma = 1 + a_{22}H + O(H^2)$ and $\gamma^q = 1 + a_{11}H + O(H^2)$. Using these approximations, we obtain

$$\begin{aligned}(1+\sigma)(1+\gamma^q) - \rho\delta\sum_{l=0}^{q-1}\gamma^l(\tfrac{2l}{q}-1) &= 4 + O(H),\\ \tfrac{\rho\delta}{q}\sum_{l=0}^{q-1}\gamma^l l - \sigma\gamma^q + 1 &= -(a_{11}+a_{22})H + O(H^2),\\ (1-\sigma)(1-\gamma^q) - \rho\delta\sum_{l=0}^{q-1}\gamma^l &= (a_{11}a_{22} - a_{12}a_{21})H^2 + O(H^3).\end{aligned} \tag{32}$$

After inserting these asymptotic expressions into (31), we obtain the following asymptotic stability conditions for (27), which coincide with the ones for (14).

$$\begin{aligned}\mathrm{tr}(\mathbf{A}) &= a_{11} + a_{22} < 0,\\ \det(\mathbf{A}) &= a_{11}a_{22} - a_{12}a_{21} > 0.\end{aligned} \tag{33}$$

Thus the SF method and the GC methods are stable for $H \to 0$ if $\mathbf{A}$ is a stable matrix.

Stability for $q \to \infty$ (fixed H)

If the multirate factor $q \to \infty$, it is necessary that $|\gamma| < 1$ such that $\gamma^q \to 0$. This means that the Euler Backward method is stable for the active part, which is the case if $a_{11} < 0$. Taking the limit $q \to \infty$, we obtain

$$\begin{aligned}(1+\sigma)(1+\gamma^q) - \rho\delta\sum_{l=0}^{q-1}\gamma^l(\tfrac{2l}{q}-1) &\to 1 + \sigma + \rho\delta\tfrac{1}{1-\gamma},\\ \tfrac{\rho\delta}{q}\sum_{l=0}^{q-1}\gamma^l l - \sigma\gamma^q + 1 &\to 1,\\ (1-\sigma)(1-\gamma^q) - \rho\delta\sum_{l=0}^{q-1}\gamma^l &\to 1 - \sigma - \rho\delta\tfrac{1}{1-\gamma}.\end{aligned} \tag{34}$$

This means that for $q \to \infty$ we have the following stability conditions

$$\begin{cases}1 + \sigma + \rho\delta\frac{1}{1-\gamma} > 0,\\ 1 - \sigma - \rho\delta\frac{1}{1-\gamma} > 0.\end{cases} \Leftrightarrow \quad |\frac{\rho\delta}{1-\gamma} + \sigma| < 1. \tag{35}$$

Because $\frac{\delta}{1-\gamma} = -\frac{a_{12}}{a_{11}}$ we get

$$|-\frac{a_{12}}{a_{11}}\rho + \sigma| < 1. \tag{36}$$

Using (17) for the SF method, condition (36) is equivalent to

$$|P_{SF}(H)| = |-\frac{a_{12}}{a_{11}}\rho + \sigma| = \frac{|1 - \frac{a_{12}a_{21}}{a_{11}}H|}{|1 - a_{22}H|} < 1.$$

If this rational function $P_{SF}(H)$ has a negative pole and $\lim_{H\to\infty}|P_{SF}(H)| = \frac{|a_{12}a_{21}|}{|a_{11}a_{22}|} < 1$, the method is unconditionally stable. Thus, if $a_{11} < 0$, $a_{22} < 0$ and $|a_{12}a_{21}| < |a_{11}a_{22}|$, the SF method is unconditionally stable for $q \to \infty$.

Using (19) for the GC methods, condition (36) is equivalent to

$$|P_{GC}(H)| = |-\frac{a_{12}}{a_{11}}\rho + \sigma| = \frac{|1 - (\frac{a_{12}a_{21}}{a_{11}} + \alpha a_{11})H|}{|1 - (\alpha a_{11} + a_{22})H + \alpha(a_{11}a_{22} - a_{12}a_{21})H^2|} < 1.$$

Table 1. Sufficient stability conditions for the SF method and the GC method

SF	GC	GC ($\alpha = 1$)
$a_{11} < 0$	$a_{11} < 0$	$a_{11} < 0$
$a_{22} < 0$	$\alpha a_{11} + a_{22} < 0$	$a_{11} + a_{22} < 0$
$\|a_{12}a_{21}\| < \|a_{11}a_{22}\|$	$-a_{11}a_{22} - 2\alpha a_{11}^2 < a_{12}a_{21}$	$-a_{11}a_{22} - 2a_{11}^2 < a_{12}a_{21}$
	$a_{12}a_{21} < a_{11}a_{22}$	$a_{12}a_{21} < a_{11}a_{22}$

Table 2. Stability of multirate methods ($H = 0.1, q = 10$)

μ	SF	GC ($\alpha = \frac{1}{q}$)	GC ($\alpha = 1$)
-10	√	√	√
-100	-	√	√
-1000	-	-	√

It can be shown that this is the case if $|\frac{a_{12}a_{21}}{a_{11}} + \alpha a_{11}| < |\alpha a_{11} + a_{22}|$, $\alpha a_{11} + a_{22} < 0$ and $\alpha(a_{11}a_{22} - a_{12}a_{21}) > 0$. Because $\alpha a_{11} + a_{22} < 0$, we find

$$\alpha a_{11} + a_{22} < \frac{a_{12}a_{21}}{a_{11}} + \alpha a_{11} < -\alpha a_{11} - a_{22}. \tag{37}$$

From the left inequality in (37) we can derive $a_{22} - \frac{a_{12}a_{21}}{a_{11}} < 0$ or

$$\frac{1}{a_{11}}(a_{11}a_{22} - a_{12}a_{21}) < 0.$$

The other inequality in (37) gives $\frac{a_{12}a_{21}}{a_{11}} < -a_{22} - 2\alpha a_{11}$ or

$$a_{12}a_{21} > -a_{11}a_{22} - 2\alpha a_{11}^2.$$

Because $\alpha > 0$, the GC method is always stable if

$$|a_{12}a_{21}| < |a_{11}a_{22}|. \tag{38}$$

4 Numerical example

Consider for $0 \geq t \geq 10$

$$\begin{pmatrix} \dot{x}_A \\ \dot{x}_L \end{pmatrix} = \begin{pmatrix} -1 & \mu \\ 1 & -1 \end{pmatrix} \begin{pmatrix} x_A \\ x_L \end{pmatrix}, \quad \begin{pmatrix} x_A(0) \\ x_L(0) \end{pmatrix} = \begin{pmatrix} 1 \\ 1 \end{pmatrix}. \tag{39}$$

For $\mu < 0$ it is a stable system with eigenvalues $-1 \pm i\sqrt{-\mu}$. The system is solved by the SF method and the GC method for $\alpha = 1$ and $\alpha = \frac{1}{q}$. Table 2 shows the stability of the numerical results for the different cases. A sufficient stability condition is $|\mu| < 1$, but for the GC methods $\mu > -1 - 2\alpha$ suffices.

5 Conclusions

We have derived stability conditions (Table 1) for the SF and GC methods if $H \to 0$ or $q \to \infty$. These results are presently be generalized to the general multi-dimensional case. The GC methods have the advantage that they do not require that $a_{22} < 0$, but only $\alpha a_{11} + a_{22} < 0$. If **A** is stable, this condition is always satisfied for $\alpha = 1$. Because for the GC methods it is sufficient if $a_{12}a_{21} > -a_{11}a_{22} - 2\alpha a_{11}^2$, large values for α are preferable.

References

1. A. Bartel, M. Günther: *A multirate W-method for electrical networks in state-space formulation*, J. of Comput. and Applied Maths., Vol. 147, pp. 411-425, 2002
2. C.W. Gear, D.R. Wells: *Multirate linear multistep methods*, BIT, 24 (1984), 484-502
3. G.R. G'omez: *Absolute stability analysis of semi-implicit multirate linear multistep methods*, PhD-thesis, Instituto Nacional de Astrofisica, Optica y Electronica, Tonantzintla, Pue, Mexico, 2002
4. A. El Guennouni, A. Verhoeven, E.J.W. ter Maten, T.G.J. Beelen: *Aspects of Multirate Time Integration Methods in Circuit Simulation Problems*, Presented at ECMI-2004, Eindhoven, The Netherlands, 21-25 June 2004
5. A. Kværnø: *Stability of multirate Runge-Kutta schemes*, The tenth Int. Conf. on Diff. Equ., Plovdiv, Bulgaria, Aug. 1999
6. S. Skelboe, P.U. Andersen: *Stability properties of backward Euler multirate formulas*, SIAM J. Sci. Stat. Comput., Vol.10-5, pp. 1000-1009, 1989

References

Stochastic Differential Algebraic Equations in Transient Noise Analysis

R. Winkler

Humboldt–Universität zu Berlin, Institut für Mathematik, 10099 Berlin,
winkler@mathematik.hu-berlin.de

Abstract In this paper we describe how stochastic differential-algebraic equations (SDAEs) arise as a mathematical model for network equations that are influenced by additional sources of Gaussian white noise. We discuss the concepts of weak and strong solutions of SDAEs and give the necessary analytical theory for the existence and uniqueness of strong solutions, provided that the systems have noise-free constraints and are uniformly of index 1. Further, we analyze discretization methods using the concept of strong convergence. Due to the differential-algebraic structure, implicit methods will be necessary. We present adaptations of known schemes for stochastic differential equations (SDEs) that are implicit in the deterministic and explicit in the stochastic part to SDAEs of index 1, in particular we discuss stochastic analogies to the the drift-implicit Euler scheme and the two-step backward differentiation formula (BDF).

1 Problem Formulation

The increasing scale of integration, high tact frequencies and low supply voltages cause smaller signal-to-noise-ratios. In several applications the noise influences the system behavior in an essentially nonlinear way such that linear noise analysis is no longer satisfactory. A possible way out is given by transient noise analysis. Here a circuit model that includes also noisy elements has to be considered and to be simulated in time-domain.

We deal with the thermal noise of resistors as well as the shot noise of semiconductors that are modeled by additional sources of additive or multiplicative Gaussian white noise currents that are shunt in parallel to the ideal, noise-free elements (see Fig. 1). Note, that modeling the internal noise of the elements as external noise sources was originally justified only for linear elements and reciprocal networks.

Nyquist's theorem (see e.g. [B96, DS98, WM98]) states that the current through an arbitrary linear resistor having a resistance R, maintained in thermal equilibrium at a temperature T, can be described as the sum of the deterministic current and a Gaussian white noise process with spectral density $S_{th} := \frac{2kT}{R}$, where k is Boltzmann's constant. Hence, the additional current is modeled as

$$I_{th} = \sigma_{th} \cdot \xi(t) = \sqrt{\frac{2kT}{R}} \cdot \xi(t) \ ,$$

Fig. 1. Thermal noise of a resistor and shot noise of a pn-junction

where $\xi(t)$ is a standard Gaussian white noise process. In [WM98] a thermo-dynamical foundation to apply this model to mildly nonlinear resistors and reciprocal networks is given.
Shot noise of pn-junctions, caused by the discrete nature of current due to the elementary charge, is also modeled by a Gaussian white noise process. Here the spectral density is proportional to the current I through the pn-junction: $S_{shot} := q_e|I|$, where q_e is the elementary charge. If the current through the pn-junction is described by a characteristic $I = g(u)$ in dependence on a voltage u, the additional current is modeled by

$$I_{shot} = \sigma_{shot}(u) \cdot \xi(t) = \sqrt{q_e|g(u)|} \cdot \xi(t),$$

where $\xi(t)$ is a standard Gaussian white noise process. For a discussion of the model assumptions we refer to [B96, DS98, WM98].

Using the charge-oriented modified nodal analysis (MNA) one formally obtains specially structured differential-algebraic equations with stochastic perturbation terms (see [W03, W02], and for the deterministic case [ET00, GF99]):

$$A\frac{d}{dt}\bar{q}(x(t)) + f(t, x(t)) + \sum_{r=1}^{m} g^r(t, x(t))\xi^r(t) = 0, \tag{1}$$

where x is the vector of unknowns consisting of the nodal potentials and the branch currents of current-controlled elements (inductances and voltage sources). The vector $\bar{q}(x)$ consists of the charges of capacitances and the fluxes of inductances. The leading constant matrix A is a singular incidence matrix which is determined by the topology of the network, $f(t, x)$ describes the impact of the static elements, $g^r(t, x)$ denotes the vector of noise intensities for the r-th noise source, and $\xi(t)$ is an m-dimensional vector of independent Gaussian white noise processes. One has to deal with a large number of equations as well as of noise sources. Compared to the other quantities the noise intensities $g^r(t, x)$ are small.

2 Solutions of SDAEs

In established numerical integrations the terms $\bar{q}(x)$ are treated as extra variables (cf. [GF99]) to guarantee charge-conservation. One can view equation (1) as a compact formulation of the larger system

$$\begin{aligned} A\tfrac{d}{dt}q(t) + f(t, x(t)) + \sum_{r=1}^{m} g^r(t, x(t))\xi^r(t) &= 0, \\ q(t) &= \bar{q}(x(t)), \end{aligned} \tag{2}$$

with unknowns (q, x). We understand (1) resp. (2) as a stochastic integral equation

$$\begin{aligned} AQ(s)|_{t_0}^{t} + \int_{t_0}^{t} f(s, X(s))ds + \sum_{r=1}^{m} \int_{t_0}^{t} g^r(s, X(s))dB^r(s) &= 0\ , \\ Q(t) &= \bar{q}(X(t)), \end{aligned} \tag{3}$$

where the second integral is an Itô-integral, and B denotes an m-dimensional Wiener process (or Brownian motion) given on the probability space $(\Omega, \mathcal{F}, P)$ with a filtration $(\mathcal{F}_t)_{t \geq t_0}$. Due to the singularity of the incidence matrix A and the special structure of the system (3), it involves constraints. We call the system (3) a stochastic differential-algebraic equation (SDAE). The solution is a stochastic process $X(t, \omega)$ depending on the time t and on the random sample ω. The value of the solution process at fixed time t is a random variable $X(t, \cdot) = X(t)$ whose argument ω is usually not written. For a fixed sample ω representing a fixed realization of the driving Wiener noise, the function $X(\cdot, \omega)$ is called a realization or a path of the solution. Due to the influence of the Gaussian white noise, typical paths of the solution are nowhere differentiable. A process is called a strong solution of (3) if it is adapted to the filtration (i.e., it does not depend on future information), and if, with probability 1, its sample paths are continuous, the integrals in (3) exist and (3) is fulfilled.

The theory of stochastic differential equations distinguishes between the concepts of weak and strong solutions. The concept of weak solutions is applied if one is interested only in the time-evaluation of

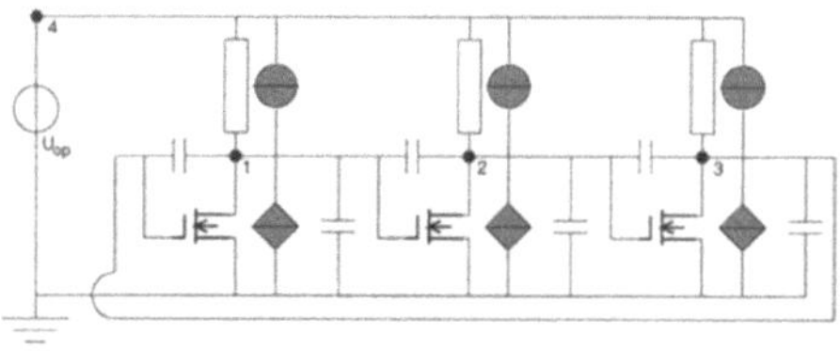

Fig. 2. Thermal noise sources in a MOSFET ring-oscillator model

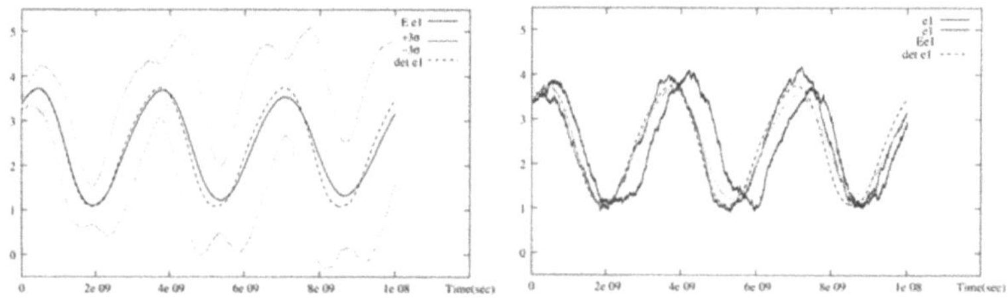

Fig. 3. Statistical parameters and solution paths for the nodal potential at node 1

the distribution of the solution. One then computes moments of the solution process like expectation and variance. The strong solution concept is applied if one is interested in solution paths.

We illustrate the output of both concepts by simulation results for a noisy MOSFET-ring-oscillator (cf. [KRS92, P00]). Only thermal noise in the MOSFETs and in the resistors is considered. The corresponding circuit diagram is given in Fig. 2. To make the differences between the solutions of the noisy and the noise-free model for this simple example more visible, we dealt with a system where the diffusion coefficients had been scaled by a factor of 1000.

We plotted values of the nodal potential at node 1 versus the time. The solution of the noise-free system is given by the dashed line. In the left of Fig. 3 we present quantities obtained from moments of the solution: the mean μ (black solid line) and the boundaries of the confidence interval $[\mu - 3\sigma, \mu + 3\sigma]$ (lightblue solid lines), where σ is the estimate for the standard deviation. The mean appears damped and differs considerably from the noiseless, deterministic solution. In the right we present two sample paths (dark solid lines) together with the mean μ (lightblue solid line). They indicate that the large deviations from the mean seen in the left picture are mainly due to phase noise.

Transient noise analysis computes paths of the solution. It allows the computation of moments in a post-processing step. The necessary numerical analysis therefore uses the concept of strong solutions and of strong convergence of approximations. Combining knowledge from the theory of stochastic differential equations and the theory of differential-algebraic equations the existence and uniqueness of a strong solution of the SDAE (3) is proved in [W03] under the following conditions, which we suppose also in the next chapter: First, assume that the deterministic MNA-system

$$\begin{aligned} A\tfrac{d}{dt}q(t) + f(t, x(t)) &= 0, \\ q(t) &= \bar{q}(x(t)), \end{aligned} \tag{4}$$

is globally an index 1 differential algebraic equation (DAE) in the sense that the constraints are regularly and globally uniquely solvable for the algebraic variables. Second, assume that the functions $f, G = (g^1, \ldots, g^r), \bar{q}$ describing the SDAE (3) are globally Lipschitz-continuous with respect to x and continuous with respect to t, and, third, assume that the SDAE (3) possesses noise-free constraints. This guarantees a solution process that is not directly affected by the white noise process, which is true for the MNA system if there are always capacitances in parallel to a noise source. This is quite restrictive in the actual noise modeling. Nevertheless, one can also handle many situations where this condition is violated. Often noisy constraints are only needed for the determination of algebraic solution components that do not interact with the dynamical ones. Future work should be dedicated to the classification of such situations.

3 Integration schemes for index 1 SDAEs

We present adaptations of known schemes for stochastic differential equations (SDEs) (cf. e.g. [Hi01]) that are implicit in the deterministic and explicit in the stochastic part to the SDAE (3). Designing the methods such that the iterates have to fulfill the constraints of the SDAE at the current time-point is the key idea to adapt known methods for SDEs to (3). This is realized by an implicit Euler or BDF-discretization of the deterministic part.

The noise densities given in Sec. 1 contain small parameters, in fact the square root of Boltzmann's constant $k = 1.3806 \times 10^{-23}$ for thermal noise and of the elementary charge $q_e = 1.602 \times 10^{-19}$ for shot noise. To exploit the smallness of the noise in the analysis of the discretization errors we express the noise densities in the form

$$G(t,x) := \epsilon \tilde{G}(t,x), \quad \epsilon \ll 1. \tag{5}$$

3.1 Drift-Implicit Euler-Maruyama Scheme

On the deterministic grid $0 = t_0 < t_1 < \ldots < t_N = t_{\text{end}}$ the drift-implicit Euler Maruyama scheme for (3) is given by

$$A\frac{\bar{q}(X_\ell) - \bar{q}(X_{\ell-1})}{h_\ell} + f(t_\ell, X_\ell) + G(t_{\ell-1}, X_{\ell-1})\frac{\Delta B_\ell}{h_\ell} = 0, \quad \ell = 1, \ldots, N, \tag{6}$$

where $h_\ell = t_\ell - t_{\ell-1}$, $\Delta B_\ell = B(t_\ell) - B(t_{\ell-1})$, and X_ℓ denotes the approximation to $X(t_\ell)$. Realizations of ΔB_ℓ are simulated as $N(0, h_\ell I)$-distributed random variables. The Jacobian of (6) is the same as in the deterministic setting. In general, the Jacobian is solution-dependent and differs from path to path. The scheme (6) for the SDAE (3) possesses the same convergence properties as the drift-implicit Euler-Maruyama scheme for SDEs (see [DW03, SD98, W03, W02]). In general, its order of strong convergence is only 1/2, i. e.,

$$\|X(t_\ell) - X_\ell\|_{L_2(\Omega)} := (E|X(t_\ell) - X_\ell|^2)^{1/2} \leq c \cdot h^{1/2}, \quad h := \max_{\ell=1,\ldots,N} h_\ell,$$

holds for the mean square norm of the global errors. For additive noise, i. e. $G(t,x) = G(t)$, the order of strong convergence is 1, for small noise (5) the error is bounded by $O(h + \epsilon^2 h^{1/2})$ (see [RW03b], or [MT97] for related results).

The smallness of the noise also allows special estimates of local error terms, which can be used to control the step-size. The local error for the Euler- Maruyama scheme applied to SDEs with small noise is analyzed in [RW03b]. As long as step-sizes with

$$h_\ell \gg \epsilon^2$$

are used, the dominating local error (per unit step) term of (6) is

$$\eta_\ell := \frac{1}{2}\|A^-(f(t_\ell, X_\ell) - f(t_{\ell-1}, X_{\ell-1}))\|_{L_2(\Omega)} = \mathcal{O}(h_\ell + \epsilon h_\ell^{1/2}), \tag{7}$$

where A^- denotes a suitable pseudo-inverse of A. For $\epsilon \to 0$ it approaches the known error estimate in the deterministic setting. If an ensemble of solution paths is computed simultaneously, the estimate η_ℓ can be approximated and may be used to control the local error corresponding to a given tolerance. This results in an adaptive step-size sequence that is uniform for all solution paths.

3.2 Higher order schemes for small noise SDAEs

Improving the (asymptotic) order of strong convergence of numerical schemes for SDEs or SDAEs would require to include more information on the driving noise process than only the increments of the Wiener process. The so-called Milstein-schemes (see [RW03a, P00]) which possess strong order 1 require derivatives of the noise densities and double stochastic integrals. In an application with a large number of small noise sources one has to pay much for a merely theoretical gain in accuracy.

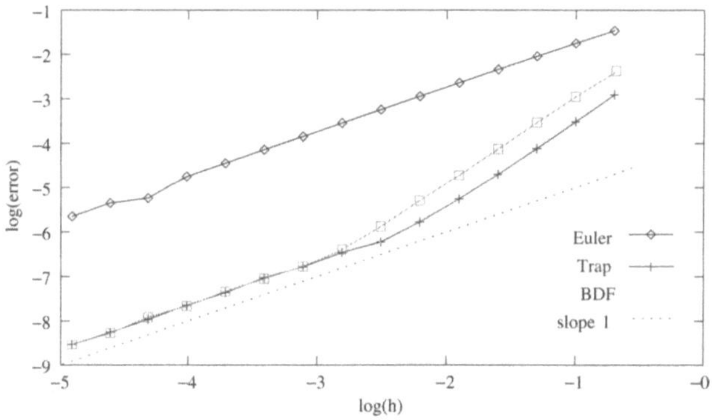

Fig. 4. Global error vs. step-size in logarithmic scale

When the noise is small, one can believe that the stochastic system, though of a completely different analytical character, has a solution that is somehow 'close' to a deterministic one. Then one can hope that for step-sizes that are not asymptotically small the error behavior is still dominated by the deterministic terms. In addition, one might expect that the stochastic scheme inherits some of the qualitative properties of the deterministic methods. Thus motivated, linear two-step Maruyama methods for SDEs have been analyzed in [BW03, BW04]. Here we present the two-step BDF-Maruyama scheme for the SDAE (3):

$$0 = A\frac{\bar{q}(X_\ell) - \frac{4}{3}\bar{q}(X_{\ell-1}) + \frac{1}{3}\bar{q}(X_{\ell-2})}{h} + \frac{2}{3}f(t_\ell, X_\ell) + G(t_{\ell-1}, X_{\ell-1})\frac{\Delta B_\ell}{h} - \frac{1}{3}G(t_{\ell-2}, X_{\ell-2})\frac{\Delta B_{\ell-1}}{h}\,.$$

For small noise the global errors of this scheme are bounded by $\mathcal{O}(h^2 + \epsilon h + \epsilon^2 h^{1/2})$. Below we illustrate this by simulation results for the scalar linear SDE

$$X(t) = 1 + \int_0^t \alpha X_\tau d\tau + \int_0^t \epsilon X_\tau dB_\tau,\ t \in [0, 1]$$

with the geometric Brownian motion $X(t) = \exp((\alpha - \frac{1}{2}\epsilon^2)t + \epsilon W(t))$ as exact solution. We have chosen the parameters $\alpha = -1,\ \epsilon = 10^{-3}$. This example is rather simple, but shows the potential of two-step schemes for small noise SDEs and SDAEs very well. The same errors can be observed for an index 1 SDAE whose inherent dynamics is described by this SDE. More experimental results are reported in [BW03, BW04]. In Fig. 4 we show the mean-square of the global errors vs. the step-size for the implicit Euler scheme, the stochastic trapezoidal rule and the two-step BDF-Maruyama-scheme for 100 computed paths in logarithmic scale with base 10. The slopes of the lines indicate the observed order of convergence of the schemes. For comparison a line with slope 1 is given. The error of the Euler scheme shows order 1 behavior in the considered range of step-sizes. The two schemes with deterministic order 2 show two different regions. For larger step-sizes the deterministic part of the error of the form $\mathcal{O}(h^2)$ dominates and leads to order 2 behavior, whereas for smaller step-sizes the error is dominated by a term $\mathcal{O}(\epsilon h)$. In both regions the errors are considerably smaller than those of the Euler-scheme. The theoretical order 1/2 of the schemes would be observed only for much smaller step-sizes.

Acknowledgement

We wish to thank Uwe Feldmann and Georg Denk (Infineon Technologies) for the excellent cooperation and two anonymous referees for their careful reading and helpful comments.

References

[B96] Blum, A.: Elektronisches Rauschen. Teubner (1996)

[BW03] Buckwar, E., Winkler, R.: Multi-step methods for SDEs and their application to problems with small noise, Preprint 03-17 Institut für Mathematik, Humboldt-Universität Berlin (2003), and submitted

[BW04] Buckwar, E., Winkler, R.: On two-step schemes for SDEs with small noise. PAMM, **4(1)**, 15–18, (2004)

[DS98] Demir, A., Sangiovanni-Vincentelli, A.: Analysis and simulation of noise in nonlinear electronic circuits and systems. Kluwer Academic Publishers (1998)

[DW03] Denk, G., Winkler, R.: Modeling and simulation of transient noise in circuit simulation. In Proceedings of 4th MATHMOD, Vienna, Feb. 5-7 (2003) and to appear in Mathematical and Computer Modelling of Dynamical Systems (MCMDS)

[ET00] Estevez Schwarz, D., Tischendorf, C.: Structural analysis for electronic circuits and consequences for MNA. Int. J. Circ. Theor. Appl., **28**, 131–162 (2000)

[GF99] Günther,F., Feldmann, U.: CAD-based electric-circuit modeling in industry I. mathematical structure and index of network equations. Surv. Math. Ind., **8**, 97–129 (1999)

[Hi01] Higham, D.J.: An algorithmic introduction to numerical simulation of stochastic differential equations. SIAM Review, **43**, 525–546 (2001)

[KRS92] Kampowsky, W., Rentrop, P. and Schmidt, W., Classification and numerical simulation of electric circuits. Surv. Math. Ind., **2**, 23–65 (1992)

[MT97] Milstein, G.N., Tretyakov, M.V.: Mean-square numerical methods for stochastic differential equations with small noise. SIAM J.Sci. Comput., **18**, 1067–1087 (1997)

[P00] Penski, C.: A new numerical method for SDEs and its application in circuit simulation. J. Comput. Appl. Math., **115**, 461–470 (2000)

[RW03a] Römisch, W., Winkler, R.: Stochastic DAEs in circuit simulation. In: Antreich,K., Bulirsch, R., Gilg, A., Rentrop, P. (eds) Mathematical Modeling, Simulation and Optimization of Integrated Circuits. Birkhäuser, 303–318 (2003)

[RW03b] Römisch, W., Winkler, R.: Stepsize control for mean-square numerical methods for stochastic differential equations with small noise, Preprint 03-8, Institut für Mathematik, Humboldt-Universität Berlin (2003) and submitted

[SD98] Schein, O., Denk, G.: Numerical solution of stochastic differential-algebraic equations with applications to transient noise simulation of microelectronic circuits. J. Comput. Appl. Math., **100**, 77–92 (1998)

[WM98] Weiß, L., Mathis, W.: A thermodynamical approach to noise in nonlinear networks. Int. J. Circ. Theor. Appl., **26**, 147–165 (1998)

[W03] Winkler, R.: Stochastic differential algebraic equations of index 1and applications in circuit simulation. J. Comp. Appl. Math., **157**, 477–505 (2003)

[W02] Winkler, R.: Stochastic DAEs in Transient Noise Simulation. In Springer Series 'Mathematics in Industry',Proceedings of 'Scientific Computing in Electrical Engineering', June, 23rd - 28th 2002, Eindhoven (2004), 408–415

Part III

Electromagnetism

Finite Element Modelling of Electrical Machines and Actuators*

D. Rodger, H.C. Lai, P.C. Coles, R.J. Hill-Cottingham, P.K. Vong, S. Viana

Department of Electronic and Electrical Engineering, University of Bath, Claverton Down, Bath, BA2 7AY, UK,
d.rodger@bath.ac.uk

Abstract Numerical models of electrical machines and actuators are becoming increasingly sophisticated. Many electrical machines move and this is taken into account using a Lagrange sliding interface. In addition to this, electrical machines are increasingly being modelled using coupled systems of equations, for example thermal and circuit effects are considered as well as the electromagnetic field equations governing the machine. In this presentation some of the methods and some case studies are described.

1 Introduction

Increasingly, electrical machines are modelled using a set of coupled equations. A familiar example may be the time transient start up of a squirrel cage induction machine connected to a non linear, speed dependent load. The machine runs at a slip which depends on the load. The efficiency of a given induction machine depends on the slip. Efficiency is important as about 50% of generated electricity in an industrialised economy is used in electric motors.

Many machines are thermally limited, so the solution of coupled electromagnetic-thermal systems of equations are required. These systems may be strongly or weakly coupled. They are said to be strongly coupled if each field has an influence on the other, for instance heating may affect the electrical resistivity and this can affect the eddy current paths and hence different parts of the device can become heated. If the heating is not sufficient to change the electrical conductivity in a significant way, the fields could be thought to be weakly coupled.

Many electrical machines are also connected to an electrical circuit, this may include non linear components. In nearly all cases, the electromagnetic-circuit system would be strongly coupled.

If a problem is weakly coupled then, as a last resort, separate codes can be used to solve each problem at each time step of, for instance, a time transient problem. If necessary, input data can be swopped between codes at the end of a time step. However, it seems unlikely that this strategy will be of much use in the case of strongly coupled problems, in which each code will have some influence on the other(s). In this case it is probable that one code should be used, solving all equations simultaneously and satisfying all the different field equations at each time step. We now present some formulations and examples.

2 3D Finite Element Formulations

2.1 Electromagnetic Equations

The non-conducting and conducting regions are often modelled using the magnetic scalar potential, ψ, and the magnetic vector potential, $\mathbf{A}$, respectively.

Non Conducting Regions

Non conducting regions are modelled using magnetic scalar potentials, either the total scalar ψ, defined as $\mathbf{H}_T = -\nabla\psi$, or the reduced scalar ϕ, defined as $\mathbf{H}_T = -\nabla\phi + \mathbf{H}_S$. Here $\mathbf{H}_T$ is the total magnetic field intensity and $\mathbf{H}_S$ is the field defined as $\nabla\times\mathbf{H}_S = \mathbf{J}_S$, where $\mathbf{J}_S$ is the source current density.

*Invited paper at SCEE-2004

Both scalars give rise to a Laplacian type equation which has to be solved.

$$\nabla \cdot \mu \nabla \psi = 0 \tag{1}$$

Voltage forced conditions can be modelled using this technique [1].

Conducting Regions

Fields in conductors can be modelled using **A**, the magnetic vector potential, and V, the electric scalar potential.
If movement is involved, the formulations are different depending on whether the moving member is smooth in the direction of motion, ie on whether the moving media cross section normal to the direction of motion is invariant as described below.

Smooth Moving Conductor Regions

If the moving conductors are smooth and the region moves at a constant velocity, the induced motional emf effect can be taken into account by including a velocity term $\mathbf{u} \times \mathbf{B}$, where $\mathbf{u}$ is the velocity (the Minkowski transformation). Using $\mathbf{B} = \nabla \times \mathbf{A}$ and $\mathbf{E} = -\frac{\partial \mathbf{A}}{\partial t} - \nabla V + \mathbf{u} \times \nabla \times \mathbf{A}$, where $\mathbf{u}$ is the material velocity, Ampère's law and the divergenceless $\mathbf{J}$ condition we have:

$$\nabla \times \frac{1}{\mu} \nabla \times \mathbf{A} = \sigma \left(-\frac{\partial \mathbf{A}}{\partial t} + \mathbf{u} \times \nabla \times \mathbf{A} - \nabla V \right) \tag{2}$$

$$\nabla \cdot \sigma \left(\frac{\partial \mathbf{A}}{\partial t} - \mathbf{u} \times \nabla \times \mathbf{A} + \nabla V \right) = 0 \tag{3}$$

Where appropriate [2], it is possible to dispense with V from the above set of equations. Substituting $V = \mathbf{A} \cdot \mathbf{u}$ in Ampère's law yields

$$\nabla \times \frac{1}{\mu} \nabla \times \mathbf{A} = \sigma \left(-\frac{\partial \mathbf{A}}{\partial t} - (\mathbf{u} \cdot \nabla)\mathbf{A} - (\mathbf{A} \cdot \nabla)\mathbf{u} - \mathbf{A} \times (\nabla \times \mathbf{u}) \right) \tag{4}$$

Now a solution of (4) involving only $\mathbf{A}$ is required, as V is specified in terms of $\mathbf{A}$.

Non Smooth Moving Conductor Regions

If the cross section of the moving object is not invariant in the direction of motion a time transient solution must be carried out. The moving object is modelled by a mesh which slides relative to the stationary parts. The independent meshes can then be coupled at their common interface using Lagrange Multipliers. Stationary and moving parts can be meshed up independently and then brought together. The distribution and density of nodes on the interface need not be the same.

If we are required to make a functional $\Pi(\phi)$ stationary subject to the constraints $C(\phi) = 0$ on a surface Γ_λ, we can introduce this constraint by forming a new functional

$$\Pi' = \Pi + \int_{\Gamma_\lambda} \lambda C(\phi) dS \tag{5}$$

where λ is a set of Lagrange Multipliers. [3],[4] Care must be taken when a nodal finite element (FE) scheme is used to solve a rotating system [5].

2.2 Coupled Equations

In addition to the above electromagnetic equations, often it is necessary to couple in the effects of other fields and external effects.

Electrical circuits:

We would usually solve coupled electrical circuits within the finite element matrix as a fully coupled problem [1]. It may in some circumstances be sufficient to run the finite element program under different conditions to build 'look up' tables which would then be used by a separate circuit solver. This strategy would become intractible if there were many separate coils, non linear effects etc.

The transient heat diffusion equation:

Here we must solve

$$\nabla \cdot \kappa \nabla T + \dot{q} = \rho C(\frac{\partial T}{\partial t} + \mathbf{u} \cdot \nabla T) \tag{6}$$

where T, $\mathbf{u}$, κ, ρ,C and $\dot{q}$ are the unknown temperature, velocity vector, thermal conductivity, material density, specific heat capacity and heat source respectively.

In electrical problems the heat source is typically the ohmic losses. A coupled problem involves the simultaneous solution of the electromagnetic equations and (6). These must both be satisfied at each time step of a transient solution. [6],[7] Usually convection and/or radiation effects are important and these can be included as surface terms in (6).

During the iterative process, the electromagnetic problem can be solved again if there are significant changes in the media electrical conductivity due to temperature rise. The temperature dependent convection coefficient can be calculated from empirical correlations such as can be found listed in [8],[9] and [10].

The electrical conductivity can also be temperature dependent and the resistivity ρ can be approximated linearly as

$$\rho(T) = \rho_o(T)[1 + \alpha(\Delta T)] \tag{7}$$

Rigid body motion:

The differential equation representing the mechanical components of a typical rotating system takes the following form,

$$\frac{d\omega_r}{dt} = \frac{1}{J}(T_e - T_l - T_f) \tag{8}$$

where J is the combined inertia of the machine rotor and connected load. T_e is the developed electromagnetic torque, T_l the load torque and T_f the friction torque. The torques T_l and T_f would usually be non linear functions.

3 Meshless Formulations

The finite element method has been successfully used to model electromagnetic systems, but its application to complex geometries still presents some difficulties often related with mesh generation. In recent years, *meshless methods*, a new numerical technique, have became popular in engineering systems modelling.

This technique eliminates the use of a mesh structure. Instead it uses only a random distribution of nodes to model the fields. Meshless methods have been applied successfully to computational mechanics, where the *mesh-free* characteristic has proved to be very useful, especially for modelling discontinuities and moving boundaries. Some of the most widely used methods are the Smooth Particle Hydrodynamics (SPH) method, the Element Free Galerkin (EFG) method, the Meshless Local Petrov-Galerkin (MLPG) method, the Point Interpolation Method (PIM), and the Reproducing Kernel Particle method (RKPM).

Owing to its attractive *mesh-free* characteristic the application of meshless methods to electromagnetic systems has been investigated [11]. In the following section a brief introduction to the Meshless Local Petrov-Galerkin (MLPG) method is presented and its main characteristics outlined.

3.1 Meshless Local Petrov Galerkin (MLPG)

Even though all meshless methods found in the literature claim to be mesh free, some of them employ a background grid in the integration process [12]. In contrast, the MLPG can be described as a truly meshless method, as it does not require any kind of mesh [13].

Consider the simple 2D magnetostatic problem described in Fig. 3.1, where a set of uniform points, $\mathbf{x} = [x, y]$, was used to discretise the domain. This problem can be represented by the following equation:

$$\nabla \cdot \frac{1}{\mu} \nabla A = 0 \tag{9}$$

The boundary conditions are given by $A = 0$ at $x = 0$ and $A = 1$ at $x = 0.1$ and $\frac{\partial A}{\partial n} = 0$ at $y = 0$ and $y = 0.1$. Here A is the z component of the magnetic vector potential.

The Meshless Local Petrov Galerkin builds the approximation by defining a small local domain around each one of the nodes and then satisfying the weak form locally. The integration points or Gauss points are then created inside this local domain and also along its boundaries. Because there is no mesh structure the *relationship* between the

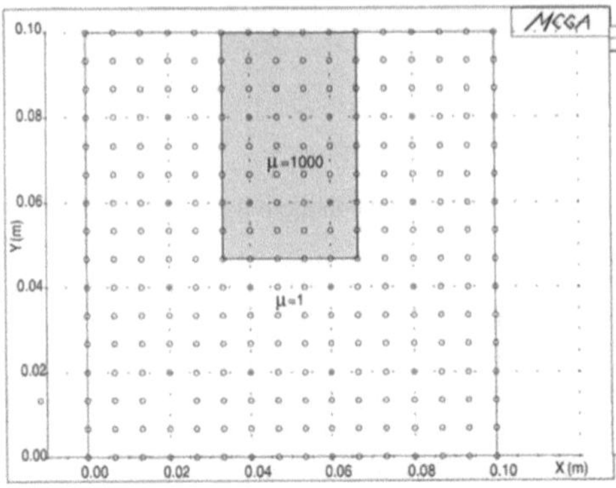

Fig. 1. Magnetostatic problem

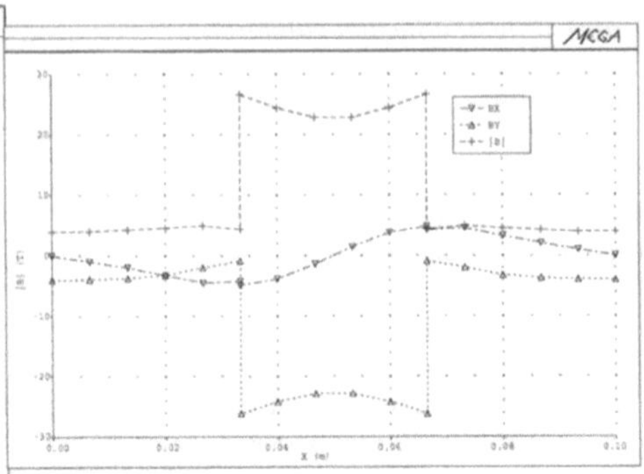

Fig. 2. The magnetic flux density computed at $y = 0.06\ m$ from $x = 0 \rightarrow x = 0.1\ m$)

nodes is determined by the influence that each node imposes on each of the other local domains, attributed by the integration points.

The Petrov-Galerkin method makes it possible to use different functions in the weak form, such as Radial Basis functions, Gaussian weight functions, Shepard functions, the Dirac Delta function and the Heaviside step function. In this work a combination of the first and the last functions was used. This results in a local system of boundary equations which simplifies the integration process, due to the fact that only the Gauss points along the local boundary are required in the process.

Finally the global system is obtained by simply evaluating the contribution of each one of the nodes. The boundary conditions are easily imposed in this method and the implementation is similar to that used in the finite element method.

The numerical procedure for implementing the method is given as:

(i) Define a finite number of nodes to describe the physical problem
(ii) Determine the local domain and its boundaries
(iii) Loop over all the nodes
  a) Create the Gauss points along the node's local boundaries
  b) Loop over the Gauss points
    - Define the influence that each of the nodes imposes on the Gauss point
    - Determine the shape function and its derivatives for those nodes
    - Evaluate the numerical boundary integral in the global system
  c) End of the Gauss points loop
(iv) End of the node loop
(v) Solve the final system

Figure 2 shows the magnetic flux density $\mathbf{B}$ and its components ($\mathbf{B}_x$ and $\mathbf{B}_y$) observed along a straight line of 16 nodes crossing the material interface at $y = 0.06m$ with $x = 0 \rightarrow x = 0.1m$). The results are similar to those found by a conventional finite element system.

4 Examples of Coupled Problems

The examples below display either strong or weak coupling.

4.1 Rigid Body Motion in a Linear Actuator

It is often difficult to model the common type of solenoidal moving plunger actuators in which the air gap varies from maximum stroke down to zero (Fig. 3). The difficulty is that the air gap is constantly varying down to zero. While remeshing techniques can be used to solve this problem, difficulties can arise and it is usually more convenient to use the method described here. The FE scheme allows two topologically unconnected 2D finite element meshes to slide over and overlap each other, whilst still coupled together in a consistent manner. Lagrange multipliers are used to connect the meshes on the sliding surfaces, while the 'shrinking air gap' region is handled using an overlapping element scheme.

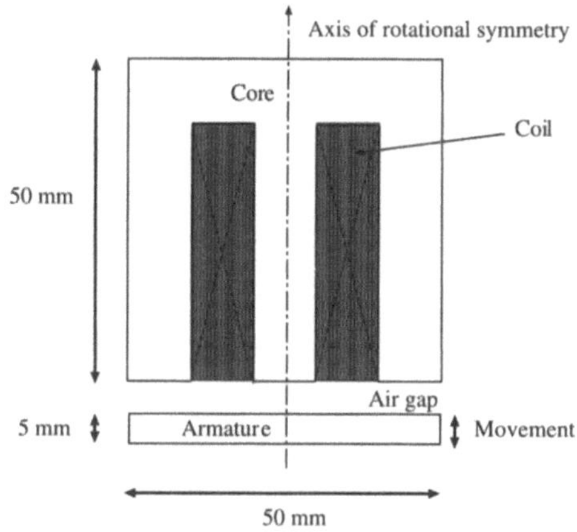

Fig. 3. A simple linear actuator

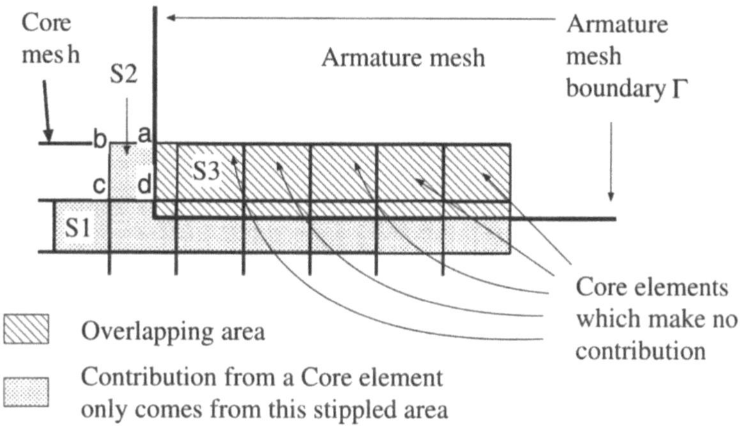

Fig. 4. Master and slave overlapping elements

The overlapping FE scheme [14] is briefly described here. In 2D, fields can be modelled using the magnetic vector potential, **A**.

If the armature of the actuator in Fig. 3 is represented by one FE mesh while the core and the surrounding air is represented by another FE mesh, then applying the usual finite element and Galerkin procedure will result in two sets of equations. There is no coupling between these two sets of equations and therefore the two meshes are still unconnected electromagnetically.

The two sets of equations can be coupled together by using the Lagrange sliding interface technique as before. It is also necessary to ensure that only elements of the armature mesh are used to model the overlapping area of the meshes. This can be achieved by using the concept of master and slave elements. In this scheme, master elements always take precedence over slave elements. The armature mesh is assigned the master mesh while the core-air mesh the slave mesh. Referring to Fig. 4, slave elements such as S1 which do not overlap with any master elements are treated in the usual manner. That is, the contribution to the system matrix from these elements are calculated over the whole element.

When a slave element, such as S2, overlaps partially with the master mesh, its contribution to the matrix will only be calculated over the portion of the element which does not overlap with the master mesh. For S2, this portion is the shaded area *abcd*. Slave elements, such as S3, that overlap totally with the master mesh become *null* elements. A *null* element makes no contribution to the system and is effectively decoupled from the model.

The equations of dynamic motion (8) are used to estimate the displacement of the armature during each time step.

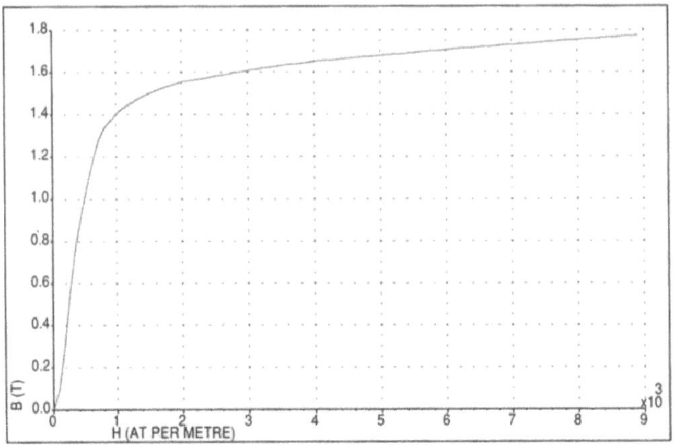

Fig. 5. Measured BH curve of armature material

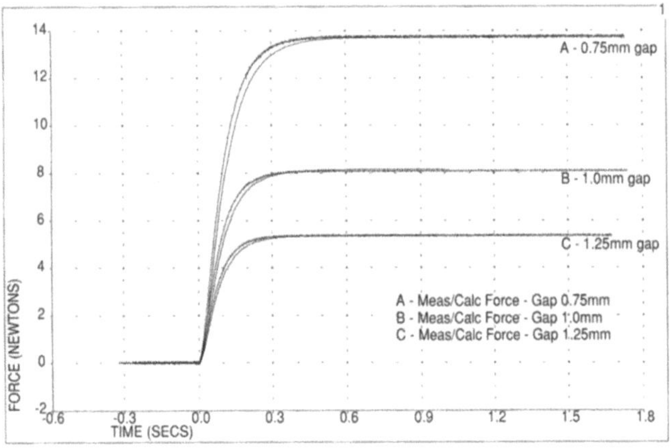

Fig. 6. Transient force at different gaps

The experimental actuator

An actuator was built to verify the FE results. It has rotational symmetry and consists of a solid steel core and a moving armature. A solenoidal type coil is fitted into the core. The dimensions of the solid core and moving armature are shown in Fig. 3. The material B-H curve is shown in Fig. 5. Force and current for the stationary actuator at a step 2V input voltage is shown in Figs. 6 and 7 respectively. Figs. 8 and 9 show the measured and calculated dynamic position and current as the device closes from three different initial air gaps.

4.2 Heating in an Induction Machine

A disk induction machine is shown in Fig. 10, dimensions are given in Table(1). The convection coefficients $\hbar_\omega$ used for the rotor are listed below [10]:

- rotating cylindrical drum

$$\hbar_\omega = \frac{\kappa(T)}{D}[0.11(0.5Re^2 + GrPr)^{0.35}] \tag{10}$$

- rotating disk in laminar flow

$$\hbar_\omega = \frac{\kappa(T)}{D}[0.36(\frac{\omega r^2}{\nu})^{0.5}] \tag{11}$$

valid for $\frac{\omega r^2}{\nu} < 2 \times 10^5$

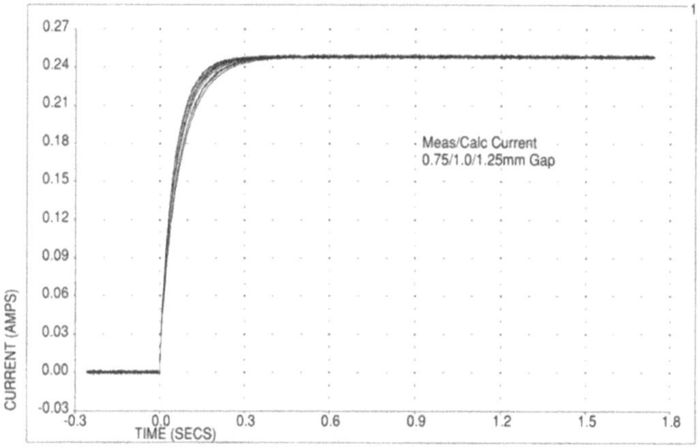

Fig. 7. Transient current at different gaps

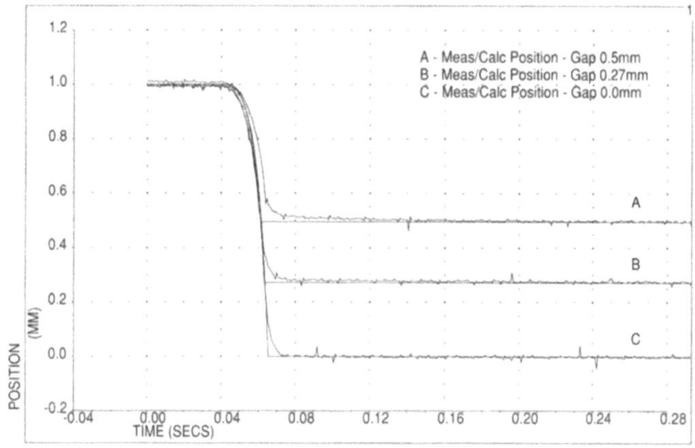

Fig. 8. Dynamic position of the armature

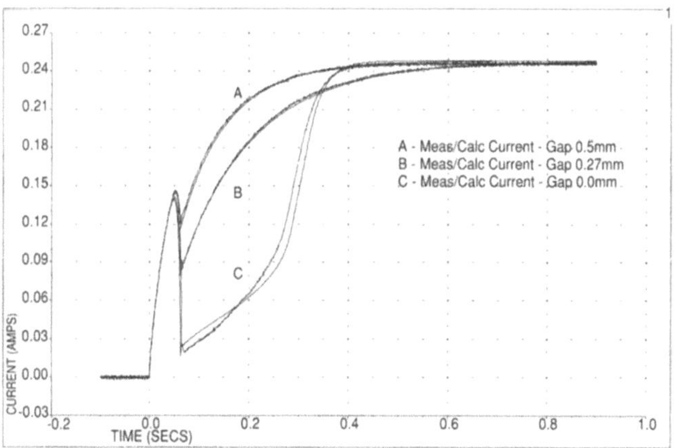

Fig. 9. Transient current as armature moves

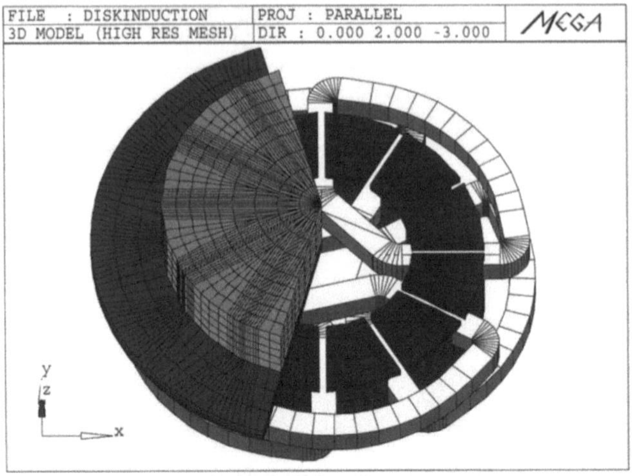

Fig. 10. Disk induction machine

Table 1. Dimensions of the disk induction machine

Machine Parameters	Dimensions
1 Stator Height	39.0 mm
2 Stator Inner Diameter	71.0 mm
3 Stator Outer Diameter	128.0 mm
4 Tooth Height	27.0 mm
5 Slot Width	14.0 mm
6 Slot Opening Width	3.0 mm
7 Air Gap	1.0 mm
8 Aluminium Disk Diameter	184.0 mm
9 Shaft Diameter	32.0 mm

- rotating disk in turbulent flow

$$\hbar_\omega = \frac{\kappa(T)}{D}[0.015(\frac{\omega r^2}{\nu})^{0.8}] \tag{12}$$

In the above D is the diameter of the cylinder/disk, r is the radius of the cylinder/disk, ω is the angular velocity and ν is the kinematic viscosity.

The induction machine is run with the aluminium disk rotating at 300 rpm at a load supplied by a DC brushless machine. The excitation current fed into the armature winding is 2.0 A and the resistance of each coil is 12.468 ohms. The main losses in the induction disk machine are the induced eddy currents in the rotating aluminium disk, the copper I^2R loss in the windings and the iron losses. The iron losses could be calculated very approximately using the following empirical formula:

$$\begin{aligned} P_h &= k_h \hat{B}^m f \text{ for hysteresis losses} \\ P_e &= k_e \hat{B}^2 f^2 \text{ for eddy current losses} \end{aligned} \tag{13}$$

where k_h and k_e are the empirical constants, $\hat{B}$ is the peak magnetic field with m = 1.76 and f is the applied excitation frequency.
The simulated and measured results for the windings and rotating aluminium disk are shown in Figs. 11 and 12.

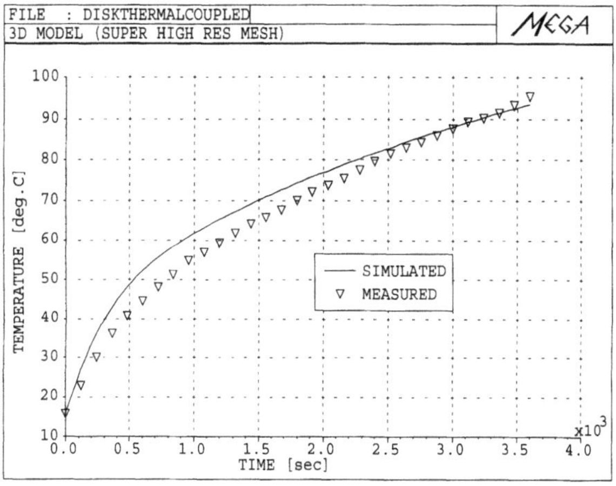

Fig. 11. Simulated and measured temperatures of the windings

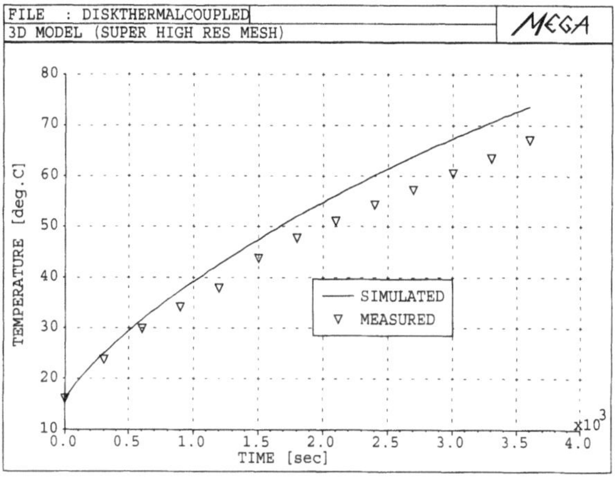

Fig. 12. Simulated and measured temperatures of the disk

References

1. P.J.Leonard and D.Rodger. "Modelling voltage forced coils using the reduced scalar potential method". *IEEE Trans. Magn.*, 28(2):1615–1617, March 1992
2. D.Rodger, N.Allen, P.C.Coles, S.Street, P.J.Leonard, and J.F. Eastham. "Finite element calculation of forces on a DC magnet moving over an iron rail". *IEEE Trans. Magn.*, 30(6):4680–4682, November 1994
3. D. Rodger, H.C. Lai, and P.J. Leonard. "Coupled elements for problems involving movement". *IEEE Trans. Magn.*, 26(2):548–550, March 1990
4. H.C.Lai, D.Rodger, and P.J.Leonard. "Coupling meshes in 3D problems involving movement". *IEEE Trans. Magn.*, 28(2):1732–1734, March 1992
5. D.Rodger, H.C.Lai, and P.J.Leonard. A comparison of finite element models for 3D rotating conductors. *IEEE Trans. Magn.*, 38(2):537–540, March 2002
6. S.H.Le G.Bisson, C.Leyden, P.J.Leonard, and D.Rodger. "Finite element analysis of transient electromagnetic heating effects in three dimensions." *IEEE Trans. Magn.*, 29(1):1102–1106, 1993
7. P.K.Vong, D.Rodger, P.C.Coles, and H.C.Lai. On modelling weakly coupled electromagnetic-thermal problems with prescribed non-linear surface heat transfer. *IEE PEMD conference April 2002 Bath IEE Conf Pub 487 pp 260-264*, 2002
8. S.W.Churchill and H.H.S.Chu. "Correlating Equations for Laminar and Turbulent Free Convection from a Horizontal Cylinder". *Int. J. Heat Mass Transfer*, 18:1049, 1975
9. S.W.Churchill and H.H.S.Chu. "Correlating Equations for Laminar and Turbulent Free Convection from a Vertical Plate". *Int. J. Heat Mass Transfer*, 18:1323, 1975

10. F.Kreith and M.S.Bohn. "Principles of Heat Transfer". *Harper and Row, Fourth Edition*, 1986
11. S.A. Viana, D. Rodger, and H.C. Lai. " Meshless local Petrov Galerkin method with radial basis functions applied to electromagnetics.". *IEE Proceedings Science, Measurement & Technology*, 151(6):449–451, November 2004
12. G.R. Liu. "Mesh free methods: Moving beyond the finite element method,". *CRC Press LLC, New York, USA*, 2003
13. S.N. Atluri and S. Shen. " The meshless local Petrov Galerkin (MLPG) method: A simple and less-costly alternative to the finite element and boundary element methods.". Comput. Modeling Eng. Sci., 3:pp 11–51, 2002
14. H.C.Lai, D.Rodger, and P.C.Coles. "A finite element scheme for colliding meshes". *IEEE Trans. Magn.*, 35(3):1362–1364, May 1999

Adaptive FEM Solver for the Computation of Electromagnetic Eigenmodes in 3D Photonic Crystal Structures

S. Burger[1,2], R. Klose[1], A. Schädle[1], F. Schmidt[1,2], and L. Zschiedrich[1,2]

[1] Konrad-Zuse-Zentrum Berlin, Takustr. 7, D-14195 Berlin, Germany, burger@zib.de
[2] JCMwave GmbH, Haarer Str.14 a, D-85640 Putzbrunn, Germany

1 Introduction

Photonic crystals (PhCs) are structures composed of different optical transparent materials with a spatially periodic arrangement of the refractive index [Joa95, Sak01]. Propagating light with a wavelength of the order of the periodicity length of the photonic crystal is significantly influenced by multiple interference effects. The most prominent effect is the opening of photonic bandgaps, in analogy to electronic bandgaps in semiconductor physics or atomic bandgaps in atom optics. Due to the fast progress in nano-fabrication technologies PhCs can be manufactured with high accuracy and with designed materials and geometrical properties. This allows for the miniaturization of optical components and a broad range of technological applications, like, e.g., in telecommunications [MBG04]. The properties of light propagating in PhCs are in general critically dependent on different system parameters, like the geometry of the device and the refractive indices of the present materials. Therefore, the design of photonic crystal devices calls for simulation tools with high accuracy, speed and reliability. In this paper we present a fast and flexible finite-element-solver for the calculation of Bloch-type eigenmodes of PhCs.

2 Light Propagation in Photonic Crystals

Light propagation in a photonic crystal is governed by Maxwell's equations with vanishing densities of free charges and currents. The dielectric coefficient $\varepsilon(\boldsymbol{x})$ and the permeablity $\mu(\boldsymbol{x})$ are real, positive and periodic, $\varepsilon(\boldsymbol{x}) = \varepsilon(\boldsymbol{x}+\boldsymbol{a})$, $\mu(\boldsymbol{x}) = \mu(\boldsymbol{x}+\boldsymbol{a})$. Here $\boldsymbol{a}$ is any elementary vector of the crystal lattice [Sak01]. For given primitive lattice vectors $\boldsymbol{a}_1$, $\boldsymbol{a}_2$ and $\boldsymbol{a}_3$ the elementary cell $\Omega \subset \mathbb{R}^3$ is defined as $\Omega = \{\boldsymbol{x} \in \mathbb{R}^3 \,|\, x = \alpha_1\boldsymbol{a}_1 + \alpha_2\boldsymbol{a}_2 + \alpha_3\boldsymbol{a}_3; 0 \le \alpha_1, \alpha_2, \alpha_3 < 1\}$. A time-harmonic ansatz with frequency ω and magnetic field $\boldsymbol{H}(\boldsymbol{x},t) = e^{-i\omega t}\boldsymbol{H}(\boldsymbol{x})$ leads to an eigenvalue equation for $\boldsymbol{H}$; additionally, the condition that $\boldsymbol{H}(\boldsymbol{x})$ is divergence-free applies:

$$\nabla \times \frac{1}{\varepsilon(\boldsymbol{x})} \nabla \times \boldsymbol{H}(\boldsymbol{x}) = \omega^2 \mu(\boldsymbol{x})\boldsymbol{H}(\boldsymbol{x}), \quad \nabla \cdot \mu(\boldsymbol{x})\boldsymbol{H}(\boldsymbol{x}) = 0, \qquad \boldsymbol{x} \in \Omega. \tag{1}$$

Similar equations are found for the electric field $\boldsymbol{E}(\boldsymbol{x},t) = e^{-i\omega t}\boldsymbol{E}(\boldsymbol{x})$:

$$\nabla \times \frac{1}{\mu(\boldsymbol{x})} \nabla \times \boldsymbol{E}(\boldsymbol{x}) = \omega^2 \varepsilon(\boldsymbol{x})\boldsymbol{E}(\boldsymbol{x}), \quad \nabla \cdot \varepsilon(\boldsymbol{x})\boldsymbol{E}(\boldsymbol{x}) = 0, \qquad \boldsymbol{x} \in \Omega. \tag{2}$$

The Bloch theorem applies for wave propagation in a periodic medium. Therefore we aim to find Bloch-type eigenmodes [Sak01] to Equations (1), defined as

$$\boldsymbol{H}(\boldsymbol{x}) = e^{i\boldsymbol{k}\cdot\boldsymbol{x}}\boldsymbol{u}(\boldsymbol{x}), \qquad \boldsymbol{u}(\boldsymbol{x}) = \boldsymbol{u}(\boldsymbol{x}+\boldsymbol{a}). \tag{3}$$

where the Bloch wavevector $\boldsymbol{k} \in \mathbb{R}^3$ is chosen from the first Brillouin zone. A similar procedure yields the Bloch-type eigenmodes to Equations (2), however, in what follows we will concentrate on Equations (1).

In order to reformulate Equations (1) and (3) we define the following functional spaces and sesquilinear forms:
(a) *The set of Bloch periodic smooth functions is defined as*

$$C_{\boldsymbol{k}}^{\infty}\left(\Omega, \mathbb{C}^d\right) = \left\{ w \in C^{\infty}(\Omega, \mathbb{C}^d) \mid w(\boldsymbol{x}+\boldsymbol{a}) = e^{i\boldsymbol{k}\cdot\boldsymbol{a}} w(\boldsymbol{x}) \right\}.$$

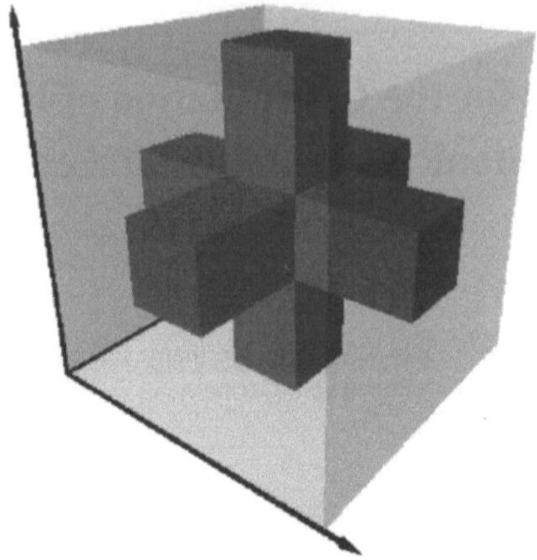

Fig. 1. Unit cell in the geometry of the *scaffold* structure. Bars with quadratic cross-sections intersect and form a 3D structure, periodic boundary conditions apply to all pairs of opposing faces

The Sobolev space $H_{\boldsymbol{k}}(\mathrm{curl})$ *is the closure of* $C^\infty_{\boldsymbol{k}}(\Omega, \mathbb{C}^3)$ *with respect to the* $H(\mathrm{curl})$*-norm. The space* $H^1_{\boldsymbol{k}}$ *is defined accordingly.*
(b) *The sesquilinear forms* $a : H_{\boldsymbol{k}}(\mathrm{curl}) \times H_{\boldsymbol{k}}(\mathrm{curl}) \to \mathbb{C}$ *and* $b : H_{\boldsymbol{k}}(\mathrm{curl}) \times H_{\boldsymbol{k}}(\mathrm{curl}) \to \mathbb{C}$ *are defined as*

$$a(\boldsymbol{w}, \boldsymbol{v}) = \int_\Omega \frac{1}{\varepsilon}(\nabla \times \boldsymbol{w}) \cdot \overline{(\nabla \times \boldsymbol{v})}\,\mathbf{d}x, \tag{4}$$

$$b(\boldsymbol{w}, \boldsymbol{v}) = \int_\Omega \mu\, \boldsymbol{w} \cdot \overline{\boldsymbol{v}}\,\mathbf{d}x. \tag{5}$$

With $\lambda \equiv \omega^2$ we get a weak formulation of Equations (1) and (3):

Problem 1. Find $\lambda \in \mathbb{R}$ and $\boldsymbol{w} \in H_{\boldsymbol{k}}(\mathrm{curl})$ such that

$$a(\boldsymbol{w}, \boldsymbol{v}) = \lambda\, b(\boldsymbol{w}, \boldsymbol{v}) \quad \forall \boldsymbol{v} \in H_{\boldsymbol{k}}(\mathrm{curl}), \tag{6}$$

under the condition that

$$b(\boldsymbol{w}, \nabla p) = 0 \qquad \forall p \in H^1_{\boldsymbol{k}}. \tag{7}$$

3 Finite Element Discretization

In order to numerically solve Problem 1 we need to discretize the corresponding functional spaces and expand the approximation of the solution in ansatz functions from these spaces [Jin93].

The Bloch periodic spaces $H_{\boldsymbol{k}}(\mathrm{curl})$ and $H^1_{\boldsymbol{k}}$ are discretized such that the corresponding edge element space $W_{h,\boldsymbol{k}} \subset H_{\boldsymbol{k}}(\mathrm{curl})$ and the Lagrange element space $V_{h,\boldsymbol{k}} \subset H^1_{\boldsymbol{k}}$ are of the same order. The finite element basis functions for $V_{h,\boldsymbol{k}}$ and $W_{h,\boldsymbol{k}}$ are denoted by φ_j, $1 \le j \le N_p$ and $\boldsymbol{\phi}_{j'}$, $1 \le j' \le N_c$. Bloch periodicity is enforced by a multiplication of basis functions associated with one of two corresponding periodic boundaries of the unit cell by the Bloch factor $\exp(i\boldsymbol{k} \cdot \boldsymbol{a}_i)$ (see Equation 3). Interior edge element functions remain unchanged.

The discretized problem corresponding to Problem 1 reads as follows:

Problem 2. Find $\lambda \in \mathbb{R}$ and $\boldsymbol{w} \in W_{h,\boldsymbol{k}}$ such that

$$a(\boldsymbol{w}, \boldsymbol{\phi}_i) = \lambda\, b(\boldsymbol{w}, \boldsymbol{\phi}_i) \quad \text{for } i = 1, ..., N_c, \tag{8}$$

under the condition that

$$b(\boldsymbol{w}, \nabla\varphi_j) = 0 \quad \text{for } j = 1, ..., N_p. \tag{9}$$

An alternative approach is discussed in [Dob01]. By expanding $\boldsymbol{w}$ in $\boldsymbol{\phi}_i$'s and inserting into Equation (8) we obtain the algebraic eigenvalue problem

$$\mathrm{Au} = \lambda \mathrm{Bu}, \tag{10}$$

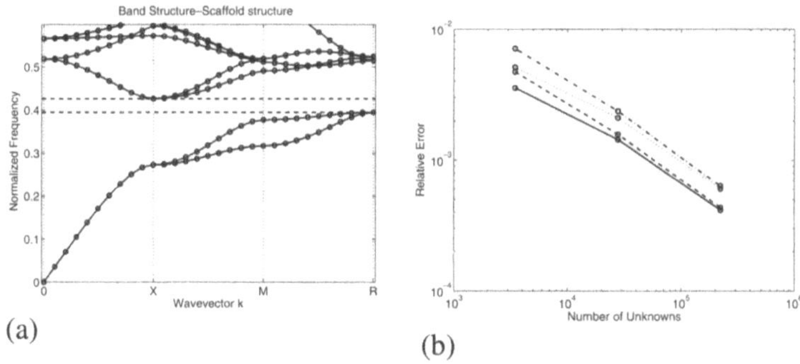

Fig. 2. (a) Band diagram for Bloch eigenmodes propagating in the scaffold structure. (b) Convergence of the first four eigenvalues at the X-point towards the eigenvalues of the quasi-exact solutions

with $A_{i,j} := a(\phi_i, \phi_j)$ and $B_{i,j} = b(\phi_i, \phi_j)$. The matrix A is hermitian, positive semidefinite and B is hermitian, positive definite.

The main advantage of the finite element method is that the matrices in Equation (10) are sparse, which is due to the locality of the chosen finite element basis functions. This allows the use of very efficient solvers.

In our realization [BKS05] we use a subspace iteration scheme similar to a method proposed by Döhler [Doe82]. This yields a solution to Equation (8) in Problem 2. In order to guarantee that Equation (9) in Problem 2 is fulfilled we add a step within the iteration scheme, which projects the iterates onto the divergence-free subspace. This is done as follows: To compute the projected, divergence-free field $\boldsymbol{w}^p$ from the (not divergence-free) iterate $\boldsymbol{w}$ we use the Helmholtz decomposition,

$$\boldsymbol{w}^p = \boldsymbol{w} + \nabla \chi_h \quad , \tag{11}$$

with the correction potential $\chi_h \in V_{h,\boldsymbol{k}}$. This leads to the following system of equations:

$$b(\boldsymbol{w} + \nabla \chi_h, \nabla \varphi_j) = 0 \quad \text{for } j = 1, ..., N_p\,, \tag{12}$$

or

$$b(\nabla \chi_h, \nabla \varphi_j) = -b(\boldsymbol{w}, \nabla \varphi_j) \quad \text{for } j = 1, ..., N_p\,. \tag{13}$$

We solve these equations by standard multi-grid algorithms [DFZ03].

For preconditioning of the algebraic problem we have implemented a multi-grid preconditioner [DFZ03] similar to the implementation in [HN02]. With this, the computational time and the memory requirements grow linearly with the number of unknowns. Furthermore, we have implemented a residuum-based error estimator [HR01] and adaptive mesh refinement for the precise determination of localized modes. As finite element (FE) ansatz functions, we typically choose edge elements (Whitney elements) of quadratic order.

4 Numerical Examples

4.1 3D scaffold structure

We illustrate the performance of our eigensolver by analyzing the convergence of the eigenvalues for an example from the literature, where the lowest eigenmodes in a 3D periodic structure (*scaffold* structure [Dob00]) are calculated. The geometry of a cubic unit cell (sidelength a) is shown in Fig. 1, it consists of bars (width $d = 0.25\,a$) of a transparent material with relative permittivity $\varepsilon_r = 13$ and a background with $\varepsilon_r = 1$ ($\varepsilon = \varepsilon_r \varepsilon_0$, ε_0: free space permittivity). For the calculation of the band structure the Bloch wave vector $\boldsymbol{k}$ is varied along symmetry lines of the Brillouin zone (*cf.* [Dob00]). The band structure which exhibits a complete bandgap around the reduced frequency of $\tilde{\omega} = \omega\, a/(2\pi\, c) = 0.4$ is shown in Fig. 2a. Table 1 shows the four lowest eigenvalues at the X-point($\boldsymbol{k} = (\pi/a, 0, 0)$) calculated on grids generated in 0, 1, resp. 2, uniform refinement steps from a coarse grid. Displayed are also the numbers of unknowns in the problem (number of ansatz functions in the finite element discretiztation) and typical computation times on a PC (intel Pentium IV, 2.5 GHz). It can be seen that the computational effort rises linearly with the number of unknowns. This behaviour is a major advantage compared to other simulation methods like plane-wave expansion methods or finite-difference time-domain methods. The convergence of the four eigenvalues towards a quasi-exact solution (obtained from a calculation on a finite-element grid with $N = 1764048$ unknowns) is shown in Fig. 2 (b).

Step	N^o DOF	CPU time [min]	$\tilde{\omega}_1$	$\tilde{\omega}_2$	$\tilde{\omega}_3$	$\tilde{\omega}_4$
0	3450	00:09.23	2.736e-01	2.740e-01	4.279e-01	4.288e-01
1	27572	01:46.33	2.730e-01	2.731e-01	4.266e-01	4.267e-01
2	220520	13:50.81	2.728e-01	2.728e-01	4.260e-01	4.260e-01

Table 1. First eigenvalues of $\boldsymbol{k} = X$-eigenmodes of the scaffold structure. Displayed are the step number, the number of degrees of freedom of the problem, the CPU time (run on a standard PC), and the reduced frequencies of the four lowest eigenmodes

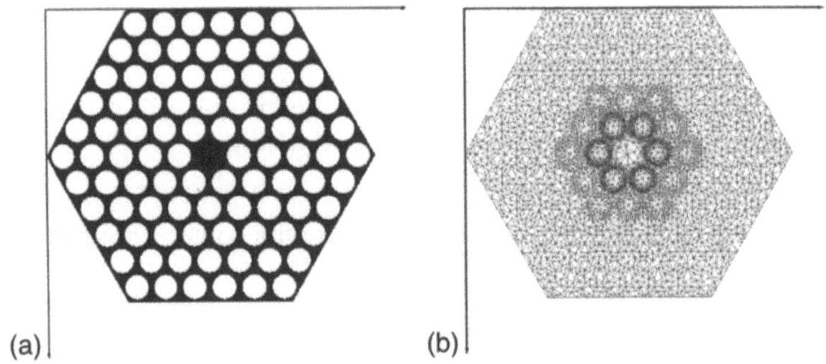

Fig. 3. Geometry (a) and coarse FE mesh (b) of a 2D photonic crystal structure with a central point defect

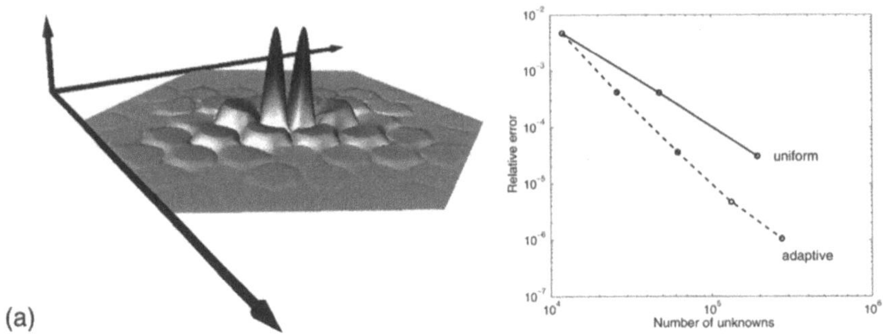

Fig. 4. (a) Distribution of the magnetic field intensity ($|\boldsymbol{H}(\boldsymbol{x})|$) for the lowest-frequency bound state at the point defect. (b) Comparison of the convergence of the eigenfrequency of the lowest frequency bound state towards a quasi-exact solution for adaptive and uniform refinement of the finite element mesh

4.2 Defect mode

Light at a frequency inside the bandgap of a photonic crystal can be "trapped" inside defects of the structure [Joa95]. This enables the construction of, e.g., waveguides (by line defects) and micro-cavities (point defects).

Figure 3 (a) shows the geometry of a 2D photonic crystal with a point defect (a missing pore in the center). It consists of a hexagonal lattice of air holes with a radius of $r = 0.4\,a$ in a material with a relative electric permittivity of $\varepsilon_r = 13$. The corresponding coarse triangular FE grid is shown in Fig. 3 (b). Fig. 4 (a) shows the modulus of the magnetic field for the lowest-frequency trapped eigenmode, computed with adaptive refinement of the FE mesh. Figure 4 (b) shows the convergence of the eigenvalue of the discrete solution, $\tilde{\omega}_d$, towards the eigenvalue of a quasi-exact solution, $\tilde{\omega} \approx 0.28272$, for adaptive grid refinement and for uniform grid refinement. Plotted is the relative error of the reduced eigenfrequency, $\Delta\tilde{\omega}/\tilde{\omega}$, where $\Delta\tilde{\omega} = |\tilde{\omega} - \tilde{\omega}_d|$. Obviously, adaptive grid refinement is especially useful when the sought solutions are geometrically localized, or when the geometry exhibits sharp features, like discontinuities in the refractive index distribution. In this example, the use of the error estimator and adaptive refinement yields an order of magnitude in the accuracy of the error for a number of unknowns of $N \sim 10^5$.

5 Conclusion

In this paper we have presented an adaptive solver for the computation of electromagnetic eigenmodes. The convergence analysis of solutions for model problems shows the efficency of the methods. Currently we are working on the implementation of transparent boundaries in our finite element code. For this we are using a new Laplace domain method which allows the treatment of inhomogeneities in the exterior domain [Sch02].

Acknowledgements

We thank P. Deuflhard for fruitful discussions, and we acknowledge support by the initiative *DFG Research Center "Mathematics for key technologies"* of the Deutsche Forschungsgemeinschaft, DFG, and by the German Federal Ministry of Education and Research, BMBF, under contract no. 13N8252 (HiPhoCs).

References

[Joa95] Joannopoulos, J. D.: Photonic Crystals – Molding the flow of light. Princeton University Press, Princeton, NJ (1995)

[Sak01] Sakoda, K.: Optical Properties of Photonic Crystals. Springer-Verlag, Berlin (2001)

[MBG04] März, R., Burger, S., Golka, S., Forchel, A., Herrmann, C., Jamois, C., Michaelis, D., Wandel, K.: Planar High Index-Contrast Photonic Crystals for Telecom Applications. In: Photonic Crystals - Advances in Design, Fabrication and Characterization, K. Busch et al. Eds., Wiley-VCH (2004)

[Jin93] Jin, J.-M.: The finite element method in electromagnetics. Wiley, New York (1993)

[Dob01] Dobson, D. C., Pasciak, J.: Analysis for an algorithm for computing electromagnetic Bloch modes using Nedelec spaces. Comp. Meth. Appl. Math. **1**, 138 (2001)

[BKS05] Burger, S., Klose, R., Schädle, A., Schmidt, F., Zschiedrich, L.: FEM modelling of 3D photonic crystals and photonic crystal waveguides. Proc. SPIE **5728**, 164 (2005)

[Doe82] Döhler, B.: A new gradient method for the simultaneous calculation of the smallest or largest eigenvalues of the general eigenvalue problem. Numer. Math. **40**, 79 (1982)

[DFZ03] Deuflhard, P., Schmidt, F., Friese, T., Zschiedrich, L.: Adaptive multigrid methods for the vectorial Maxwell eigenvalue problem for optical waveguide design. In: Mathematics - Key Technologies for the Future, Springer-Verlag, Berlin (2003)

[HN02] Hiptmair, R., Neymeyr, K.: Multilevel method for mixed eigenproblems. SIAM J. Sci. Comp. **23**, 2141 (2002)

[HR01] Heuveline, V., Rannacher, R.: A posteriori error control for finite element approximations of elliptic eigenvalue problems. J. Adv. Comp. Math. **15**, 107 (2001)

[Dob00] Dobson, D. C., Gopalakrishnan, J., Pasciak, J. E.: An efficient method for band structure calculations in 3D photonic crystals. J. Comp. Phys. **161**, 668 (2000)

[Sch02] Schmidt, F.: Solution of Interior-Exterior Helmholtz-Type Problems Based on the Pole Condition Concept – Theory and Algorithms. Habilitation thesis, Free University, Berlin (2002)

COLLGUN: a 3D FE Simulator for the Design of TWTs Electron Guns and Multistage Collectors

S. Coco[1], S. Corsaro[1], A. Laudani[1], G. Pollicino[1], R. Dionisio[2], and R. Martorana[2]

[1] Dipartimento di Ingegneria Elettrica Elettronica e dei Sistemi, University of Catania - Viale A. Doria 6, I-95125 Catania, coco@diees.unict.it

[2] Galileo Avionica, Via Villagrazia 79, Palermo, roberto.dionisio@galileoavionica.it

Abstract In this paper a new simulator for the design of Traveling Wave Tubes (TWT) electron guns and multistage collectors is presented. The simulator is based on the 3-D FE discretization of the Poisson equation combined with a particle model for the solution of the Vlasov equation in the space charge limited regime.

1 Introduction

Traveling Wave Tube (TWT) are electronic vacuum devices used for high-power amplification of RF signals. Three main regions are usually individuated in a TWT, as shown in Fig. 1 from [ALMP01]: the electron gun, where the electron beam is generated, consisting of a cathode, a focusing electrode, one or two grids and an anode; a slow wave helicoidal structure, where the beam interacts with the low-level input RF signal; and a depressed collector, consisting of several electrodes suitably voltaged in such a way to slow down the electrons after their interaction with the RF signal [Gil94]. In this way the spent beam residual energy is recovered and overall efficiency is increased. Periodic permanent magnets (PPM) structures are also employed to keep the electron beam laminar. Electron guns and collectors are also used in a wide variety of vacuum devices including klystrons and particle accelerators. In the TWT design process the geometries of electron guns and collectors are critical issues in order to achieve optimal overall efficiency. In fact, usage of more complex geometries and insertion of control grids allow the designer to obtain a better performance. For this reason dedicated tools are required to test innovative geometries for collectors and guns addressing optimal placement of control grids and electrodes [SWKR99, Pet02, KSCP95]. Numerical analysis tools based on the Finite Element Method (FE) are widely used for general electromagnetic analysis, since they possess several advantages in comparison with other numerical techniques; such advantages include flexibility in the treatment of realistic geometries, easy utilization of irregular meshes and/or mesh adaptation in the discretization process and capability to easily manage material non-homogeneities [CEL01, CoL02].

In this paper we present an innovative dedicated 3-D FE simulator (COLLGUN) expressly conceived for the design of TWT electron guns and multistage collectors, developed at the University of Catania in collaboration with Galileo Avionica. The complete simulator consists of three main modules: a fully 3-D FE mesh generator, a 3-D FE Vlasov solver, including space-charge effects with an integrated electron trajectory tracer, and a graphic post-processing module for result restitution.

In COLLGUN the electromagnetic analysis of the TWT device is aimed at the tracing of electron trajectories and is based on the iterative solution of a 3-D steady-state Vlasov-Poisson electromechanical problem. In the iterative scheme three main phases are followed: the solution of the electromagnetic problem, the integration of the mechanical equations for the tracing of the electron trajectories and the computation of space charge distribution. The above steps are repeated until the "distance" between two consecutive solutions is less than a prescribed end-iteration tolerance.

Some results concerning the analysis of an electron gun and a multistage depressed collector are presented, showing good agreement with available experimental data.

The paper is structured as follows. In Sect. 2 the architecture of the COLLGUN simulator is described. In Sect. 3 some examples of simulations are given.

2 The COLLGUN simulator

The COLLGUN simulator is written in C++ and has been developed under the MS-WINDOWS OS environment. The COLLGUN structure follows the classical scheme of FE simulators in which three main modules are present. The

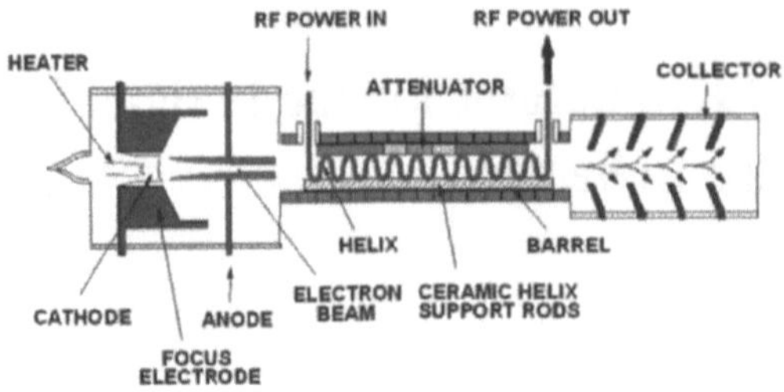

Fig. 1. Helix TWT schematic

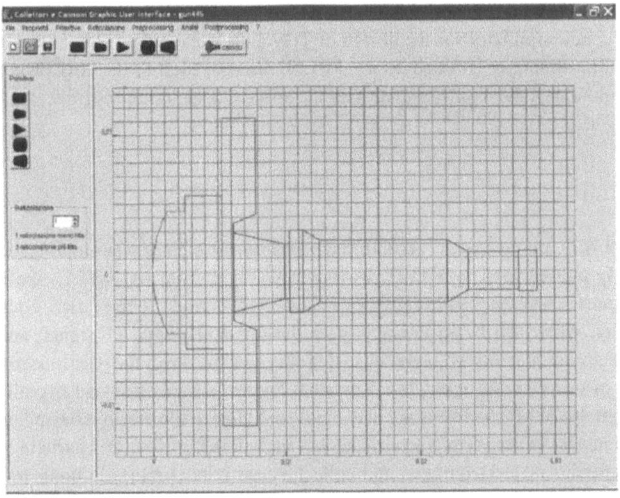

Fig. 2. GUI main window

first one groups preprocessing functions (such as the construction of the geometrical model and FE mesh generation). The second module is devoted to the processing functions (solution of the mathematical model). The last module is devoted to post-processing (such as further elaboration of results, analysis plots, etc.). The tool has been specifically conceived to provide the TWT designer with an easy-to-use environment through a friendly CAD-based Graphical User Interface (GUI). Differently from other interfaces dedicated only to post-processing, the GUI of COLLGUN aims at facilitating the management of all the various aspects of a simulation session. In fact all the simulator functions are interactively executed by using the GUI window, where several menus foresee all the actions necessary to create, modify and simulate the device. The COLLGUN GUI is based on a window presentation supported by suitable pop-up menus. The main window, see Fig. 2, contains a toolbar displaying the accelerator buttons and a main view visualizing the device geometry structure in 2-D space. The GUI also includes visualization tools, together with some specific pre/post-processing functions related to TWT design.

2.1 Preprocessing: Construction of the FE Model

The developed preprocessor has been especially tailored in order to facilitate the input of geometrical data and boundary conditions. The philosophy followed in COLLGUN for the construction of the geometrical and functional model considers the device (collector/gun) as the union of blocks of simpler shape (primitives) representing the various regions such as electrodes, cathode, grids, etc. These primitive are quite common shapes (cylinders, cones, spikes, cut off cone, etc.), which cover a large number of geometries and other nonaxisymmetrical blocks like the gridded cathode shown in Fig. 3, etc. The specification of the primitives includes assignment of boundary conditions. It is worth noticing that during the creation of a new gun (or collector) the designer is assisted by input validation functions. When this formal description is complete, the specified primitives are automatically assembled together and the whole FE mesh is generated.

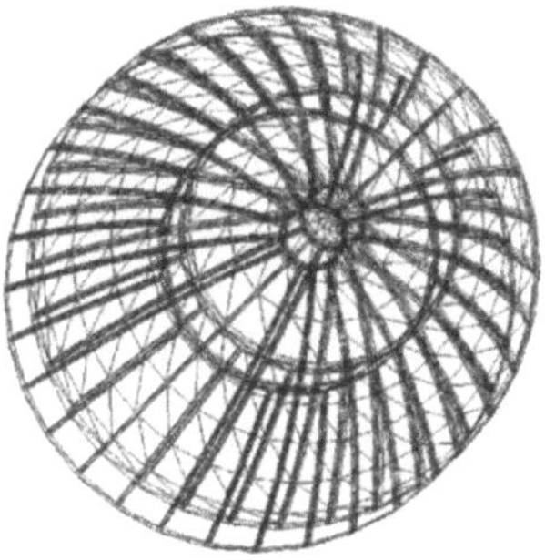

Fig. 3. Example of discretization of the shadow and control grids

2.2 Processing 3D FE Steady-State Electron Beam Analysis

Starting from an available FE geometrical and functional device model, the COLLGUN processor module performs a fully 3-D steady-state solution of the coupled electromechanical problem inside the device region, assuming a macro-particle model for the electron beam. In this approach, the steady-state spatial distribution of electric charge in the collector is governed by the following 3-D Vlasov equation for the scalar potential ϕ, associated with the relativistic dynamic equations of electrons:

$$\nabla^2 \phi + \int \int \int_{\Omega} f_P d\Omega = 0 \tag{1}$$

$$\frac{d\mathbf{p}}{dt} = e \cdot (\mathbf{E} + \mathbf{v} \times \mathbf{B}) \tag{2}$$

where f_P is an unknown function describing the space charge distribution, and e, $\mathbf{p}$ and $\mathbf{v}$ are the electron charge, impulse and velocity respectively. In the 3-D FE numerical solution of the above coupled problem, the unknown potential is approximated by using Lagrangian interpolating polynomials at nodal points (vertices of tetrahedra). The resulting discretized problem consists of two systems of equations: the first is an FE linear algebraic system, concerning the spatial distribution of the unknown potential values, the other regards the positions at a certain time of all the discrete particles used in the model. The strategy adopted for the solution of the complete set of equations is based on an iterative scheme, which alternates the solution of the FE algebraic system with the integration of the dynamical equation in time. At each step, from the computed trajectories a new estimate of the space charge distribution is evaluated and then used to perform a new Poisson-solver step. These two steps are iteratively repeated until convergence is reached, when the distance between two consecutive solutions is less than a user-specified end-iteration tolerance. It is worth noticing that the presence of an externally assigned focusing magnetic field can also be considered in the integration of the dynamical equation. Two aspects are fundamental for the convergence of the iterative procedure: a congruent redistribution of the 3-D space charge density and an accurate representation of the input beam. This second aspect plays a more relevant role in the simulation of electron guns, since in this case, an appropriate model must be considered for the electron emission, differently from collector analysis, for which the spatial and energetic distribution of the spent beam is assumed known. The model of cathodic emission adopted is based on the Child's law, suitably modified in order to take into account the cathode geometry. In fact, the developed model includes corrections for cathode geometric shape (spherical), relativistic beams and allows an accurate prediction of emission currents from the knowledge of the field distribution near the cathode.

2.3 Postprocessing

The post-processing functions presently implemented allow computation and visualization of all the global and local quantities commonly needed to perform accurate 3-D collector and electron gun design. The GUI allows an easy management of postprocessing functions including graphical output. The device geometries and discretization can be viewed and printed in 2-D sections and in 3-D axonometric views from any angle. Zoom in/out and rotation functions are available to facilitate detail visualization. Upon completion of the simulation task, the COLLGUN GUI allows the user to visualize and print several results useful to the designer such as trajectories in a 3-D space (showing also only the trajectories pertaining to each electrode), contour lines and color maps in any plane selected by the user, V-I diagrams, on axis magnetic field profiles, cathode emission densities, etc. In addition all the global quantities of interest for the designer (total current, current and power recovered for each stage, power globally recovered, collector efficiency, cathode loading, perveance, etc.) are also available in a text output file.

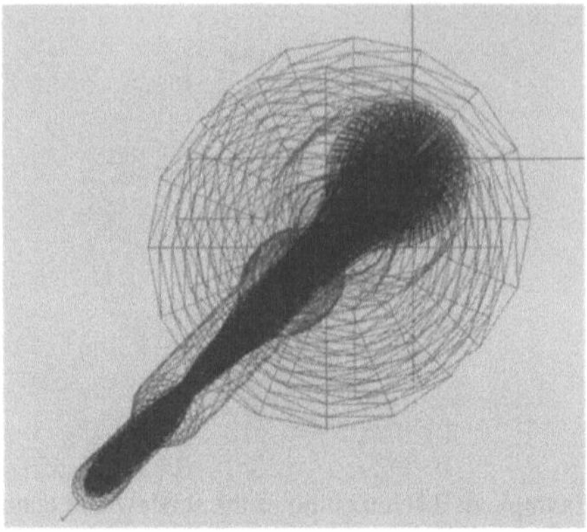

Fig. 4. 3-D plot of emitted electron trajectories

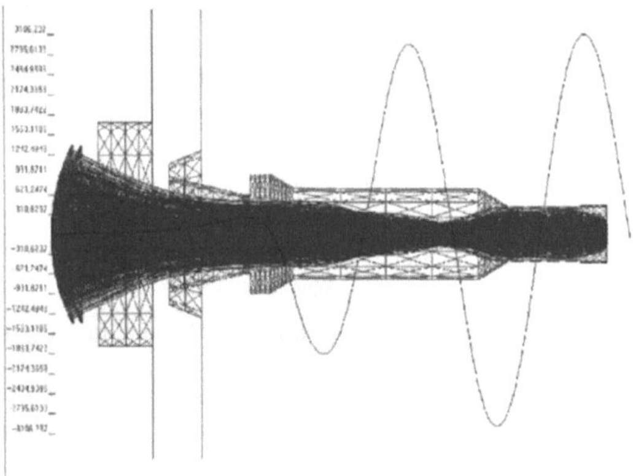

Fig. 5. 2-D projection of trajectories and on-axis profile of the focusing magnetic field

3 Examples of TWT simulations

Several examples of 3-D simulations of electron guns and multistage depressed collectors have been performed to check the COLLGUN functions; the results of the tests based on geometries available from Galileo Avionica are in excellent agreement with measured data. Hereafter some results concerning a sample grid electron gun and a two-stage depressed collector are reported.

3.1 Analysis of a gridded electron gun

In this paragraph the analysis of a grid electron gun for which measured data are available from Galileo Avionica is illustrated. The 2-D longitudinal section view of the electron gun is shown in Fig. 2. This gun has a grid with the complex geometry shown in Fig. 3. An axisymmetric magnetic field is applied in order to focalize and to maintain the generated beam laminar. The cathode, control grid and anode voltages were settled to 0V, 304V and 15kV respectively. For the analysis an irregular mesh consisting of 79952 first order tetrahedra and 18337 nodes was generated by the preprocessing module; a more refined mesh was used in the discretization of the inter-grid region. For this discretization five iteration were needed to solve the Vlasov-Poisson system, using an end-iteration tolerance of 0.1%. Figure 4 shows a 3-D plot of the emitted beam consisting of the 976 macro-electron trajectories generated by COLLGUN. In Fig. 5 the trajectories in the yz plane with an applied focusing magnetic field are shown. The comparison between simulation results and measured parameters are reported in Table 1. Total cathode current is accurately represented in the simulation. As far as intercepted current is considered, simulation results exhibit an absolute error of the same order of the cathode current, even if they are referred to a vanishing quantity (approching to zero). Neverthless this result is

Table 1. Comparison between experimental values and simulated results

	Experimental	Simulated
Cathode current [A]	2.504	2.509
Intercepted current [A]	0.002	0.005

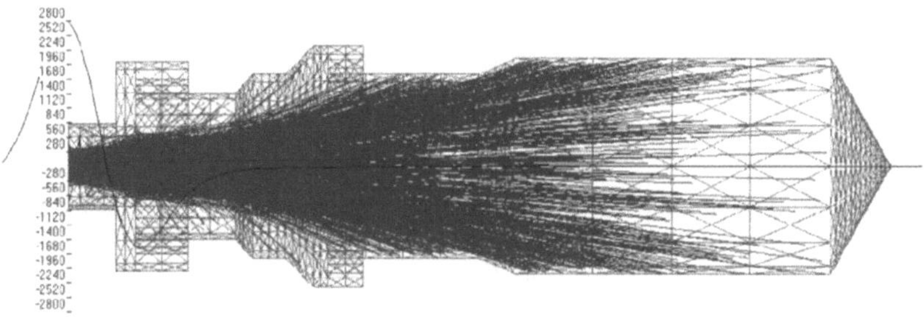

Fig. 6. 2-D projection of trajectories and on-axis profile of the focusing magnetic field

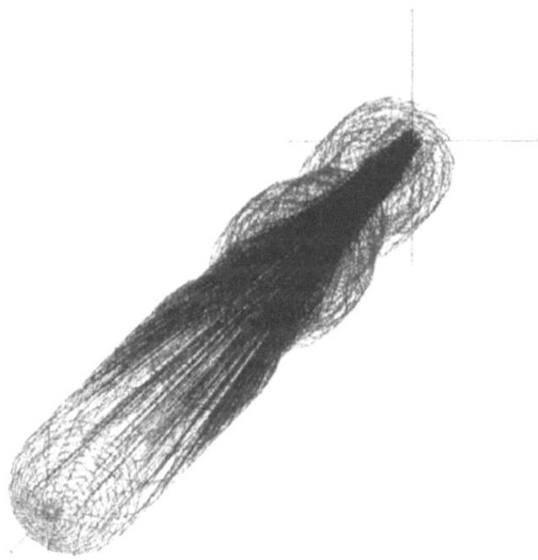

Fig. 7. 3-D plot of electron trajectories

reported because it is significant to ascertain correct operation of the device (very low intercepted current). The CPU time required for the whole simulation was about 10 min, using a PC Pentium IV at 2.4 GHz with 1GB RAM.

3.2 Analysis of a multistage depressed collector

In this paragraph the computation of electron trajectories inside a TWT collector is presented. The analyzed collector is an axisymmetric two-stage depressed for which geometrical data are available from Galileo Avionica. For the considered collector a focusing magnetic field is applied in order to reduce backstreaming current. For this simulation an irregular mesh of 28528 tetrahedra and 6257 nodes was generated, using a more refined mesh in the inter-electrode regions. The beam entering the collector has a reference voltage of 15 kV, a current of 1.93 A and a radius of 1.33 mm. The cross section of the electron beam was modelled assuming that the beam total current is radially distributed into 80 concetrical rings, each one carrying 10 macro-electrons. Figure 6 shows the trajectories landing on the first electrode (first stage), carrying a total current of 180 mA, and those landing on the second electrode, carrying 1.74 A. The on-axis profile of the focusing magnetic field applied is also shown. In Fig. 7 all the trajectories are shown in a 3D space. Table 2 summarizes the simulation results for the considered collector. In this case the CPU time required for the whole simulation was about 8 minutes. Other collector geometries, not reported here, have been analysed, all of which showed results in excellent agreement (within a few percent) with measurement data from Galileo Avionica.

Table 2. Results of the simulation of the two stages collectors

Spent Beam current [A]	1.93
Spent beam power [W]	24165
Recovered Power [W]	11474
Efficiency [%]	47.48
Backstreaming current [mA]	9.64
Drift Current [mA]	2.41
1st stage current [A]	0.18075
2nd stage current [A]	1.74

References

[Gil94] Gilmour, A. S. Jr.: Principles of Traveling Wave Tubes, Artech House fnc., Norwood (1994)

[ALMP01] Abrams, R.H., Levush, B., Mondelli, A.A., Parker, R.K.: Vacuum for the 21st Century Electronics, IEEE Microwave Magazine, Vol. 2, Issue 3, 61-72 (2001)

[SWKR99] Staats J., Weiland T., Kostial S., Richter A.:Tracking of electron beams with numerically determined space charge forces, Proceedings of the 1999 Particle Accelarator Conference, pp. 2740-2742, New York (1999)

[CoL02] Coco, S. and Laudani, A.: An Iterative approach for the FE solution of 3D Vlasov-Poisson problems in TWT collectors, 10th International IGTE Symposium on Numerical Field Calculation in Electrical Engineering, Graz (Austria), 16-19 settembre 2002

[CEL01] Coco, S., Emma, F., Laudani, A., Pulvirenti, S. and Sergi, M.: COCA: a novel 3D FE Simulator for the Design of TWTs Multistage Collectors. IEEE Transactions on Electron Devices, Vol. 48, Number 1, 24-31 (2001)

[Pet02] Petillo, J. et alii: The MICHELLE Three-Dimensional Electron Gun and Collector Modeling Tool: Theory and Design, IEEE Transactions on Plasma Science, vol. 30, n. 3, 1238-1264 (2002)

[KSCP95] Kumar L., Spatke P.,Carter R. G and Perring D.: Three-dimensional simulation of multistage depressed collectors on micro-computers. IEEE Trans. Electron Devices, vol. ED-42, pp. 1163-1173, (1995)

A New Thin-Solenoid Model for Accurate 3-D Representation of Focusing Axisymmetric Magnetic Fields in TWTs

S. Coco, A. Laudani and G. Pollicino

Dipartimento di Ingegneria Elettrica, Elettronica e dei Sistemi, University of Catania, Viale A. Doria 6, Catania I-95125, Italy, coco@diees.unict.it

Abstract In this paper an iterative procedure is presented for the computation of equivalent source representations of focusing axisymmetric magnetic fields inside Traveling Wave Tubes (TWT). The procedure uses thin solenoid pairs as equivalent sources and solves iteratively a sequence of inverse problems until user defined end-iteration tolerance is achieved. The adopted approach is accurate, robust and allows us to obtain a very accurate representation of real complicated shape fields by using few thin solenoid pairs. In order to illustrate the effectiveness and the advantages of the procedure, several examples of representation of field profiles are also given.

1 Introduction

Travelling Wave Tubes (TWTs) are vacuum electronic devices used as high-power high-frequency amplifiers for various applications such as telecommunication and radar systems. A TWT consists of an electron gun, where the electron beam is generated, a slow wave structure (SWS), where the RF signal is amplified, and a collector region, where the spent beam energy is recovered by slowing down the electrons. TWTs signal amplification is based on the interaction between the relativistic electron beam and the input RF signal, taking place in the aforementioned helicoidal slow wave structure. In order to improve TWTs performance, avoiding the spreading of the electron beam, magnetic focusing systems are used both in the electron gun region and in the SWS and collector regions. The focusing field is usually generated by means of permanent magnets suitably positioned along the axisymmetry axis. In the numerical analysis of TWTs accurate three-dimensional (3-D) representations of the focusing axisymmetric magnetic field are required in order to precisely compute electron trajectories [CEL01]. In literature 3-D magnetic field representations are generally built from experimentally measured 1-D axis values [Vau72, Vau74, Sta79, Jac99, CL02]. Various models are then used in order to obtain an approximate 3-D field representation by minimizing the error with respect to experimental on-axis values. Equivalent ideal loop sources are commonly used for this purpose but a single loop source is intrinsically different from a permanent magnet source (PMS) and consequently a non-realistic high number of loops often results from ideal loop procedures, when accurate representations are required. Furthermore the off-axis components of the magnetic field representations based on ideal loop models may not be close to those coming out from real PMS. In literature [Vau72] it has been pointed out that considerable errors may arise in off axis values when models not adherent with real sources are used, even if on-axis approximations are satisfactory. In this paper the authors present a new 3-D magnetic field representation procedure based on thin solenoid pairs (TSP) sources (Fig. 1), more adherent to the real behaviour of permanent magnets. In this way the model allows us to calculate very accurate representations. From this model an iterative procedure based on the minimization of the representation error is derived to calculate the parameters of the TSP model. In the procedure local and global inverse problems are iteratively solved in sequence until the error in the representation is lower than a prescribed tolerance. The adopted approach allows us to achieve a very accurate representation of complicated shape fields by using only few equivalent sources, so reducing the overall computational effort. In addition the procedure is robust and convergence has been observed in all the cases examined. The paper is structured as follows: in Section II the TSP model is illustrated; in Section III the developed procedure for the calculation of TSP model representation is presented; in Section IV several examples of application are shown; the authors' conclusions follow in Section V.

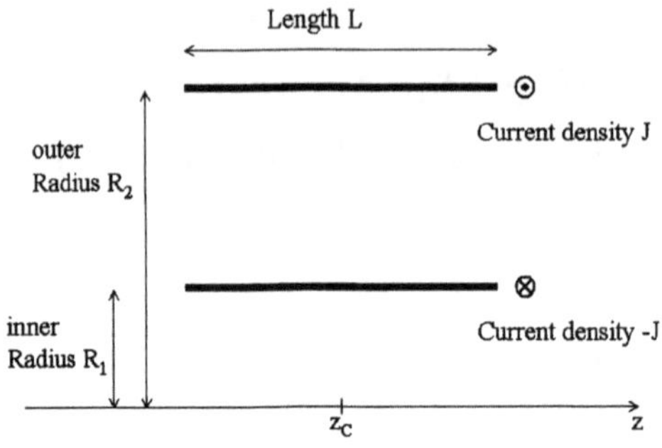

Fig. 1. Model of a permanent magnet by using a couple of thin solenoids

2 The thin solenoid pair model

The geometrical and physical configuration of a thin solenoid pair is represented in Fig. 1. The on-axis magnetic field component B_z of a single thin solenoid pair is given by the following formulas

$$B_z(z) = M \cdot \frac{(z - z_C + \frac{L}{2})}{\sqrt{R_1^2 + (z - z_C + \frac{L}{2})^2}} - M \cdot \frac{(z - z_C - \frac{L}{2})}{\sqrt{R_1^2 + (z - z_C - \frac{L}{2})^2}}$$
$$-M \cdot \frac{(z - z_C + \frac{L}{2})}{\sqrt{R_2^2 + (z - z_C + \frac{L}{2})^2}} - M \cdot \frac{(z - z_C - \frac{L}{2})}{\sqrt{R_2^2 + (z - z_C - \frac{L}{2})^2}} \quad (1)$$

where $M = \frac{\mu_0 \cdot J}{2}$, and J is the linear current density of each thin solenoid, z_C is the z coordinate of thin solenoid pair center, R_1 and R_2 are the radii of the inner and outer thin solenoid respectively. The above model is usually used to represent a cylindrical permanent magnet of length L, inner radius R_1 and outer radius R_2 [Jac99]. The overall on-axis magnetic field due to a set of N thin solenoid pairs is expressed by adding all the TSP contributions as follows

$$B_z(z) = \sum_{i=1}^{N} M_i \cdot \left\{ \frac{(z - z_{C_i} + \frac{L_i}{2})}{\sqrt{R_{i1}^2 + (z - z_{C_i} + \frac{L_i}{2})^2}} - \frac{(z - z_{C_i} - \frac{L_i}{2})}{\sqrt{R_{i1}^2 + (z - z_{C_i} - \frac{L_i}{2})^2}} \right\}$$
$$-\sum_{i=1}^{N} M_i \cdot \left\{ \frac{(z - z_{C_i} + \frac{L_i}{2})}{\sqrt{R_{i2}^2 + (z - z_{C_i} + \frac{L_i}{2})^2}} - \frac{(z - z_{C_i} - \frac{L_i}{2})}{\sqrt{R_{i2}^2 + (z - z_{C_i} - \frac{L_i}{2})^2}} \right\} \quad (2)$$

where M_i, z_{C_i}, R_{i1} and R_{i2} refer to the ith TSP in the summation.

3 The TSP iterative procedure for the 3-D magnetic field representation

The input data of the TSP procedure are the measured magnetic field values along the z-directed symmetry axis in the interval (z_i, z_f). The outputs of the TSP procedure are the parameters of the thin solenoids: that is the final number of the pairs, their axial positions, their geometrical data (radii and length) and the currents carried by each solenoid. At the beginning the procedure builds a tentative 3-D representation of the magnetic field by using a minimum number of TSP (even a single pair) positioned on the z-directed symmetry axis in correspondence with the maximum value of the experimentally known axial magnetic field. This follows a general rule also adopted for determining the position of successively added pairs, which corresponds to the z-axis point where the difference between the experimental and computed profiles (*differential profile*) exhibits its maximum value. The initial tentative value of the parameters of each inserted TSP are individually computed and assigned by solving a local inverse problem according to the formula:

$$R = |z_1 - z_2| \cdot \sqrt{S + \sqrt[3]{S \cdot (1+S)^2} + \sqrt[3]{S^2 \cdot (1+S)}} \quad (3)$$

where

$$S = \frac{1}{\frac{B_1^2}{B_2^2} - 1} \quad (4)$$

and B_1 and B_2 are the measured values of the magnetic field at points z_1 (where an extremal value occurs) and z_2 separated from z_1 by a user chosen z-axis discretization interval. In particular the inner radius is chosen equal to R, the outer twice R, while the length is assigned equal to four time R.
At this point all the parameters of the tentative representation are adjusted by following a constrained minimization of the overall representation error e(B) evaluated according to the expression:

$$e(B) = \frac{\| \Delta B \|}{\| B \|} \tag{5}$$

where

$$\| \Delta B \| = \int_{z_i}^{z_f} | B_{measured} - B_{computed} |^2 \, dz \tag{6}$$

$$\| B \| = \int_{z_i}^{z_f} | B_{measured} |^2 \, dz \tag{7}$$

and z_i and z_f are the extremes of the z-axis interval.

In particular the error minimization procedure tries to find iteratively better estimates for both the radii and length of the last inserted TSP in order to reduce the representation error. For this purpose only a set of K points (typically 12-16) around the axial position of the last inserted sources are used in a least square minimization algorithm. Diversely from the radii, which are determined for each TSP independently of the others, the currents to be assigned to each equivalent source are obtained all together as solutions of a global inverse problem for the on axis magnetic field. In fact the field B along the z-axis is related to the thin solenoid pairs parameters by expression 2. When this expression is evaluated at N points z_j on the z-axis, coincident with the positions of the center of the N TSPs, the following linear system is obtained, where the only unknowns are the coefficient M_i, directly related to currents.

$$B_j = \sum_{i=1}^{N} C_{ij} \cdot M_i \tag{8}$$

where $M_i = \frac{\mu \cdot J}{2}$ and the coefficients C_{ij} are

$$C_{ij} = \frac{z_j - z_i + \frac{L_i}{2}}{\sqrt{R_{1i}^2 + (z_j - z_i + \frac{L_i}{2})^2}} - \frac{z_j - z_i - \frac{L_i}{2}}{\sqrt{R_{1i}^2 + (z_j - z_i - \frac{L_i}{2})^2}} - \frac{z_j - z_i + \frac{L_i}{2}}{\sqrt{R_{2i}^2 + (z_j - z_i + \frac{L_i}{2})^2}} + \frac{z_j - z_i - \frac{L_i}{2}}{\sqrt{R_{2i}^2 + (z_j - z_i - \frac{L_i}{2})^2}} \tag{9}$$

The unknown coefficients M_i are easily found by solving the above linear problem.

After this phase, if the error still exceeds the predetermined tolerance, the procedure restarts building a more complex tentative TSP representation, obtained by adding one more TSP, positioned according to the previously discussed rule. If even the modified representation is not adequate other TSP are added one at a time in a similar way until the prescribed tolerance is reached.

It is worth noticing that for a desired target accuracy the minimization process allows us to achieve an optimal solution with respect to the number of TSPs used. In addition, direct calculation of the source currents makes the procedure very robust allowing a solution to be obtained in all the analyzed cases.

4 Examples of application

The TSP procedure has been tested by using several on-axis experimental magnetic field data available by TWT manufacturers. Hereafter three examples of application are illustrated in order to show the achievable degree of accuracy. The first example regards the 3-D representation of an on-axis measured magnetic field curve which exhibits emphasized slope variations. The z-axis total length l=0.08 m has been subdivided into 800 discretization intervals and the target tolerance was fixed to 0.1% as specified in [CL02]. The experiments have shown that the procedure is able to achieve the target representation error by using only 4 thin solenoid pairs, a lower number with respect to ideal loop source models (10 ideal coils)[CL02], thus obtaining a remarkable reduction of computational effort. Figure 2 shows the simulated and measured on-axis magnetic fields and their difference. The TSP parameters (z_C, R_1, R_2, L and M) for this representation are summarized in table 1.

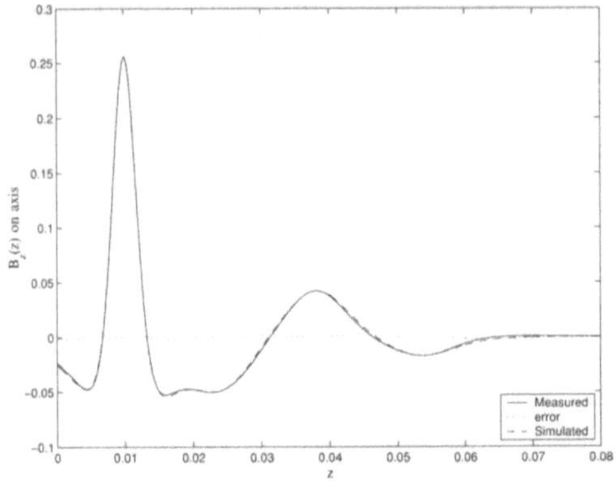

Fig. 2. On-axis profile of the magnetic field for the first example

Table 1. TSP parameters for the first example

$z_C[m]$	$R_1[m]$	$R_2[m]$	$L[m]$	$M[H \cdot A/m^2]$
1.0000e-02	3.3136048e-003	5.9969731e-003	1.7169790e-03	-1.1642526
3.8000e-02	8.2584476e-003	1.6735344e-002	6.6755083e-03	-0.10463623
2.4300e-02	5.7789974e-003	2.2482202e-002	6.6118100e-03	0.042748410
5.3600e-02	8.9701603e-003	1.9385739e-002	7.3357048e-03	0.26056058

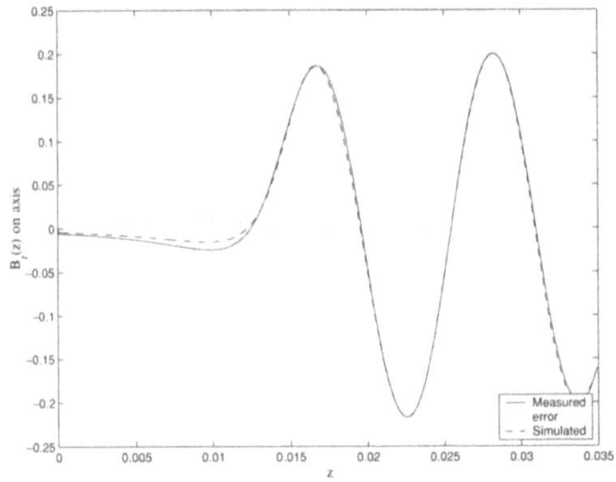

Fig. 3. On-axis profile of the magnetic field for the second example

Table 2. TSP parameters for the second example

$z_C[m]$	$R_1[m]$	$R_2[m]$	$L[m]$	$M[H \cdot A/m^2]$
2.2540e-02	2.7133626e-03	5.4060805e-03	4.2544959e-03	0.35228756
3.3670e-02	2.9002675e-03	4.9427654e-03	4.3723550e-03	0.45541084
1.6660e-02	2.5553433e-03	8.2093684e-03	3.4851959e-03	-0.20566813
2.8210e-02	3.3791592e-03	4.2215791e-03	3.5500570e-03	-0.80645423

The second example regards a typical magnetic field profile adopted in the focalization of the beam of an electron gun. The z-axis total length l = 0.035 m has been subdivided into 500 discretization intervals. The target tolerance was chosen as 0.5% and the number of equivalent sources coming out from the iterative procedure for this example is 4. Figure 3 shows the simulated and measured on-axis magnetic fields. Table 2 summarizes the resulting thin solenoid pairs parameters (z_C, R_1, R_2, L and M) for this example.

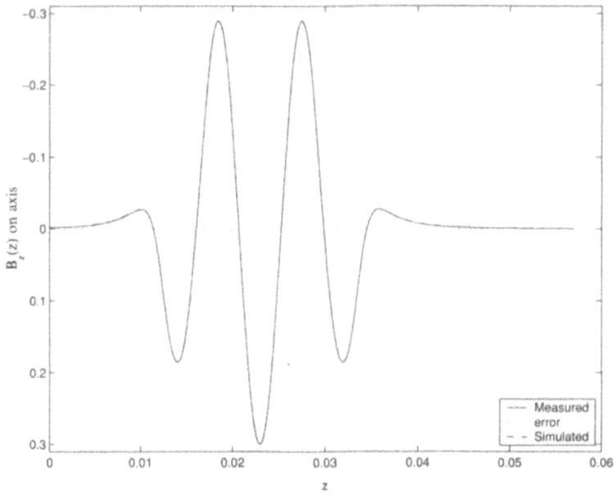

Fig. 4. On-axis profile of the magnetic field for the 3rd example

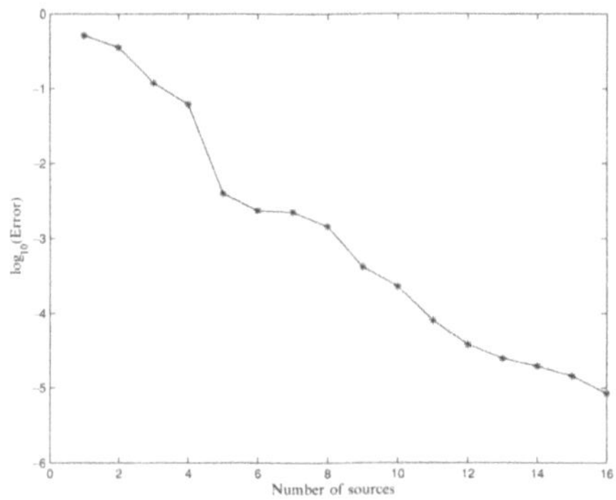

Fig. 5. Monotonic behavior of the representation error

Table 3. TSP parameters for the third example

$z_C[m]$	$R_1[m]$	$R_2[m]$	$L[m]$	$M[H \cdot A/m^2]$
2.30e-02	2.4667401e-03	2.5073321e-03	3.4908720e-03	-17.708462
1.84e-02	2.3315473e-03	2.5022015e-03	3.0303974e-03	3.9044845
2.76e-02	1.9167791e-03	3.9493502e-03	3.0508228e-03	0.39466338
3.20e-02	1.6986424e-03	4.4763201e-03	2.8323448e-03	-0.21892332
1.38e-02	1.8448416e-03	3.3865365e-03	2.8589395e-03	-0.32496537

The last example shows the performance of the algorithm in the representation of a magnetic field profile due to a short regular PPM (Periodic Permanent Magnets) structure. Even for this example the total length of the z-axis l = 0.058 m has been divided in 500 intervals and the target tolerance was chosen as 0.5%. The number of thin solenoid pairs coming out from procedure was 5 exactly as the numbers of PPM sources; their parameters are summarized in table 3. Furthermore in order to assess the robustness of the TSP procedure with respect to convergence a second test was performed assigning an end tolerance of 0.001%. In Fig. 5 the $log_{10}(Error)$ (the representation error) is plotted in order to demonstrate the monotonic decreasing behaviour of the representation error. Figure 4 shows the perfect matching between the measured and simulated magnetic profile achieved for this example as expected.

5 Conclusions

The iterative procedure presented gives accurate representations of the 3-D focusing axisymmetric magnetic field in TWTs devices by using a minimum number of TSP sources. The parameters of the TSP model (axial position, radii, length and currents) are obtained by minimizing the representation error, assigning a reasonable end iteration tolerance (in the range 0.1-1%). The resulting number of TSP sources gives a realistic estimate of the number of permanent magnets to be used for the focalization system. Furthermore the 3-D field direct computation by using analytical formulae of source model more adherent to PPM gives advantages of high accuracy even for off-axis values and consistent reduction of computational effort with respect to the case of numerical calculations performed by means of dedicated FE tools.

References

[Vau72] Vaughan, J. R. M.: Representation of axisymmetric magnetic fields in computer programs. IEEE Trans. Electron Devices, vol. **ED-19**, pp.144-151, (1972)

[Vau74] Vaughan, J. R. M.: Methods of finding the parameters of ideal current loops for computer simulation of magnetic fields. IEEE Trans. Electron Devices,vol. **ED-21**, pp. 310-312,(1974)

[Sta79] Stankiewicz, N.: A matrix solution for the simulation of magnetic fields with ideal current loops. IEEE Trans. Electron Devices. vol. **ED-26**, pp.1598-1601, (1979)

[Jac99] Jackson, R. H.: Off-axis expansion solution of Laplace's equation: Application to accurate and rapid calculation of coil magnetic fields. IEEE Trans. Electron Devices, vol. **46**, pp. 1050-1062, (1999)

[CL02] Coco, S. and Laudani, A. : An Iterative Approach for the 3D Representation of Focusing Axisymmetric Magnetic Fields in TWT Collectors. IEEE Trans. on Magnetics, vol. **38**, no. **2**, pp. 1137-1140, (2002)

[CEL01] Coco, S., Emma, F., Laudani, A., Pulvirenti, S. and Sergi, M. : COCA: a novel 3D FE Simulator for the Design of TWTs Multistage Collectors. IEEE Trans. on Electron Devices, Vol. 48, No 01, pp. 24-31, (2001)

Hybridised PTD/AWE for Modelling Wide-Band Electromagnetic Wave Scattering

M. Condon, C. Brennan and E. Dautbegovic

Research Institute for Networks and Communications Engineering, School of Electronic Engineering, Dublin City University, Ireland

Abstract A hybridised Physical Theory of Diffraction (PTD) / Asymptotic Waveform Evaluation (AWE) technique is presented for the efficient solution of electromagnetic wave scattering problems over a wide frequency band. The scatterer is discretised using Rao Wilton Glisson (RWG) basis functions. Regions of the scatterer where the PTD solution is deemed accurate are identified and the corresponding basis coefficient amplitudes are computed using this asymptotic technique. A revised matrix equation is then formed for the surface currents over the remainder of the structure. The AWE technique is used to efficiently solve this matrix equation over a wide frequency band.

1 Introduction

Many practical electromagnetics problems require the solution of a large-scale linear system over a wide frequency band. Some examples include the analysis of antennas, the computation of radar cross sections, and the calculation of scattering from perfectly and imperfectly conducting objects. There exist several candidate formulations for such problems. The integral equation is a compact formulation offering a full-wave solution. It proceeds by expressing the scattered field in terms of an integral involving surface currents and an appropriate Green's function and then enforcing an appropriate boundary condition to yield an integral equation. The Method of Moments (MoM) [1] is frequently employed to solve such an integral equation formulation. A major drawback of the MoM is the necessity to employ large numbers of basis functions to adequately capture the oscillatory nature of the unknown surface currents. As a consequence, the sequential specification and solution of the associated matrix equations over a wide band of frequencies is a computationally intensive process. An alternative is to use Asymptotic Waveform Evaluation (AWE)[2]. AWE involves expanding the unknown basis function coefficient vector in a Taylor series around a central frequency. Derivatives (with respect to the wave-number) of the impedance matrix and incident field vector are used to evaluate the coefficients of this expansion. From this Taylor expansion a more accurate rational function expression can be computed.

For large scatterers the necessity to compute and store the inverse and derivatives of the impedance matrix at the expansion frequency places onerous computational requirements on the AWE. In this paper, a hybridised approach is proposed. For a smooth scatterer it is possible to get a reliable estimate of the surface current in regions away from edges and corners by employing an asymptotic solution. The PTD solution that we employ supplements the Physical Optics current with edge waves based on Sommerfeld's solution for a perfectly conducting half-plane excited by a plane wave. For a flat polygon we employ this solution in the central portion of the scatterer, restricting the MoM description to the currents at the edges and corners where the asymptotic solution is less valid. In this fashion we can reduce the storage requirements of the AWE as the problem size grows.

2 Method of moments

The MoM proceeds by expanding the unknown surface current at each point on the scatterer surface $\mathbf{r}$ in terms of a set of N basis functions.

$$\mathbf{J}(k, \mathbf{r}) = \sum_{i=1}^{N} I_i(k)\mathbf{f}_i(\mathbf{r}) \tag{1}$$

We have made explicit the dependence on both wave-number and position of the surface currents. The basis functions $\mathbf{f}_i$ are frequency independent, assuming that we have sufficient quantity of them to adequately capture the oscillatory nature of the surface current. Typically one needs on the order of 10 basis functions per wavelength. In this paper we choose the sub-domain basis functions described in [3]. Inserting (1) into the Electric Field Integral Equation (EFIE) and testing with suitably chosen testing functions yields a matrix equation

$$\mathbf{Z}(k)\mathbf{I}(k) = \mathbf{V}(k) \tag{2}$$

$\mathbf{Z}$ contains information about the interaction between pairs of basis functions and is referred to as the impedance matrix. $\mathbf{V}$ is a vector containing information about the incident field while the vector $\mathbf{I}$ holds the unknown basis coefficients $I_i(k)$. Reference [3] gives explicit formulae for the quantities in equation (2). The solution of (2) yields the unknown amplitudes I_i at the specified frequency from which can be calculated the scattered and total electromagnetic fields at any point. Repeated formation and solution of equation (2) over a range of frequencies is a time-consuming process.

Instead AWE prescribes solving equation (2) at a central frequency f_0 as well as computing derivatives of $\mathbf{Z}(k)$ and $\mathbf{V}(k)$ with respect to the wave-number k at this frequency. These vectors and matrices can be used to form rational approximations to $\mathbf{I}(k)$ at many frequencies.

$$\bar{\mathbf{I}}(k) = \frac{\sum_{j=0}^{m} \mathbf{a}_j (k - k_0)^j}{\sum_{j=0}^{n} \mathbf{b}_j (k - k_0)^j} \tag{3}$$

k_0 is the wave-number at the frequency f_0 while $\mathbf{a}_j$ and $\mathbf{b}_j$ are vectors of size N representing the coefficients of the rational expansion. These coefficients are determined by equating the expression in equation (3) with a Taylor series expansion of the form

$$\bar{\mathbf{I}}(k) = \sum_{j=0}^{n+m+1} \mathbf{m}_j (k - k_0)^j \tag{4}$$

where

$$\mathbf{m}_j = \mathbf{Z}^{-1}(k_0) \left(\frac{\mathbf{V}^{(j)}(k_0)}{j!} - \sum_{q=0}^{j} \frac{(1 - \delta_{q0})\, \mathbf{Z}^{(q)}(k_0)\mathbf{m}_{j-q}}{q!} \right) \tag{5}$$

Explicit expressions for the derivatives of the impedance matrix can be found in [4]. However, implementation of AWE requires the explicit computation and storage of the inverse of the impedance matrix as well as several derivatives. These requirements grow as the scatterer size increases, placing a natural limit on the size of problem that can be thus analysed.

3 Physical Theory of Diffraction

An alternative approach is to use approximate surface currents predicted by a high frequency solution. For scattering from a large polygon a suitable asymptotic solution is the Physical Optics current along with fringe waves produced by the diffracting edges. Referring to Fig. (1) the Physical Optics current is given by

$$\mathbf{J}^{PO}(\mathbf{r}) = 2\hat{n} \times \mathbf{H}^i(\mathbf{r}) \tag{6}$$

In addition each edge produces a fringe or edge wave. The total fringe current due to an edge is given by

$$\mathbf{J}^{edge} = J_l^{edge}\hat{l} + J_m^{edge}\hat{m} \tag{7}$$

The l coefficient is given by

$$J_l^{edge} = -\frac{4}{\sqrt{\pi}} \hat{e}_{i\|} \cdot \mathbf{H}^i(\mathbf{r}_d)\, \mathcal{F}\left(\psi \cos\frac{\phi^i}{2}\right) \exp(\jmath\kappa) \sin\alpha^i \tag{8}$$

while the m coefficient is given by

$$\begin{aligned} J_m^{edge} = {} & \tfrac{4}{\sqrt{\pi}} \left(\cos\alpha_i \cos\phi \hat{e}_{\|}^i + \sin\phi_i \hat{e}_{\perp}^i\right) \cdot \mathbf{H}^i(\mathbf{r}_d)\, \mathcal{F}\left(\psi \cos\tfrac{\phi^i}{2}\right) \exp(\jmath\kappa) \\ & - \tfrac{4}{\sqrt{\pi}} \left(\sin\tfrac{\phi_i}{2} \hat{e}_{\perp}^i + \cos\tfrac{\phi_i}{2} \cos\alpha_i \hat{e}_{\|}^i\right) \cdot \mathbf{H}^i(\mathbf{r}_d)\, \tfrac{e^{-\jmath(ks + \pi/4)}}{\psi} \end{aligned} \tag{9}$$

In both the equations above we have used the expressions

$$\psi = \sqrt{2ks} \sin\alpha^i \tag{10}$$

$$\kappa = ks\left(\sin^2\alpha_i \cos\phi_i - \cos^2\alpha_i\right) + \pi/4 \tag{11}$$

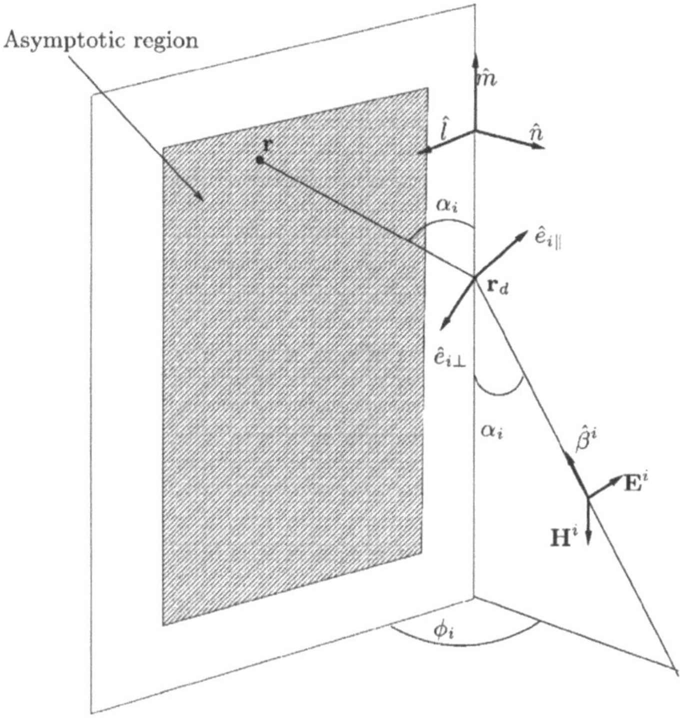

Fig. 1. Geometry for problem

The Fresnel function is defined by

$$\mathcal{F}(x) = \int_x^\infty \exp\left(-\jmath t^2\right) dt \tag{12}$$

while the unit vectors $e^i_\parallel$ and $e^i_\perp$ are given by

$$e^i_\perp = -\frac{\beta^i \times \hat{m}}{|\beta^i \times \hat{m}|}$$
$$e^i_\parallel = \beta^i \times e^i_\perp$$

The diffracting point $\mathbf{r}_d$ can be determined by observing that it must satisfy the Keller condition and s is the distance from this diffracting point to $\mathbf{r}$. For a polygon with N_e edges there is a diffraction contribution from each edge and the current is given by

$$\mathbf{J}^{PTD}(\mathbf{r}) = \mathbf{J}^{PO}(\mathbf{r}) + \sum_{n=1}^{N_e} \mathbf{J}^{edge,n}(\mathbf{r}) \tag{13}$$

The principal drawback of such an asymptotic solution is the failure to rigorously include higher order scattering effects, such as corner diffraction or repeated edge diffraction. Such effects can only be rigorously accounted for using a full-wave technique. However for a smooth scatterer we expect it to yield reasonable results away from the corners and edges.

4 Hybrid Method

We have investigated a hybrid method which proposes to use each formulation in the region for which it is best suited. To achieve this we split the scatter into two regions (See Fig. (1)). Region 1 is called the method of moments region, where we will use a matrix equation to compute the currents. It contains N_{MOM} basis functions. Region 2 is the asymptotic region where we shall approximate the surface current using the Physical Optics and fringe wave currents.

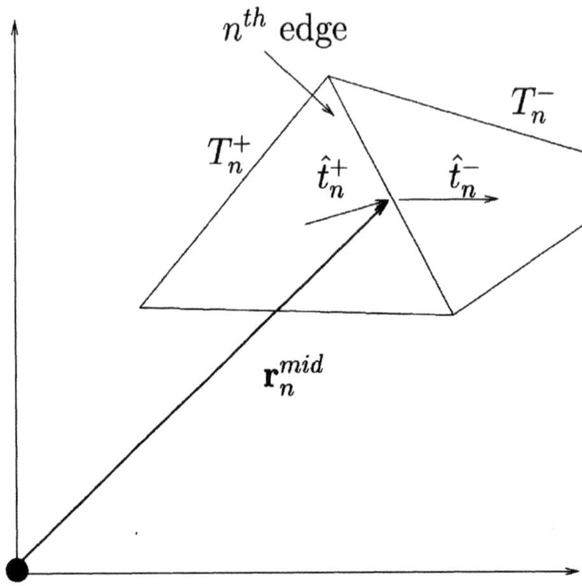

Fig. 2. Geometry for computing current coefficient values for region 2

It contains N_{PTD} basis functions. Obviously N_{MOM} and N_{PTD} sum to give us N the total number of basis functions. By re-arranging the ordering of our basis functions if necessary we can express the matrix equation (2) as

$$\mathbf{Z} = \begin{bmatrix} \mathbf{Z}^{(1,1)} & \mathbf{Z}^{(1,2)} \\ \mathbf{Z}^{(2,1)} & \mathbf{Z}^{(2,2)} \end{bmatrix} \tag{14}$$

where $Z^{(1,1)}$ is a matrix of order N_{MOM} containing interactions between the basis functions in region 1 while while $Z^{(2,2)}$ is a matrix of order N_{PTD} containing interactions between basis functions in region 2. $Z^{(1,2)}$ and $Z^{(2,1)}$ are matrices representing coupling interactions between the regions. In addition we split the incident vector

$$\mathbf{V} = \begin{bmatrix} \mathbf{V}^{(1)} \\ \mathbf{V}^{(2)} \end{bmatrix} \tag{15}$$

and the current amplitude vector

$$\mathbf{I} = \begin{bmatrix} \mathbf{I}^{(1)} \\ \mathbf{I}^{(2)} \end{bmatrix} \tag{16}$$

If a solution for $\mathbf{I}^{(2)}$ is available one can write

$$\mathbf{Z}^{(1,1)}\mathbf{I}^{(1)} = \mathbf{V}^{(1)} - \mathbf{Z}^{(1,2)}\mathbf{I}^{(2)} \tag{17}$$

$$= \hat{\mathbf{V}}^{(1)} \tag{18}$$

We compute an estimate for $\mathbf{I}_2$ using the technique outlined in [5]. We note that for a point in region 2 the PTD solution can be expressed in terms of basis functions as

$$\mathbf{J}^{PTD}(\mathbf{r}) = \sum_{n=1}^{N_{PTD}} I_n \mathbf{f}_n(\mathbf{r}) \tag{19}$$

We introduce the unit vectors $\hat{t}_n^{\pm}$ as indicated in Fig. (2). These unit vectors are perpendicular to the nth edge and the point $\mathbf{r}_n^{mid}$ which is the centre of the nth edge. Using the fact that $\mathbf{f}_n$ has a normal component of unity across the nth edge and that the normal component vanishes across other edges it is possible to write expressions for the coefficients of the basis functions in region 2 as

$$I_n = \frac{1}{2}\left(\hat{t}_n^+ + \hat{t}_n^-\right) \cdot \mathbf{J}^{PTD}\left(\mathbf{r}_n^{mid}\right) \tag{20}$$

Employing this approximation leaves the lower order matrix equation (18) to be solved numerically. One can employ AWE to efficiently solve this equation over a range of frequencies.

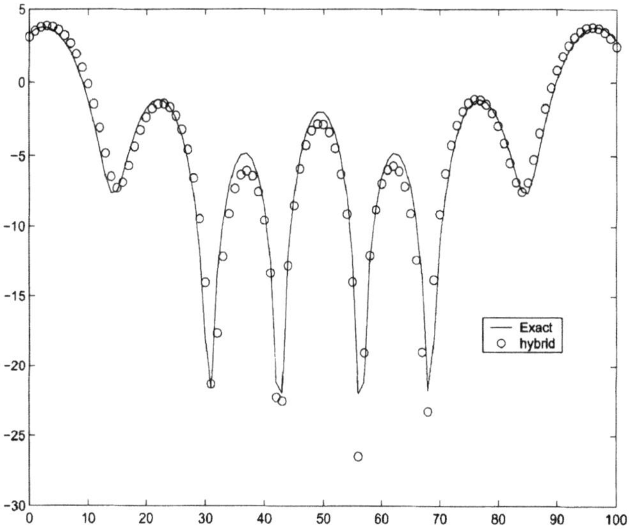

Fig. 3. Total fields 1 wavelength behind 3 wavelength metallic plate

5 Results

The first example involves scattering from a square metallic plate of side length $3cm$ (or 3λ at the expansion frequency $30GHz$) lying in the xy plane with $z = 0$. The southwest corner of the plate is at $(-0.015, -0.015, 0)$ while the northwest corner is at $(0.015, 0.015, 0)$. The incident wave is normally incident from the $+z$ direction and the incident **E** field is in the x direction. The plate was divided initially into a coarse grid of 8 by 8 squares. The outer 28 squares were deemed to be region 1, while the inner 36 squares were region 2. These coarse squares were then further discretised and a total of 1680 basis functions defined. Of these 708 were in region 1, with 972 in region 2. Region 2 coefficients were computed using PTD while the currents in region 1 were computed using a 7th order AWE. The AWE solution was thus confined to a strip of width $0.375cm$ around the edge of the plate, where we expect the PTD solution to perform poorly. Obviously a trade off exists between the size of the AWE region and the computational resource required by it. Numerical experimentation showed that, for this problem, this was the minimum size for region 1 such the resultant average error along a test line was less than 1dB (see below). Unfortunately it is hard to state any general rules that can inform the choice of the physical extent of these regions. The expansion frequency was $30GHz$ and a Pade expansion was used over the range $25GHz$ to $35Ghz$. Total fields were computed computed along a line in the shadow region directly behind the scatterer. The line ran in the x direction from $(-0.045, 0, -0.01)$ to $(0.045, 0, -0.01)$. The average error in dB ranged from $0.2dB$ at the expansion frequency to the worst value of $0.9dB$ at $35GHz$. Figure (3) show the total fields at $35GHz$ along the trial line. The computation time using a 3.2GHz processor was 32954 seconds to make and store the AWE matrices for the exact solution. In contrast the hybrid solution required 5853 seconds.

6 Conclusions

A hybridised Physical Theory of Diffraction (PTD) / Asymptotic Waveform Evaluation (AWE) technique has been presented for the efficient solution of electromagnetic wave scattering problems over a wide frequency band. The scatterer is discretised using Rao Wilton Glisson basis functions. Regions of the scatterer where the PTD solution is deemed accurate are identified and the corresponding basis coefficient amplitudes are computed using the asymptotic solution. A method of moments technique is used to compute the surface currents over the remainder of the structure. The AWE technique is used to efficiently solve this matrix equation over a wide frequency band.

References

1. Harrington R.F. 'Field computation by Moment Methods' New York Macmillan, 1968
2. Reddy C.J., Deshpande, M.D., Cockrell C.R. and Beck F.B.: 'Fast RCS Computation over a Frequency Band Using Method of Moments in conjunction with Asymptotic Waveform Evaluation Technique', IEEE Trans. on Antennas and Propagation, Vol. 46, No. 8, Aug. 1998

3. Rao S., Wilton D. and Glisson A., 'Electromagnetic scattering by surfaces of arbitrary shape' IEEE Trans. Ant. Prop. 30 pp. 409-418, 1982
4. Cockrell C.R., and Beck F. B. 'Asymptotic Waveform Evaluation (AWE) technique for frequency domain electromagnetic analysis' NASA technical memorandum 110292
5. Jakobus U. and Landstorfer F. 'Improved PO-MoM hybrid formulation for scattering from Three-Dimensional Perfectly Conducting bodies of arbitrary shape', IEEE Trans. Ant. Prop. Vol. 43 No. 2 February 1995 pp. 162-169

Transverse Electric Plane Wave Scattering by Two Infinitely Long Conducting Elliptic Cylinders: Iterative Solution

A-K. Hamid[1] and Q. Nasir[2]

[1] Department of Electrical/Electronics/ and Computer Engineering, University of Sharjah, P.O. Box 27272, Sharjah, United Arab Emirates, akhamid@sharjah.ac.ae
[2] Department of Electrical/Electronics/and Computer Engineering, University of Sharjah, P.O. Box 27272, Sharjah, United Arab Emirates, nasir@sharjah.ac.ae

Abstract An analytic solution to the problem of a TE polarized plane electromagnetic wave scattering by two infinitely long conducting elliptic cylinders is presented using an iterative procedure to account for the multiple scattered field between the cylinders. To compute the higher order terms of the scattered fields, the translation addition theorem for Mathieu functions is implemented to express the field scattered by one cylinder in terms of the elliptic coordinate system of the other cylinder in order to impose the boundary conditions. Scattered field coefficients of various scattered orders are obtained and written in matrix form. Numerical results are obtained for the scattered field in the far zone for different axial ratio, electrical separation distance and angles of incidence.

Key words: TE scattering, Conducting elliptic cylinders, Iterative solutions, Mathieu functions

1 Introduction

The multiple scattering of a TE polarized plane electromagnetic wave by a system of infinitely long conducting elliptic cylinders is important in a variety of practical applications. For example, the solution may be used to study the scattering by complex bodies modeled by a collection of cylinders, prediction of radiation from elliptical reflector antennas, and to check the accuracy of the results of numerical and approximate methods. Exact analytic solutions of the problem of scattering by a system of N conducting elliptic cylinders have been formulated using the translation addition theorem for Mathieu functions to enforce the boundary conditions [1, 2]. The required computer time and memory to invert the resulting system of matrix increase rapidly with the number of cylinders. In addition, numerical results for certain cylinder dimensions, electrical separations and angles of incidence are difficult to obtain by this analytical method may be due to the associated ill-condition system matrices.

In the present paper an iterative procedure which was employed for the TM case [3] is extended to the TE scattering by an arbitrary oriented two infinitely long conducting elliptic cylinders. This approach requires the solution of the scattered field by each cylinder, assumed to be alone in the incident field that acts as an incident field on the other cylinder. Therefore, the first order scattered field results from the excitation of each cylinder by the incident field only, while the second order scattered field results from the excitation of each cylinder by the first order scattered field. Hence, this iterative procedure continues until the solution convergence. One of the advantages of the iterative procedure is that the proposed solution does not require matrix inversion and therefore the desired scattered field coefficients are obtained after each iteration and used in the subsequent iteration.

The solution of the electromagnetic scattering by a system of N infinitely long conducting cylinders has received little attention in the literature due to the complexity of computing Mathieu functions of higher orders and its associated translation addition theorem. Recently, there have been many studies on the multiple scattering elliptic cylinders [1]-[4], and circular cylinders [5]-[7] using different techniques. Numerical results showing the number of scattered fields are plotted for the normalized echo pattern width with various electrical separations, sizes, angles of incidence, and also compared with published results to demonstrate the efficiency of the method [2].

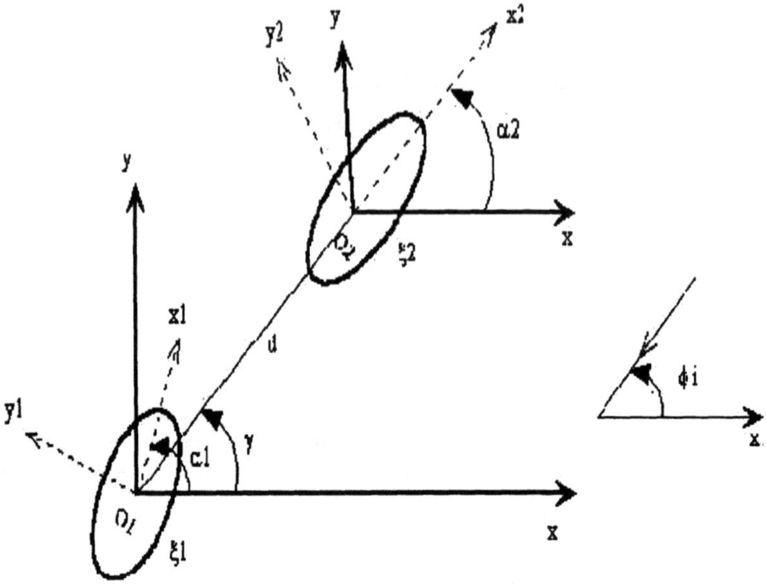

Fig. 1. Scattering geometry of two conducting elliptic cylinders

2 Formulation

Figure 1 shows the scattering geometry of two infinitely long conducting elliptic cylinders with different cross section and arbitrary orientation. The center axes of the two cylinders are assumed to be parallel to the z-axes. The first cylinder is located at the origin o_1 while the second cylinder is located at the polar coordinate point (d,γ) with respect to the global coordinate system (x,y,z). The major axes of the cylinders are a_1 and a_2 while the minor axes are b_1 and b_2 respectively, and each cylinder's local coordinate system makes angle for the first cylinders and for the second cylinder with its global coordinate system.

Consider elliptic coordinate systems u,ν, and z such that

$$x = F\cosh u \cos\nu, y = F\sinh h \sin\nu, z = z \tag{1}$$

where F is the semifocal length,$0 \leq u < \infty$,$0 \leq \nu < 2\pi$ and $-\infty \leq z < \infty$. It is usually convenient to introduce

$$\zeta = \cosh u, \eta = \cos\nu \tag{2}$$

with $1 \leq \zeta \leq \infty$ and $-1 \leq \eta \leq 1$.

Consider the case of a linearly polarized electromagnetic plane wave incident on the two infinitely long conducting elliptic cylinders at an angle ϕ_i with respect to the positive axis x , as shown in Fig. 1, with e^{jwt} time dependence suppressed. The magnetic field component of the TE polarized plane wave of amplitude H_0 is given by

$$H_z^i = H_0 e^{jk\rho\cos(\phi-\phi_i)} \tag{3}$$

where k is the wave number in free space. The incident magnetic field may be expressed in terms of Mathieu functions about the origins o_1 and o_2 as follows:

$$H_{1z}^i = \sum_{m=0}^{\infty} A_{1em} R_{em}^{(1)}(c_1,\zeta_1) S_{em}(c_1,\eta_1) + \sum_{m=1}^{\infty} A_{1om} R_{om}^{(1)}(c_1,\zeta_1) S_{om}(c_1,\eta_1) \tag{4}$$

$$H_{2z}^i = \sum_{m=0}^{\infty} A_{2em} R_{em}^{(1)}(c_2,\zeta_2) S_{em}(c_2,\eta_2) + \sum_{m=1}^{\infty} A_{2om} R_{om}^{(1)}(c_2,\zeta_2) S_{om}(c_2,\eta_2) \tag{5}$$

where

$$A_{1\begin{smallmatrix}em\\om\end{smallmatrix}} = H_0 j^m \frac{\sqrt{8\pi}}{N_{\begin{smallmatrix}em\\om\end{smallmatrix}}(c_1)} S_{\begin{smallmatrix}em\\om\end{smallmatrix}}(c_1, \cos\phi_i^1) \tag{6}$$

$$A_{\substack{2em\\om}} = H_0 j^m \frac{\sqrt{8\pi}}{N_{\substack{em\\om}}(c_2)} S_{\substack{em\\om}}(c_2, \cos\phi_i^2) e^{jkd\cos(\gamma-\phi_i)} \tag{7}$$

$$N_{\substack{em\\om}}(c_1) = \int_0^{2\pi} [S_{\substack{em\\om}}(c_1, \eta_1)]^2 dv \tag{8}$$

$$N_{\substack{em\\om}}(c_2) = \int_0^{2\pi} [S_{\substack{em\\om}}(c_2, \eta_2)]^2 dv \tag{9}$$

$$\phi_i^1 = \phi_i - \alpha_1 \quad , \quad \phi_i^2 = \phi_i - \alpha_2 \tag{10}$$

and $c_1 = kF_1$,$c_2 = kF_2$,S_{em} and S_{om} are the even and odd angular Mathieu functions of order m, respectively,$R_{em}^{(1)}$ and $R_{om}^{(1)}$ are the even and odd radial Mathieu functions of the first kind, and $N_{em}$and N_{om} are the even and odd normalized functions.

The scattered magnetic field from the conducting elliptic cylinders can also be expressed in terms of Mathieu functions as

$$H_{1z}^s = \sum_{m=0}^{\infty} B_{em} R_{em}^{(4)}(c_1, \zeta_1) S_{em}(c_1, \eta_1) + \sum_{m=1}^{\infty} B_{om} R_{om}^{(4)}(c_1, \zeta_1) S_{om}(c_1, \eta_1) \tag{11}$$

$$H_{2z}^s = \sum_{m=0}^{\infty} C_{em} R_{em}^{(4)}(c_2, \zeta_2) S_{em}(c_2, \eta_2) + \sum_{m=1}^{\infty} C_{om} R_{om}^{(4)}(c_2, \zeta_2) S_{om}(c_2, \eta_2) \tag{12}$$

where B_{em} , C_{em} , B_{om}, and C_{om} are the unknown even and odd scattered field expansion coefficients and $R_{em}^{(4)}$ and $R_{om}^{(4)}$ are the even and odd Mathieu functions of the fourth kind.

3 First Order Scattered Fields by Cylinders

The first order scattered field results from the separate excitation of each cylinder by the incident plane wave alone. The boundary condition at the surface of first cylinder requires the tangential components of the total electric field to vanish ($E_{1\eta}^i + E_{1\eta}^s = 0$), i.e.,

$$\begin{aligned}
&\sum_{m=0}^{\infty} A_{1em} R_{em}^{(1)'}(c_1, \zeta_1) S_{em}(c_1, \eta_1) \\
&\quad + \sum_{m=1}^{\infty} A_{1om} R_{om}^{(1)'}(c_1, \zeta_1) S_{om}(c_1, \eta_1) \\
&\quad + \sum_{m=0}^{\infty} B_{em}^1 R_{em}^{(4)'}(c_1, \zeta_1) S_{em}(c_1, \eta_1) \\
&\quad + \sum_{m=1}^{\infty} B_{om}^1 R_{om}^{(4)'}(c_1, \zeta_1) S_{om}(c_1, \eta_1) = 0
\end{aligned} \tag{13}$$

where B_{em}^1 and $B_{om}^1$are the first order scattered field expansion coefficients of the first cylinder. A similar equation may be written corresponds to the second cylinder. Using the orthogonality properties of the angular Mathieu function yields the first order scattered field coefficients, which may be written for each cylinder in matrix form as

$$\begin{bmatrix} B_{em}^1 \\ B_{om}^1 \end{bmatrix} = \begin{bmatrix} Q_{enm}^{11} & 0 \\ 0 & Q_{onm}^{11} \end{bmatrix} \begin{bmatrix} A_{1em} \\ A_{1om} \end{bmatrix} \tag{14}$$

$$\begin{bmatrix} C_{em}^1 \\ C_{om}^1 \end{bmatrix} = \begin{bmatrix} Q_{enm}^{22} & 0 \\ 0 & Q_{onm}^{22} \end{bmatrix} \begin{bmatrix} A_{2em} \\ A_{2om} \end{bmatrix} \tag{15}$$

where C^1_{em} and C^1_{om} are the first order scattered field coefficients of the second cylinder, and

$$Q^{11}_{enm} = \frac{R^{(1)'}_{en}(c_1,\zeta_1)}{R^{(4)'}_{en}(c_1,\zeta_1)}, \quad Q^{11}_{onm} = \frac{R^{(1)'}_{on}(c_1,\zeta_1)}{R^{(4)'}_{on}(c_1,\zeta_1)}, \tag{16}$$

$$= 0, n \neq m, \qquad\qquad = 0, n \neq m$$

Similar equations may be written correspond to Q^{22}_{enm} and Q^{22}_{onm} .

4 Higher Order Scattered Fields by Cylinders

The second order scattered field results from the excitation of each cylinder by the scattered field from the other cylinder due to the initial incident field. The boundary condition at the surface of first cylinder requires the tangential components of the total electric field to be zero, i.e.,

$$\sum_{m=0}^{\infty} C^1_{enm} R^{(4)'}_{em}(c_2,\zeta_2) S_{em}(c_2,\eta_2) + \sum_{m=0}^{\infty} C^1_{onm} R^{(4)'}_{om}(c_2,\zeta_2) S_{om}(c_2,\eta_2) + \sum_{m=0}^{\infty} B^2_{em} R^{(4)'}_{em}(c_1,\zeta_1) S_{em}(c_1,\eta_1) + \sum_{m=0}^{\infty} B^2_{om} R^{(4)'}_{om}(c_1,\zeta_1) S_{om}(c_1,\eta_1) = 0 \tag{17}$$

where B^2_{em} and B^2_{om} are the second order scattered field expansion coefficients of the first cylinder. To enforce the boundary condition, the first order scattered field from the second cylinder must be expressed in terms of the coordinate systems of the first cylinder by using the addition theorem of the Mathieu functions [1]-[3]. Thus, the second order scattered field coefficients which may be written for each cylinder in matrix form as

$$\begin{bmatrix} B^2_{em} \\ B^2_{om} \end{bmatrix} = \begin{bmatrix} Q^{11}_{enm} & 0 \\ 0 & Q^{11}_{onm} \end{bmatrix} \begin{bmatrix} Q^{12}_{eenm} & Q^{12}_{eonm} \\ Q^{12}_{oenm} & Q^{12}_{oonm} \end{bmatrix} \begin{bmatrix} C^1_{1em} \\ C^1_{1om} \end{bmatrix} \tag{18}$$

$$\begin{bmatrix} C^2_{em} \\ C^2_{om} \end{bmatrix} = \begin{bmatrix} Q^{22}_{enm} & 0 \\ 0 & Q^{22}_{onm} \end{bmatrix} \begin{bmatrix} Q^{21}_{eenm} & Q^{21}_{eonm} \\ Q^{21}_{oenm} & Q^{21}_{oonm} \end{bmatrix} \begin{bmatrix} B^1_{1em} \\ B^1_{1om} \end{bmatrix} \tag{19}$$

where C^2_{em} and C^2_{om} are the second order scattered field expansion coefficients of the second cylinder, and

$$Q^{12}_{eenm} = WE^{2\to1}_{enm}, \; Q^{12}_{eonm} = WE^{2\to1}_{onm}, \; Q^{12}_{oenm} = WE^{2\to1}_{enm}, \; Q^{12}_{oonm} = WE^{2\to1}_{onm} \tag{20}$$

are defined in [1]-[3]. Similar equations may be written correspond to Q^{21}_{eenm},Q^{21}_{eonm},Q^{21}_{oenm}, and Q^{21}_{oonm}.

To obtain a general solution, we solve for the higher order scattered field which are sensitive to the electrical size , separation distance between the cylindrers and the angles of incidence. The general expression for the kth order scattered field coefficients may be written as

$$\begin{bmatrix} B^k_{em} \\ B^k_{om} \end{bmatrix} = \begin{bmatrix} Q^{11}_{enm} & 0 \\ 0 & Q^{11}_{onm} \end{bmatrix} \begin{bmatrix} Q^{12}_{eenm} & Q^{12}_{eonm} \\ Q^{12}_{oenm} & Q^{12}_{oonm} \end{bmatrix} \begin{bmatrix} C^{k-1}_{1em} \\ C^{k-1}_{1om} \end{bmatrix}, \quad k = 2, 3 \cdots \tag{21}$$

$$\begin{bmatrix} C^k_{em} \\ C^k_{om} \end{bmatrix} = \begin{bmatrix} Q^{22}_{enm} & 0 \\ 0 & Q^{22}_{onm} \end{bmatrix} \begin{bmatrix} Q^{21}_{eenm} & Q^{21}_{eonm} \\ Q^{21}_{oenm} & Q^{21}_{oonm} \end{bmatrix} \begin{bmatrix} B^{k-1}_{1em} \\ B^{k-1}_{1om} \end{bmatrix}, \quad k = 2, 3 \cdots \tag{22}$$

It should be noted that the matrices in equations 21 and 22 are computed once (i.e. k=2) for the electrical size and electrical separations considered and used for the subsequent iterations (i.e. k = $3, 4, \ldots$).

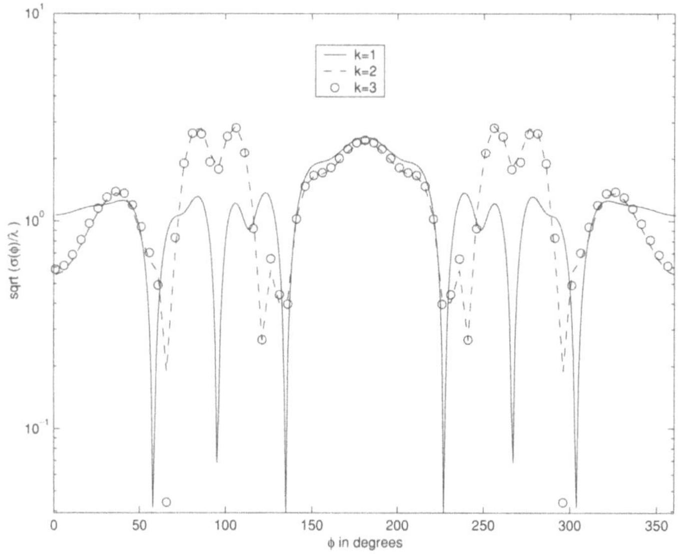

Fig. 2. Normalized echo pattern width versus the scattering angle ϕ for two identical elliptic cylinders with $ka = 5.0, kb = 2.5, kd = 10, \alpha_1 = \alpha_2 = 0^o, \phi_i = 0^o, \gamma = 0^o$

Once the scattered field coefficients are determined, the total scattered field from the cylinderss due to the kth order scattered field can be written as

$$H_z^s = \left(\frac{j}{k\rho}\right)^{0.5} e^{-jk\rho} \left\{ \sum_{k=1,2,\cdots} \left\{ \sum_{m=0}^{\infty} j^m B_{em}^k S_{em}(c_1, \cos(\phi - \alpha_1)) + \sum_{m=1}^{\infty} j^m B_{om}^k S_{om}(c_1, \cos(\phi - \alpha_1)) + e^{jkd\cos(\gamma - \phi)} \left\{ \sum_{m=0}^{\infty} j^m C_{em}^k S_{em}(c_2, \cos(\phi - \alpha_2)) + \sum_{m=1}^{\infty} j^m C_{om}^k S_{om}(c_2, \cos(\phi - \alpha_2)) \right\} \right\} \right\} \quad (23)$$

Far field data are usually expressed in terms of the scattering cross section per unit length, i.e., the echo width. For the TE polarization case it is defined as

$$\sigma_{TE} = 2\pi\rho \lim_{\rho \to \infty} \frac{|H_z^s|^2}{|H_z^i|^2} \quad (24)$$

5 Numerical Results

In order to solve for the unknown scattered field coefficients, the infinite series are first truncated to include only the first N terms, where N in general, is a suitable truncation number proportional to the cylinders electrical size. In the computation, the value of N has been chosen to impose a convergence condition that provides solution accuracy with at least four significant figures. To check the accuracy of our computer program, we recomputed first the results given in references [2] and we obtained complete agreement between both methods ,by only implementing in some cases the first order scattered field using the iterative solution. Figure 2 shows the numerical result of the normalized echo width pattern $\sqrt{\sigma/\lambda}$ versus the scattering angle ϕ for two identical conducting elliptic cylinders with electrical major axes $ka = 5$ and electrical minor axes $kb = 2.5$. The electrical separation distance between the center of the cylinders is assumed to be $kd = 10$ (touching) and at an angle of incidence $\phi_i = 0^o$ (endfire incidence). It can be

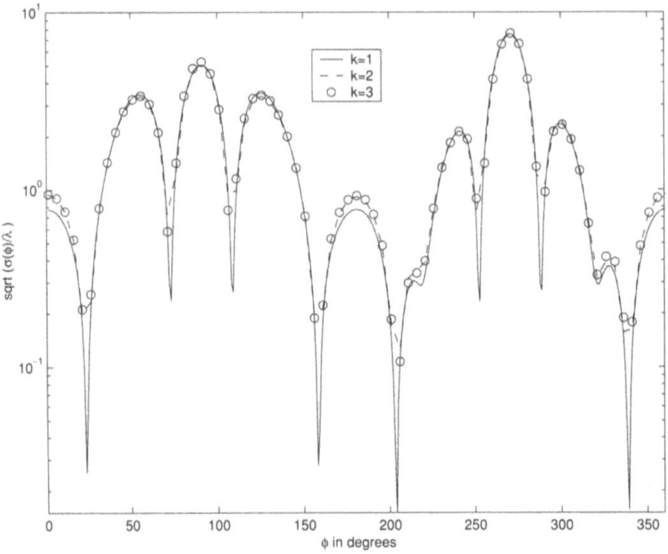

Fig. 3. Normalized echo pattern width versus the scattering angle ϕ for two identical elliptic cylinders with $ka = 5.0, kb = 2.5, kd = 10, \alpha_1 = \alpha_2 = 0^o, \phi_i = 90^o, \gamma = 0^o$

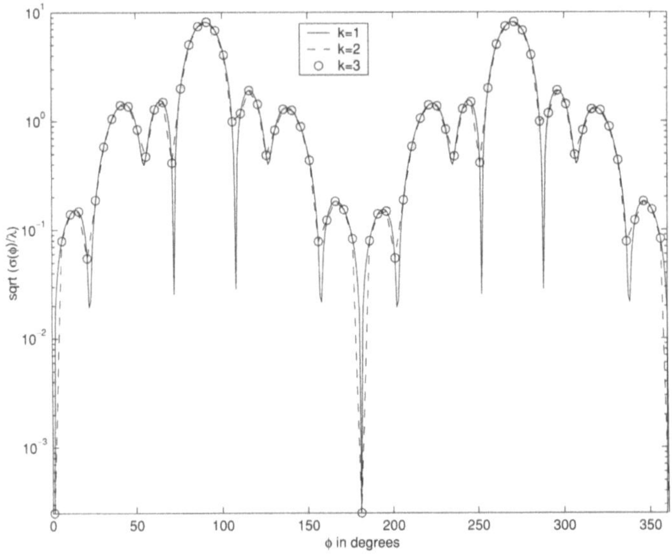

Fig. 4. Normalized echo pattern width versus the scattering angle ϕ for two identical strips with $ka = 5.0, kb = 0.0, kd = 10, \alpha_1 = \alpha_2 = 0^o, \phi_i = 90^o, \gamma = 0^o$

seen that the results of the first order field scattered order (k=1) presented by solid line is not satisfactory since kd is not large when it is compared with cylinders dimensions [2]. This is because the first order scattered field does not take into account the interaction between the cylinders and hence k=1 represents the sum of the scattered field due to the incident field only.The significance of the multiple scattered fields can be seen in the second order scattered term (k=2) which includes the scattered fields due to the plane wave incidence plus the scattered fields due to the first order scattered field due to the incident field on each cylinder. However, the results show that three scattered field orders are needed to obtain convergent solution. The results also show that only $k \approx 2$ is needed for scattering angles higher than 140 degrees and less than 240 degrees. Figure 3 has the same electrical dimensions of Fig. 2 except with $\phi_i = 90^o$(broadside incidence).

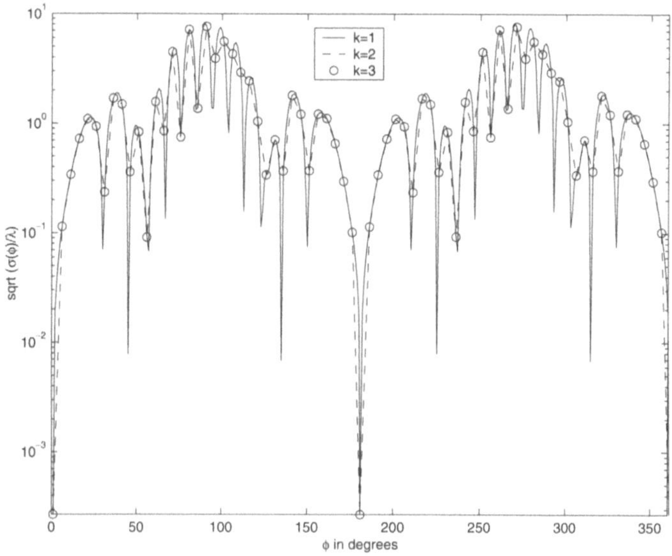

Fig. 5. Normalized echo pattern width versus the scattering angle ϕ for two identical strips with $ka = 5.0, kb = 0.0, kd = 40, \alpha_1 = \alpha_2 = 0^o, \phi_i = 90^o, \gamma = 0^o$

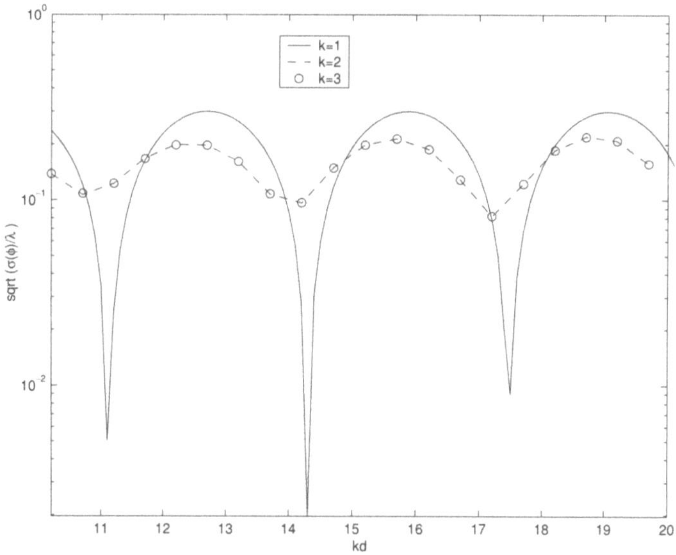

Fig. 6. Normalized backscattering cross section versus the electrical length kd for two identical elliptic cylinders with $ka = \pi, kb = \pi/4, \alpha_1 = \alpha_2 = 0^o, \phi_i = 0^o, \gamma = 0^o$

It can be seen that the shape of the echo pattern changes significantly by changing the incident angle and the location of peak values is shifted from 180 degrees to 90 degrees.

Figure 4 shows the numerical result of the normalized echo width pattern versus the scattering angle ϕ for two identical identical strips with electrical major axes $ka = 5$ and electrical minor axes $kb = 0$. The electrical separation between the center of the strips is assumed to be $kd = 10$ and at an angle of incidence $\phi_i = 90^o$. Figure 5 has the same electrical dimensions of Fig. 4 except more interaction between cylinders in case of $kd = 40$. *It can be noticed that there is an abrupt change in the echo pattern as the scattering angles varies if incidence angle is* $\phi_i = 0^o$ *while the change is smoother if the incidence angle of is changed to* $\phi_i = 90^o$

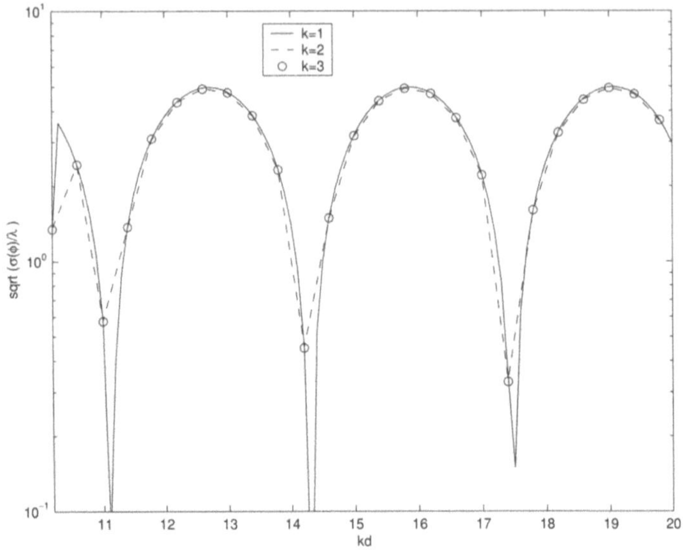

Fig. 7. Normalized backscattering cross section versus the electrical length kd for two identical elliptic cylinders with $ka = \pi, kb = \pi/4, \alpha_1 = \alpha_2 = 0^o, \phi_i = 90^o, \gamma = 0^o$

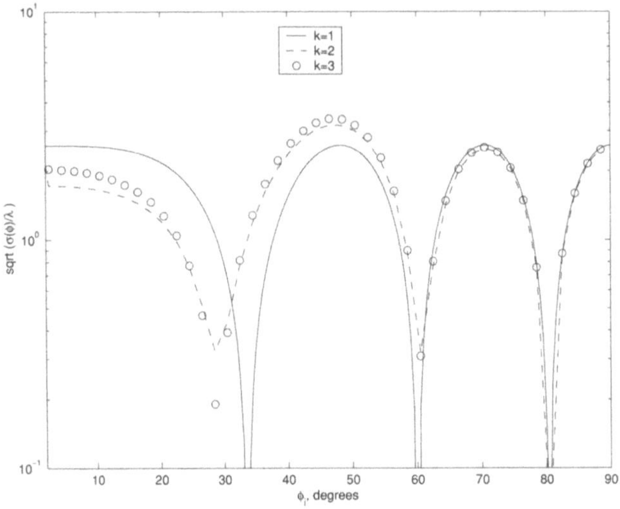

Fig. 8. Normalized backscattering cross section versus the incident angleϕ_i for two identical elliptic cylinders with $ka = \pi, kb = \pi/2, kd = 3\pi, \alpha_1 = \alpha_2 = 0^o, \phi_i = 0^o, \gamma = 0^o$

Figure 6 shows the numerical result of the normalized backscattering cross section versus kd for an elliptic cylinder with $ka = \pi, kb = \pi/4$ and at an angle of incidence $\phi_i = 0^o$. It can be seen that three orders scattered fields required to obtain satisfactory solutions. Figure 7 is similar to Fig. 6 except $\phi_i = 90^o$. Figure 8 shows the numerical result of the normalized backscattering cross section versus versus the incident angle ϕ_i for two identical elliptic cylinders with $ka = \pi, kb = \pi/2$,and $kd = 3\pi$. Again scattering order $k = 3$ is needed to have convergent operation especially at incident angles less than 60°. It is noticed that the normalized backscattering behaviour is sinusodialy when the incident angle $\phi_i > 45^o$ with decreasing maximum peaks.

6 Conclusions

We have investigated the problem of multiply field scattered due to a TE polarized plane electromagnetic wave incident on arbitrary oriented two perfectly conducting elliptic cylinders. An iterative procedure was presented for the first time, for TE case, in elliptic coordinate systems and the boundary conditions were implemented using the translation addition theorem. The numerical results indicated that the number of multiple scattered fields depends on the shape and electrical size of the scatterers, electrical separations and incident angles. A potential advangane of using the iterative solution is that of saving computer time and memory by avoiding the inversion of system matrix.

Acknowledgement. The authors wish to acknowledge the support provided by the University of Sharjah and United Arab Emirates University, U.A.E.

References

[1] Sebak A. : Transverse magnetic scattering by parallel conducting elliptic cylinders. Can. J. Phys., **69**, 1233–1241(1991)

[2] Sebak A. and Antar A. : Multiple scattering by parallel conducting elliptic cylinders: TE case.IEE Proc.-Microw. Antennas Propag., **142**, 178–182(1995)

[3] Hamid A-K.and Hussein M.I.:Iterative solution to the electromagnetic plane wave scattering by two parallel conducting elliptic cylinders . J. of Electromagnetic Waves and Applications, **17**, 813–828(2003)

[4] Hamid A-K. :Iterative Solution to the TM Scattering by Two Infinitley Long Lossy Dielelctric Elliptic Cylinders. J. of Electromagnetic Waves and Applications, **18**, 529–546(2004)

[5] Hongo K. :Multiple scattering by two conducting circular cylinders. IEEE Trans. Antennas Propagat., **26**, 784–751(1978)

[6] Ragheb H. and Hamid M. : Scattering by N parallel conducting circular cylinders.Int. J. of Electronics, **59**, 407–421(1985)

[7] Elsherbeni A.Z. and Hamid M. : Scattering by parallel conducting circular cylinders . IEEE Trans. Antennas Propagat., **35**, 335–358(1987)

Simulation of Microwave and Semiconductor Laser Structures Including PML: Computation of the Eigen Mode Problem, the Boundary Value Problem, and the Scattering Matrix

G. Hebermehl[1], J. Schefter[1], R. Schlundt[1], T. Tischler[2], H. Zscheile[2] and W. Heinrich[2]

[1] Weierstrass Institute for Applied Analysis and Stochastics, Mohrenstr. 39, 10117 Berlin, Germany, {hebermehl,schlundt}@wias-berlin.de
[2] Ferdinand-Braun-Institut für Höchstfrequenztechnik, Gustav-Kirchhoff-Str. 4, 12489 Berlin, Germany, w.heinrich@ieee.org

1 Introduction

Today, electromagnetic simulation forms an indispensable part in the development of microwave circuits as well as in diode laser design. Since the simulation methods are computationally too expensive to handle complete microwave circuits, analysis has to concentrate on critical parts, such as transmission-line discontinuities and junctions. These elements can be represented by the basic description shown in Fig. 1: a structure of arbitrary geometry which is connected to the remaining circuit by transmission lines. The passive structure (e.g. coplanar waveguide, coupled spiral inductors, via hole, impedance step) forms the central part of the problem. Short transmission line sections are attached to it in order to describe its interaction with other circuit elements.

2 Scattering Matrix

The aim consists in the computation of the scattering matrix, which describes the structure in terms of the wave modes on the transmission line sections at the ports. The wave-mode quantities are derived by assuming the transmission-line sections to be infinitely long and longitudinally homogeneous. The generalized scattering matrix is defined as follows:

$$S = (S_{\rho,\sigma}), \quad \rho,\sigma = 1(1)m_s, \quad \text{with} \quad m_s = \sum_{p=1}^{\overline{p}} m^{(p)}, \quad \rho = l + \sum_{q=1}^{p-1} m^{(q)}. \tag{1}$$

$m^{(p)}$ denotes the number of modes which have to be taken into account at the port p. $\overline{p}$ is the number of ports. The modes on a port p are numbered with l, $l = 1(1)m^{(p)}$. That means, the dimension m_s of this matrix is determined by the total number of modes at all ports.

The computation of the scattering matrix is outlined as follows. The scattering matrix can be extracted from the orthogonal decomposition of the electric field into a sum of mode fields [4]. This has to be done at a pair of neighboring cross-sectional planes z_p and $z_{p+\Delta p}$ on each waveguide for a number of linearly independent excitations. The electric fields at the planes z_p and $z_{p+\Delta p}$ are calculated solving an eigenvalue problem for the infinitely long waveguide (see section 5) and a boundary value problem for the 3D structure (see section 3), respectively.

The computation of the scattering matrix is based on the orthogonality relation for the electric and magnetic fields of different modes

$$\int_{\Omega} (\boldsymbol{E}_{t,l}(z) \times \boldsymbol{H}_{t,m}(z)) \cdot d\boldsymbol{\Omega} = \eta_m \delta_{l,m}, \tag{2}$$

where $\delta_{l,m}$ is the Kronecker symbol. $\boldsymbol{H}_{t,m}$ denotes the transverse magnetic mode fields.

In the case of degenerate modes, i.e., the algebraic multiplicity of the corresponding eigenvalues is larger than unity, we have to use first (2) in order to orthogonalize the modes (see [12]).

For sake of simplicity we assume the cross section is located on the left-handed (x,y)-plane of the enclosure (see Fig. 1). We consider all exciting modes with amplitudes a_l in positive z-direction and all outgoing modes with amplitudes b_l in negative z-direction. Then the transverse mode field at a cross-sectional plane z is given by

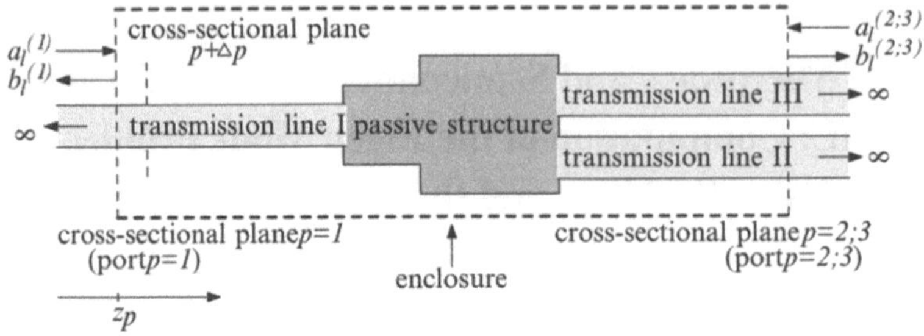

Fig. 1. The basic structure under investigation

$$\boldsymbol{E}_t(z) = \sum_{l=1}^{m^{(p)}} a_l \boldsymbol{E}_{t,l} e^{-\jmath k_{z_l} z} + \sum_{l=1}^{m^{(p)}} b_l \boldsymbol{E}_{t,l} e^{+\jmath k_{z_l} z} = \sum_{l=1}^{m^{(p)}} w_l(z) \boldsymbol{E}_{t,l} \tag{3}$$

with

$$w_l(z) = a_l e^{-\jmath k_{z_l} z} + b_l e^{+\jmath k_{z_l} z} = \tilde{a}_l(z) + \tilde{b}_l(z), \tag{4}$$

where k_{z_l} is the propagation constant. The application of (3) with (4) at a pair of neighboring cross-sectional planes z_p and $z_{p+\Delta p}$ gives because of $\boldsymbol{H}_{t,m}(z_{p+\Delta p}) = \boldsymbol{H}_{t,m}(z_p)$:

$$\begin{aligned} \frac{1}{\eta_m} \int_\Omega (\boldsymbol{E}_t(z_p) \times \boldsymbol{H}_{t,m}(z_p) \cdot d\boldsymbol{\Omega} &= \tilde{a}_m^{(p)} + \tilde{b}_m^{(p)} &&= w_m^{(p)}, \\ \frac{1}{\eta_m} \int_\Omega (\boldsymbol{E}_t(z_{p+\Delta p}) \times \boldsymbol{H}_{t,m}(z_p) \cdot d\boldsymbol{\Omega} &= \tilde{a}_m^{(p+\Delta p)} + \tilde{b}_m^{(p+\Delta p)} &&= w_m^{(p+\Delta p)}. \end{aligned} \tag{5}$$

We get $\boldsymbol{E}_t(z_p)$ solving eigenvalue problems for the transmission lines (see section 5). $H_{t,m}(z_p)$ can be computed from the known electric field $E_{t,m}$ of mode m (see [12]). The values of the weighted mode amplitude sums $w_m^{(p)}$ are given (see the discussion to follow). Thus, the normalization constant η_m can be computed by evaluating the orthogonality relation in the first equation of (5). $\boldsymbol{E}_t(z_{p+\Delta p})$ is computed solving boundary value problems for the discontinuity (see section 3). Thus, the weighted mode amplitude sums $w_m^{(p+\Delta p)}$ can be calculated by using the second equation of (5).

By using (see (4))

$$\tilde{a}_m^{(p+\Delta p)} = \tilde{a}_m^{(p)} e^{-\jmath k_{z_l}^{(p)} \Delta z_p}, \quad \tilde{b}_m^{(p+\Delta p)} = \tilde{b}_m^{(p)} e^{+\jmath k_{z_l}^{(p)} \Delta z_p}, \tag{6}$$

we eliminate $\tilde{a}_m^{(p+\Delta p)}$ and $\tilde{b}_m^{(p+\Delta p)}$ in (5), and obtain

$$\tilde{a}_m^{(p)} = \frac{w_m^{(p)} e^{+\jmath k_{zm}^{(p)} \Delta z_p} - w_m^{(p+\Delta p)}}{e^{+\jmath k_{zm}^{(p)} \Delta z_p} - e^{-\jmath k_{zm}^{(p)} \Delta z_p}}, \quad \tilde{b}_m^{(p)} = \frac{w_m^{(p+\Delta p)} - w_m^{(p)} e^{-\jmath k_{zm}^{(p)} \Delta z_p}}{e^{+\jmath k_{zm}^{(p)} \Delta z_p} - e^{-\jmath k_{zm}^{(p)} \Delta z_p}}. \tag{7}$$

By using (7) reflection coefficients

$$r_m^{(p)} = \frac{\tilde{b}_m^{(p)}}{\tilde{a}_m^{(p)}} = \frac{e^{-\jmath k_{zm}^{(p)} \Delta z_p} - \frac{w_m^{(p+\Delta p)}}{w_m^{(p)}}}{\frac{w_m^{(p+\Delta p)}}{w_m^{(p)}} - e^{+\jmath k_{zm}^{(p)} \Delta z_p}} \tag{8}$$

are computed for all modes $\rho = 1(1)m_s$ and all excitations $\nu = 1(1)m_s$.

The values $w_m^{(p)}$ are given, and than we form the vectors

$$\bar{\boldsymbol{w}}_\nu = (\bar{w}_{1,\nu}, \ldots, \bar{w}_{\rho,\nu}, \ldots, \bar{w}_{m_s,\nu})^T, \quad \nu = 1(1)m_s. \tag{9}$$

The vectors (9) have to be linear independent. That is achieved here by choosing the components of $\bar{\boldsymbol{w}}_\nu$ in the following way:

$$\bar{w}_{\rho,\nu} = \begin{cases} |w_m^{(p)}| \text{ for } 1 \le \rho \le m_s + 1 - \nu \\ -|w_m^{(p)}| \text{ for } m_s + 2 - \nu \le \rho \le m_s \end{cases}, \quad \rho = m + \sum_{q=1}^{p-1} m^{(q)}, \tag{10}$$

with

$$w_m^{(p)} = 1.0, \quad m = 1(1)m^{(p)}, \quad p = 1(1)\bar{p}. \tag{11}$$

With this choice of $\bar{\boldsymbol{w}}_\nu$ (see (9), (10), and (11)) the vectors $\bar{\boldsymbol{r}}_\nu$, $\bar{\boldsymbol{a}}_\nu$ and $\bar{\boldsymbol{b}}_\nu$ are built up analogously (see (7) and (8)):

$$\begin{aligned}\bar{\boldsymbol{r}}_\nu &= (\bar{r}_{1,\nu},\ldots,\bar{r}_{\rho,\nu},\ldots,\bar{r}_{m_s,\nu})^T, & \bar{r}_{\rho,\nu} &= r_m^{(p)},\\ \bar{\boldsymbol{a}}_\nu &= (\bar{a}_{1,\nu},\ldots,\bar{a}_{\rho,\nu},\ldots,\bar{a}_{m_s,\nu})^T, & \bar{a}_{\rho,\nu} &= \tilde{a}_m^{(p)},\\ \bar{\boldsymbol{b}}_\nu &= (\bar{b}_{1,\nu},\ldots,\bar{b}_{\rho,\nu},\ldots,\bar{b}_{m_s,\nu})^T, & \bar{b}_{\rho,\nu} &= \tilde{b}_m^{(p)}.\end{aligned} \tag{12}$$

The relation between (ρ,ν) on the one hand and $(m,(p))$ on the other hand is given by (10) and (11). The choice of $\bar{\boldsymbol{w}}_\nu$ and the relations between the indices are demonstrated by a small example in [12].

That means, we have to solve m_s boundary value problems with the boundary conditions (see sections 3 and 6)

$$\boldsymbol{E}_{t,\nu} = \sum_{\rho=1}^{m_s} \bar{w}_{\rho,\nu}\boldsymbol{E}_{t,l}(z_p), \quad \rho = l + \sum_{q=1}^{p-1} m^{(q)}, \quad p = 1(1)\bar{p}, \quad \nu = 1(1)m_s, \tag{13}$$

in order to compute $w_m^{(p+\Delta p)}$.

The scattering matrix S (see (1)) is defined by

$$\bar{\boldsymbol{b}}_\nu = S\bar{\boldsymbol{a}}_\nu, \quad \nu = 1(1)m_s, \tag{14}$$

or (see (12))

$$\bar{b}_{\rho,\nu} = \sum_{\sigma=1}^{m_s} S_{\rho,\sigma}\cdot\bar{a}_{\sigma,\nu}, \quad \rho,\nu = 1(1)m_s. \tag{15}$$

Because of (5) and (8) we have

$$\begin{aligned}\bar{w}_{\rho,\nu} &= \bar{a}_{\rho,\nu} + \bar{b}_{\rho,\nu},\\ 0 &= \bar{r}_{\rho,\nu}\bar{a}_{\rho,\nu} - \bar{b}_{\rho,\nu},\end{aligned} \quad \text{or} \quad \begin{aligned}\bar{a}_{\rho,\nu}(1+\bar{r}_{\rho,\nu}) &= \bar{w}_{\rho,\nu},\\ \bar{b}_{\rho,\nu}(1+\bar{r}_{\rho,\nu}) &= \bar{r}_{\rho,\nu}\bar{w}_{\rho,\nu},\end{aligned} \quad \rho,\nu = 1(1)m_s. \tag{16}$$

Multiplying Eq. (15) with the product $\prod_{\mu=1}^{m_s}(1+\bar{r}_{\mu,\nu})$ gives

$$\bar{b}_{\rho,\nu}\prod_{\mu=1}^{m_s}(1+\bar{r}_{\mu,\nu}) = \sum_{\sigma=1}^{m_s} S_{\rho,\sigma}\bar{a}_{\sigma,\nu}\prod_{\mu=1}^{m_s}(1+\bar{r}_{\mu,\nu}), \quad \rho,\nu = 1(1)m_s. \tag{17}$$

Substitution of (16) into the relation (17) gives

$$R_{\rho,\nu} = \sum_{\sigma=1}^{m_s} S_{\rho,\sigma}W_{\sigma,\nu} \quad \text{or} \quad R = SW \tag{18}$$

with

$$W_{\rho,\nu} = \bar{w}_{\rho,\nu}\prod_{\mu=1,\mu\neq\rho}^{m_s}(1+\bar{r}_{\mu,\nu}), \quad R_{\rho,\nu} = \bar{r}_{\rho,\nu}W_{\rho,\nu}. \tag{19}$$

That means, we have to solve m_s linear algebraic equations in order to compute the $(m_s)^2$ coefficients of S:

$$W^T(S_{\rho,1},\ldots,S_{\rho,m_s})^T = (R_{\rho,1},\ldots,R_{\rho,m_s})^T, \quad \rho = 1(1)m_s. \tag{20}$$

3 Boundary Value Problem

A three-dimensional boundary value problem can be formulated using the integral form of Maxwell's equations in the frequency domain [1] in order to compute the electromagnetic field within the structure of interest:

$$\oint_{\partial\Omega}\boldsymbol{H}\cdot d\boldsymbol{s} = \int_\Omega \jmath\omega[\epsilon]\boldsymbol{E}\cdot d\boldsymbol{\Omega}, \quad \oint_{\cup\Omega}([\epsilon]\boldsymbol{E})\cdot d\boldsymbol{\Omega} = 0, \tag{21}$$

$$\oint_{\partial\Omega}\boldsymbol{E}\cdot d\boldsymbol{s} = -\int_\Omega \jmath\omega[\mu]\boldsymbol{H}\cdot d\boldsymbol{\Omega}, \quad \oint_{\cup\Omega}([\mu]\boldsymbol{H})\cdot d\boldsymbol{\Omega} = 0, \tag{22}$$

$$\boldsymbol{D} = [\epsilon]\boldsymbol{E}, \quad \boldsymbol{B} = [\mu]\boldsymbol{H}. \tag{23}$$

The electric and magnetic flux densities $\boldsymbol{D}$ and $\boldsymbol{B}$ are complex functions of the spatial coordinates. $\omega = 2\pi f$ is the angular frequency of the sinusoidal excitation, and $\jmath^2 = -1$. f denotes the frequency. In the left-hand sides of formulae (21) and (22) Ω is an open surface surrounded by a closed contour $\partial\Omega$. The direction of the element $d\boldsymbol{s}$ of the contour

$\partial\Omega$ is determined according to a right-hand system. In the right-hand sides of (21) and (22) $\cup\Omega$ is a closed surface with an interior volume. The complex electric permittivity $[\epsilon]$ and the magnetic permeability $[\mu]$ are diagonal tensors.

At the ports p the transverse electric field $\boldsymbol{E}_t(z_p)$ is given by superposing weighted transmission line modes $\boldsymbol{E}_{t,l}(z_p)$ (see (3)):

$$\boldsymbol{E}_t(z_p) = \sum_{l=1}^{m^{(p)}} w_l(z_p)\boldsymbol{E}_{t,l}(z_p). \tag{24}$$

The transverse electric mode fields have to be computed solving an eigenvalue problem for the transmission lines (see section 5). All other parts of the surface of the computation domain are assumed to be an electric or a magnetic wall:

$$\boldsymbol{E} \times \boldsymbol{n} = 0 \quad \text{or} \quad \boldsymbol{H} \times \boldsymbol{n} = 0. \tag{25}$$

The simulation of open-region problems usually requires absorbing boundary conditions to properly truncate the computational domain. Perfectly matched layers (PML) are absorption layers. The PML was introduced by Berenger [2] using artificial electric and magnetic conductivities κ_e and κ_m, respectively, and splitting the electromagnetic field components (split-field formulation). The PML was later shown to be equivalent to a complex coordinate stretching of the coordinate space (coordinate stretching formulation, [3]) and to the uniaxial Maxwellian PML formulation [15].

Using the uniaxial PML formulation the original form of Maxwell's equations is retained. That means, we could easily implement the PML into an existing code. A complex permittivity $[\epsilon]$ and a complex permeability $[\mu]$ diagonal tensor are introduced (see (23), (28), and (29)), resulting in a reflection-free interface between the computational area and the lossy PML region:

$$[\epsilon] = (\epsilon)[\Lambda^{(\epsilon)}], \quad [\mu] = (\mu)[\Lambda^{(\mu)}] \tag{26}$$

with

$$(\epsilon) = \mathrm{diag}(\epsilon_x, \epsilon_y, \epsilon_z), \quad (\mu) = \mathrm{diag}(\mu_x, \mu_y, \mu_z). \tag{27}$$

$[\Lambda^{(\epsilon)}]$ and $[\Lambda^{(\mu)}]$ are defined for a PML in x-, y-, or z-direction in the following way ($\nu \in \{\epsilon, \mu\}$):

$$[\Lambda^{(\nu)}] = \left\{ \begin{array}{l} [\Lambda^{(\nu)}]_x = \mathrm{diag}(\frac{1}{\lambda_\nu}, \lambda_\nu, \lambda_\nu) \\ [\Lambda^{(\nu)}]_y = \mathrm{diag}(\lambda_\nu, \frac{1}{\lambda_\nu}, \lambda_\nu) \\ [\Lambda^{(\nu)}]_z = \mathrm{diag}(\lambda_\nu, \lambda_\nu, \frac{1}{\lambda_\nu}) \end{array} \right\} \quad \text{with } \lambda_\nu = 1 - \jmath\frac{\kappa_\nu}{\nu_0\omega}. \tag{28}$$

That means, we get for an overlapping region in x-, y-, and z-direction:

$$[\epsilon] = (\epsilon)[\Lambda^{(\epsilon)}]_x[\Lambda^{(\epsilon)}]_y[\Lambda^{(\epsilon)}]_z \quad \text{and} \quad [\mu] = (\mu)[\Lambda^{(\mu)}]_x[\Lambda^{(\mu)}]_y[\Lambda^{(\mu)}]_z. \tag{29}$$

The quantities ϵ_0 and μ_0 denote the permittivity and the permeability for a vacuum, κ_ϵ and κ_μ the electric and magnetic (introduced for PML) conductivity, respectively. The lossfree and the lossy case are special variants of (28).

The conductivities have to fulfill the relation

$$\frac{\kappa_\epsilon}{\epsilon_0} = \frac{\kappa_\mu}{\mu_0}. \tag{30}$$

There is always an electric or magnetic wall (see (25)) behind the PML. On the one hand, the PML allows computing the leakage due to radiation effects, on the other hand, the PML can be used to suppress the influence of the boundary on the electric behavior of the structure.

4 Maxwellian Grid Equations

Maxwellian grid equations are formulated for staggered nonequidistant rectangular grids [1, 20, 9] and for tetrahedral nets with corresponding dual Voronoi cells using the Finite Integration Technique with lowest order integration formulae:

$$\oint_{\partial\Omega} \boldsymbol{f} \cdot d\boldsymbol{s} \approx \sum(\pm f_i s_i), \quad \int_{\Omega} \boldsymbol{f} \cdot d\boldsymbol{\Omega} \approx f\Omega. \tag{31}$$

4.1 Staggered Nonequidistant Rectangular Grids

The use of rectangular grids is the standard approach. In general, it is very well adapted to planar microwave structures, since most circuits have a basically rectangular geometry. Using (31) Eqs. (21,22) are transformed into a set of grid equations:

$$A^T D_{s/\mu} \boldsymbol{b} = \jmath\omega\epsilon_0\mu_0 D_{A_\epsilon} \boldsymbol{e}, \qquad B D_{A_\epsilon} \boldsymbol{e} = 0, \tag{32}$$

$$A D_s \boldsymbol{e} = -\jmath\omega D_A \boldsymbol{b}, \qquad \tilde{B} D_A \boldsymbol{b} = 0. \tag{33}$$

The vectors $\boldsymbol{e}$ and $\boldsymbol{b}$ contain the components of the electric field intensity and the magnetic flux density of the elementary cells, respectively. The diagonal matrices $D_{s/\mu}$, D_{A_ϵ}, D_s, and D_A contain the information on cell dimension and material. A, B, and $\tilde{B}$ represent the integrals. A is a singular matrix. B and $\tilde{B}$ are rectangular matrices. A, B, and $\tilde{B}$ are sparse and contain the values $1, -1$, and 0 only. An explicit derivation and a discussion of the properties of (32) and (33) can be found in [10].

By eliminating the components of the magnetic flux density from the two equations on the left-hand sides of (32) and (33), we obtain the system of linear algebraic equations

$$(A^T D_{s/\mu} D_A^{-1} A D_s - k_0^2 D_{A_\epsilon})\boldsymbol{e} = 0, \quad k_0 = \omega\sqrt{\epsilon_0\mu_0}, \tag{34}$$

which have to be solved using the boundary conditions (13) and (25), possibly supplemented by PML. k_0 denotes the wavenumber in vacuum.

4.2 Tetrahedral Grids and Voronoi Cells

Using rectangular grids a mesh refinement in one point results in an accumulation of small elementary cells in all coordinate directions, although the refinement is needed only in inner regions. In addition, rectangular grids are not well suited for treatment of curved and non-rectangular structures. A finite-volume method, which uses tetrahedral nets with corresponding Voronoi cells for the three-dimensional boundary value problem, reduces the number of elementary cells by local grid refinement and improves the description of curved structures. The primary grid is formed by tetrahedra and the dual grid by the corresponding Voronoi cells [13].

We consider a tetrahedron $ABCD$ with the internal edge AB (see Fig. 2) and the neighbouring elements, which share the edge AB with it. The electric field intensity components are located at the centers of the edges of the tetrahedra, and the magnetic flux density components are normal to the circumcenters of the triangular faces. The Voronoi cells are polytopes. We use the notations given in Table 1 with $X, Y, Z, W \in \{A, B, C, D\}$, where X, Y, Z, W are different from each other, in order to develop the grid equations for tetrahedral nets. E_{XY} and B_{XYZ} satisfy

$$\begin{aligned} E_{XY} &= -E_{YX}, \\ B_{XYZ} &= B_{YZX} = B_{ZXY} = -B_{YXZ} = -B_{XZY} - -B_{ZYX}, \end{aligned} \tag{35}$$

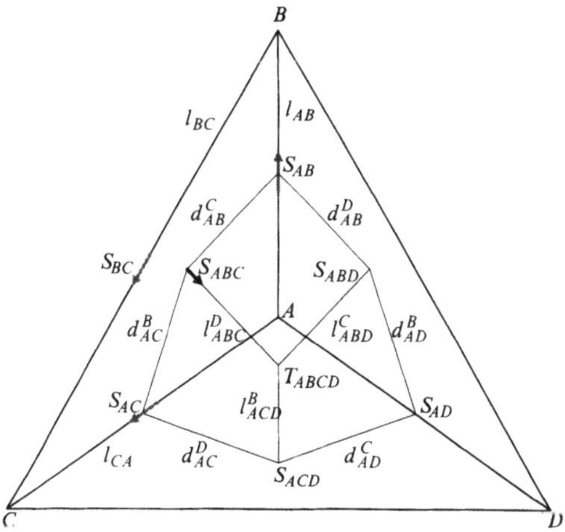

Fig. 2. Tetrahedron with partial areas of the Voronoi cell faces related to node A

Table 1. Notations

X, Y, Z, W	nodes	l_{XY}	distance of X to Y
XY	edge between X and Y	l_{XYZ}^{W}	distance of T_{XYZW} to XYZ
XYZ	triangle	d_{XY}^{Z}	distance of S_{XYZ} to XY
$XYZW$	tetrahedron	a_{XYZ}	area of XYZ
S_{XY}	center of XY	μ_{XYZW}	permeability in $XYZW$
S_{XYZ}	circumcenter of XYZ	ϵ_{XYZW}	permittivity in $XYZW$
T_{XYZW}	circumcenter of $XYZW$		
E_{XY}	magnitude of the electric field on S_{XY}		
B_{XYZ}	magnitude of the magnetic flux density on S_{XYZ}		

respectively. The PML boundary conditions are not implemented for tetrahedral grids, i.e. one has (see (26)-(28)),

$$\mu_x = \mu_y = \mu_z = \mu_{XYZW}, \quad \epsilon_x = \epsilon_y = \epsilon_z = \epsilon_{XYZW}. \tag{36}$$

Using a finite volume approach with the lowest-order integration formulae (31), Eqs. (21) and (22) are transformed into a set of grid equations.

Taking into account the constitutive relations (23) the first equation of (21) is discretized on the dual grid. The internal edge AB is orthogonal to the corresponding Voronoi cell face over which we have to integrate (see Fig. 2). The closed integration path $\partial\Omega$ (see (21) and (31)) consists of the edges with length $s_i = l_{XYZ}^{W}$, and is the polygon around the periphery of the mentioned Voronoi cell face. The vertices of the polygon are the circumcenters of the tetrahedra which share the edge AB with the tetrahedron $ABCD$. $f_i = B_{XYZ}$ denotes the function values on S_{XYZ}. Ω is the area of the Voronoi cell face. $f = E_{AB}$ denotes the function value on the center S_{AB}. Thus, the discretized equation takes the form:

$$\begin{aligned}&\sum\nolimits_{CD} \tfrac{1}{\mu_{ABCD}} \left[l_{ABC}^{D}\, B_{ABC} + l_{ABD}^{C}\, B_{ABD}\right] \\ &= \jmath\omega \left[\sum\nolimits_{CD} \tfrac{1}{2}\epsilon_{ABCD} \left(d_{AB}^{C}\, l_{ABC}^{D} + d_{AB}^{D}\, l_{ABD}^{C}\right)\right] E_{AB}\end{aligned} \tag{37}$$

where the sum is over those tetrahedra $ABCD$ which share the edge AB.

The first equation of (22) is discretized using (31) on the primary grid. We have to integrate over the triangle ABC. This yields the following form:

$$l_{AB}\, E_{AB} + l_{BC}\, E_{BC} + l_{CA}\, E_{CA} = -\jmath\omega a_{ABC}\, B_{ABC}. \tag{38}$$

Now we address the first of the surface integrals (second equation of (21)) reverting to the dual grid. Here, $\cup\Omega$ is a closed surface with an interior volume. The discretization formula (39), with a form similar to the right-hand side of (37) is obtained, except for the additional outer summation taken over all the nodes B neighboring A (in the primary grid). For our final integral equation (second equation of (22)) the primary grid is used again, but now the integration is over the surface of the tetrahedron $ABCD$. As a consequence, the discretized form (40) can be deduced:

$$\sum_{B} \left(\left[\sum_{CD} \frac{1}{2}\epsilon_{ABCD} \left(d_{AB}^{C}\, l_{ABC}^{D} + d_{AB}^{D}\, l_{ABD}^{C}\right) \right] E_{AB} \right) = 0, \tag{39}$$

$$-a_{ABC}\, B_{ABC} - a_{ACD}\, B_{ACD} + a_{ABD}\, B_{ABD} + a_{BCD}\, B_{BCD} = 0. \tag{40}$$

Substituting the components of the magnetic flux density in (37), (38) the number of unknowns in this system can be reduced by a factor of two:

$$\begin{aligned}&\sum\nolimits_{CD} \tfrac{1}{\mu_{ABCD}} \left[\left(\tfrac{l_{ABC}^{D}}{a_{ABC}} + \tfrac{l_{ABD}^{C}}{a_{ABD}} \right) l_{AB}\, E_{AB} + \tfrac{l_{ABC}^{D}\, l_{BC}}{a_{ABC}}\, E_{BC} \right. \\ &\left. + \tfrac{l_{ABC}^{D}\, l_{CA}}{a_{ABC}}\, E_{CA} + \tfrac{l_{ABD}^{C}\, l_{BD}}{a_{ABD}}\, E_{BD} + \tfrac{l_{ABD}^{C}\, l_{DA}}{a_{ABD}}\, E_{DA} \right] \\ &= \tfrac{\omega^2}{2} \left[\sum\nolimits_{CD} \epsilon_{ABCD} \left(d_{AB}^{C}\, l_{ABC}^{D} + d_{AB}^{D}\, l_{ABD}^{C}\right)\right] E_{AB}.\end{aligned} \tag{41}$$

Here, summation is taken over these tetrahedra $ABCD$, which possess the common edge AB.

The method requires a triangulation of the domain in tetrahedra. Thus, triangulation algorithms and grid management are of major importance in the numerical simulation.

Using the grid management interface of the software package pdelib [6], the meshing algorithm COG [17], [18] has been applied.

Based on the octree decomposition technique the software package COG for grid generation and geometry description allows to generate tetrahedral Delaunay meshes [8] with local and anisotropic refinement for arbitrary geometries.

A tetrahedral triangulation is roughly spoken a Delaunay triangulation if the circumsphere of each tetrahedron does not contain any vertices of the grid. COG generates - regardless rounding errors - accurate representations of vertices, edges and planar areas at the inner material interfaces and the boundaries of the structures for triangular and rectangular geometries and for geometries which results from its by coordinate transformations. Near curved boundaries special coordinate systems are used which are adapted at a sufficiently large distance to the usual cartesian coordinate system.

Especially, if the circumcenter of a tetrahedron is located within the tetrahedron, we have a clear physical interpretation. The restriction that the circumcenter of a tetrahedron is located within the tetrahedron can not fulfilled in general by a mesh generator. Thus, it can be that the circumcenter of any tetrahedron of the generated Delaunay triangulation is located outside of the tetrahedron, but COG avoids the case that this will be for tetrahedra which are located at inner material interfaces and boundaries. There are no negative distances between two circumcenters. Thus, apart from the physical interpretation the deduced grid equations can be applied using the mentioned properties of COG.

As an example we have simulated a junction of a microstrip line with a coaxial line (see Figs. 3, 4, 5). The structure is symmetric. Thus, only the right half is discretized.

For comparison the structure is subdivided in nonequidistant rectangular three-dimensional elementary cells on the one hand and in tetrahedra on the other hand. In case of rectangular grids, the order of the system of linear algebraic

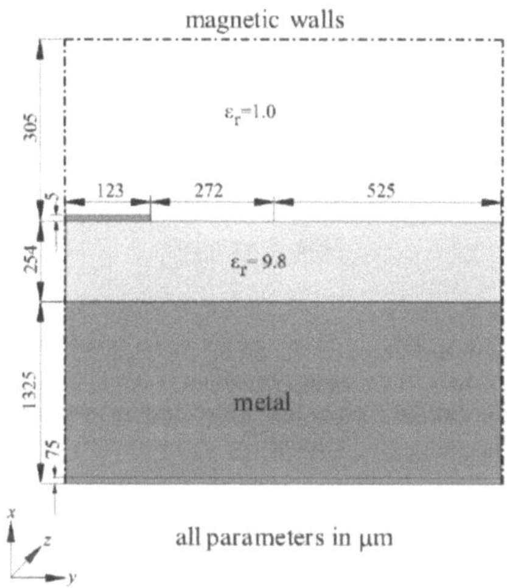

Fig. 3. xy-plane

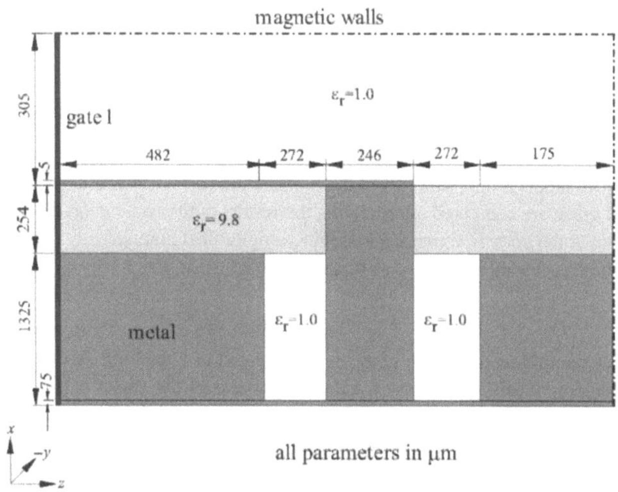

Fig. 4. xz-plane

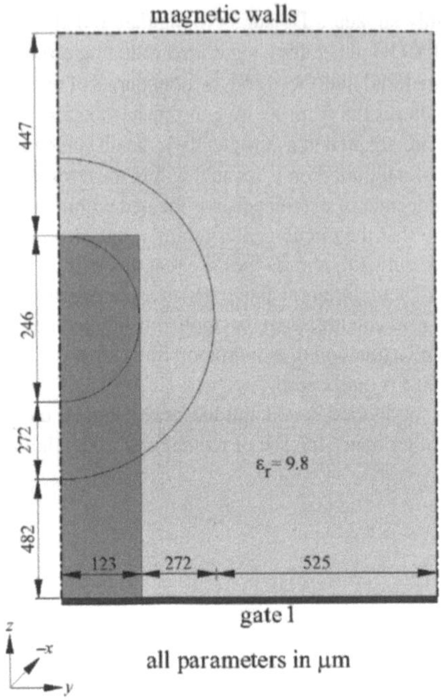

Fig. 5. yz-plane

equations (see section 6), which corresponds to the boundary value problem (see section 3), is $n = 3n_x n_y n_z = 163\,944$. $n_x n_y n_z$ is the number of cells of the structure which is assumed to be a parallelepiped. We need a high mesh refinement near the microstrip and the coaxial line which results in an accumulation of elementary cells in all coordinate directions even though the refinement is not necessary in order to approximate the solution with the required accuracy.

The tetrahedral grid consists of $n_n = 11\,368$ nodes, $n_t = 58\,742$ tetrahedra, and $n_p = 11\,446$ peripheral cell faces. The order of the corresponding system of linear algebraic equations is less than the number of edges:

$$n = n_n + n_t + n_p/2 - 1 = 75\,832. \tag{42}$$

The disadvantage of rectangular grids, the accumulation of elementary cells in all coordinate directions, is avoided here. Curved boundaries are better approximated.

5 Eigenvalue Problem Including PML

For the eigenvalue problem, we refer to the rectangular grid [4].

The transverse electric mode fields (see (24)) at the ports of the three-dimensional structure, which is discretized by means of tetrahedral grids, are computed interpolating the results of the rectangular discretization.

The field distribution at the ports is computed assuming longitudinal homogeneity for the transmission line structure. Thus, any field can be expanded into a sum of so-called modal fields which vary exponentially in the longitudinal direction:

$$\boldsymbol{E}(x, y, z \pm 2h) = \boldsymbol{E}(x, y, z)e^{\mp \jmath k_z 2h}. \tag{43}$$

k_z is the propagation constant. $2h$ is the length of an elementary cell in z-direction. We consider the field components in three consecutive elementary cells. The electric field components of the vector $\boldsymbol{e}$ (see (34)) $E_{x_{i,j,k+1}}$, $E_{x_{i,j,k-1}}$, $E_{y_{i,j,k+1}}$, $E_{y_{i,j,k-1}}$, $E_{z_{i,j,k-1}}$, $E_{z_{i+1,j,k-1}}$, and $E_{z_{i,j+1,k-1}}$ are expressed by the values of cell k using ansatz (43). The longitudinal electric field components E_z can be eliminated by means of the electric-field divergence equation $BD_{A_\epsilon}\boldsymbol{e} = 0$ (see (32)). Thus, we get an eigenvalue problem for the transverse electric fields $\boldsymbol{y} = \boldsymbol{E}_{t,l}(z_p)$, $l = 1(1)m^{(p)}$, (see (24)) on the transmission line region:

$$G\boldsymbol{y} = \gamma\boldsymbol{y}, \;\; \gamma = e^{-\jmath k_z 2h} + e^{+\jmath k_z 2h} - 2 = -4\sin^2(hk_z). \tag{44}$$

The problem of the transmission line region is reduced to a two-dimensional problem. A detailed derivation of the eigenvalue problem can be found in [10], [11]. The eigenvalue problem has to be solved for each port z_p, $p = 1(1)\bar{p}$, (see (1)). The sparse matrix G is general complex. The order of G is $n = 2n_xn_y - n_b$. n_xn_y is the number of elementary cells at the port. The size n_b depends on the number of cells with perfectly conducting material. The solutions of the eigenvalue problem correspond to the propagation constants of the modes. Using a conformal mapping it can be shown that the eigenvalues corresponding to the few interesting modes of smallest attenuation are located in a region bounded by two parabolas. The modes are found solving a controlled sequence of eigenvalue problems of modified matrices [12] applying the invert mode of the Arnoldi iteration with shifts.

The m_s (see (1)) eigenvectors (see (13)) determine the number of right-hand sides of the system of linear algebraic equations (see (48)).

The PML influences the mode spectrum. The absorption inside the PML operates through conductive losses, so that an exponential decay of the fields inside the PML is obtained. The PML achieves a refectionless absorption if the mesh discretization size goes to zero. Caused by the finite mesh size in the finite simulation domain spurious modes are generated due to the electric and magnetic walls behind the absorbing layers. The PML shifts these modes inside the region of propagating modes. We want to distinguish the spurious modes from the desired ones. As a result of our numerical calculations we found that examination of the eigenfunctions provides a useful criterion to select the modes of interest. While the field of guided modes is concentrated around the waveguide structure, the parasitic box modes exhibit a strong field accumulation inside the PML area. Thus, modes that are related to the PML boundary can be detected, using the PPP criterion (Power Part in PML) which is based on the comparison between the power concentration inside the PML region to the whole computational domain [19].

This method, developed initially for a reliable calculation of all interesting complex eigenvalues of microwave structures, was expanded then to meet the special requirements of optoelectronic structure calculations. Relatively large cross sections and highest frequencies (i.e., small wavelengths) yield increased dimensions for the eigenvalue problems. Using the results of a coarse grid calculation within the final fine grid reduces the numerical efforts significantly. A laser application can be found in [12]. A self aligned stripe (SAS) laser with a discretized large cross section of (4050×7750) nm is investigated there. Thin layers of 100 nm with complex material properties have to be taken into account. The frequency is fixed to $300 * 10^{12}$ Hz, which corresponds to a vacuum wavelengths of 1000 nm. A graded mesh of 121 times 127 elementary cells, including 10-cell PML regions, is used as a coarse grid in order to find approximately the location of the guided mode. A sequence of 84 eigenvalue problems have been used to cover the long small region in the complex plane. The circle that contains the guided mode is known after this step. A graded mesh of 283 times 345 elementary cells, including including 10-cell PML regions, is used as a fine grid in order to find the accurate value of the guided mode in the reduced region. The computational time is reduced by a factor of 10 using a coarse and a fine grid.

The use of two levels of parallelization results in an additional speedup in terms of computation time.

6 Systems of Linear Algebraic Equations Including PML

All boundary conditions are known after the computation of the eigen mode problem, and the systems of linear algebraic equations can be solved.

Besides the locations and values of the entries, the matrix representations of (37) - (41) have the same structure as (32) - (34). Thus, we refer to (34) for the solution of the linear algebraic equations.

Multiplying (34) by $D_s^{1/2}$ yields a symmetric form of linear algebraic equations:

$$\bar{A}\boldsymbol{x} = 0, \quad \bar{A} = (D_s^{1/2} A^T D_{s/\tilde{\mu}} D_A^{-1} A D_s^{1/2} - k_0^2 D_{A_{\tilde{\epsilon}}}) \tag{45}$$

with $\boldsymbol{x} = D_s^{1/2}\boldsymbol{e}$. Moreover, the gradient of the electric field divergence

$$[\epsilon]\nabla([\epsilon]^{-2}\nabla \cdot [\epsilon]\boldsymbol{E}) = 0 \tag{46}$$

is used. It can be written as matrix equation

$$\bar{B}\boldsymbol{x} = 0, \quad \bar{B} = D_s^{-1/2} D_{A_{\tilde{\epsilon}}} B^T D_{V_{\tilde{\epsilon}\tilde{\epsilon}}}^{-1} B D_{A_{\tilde{\epsilon}}} D_s^{-1/2}. \tag{47}$$

The diagonal matrix $D_{V_{\tilde{\epsilon}\tilde{\epsilon}}}$ is a volume matrix for the 8 partial volumes of the dual elementary cell. In case of tetrahedral grids, the gradient of the divergence at an internal point is obtained considering the partial volumes of the appropriate Voronoi cell.

Taking into account the boundary conditions (13) and (25), Eqs. (45) and (47) yield the form $\hat{A}\boldsymbol{x} = \boldsymbol{b}$ and $\hat{B}\boldsymbol{x} = 0$, respectively, and

$$(\hat{A} + \hat{B})\boldsymbol{x} = \boldsymbol{b}, \quad \hat{A} + \hat{B} \text{ complex indefinite symmetric,} \tag{48}$$

can be solved faster than $\hat{A}\boldsymbol{x} = \boldsymbol{b}$.

Independent set orderings [14], Jacobi and SSOR preconditioning using Eisenstat's trick [5] are applied to accelerate the speed of convergence of the used block Krylov subspace method [7, 16] for the system of linear algebraic equations (48) that has to be solved with the same coefficient matrix, but m_s (see (1)) right-hand sides.

The permutations P_i transform the matrices A_i with $A_0 = \hat{A} + \hat{B}$ in the form

$$A_i \quad \longrightarrow \quad P_i A_i P_i^T = \begin{pmatrix} D_i & E_i^T \\ E_i & H_i \end{pmatrix}, \tag{49}$$

where D_i is a diagonal, E_i, and H_i are sparse matrices. Using the factorized form of (49) we get a system of linear equations

$$\begin{pmatrix} I_i & 0 \\ E_i D_i^{-1} & I_i \end{pmatrix} \begin{pmatrix} D_i & E_i^T \\ 0 & H_i - E_i D_i^{-1} E_i^T \end{pmatrix} \begin{pmatrix} \boldsymbol{y}_{i,1} \\ \boldsymbol{y}_{i,2} \end{pmatrix} = \begin{pmatrix} \boldsymbol{c}_{i,1} \\ \boldsymbol{c}_{i,2} \end{pmatrix} \tag{50}$$

with $\boldsymbol{y}_i = P_i \boldsymbol{x}_i = (\boldsymbol{y}_{i,1}, \boldsymbol{y}_{i,2})^T$ and $\boldsymbol{c}_i = P_i \boldsymbol{b}_i = (\boldsymbol{c}_{i,1}, \boldsymbol{c}_{i,2})^T$. The algorithm for solving Eq. (48) is described in the following:

(i) Set $A_0 = \hat{A} + \hat{B}$, $\boldsymbol{x}_0 = \boldsymbol{x}$, $\boldsymbol{b}_0 = \boldsymbol{b}$
(ii) Forward substitution: $i = 0, \ldots, lev - 1$
  a) Compute P_i: $P_i A_i P_i^T$, $\boldsymbol{y}_i = P_i \boldsymbol{x}_i$, $\boldsymbol{c}_i = P_i \boldsymbol{b}_i$
  b) Compute $\boldsymbol{x}_{i+1} = \boldsymbol{y}_{i,2}$, $\boldsymbol{b}_{i+1} = \boldsymbol{c}_{i,2} - E_i D_i^{-1} \boldsymbol{c}_{i,1}$
  c) Compute $A_{i+1} = H_i - E_i D_i^{-1} E_i^T$
(iii) Solve $A_{lev} \boldsymbol{x}_{lev} = \boldsymbol{b}_{lev}$ for $\boldsymbol{x}_{lev}$
(iv) Backward substitution: $i = lev - 1, \ldots, 0$
  a) Compute $\boldsymbol{y}_{i,2} = \boldsymbol{x}_{i+1}$, $\boldsymbol{y}_{i,1} = D_i^{-1}(\boldsymbol{c}_{i,1} - E_i^T \boldsymbol{y}_{i,2})$
  b) Compute $\boldsymbol{x}_i = P_i^T \boldsymbol{y}_i$

In comparison to the simple lossy case the number of iterations of Krylov subspace methods increases significantly if the structure contains a PML. In this case, among others, the speed of convergence depends on the relations of the edge lengths in an elementary cell of the nonequidistant rectangular. The best results can be obtained using nearly cubic cells. Moreover, overlapping conditions at the corner regions of the computational domain cause an increase of the magnitude of the corresponding off-diagonal elements in comparison to the diagonal of the coefficient matrix. This deteriorates the properties of the matrix. Thus, overlapping PML should be avoided.

The PML layers, which form the absorbing boundary condition, have a significant influence on computational efforts, which is demonstrated in Table 2 for a quasi-TEM waveguide (in Table 2, ω denotes the relaxation parameter of the Krylov subspace method). A nonequidistant mesh of $27 * 24 * 21$ elementary cells including graded PML regions is used, that means the order of the system of linear algebraic equations is 40 824. The structure is symmetric with respect to the (x, z)-plane. Here, a magnetic wall is used, all other parts of the surface are assumed to be electric walls covered by PML. The longitudinal z-PML region consists of 10 layers, the lateral (x, y)-PML's of 5 layers. The number of iterations also depends on the frequency f and the relaxation parameter ω.

Table 2. Influence of the PML layers on computational efforts

	Number of Iteration								
	$\omega = 1.00$			$\omega = 1.30$			$\omega = 1.58$		
f/GHz Structure	10	50	100	10	50	100	10	50	100
no PML	63	72	127	51	58	104	45	53	91
z-PML	649	647	716	501	518	591	431	452	543
yz-PML	13 912	27 924	32 298	13 501	29 077	45 371	16 457	44 824	104 642
xyz-PML	12 307	44 723	213 358	11 475	55 221	322 155	15 983	111 965	$>10^6$
xyz-PML (nonov.)	628	591	742	527	479	609	493	436	624

References

[1] Beilenhoff, K., Heinrich, W., Hartnagel, H.L.: Improved Finite-Difference Formulation in Frequency Domain for Three-Dimensional Scattering Problems. IEEE Transactions on Microwave Theory and Techniques, **40**, 540–546 (1992)

[2] Berenger, J.P.: A perfectly matched layer for the absorption of electromagnetic waves. J. Comput. Phys., **114**, 185–200 (1994)

[3] Chew, W.C., Weedon, W.: A 3D perfectly matched medium from modified Maxwell's equation with stretched coordinates. Microwave Opt. Technol. Lett. **7**, 599–604 (1994)

[4] Christ, A., Hartnagel, H.L.: Three-Dimensional Finite-Difference Method for the Analysis of Microwave-Device Embedding. IEEE Transactions on Microwave Theory and Techniques, **35**, 688–696 (1987)

[5] Eisenstat, S.C.: Efficient implementation of a class of preconditioned conjugate gradient methods. SIAM J. Sci Statist. Comput. **2**, 1–4 (1981)

[6] Fuhrmann, J., Langmach, H., Schmelzer, I., Uhle, M.: Grid management in pdelib: sxgrid. http://www.wias-berlin.de/pdelib/documentation/pdelib-1.14.html, sxgrid.ps, 1–29 (2001)

[7] Freund, R.W., Malhotra, W: A Block-QMR Algorithm for Non-Hermitian Linear Systems with Multiple Right-Hand Sides. Linear Algebra and Its Applications, **254**, 119–157 (1997)

[8] George, P.-L., Borouchaki, H.: Delaunay Triangulation and Meshing. Editions Hermes, Paris (1998)

[9] Hebermehl, G., Schlundt, R., Zscheile, H., Heinrich, W.: Improved Numerical Methods for the Simulation of Microwave Circuits. Surveys on Mathematics for Industry, **9**, 117–129 (1999)

[10] Hebermehl, G., Schlundt, R.: Simulation of Monolithic Microwave Integrated Circuits. Weierstrass Institute for Applied Analysis and Stochastics, http://www.wias-berlin.de/publications/preprints/235/, Preprint **235**, 1–37 (1996)

[11] Hebermehl, G., Schlundt, R.: Improved Numerical Solutions for the Simulation for Monolithic Microwave Integrated Circuits. Weierstrass Institute for Applied Analysis and Stochastics, http://www.wias-berlin.de/publications/preprints/236/, Preprint **236**, 1–43 (1996)

[12] Hebermehl, G., Hübner, F.-K., Schlundt, R., Tischler, Th., Zscheile, H., Heinrich, W.: Simulation of Microwave and Semiconductor Laser Structures Including Absorbing Boundary Conditions. In: Bänsch, E. (ed) Challenges in Scientific Computing - CISC2002. Lecture Notes in Computational Science and Engineering, **35**, 131–159, Springer Berlin Heidelberg New York (2003)

[13] Hebermehl, G., Schefter, J., Schlundt, R., Tischler, Th., Zscheile, H., Heinrich, W.: Simulation of Microwave Circuits and Laser Structures Including PML by means of FIT. Advances in Radio Science, **2**, 107–112 (2004)

[14] Saad, Y.: Iterative methods for sparse linear systems. PWS Publishing Company (1996)

[15] Sacks, Z.S., Kingsland, D.M., Lee, R., Lee, J.-F.: A Perfectly Matched Anisotropic Absorber for Use as an Absorbing Boundary Condition. IEEE Transactions on Antennas and Propagation, **43**, 1460–1463 (1995)

[16] Schlundt, R., Hebermehl, G., Hübner, F.-K., Zscheile, H., Heinrich, W.: Iterative Solution of Systems of Linear Equations in Microwave Circuits Using a Block Quasi-Minimal Residual Algorithm In: van Rienen, U., Günther, M., Hecht, D. (ed) Scientific Computing in Electrical Engineering. Lecture Notes in Computational Science and Engineering, **18**, 325–333, Springer Berlin Heidelberg New York (2001)

[17] Schmelzer, I.: Grid Generation and Geometry Description with COG. Proceedings of contributed papers and posters, ALGORITM 2000, 15th Conference on Scientific Computing, Vysoké Tatry - Podbanské, Slovakia, September 10 - 15, 2000, Ed. A. Handlovičová, M. Kormorníková, K. Mikula, D. Ševčovič, Slovak University of Technology, Bratislava, 399–405

[18] Schmelzer, I.: COG, http://www.wias-berlin.de/software/cog/index.html

[19] Tischler, Th., Heinrich, W.: Perfectly Matched Layer as Lateral Boundary in Finite-Difference Transmission-Line Analysis. IEEE Transactions on Microwave Theory and Techniques, **48**, 2249–2253 (2000)

[20] Weiland, T.: A discretization method for the solution of Maxwell's equations for six-component fields. Electronics and Communication (AEÜ), **31**, 116–120 (1977)

Solving of an Electric Arc Motion in a Vacuum Interrupter

P. Kacor and D. Raschka

VSB - Technical University of Ostrava, Faculty of Electrical Engineering, Department of Electrical Machines and Apparatuses, Ostrava, Czech Republic, {petr.kacor,david.raschka.fei}@vsb.cz

Abstract The paper describes using of the numerical modeling for solving of problems related to design and optimalization of electric apparatus, in this case of the vacuum interrupter. The method of simplified numerical model creation is described and also obtained results, voltage drop on interrupter, distribution of current density, electromagnetic induction and deformation of the electric arc, are presented.

1 Introduction

The vacuum interrupter is one of the main switching elements in a high voltage distribution networks today. Main advantages of vacuum interrupter are very easy construction, long working life and operational reliability. An idea of vacuum breaking is relatively old. The first experiments have started about 80 years ago, but there were problems with joining tight and contact materials degradation mainly. So the first types of vacuum interrupters have been able to work since 6th decade of 20th century.

Generally, the vacuum interrupter consists of two contacts pairs inside of ceramic vacuum container, see Fig. 1. The pressure of internal space (1) is 0,001 Pa approximately. One of the contacts (2) is stable and fixed to the insulating ceramic enclosure (6) by the metal cap. Due to elastic bellows (4) the second contact (3) is able to move without vacuum lost. In ordinary conditions of circuit the interrupter is switch-on and there is a frontal coupling of contacts. An electric arc begins to burn inside of interrupter by movement of moving contact and the electric arc is formed by small amount of metal vapors. The metal vapors leave gradually the inner contacts space and condensate on the contacts and on an inner shielding cover (5). So the shielding cover (5) prevents condensation of arc particles on the inner areas of insulating enclosure (6). In switch-off situation of interrupter the dielectric strength is secured by the contacts distance and insulating enclosure. The electric arc in vacuum occurs, from simplifying view, in two forms. The first form is a diffuse mode; the second form is a constricted mode. The diffuse mode exists in value of electric current up to a few kA. The cathode spot is a main source of arc particles. There is only one cathode spot but there may be several cathode spots with irregular motion on the cathode surface. The cathode spot is a very small section capable of emitting a current about hundred amperes. There is a high rate of current density too, about 1000 A/m2. Globally neutral plasma has usually a conical shape and arc particles are diffused on the large surface of the opposite contact (anode), Fig. 2 left. With the increasing of electric current value (over 10kA) cathode spots begin to become one together by the acting of electromagnetic forces. The original conical shape of plasma arc is transformed to a cylindrical. This situation leads to the formation of a positive anode voltage. The energy received by the anode increases and tends to be concentrated on a reduced area. The anode heats up and starts to emit neutral particles that are ionized by the incident electrons, anode spot comes into existence. There is a high pressure electric arc between electrodes (contacts). In this case is very hard to break of flowing electric current without other steps, Fig. 2 right.

2 Breaking techniques of the arc in vacuum

The main problem of vacuum interrupter design is protection against the overheating of contact surfaces in area of arc root or beginning of anode spot. During a long time of vacuum breaker designing were developed the two techniques of interrupting where operation of electromagnetic fields is used.

The first technique uses a fast circular movement of constricted arc. The energy of arc is distributed onto large parts of contacts in this case and the overheating of contact surfaces is limited. The circular movement is obtained through

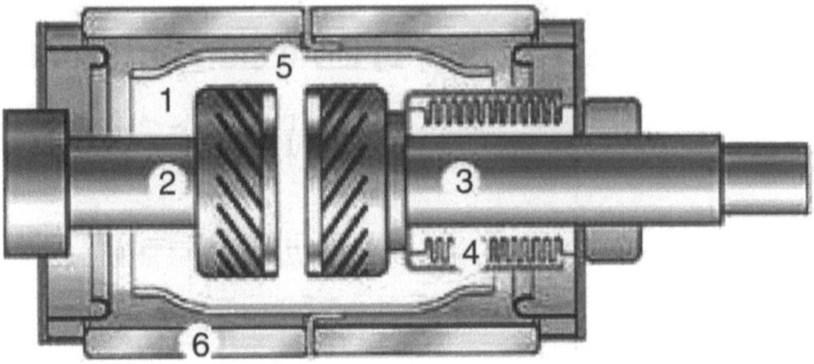

Fig. 1. Longitudinal section of vacuum interrupter (ABB construction)

Fig. 2. (left) Diffuse mode of the arc, (right) Contracted mode of the arc

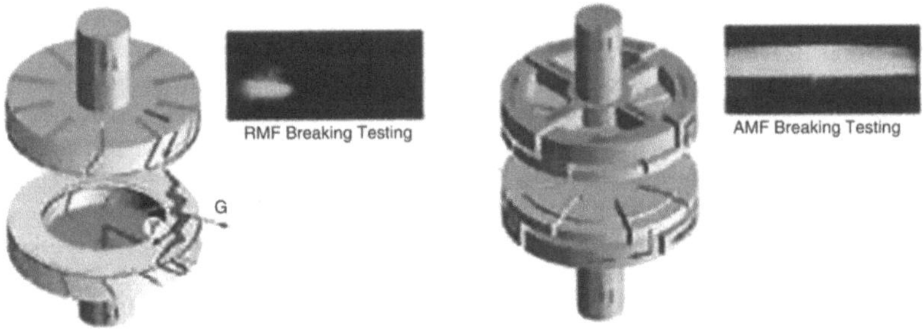

Fig. 3. The techniques of RMF - radial mag. field and of AMF - axial mag. field breaking

application of radial magnetic field (RMF) in the arc zone. Contact has got a special spiral or cup shape, Fig. 3 left. The second technique prevents the arc transformation into constricted mode by the help of axial acting of magnetic field (AMF) in the arc zone. The electric arc is immobile and during its burning is extended on the large surfaces of contacts. The contact shape is more complex and looks like, mostly, a parallel combination of one-turn coils, Fig. 3 right.

3 Creating of the interrupter model

We would like to present one of the possibilities how to solve force interaction between electric arc and current conducting way of vacuum interrupter contacts by the help of Finite Element Method (ANSYS). There was chosen interrupter with radial magnetic field breaking technique (RMF) for solution. Generally, the electric arc (plasma) is described by equations summarized on Fig. 4.

$$\bar{E}_0 = -\,grad v_1 \qquad \text{(eq.1)}$$

$$div E_0 = \frac{e}{\varepsilon_0}(n_i - n_e) \qquad \text{(eq.2)}$$

$$div(j_i + j_e) = e\left(\frac{\partial n_i}{\partial_e} - \frac{\partial n_e}{\partial t}\right) \qquad \text{(eq.3)}$$

$$E_0(j_i + j_e) = -\,div\,(\lambda.gradT) + \xi_{xar}(T) + s.c, \bar{v},\ grad\ T + $$
$$+ eU_i\left(\frac{\partial n_e}{\partial_t} - \frac{1}{e}\,divj_e\right) - s.c,\ \frac{dT}{dt} \qquad \text{(eq.4)}$$

$$j_i = en_i\mu_i\bar{E}_0 - eD_i\,grand\,(n_i) + 2en_i\mu_i^2(\bar{E}_0\bar{B}) + en_i\bar{v}_i \qquad \text{(eq.5)}$$

$$j_e = en_e\mu_e\bar{E}_0 + eD_e grand(n_e) - 2en_e\mu_e^2\,(\bar{E}_0\bar{B}) - en_e\bar{v}_i \qquad \text{(eq.6)}$$

$$n_i.n_e = K_s.\ pT^{0.5}.exp\left(-\frac{eU_i}{kT}\right) \qquad \text{(eq.7)}$$

Fig. 4. Equations defining the electric arc

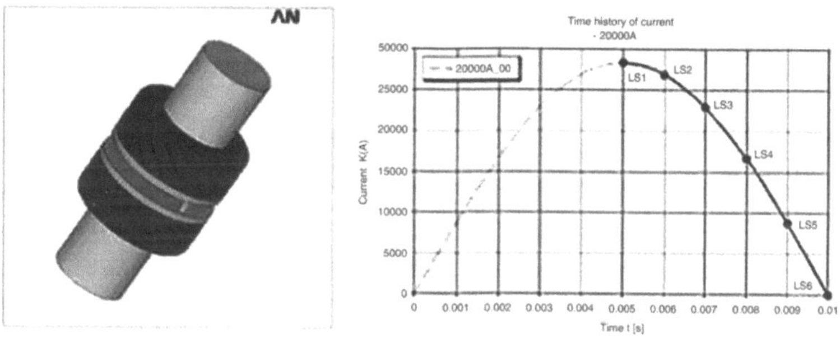

Fig. 5. (left) FEM model of vacuum interrupter, (right) Time history of the current

As you can see, without any simplification of input conditions is very hard to obtain any analytical solution of these equations. Therefore, in the first point of the arc movement analysis the thermal or fluid dynamic relationships are fully neglected. The electric arc burns in a constricted form with rapid circular movement around contact surfaces in this type of interrupter. So, the first approximation the constricted arc can be compared to a cylindrical conductor through which a current flows, the direction of which is parallel to the axis of the contacts.

Simplified numerical model of vacuum interrupter, Fig. 5 left, with RMF breaking technique contains the one pair of contacts and the electric arc model. This picture does not show surrounded air but the main interrupter geometry only. However, the model has got relative accurate geometry with a sufficient number of elements.

4 Analysis and Results

For the solution of force interaction is used advantage of coupled field analysis. The beginning of contact moving was neglected, the contacts are fixed. Solution cycle started with choosing of the arc position and applying the first value of electric current from the Fig. 5 right (LS1). After that there was solved electrical model of interrupter. The current density is provided by solution of the electric model and it serves as an initial condition of electromagnetic model

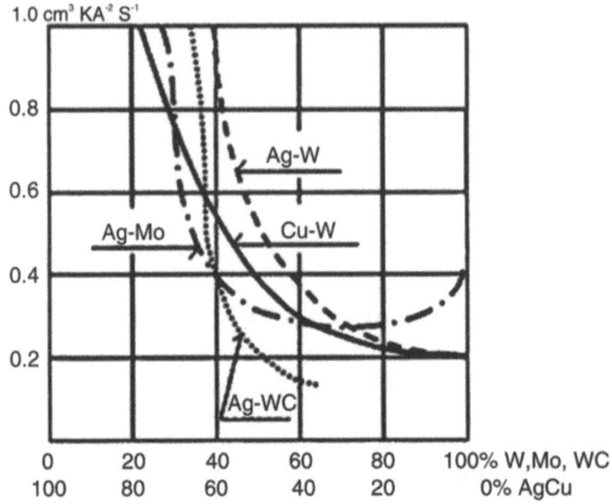

Fig. 6. Weight decreasing of contact material in dependence of time and flowing current

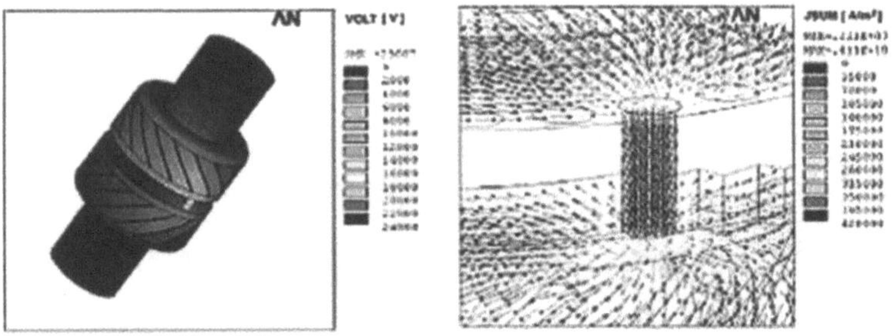

Fig. 7. (left) Voltage drop on interrupter, (right) Distribution of current density

solution. It means we need to load the distribution of current density for the further electromagnetic solution. After finding out the value of radial electromagnetic force in arc, we used the graphical dependence of weight decreasing of contact material, Fig. 6, and general equation of uniformly slowed-down motion for the next prediction of the electric arc position.

Of course, we know that the arc make about one and half turn around contact boundary during breaking process from laboratory breaking test. Solving cycle continues with a new position of the electric arc as long time as the electric current falls to zero (LS6).

There is a uniform voltage drop on the Fig. 7 left, and vector display of flowing current on the Fig. 7 right. As we can see, the current lines come out from massive part of contact and they are pointed on the little area of contacts. Current lines flow through volume of arc model and there is the maximal value of current density. Arc volume represents the smallest cross-section part of current conducting way of interrupter.

There is result of electromagnetic model solution in the first load step (LS1) on the Fig. 8 left. As we must have guessed from the distribution of current density, we can find there a maximal value of electromagnetic induction in area of the electric arc, which is about B=3T. As a source of arc movement the radial force effect is created by interaction between flowing current through electric arc and magnetic field excited by current conducting way of interrupter. There is solution of structural model of the electric arc on the Fig. 8 right, too. We are not able to describe structural material constant of arc correctly, so the results of the arc deformation is not absolutely correct. But we can use it for complex view in the course of arc breaking process.

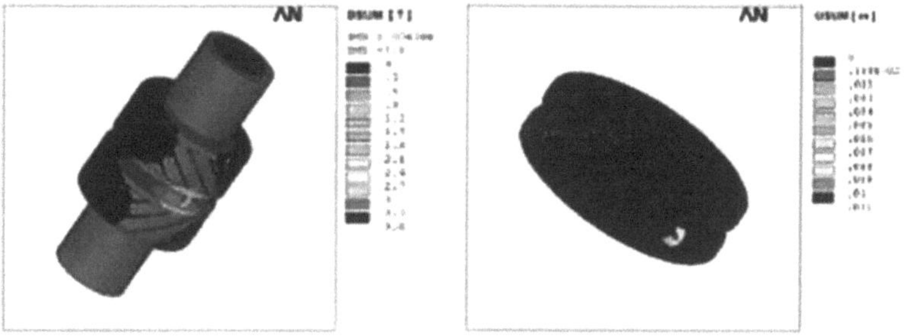

Fig. 8. (left) Electromagnetic induction, (right) Deformation of the electric arc

5 Conclusion

The results of this analysis show the next FEM using for the electromagnetic field solution in construction and electrical apparatus optimalization. Performed analysis and selected method is not able to respect all processes running in a breaking cycle of vacuum interrupter. It is able to reply for a question of optimal shape configuration of the vacuum interrupter current conducting way and optimal slant angle of contacts cup shape.

References

[1] Bozak P: Simulation of an Electric Arc in Vacuum Interrupter, diploma thesis, VSB-TU Ostrava, 2003

[2] Kacor P: Numerical Solution of Electrodynamics Forces on Current Conducting Way of Switching Apparatus, Ph.D. thesis, VSB-TU Ostrava, 2003

[3] Ansys On-line Help release ANSYS 5.7., Sequential Coupled-Field Analysis

[4] Electromagnetic Fields Analysis Guide release 5.5

[5] Technical literature of SIEMENS, `http://www.siemens.com`

[6] Technical literature of ABB, `http://www.abb.com`

[7] Technical literature of SCHNEIDER-ELECTRIC, `http://www.schneider-electric.com`

[8] Laboratory of Physical Field Simulation, `http://fei.vsb.cz/kat453/www453/soubory/MFP_LAB/cz/cz_index.htm`

Analysis of Eddy Currents in a Gradient Coil

J. M. B. Kroot

Eindhoven University of Technology – P.O.Box 513; 5600 MB Eindhoven, The Netherlands

Abstract To model the z-coil of an MRI-scanner, a set of circular loops of strips is shown in [4] to be a good approximation. This ring model yields a current distribution that only depends on the axial direction. In order to take the dependence of the tangential direction into account, we introduce rectangular pieces of copper (called islands) in between the rings. In this paper the current distribution in a set of rings and islands, driven by an external applied source current is investigated. The source, and all excited fields, are time harmonic, and the frequency is low enough to allow for a quasi-static approximation. Due to induction eddy currents occur which form the so-called edge-effect. The edge-effect depends on the applied frequency and the distances between the strips, and causes higher impedances. From the Maxwell equations, an integral equation for the current distribution in the strips is derived. The Galerkin method is applied, using global basis functions to solve this integral equation. Using Legendre polynomials for the axial direction turns out to be an appropriate choice. It provides a fast convergence, so only a very small number of Legendre polynomials is needed.

1 Introduction

Magnetic Resonance Imaging (MRI) is an imaging technique that plays an important role in the medical community. It provides images of cross-sections of a body, taken from any angle [1]. The selection of a slice is realized by the so-called gradient coils. A gradient coil consists of copper strips wrapped around a cylinder. Due to mutual magnetic coupling, the current is not uniformly distributed and eddy currents arise which affect the quality of the image. For analysis and design of gradient coils, finite element packages are used. However, these packages cannot sufficiently describe the characteristics that give insight in the qualitative behaviour of the distribution of the currents, relating the geometry to typical parameters like edge effects, mutual coupling and heat dissipation. One of the reasons is that the coils are large, but very thin, such that numerical simulations become inaccurate and inefficient.

In this paper we focus on the z-coil, which has the function to create a gradient in the magnetic field in the axial (z-) direction of the scanner. In [2] and [3], a parallel set of conducting strips is used to model the z-coil. In [4], the z-coil has been modelled as a set of rings. The current distribution is in that case independent of the tangential direction (i.e. an axi-symmetric solution). However, in a z-coil embedded in a system of more coils and magnets, extra eddy currents are present, making the distribution of the current in the z-coil non-symmetric. In order to obtain a dependence of the tangential direction, we make use of so-called islands. These islands are thin pieces of (copper) strips situated between the rings on the same cylindrical surface. The current through the rings now induces eddy currents in the islands and vice versa.

The overall aim is the calculation of the electric current distribution in a set of rings and islands. The system is driven by a source current, which changes harmonically in time, with a low frequency (in the order of kHz). In the mathematical analysis an integral equation is derived for the current distribution and the Galerkin method is used to solve this equation. The most important issue is the choice of the basis functions. Numerical implementation is needed to determine the coefficients for the basis functions.

2 Model definition

For the model, we consider a set of N_r coaxial circular strips, or rings, and N_i rectangular pieces of strips, called islands. All these conductors are on the same imaginary cylinder S_c, defined as

$$S_c = \{(r, \varphi, z) \in \mathbb{R}^3 | r = R\}. \tag{1}$$

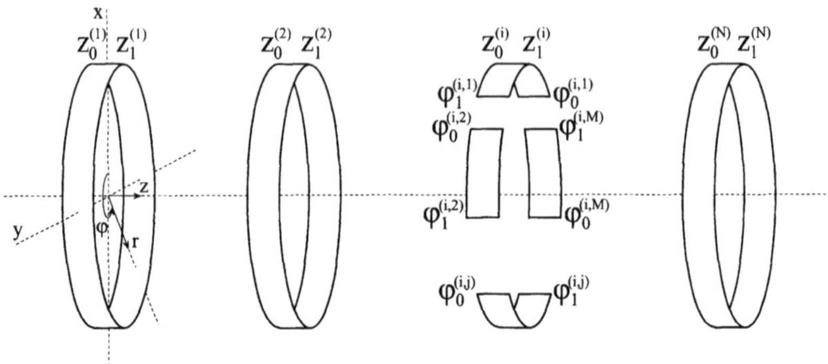

Fig. 1. The geometry of the model

The geometry is depicted in Fig. 1. We use cylindrical coordinates (r, φ, z), where the z-axis coincides with the central axis of the cylinder. All rings and islands have thickness h and are of uniform width.

A source current is applied to the rings, which is time-harmonic at frequency ω. The total current has a distribution $\mathbf{J}(r,\varphi,z,t) = \mathrm{Re}(\mathbf{J}(r,\varphi,z)e^{-i\omega t})$. The penetration depth δ, defined as $\delta = \sqrt{2/\mu\sigma\omega}$, is much larger than the thickness of the conductor, for frequencies $\omega < 10^3$ rad/s. So, the current density is almost distributed uniformly throughout the thickness of the conductor. Consequently, we can assume that the conductors are infinitely thin, and replace the current density $\mathbf{J}$ (in A/m^2) by the current per unit of length $\mathbf{j}$ (in A/m), such that $\mathbf{j} = h\mathbf{J}$. From now on, the current distribution in the conductors is independent of r, $\mathbf{j} = \mathbf{j}(\varphi, z)$, and has no component in the r-direction.

The strips occupy the surface $S_\cup = S_r + S_i$ in space, described in cylindrical coordinates by (see Fig. 1)

$$S_r = \sum_{n=1}^{N_r} S_n^{(r)}, \qquad S_n^{(r)} = \{(r,\varphi,z) \in \mathbb{R}^3 | r = R, z \in [z_0^{(n)}, z_1^{(n)}]\}, \tag{2}$$

$$S_i = \sum_{n=1}^{N_i} S_{n,p}^{(i)}, \qquad S_{n,p}^{(i)} = \{(r,\varphi,z) \in \mathbb{R}^3 | r = R, \varphi \in [\varphi_0^{(n,p)}, \varphi_1^{(n,p)}], z \in [z_0^{(n)}, z_1^{(n)}]\}, \tag{3}$$

such that S_r is the unified surface of the rings, S_i is the unified surface of the islands, and $S_\cup \subset S_c$. By G^-, we indicate the inner region of the cylinder, and by G^+ the outer region:

$$G^- = \{(r,\varphi,z) \in \mathbb{R}^3 | 0 \le r < R\}, \tag{4}$$
$$G^+ = \{(r,\varphi,z) \in \mathbb{R}^3 | r > R\}. \tag{5}$$

To obtain the mathematical description of the problem, we use the Maxwell's theory applied to the geometry of the model. The set of equations is reduced by using the following assumptions:

(i) The strips are isotropic homogeneous non-polarizable and non-magnetizable conductors (copper).
(ii) The current distribution is time-harmonic and with that also the magnetic field and the electric field are.
(iii) The frequency is low (in the order of kHz), such that we can neglect the displacement current in Ampère's law, i.e. we may use a quasi-static approach.
(iv) The strips are negligibly thin.
(v) The conductors are rigid; magneto-mechanical influences (vibrations) are not considered.

The set of equations, valid in both G^- and G^+, become

$$\nabla \times \mathbf{E} = i\omega\mu\mathbf{H}\,, \quad \nabla \times \mathbf{H} = \mathbf{0}\,, \quad \nabla \cdot \mathbf{E} = \nabla \cdot \mathbf{H} = 0\,. \tag{6}$$

Denoting the jump across $S_\cup$ by $[\![\ ]\!]$, we can write the boundary conditions as

$$[\![\mathbf{E} \times \mathbf{n}]\!] = \mathbf{0}\,, \ [\![\mathbf{H} \cdot \mathbf{n}]\!] = 0\,, \ [\![\mathbf{E} \cdot \mathbf{n}]\!] = Q_s\,, \ [\![\mathbf{H} \times \mathbf{n}]\!] = -\mathbf{j}\,, \tag{7}$$

where $\mathbf{n} = \mathbf{e}_r$, $\mathbf{j}$ is the surface current and Q_s is the surface charge. For our purposes $Q_s = 0$, because a jump over a negligibly thin conductor experiences no surface charge. In the sheets, we have

$$\nabla \cdot \mathbf{j} = 0, \quad \mathbf{j} = \mathbf{j}^s + \sigma h \mathbf{E}, \tag{8}$$

where $\mathbf{j}^s$ is the prescribed source current. The total current consists of the source current $\mathbf{j}^s$ and the induced eddy current $\mathbf{j}^e$, so $\mathbf{j} = \mathbf{j}^s + \mathbf{j}^e$. Furthermore, the normal component of the current on the edges has to be zero, i.e. $j_\varphi(\varphi_e, z) = j_z(\varphi, z_e) = 0$, where φ_e and z_e are the values of φ and z on the edges, respectively. Finally, at infinity, we require $|\mathbf{H}| \to 0$.

Using a vector potential $\mathbf{A}$, defined by $\mathbf{B} = \nabla \times \mathbf{A}$, and a scalar potential Φ, we can write

$$\mathbf{E} = -\frac{\partial \mathbf{A}}{\partial t} - \nabla \Phi = i\omega \mathbf{A} - \nabla \Phi. \tag{9}$$

The scalar potential Φ must satisfy the Laplace equation (which follows from Gauss' law and the Coulomb gauge) and must vanish at infinity. It will therefore be identical to zero.

For the dimension analysis, the distances are scaled by the radius of the cylinder, and the current is scaled by the average current through all rings. The Ohm's law (see (8)) in dimensionless form then becomes

$$i\kappa A_\varphi(1, \varphi, z) = j_\varphi(\varphi, z) - j_\varphi^s(\varphi, z), \tag{10}$$

$$i\kappa A_z(1, \varphi, z) = j_z(\varphi, z) - j_z^s(\varphi, z), \tag{11}$$

where

$$\kappa = h\sigma\mu\omega R. \tag{12}$$

The vector potential $\mathbf{A}$ can be written in an integral form, which follows from the Maxwell equations (6) and the boundary conditions (7). We obtain

$$A_\varphi(1, \varphi, z) = \frac{1}{4\pi} \int_{S_\cup} \frac{\cos(\varphi - \theta) j_\varphi(\theta, \zeta)}{\sqrt{(z-\zeta)^2 + 4\sin^2(\frac{\varphi-\theta}{2})}} \, \mathrm{d}\theta \, \mathrm{d}\zeta, \tag{13}$$

$$A_z(1, \varphi, z) = \frac{1}{4\pi} \int_{S_\cup} \frac{j_z(\theta, \zeta)}{\sqrt{(z-\zeta)^2 + 4\sin^2(\frac{\varphi-\theta}{2})}} \, \mathrm{d}\theta \, \mathrm{d}\zeta. \tag{14}$$

3 Solution procedure

In this section, we explain how we solve $j_\varphi(\varphi, z)$ from (10) and (13). Note that $j_z(\varphi, z)$ follows automatically, because the current is divergence free. The Galerkin method is applied, for which we have to choose appropriate basis functions. Therefore, we first investigate the behaviour of the kernel function in (13). This function has a singularity in the point $(\varphi, z) = (0, 0)$. In abstract form, (10) is written as $\mathsf{K} j_\varphi - i\epsilon j_\varphi = -i\epsilon j_\varphi^s$, where

$$\mathsf{K} j_\varphi(\varphi, z) = \int_{S_\cup} \mathcal{K}_\varphi(\varphi - \theta, z - \zeta) j_\varphi(\theta, \zeta) \, \mathrm{d}\theta \, \mathrm{d}\zeta, \qquad \epsilon = \frac{1}{\kappa}, \tag{15}$$

and the kernel function $\mathcal{K}_\varphi(\varphi, z)$ can be expressed by a Fourier series [5]

$$\begin{aligned} \mathcal{K}_\varphi(\varphi, z) &= -\frac{\cos(\varphi)}{4\pi\sqrt{z^2 + 4\sin^2(\frac{\varphi}{2})}} \\ &= -\frac{1}{4\pi^2}\Big(Q_{\frac{1}{2}}(\chi) + \sum_{k=1}^{\infty} \cos(k\varphi)\big(Q_{k-\frac{3}{2}}(\chi) + Q_{k+\frac{1}{2}}(\chi)\big)\Big). \end{aligned} \tag{16}$$

Here, $Q_{k-1/2}$ is the Legendre function of the second kind of half-integer degree [6], and $\chi = (2 + z^2)/2$. We can now determine the behaviour of each term in the series of (16) around the point $z = 0$. We find

$$Q_{m-\frac{1}{2}}(\chi) \approx \frac{1}{2}\Big(-2\gamma + \ln 4 - 2\Psi^{(0)}\big(\frac{2m+1}{2}\big) - 2\ln|z|\Big) + \mathcal{O}(z), \tag{17}$$

for $m \geq 0$, establishing that the singularity is logarithmic in the z-direction. Here, γ is Euler's constant and $\Psi^{(0)}$ is the polygamma function.

The basis functions we use, are global, i.e. they are valid on the complete rings/islands. Due to the logarithmic singularity of the kernel function, we choose Legendre polynomials of the first kind in the z-direction. We then obtain analytical solutions for the integrals as demonstrated in (28). In φ-direction, we need to use 2π-periodic functions. For the inner products to be computed in the Galerkin method [7], we distinguish the following situations:

(i) Inner products of basis functions of rings mutually.
(ii) Inner products of basis functions of a ring and an island.
(iii) Inner products of basis functions of islands mutually.

Every situation starts with the same basic idea: We consider the current distribution on each ring and each island as a Fourier series in the φ-direction, after which we can focus on the projections on the basis functions $\cos(n\varphi)$ and $\sin(n\varphi)$, and use the fact that these functions are orthogonal. The current distribution can be expressed as

$$\begin{aligned} j_\varphi &= \Pi_0 j_\varphi + \sum_{n=1}^{\infty} \Pi_n^{(1)} j_\varphi \cos(n\varphi) + \sum_{n=1}^{\infty} \Pi_n^{(2)} j_\varphi \sin(n\varphi) \\ &= j_0(z) + \sum_{n=1}^{\infty} j_n^{(1)}(z)\cos(n\varphi) + \sum_{n=1}^{\infty} j_n^{(2)}(z)\sin(n\varphi), \end{aligned} \tag{18}$$

where Π_0 is the projection operator on the constant function, $\Pi_n^{(1)}$ is the projection operator on $\cos(n\varphi)$, and $\Pi_n^{(2)}$ is the projection operator on $\sin(n\varphi)$. For the functions in the z-direction we can read the Legendre polynomials P_n. The operator K is bounded, so we can write

$$\mathsf{K} j_\varphi = \Pi_0 \mathsf{K} j_\varphi + \sum_{n=1}^{\infty} \Pi_n^{(1)} \mathsf{K} j_\varphi \cos(n\varphi) + \sum_{n=1}^{\infty} \Pi_n^{(2)} \mathsf{K} j_\varphi \sin(n\varphi). \tag{19}$$

If we use the Fourier cosine series of (16), and define

$$k_0(z) = \frac{1}{2\pi}\int_{-\pi}^{\pi} \mathcal{K}(\varphi, z)\,\mathrm{d}\varphi = \frac{1}{4\pi^2} Q_{\frac{1}{2}}(\chi), \tag{20}$$

$$k_n(z) = \frac{1}{\pi}\int_{-\pi}^{\pi} \mathcal{K}(\varphi, z)\cos(n\varphi)\,\mathrm{d}\varphi = \frac{1}{4\pi^2}\left(Q_{n-\frac{3}{2}}(\chi) + Q_{n+\frac{1}{2}}(\chi)\right), \tag{21}$$

then (19) yields

$$\mathsf{K} j_\varphi = (k_0 * j_0) + \sum_{n=1}^{\infty} (k_n * j_n^{(1)})\cos(n\varphi) + \sum_{n=1}^{\infty} (k_n * j_n^{(2)})\sin(n\varphi), \tag{22}$$

with

$$\begin{aligned} (k_0 * j_0) &= \int_{S_\cup} \mathcal{K}(\varphi, z-\zeta) j_0(\zeta)\,\mathrm{d}\varphi\,\mathrm{d}\zeta, \\ (k_n * j_n^{(1)}) &= \int_{S_\cup} \mathcal{K}(\varphi, z-\zeta) j_n^{(1)}(\zeta)\cos(n\varphi)\,\mathrm{d}\varphi\,\mathrm{d}\zeta, \\ (k_n * j_n^{(2)}) &= \int_{S_\cup} \mathcal{K}(\varphi, z-\zeta) j_n^{(2)}(\zeta)\cos(n\varphi)\,\mathrm{d}\varphi\,\mathrm{d}\zeta. \end{aligned} \tag{23}$$

Now consider two rings, r_1 and r_2, situated at $[z_0^{(r_1)}, z_1^{(r_1)}]$ and $[z_0^{(r_2)}, z_1^{(r_2)}]$, respectively. Define $B_{nk}^{(r_1)}$ as a basis function on r_1 and $B_{n'k'}^{(r_2)}$ as a basis function on r_2, where $n, n' > 0$ correspond with the order of the cosine function in the φ-direction and k, k' correspond with the order of the Legendre polynomial in the z-direction. We remark that results for $n, n' = 0$ and for the sine function follow analogously. The inner product to be computed in the Galerkin method becomes

$$(\mathsf{K} B_{nk}^{(r_1)}, B_{n'k'}^{(r_2)}) = \pi\delta_{n'n}\int_{z_0^{(r_2)}}^{z_1^{(r_2)}} (k_n * P_k) P_{k'}\,\mathrm{d}z. \tag{24}$$

On the island i_1, positioned at $[z_0^{(i_1)}, z_1^{(i_1)}]$, we denote a basis function as $B_{nk}^{(i_1)}$, where n is the order in the φ-direction and k is the order in the z-direction. Each separate basis function satisfies the condition that the normal component is zero at the edges. In order to get 2π periodic functions, we expand $B_{nk}^{(i_1)}$ as follows:

$$B_{nk}^{(i_1)} = \left(\alpha_{n0} + \sum_{m=1}^{\infty} \alpha_{nm}\cos(m\varphi) + \sum_{m=1}^{\infty} \beta_{nm}\sin(m\varphi)\right) P_k, \tag{25}$$

in which the coefficients are known. The inner products $(\mathsf{K}B_{nk}^{(r_1)}, B_{n'k'}^{(i_1)})$ become

$$(\mathsf{K}B_{nk}^{(r_1)}, B_{n'k'}^{(i_1)}) = \pi\alpha_{n'n} \int_{z_0^{(i_1)}}^{z_1^{(i_1)}} (k_n * P_k)P_{k'}\,\mathrm{d}z. \tag{26}$$

For two islands i_1 and i_2 we obtain

$$(\mathsf{K}B_{nk}^{(i_1)}, B_{n'k'}^{(i_2)}) = 2\pi\alpha_{n0}\alpha_{n'0} \int_{z_0^{(i_2)}}^{z_1^{(i_2)}} (k_0 * P_k)P_{k'}\,\mathrm{d}z$$
$$+\pi \sum_{m=1}^{\infty} (\alpha_{nm}\alpha_{n'm} + \beta_{nm}\beta_{n'm}) \int_{z_0^{(i_2)}}^{z_1^{(i_2)}} (k_m * P_k)P_{k'}\,\mathrm{d}z. \tag{27}$$

In numerical computations, we can truncate the series at $m = M$, with M sufficiently large. The integrals in (24) and (27) have a singular integrand when the two rings/islands coincide. In that case, we split off the logarithmic part, and use

$$\int_{-1}^{1}\int_{-1}^{1} P_k(z)P_{k'}(z)\log|z-\zeta|\,\mathrm{d}\zeta\,\mathrm{d}z \tag{28}$$
$$= \begin{cases} \dfrac{8}{(k+k')(k+k'+2)[(k-k')^2-1]}, & \text{if } k+k'>0 \text{ even}, \\ 0, & \text{if } k+k' \text{ odd}, \\ 4\log 2 - 6, & \text{if } k=k'=0. \end{cases}$$

The remaining part is regular and can therefore be solved numerically.

4 Results

Considered are two rings and one island, placed on a cylinder with radius $R = 0.35$m. The rings have a width of 4 cm and carry a source current of 600 A. The island is placed in between the two rings, has a width of 2 cm, a length of 55 cm, and its center is defined at $(\varphi, z) = (0, 0)$.

If the current through the rings is in phase, then we observe two eddies in the island and an edge-effect in the rings towards the outside of the system. This is visualized in Fig. 2 (a), where the amplitude of the tangential component of the current density $|j_\varphi|$ is plotted along the line $\varphi = 0$. Note that the edge-effects become stronger if the frequency is increased.

If the current through the rings is in anti-phase, then we observe one eddy in the island and an edge-effect in the rings towards the center of the system. This is visualized in Fig. 2 (b). Note that the current density in the island is stronger than in the previous case.

The consistency of the method is checked by comparison with a configuration of two rings only. The method for a coil modeled as circular loops of strips is described in [4]. In Fig. 2 (c), the current distribution is shown for the configuration similar to the first example, but without island. In Fig. 2 (d), the current distribution is shown for the configuration similar to the second example, but without island. We observe that in both cases, the eddies in the island hardly affect the current in the rings.

5 Conclusions

In this paper, we have modeled a z-coil of an MRI-scanner by a set of rings and islands. The model is an extension of the existing model for rings only, in order to take into account the currents in axial direction. The resulting program is a handy tool for the design of gradient coils. The simulations are used for instance to investigate the presence of islands in a coil.

An appropriate choice for the basis functions in the axial direction is the use of Legendre polynomials. Integrals containing Legendre polynomials and a logarithmic function can be computed analytically and show a fast convergence; less than ten polynomials are needed only for an accurate approximation. The results are consistent with the ones for rings only.

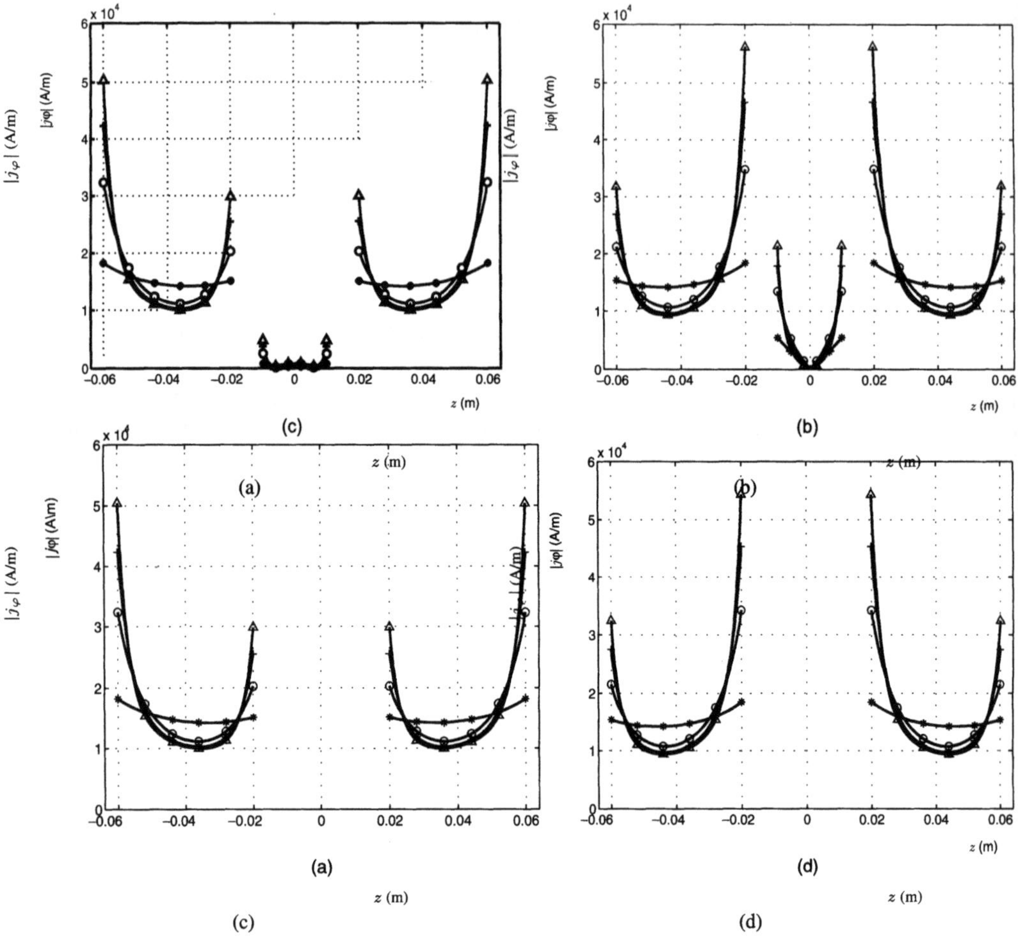

Fig. 2. Amplitude of the current density along the line $\varphi = 0$, at frequencies $f = 100$ Hz ($*$), $f = 400$ Hz ($\circ$), $f = 700$ Hz ($+$), $f = 1000$ Hz ($\triangle$). (a) Two rings, one island, sources in phase; (b) Two rings, one island, sources in anti-phase; (c) Two rings, sources in phase; (d) Two rings, sources in anti-phase

References

1. M.T. Vlaardingerbroek and J.A. den Boer, *Magnetic Resonance Imaging.* Berlin: Springer (1999)
2. T. Ulicevic, *Skin effect in a gradient coil.* Eindhoven: Final report of the postgraduate program Mathematics for Industry, Eindhoven University of Technology (2001)
3. T. Ulicevic, J.M.B. Kroot, S.J.L. van Eijndhoven, A.A.F van de Ven *Current distribution in a parallel set of conducting strips.* Journal of Engineering Mathematics, Vol 51, pp 381-400 (2005)
4. J.M.B. Kroot, *Current Distribution in a Gradient Coil, Modeled as Circular Loops of Strips.* Eindhoven: Final report of the postgraduate program Mathematics for Industry, Eindhoven University of Technology (2002)
5. H. S. Cohl, J.E. Tohline, *A compact cylindrical Green's function expansion for the solution of potential problems.* The astrophysical journal, Vol 257, pp 86-101 (1999)
6. M. Abramowitz and I.A. Stegun, *Handbook of Mathematical Functions.* New York: Dover (1972)
7. R. F. Harrington, *Field Computation by Moment Methods.* London: Macmillan (1968)

An Integration of Optimal Topology and Shape Design for Magnetostatics *

D. Lukáš

SFB F013 "Numerical and Symbolic Scientific Computing", University Linz, Altenberger Strasse 69, A-4040 Linz, Austria, dalibor.lukas@vsb.cz, http://lukas.am.vsb.cz

Abstract Topology optimization searches for an optimal distribution of material and void without any restrictions on the structure of the design geometry. Shape optimization tunes the shape of the geometry, while the topology is fixed. In this paper we proceed sequentially with the optimal topology and shape design so that a coarsely optimized topology is the initial guess for the following shape optimization. In between we identify the topology by hand and approximate it by piecewise Bézier shapes by means of the least square method. For the topology optimization we use the steepest descent method, while a quasi–Newton method and multilevel techniques are used for the shape optimization. We apply the machinery to optimal design of a direct electric current electromagnet. The resulting optimal design corresponds to physical experiments.

1 Introduction

In the process of development of industrial components one looks for the parameters to be optimal subject to a proper criterion. The geometry is usually crucial as far as the design of electromagnetic components is concerned. We can employ topology optimization, cf. [Ben95], to find an optimal distribution of the material without any preliminary knowledge. Shape optimization, cf. [HN97, Luk04], is used to tune shapes of a known initial design. While in the structural mechanics topology optimization results in rather complicated structures the shapes of which are not needed to be then optimized, in magnetostatics we end up with simple topologies which, however, serve as very good initial points for the further shape optimization. The idea here is to couple them sequentially.

In [Cea00] a connection between topological and shape gradient is shown and applied in structural mechanics. They proceed shape and topology optimization simultaneously so that at one optimization step both the shape and topology gradient are calculated. Then shapes are displaced and the elements with great values of the topology gradient are removed, while introducing the natural boundary condition along the new parts, e.g. a hole. Here we are rather motivated by the approach in [OBR91, TCh01], where they apply a similar algorithm as we do to structural mechanics, however, using re-meshing in a CAD software environment, which was computationally very expensive. Our aim here is to make the algorithm fast. Therefore, we additionally employ semianalytical sensitivity analysis and a multilevel method.

2 Topology Optimization for Magnetostatics

Let us consider a fixed computational domain $\Omega \subset \mathbf{R}^d$, where $d = 2, 3$. Let $\Omega_{\mathrm{d}} \subset \Omega$ be the subdomain where the designed structure can arise. The set of admissible material distributions is denoted by $\mathcal{Q} := \{\rho \in L^2(\Omega_{\mathrm{d}}) \mid 0 \le \rho \le 1\}$. We penalize the intermediate values by

$$\widetilde{\rho}_p(\rho) := \frac{1}{2}\left(1 + \frac{1}{\arctan(p)}\arctan(p(2\rho - 1))\right),$$

*This research has been supported by the Austrian Science Fund FWF within the SFB "Numerical and Symbolic Scientific Computing" under the grant SFB F013, subproject F1309.

where $p := 100$ is typically good enough. Further, we consider the following linear magnetic reluctivity:

$$\nu(\widetilde{\rho}) := \begin{cases} \nu_0 + (\nu_1 - \nu_0)\widetilde{\rho}, & \text{in } \Omega_\mathrm{d} \\ \nu_0, & \text{otherwise} \end{cases},$$

where ν_0, ν_1 are the reluctivities of the air and ferromagnetics, respectively. Finally, let $\mathcal{I} : \mathbf{L}^2(\Omega) \to \mathbf{R}$ be a cost functional, possibly involving penalization of state constraints. Given a maximal volume $V_{\max}$ of the designed structure, the 3D topology optimization problem governed by the linear magnetostatics then reads as follows:

$$\begin{cases} \min_{\rho \in \mathcal{Q}} \mathcal{I}(\mathbf{curl}(\mathbf{u})) \\ \text{w.r.t.} \\ \int_{\Omega_\mathrm{d}} \widetilde{\rho}(\rho)\, d\mathbf{x} \le V_{\max} \\ \int_{\Omega} \nu\left(\widetilde{\rho}(\rho)\right) \mathbf{curl}(\mathbf{u}) \cdot \mathbf{curl}(\mathbf{v})\, d\mathbf{x} = \int_{\Omega} \mathbf{J} \cdot \mathbf{v}\, d\mathbf{x} \text{ in } \mathbf{H}_{0,\perp}(\mathbf{curl}; \Omega), \end{cases} \tag{1}$$

where $\mathbf{J} \in \mathbf{L}^2(\Omega)$ is a divergence–free current density and where

$$\mathbf{H}_{0,\perp}(\mathbf{curl}; \Omega) := \{\boldsymbol{v} \in \mathbf{H}_0(\mathbf{curl}; \Omega) \mid \forall p \in H_0^1(\Omega) : \int_\Omega \mathbf{grad}(p) \cdot \boldsymbol{v}\, d\mathbf{x} = 0\},$$

$$\mathbf{H}_0(\mathbf{curl}; \Omega) := \{\boldsymbol{v} \in \mathbf{L}^2(\Omega) \mid \mathbf{curl}(\boldsymbol{v}) \in L^2(\Omega)\}.$$

Note that the 2-dimensional (2D) reduced magnetostatic problem leads to the Poisson equation.

Concerning the numerical solution, the 3-dimensional (3D) problem is discretized by the finite element method using the lowest order edge Nédélec elements on tetrahedra, while we use the lowest order nodal Langrange elements on triangles in case of the 2D reduced problem. The design material distribution is elementwise constant. Note that in the 3D case we do not solve the mixed formulation in $\mathbf{H}_{0,\perp}(\mathbf{curl}; \Omega)$ but rather a non-mixed one in $\mathbf{H}_0(\mathbf{curl}; \Omega)$ while we add the regularization term $\varepsilon \int_\Omega \mathbf{u} \cdot \mathbf{v}\, d\mathbf{x}$ to the bilinear form. In the optimization process we always choose the initial value of ρ to be 0.5.

3 Piecewise Smooth Approximation of Shapes

We will use the optimal topology design as the initial guess for the shape optimization. The first step towards a fully automatic procedure is a shape identification, which we are doing by hand for the moment. The second step we are treating now is a piecewise smooth approximation of the shapes by Bézier curves or patches. Let $\rho^{\mathrm{opt}} \in \mathcal{Q}$ be an optimized discretized material distribution. Recall that it is not a strictly 0-1 function. Let $\boldsymbol{p}_1, \dots, \boldsymbol{p}_n$ denote vectors of Bézier parameters of the shapes $\alpha_1(\boldsymbol{p}_1), \dots, \alpha_n(\boldsymbol{p}_n)$ which form the air and ferromagnetic subdomains $\Omega_0(\alpha_1, \dots, \alpha_n)$ and $\Omega_1(\alpha_1, \dots, \alpha_n)$, respectively, i.e. $\Omega_1 \subset \Omega_\mathrm{d}$, $\overline{\Omega} = \overline{\Omega_0} \cup \overline{\Omega_1}$ and $\Omega_0 \cap \Omega_1 = \emptyset$. Let further $\underline{\mathbf{p}_i}$ and $\overline{\mathbf{p}_i}$ denote the lower and upper bounds, respectively, and let $\mathcal{P} := \left\{(\boldsymbol{p}_1, \dots, \boldsymbol{p}_n) \mid \underline{\boldsymbol{p}_i} \le \boldsymbol{p}_i \le \overline{\boldsymbol{p}_i} \text{ for } i = 1, \dots, n\right\}$ be the set of admissible Bézier parameters. We solve the following least square fitting problem:

$$\min_{(\boldsymbol{p}_1, \dots, \boldsymbol{p}_n) \in \mathcal{P}} \int_{\Omega_\mathrm{d}} \left(\rho^{\mathrm{opt}} - \chi(\Omega_1(\alpha_1(\boldsymbol{p}_1), \dots, \alpha_n(\boldsymbol{p}_n)))\right)^2 d\mathbf{x}, \tag{2}$$

where $\chi(\Omega_1)$ is the characteristic function of Ω_1.

When solving (2) numerically, one encounters the problem of intersection of the Bézier shapes with the mesh on which ρ^{opt} is elementwise constant. In order to avoid it we use the property that the Bézier control polygon converges quite fast to the shape under the refinement procedure, which is in 2D as follows:

$$\begin{aligned} \left[\boldsymbol{p}_i^{k+1}\right]_0 &:= \left[\boldsymbol{p}_i^k\right]_0 \\ \left[\boldsymbol{p}_i^{k+1}\right]_j &:= \tfrac{j-1}{m_i+1}\left[\boldsymbol{p}_i^k\right]_{j-1} + \tfrac{n-j}{m_i+1}\left[\boldsymbol{p}_i^k\right]_j, \; j = 2, \dots, m_i \\ \left[\boldsymbol{p}_i^{k+1}\right]_{m_i+1} &:= \left[\boldsymbol{p}_i^k\right]_{m_i} \end{aligned} \tag{3}$$

where $\boldsymbol{p}_i^0 := \boldsymbol{p}_i$, see also Fig. 1. Note that in 3D one uses a similar procedure provided a tensor-product grid of Bézier control nodes. Then the integration in (2) is replaced by a sum over the elements and we deal with intersecting the mesh with a polygon. Note that our least square functional is not twice differentiable whenever a shape touches the grid. This is still acceptable for the quasi-Newton optimization method that we apply.

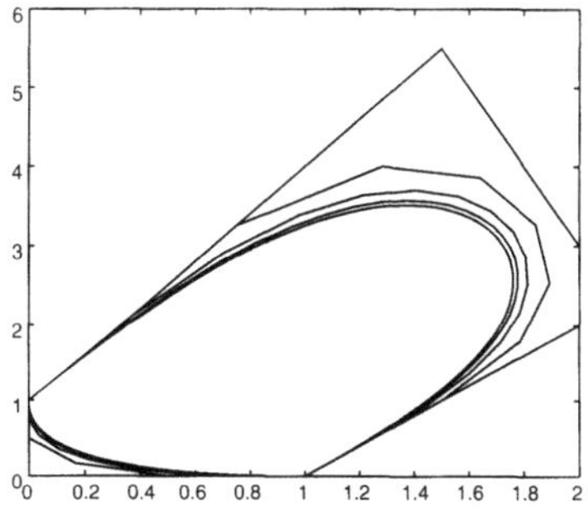

Fig. 1. Approximation of Bézier shapes by the refined control polygon

4 Multilevel Shape Optimization for Magnetostatics

With the notation of Sect. 2, the shape optimization problem under consideration is as follows:

$$\left\{\begin{array}{l} \min\limits_{(\boldsymbol{p}_1,\dots,\boldsymbol{p}_n)\in\mathcal{P}} \mathcal{I}(\mathbf{curl}(\mathbf{u})) \\ \text{w.r.t.} \\ \int\limits_{\Omega_1(\alpha_1(\boldsymbol{p}_1),\dots,\alpha_n(\boldsymbol{p}_n))} d\mathbf{x} \le V_{\max} \\ \sum\limits_{i=0}^{1} \int\limits_{\Omega_i(\alpha_1(\boldsymbol{p}_1),\dots,\alpha_n(\boldsymbol{p}_n))} \nu_i \mathbf{curl}(\mathbf{u})\cdot\mathbf{curl}(\mathbf{v})\,d\mathbf{x} = \int\limits_{\Omega} \mathbf{J}\cdot\mathbf{v}\,d\mathbf{x} \text{ in } \mathbf{H}_{0,\perp}(\mathbf{curl};\Omega). \end{array}\right. \tag{4}$$

Again, we use the regularization and the 2D reduction as in Sect. 2

Concerning the discretization, we have to take special care of how the shape enters the bilinear form in order not to change the topology of the mesh. We use two approaches here. First, the control design nodes interpolate the Bézier shape and the remaining grid nodes displacements are given by solving an auxiliary discretized linear elasticity problem with the nonzero Dirichlet boundary condition along the design shape. The drawback is that on fine meshes some elements may flip whenever the shape changes significantly. Another approach is to use (3) again and intersect the refined Bézier control polygon with the mesh so that the design interface goes across some elements. This brings a little nonsmoothness, which is still acceptable for a quasi-Newton optimization method we use. Moreover, assembling the bilinear form takes much longer. On the other hand, the design change is not limited by the finesty of the grid.

Perhaps, the main reason for solving the coarse topology optimization as a preprocessing is that we get rid of a large number of design variables in cases of fine discretized topology optimization. Once we have a good initial shape design, we will proceed the shape optimization in a multilevel way in order to speed up the algorithm as much as possible. We propose to couple the outer quasi-Newton method with the nested conjugate gradient method preconditioned by a geometric multigrid (PCG), as depicted in Algorithm 1, in which $\mathbf{A}^l(\boldsymbol{p}_1,\dots,\boldsymbol{p}_n)$ denotes the reluctivity matrix assembled at the l-th level.

Algorithm 1 Newton iterations coupled with nested multigrid PCG

Given $\boldsymbol{p}_1^{\text{init}},\dots,\boldsymbol{p}_n^{\text{init}}$
Discretize at the first level $\longrightarrow h^1, \mathbf{A}^1(\boldsymbol{p}_1^{\text{init}},\dots,\boldsymbol{p}_n^{\text{init}})$
Solve by a quasi-Newton method and the nested direct solver $\longrightarrow \boldsymbol{p}_1^1,\dots,\boldsymbol{p}_n^1$
Store the first level preconditioner $\mathbf{C}^1 := \left[\mathbf{A}^1(\boldsymbol{p}_1^1,\dots,\boldsymbol{p}_n^1)\right]^{-1}$
for $l = 2,\dots$ **do**
  Refine $h^{l-1} \longrightarrow h^l$
  Prolong $\boldsymbol{p}_1^{l-1},\dots,\boldsymbol{p}_n^{l-1} \longrightarrow \boldsymbol{p}_1^{l,\text{init}},\dots,\boldsymbol{p}_n^{l,\text{init}}$
  Solve by a quasi-Newton method and the nested multigrid solver $\longrightarrow \boldsymbol{p}_1^l,\dots,\boldsymbol{p}_n^l$
  Store the l–th level preconditioner $\mathbf{C}^l$
end for

5 An Application

We consider a direct electric current (DC) electromagnet, see Fig. 2. The electromagnets are used for measurements of Kerr magnetooptic effects, cf. [ZK97]. They require the magnetic field among the pole heads as homogeneous, i.e. as constant as possible. Let us note that the magnetooptic effects are investigated for applications in high capacity data storage media, like development of new media materials for magnetic or compact discs recording. Let us also note that the electromagnets have been developed at the Institute of Physics, Technical University of Ostrava, Czech Republic, see [Pos02]. A number of instances have been delivered to laboratories in France, Canada or Japan.

Our aim is to improve the current geometries of the electromagnets in order to be better suited for measurements of the Kerr effect. The generated magnetic field should be strong and homogeneous enough. Unfortunately, these assumptions are contradictory and we have to balance them. The cost functional reads as follows:

$$\mathcal{I}(\mathbf{curl}(\boldsymbol{u})) := \int_{\Omega_{\mathrm{m}}} \|\mathbf{curl}(\mathbf{u}) - B_{\mathrm{m}}^{\mathrm{avg}}\mathbf{n}_{\mathrm{m}}\|^2 + 10^6 \left(\min\{0, B_{\mathrm{m}}^{\mathrm{avg}} - B^{\mathrm{min}}\}\right)^2,$$

where $\Omega_{\mathrm{m}} \subset \Omega$ is the subdomain where the magnetic field should be homogeneous, $B_{\mathrm{m}}^{\mathrm{avg}}$ is the mean value over Ω_{m} of the magnetic flux density component in the direction $\boldsymbol{n}_{\mathrm{m}} := (0,1)$ and $B^{\mathrm{min}} := 0.12\,[\mathrm{T}]$ is the minimal required magnitude. There are 600 turns pumped by the current of $5\,[\mathrm{A}]$. We use the linearized value of the relative permeability of the ferromagnetics, which is 5100. Some results were already presented in [Luk01].

6 Numerical Results

We present numerical results for our application in 2D. For simplicity we consider only two coils to be active and take, due to the symmetry, a quarter of the domain, see Fig. 3 (**a**). Given the initial design $\rho^{\mathrm{init}} := 0.5$ in Ω_{d} we start with the topology optimization. Concerning (1), we choose $V_{\max} := 0.0155\,[\mathrm{m}^2]$ and $p := 100$. A coarse optimized topology design is depicted in Fig. 3 (**b**). There are 861 design, 1105 state variables and the optimization was done in 7 steepest descent iterations which took 2.5 seconds, when using the adjoint method for the sensitivity analysis.

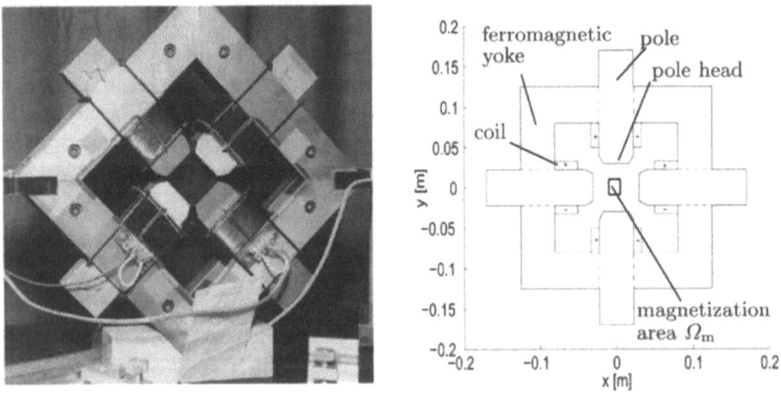

Fig. 2. An electromagnet of the Maltese Cross geometry

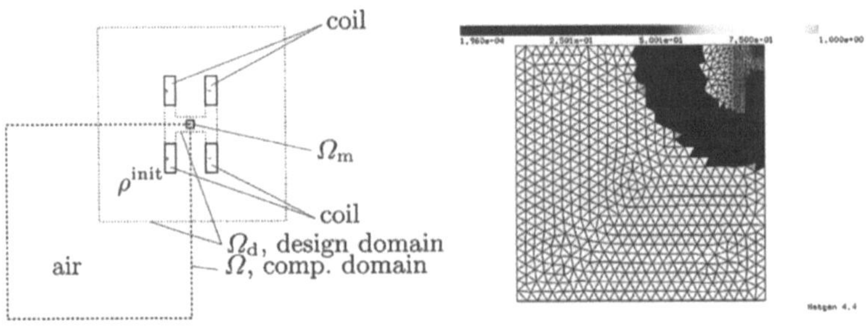

Fig. 3. Topology optimization: (**a**) initial design; (**b**) coarsely optimized design ρ^{opt}

The second part of the computation is the shape approximation. Here we refer to Fig. 4. We are looking for three Bézier curves that fit the optimized topology. Here we have 19 design parameters in total and solving the least square problem (2) was finished in 8 quasi-Newton iterations which took 26 seconds, when using the numerical differentiation.

Finally, we used the smooth shape design as the initial guess for the shape optimization (4). In Tables 1 and 2 there are parameters of the computation when using the mesh deformation and the so-called shape-across-elements approach, respectively. In the first case the multigrid acts very efficiently, however, on the finest level we end up with the design almost the same as the very initial one $\boldsymbol{p}_1^{1,\mathrm{init}}, \ldots, \boldsymbol{p}_n^{1,\mathrm{init}}$. This is due to that the mesh deformation is very limited at the finest mesh. In the second approach we observed a significant improvement of the shape in terms of the cost functional, however, the multigrid preconditioner is by far not efficient, see Table 2, due to the reluctivity being jumping within some elements. The final optimized geometry calculated by the second approach is depicted in Fig. 5

Table 1. Multilevel shape optimization using the mesh deformation approach

level	design variables	outer Newton iterations	state variables	nested CG iterations	total time
1	19	7	1098		27s
2	40	8	4240	3	3min 9s
3	82	8	16659	4–5	29min 14s
4	166	8	66037	4–5	3h 37min 42s

Table 2. Multilevel shape optimization using the shape-across-elements approach

level	design variables	outer Newton iterations	state variables	nested CG iterations	total time
1	19	14	1098		4min 32s
2	40	6	4240	11–14	26min 37s
3	82	8	16659	21–26	3h 20min 15s

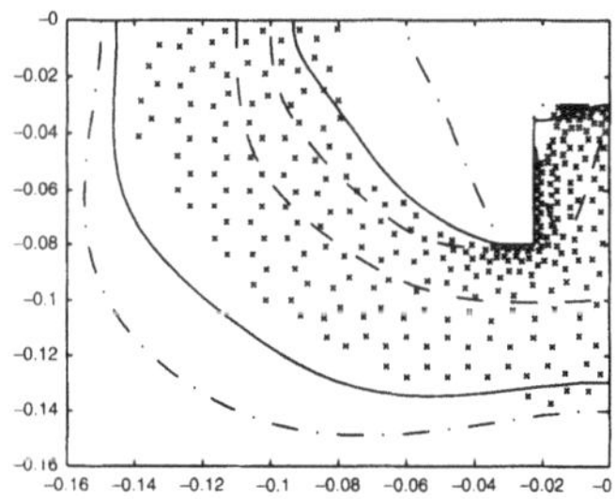

Fig. 4. Shape approximation: dashed line – lower bound; dash-and-dot line – upper bound; solid line – optimal shape approximation; crosses – mid-points of the elements with $\rho^{\mathrm{opt}} \geq 0.5$

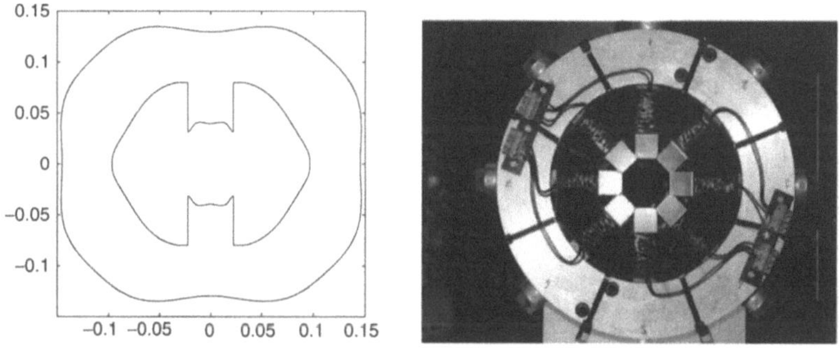

Fig. 5. Multilevel shape optimization: **(a)** optimized geometry; **(b)** the O-Ring electromagnet

(**a**). We can see that the result is in a good correspondance with the so-called O-Ring electromagnet which was already designed and manufactured by physicists.

7 Conclusion

This paper presented a method which sequentially combines topology and shape optimization. First, we solved a coarsely discretized topology optimization problem. Then we approximated some chosen interfaces by Bézier shapes. Finally, we proceeded with shape optimization in a multilevel way. We also discussed two different shape-to-state mappings. We applied the method to a 2D optimal shape design of a DC electromagnet. Without the multilevel procedure, we can get already fine optimized geometries in minutes. However, as we aim at large-scale discretizations, it still remains to analyze and improve the multigrid convergence.

References

[Ben95] Bendsøe, M.P.: Optimization of Structural Topology, Shape and Material. Springer, Berlin, Heidelberg (1995)

[Cea00] Céa, J., Garreau, S., Guillaume, P., Masmoudi, M.: The shape and topological optimizations connection. Comput. Methods Appl. Mech. Eng. **188**, 713–726 (2000)

[HN97] Haslinger J., Neittaanmäki P.: Finite Element Approximation for Optimal Shape, Material and Topology Design. Wiley, Chinchester (1997)

[Luk01] Lukáš, D.: Shape optimization of homogeneous electromagnets. In: van Rienen, U., Günther, M., Hecht, D. (eds.) Scientific Computing in Electrical Engineering 2000, Lect. Notes Comp. Sci. Engrg. **18**, pp. 145–152 (2001)

[Luk04] Lukáš, D.: On solution to an optimal shape design problem in 3-dimensional magnetostatics. Appl. Math. **49**:5, 24 pp. (2004)

[OBR91] Olhoff, N., Bendsøe, M.P., Rasmussen, J.: On CAD-integrated structural topology and design optimization. Comp. Meth. Appl. Mech. Eng. **89**, 259–279 (1991)

[Pos02] Postava, K., Hrabovský, D., Pištora, J., Fert, A.R., Višňovský, Š., Yamaguchi, T.: Anisotropy of quadratic magneto-optic effects in reflection. J. Appl. Phys. **91**, 7293–7295 (2002)

[TCh01] Tang, P.-S., Chang, K.-H.: Integration of topology and shape optimization for design of structural components. Struct. Multidisc. Optim. **22**, 65–82 (2001)

[ZK97] Zvedin, A.K., Kotov, V.A.: Modern Magnetooptics and Magnetooptical Materials. Institute of Physics Publishing Bristol and Philadelphia (1997)

Numerical Computation of Magnetic Field and Inductivity of Power Reactor with Respect of Real Magnetic Properties of Iron Core

M. Marek

VSB - Technical University of Ostrava, Faculty of Electrical Engineering, Department of Electrical Machines and Apparatuses, Ostrava, Czech Republic, martin.marek@vsb.cz

Abstract In this paper introduced electromagnetic analysis presents just a part of all analyses which have been performed within the solving of plasmatron power system project. This system is the significant part of plasma technology which is designed for coal-energy blocks smelting and their stabilization. In recent years is this technology developed and realized with the firm ORGREZ Corp in conjunction with VSB - Technical University of Ostrava. To get the complex view about the solving problem it is necessary to describe basic principles and some parts of plasma technology.

1 Plasma technology

The mentioned plasma technology is designed for classical energy blocks containing boilers for coal fuel burning (power station, heat station). At the present time for energy blocks smelting and stabilization is mostly used the secondary fuel (gas, black oil). Prices of these raw materials are continually on the rise and the ultimate reserves are continually less. On this account the plasma technology deriving benefit from the fourth state of substance can bring the significant economic profit. The structural element of this technology is the power generator of low-temperature plasma plasmatron GNP320. By generator produced plasma effects in the thermochemical chamber on the dust coal-air mixture. The mixture input into the chamber is realized by the help of powder-conduit. Thanks to the acting plasma high temperature on the mixture come to the forcefully thermochemical break up of coal elements, to thermochemical reaction and subsequently to the aero-mixture burning in the area of burner mouth into the burning chamber. The fundamental process schema of coal powder burning initialization is shown on the figure 1. The more detailed description of burning technology can be found in the literature [1], [2].

The main benefit of this technology - unlike other smelting methods and stabilization methods of coal energy-blocks which are nowadays used, is to eliminate the use and consumption of others secondary fuels (gas, black oil). The plasma acts here as a starting heat source and is extracted directly from electrical energy and compressed air. By plasmatron GNP320 generated plasma is shown on the figure 2 (left), the working principle of this generator is demonstrated on the right part of the figure 2. The d.c. arc acts here as plasma source and is high stabilized by the loaded compressed air which is burning between two cylindrical electrodes. One part of so burning arc is carried away out of plasmatron through the positive electrode. The total plasma power is adequate to the arc-drop voltage size and to the size of arc flowing current. For the plasmatron GNP320 is this power regulable in the range of 150-320kW by the operating current extent of 370-750A.

1.1 Power electric supply of plasmatron GNP320

The basic requirement on the power supply system for plasmatrons GNP320 (electrical element with the d.c. arc character) is to ensure the supply of stabilized direct current with a little wave. This requirement is the fundamental prerequisite for right working of plasma generator and for the stable burning process of above mentioned voltaic arc inside the plasmatron. At the plasmatron use is various plasma power mode required. This power mode depends on combustion burner working mode. On this account it is important to ensure the above mentioned fundamental prerequisite by the whole working power range. For this need has been the power supply system developed. The simplified scheme of this system is shown on the fig. 3. Basic elements of the circuit are supply power transformer (VN side selectable, NN side 3×400V), controlled three-phase rectifier, starting resistor, smoothing inductor, plasmatron, HV-hf ionizing ignition unit.

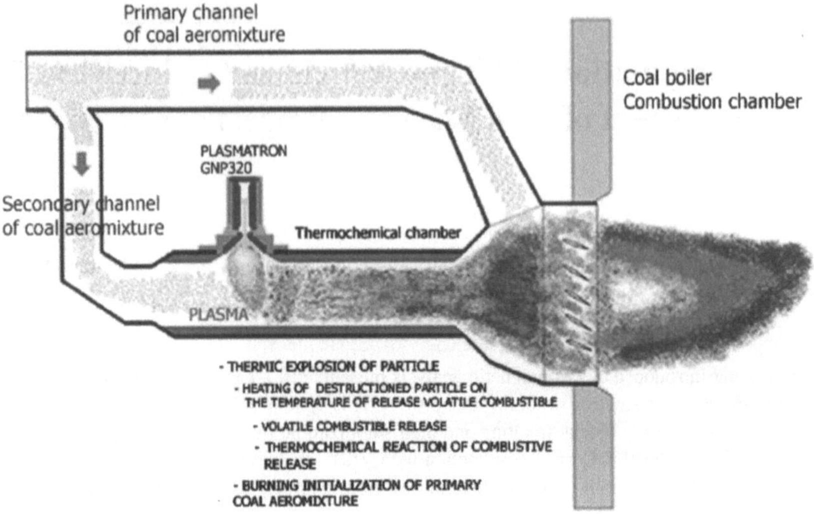

Fig. 1. Initialization principle of coal mixture burning by the help of plasma

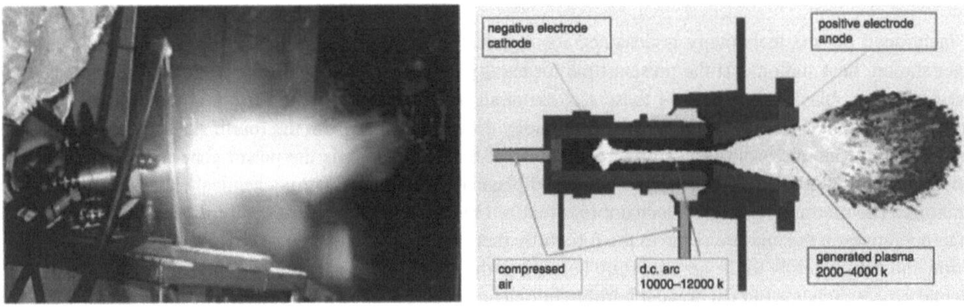

Fig. 2. Generator of low-temperature plasma plasmatron GNP320

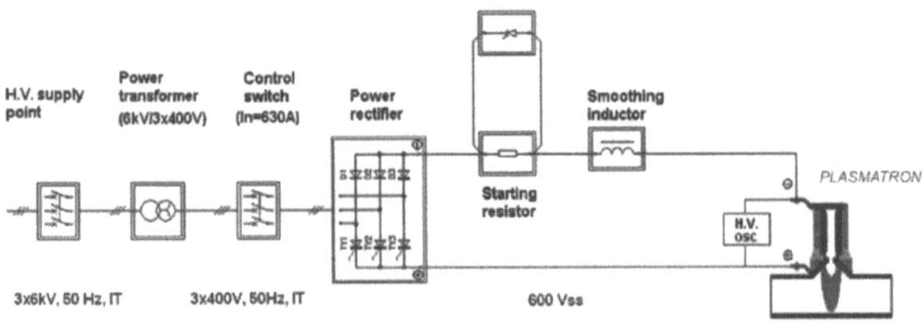

Fig. 3. Block diagram of plasmatron GNP320 power supply system

On the ground of the next description and electrotechnical analyses it is important to mention that the plasmatron is a non-linear electrical element with the own VA characteristic. This characteristic is given by the general VA characteristic with specific parameters and depends on the setting size of loading air pressure. For the clearness are these characteristics shown on the figure 4. Pursuant to this picture it is evident that the plasmatron power regulation is not possible through the voltage regulation.

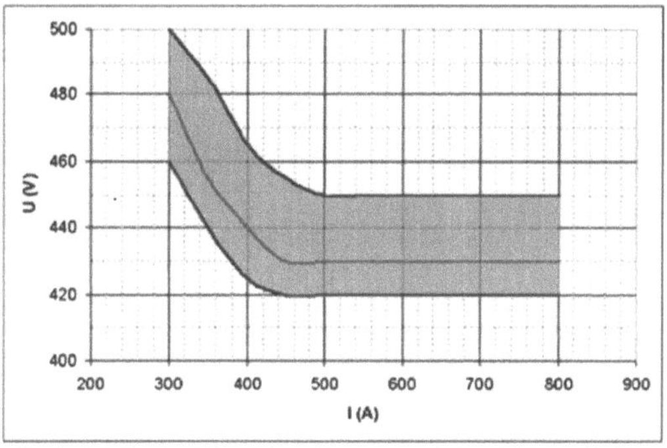

Fig. 4. VA characteristics of plasmatron GNP320 (various pressure relations)

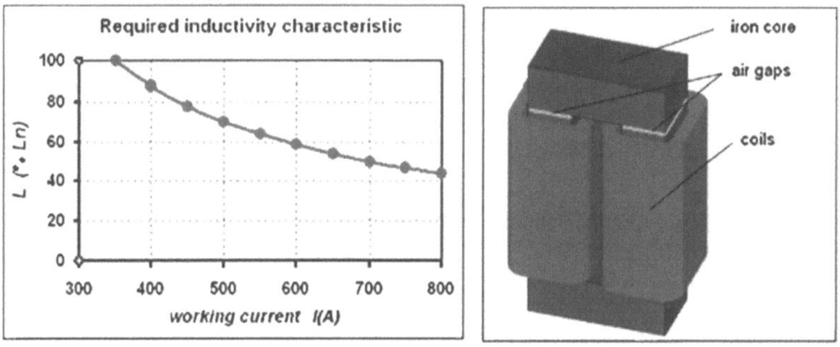

Fig. 5. (left) Required percent inductivity size in dependence on working current, (right) Designed inductor construction

1.2 Smoothing inductor

The power smoothing inductor is as electrical element in the power supply circuit which ensures the needed stabilization and current extermination. This element has to ensure the adequate current smoothing in the whole range of plasmatrons working current (In=350-750A) by the correspondent arc-drop voltage (Upl=400-460V). For the own power smoothing inductor project and for the determination of needed parameters is important to take into account mentioned VA characteristics and the defined supply voltage. The supply voltage is given by the secondary side of transformer and by the rectifier design. By the analysis of these data we get the minimal needed inductance size depending on working current size (for set specific acceptable value of current waviness). The given inductivity dependence on working current is shown on the figure 5 (left) in percentage. The nominal inductivity value of inductor is shown on the figure 5. in the form of 100 (%)Ln and it has been determined for minimal working current of 350A. For these parameters the inductor with copper winding and with split magnetic core which includes air gaps has been projected. The basic concept of inductor design is shown on the figure 5 (right). Particular parameters have been determined and optimized pursuant to empirical relations by the use of BH and magnetization characteristics of electrical sheets metal. These sheets have been chosen for magnetic core construction and their characteristics have been discovered by measuring.

1.3 Magnetic characteristics measuring of sheets metal of magnetic core

Magnetic characteristics of electrotechnical sheets metal chosen type, which are designed for core production, have been determined on belt samples made of this material - electrical sheet metal 0,35mm thick. Magnetic characteristics measuring has been performed by the help of gauging system REMACOMP, which is designed for BH characteristics measuring and magnetization characteristics measuring of magnetically soft materials in dynamically magnetic fields with frequency range of 1 Hz 10 kHz. Following gauging extenders have been used for measuring a small Epstein frame and SST yoke. Sizes of the set of sheets metal samples satisfied the standard size of this frame (280×30mm). The design of gauging system REMACOMP and of gauging extenders is shown on figure 6. Magnetic characteristics measuring of sheets metal has been made for various frequency 1-150Hz and for various parameters setting of exciting magnetic field. Some of measuring results are shown on figure 7. In this way have been determined real magnetic

Fig. 6. REMACOMP Gauging system for magnetic characteristics measuring of constructional materials in dynamic fields 1Hz 10 kHz. Epstein frame (in the middle). SST yoke (right)

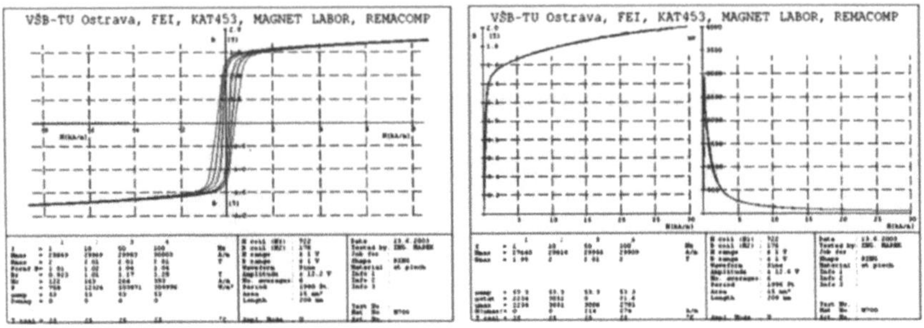

Fig. 7. Chosen BH and magnetization characteristics measured onto electrotechnical sheets metal samples of core material for the frequency of 1,10,50,100 Hz

characteristics of chosen core material and have been guaranteed exact input data for the inductor design and also for next computations.

2 Numerical computation of electromagnetic field of inductor

The aim of this computation part has been the check on exactness of made empirical inductor design. By this empirical design were used the exact material characteristics but the own computation resulted from empirical founded relations. On this account this computation had to be loaded in error. This error is given by the inaccurate determination of magnetic flux lay-out in the area of working air gaps, by the no homogeneity of magnetic field lay-out in the core, by leakage flux, etc. On this account has been produced 3D FEM inductor model, whose geometry has complied with in empirical design given parameters. In this model has been thought also existing magnetic characteristics of electrotechnics sheets metal of core material which have been determined by measuring (magnetization characteristic f=150Hz). With in this way produced 3D FEM model have been performed two types of analyses: a) Computation of magnetic field lay-out of inductor. b) Computation of inductor inductivity.

2.1 Computation of magnetic field lay-out of inductor

The main aim of this computation has been the check on magnetic field lay-out of inductor to the whole extent of working current. The check has been target on the magnetic saturation of sheets metal which are forming the magnetic core and then on the magnetic field lay-out in the operating air gap. The own check computation has been realized on the produced model (see figure 8) where have been performed a number of stationary electromagnetic analyses. For every partial analysis has been set the specific value of load current (see figure 8 right). The set value of operating current, which is vector-distributed by the volume winding, made load conditions for one computational step. For in this way loaded model has been searched for final magnetic field lay-out by the help of the scalar potential.

The mathematical calculations procedure of scheduled task results from Maxwells equations and from material equation. In this case it applies to the computation of magnetic field lay-out which is excited by the load current. The computing procedure is here the 3D static magnetic analysis by the use of scalar potential and GSP method (General scalar potential). Reduced Maxwells equations for magnetic field:

$$\nabla \times \{H\} = \{J_s\}$$
$$\nabla . \{B\} = 0 \qquad (1)$$

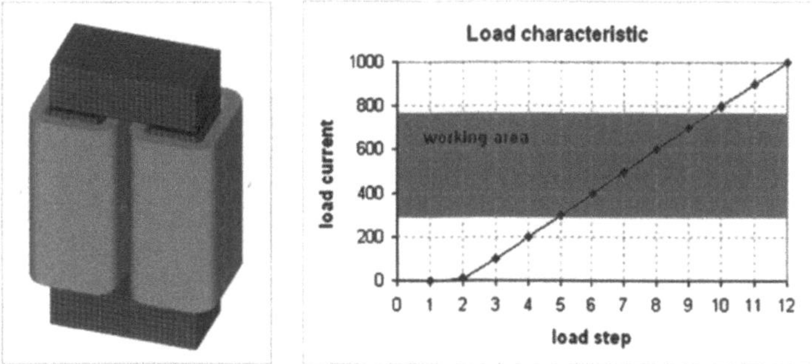

Fig. 8. FEM inductor model and Load characteristic

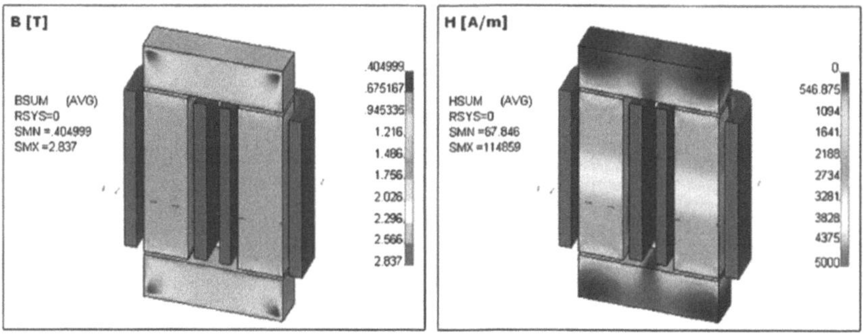

Fig. 9. Computed magnetic field lay-out in the core (I=800A)

Constitutive relation:

$$\{B\} = [\mu].\{H\} + \mu_0.\{M_0\} \tag{2}$$

The solving region includes subregions: Ω_0 is free space region (vacuum) and Ω_1 is non-conducting permeable region. A solution is sought which satisfies equations eg.1 and eg.2 in the following form:

$$\begin{aligned}&\{H\} = \{H_g\} - \nabla\phi_g\\&\nabla.[\mu]\nabla\phi_g - \nabla.[\mu]\{H_g\} - \nabla.[\mu_0]\{M_0\} = \{0\}\end{aligned} \tag{3}$$

Where $\{H_g\}$ is preliminary or guess magnetic field and ϕ_g is generalized potential. Solution of Hg relates to the field development depending on Biot-Savart field which is a function of current source. The main part of the numerical computation of magnetic field lay-out is then the development of Biot-Savart field. For this field is valid:

$$\{H_S\} = \frac{1}{4.\pi}\int_{volc}\frac{\{J_S\}\times\{r\}}{\left|\{r\}\right|^3}\times d(volc) \tag{4}$$

Where $\{J_s\}$ is current source density vector at d(volc), $\boldsymbol{r}$ is position vector from current source to node point and (volc) is volume of current source.

The own computing strategy by the GSP method includes three steps:

The fist step: The fields solution procedure in the iron region: $\{H_g\} = \{H_s\}$ in region Ω_1 subject to: $\{n\} \cdot [\mu] \cdot (\{H_g\} - \nabla \cdot \phi_g) = 0$ on boundary S1 (where S1 is the surface on the iron air interface). The resulting field in this region is: $\{H_1\} = \{H_s\} - \nabla \cdot \phi_g$

The second step: The fields solution procedure in the air region: $\{H_g\} = \{H_s\}$ in region Ω_0 subject to: $\{n\} \times \{H_g\} = \{n\} \times \{H_1\}$ on boundary S1. The resulting field in this region is: $\{H_0\} = \{H_s\} - \nabla \cdot \phi_g$ in region Ω_0.

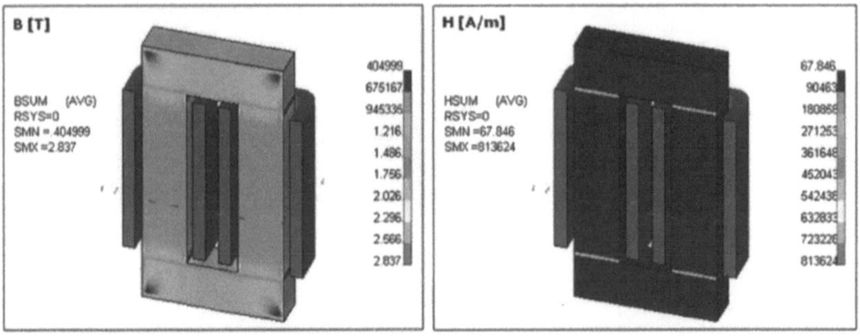

Fig. 10. Computed magnetic field lay-out in the core and in the air gap (I=800A)

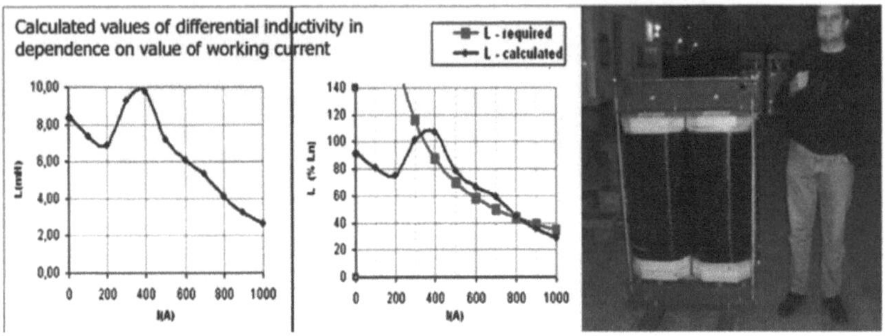

Fig. 11. (left) Computed inductance, (middle) Comparing of inductance computed values with required percent values (right) The real inductor design

The third step: Uses the fields calculated on the first two steps at the preliminary field for equations (eg.3). $\{H_g\} = \{H_1\}$ in region Ω_1, $\{H_g\} = \{H_0\}$ in region Ω_0. The total field in all regions: $\{H\} = \{H_b\} - \nabla \cdot \phi_g$ in region Ω.

The detailed description of this static analysis is explained in the literature see [4, 5]. The practical implementation of this static analysis has been made by the help of the software ANSYS. The next part of this entry describes some results of the performed analysis for one load step which corresponds to the operating current of I=800A.

2.2 Inductance computation of inductor

The inductance computation of an inductor closely connects with the foregoing chapter with calculations. The main aim of these calculations is to determine the self inductance of an inductor for the definite current value which is flowing through the winding. The inductance value of the inductor, which has been calculated on the model in dependence on the operating current size, has been compared with the required inductance procedure. The empirical design has been taken as correct provided that the by model computed inductance was higher then the required inductance. Therefore this calculation has been thought as the main check computation of inductors design accuracy. The self inductance computation has been made for identical load steps as by the magnetic field computation, what means for the identical operating current value. The inductance for the given load step has been computed from already computed database of electromagnetic calculation for the same load step. The inductance computation results from the definition of total flux linkage with coil winding with N turns and current size I. The relationship between flux linkage and current can be described by secant inductance matrix [Ls] in the basic form:

$$\{\Phi\} = \left[L_S\left(t, \{I\}\right)\right].\{I\} + \{\Phi_0\} \tag{5}$$

Where $\{\Phi\}$ is vector of coil flux linkages, $\{I\}$- vector of coil currents, $\{\Phi_0\}$ is vector of flux linkages for zero coil currents. In the time invariant non-linear case:

$$\{U\} = \left(\frac{d[L_S]}{d\{I\}}\{I\} + [L_S]\right)\frac{\partial}{\partial t}\{I\} = \left[L_d\{I\}\right]\frac{\partial}{\partial t}\{I\} \tag{6}$$

Where $[L_d]$ is differential inductance matrix. On the figure 11 you can see the size progression of inductor inductivity Ln in dependence on the operating current size. The operating current has been obtained in single steps by the help of individual electromagnetic analyses. On the figure is further shown the comparing of inductance computed values with required values.

Conclusion

The content of this entry presents the practically made design way of electromagnetic apparatus. By this project solution have been used two elements which assure the high accuracy of the design, magnetic measuring and magnetic fields lay-out computation by the help of FEM methods. Performed analyses and calculations which are just partially presented in this entry markedly achieved a development of the whole plasmatrons GNP320 supply system and also achieved the real plasmatron practical use (fig. 2). The advantage of these computation ways (magnetic fields lay-out computation and inductance computation) is the possibility of subsequent results use for example for the dynamic power analysis of signals.

References

[1] Marek, M., Starek, K., Maly, R.: Plasma technology in the area of coal energy blocks. CEEERES03, Prague 2003
[2] Maly, R., Starek, K., Marek, M., ORGREZ a.s.: Influence of plasma technology at starting and stabilization of coal blocks destined for coal burning, Congress of industrial technology and living environment, Kosice 2002
[3] Marek, M.: Lay-out simulation of electroheat field at d.c. arc. 11th ANSYS Users meeting, Znojmo 2003,CZ
[4] ANSYS - Electromagnetic Fields Analysis Guide
[5] Polak, J.: Variation principles and methods Theory of electro/magnetic field, ACADEMIA, Prague 1988

Calculation of 3D Space-Charge Fields of Bunches of Charged Particles by Fast Summation*

G. Pöplau[1], D. Potts[2], and U. van Rienen[1]

[1] Institute of General Electrical Engineering, Rostock University, D-18051 Rostock, Germany, {gisela.poeplau,ursula.van-rienen}@etechnik.uni-rostock.de
[2] Institute of Mathematics, University of Lübeck, D- Lübeck, Germany, potts@math.uni-luebeck.de

Abstract The fast calculation of space-charge fields of bunches of charged particles in three dimensional space is a demanding problem in accelerator design. Since particles of equal charge repel each other due to space-charge forces, it is difficult to pack a high charge in a small volume. For this reason, the calculation of space-charge forces is an important part of the simulation of the behaviour of charged particles in these machines. As the quality of the charged particle bunches increases, so do the requirements for the numerical space-charge calculations.

In this paper we develop a new fast summation algorithm for the determination of the electric field generated by N charged particles. Applying the nonequidistant Fast Fourier Transform (NFFT) the fast summation requires only $\mathcal{O}(N \log N)$ operations. The numerical test cases confirm this behaviour.

1 Introduction

Recent developments in the field of charged particle accelerator research make high demands on numerical simulations. Among the simulation problems of particle dynamics is the three dimensional calculation of Coulomb repulsion, so-called space-charge fields, of bunches containing millions of particles.

Widely used methods for the calculation of these space-charge fields are the particle-mesh method and the particle-particle method [6]. The particle-mesh method, based on solving Poisson's equation for the electrostatic potential, is typically much faster than the particle-particle method. Furthermore, it provides better numerical results for sufficiently "smooth" distributed particles. Progress in the particle-mesh method has been achieved with the construction of non-equispaced adaptive grids and the development of multigrid Poisson solvers for grids with large aspect ratio [8, 9]. The computational effort of the resulting algorithm scales linearly with the number of particles for a wide range of particle distributions (see [15] for numerical tests).

Although the particle-mesh method provides good results for most real life simulations [8], it is on the edge of the requirements for the simulation of very short bunches present in rf-photoguns based on femtosecond excitation lasers. Also problematic are simulations of high peak current bunches with a long tail as present after the compression stage (first bunch compressor) of the Tesla Test Facility (TTF), a novel linear accelerator recently under development and construction at DESY in Hamburg [1]. In both cases, the main difficulty is the fact that to keep computational costs and memory consumption at an acceptable level, a very high aspect-ratio mesh needs to be constructed resulting in the degradation of the convergence behaviour of the Poisson solver [9].

Motivated by the above-mentioned problems with the particle-mesh method we deal in this paper with the development of a new fast calculation technique for the particle-particle model. The particle-particle method calculates the self-induced field $\boldsymbol{E}$ generated by N charged particles with the superposition principal. Let the ℓ-th particle have the charge q_ℓ and the position $\boldsymbol{r}_\ell$ ($\ell = 1, \ldots, N$) and let ε_0 denote the dielectric constant and $\|\cdot\|$ the Euclidean norm in $\mathbb{R}^3$, then

$$\boldsymbol{E}(\boldsymbol{r}) = \frac{1}{4\pi\varepsilon_0} \sum_{\ell=1}^{N} q_\ell \frac{\boldsymbol{r} - \boldsymbol{r}_\ell}{\|\boldsymbol{r} - \boldsymbol{r}_\ell\|^3}, \qquad \boldsymbol{r}, \boldsymbol{r}_\ell \in \mathbb{R}^3,\ \boldsymbol{r} \neq \boldsymbol{r}_\ell,\ \ell = 1, \ldots, N. \tag{1}$$

The direct summation which requires $\mathcal{O}(N^2)$ operations is either very time consuming or causes large simulation errors due to the restricted number of particles. This essentially eliminates its applicability to real life simulations, unless the computation is the restriction to 2 D models [14]. In order to make large scale problems tractable it is essential to compute these interactions efficiently. A number of algorithms have been proposed for this purpose. The

*supported by a research grant from DESY, Hamburg

fast multipole method (FMM) has been one of the most successful, especially for nonuniform particle distributions (see [16] and references therein). Our new method is fully 3 D and based on the nonequidistant Fast Fourier Transform (NFFT) [7], hereby reducing the computational from $\mathcal{O}(N^2)$ to $\mathcal{O}(N \log N)$. Although this is still slower compared to the best particle-mesh methods, it could prove to be advantageous for ultra-short and 'TTF'-like bunches because no mesh needs to be constructed.

In the next chapter we develop the main principles of the fast summation by NFFT and present an algorithm for the computation of (1). Finally the numerical experiments in section 3 show that the fast summation technique provides the values for the field with an acceptable numerical error in much shorter simulation time compared to direct methods.

2 Fast Summation at Nonequispaced Knots by NFFTs

The fast computation of special structured discrete sums similar to (1) is a frequently appearing task in the study of particle models [3, 4, 16]. The new fast summation technique we develop in this paper is based on a method first presented in [10].

The fast computation of $\boldsymbol{E}$ at the positions $\boldsymbol{r}_j$ $(j = 1, \ldots, N)$ is performed for the two sums

$$\boldsymbol{E}(\boldsymbol{r}_j) = \frac{1}{4\pi\varepsilon_0} \left(\boldsymbol{r}_j \sum_{\substack{\ell=1 \\ j\neq\ell}}^{N} \frac{q_\ell}{\|\boldsymbol{r}_j - \boldsymbol{r}_\ell\|^3} - \sum_{\substack{\ell=1 \\ j\neq\ell}}^{N} q_\ell \frac{\boldsymbol{r}_\ell}{\|\boldsymbol{r}_j - \boldsymbol{r}_\ell\|^3} \right) \tag{2}$$

in the following way: As suggested in [10] we use a separation of the knots $\boldsymbol{r}_j$ and $\boldsymbol{r}_\ell$ by Fourier expansions. More precisely, we split the function $1/\|\boldsymbol{x}\|^3$ into the sum $1/\|\boldsymbol{x}\|^3 \approx \mathcal{K}_{\rm NE} + \mathcal{K}_{\rm R}$. Thereby the function $\mathcal{K}_{\rm NE}$ is supposed to have small support with $\operatorname{supp}\mathcal{K}_{\rm NE} = \{x \in \mathbb{R}^3;\ \|x\| \le \varepsilon_I\}$. It can be considered as the near field approximation of $1/\|\boldsymbol{x}\|^3$. Further the function $\mathcal{K}_{\rm R}$ is chosen as a smooth 1–periodic function also referred to as the regularisation of $1/\|\boldsymbol{x}\|^3$. The construction of $\mathcal{K}_{\rm R}$ is somewhat technical so we don't give it at this place. It is needed for the computation of the discrete Fourier coefficients $b_{\boldsymbol{k}}$ defined by

$$b_{\boldsymbol{k}} := \frac{1}{n^3} \sum_{\boldsymbol{j} \in I_n} \mathcal{K}_{\rm R}(\boldsymbol{j}/n)\, \mathrm{e}^{-2\pi \mathrm{i} \boldsymbol{j}\boldsymbol{k}/n} \tag{3}$$

where $\boldsymbol{k}$ runs over the finite index set $I_n := \{-n/2, \ldots, n/2-1\}^3$. A detailed description can be found in [10] for the one dimensional case which can be straightforward applied to the three dimensional problem.

Next, we approximate the smooth function $\mathcal{K}_{\rm R}$ by the discrete finite Fourier sum $\mathcal{K}_{\rm RF}$ given by

$$\mathcal{K}_{\rm R} \approx \mathcal{K}_{\rm RF} = \sum_{\boldsymbol{k} \in I_n} b_{\boldsymbol{k}}\, \mathrm{e}^{2\pi \mathrm{i} \boldsymbol{k} \cdot} . \tag{4}$$

Then, $1/\|\boldsymbol{x}\|^3$ is replaced by $1/\|\boldsymbol{x}\|^3 \approx \mathcal{K}_{\rm RF} + \mathcal{K}_{\rm NE}$. Using the outstanding property $\mathrm{e}^{2\pi\mathrm{i}(\boldsymbol{r}_j - \boldsymbol{r}_\ell)} = \mathrm{e}^{2\pi\mathrm{i}\boldsymbol{r}_j}\, \mathrm{e}^{-2\pi\mathrm{i}\boldsymbol{r}_\ell}$, we obtain the desired separation of $\boldsymbol{r}_j$ and $\boldsymbol{r}_\ell$ by

$$\frac{1}{\|\boldsymbol{r}_j - \boldsymbol{r}_\ell\|^3} \approx \sum_{\boldsymbol{k} \in I_n} b_{\boldsymbol{k}}\, \mathrm{e}^{2\pi\mathrm{i}\boldsymbol{k}\boldsymbol{r}_j}\, \mathrm{e}^{-2\pi\mathrm{i}\boldsymbol{k}\boldsymbol{r}_\ell} + \mathcal{K}_{\rm NE}(\boldsymbol{r}_j - \boldsymbol{r}_\ell)$$

and finally

$$\begin{aligned} \hat{\alpha}_j &:= \sum_{\substack{\ell=1 \\ j\neq\ell}}^{N} \frac{\alpha_\ell}{\|\boldsymbol{r}_j - \boldsymbol{r}_\ell\|^3} \\ &\approx \sum_{\boldsymbol{k} \in I_n} b_{\boldsymbol{k}} \left(\sum_{\ell=1}^{N} \alpha_\ell\, \mathrm{e}^{-2\pi\mathrm{i}\boldsymbol{k}\boldsymbol{r}_\ell} \right) \mathrm{e}^{2\pi\mathrm{i}\boldsymbol{k}\boldsymbol{r}_j} + \sum_{\substack{\ell=1 \\ j\neq\ell}}^{N} \alpha_\ell \mathcal{K}_{\rm NE}(\boldsymbol{r}_j - \boldsymbol{r}_\ell) - \alpha_j \sum_{\boldsymbol{k} \in I_n} b_{\boldsymbol{k}} . \end{aligned} \tag{5}$$

The expression in the inner brackets can be computed by a multivariate $\mathrm{NFFT}^{\mathrm{T}}(n)$, where $\mathrm{NFFT}^{\mathrm{T}}$ denotes the transposed version of the NFFT [7]. This is followed by n^3 multiplications with $b_{\boldsymbol{k}}$ and completed by a multivariate NFFT(n) to compute the outer sum with the complex exponentials. By construction the function $\mathcal{K}_{\rm NE}$ has a small support such that the summation can be done very efficiently. The approximation of (5) is used in (2) with $\alpha_\ell = q_\ell$ and $\alpha_\ell = q_\ell \boldsymbol{r}_\ell$, respectively. Applying the recently developed fast Fourier transform for nonequispaced data (NFFT) (see

[12] and references therein), we come up with a fast summation algorithm. This NFFT summation requires for "sufficiently uniformly distributed" points $\boldsymbol{r}_\ell$ only $\mathcal{O}(N \log N)$ arithmetic operations and can be simply implemented using the public domain NFFT toolbox (see e.g. [7]). Note that the NFFT itself is based on the approximation of functions by translates of one function, which is taken as a Kaiser–Bessel function in our numerical computations. In summary we obtain the following

Algorithm:

Precomputation:

i) Computation of $(b_{\boldsymbol{k}})_{\boldsymbol{k}\in I_n}$ by (3).

ii) Computation of $K_{\mathrm{NE}}(\boldsymbol{r}_j - \boldsymbol{r}_\ell)$ for all $(j = 1, \ldots, N)$ and $\ell \in I_{\varepsilon_{\mathrm{I}}}^{\mathrm{NE}}(j)$, where $I_{\varepsilon_{\mathrm{I}}}^{\mathrm{NE}}(j) := \{\ell \in \{1, \ldots, N\} : \|\boldsymbol{r}_j - \boldsymbol{r}_\ell\| < \varepsilon_{\mathrm{I}}\}$.

1. For $\boldsymbol{k} \in I_n$ compute by four multivariate $\mathrm{NFFT}^{\mathrm{T}}(n)$s

$$\hat{q}_{\boldsymbol{k}} := \sum_{\ell=1}^{N} q_\ell \,\mathrm{e}^{-2\pi \mathrm{i} \boldsymbol{k} \boldsymbol{r}_\ell}, \quad \hat{\boldsymbol{r}}_{\boldsymbol{k}} := \sum_{\ell=1}^{N} q_\ell \boldsymbol{r}_\ell \,\mathrm{e}^{-2\pi \mathrm{i} \boldsymbol{k} \boldsymbol{r}_\ell}.$$

2. For $\boldsymbol{k} \in I_n$ compute the products $d_{\boldsymbol{k}} := \hat{q}_{\boldsymbol{k}} b_{\boldsymbol{k}} \in \mathbb{C}$.
3. For $j = 1, \ldots, N$ compute by a multivariate NFFT(n)

$$f_{\mathrm{RF}}(\boldsymbol{r}_j) := \boldsymbol{r}_j \sum_{\boldsymbol{k}\in I_n} d_{\boldsymbol{k}} \,\mathrm{e}^{2\pi \mathrm{i} \boldsymbol{k} \boldsymbol{r}_j}.$$

4. For $\boldsymbol{k} \in I_n$ compute the products $\boldsymbol{d}_{\boldsymbol{k}} := \hat{\boldsymbol{r}}_{\boldsymbol{k}} b_{\boldsymbol{k}} \in \mathbb{C}^3$.
5. For $j = 1, \ldots, N$ compute by three multivariate NFFTs(n)

$$f_{\mathrm{RF}}(\boldsymbol{r}_j) := f_{\mathrm{RF}}(\boldsymbol{r}_j) - \sum_{\boldsymbol{k}\in I_n} \boldsymbol{d}_{\boldsymbol{k}} \,\mathrm{e}^{2\pi \mathrm{i} \boldsymbol{k} \boldsymbol{r}_j}.$$

6. For $j = 1, \ldots, N$ compute the near field sums

$$f_{\mathrm{NE}}(\boldsymbol{r}_j) = \boldsymbol{r}_j \sum_{\ell \in I_{\varepsilon_{\mathrm{I}}}^{\mathrm{NE}}(j)} q_\ell \mathcal{K}_{\mathrm{NE}}(\boldsymbol{r}_j - \boldsymbol{r}_\ell) - \sum_{\ell \in I_{\varepsilon_{\mathrm{I}}}^{\mathrm{NE}}(j)} q_\ell \boldsymbol{r}_\ell \mathcal{K}_{\mathrm{NE}}(\boldsymbol{r}_j - \boldsymbol{r}_\ell).$$

7. For $j = 1, \ldots, N$ compute the near field corrections

$$\tilde{E}(\boldsymbol{r}_j) = \frac{1}{4\pi\varepsilon_0} \left(f_{\mathrm{NE}}(\boldsymbol{r}_j) + f_{\mathrm{RF}}(\boldsymbol{r}_j) \right).$$

Note, that usually the field values are requested at same the locations as the location of the particles. But the algorithm can also evaluate field values at other points the number of which has not to be in coincidence with the number of particles (see [10]).

3 Numerical Results

The algorithms for the fast summation have been implemented in C and tested on an AMD Atlon xp1800+ 512MB RAM, SuSe-Linux 8.0 using double precision arithmetic. Throughout our experiments we have applied the NFFT/$\mathrm{NFFT}^{\mathrm{T}}$ package [7] with Kaiser–Bessel functions, oversampling factor $\rho = 2$ and several bandwidth parameters n which will be specified in the examples. Further the NFFT/$\mathrm{NFFT}^{\mathrm{T}}$ algorithms require the parameters p (guarantees the smoothness of $\mathcal{K}_{\mathrm{R}}$ up to the derivative of order $p-1$) and m (controls the accuracy of interpolation by the Kaiser–Bessel functions) which are for the fast summation chosen as $p = 2$ and $m = 2$. Note, that the fast summation method suggested in (5) was first proposed for the univariate case in [10] and for the bivariate case in [11] (see also [2]). There error estimates are proved to obtain clues about the choice of the involved parameters. For a numerical comparison with the fast multipole method in 2D see [11]. With the algorithm for the calculation of the electric field we extend these methods to $\mathbb{R}^3$.

As numerical test we used a spherical bunch uniformly filled with charged particles. The total charge of the sphere has been kept with $Q = -1$ nC. Thus the particles are assumed to posses the charge $q_i = q = -1/N$ nC $(i = 1, \ldots, N)$, where N denotes the number of particles in the sphere. These particles are also regarded as macro-particles representing the distribution of all particles (for instance electrons) in a bunch. The uniform particle distributions have been generated with the tracking code GPT (General Particle Tracer) [13] by means of Hammersley sequences [5].

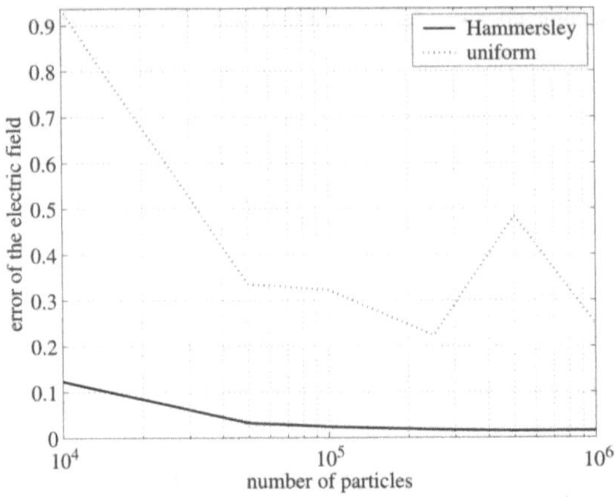

Fig. 1. The error $\mathcal{E}_2$(theo,fast)) (see equation (3)) of the electric field for particle distributions generated by Hammersley sequences and by straightforward computed random numbers, respectively

These sequences provide pseudo random numbers such that distance between two particles does not become too small. The advantage of such generated distributions is represented in Fig. 1 where the numerical error is compared to particle distributions generated with straightforward computed random numbers.

The fast summation technique is not restricted to the calculation of the discrete sum (1) but can be applied to a great variety of discrete sums appearing in the study of particle models. In order to demonstrate the efficiency of our new method with a more simple discrete sum we start with the calculation of the potential φ caused by N charged particles with charge q given by

$$\varphi(\boldsymbol{r}_j) = \frac{1}{4\pi\varepsilon_0} \sum_{\substack{\ell=1 \\ j\neq\ell}}^{N} \frac{q}{\|\boldsymbol{r}_j - \boldsymbol{r}_\ell\|}, \qquad (\boldsymbol{r}_j \in \mathbb{R}^3).$$

The fast summation strategy described in section 2 can be easily adapted to the above discrete sum. Since a sphere uniformly filled with an increasing number of particles of equal charge gets more and more close to a sphere with charge $Q = \sum_{i=1}^{N} q$, we compare the results of the summation to the analytically known potential given by

$$\varphi_{\text{theo}}(\boldsymbol{r}_j) = \frac{Q}{4\pi\varepsilon_0}\left(\frac{3}{2} - \frac{\|\boldsymbol{r}_j\|}{2R^2}\right), \qquad \|\boldsymbol{r}_j\| \leq R,$$

where R denotes the radius of the sphere.

We have investigated the numerical error

$$\mathcal{E}_2(\text{a,b}) = \left(\sum_{k=1}^{N} |\varphi_{\text{a}}(\boldsymbol{r}_k) - \varphi_{\text{b}}(\boldsymbol{r}_k)|^2\right)^{1/2} \left(\sum_{k=1}^{N} |\varphi_{\text{a}}(\boldsymbol{r}_k)|^2\right)^{-1/2},$$

where a and b represent the different techniques for the computation either of the potential or the electric field (slow: straightforward summation, fast: fast summation, theo: analytical solutions). Similarly the computational time for the straightforward summation and for the fast summation based on (5) is denoted by t_{slow} and t_{fast}, respectively.

The numerical experiments documented in Table 1 show that we obtain with our fast algorithm the same errors as with the straightforward (slow) summation but with an numerical effort of only $\mathcal{O}(N \log N)$. Hereby the parameters of the NFFT are chosen such that the approximation error is less than the simulation error. Depending on the number of particles the Fourier sum (4) has been computed as NFFT(n) with $n = 32$, $n = 64$ and $n = 128$, respectively (see Tables 1 and 2). The star * means that the running time of the direct evaluation is obtained by extrapolation. Note that the straightforward evaluation of the potential φ with $N = 5 \cdot 10^6$ requires more that 10 days (see Fig. 2). Finally we test the algorithm for the computation of the electrostatic field suggested in section 2. It is well known that the field of a charged sphere is given by

$$\boldsymbol{E}_{\text{theo}}(\boldsymbol{r}_j) = \frac{Q}{4\pi\varepsilon_0}\left(\frac{\boldsymbol{r}_j}{R^3}\right), \qquad \|\boldsymbol{r}_j\| \leq R.$$

N	n	t_{slow}	t_{fast}	$\mathcal{E}_2$(theo,slow)	$\mathcal{E}_2$(theo,fast)	$\mathcal{E}_2$(slow,fast)
10000	64	6.680e+00	2.310e+00	2.586e-03	2.544e-03	1.206e-04
50000	64	1.777e+02	7.140e+00	1.018e-03	9.755e-04	9.654e-05
100000	64	7.092e+02	1.770e+01	5.630e-04	5.283e-04	1.002e-04
250000	128	4.470e+03	4.821e+01	2.952e-04	2.584e-04	1.125e-04
500000	128	1.756e+04	8.951e+01	2.043e-04	1.647e-04	1.103e-04
1000000	128	7.024e+04*	2.257e+02		1.079e-04	

Table 1. Computational time and the error $\mathcal{E}_2$ for the potential φ, *estimated

N	n	t_{slow}	t_{fast}	$\widetilde{\mathcal{E}}_2$(theo,slow)	$\widetilde{\mathcal{E}}_2$(theo,fast)	$\widetilde{\mathcal{E}}_2$(slow,fast)
10000	32	7.580e+00	3.680e+00	1.232e-01	1.232e-01	1.068e-03
50000	64	1.930e+02	2.185e+01	3.204e-02	3.205e-02	5.765e-04
100000	64	7.710e+02	4.810e+01	2.393e-02	2.394e-02	4.662e-04
250000	128	5.781e+03	1.635e+02	1.716e-02	1.718e-02	5.462e-04
500000	128	2.312e+04*	2.699e+02		1.446e-02	
1000000	128	9.245e+04*	6.031e+02		1.468e-02	

Table 2. Computational time and the error $\widetilde{\mathcal{E}}_2$ for the electric field $\boldsymbol{E}$, *estimated

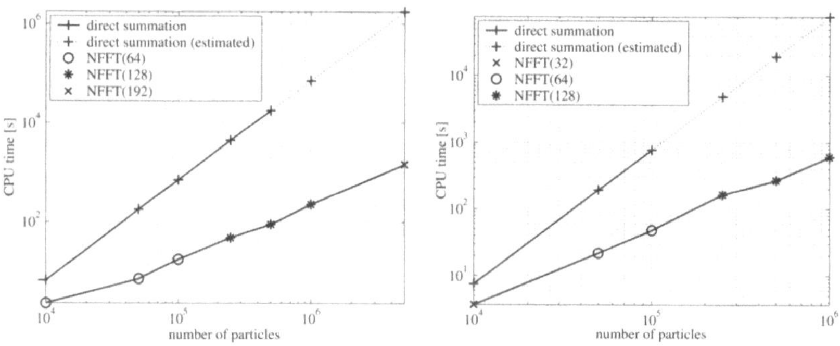

Fig. 2. Performance of the fast NFFT-algorithm compared to the direct summation: computation of φ (left), computation of $\boldsymbol{E}$ (right)

Here we consider the error

$$\widetilde{\mathcal{E}}_2(\text{a,b}) = \left(\sum_{k=1}^{N} \|\boldsymbol{E}_{\text{a}}(\boldsymbol{r}_k) - \boldsymbol{E}_{\text{b}}(\boldsymbol{r}_k)\|^2 \right)^{1/2} \left(\sum_{k=1}^{N} \|\boldsymbol{E}_{\text{a}}(\boldsymbol{r}_k)\|^2 \right)^{-1/2} .$$

Table 2 represents the results of the related numerical simulations. Figure 2 compares the performance of the fast summation algorithm with NFFT to the direct slow summation. It shows that the NFFT summation scales with $\mathcal{O}(N \log N)$. The numerical errors are acceptable (see Table 1 and Table 2). Hence this new summation technique enables the computation of fully 3 D particle-particle interactions in real life applications.

4 Acknowledgment

The authors like to thank the physicists Marieke de Loos (Pulsar Physics, the Netherlands) and Bas van der Geer (TU Eindhoven, the Netherlands) for supporting the work with both a lot of helpful comments related to accelerator physics and a lot of space-charge simulations performed with the tracking code GPT (General Particle Tracer) [13].

References

1. DESY, Hamburg, Germany, `tesla.desy.de`. *DESY-TTF*
2. M. Fenn and G. Steidl. Fast NFFT based summation of radial functions. *Sampling Theory in Signal and Image Processing*, 3:1–28, 2004
3. L. Greengard. *The Rapid Evaluation of Potential Fields in Particle Systems*. MIT Press, Cambridge, 1988
4. L. Greengard and V. Rokhlin. A fast algorithm for particle simulations. *J. Comput. Phys.*, 73:325–348, 1987
5. J. M. Hammersley. Monte carlo methods for solving multivariable problems. *Proceedings of the New York Academy of Science*, 86:844–874, 1960
6. R. Hockney and J. Eastwood. *Computer Simulation Using Particles*. Institut of Physics Publishing, Bristol, 1992
7. S. Kunis and D. Potts. NFFT, Softwarepackage, C subroutine library. `http://www.math.uni-luebeck.de/potts/nfft`, 2002–2004
8. G. Pöplau, U. van Rienen, S. van der Geer, and M. de Loos. Fast calculation of space charge in beam line tracking by multigrid techniques. In W. Schilders, E. ter Marten, and S. Houben, editors, *Scientific Computing in Electrical Engineering*, number 4 in Mathematics in Industry, pages 329–336, Berlin, 2004. Springer-Verlag
9. G. Pöplau, U. van Rienen, S. van der Geer, and M. de Loos. Multigrid algorithms for the fast calculation of space-charge effects in accelerator design. *IEEE Transactions on Magnetics*, 40(2):714–717, 2004
10. D. Potts and G. Steidl. Fast summation at nonequispaced knots by NFFTs. SIAM *J. Sci. Comput.*, 24:2013–2037, 2003
11. D. Potts, G. Steidl, and A. Nieslony. Fast convolution with radial kernels at nonequispaced knots. *Numer. Math.*, 98:329–351, 2004
12. D. Potts, G. Steidl, and M. Tasche. Fast Fourier transforms for nonequispaced data: A tutorial. In J. J. Benedetto and P. J. S. G. Ferreira, editors, *Modern Sampling Theory: Mathematics and Applications*, pages 247–270, Boston, 2001. Birkhäuser.
13. Pulsar Physics, De Bongerd 23, 3762 XA Soest, The Netherlands, `www.pulsar.nl/gpt`. *General Particle Tracer (GPT)*, release 2.60 edition
14. S. van der Geer and M. de Loos. *The General Particle Tracer Code. Design, implementation and application.* PhD thesis, TU Eindhoven, 2001.
15. S. van der Geer, M. de Loos, O. Luiten, G. Pöplau, and U. van Rienen. 3D space-charge model for GPT simulations of high-brightness electron bunches. In M. Berz and K. Makino, editors, *Computational Accelerator Physics 2002*, number 175 in Institute of Physics Conference Series, pages 101–110, Bristol and Philadelphia, 2005. Institute of Physics Publishing
16. L. Ying, G. Biros, and D. Zorin. A kernel-independent adaptive fast multipole method in two and three dimensions. *J. Comput. Pysics*, to appear

Comparison of the A, V-formulation and Hiptmair's Smoother

B. Weiß and O. Bíró

Institute of Fundamentals and Theorie of Electrical Engineering, IGTE, Graz University of Technology
Kopernikusgasse 24, Graz, AUSTRIA, {bernhard.weiss,biro}@TUGraz.at

1 Introduction

In the numerical analysis of three dimensional electric and magnetic field problems, the finite element method (FEM) is very common. In the electrostatic case nodal finite elements are used, while for magnetostatic or eddy-current problems edge elements proved their worth. The basis functions of these edge elements introduced by Nédélec in [5] are vector functions. One of their advantages is that edge elements enforce the tangential continuity of the vector fields only, but not that of the normal component. Because of this, only the essential continuity properties of the electric field intensity $\boldsymbol{E}$ are fulfilled when using edge elements.

To describe the eddy current problem, various formulations have been introduced. The Maxwell equations are used as the starting point. From this one can derive differential equations for e.g. the electric field intensity $\boldsymbol{E}$ or the magnetic field intensity $\boldsymbol{H}$. Nevertheless the use of scalar or vector potential functions are more common to describe electric and magnetic fields [2].

However, different formulations of the problem lead to different systems of equations. For time harmonic eddy current problems, the resulting sparse system matrices are complex symmetric. In this paper two different formulations, namely the $\boldsymbol{A}^*$-formulation and the $\boldsymbol{A}, V$-formulation are used.

A main field of computational electromagnetics is the fast solution of the resulting system of equations. One of the fastest solvers is the multigrid (MG) algorithm (e.g. [7]). While multigrid is straightforward for nodal elements, more difficulties arise when using edge elements. This is due to the non trivial kernel of the curl operator. Two common approaches to solve this problem are the hybrid smoother of Hiptmair [4] and the block smoother of Arnold, Falk and Winther [1]. It should be mentioned that these smoothers can also be used very effectively as preconditioners for Krylov subspace methods ([6], [9]).

Instead of these more sophisticated smoothers, a standard Gauss-Seidel (GS) smoother can be used with a formulation combining the vector potential, $\boldsymbol{A}$, with the scalar potential, V [9]. The aim of this paper is to compare the multigrid algorithm of an $\boldsymbol{A}, V$-formulation to the multigrid algorithm of a vector formulation with the hybride smoother. It will be shown, that these two methods are mathematically identical.

The paper is structured as follows: In the next section, the $\boldsymbol{A}^*$-formulation and the $\boldsymbol{A}, V$-formulation are presented. The different properties of these formulations will be discussed. Section 3 describes the multigrid smoothers and shows the similarities of the hybrid smoother and the $\boldsymbol{A}, V$-formulation. In section 4, numerical examples will show the properties of the different methods. Finally, the conclusions are presented in section 5.

2 Finite Element Formulations

An eddy current problem involves two regions: a conducting region Ω_c with an unknown current density distribution and a nonconducting region Ω_n with a given source current density $\boldsymbol{J}_0$.

In the nonconducting region, the magnetic vector potential $\boldsymbol{A}$ is defined as usual by $\boldsymbol{B} = \mathrm{curl}\boldsymbol{A}$. Neglecting the dielectric displacement leads to the differential equation $\mathrm{curl}(\nu \mathrm{curl}\boldsymbol{A}) = \boldsymbol{J}_0$ in Ω_n, where $\nu = \nu(\boldsymbol{x})$ is the reluctivity of the material.

Since the resulting system of equations is singular, it is essential to ensure that the right hand side is in the range of the system matrix. This can be done by introducing an electric vector potential $\boldsymbol{T}_0$ instead of using $\boldsymbol{J}_0$ [2]. Since the divergence of the given source current density $\boldsymbol{J}_0$ is zero $\boldsymbol{J}_0$, can be described as $\mathrm{curl}\boldsymbol{T}_0 = \boldsymbol{J}_0$, leading to following differential equation:

$$\mathrm{curl}(\nu \mathrm{curl}\boldsymbol{A}) = \mathrm{curl}\boldsymbol{T}_0 \qquad \text{in } \Omega_n \,. \tag{1}$$

In the conducting region, two main formulations will be presented.

2.1 The A^*-formulation

One possibility to describe the eddy current field is by means of a modified vector potential $\boldsymbol{A}^*$ where $\boldsymbol{B} = \mathrm{curl}\boldsymbol{A}^*$ and $\boldsymbol{E} = -j\omega\boldsymbol{A}^*$. This formulation leads to the following complex differential equation

$$\mathrm{curl}(\nu \mathrm{curl}\boldsymbol{A}^*) + j\omega\sigma\boldsymbol{A}^* = \boldsymbol{0} \qquad \text{in } \Omega_c \tag{2}$$

where $\sigma = \sigma(\boldsymbol{x})$ is the conductivity in the conducting region. The modified vector potential will be approximated by edge basis functions $\boldsymbol{N}_i$. Using the boundary conditions $\boldsymbol{A}^* \times \boldsymbol{n} = \boldsymbol{0}$ or $\nu\mathrm{curl}\boldsymbol{A}^* = \boldsymbol{T}_0 \times \boldsymbol{n}$ and applying Galerkin techniques to (2) results in the system of equations:

$$(\mathrm{curl}\boldsymbol{N}_i, \nu\mathrm{curl}\boldsymbol{A}_h^*) + j\omega\,(\boldsymbol{N}_i, \sigma\boldsymbol{A}_h^*) = (\mathrm{curl}\boldsymbol{N}_i, \boldsymbol{T}_0) \tag{3}$$

for $i = 1, 2, \ldots, n_e$ with $\boldsymbol{A}_h^* = \sum_{i=1}^{n_e} a_i^* \boldsymbol{N}_i$ being the edge element discretization of $\boldsymbol{A}^*$, n_e the number of edges and $(\boldsymbol{a}, \boldsymbol{b}) = \int_{\Omega_c} \boldsymbol{a} \cdot \boldsymbol{b} d\Omega$. For the term on right hand side use hase been made of the fact that the basis functions $\boldsymbol{N}_i$ satisfy the homogeneous Dirichlet boundary conditions as well as of the given source current density $\boldsymbol{J}_0$ being zero in Ω_C. This regular system of equations can be written as $K\boldsymbol{a}^* = \boldsymbol{f}$ where

$$k_{ij} = \int_{\Omega_c} \mathrm{curl}\boldsymbol{N}_i \cdot \nu\mathrm{curl}\boldsymbol{N}_j \; d\Omega + j\omega \int_{\Omega_c} \boldsymbol{N}_i \cdot \sigma\boldsymbol{N}_j \; d\Omega \tag{4}$$

are the entries of the matrix K and $\boldsymbol{f}$ is the right hand side:

$$f_i = \int_{\Omega_c} \mathrm{curl}\boldsymbol{N}_i \cdot \boldsymbol{T}_0 \; d\Omega \,. \tag{5}$$

2.2 The A, V-formulation

Another formulation for eddy current fields is the $\boldsymbol{A},\ V$-formulation. In this case an additional modified electric scalar potential V is introduced as

$$\boldsymbol{E} = -j\omega\boldsymbol{A} - j\omega\mathrm{grad}V \,. \tag{6}$$

Since there are four unknowns, $\boldsymbol{A}$ and V, an additional equation has to be used. A common way is to use the divergence free property of the current density, $\mathrm{div}\boldsymbol{J} = 0$. This leads to the following system of differential equations:

$$\mathrm{curl}(\nu\mathrm{curl}\boldsymbol{A}) + j\omega\sigma\boldsymbol{A} + j\omega\sigma\mathrm{grad}V = \boldsymbol{0} \qquad \text{in } \Omega_c \,, \tag{7}$$

$$-\mathrm{div}\,(j\omega\sigma\boldsymbol{A} + j\omega\sigma\mathrm{grad}V) = 0 \qquad \text{in } \Omega_c \,. \tag{8}$$

In addition to the boundary conditions in section 2.1, the boundary conditions on the scalar potential are introduce as $\boldsymbol{n} \cdot (-j\omega\sigma\boldsymbol{A} - j\omega\sigma\mathrm{grad}V) = 0$ or $V = V_0 = \text{constant}$. Using edge basis functions $\boldsymbol{N}_{i1}$ for the vector potential with $i1 = 1, 2, \ldots, n_e$, and the nodal basis functions N_{i2} with $i2 = 1, 2, \ldots, n_n$ for the scalar potential where n_n is the number of nodes, the Galerkin equations can be written as

$$(\mathrm{curl}\boldsymbol{N}_{i1}, \nu\mathrm{curl}\boldsymbol{A}_h) + j\omega\,(\boldsymbol{N}_{i1}, \sigma\boldsymbol{A}_h) + j\omega\,(\boldsymbol{N}_{i1}, \sigma\mathrm{grad}V_h) = \text{r.h.s.} \tag{9}$$

$$j\omega\,(\mathrm{grad}N_{i2}, \sigma\boldsymbol{A}_h) + j\omega\,(\mathrm{grad}N_{i2}, \sigma\mathrm{grad}V_h) = 0 \tag{10}$$

where r.h.s. $= (\mathrm{curl}\boldsymbol{N}_{i1}, \boldsymbol{T}_0)$ and V_h is the approximation of the scalar potential. The resulting system of equations is singular since the gradient of the nodal finite element space is included in the edge element space. Using matrices the equation system can be written as

$$\begin{bmatrix} K & C \\ C^T & B \end{bmatrix} \begin{Bmatrix} \boldsymbol{a} \\ \boldsymbol{v} \end{Bmatrix} = \begin{Bmatrix} \boldsymbol{f} \\ \boldsymbol{0} \end{Bmatrix} \tag{11}$$

where the matrix K and $\boldsymbol{f}$ have already been described in section 2.1. Writing b_{ij} for the entries in matrix B and c_{ij} for the entries of C, these matrices can be calculated as

$$b_{ij} = j\omega \int_{\Omega_c} \mathrm{grad}N_i \cdot \sigma\mathrm{grad}N_j \; d\Omega \qquad \text{and} \tag{12}$$

$$c_{ij} = j\omega \int_{\Omega_c} \boldsymbol{N}_i \cdot \sigma\mathrm{grad}N_j \; d\Omega \,. \tag{13}$$

3 Multigrid Smoothers

A fast method for solving the resulting systems of complex equations is the multigrid method. In this work, a geometric multigrid is used. The two main parts which influence the convergence are the smoothing operator as well as the restriction and prolongation operators.

For the discretization of the scalar Laplace-Poisson equation with nodal elements, the smoothing iterations are usually carried out by methods like Jacobi or Gauss-Seidel. Using edge elements, the smoothing operator has to be adjusted to the finite element formulation.

In case of the $\boldsymbol{A}^*$-formulation attention must be paid to the kernel of the curl operator. To solve this problem, Hiptmair suggested a hybrid smoother. The idea is based on a Helmholtz decomposition and from this follows that a second smoothing step has to be carried out in the space of scalar functions corresponding to the nodes. For more details see [4].

Here we want to emphasize the properties of the edge basis functions $\boldsymbol{N}_j$ and the scalar basis functions N_i. It is well known that the gradient of a scalar basis function can be described as the weighted sum of edge basis functions:

$$\mathrm{grad} N_i = \sum_{j=1}^{n_e} g_{ij} \boldsymbol{N}_j \,. \tag{14}$$

The weighting values g_{ij} are -1 for edges j with the starting node i, 1 for edges j with the ending node i and elsewhere 0. The n_n by n_e matrix G with the entries g_{ij} is the incidence matrix multiplied by -1.

Using this notation and denoting the iteration matrix of a matrix X as W_X, one iteration step of the hybrid smoother can be written as:

$$\tilde{\boldsymbol{a}}^*_{m+1} = \boldsymbol{a}^*_m + W_K^{-1}\left(\boldsymbol{f} - K\boldsymbol{a}^*_m\right) \tag{15}$$

$$\boldsymbol{d} = \boldsymbol{f} - K\tilde{\boldsymbol{a}}^*_{m+1} \tag{16}$$

$$\boldsymbol{v}_{m+1} = W^{-1}_{GKG^T}\left(G\boldsymbol{d}\right) \tag{17}$$

$$\boldsymbol{a}^*_{m+1} = \tilde{\boldsymbol{a}}^*_{m+1} + G^T\boldsymbol{v}_{m+1} \tag{18}$$

On the other hand, a simple Gauss-Seidel smoother can be used for the $\boldsymbol{A}, V$-formulation [8]. Splitting up the two systems of equations of (11) one iteration step can be written as

$$\boldsymbol{a}_{m+1} = \boldsymbol{a}_m + W_K^{-1}\left(\boldsymbol{f} - K\boldsymbol{a}_m - C\boldsymbol{v}_m\right) \tag{19}$$

$$\boldsymbol{v}_{m+1} = \boldsymbol{v}_m + W_B^{-1}\left(-C^T\boldsymbol{a}_{m+1} - B\boldsymbol{v}_m\right) . \tag{20}$$

The discrete version of (6) can now be written as $\boldsymbol{e} = j\omega(-\boldsymbol{a} - G^T\boldsymbol{v}) = -j\omega\boldsymbol{a}^*$ with $\boldsymbol{e}$ being the discretization of the electric field intensity $\boldsymbol{E}$. From this follows that $\boldsymbol{a}^* = \boldsymbol{a} + G^T\boldsymbol{v}$.

3.1 Comparison

First of all, to analyze these two previous algorithms, the multiplication of the system matrix K with the incidence matrix G will be carried out. Thereby, the properties (14) will be used. One entry $\tilde{c}_{ij}$ of the resulting matrix $\tilde{C} = KG^T$ can be written as

$$\begin{aligned}\tilde{c}_{ij} &= \sum_{k=1}^{n_e}\left(\int_{\Omega_c} \mathrm{curl}\boldsymbol{N}_i \cdot \nu\mathrm{curl}\boldsymbol{N}_k \, d\Omega + j\omega\int_{\Omega_c} \boldsymbol{N}_i \cdot \sigma\boldsymbol{N}_k \, d\Omega\right) g_{jk} \\ &= \int_{\Omega_c} \mathrm{curl}\boldsymbol{N}_i \cdot \nu\mathrm{curl}\sum_{k=1}^{n_e}\left(\boldsymbol{N}_k g_{jk}\right) d\Omega + j\omega\int_{\Omega_c} \boldsymbol{N}_i \cdot \sigma\sum_{k=1}^{n_e}\left(\boldsymbol{N}_k g_{jk}\right) d\Omega \\ &= \int_{\Omega_c} \mathrm{curl}\boldsymbol{N}_i \cdot \nu\mathrm{curl}(\mathrm{grad}N_j) \, d\Omega + j\omega\int_{\Omega_c} \boldsymbol{N}_i \cdot \sigma\mathrm{grad}N_j \, d\Omega \,. \end{aligned} \tag{21}$$

Since the curl of a gradient function equals zero, $\tilde{c}_{ij}$ can now be written as

$$\tilde{c}_{ij} = j\omega\int_{\Omega_c} \boldsymbol{N}_i \cdot \sigma\mathrm{grad}N_j \, d\Omega \,. \tag{22}$$

Comparing this result with (13) it can be seen that $KG^T = C$.

Hence the multiplication of the incidence matrix with the system matrix results in $GK = C^T$. Using the previous steps, the product GKG^T can be seen to satisfy $B = GKG^T$. With these results and substituting $\boldsymbol{a} + G^T\boldsymbol{v}$ for $\boldsymbol{a}^*$ the hybrid smoother (15) can now be written as

$$\tilde{\boldsymbol{a}}^*_{m+1} = \boldsymbol{a}_m + G^T\boldsymbol{v}_m + W_K^{-1}\left(\boldsymbol{f} - K\boldsymbol{a}_m - \underbrace{KG^T}_{C}\boldsymbol{v}_m\right) \tag{23}$$

$$= \boldsymbol{a}_m + W_K^{-1}\left(\boldsymbol{f} - K\boldsymbol{a}_m - C\boldsymbol{v}_m\right) + G^T\boldsymbol{v}_m \tag{24}$$

$$= \boldsymbol{a}_{m+1} + G^T\boldsymbol{v}_m \tag{25}$$

$$\boldsymbol{v}_{m+1} = W_{GKG^T}^{-1}\left(G\boldsymbol{f} - \underbrace{GK}_{C^T}\boldsymbol{a}_{m+1} - \underbrace{GKG^T}_{B}\boldsymbol{v}_m\right) \tag{26}$$

$$= W_B^{-1}\left(-C^T\boldsymbol{a}_{m+1} - B\boldsymbol{v}_m\right) \tag{27}$$

$$\boldsymbol{a}^*_{m+1} = \tilde{\boldsymbol{a}}^*_{m+1} + G^T\boldsymbol{v}_{m+1}\,. \tag{28}$$

It should be mentioned that $G\boldsymbol{f} = 0$ has been assumed in (26). This of course is only true for a right hand side which is described as the curl of a vector field as written in section 2.

Since $\boldsymbol{a}^* = \boldsymbol{a} + G^T\boldsymbol{v}$, the Gauss-Seidel smoother for the $\boldsymbol{A}$, V-formulation (19) produces the same results as the hybrid smoother in (23). It can hence be seen that the two algorithms are equivalent.

To sum it up, the pros and cons of the different algorithms can be described as follows: In case of the hybrid smoother, only edge basis functions are used. Less memory is used since the matrix C is not stored. On the other hand the incidence matrix has to be constructed. But the more time consuming task is the calculation of $B = GKG^T$ which increases with n_e^2.

For the GS-smoother with the $\boldsymbol{A}$, V-formulation, additional nodal basis functions have to be introduced and more memory is used because of the additional C-matrix. The advantage of this algorithm is the short assembly time for the matrices B and C.

Though the equivalence of the hybrid smoother applied to the $\boldsymbol{A}^*$-formulation and the classical Gauss-Seidel smoother for the $\boldsymbol{A}$, V-formulation has been shown only for the conducting region, it is also true for problems containing both conducting and non conducting regions. The incidence matrix has to cover the conducting region and at least the finite elements in the non conducting regions which border on the interface between conducting and non conducting region. This has to be done to ensure (14) also for nodes on the interface.

4 Numerical example

To show the properties of these different formulations and MG-smoothers, TEAM problem No. 7 has been analyzed. It consists of a conducting plate with a hole [3]. The geometry of the conducting region with its discretization on the coarse grid and the coil can be seen in Fig. 1. For the discretization, hexahedral elements of second order are used. They consist of 20 nodes and 36 edges.

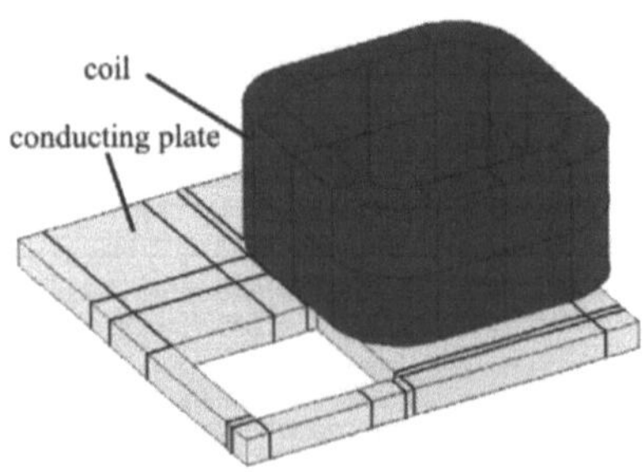

Fig. 1. Geometry of TEAM Problem No. 7

Table 1. Multigrid iterations

Degrees of freedom	Multigrid levels	No. of MG iterations: Hybrid smoother ($\boldsymbol{A}^*$-formulation)	SGS smoother ($\boldsymbol{A}$, V-formulation)
94102	3	111	111
325830	4	117	117
465246	4	142	142
782432	5	151	151

Table 2. Multigrid iterations and iteration time with 6 smoothing steps

Degrees of freedom	No. of MG iterations			Iteration time in sec		
	Hybrid smoother		SGS smoother	Hybrid smoother		SGS smoother
	extern	intern		extern	intern	
94102	21	21	21	344.1	244.4	225.1
325830	23	23	23	1601	1032	998
465246	27	28	28	2269	1487	1450
782432	27	29	29	4171	2798	2659

The problem has been calculated for different discretizations with the coarse grid remaining the same. The fine grid is achieved by subdividing individually the coarse grid elements. All calculations where done on a Windows 2000 PC with 1100MHz and 1.5GB RAM.

For the MG algorithm a standard V-cycle has been chosen. Instead of a GS iteration step, a symmetric Gauss-Seidel (SGS) iteration has been used in the algorithm of the hybrid smoother and as a smoother for the $\boldsymbol{A}$, V-formulation. Even though the comparison in section 3.1 is only true for a forward Gauss-Seidel algorithm, experiences have shown that the properties are similar even for SGS iterations. In addition, the number of MG iterations is lower when using smoothers with SGS iterations. In Table 1, the numbers of MG iterations for different discretizations and different smoothers are shown. It can be seen that even if the number of degrees of freedom is increasing rapidly, the number of iterations is only slightly changing. Furthermore, the number of iterations is the same for both smoothers.

By increasing the number of smoothing iterations, the number of MG cycles can be reduced. From experience it can be said that using about 6-8 SGS smoothing steps results in an optimal iteration time. If more iteration steps are used, the iteration time will not decrease any more because the gain of the iteration time will be overrun by the - especially on the finest grid - time consuming smoothing steps. In case of the hybrid smoother more smoothing steps can be implemented in two different ways: On the one hand, the whole algorithm (15) can be repeated several times, on the other hand only the internal GS (or SGS)-iterations (15) and (17) can be repeated. The first one will be denoted as external, the latter one as internal.

More significant than the number of iterations is the iteration time, since the problem should be solved as fast as possible. In Table 2, the properties of the different algorithms for 6 smoothing steps can be seen. Even though the number of iterations is lower for the external hybrid smoother, the iteration time is much higher. This is due to the fact that Gd and (18) have to be carried out 6 times for the external smoother, while for the internal algorithm it has to be done only once.

5 Conclusion

In this paper we have compared two different smoothers for different formulations of time harmonic eddy current problems. It turned out that the hybrid smoother for the $\boldsymbol{A}^*$-formulation is equivalent to the Gauss-Seidel smoother for the $\boldsymbol{A}$, V-formulation. The only difference is in the implementation. In one case, the incidence matrix and the matrix product $B = GKG^T$ has to be calculated and in the other, nodal basis functions have to be introduced.

A numerical example has illustrated the properties of these smoothers. Finally, the use of several smoothing steps has been discussed.

References

1. D. N. Arnold, R. S. Falk, and R. Winther. Multigrid in h(div) and h(curl). *Numer. Math.*, 85(2):197–217, 2000
2. O. Bíró. Edge element formulations of eddy current problems. *Computer methods in applied mechanics and engineering*, 169:391–405, 1999
3. K. Fujiwara and T. Nakata. Results for benchmark problem 7. *Compel*, 9(3):137–154, 1990
4. R. Hiptmair. Multigrid method for Maxwell's equations. *SIAM J. Numer. Anal.*, 36:204–225, 1999
5. J.C. Nédélec. Mixed finite elements in $\mathbf{R}^3$. *Numer. Math.*, 35:315–341, 1980
6. R. Perrussel, L. Nicolas, and F. Musy. An efficient preconditioner for linear systems issued from the finite element method for scattering problems. *IEEE Trans. Magn.*, 40(2):1080–1083, March 2004
7. U. Trottenberg, C. Oosterlee, and A. Schller. *Multigrid.* Academic Press, London, 2001
8. B. Weiß and O. Bíró. Multigrid for transient 3d eddy current analysis. *Compel*, 22(3):779–788, 2003
9. B. Weiß and O. Bíró. On the convergence of transient eddy-current problems. *IEEE Trans. Magn.*, 40(2):957–960, March 2004

Iterative Solution of Field Problems with a Varying Physical Parameter

A. G. Tijhuis[1], M. C. van Beurden[1] and A. P. M. Zwamborn[2]

[1] Faculty of Electrical Engineering, Eindhoven University of Technology – P.O. Box 513, 5600 MB Eindhoven, the Netherlands
[2] TNO Physics and Electronics Laboratory – P.O. Box 96864, 2509 JG 's-Gravenhage, the Netherlands

Abstract In this paper, linear field problems with a varying physical parameter are solved with the conjugate gradient method and a dedicated extrapolation procedure for generating the initial estimate. The scheme is formulated in detail, and its application to three-dimensional scattering problems for a rectangular conducting plate and an inhomogeneous, dispersive dielectric body are discussed.

1 Introduction

In modern society different trends are recognized in the usage of the available electromagnetic spectrum. One can think of wireless communication or transport of (digital) information. The density of such applications is increasing rapidly. Obtaining electromagnetic compatibility and/or reducing electromagnetic interference sometimes seems to be an impossible task. Another trend is found in electromagnetic inverse scattering and profiling. For example, this development is used in the detection and classification of land mines and other unexploded ordnance. Regarding electromagnetic inversion, one can also think of medical applications such as tomography or the detection of defects in metallic heart valves. Finally, we would like to mention the problem of electromagnetic coupling into humans in the area of clinical hyperthermia or non-ionizing radiation hazards analysis. In these applications, a rigorous electromagnetic analysis is indispensable.

The focus of this chapter is found in computational tools in the field of electromagnetic analysis and design. One can identify the roadmap "going from engineering electromagnetics to electromagnetic engineering". One of the extensively used and most versatile methods is the integral equation technique. It takes into account that the irradiated object is present in free space and that it manifests itself through the presence of secondary sources or contrast sources. Integral equations can be solved by the method of moments [1]. This leads to a system of linear algebraic equations.

To solve this system, one can use direct discrete numerical solution methods, such as Gaussian Elimination or Singular Value Decomposition, or suitable iterative techniques such as a conjugate gradient (CG) method. An overview of numerical solution methods for linear systems of equations can be found in the book by Golub and Van Loan [2]. In electromagnetic scattering and coupling problems, integral equations are often solved by using a Fast Fourier Transform to compute the spatial convolution of the integral operator and a conjugate gradient iterative scheme. This so called CGFFT method has been used successfully for many electromagnetic scattering and coupling problems [3, 4, 5, 6, 7, 8, 9].

In analysis or design procedures, the engineer has the freedom to change one or more parameters to obtain an optimal design with respect to performance and costs. This means that he or she will need to consider the determination of electromagnetic fields for a (large) number of values of a physical parameter. In this chapter we present a strong approach for this type of problem, which utilizes the iterative schemes mentioned above. We restrict ourselves to the case where the linear system originates from one or more integral equations. We apply an iterative procedure based on the minimization of an integrated squared error, and start this procedure from an initial estimate that is a linear combination of the last few "final" results. When the coefficients in this extrapolation are determined by minimizing the integrated squared error for the actual value of the parameter, the built-in orthogonality in this type of scheme ensures that only a few iteration steps are required to obtain the solution. The success of this strategy has been demonstrated before [10, 11, 12]. However, it has not been applied to 3-D problems.

The outline of the chapter is as follows. In Sect. 2, the method of solution is discussed. Special attention is given to the iterative procedure and the implementation of a relevant initial estimate based on the previous solutions. Explicit examples are discussed in Sect. 4 and Sect. 5 presents the conclusions.

2 Method of Solution

In the computational modeling of electromagnetic fields for practical applications, typically a large system of linear equations must be solved. This system originates from spatially discretizing Maxwell's differential equations (in "finite" or "local" techniques) or equivalent integral equations (in "global" techniques). In formal notation, such a system can be written as

$$L(p)\,u(p) = f(p), \tag{1}$$

where

$$\begin{aligned} L(p) &= \text{a linear operator,} \\ u(p) &= \text{the unknown field,} \\ f(p) &= \text{the forcing function,} \\ p &= \text{a physical parameter.} \end{aligned}$$

The operator $L(p)$ originates from discretizing its counterpart in the continuous equation, $u(p)$ is a discretized field and $f(p)$ corresponds to an impressed source or an incident field. We are interested in the situation where this problem must be solved for a large number of sampled values of the parameter p, e.g., $p_m = p_0 + m\Delta p$, with $m = 0, 1, \ldots, M$.

To solve the system of equations (1), we use the conjugate gradient method. This algorithm is described in detail by Van den Berg [3, 4]. Further, we organize the space discretization such that the convolution structure of the continous equation is preserved. In that case, the matrix-vector products in the CG algorithm can be evaluated by FFT operations, which considerably improves the speed of the so-called CGFFT algorithm [3, 4, 5, 6, 7, 8, 9].

In many applications of the conjugate gradient method, a simple initial estimate is used. Typically, the scheme is started from an initial vector $u^{(0)} = 0$. Depending on the nature of the problem at hand, we can also start from an incident field or from the Kirchhoff approximation to an unknown surface current.

Our choice of the initial estimate is inspired by the fact that $u(p)$ depends in a well-behaved manner on the parameter p. Therefore, it should be possible to extrapolate, by choosing

$$u^{(0)}(p_m) = \sum_{k=1}^{K} \gamma_k\, u(p_{m-k}). \tag{2}$$

The interpretation of the conjugate gradient scheme given above suggested that the values $\{\gamma_k \mid k = 1, \ldots, K\}$ should be found by minimizing the squared error

$$< L(p_m)u^{(0)}(p_m) - f(p_m) \mid L(p_m)u^{(0)}(p_m) - f(p_m) > . \tag{3}$$

where we have defined the inner product as

$$< g \mid h >= \sum_j g_j^* h_j, \tag{4}$$

where g_j and h_j denote the components of g and h, and where the asterisk denotes complex conjugation.

Because of the built-in orthogonality of the conjugate gradient method, we are then certain that this procedure must start its search for components of $f(p_m)$ outside the space spanned by the "previous" functions $\{Lu(p_{m-k}) \mid k = 1, \ldots, K\}$.

The coefficients γ_k that minimize the squared error in (3) can be found from the system of linear equations

$$\begin{aligned} \sum_{k'=1}^{K} &< L(p_m)u(p_{m-k}) \mid L(p_m)u(p_{m-k'}) > \gamma_{k'} \\ &=< L(p_m)u(p_{m-k}) \mid f(p_m) >, \end{aligned} \tag{5}$$

with $k = 1, \ldots, K$. Typically, we choose $K = 2$ (linear extrapolation) or $K = 3$ (quadratic extrapolation). For larger values of K, the basis vectors $L(p_m)u(p_{m-n})$ with $n = 1, \ldots, K$ become almost linearly dependent, and therefore the coefficients $\{\gamma_k\}$ can no longer be resolved from (5).

3 Scattering by Three-Dimensional Objects

To illustrate our approach, we have extended existing implementations of the CGFFT procedure for two three-dimensional objects that have become standards in the literature. In both cases, no special precautions were taken to enhance the discretization, which is first-order accurate as a function of the mesh size.

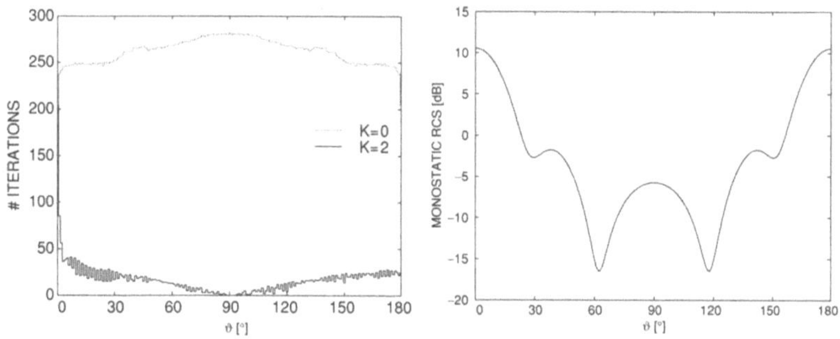

Fig. 1. Marching-on-in-angle version of the CGFFT method for a flat plate. (a) Number of iterations required to reach a relative error of 10^{-3} versus angle of incidence using zero (gray line) and two previous results (black line) as an initial estimate. (b) Monostatic radar cross section versus angle of incidence

3.1 Scattering by a Flat Plate

The first example is a flat, rectangular plate in free space located at $0 < x < a$, $0 < y < b$ and $z = 0$. For this problem, we solve the well-known electric-field integral equation

$$\left[\nabla_{\mathrm{T}}\nabla_{\mathrm{T}} \cdot -\frac{s^2}{c_0^2}\right] \int_0^a dx' \int_0^b dy' \,\frac{\exp(-sR/c_0)}{4\pi R}\, \boldsymbol{J}_{\mathrm{S}}(\boldsymbol{r}'_{\mathrm{T}}, s) = -s\varepsilon_0\, \boldsymbol{E}^{\mathrm{i}}_{\mathrm{T}}(\boldsymbol{r}_{\mathrm{T}}, s), \tag{6}$$

where s is a complex frequency, $R = |\boldsymbol{r}_{\mathrm{T}} - \boldsymbol{r}'_{\mathrm{T}}|$, and where the subscript $_{\mathrm{T}}$ stands for a transverse component. The unknown surface current $\boldsymbol{J}_{\mathrm{S}}(\boldsymbol{r}_{\mathrm{T}}, s)$ is approximated by rooftop functions, and we use a weak formulation of (6), weighted by the same rooftop functions [13]. In the resulting discretized form, the convolution symmetry is preserved, so that the matrix-vector products in the conjugate gradient procedure can be evaluated with the aid of two-dimensional FFT operations.

In particular, we have computed the monostatic radar cross section of a $\lambda \times \lambda$ plate for the special case $s = j\omega$. A plane wave is incident on the plate at an angle ϑ with respect to the z-axis and an angle $\phi = 90°$ with respect to the x-axis. The incident plane wave is x-polarized. The discretized plate has a mesh of 31×31 points. Figure 1(a) shows the number of iterations for increasing ϑ. In the generic formulation of Section 2, this means that $p = \vartheta$. The gray line represents starting from a zero initial estimate, and the black line is for two previous results in the initial estimate, i.e. $K = 2$. Figure 1 presents the monostatic radar cross section of the $\lambda \times \lambda$ plate in the plane $\phi = 90°$.

Another result for the plate concerns marching on in length. Now, the parameter p represents the length of the plate in the x-direction. The idea was inspired by the shape sensitivity analysis in [14, 15]. Here, we start from a $\lambda \times \lambda$ plate and we increase the length of the plate in 100 steps to a $2\lambda \times \lambda$ plate. We used a fixed spatial discretization of 62×31 mesh points. The number of iterations required to reach a relative error of 10^{-3} versus the length of the plate is shown in Fig. 2. In the computations leading to Figs. 1 and 2, it turned out that extrapolation with $K = 2$ was in fact more efficient than extrapolation with $K = 3$.

3.2 Scattering by an Inhomogeneous Dielectric Cube

The second example is an inhomogeneous dielectric cube, again in free space. We formulate the scattering problem as a domain integral equation over the object domain $\mathcal{D}$ as

$$\boldsymbol{E}^{i}(\boldsymbol{r}, s) = \frac{\boldsymbol{D}(\boldsymbol{r}, s)}{\varepsilon(\boldsymbol{r}, s)} + \left(\frac{s^2}{c_0} - \nabla\nabla\cdot\right) \boldsymbol{A}(\boldsymbol{r}, s), \tag{7}$$

where s is a complex frequency and where the vector potential $\boldsymbol{A}(\boldsymbol{r}, s)$ is given by

$$\boldsymbol{A}(\boldsymbol{r}, s) = \frac{1}{\varepsilon_0} \iiint_{\mathcal{D}} \frac{\exp(-sR/c_0)}{4\pi R}\, \frac{\varepsilon(\boldsymbol{r}, s) - \varepsilon_0}{\varepsilon(\boldsymbol{r}, s)}\, \boldsymbol{D}(\boldsymbol{r}, s)\, dV', \tag{8}$$

where $R = |\boldsymbol{r} - \boldsymbol{r}'|$. We take the contrast function in (8) constant in each rectangular subdomain in the space discretization. Like the current in the plate problem, the dielectric displacement $\boldsymbol{D}(\boldsymbol{r}, s)$ is approximated by an expansion

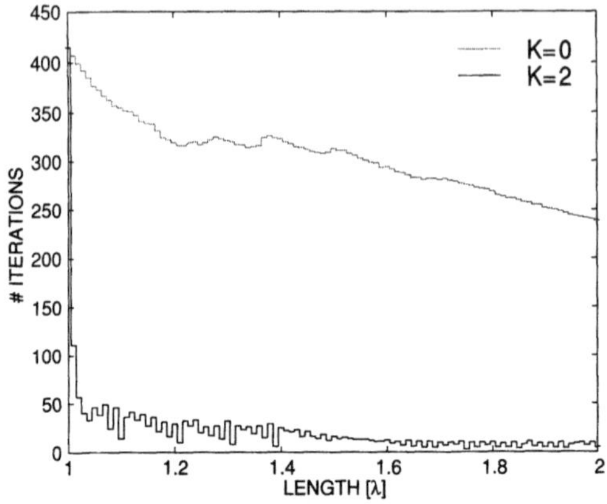

Fig. 2. Number of iterations required to reach a relative error of 10^{-3} versus length of the plate for the marching-on-in-length version of the CGFFT method for a flat plate using zero (gray line) and two previous results (black line) as an initial estimate

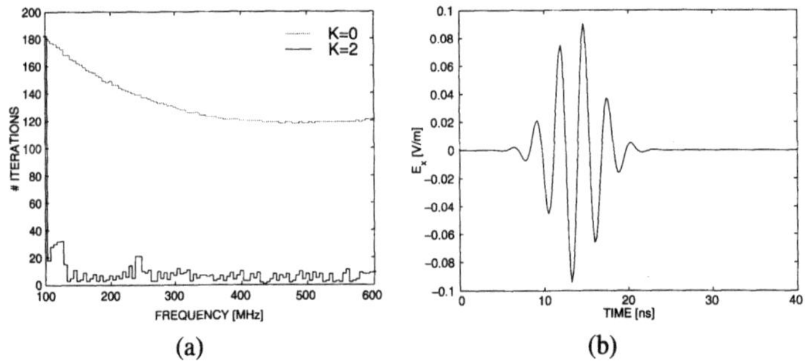

Fig. 3. Marching-on-in-frequency version of the CGFFT method for an inhomogeneous dielectric cube. (a) Number of iterations required to reach a relative error of 10^{-3} versus frequency for the using zero (gray line) and two previous results (black line) as an initial estimate. (b) Time domain signal at the center of the muscle cube for an incident x-polarized wave of 1 V/m

that is piecewise linear in the longitudinal direction and constant in the transverse directions. The Green's function is replaced by a weak form, and the result is weighted by testing functions that are identical to the expansion functions. Again, the space discretization preserves the convolution symmetry of the continuous form of the integral equation given in (7) and (8). More details can be found in the papers by Zwamborn and Van den Berg [16, 17].

As an illustration, we have modeled a cube of muscle tissue centered inside a cube of fat tissue. The incident field is x-polarized with propagation vector parallel to the z-axis and a strength of 1 V/m. The dispersive tissues are modeled using a Debye model [18] and the dimensions of the inner and outer cubes are 0.14 m and 0.30 m, respectively. The discretized object has $30 \times 30 \times 30$ mesh points. The field is computed in the middle of the muscle cube for real-valued frequencies $f = \omega/2\pi = -js/2\pi$ of 100 to 600 MHz and then converted to a time domain signal. In this case, we vary $p = \omega$. The number of iterations needed is shown in Fig. 3(a), where the gray line is for a zero initial estimate, and the black line for minimization using two previous results. Again, using $K = 2$ in the extrapolation procedure led to the most rapid convergence. The time signal, shown in Fig. 3(b), is computed by an FFT using the waveform $\exp\left[-(t-\tau)^2/(2T^2)\right]\sin(\omega_0 t)$, where $\tau = 14$ ns, $T = 2.75$ ns and $\omega_0/2\pi = 450$ MHz.

4 Conclusions

In this chapter, we have extended the conventional conjugate gradient method with a dedicated extrapolation procedure that considerably enhances the speed of convergence. Although the procedure has already been demonstrated successfully for a range of applications, including transient scattering, radar cross section computations and inverse profiling, until now no applications to three-dimensional configurations have been reported. In the present chapter, this gap has been filled.

References

1. Harrington, R.F: Field computation by Moment Methods. Macmillan, New York (1968)
2. Golub, G.H., and Van Loan, C.F.: Matrix Computations (Third Edition). The Johns Hopkins University Press, Baltimore (1996)
3. Van den Berg, P.M.: Iterative computational techniques in scattering based upon the integrated square error criterion. IEEE Trans. Antennas Propagat. **32** (1984) 1063–1071
4. Van den Berg, P.M.: Iterative schemes based on the minimization of the error in field problems. Electromagnetics **5** (1985) 237–262
5. Sarkar, T.K., Arvas, E. and Rao, S.M.: Application of the fast Fourier transform and the conjugate gradient method for the solution of electromagnetic radiation from electrically large and small conducting bodies. Electromagnetics **5** (1985) 99–122
6. Bokhari, S.A., and Balakrishnan, N.A.: A method to extend the spectral iteration technique. IEEE Trans. Antennas Propagat. **34** (1986) 51–57
7. Sarkar, T.K., Arvas, E. and Rao, S.M.: Application of the FFT and the conjugate gradient method for the solution of electromagnetic radiation from electrically large and small conducting bodies. IEEE Trans. Antennas Propagat. **34** (1986) 635–640
8. Yuan Zhuang, Ke-Li Wu, Chen Wu and Litva, J.: A combined full-wave CGFFT method for rigorous analysis of large microstrip antenna arrays. IEEE Trans. Antennas Propagat. **44** (1996) 102–109
9. Basterrechea, J. and Catedra, M.F.: Computatation of microstrip S-parameter using a CG-FFT scheme, IEEE Trans. Microwave Theory Tech. **42** (1994) 234–240
10. Tijhuis, A.G., Peng, Z.Q. and Rubio Bretones, A.: Transient excitation of a straight thin wire segment: a new look at an old problem. IEEE Trans. Antennas Propagat. **40** (1994) 1132–1146
11. Tijhuis, A.G., and Peng, Z.Q.: Marching-on-in-fequency method for solving integral equations in transient electromagnetic scattering. IEE Proc. H **138** (1991) 347–355
12. Peng, Z.Q. and Tijhuis, A.G.: Transient scattering by a lossy dielectric cylinder: marching-on-in-frequency approach. J. Electromagn. Waves Applicat. **7** (1993) 739–763
13. Zwamborn, A.P.M. and Van den Berg, P.M.: The weak form of the conjugate gradient method for plate problems. IEEE Trans. Antennas Propagat. **39** (1991) 224–228
14. Ureel, J. and De Zutter, D.: Shape sensitivities of capacitances of planar conducting surfaces using the method of moments. IEEE Trans. Microwave Theory Tech. **44** (1996) 198–207
15. Ureel, J. and De Zutter, D.: A new method for obtaining shape sensitivities of planar microstrip structures by a full-wave analysis, IEEE Trans. Microwave Theory Tech. **44** (1996) 249–260
16. Zwamborn, A.P.M. and Van den Berg, P.M.: The three-dimensional weak form of the conjugate gradient FFT method for solving scattering problems. IEEE Trans. Microwave Theory Tech. **40** (1992) 1757–1766
17. Zwamborn, A.P.M. and Van den Berg, P.M.: Computation of electromagnetic fields inside strongly inhomogeneous objects by the weak conjugate gradient FFT method. JOSA A **11** (1994) 1414–1421
18. Lepelaars, E.S.A.M.: Electromagnetic pulse distortion in living tissue. Med. Biol. Eng. Comput. **34** (1996) 213–220

Part IV

General Mathematical and Computational Methods

Time Integration Methods for Coupled Equations*

A. Kværnø

Department of Mathematical Sciences, Norwegian University of Science and Technology, N-7491 Trondheim, Norway, anne@math.ntnu.no

Abstract In this paper we discuss time integration methods designed for solving stiff-nonstiff problems. A tool for analysing the effect of using stepsizes larger than the time scale of the stiff subsystem is presented.

1 Introduction

In many applications we have to deal with time integration of coupled systems, with subsystems of different time scales. Over the years, several approaches have been developed to exploit the particular properties of each subsystem, like multirate methods, implicit-explicit methods and splitting methods. More recently, also exponential integrators are enjoying a renaissance. Most of these methods are well understood in terms of classical local error / order analysis. However, the desired modus operandi often gives stepsizes larger than the time scales of the rapid subsystems. In this case, the classical order analysis is of limited, although important, relevance.

The problem can be illustrated by the following simple example: Consider the equation

$$y' = \lambda y + y + \mathrm{e}^t, \qquad y(0) = 1, \qquad \mathrm{Re}(\lambda) << 0.$$

The linear term λy represents the fast subsystem, while $y + \mathrm{e}^t$ is the slow one. The problem is solved by two different explicit exponential integrators, both of order 3. Exponential integrators work such that the fast linear part is integrated exactly. Figure 1 shows the relative error after one step, using different stepsizes. The local error is measured for two values of t, at $t = 0$ where the solution is dominated by its transient, and at $t = 0.5$, in which the transient is completely damped.

From these pictures, we can draw several conclusions. First, even if the two methods are both of classical order 3, they behave quite differently, in the nonstiff regime (for which λh is small) as well as in the stiff. We also observe that the error depends not only on the stepsize h, but also of the stiffness parameter λ and of the initial values. Unfortunately, this behaviour can not be completely understood by a classical local error analysis, neither by a standard stability analysis.

In this paper, we will first describe two different strategies for solving stiff-nonstiff problems. In Sect. 3 an alternative local error analysis is presented, although details are only given for the linear problems. A simple numerical test verifies the theoretical results.

2 Stiff-nonstiff problems

Given the problem

$$y' = f_S(t,y) + f_N(t,y), \qquad y(t_0) = y_0, \tag{1}$$

where f_S corresponds to the stiff term and f_N to the nonstiff. Such problems arise frequently from discretization of partial differential equations (PDEs) of advection-diffusion-reaction type, see e.g. [7]. In this paper we will put emphasis on semilinear problems

$$y' = Ly + f_N(t,y) \qquad y(t_0) = y_0, \tag{2}$$

coming from e.g. the discretization of semilinear parabolic equations or the Schrödinger equation. In the following, we will present two different strategies for solving such problems.

*Invited paper at SCEE-2004

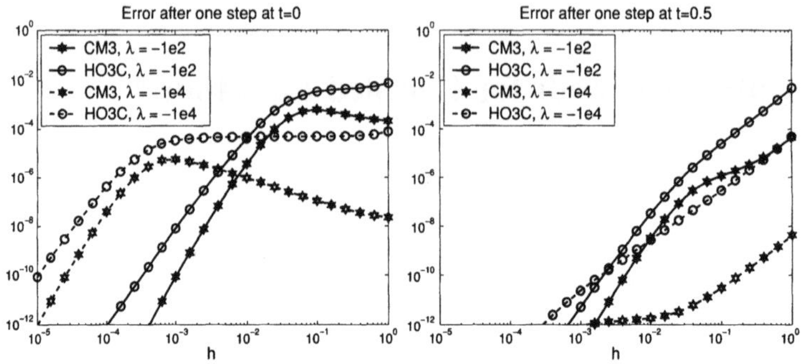

Fig. 1. Local error for two exponential integrators

Table 1. IMEX3: A third order, L-stable IMEX-RK method

$$\begin{array}{c|ccccc} 0 & 0 & 0 & 0 & 0 & 0 \\ \frac{1}{2} & 0 & \frac{1}{2} & 0 & 0 & 0 \\ \frac{2}{3} & 0 & \frac{1}{6} & \frac{1}{2} & 0 & 0 \\ \frac{1}{2} & 0 & -\frac{1}{2} & \frac{1}{2} & \frac{1}{2} & 0 \\ 1 & 0 & \frac{3}{2} & -\frac{3}{2} & \frac{1}{2} & \frac{1}{2} \\ \hline 1 & 0 & \frac{3}{2} & -\frac{3}{2} & \frac{1}{2} & \frac{1}{2} \end{array}, \qquad \begin{array}{c|ccccc} 0 & 0 & 0 & 0 & 0 & 0 \\ \frac{1}{2} & \frac{1}{2} & 0 & 0 & 0 & 0 \\ \frac{2}{3} & \frac{11}{18} & \frac{1}{18} & 0 & 0 & 0 \\ \frac{1}{2} & \frac{5}{6} & -\frac{5}{6} & \frac{1}{2} & 0 & 0 \\ 1 & \frac{1}{4} & \frac{7}{4} & \frac{3}{4} & -\frac{7}{4} & 0 \\ \hline 1 & \frac{1}{4} & \frac{7}{4} & \frac{3}{4} & -\frac{7}{4} & 0 \end{array}$$

2.1 Implicit-explicit Runge-Kutta methods

The strategy of applying an implicit scheme for f_S and an explicit one for f_N is the idea behind implicit-explicit (IMEX) methods. Multistep methods as well as one-step methods have been constructed this way. In this paper, we restrict ourself to IMEX Runge-Kutta (IMEX-RK) schemes as defined in [1, 10]. One step of an s-stage IMEX-RK scheme applied to (1) is given by

$$\begin{aligned} Y_1 &= y_0, \\ Y_i &= y_0 + h\sum_{j=1}^{i} a_{ij} f_S(t_0 + c_j h, Y_j) + h\sum_{j=1}^{i-1} \hat{a}_{ij} f_N(t_0 + c_j h, Y_j), \quad i = 2, \cdots, s, \\ y_1 &= y_0 + h\sum_{i=1}^{s} b_i f_S(t_0 + c_i h, Y_i) + h\sum_{i=1}^{s} \hat{b}_i f_N(t_0 + c_i h, Y_i), \end{aligned} \tag{3}$$

where the coefficients are given in the following tableaux

$$\begin{array}{c|cccccc} 0 & 0 & 0 & 0 & \cdots & 0 \\ c_2 & a_{21} & a_{22} & 0 & \cdots & 0 \\ c_3 & a_{31} & a_{32} & a_{33} & \cdots & 0 \\ \vdots & \vdots & \vdots & \vdots & \ddots & \vdots \\ c_s & a_{s1} & a_{s2} & a_{s3} & \cdots & a_{ss} \\ \hline & b_1 & b_2 & b_3 & \cdots & b_s \end{array}, \qquad \begin{array}{c|cccccc} 0 & 0 & 0 & 0 & \cdots & 0 \\ c_2 & \hat{a}_{21} & 0 & 0 & \cdots & 0 \\ c_3 & \hat{a}_{31} & \hat{a}_{32} & 0 & \cdots & 0 \\ \vdots & \vdots & \vdots & \vdots & \ddots & \vdots \\ c_s & \hat{a}_{s1} & \hat{a}_{s2} & \hat{a}_{s3} & \cdots & 0 \\ \hline & \hat{b}_1 & \hat{b}_2 & \hat{b}_3 & \cdots & \hat{b}_s \end{array}$$

or in short form as

$$\begin{array}{c|c} c & A \\ \hline & b^T \end{array}, \qquad \begin{array}{c|c} c & \hat{A} \\ \hline & \hat{b}^T \end{array}.$$

A third order, L-stable scheme, proposed in [1] is given in Table 1. Sometimes it might be useful to write IMEX-RK methods applied to the semilinear problem (2) as

$$Y = (I_{ms} - hA \otimes L)^{-1}(\ _s \otimes y_0) + (I_{ms} - hA \otimes L)^{-1} f_N(t_0 + ch, Y),$$
$$y_1 = r(hL)y_0 + h(\hat{b}^T \otimes I_m + (b^T \otimes L)(I_{ms} - hA \otimes L)^{-1}) f_N(t_0 + ch, Y), \tag{4}$$

where $Y = [Y_1^T, \cdots, Y_s^T]^T$, $f_N(t_0 + ch, Y) = [f(t_0 + c_1 h, Y_1)^T, \cdots, f(t_0 + c_s h, Y_s)^T]^T$, $\ = [1, 1, \cdots, 1]^T$, s refers to the number of stages and m to the dimension of the problem (1). Further, $r(z) = 1 + zb^T(I_s - zA)^{-1}\ _s$ is the stability function for the implicit method.

2.2 Exponential integrators

Exponential integrators are mostly constructed to solve problems of the form (2). The idea behind these integrators dates back to the sixties, but has not been considered practical since the schemes involve computation of matrix exponential functions. Using modern techniques, such functions can now be computed quite efficiently, see [13] and [9]. The latter pays particular attention to stable computations of the exponential function of which exponential integrators are composed. Today exponential integrators are enjoying a renaissance, numerical comparisons reveal several examples where they outperform standard integrators. A nice introduction to the idea of exponential integrators can be found in [3, 6], see also the review paper [12].

In the presentation of exponential integrators, we will frequently use the following function

$$\phi_q(hL) = \frac{1}{(q-1)!}\frac{1}{h^q} \mathrm{e}^{hL} \int_0^h \mathrm{e}^{-\tau L} \tau^{q-1} d\tau,$$

or

$$\phi_q(z) = \frac{1}{z^q}\left(\mathrm{e}^z - \sum_{j=0}^{q-1} \frac{z^j}{j!}\right), \qquad q = 1, 2, \cdots. \tag{5}$$

Note that ϕ_q is an analytic function of z and $\phi_q(0) = \frac{1}{q!}$.

Exponential integrators can be considered as approximations of the variation-of-constants formula, which gives the exact solution of (2) as

$$y(t_0 + h) = \mathrm{e}^{hL} y_0 + \mathrm{e}^{hL} \int_0^h \mathrm{e}^{-\tau L} f_N(t_0 + \tau, y(t_0 + \tau)) d\tau. \tag{6}$$

A first order method can be derived by using $f_N \approx f_N(t_0, y_0)$. Inserting this into (6) gives an exponential forward Euler method

$$y_1^{fe} = \mathrm{e}^{hL} y_0 + h\phi_1(hL) f(t_0, y_0). \tag{7}$$

This result can be improved by using

$$f_N \approx f_N(t_0, y_0) + \frac{t - t_0}{h}(f_N(t_0 + h, y_1^{fe}) - f_N(t_0, y_0))$$

which, when inserted into (6) gives

$$\begin{aligned} y_1^{imp} &= \mathrm{e}^{hL} y_0 + h\phi_1(hL) f_N(t_0, y_0) \\ &+ h\phi_2(hL)(f_N(t_0 + h, y_1^{fe}) - f_N(t_0, y_0)). \end{aligned} \tag{8}$$

In general, explicit exponential Runge-Kutta integrators are given by

$$Y_i = \mathrm{e}^{c_i hL} y_0 + h \sum_{j=1}^{i-1} a_{ij}(hL) f_N(t_0 + c_j h, Y_j), \quad i = 1, 2, \cdots, s,$$
$$y_1 = \mathrm{e}^{hL} y_0 + h \sum_{i=1}^{s} b_i(hL) f_N(t_0 + c_i h, Y_i). \tag{9}$$

where the method coefficients are exponential functions evaluated at hL. Table 2 presents two third order exponential RK methods. The first, called CM3, was proposed by Cox and Matthew in [3]. The second, HO3C, was presented in a talk by Hochbruck and Ostermann, [5].

By comparing (4) and (9) we observe that both IMEX-RK and explicit exponential RK methods can be written in the general form

Table 2. Exponential Runge-Kutta methods of order 3

Cox and Matthew: CM3

$$
\begin{array}{c|c|ccc}
0 & 1 & 0 & & \\
\frac{1}{2} & e^{z/2} & \frac{1}{2}\phi_1(\frac{z}{2}) & & \\
1 & e^{z} & -\phi_1(z) & 2\phi_1(z) & \\
\hline
0 & e^{z} & \phi_1(z) - 3\phi_2(z) + 4\phi_3(z) & 4\phi_2(z) - 8\phi_3(z) & -\phi_2(z) + 4\phi_3(z)
\end{array}
$$

Hochbruck and Ostermann: HO3C

$$
\begin{array}{c|c|ccc}
0 & 1 & 0 & & \\
\frac{1}{3} & e^{z/3} & \frac{1}{3}\phi_1(\frac{z}{3}) & & \\
\frac{2}{3} & e^{2z/3} & 0 & \frac{2}{3}\phi_1(\frac{2z}{3}) & \\
\hline
 & e^{z} & \phi_1(z) - \frac{3}{2}\phi_2(z) & 0 & \frac{3}{2}\phi_2(z)
\end{array}
$$

$$
\begin{aligned}
Y_i &= \chi_i(hL)y_0 + h\sum_{j=1}^{i-1} \alpha_{ij}(hL) f_N(t_0 + c_j h, Y_j), \quad i = 1, 2, \cdots, s, \\
y_1 &= r(hL)y_0 + h\sum_{i=1}^{s} \beta_i(hL) f_N(t_0 + c_i h, Y_i).
\end{aligned} \tag{10}
$$

where the coefficients are given in the tableau

$$
\begin{array}{c|c|ccccc}
0 & \chi_1(z) & & & & & \\
c_2 & \chi_2(z) & \alpha_{21}(z) & & & & \\
c_3 & \chi_3(z) & \alpha_{31}(z) & a_{32}(z) & & & \\
\vdots & \vdots & \vdots & \vdots & \ddots & & \\
c_s & \chi_s(z) & \alpha_{s1}(z) & \alpha_{s2}(z) & \cdots & \alpha_{s,s-1}(z) & \\
\hline
 & r(z) & \beta_1(z) & \beta_2(z) & \cdots & \beta_{s-1}(z) & \beta_s(z)
\end{array}
\qquad \text{or short as} \qquad
\begin{array}{c|c|c}
c & \chi(z) & \mathcal{A}(z) \\
\hline
 & r(z) & \beta^T(z)
\end{array}.
$$

The coefficients are either exponential or rational functions evaluated in hL. Other methods might fit into this formulation as well. For the two schemes in question, the coefficients are given by

$$
\begin{array}{rr}
\text{IMEX} & \text{ExpRK} \\
\chi(z) = (I_s - zA)^{-1} \quad {}_s & \chi(z) = e^{cz} \\
r(z) = 1 + zb^T(I_s - zA)^{-1} \quad {}_s & r(z) = e^{z} \\
\mathcal{A}(z) = (I_s - zA)^{-1}\hat{A} & \mathcal{A}(z) = A(z) \\
\beta^T(z) = zb^T(I_s - zA)^{-1}\hat{A} + \hat{b}^T , & \beta^T(z) = b(z)^T .
\end{array}
$$

This general formulation will be used in the local error analysis of the next section.

3 Local error analysis

In this section, we illustrate the error behaviour of the methods by studying a single Fourier component of a linear problem, represented by

$$
y' = \lambda y + f(t), \qquad y(t_0) = y_0, \qquad \lambda \in \mathbb{C}^-, \quad |\lambda| >> 1. \tag{11}
$$

The exact solution of (11) is given by

$$
y(t_0 + h) = e^{\lambda h} y_0 + e^{\lambda h} \int_0^h e^{-\lambda\tau} f(t_0 + \tau) d\tau.
$$

Table 3. Weight functions for the linear problem

q	$\phi_q(z)$	$\psi_q(z)$		
		IMEX3	CM3	HO3C
0	e^z	$\frac{8(z^3-6z+6)}{3(z-2)^4}$	e^z	e^z
1	$\frac{e^z-1}{z}$	$\frac{-3z^3+32z^2-72z+48}{3(z-2)^4}$	$\frac{e^z-1}{z}$	$\frac{e^z-1}{z}$
2	$\frac{e^z-1-z}{z^2}$	$\frac{-25z^2+84z-72}{18(z-2)^3}$	$\frac{e^z-1-z}{z^2}$	$\frac{e^z-1-z}{z^2}$
3	$\frac{e^z-1-z-\frac{z^2}{2}}{z^3}$	$\frac{-27z^2+100z-96}{72(z-2)^3}$	$\frac{e^z-1-z-\frac{z^2}{2}}{z^3}$	$\frac{e^z-1-z}{3z^2}$
4	$\frac{e^z-1-z-\frac{z^2}{2}-\frac{z^3}{6}}{z^4}$	$\frac{-89z^2+364z-384}{1296(z-2)^3}$	$\frac{(6-z)e^z-6-5z-2z^2}{12z^3}$	$\frac{2(e^z-1-z)}{27z^2}$

Taking the Taylor expansion of $f(t_0 + \tau)$ around $\tau = 0$ and integrating each term separately gives

$$y(t_0 + h) = e^z y_0 + \sum_{q=1}^{\infty} \phi_q(z) h^q f^{(q-1)}(t_0), \qquad z = \lambda h, \tag{12}$$

and $\phi_q(z)$ is given by (5). In the case of $\text{Re}(\lambda) << 0$, the solution will be attracted to a slow solution manifold, given by

$$y_s(t) = -\sum_{q=1}^{\infty} \frac{f^{(q-1)}(t)}{\lambda^q}. \tag{13}$$

A series similar to (12) can be derived for the numerical solution. Applying (10) on (11) and replacing f by its Taylor expansion gives

$$y_1 = r(z)y_0 + h\sum_{i=1}^{s} \beta_i(z) f(t_0 + c_i h) = r(z)y_0 + \sum_{q=1}^{\infty} \psi_q(z) h^q f^{(q-1)}(t_0), \tag{14}$$

where

$$\psi_q(z) = \frac{1}{(q-1)!} \sum_{i=1}^{s} \beta_i(z) c_i^{q-1}.$$

For convenience we will use the notation $\phi_0(z) = e^z$ and $\psi_0(z) = r(z)$. Table 3 lists the functions ϕ_q as well as ψ_q for the methods given in Table 1 and 2.
The local truncation error is given by

$$y(t_0 + h) - y_1 = \mathcal{E}_0 y_0 + \sum_{q=1}^{\infty} \mathcal{E}_q(z) h^q f^{(q-1)}(t_0), \tag{15}$$

where the error functions $\mathcal{E}_q$ are given by

$$\mathcal{E}_q(z) = \phi_q(z) - \psi_q(z).$$

Obviously, the error is of order $p + 1$ independent of the stiffness parameter λ if

$$\mathcal{E}_q(z) = 0, \qquad q = 1, 2, \cdots p,$$

and for the three methods under consideration

$$p^{\text{IMEX3}} = 0, \qquad p^{\text{CM3}} = 3 \quad \text{and} \qquad p^{\text{HO3C}} = 2.$$

Only the IMEX method has a local error depending on the initial value y_0. IMEX methods approximate the exponential e^z by a rational function $r(z)$, thus $\mathcal{E}_0 \sim \lambda^{\rho+1} h^{\rho+1}$ for some ρ. This term usually dominates the error when λ is large. However, if the initial value is on the smooth manifold, then y_0 in (15) can be replaced by $y_s(t_0)$ given in (13), thus

$$y(t_0 + h) - y_1 = \sum_{q=1}^{\infty} \tilde{\mathcal{E}}_q(z) h^q f^{(q-1)}(t_0),$$

with $\tilde{\mathcal{E}}_q = \mathcal{E}_q - \mathcal{E}_0 / z^q$. For IMEX3 these terms are

$$\tilde{\mathcal{E}}_1 = 0, \quad \tilde{\mathcal{E}}_2 = \frac{7z^3 - 8z^2}{18(z-2)^4}, \quad \tilde{\mathcal{E}}_3 = \frac{-9z^3 + 62z^2 - 64z}{72(z-2)^4}, \quad \dots$$

giving $\tilde{p}^{\text{IMEX3}} = 2$. In the following, we will use the term IMEX3(s) to denote the situation when the initial value is on the slow manifold.

Examination of the error functions in the extreme cases, like the nonstiff, the strongly damped and the highly oscillatory case, gives further insight into the behaviour of the local error.

The nonstiff case

This situation is characterised by $|z|$ small and the error functions can be studied in terms of their series expansions. For the methods in question, the dominant terms of $\mathcal{E}_q$ are given by

q	IMEX3	IMEX3(s)	CM3	HO3C
0	$\frac{1}{48}z^4 + \mathcal{O}(z^5)$	0	0	0
1	$\frac{1}{48}z^3 + \mathcal{O}(z^4)$	0	0	0
2	$-\frac{1}{144}z^2 + \mathcal{O}(z^3)$	$-\frac{1}{36}z^2 + \mathcal{O}(z^3)$	0	0
3	$-\frac{5}{144}z + \mathcal{O}(z^2)$	$-\frac{1}{18}z + \mathcal{O}(z^2)$	0	$-\frac{1}{72}z + \mathcal{O}(z^2)$
4	$\frac{1}{216} + \mathcal{O}(z)$	$-\frac{7}{432} + \mathcal{O}(z)$	$\frac{1}{720} + \mathcal{O}(z)$	$\frac{1}{216} + \mathcal{O}(z)$

By inserting this into (15), keeping in mind that $z = \lambda h$, we obtain the following expressions for the local error:

$$y(x_0 + h) - y_1 = \begin{cases} \left(\frac{\lambda^4}{48}y_0 + \frac{\lambda^3}{48}f - \frac{\lambda^2}{144}f' - \frac{5\lambda}{144}f'' + \frac{1}{216}f'''\right)h^4 + \mathcal{O}(h^5) & \text{for IMEX3} \\ \left(-\frac{\lambda^2}{36}f' - \frac{\lambda}{18}f'' - \frac{7}{432}f'''\right)h^4 + \mathcal{O}(h^5) & \text{for IMEX3(s)} \\ \left(\frac{\lambda}{720}f''' - \frac{1}{2880}f^{(4)}(t_0)\right)h^5 + \mathcal{O}(h^6) & \text{for CM3} \\ \left(-\frac{\lambda}{72}f'' + \frac{1}{216}f'''\right)h^4 + \mathcal{O}(h^5) & \text{for HO3C} \end{cases}$$

The error terms all depend on some power of λ. Since $|\lambda| >> 1$ by assumption, we will prefer this power to be as small as possible. In this sense the exponential integrators outperform the IMEX method in the transient case. The situation improves significantly in the slow case, but still the error is $\sim \lambda^2$ for the IMEX method while it is $\sim \lambda$ for the exponential methods. The order of the local error of CM3 is one more than expected, and the error constants are about 1/10 of those for HO3C. However, the higher order only occurs in the linear case, for a nonlinear problem the order reduces to 4.

Rapid decay

In this case we assume $\text{Re}(z) << 0$, such that all transients represented by exponential functions are completely damped. In this case it makes sense to write the error functions as inverse power series of z. The dominant terms of $\mathcal{E}_q$ are given by

q	IMEX3	IMEX3 (s)	CM3	HO3C
0	$-\frac{8}{3z} + \mathcal{O}(\frac{1}{z^2})$	0	0	0
1	$-\frac{8}{3z^2} + \mathcal{O}(\frac{1}{z^3})$	0	0	0
2	$\frac{7}{18z} + \mathcal{O}(\frac{1}{z^2})$	$\frac{7}{18z} + \mathcal{O}(\frac{1}{z^2})$	0	0
3	$-\frac{1}{8z} + \mathcal{O}(\frac{1}{z^2})$	$-\frac{1}{8z} + \mathcal{O}(\frac{1}{z^2})$	0	$-\frac{1}{6z} + \mathcal{O}(\frac{1}{z^2})$
4	$-\frac{127}{1296z} + \mathcal{O}(\frac{1}{z^2})$	$-\frac{127}{1296z} + \mathcal{O}(\frac{1}{z^2})$	$-\frac{1}{12z^2} + \mathcal{O}(\frac{1}{z^3})$	$-\frac{5}{54z} + \mathcal{O}(\frac{1}{z^2})$

The local truncation error behaves as

$$y(t_0 + h) - y_1 = \begin{cases} \left(-\frac{8}{3\lambda}y_0 - \frac{8}{3\lambda^2}f\right)\frac{1}{h} + \mathcal{O}\left(\frac{1}{\lambda^2 h^2} + \frac{h}{\lambda}\right) & \text{for IMEX3} \\ \frac{7}{8\lambda}hf' + \mathcal{O}\left(\frac{1}{\lambda^2} + \frac{h^2}{\lambda}\right) & \text{for IMEX3 (s)} \\ -\frac{1}{12\lambda^2}h^2 f''' + \mathcal{O}\left(\frac{h}{\lambda^3} + \frac{h^3}{\lambda^2}\right) & \text{for CM3} \\ -\frac{1}{6\lambda}h^2 f'' + \mathcal{O}\left(\frac{h^3}{\lambda} + \frac{h}{\lambda^2}\right) & \text{for HO3C} \end{cases}$$

In the transient case the error of IMEX3 goes as $\sim 1/h$. The error increases as the stepsize decreases! This phenomenon is known from the literature as "the hump". The situation is improved in the slow case, but still the IMEX method has a local error of one order less than the two exponential RK-methods. For very stiff problem, the $1/\lambda^2$ behaviour of CM3 is an advantage.

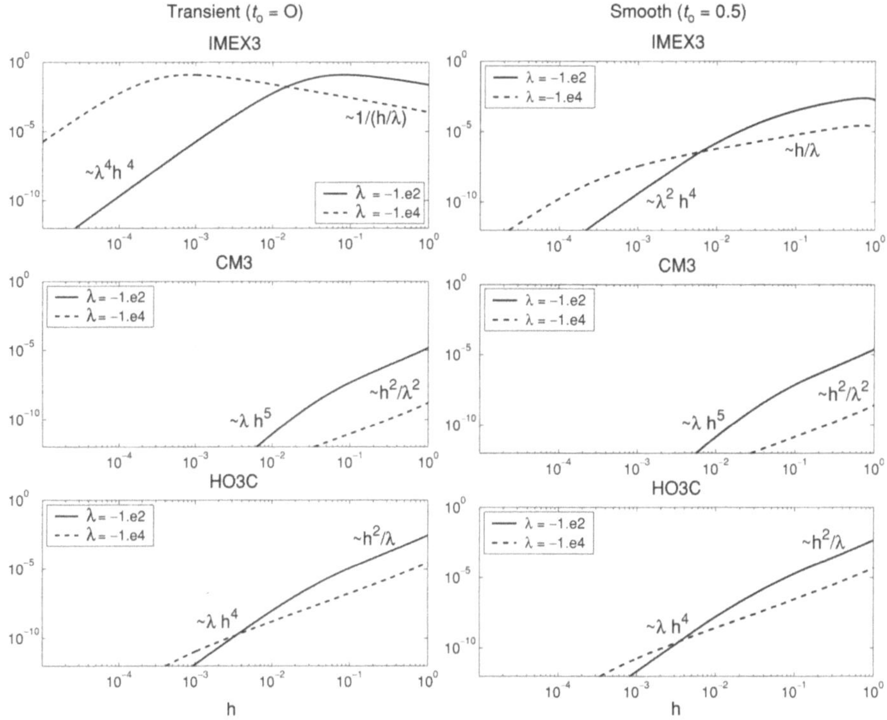

Fig. 2. Local error in the rapid decay case

Rapid oscillations

In this situation, we assume $|z|$ large and λ purely imaginary. The IMEX3 method is not constructed for solving oscillatory problems, so its behaviour is not considered here. For the two exponential RK-methods, the exponentials will represent rapid oscillations in the error functions which are dominated by the terms:

q	CM3	HO3C
3	0	$-\frac{1}{6z}+\mathcal{O}(\frac{1}{z^2})$.
4	$\frac{e^z-1}{12z^2}+\mathcal{O}(\frac{1}{z^3})$	$-\frac{5}{54z}+\mathcal{O}(\frac{1}{z^2})$

The absolute value of the local truncation error is

$$|y(t_0+h)-y_1| = \begin{cases} \frac{1}{12|\lambda|^2} M h^2 f'''(t_0) + \mathcal{O}(\frac{h^3}{|\lambda|^2}+\frac{h}{|\lambda|^3}),\ M\in[0,2] & \text{for CM3} \\ \frac{1}{6|\lambda|} h^2 f''(t_0) + \mathcal{O}(\frac{h^3}{|\lambda|}+\frac{h}{|\lambda|^2}) & \text{for HO3C} \end{cases}.$$

Both methods have local errors of order 2. But again, the $1/\lambda^2$ term for the CM3 method results in very small errors for stiff systems.

The theoretical results can be verified by the following example:

Example 1. Consider the equation

$$y' = \lambda y + e^t, \qquad y(0) = y_0,$$

with exact solution

$$y(t) = e^{\lambda t} y_0 + \frac{e^{\lambda t} - e^t}{\lambda - 1}.$$

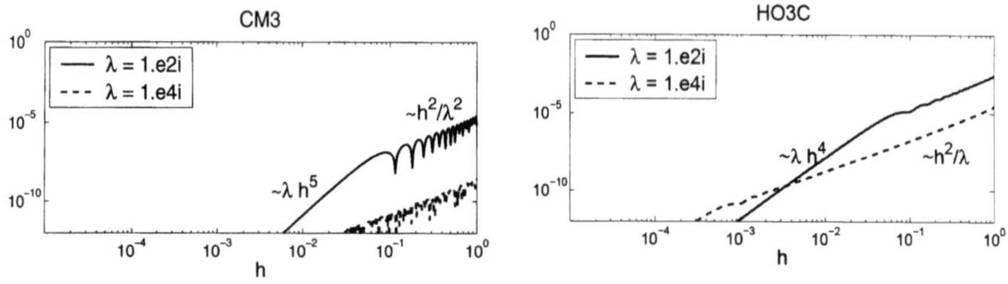

Fig. 3. Local error in the rapid oscillation case

Figure 2 shows the local error in the rapid decay case, both in the transient and the slow case. Figure 3 shows the local error in the highly oscillatory case. Both verify the theoretical results.

A analysis of the error behaviour of exponential RK methods applied to the nonlinear equation

$$y' = \lambda y + f(t, y), \qquad y(t_0) = y_0$$

can be found in [11], using B-series and rooted trees. These results can be extended to the IMEX-RK methods. As expected from the example in the introduction and Fig. 1 the error depends on the initial value y_0 also for the exponential RK methods, and we get a completely different error behaviour depending on whether the initial value is on the smooth manifold or not. The results for the present methods are shortly summarised in the following, where we have used the notation METHOD(s) for the case of initial value on the smooth manifold:

For the nonstiff case get

$$y(t_0 + h) - y_1 = \begin{cases} \mathcal{O}(\lambda^4 h^4) \;/\; \mathcal{O}(\lambda^2 h^4) & \text{for IMEX / IMEX(s),} \\ \mathcal{O}(h^4 + \lambda^4 h^5) \;/\; \mathcal{O}(h^4 + \lambda^2 h^5) & \text{for CM3 / CM3(s),} \\ \mathcal{O}(\lambda^3 h^4) \;/\; \mathcal{O}(\lambda h^4) & \text{for HO3C / HO3C(s),} \end{cases}$$

and in the rapid decay case the results become

$$y(t_0 + h) - y_1 = \begin{cases} \mathcal{O}(\frac{1}{\lambda h}) \;/\; \mathcal{O}(\frac{h}{\lambda}) & \text{for IMEX / IMEX(s),} \\ \mathcal{O}(\frac{1}{\lambda^2 h}) \;/\; \mathcal{O}(\frac{1}{\lambda^3} + \frac{h^2}{\lambda^2}) & \text{for CM3 / CM3(s),} \\ \mathcal{O}(\frac{1}{\lambda}) \;/\; \mathcal{O}(\frac{h^2}{\lambda}) & \text{for HO3C / HO3C(s).} \end{cases}$$

The investigation of a single Fourier mode, linear or nonlinear, will certainly not necessarily give a representative solution of more complex equations. But it is a quite straightforward tool to reveal certain characteristic properties of a method. Different approaches to error analysis can be found in [6, 2].

4 Remarks

The main practical question will probably be how the exponential RK methods compare when applied to space discretized partial differential equations, like (2). Several papers have appeared recently, comparing different exponential integrators applied to certain test problems. But to my knowledge, no direct efficiency comparisons between exponential RK methods and IMEX-RK methods has so far been published. However, Kassam [8] has in his thesis performed numerous experiments comparing different methods for time-stepping of partial differential equations, among them an exponential RK method and a multistep IMEX method, both of order 4. Kassam observe that among these methods, the exponential RK methods is far the most stable and is able to solve problems for which the IMEX method fails. What efficiency concerns, the picture is less clear. For problems in 1D or in the case for which L is diagonal, generally the result of a Fourier spectral discretization, the exponential RK method seems to be superior both in accuracy and efficiency. When L is a full matrix, this is no longer necessarily the case. Most exponential integrators are implemented in a fixed stepsize regime, this allows for an explicit implementation once the appropriate preprocessing has been done. Kassam reports that the startup time for the exponential integrators can be up to 10 times larger than that of the multistep IMEX method. It is then an open question whether exponential integrators will be efficient when variable stepsizes are allowed when L is a full matrix. However, the efficiency might be improved by Krylov approximations for the exponential functions as proposed by [4].

As a conclusion, carefully constructed exponential RK methods can be a promising candidate for time-stepping of some PDEs, due to their improved error behaviour and stability properties. But to be efficient on a broad range of problems, there are still unsolved implementation issues.

References

1. U.M. Ascher, S.J. Ruuth, and R.J. Spiteri. Implicit-explicit Runge-Kutta methods for time-dependent partial differential equations. *App. Numer. Math.*, 25:151–167, 1997
2. H. Berland, B. Owren, and B. Skaflestad. B-series and order conditions for exponential integrators. Technical report, NTNU, 2004
3. S.M. Cox and P.C. Matthews. Exponential time differencing for stiff systems. *J. of Comp. Phys.*, 176:430–455, 2002
4. M. Hochbruck, C. Lubich, and H. Selhofer. Exponential integrators for large systems of differential equations. *SIAM J. Sci. Comput.*, 19(5):1552–1574, 1998
5. M. Hochbruck and A. Ostermann. Exponential integrators for semilinear parabolic problems. Numdiff 10, Halle, September 2003
6. M. Hochbruck and A. Ostermann. Explicit exponential Runge-Kutta for semilinear parabolic equations. 2004
7. W. Hundsdorfer and J. G. Verwer. *Numerical solution of time-dependent advection-diffusion-reaction equations*. Springer series in computaional mathematics. Springer, Berlin, 2003
8. A.-K. Kassam. *High order timestepping for stiff semilinear partial differential equations*. PhD thesis, University of Oxford, 2004
9. A.-K. Kassam and L.N. Trefethen. Fourth-order time-stepping for stiff pdes. *SIAM J. Sci. Comput.*, 26(4):1214–1233, 2005
10. C.A. Kennedy and M.H. Carpenter. Additive Runge-Kutta schemes for convection-diffusion-reaction equations. *App. Numer. Math.*, 44:139–181, 2003
11. A. Kværnø. Error behaviour of exponential Runge-Kutta methods. In preparation
12. B.V. Minchev and W.M. Wright. A review of exponential integrators for semilinear problems. Technical report, NTNU, 2005
13. C. Moler and C. van Loan. Nineteen dubious ways to compute the exponential of a matrix, twenty-five years later. *SIAM Review*, 45(1):3–49, 2003

Two-Band Quantum Models for Semiconductors Arising from the Bloch Envelope Theory

G. Alì[1], G. Frosali[2], and O. Morandi[3]

[1] Istituto per le Applicazioni del Calcolo "M. Picone", CNR - Via P. Castellino 111, I-80131 Napoli, Italy, and INFN-Gruppo c. Cosenza, g.ali@iac.cnr.it
[2] Dipartimento di Matematica Applicata "G.Sansone", Università di Firenze - Via S.Marta 3, I-50139 Firenze, Italy, giovanni.frosali@unifi.it
[3] Dipartimento di Elettronica e Telecomunicazioni, Università di Firenze - Via S.Marta 3, I-50139 Firenze, Italy, omar.morandi@unifi.it

1 Introduction

In the recent past, the interest in quantum hydrodynamic models for semiconductors has increased considerably. In fact, classical fluid dynamical models fail to be adequate for new generations of semiconductor devices, where quantum effects tend to become not negligible or even dominant (see [6] and the references therein). This paper is particularly devoted to multi-band quantum models, introduced to describe the Resonant Interband Tunneling Diode (RITD) [5, 11]. In section 2 we review briefly two-band quantum models for semiconductors arising from the Bloch envelope theory [7, 10]. In section 3 we present a new Madelung-like hydrodynamical formulation for the previous models, based on a suitable definition of osmotic and current velocities. This method has been applied in [1] to the Kane model. We conclude this paper with a thorough physical discussion of the models, with some numerical experiments showing the different description of the interband resonant tunneling of the previous models.

2 The envelope function models

In quantum mechanics the motion of an electron is described by a quantum Hamiltonian operator, which governs the evolution of a wave function Ψ, whose modulus $n(x,t) := |\Psi(x,t)|^2$ represents the probability density of finding the electron at the position x and time t. Since we are interested in modeling multi-band quantum effects, it is necessary to introduce quantum densities for each band, with a possibly clear physical meaning. Then, the tunneling process will be described by an operator which is non-diagonal with respect to the band index. In view of this, the effective mass formalism, and in particular the $\mathbf{k}\cdot\mathbf{p}$ envelope function method, seems to be the natural framework for multi-band analysis [12].

The basic idea behind $\mathbf{k}\cdot\mathbf{p}$ theory is that we do not need to calculate the evolution of the full wave function to obtain the trajectory of an electron through the crystal but it is sufficient to calculate the evolution of the so-called envelope function, a smooth function which is obtained by replacing Ψ by its average in each primitive cell. So, the microscopic structure of the full wave function is not relevant.

The application of the $\mathbf{k}\cdot\mathbf{p}$ theory gives rise to models which differ both by the choice of the set of basis functions, and by the approximation procedure. Generally speaking, $\mathbf{k}\cdot\mathbf{p}$ models arise from perturbative methods where the crystal momentum $\mathbf{k}$ is considered a "small" quantity. Typically, this analysis applies when $k=0$ is a stationary point (the Γ point) for the dispersion relation of the band, since the momentum of the electron is localized around this point. Different sets of basis functions are generated by using suitable projection operators on the Bloch basis in such a way that the new basis elements "approach" the original Bloch waves as $|k|$ tends to zero. The most common choice of $\mathbf{k}\cdot\mathbf{p}$ basis was proposed by Kane (in its pioneering paper of 1956 [7]) with the aim of approximating the periodic part of each Bloch function by its value at the Γ point. In spite of its simplicity, this choice fails to give an adequate physical interpretation of tunneling phenomena. To overcome this difficulty, other choices have been proposed in literature [9]. In particular, in this paper, we refer to the Multi-band Envelope Function (MEF) model [10].

Since every $\mathbf{k}\cdot\mathbf{p}$ model involves a full coupling among all unperturbed bands, to retain only those terms which are well localized into the bands of interest, a cutoff is employed in the expansion of the solution Ψ. In semiconductor devices, the current is mainly generated by transport of electrons in conduction band and in light hole valence band, thus it is customary to approximate the wave function Ψ solely by its conduction and valence components, denoted here ψ_c and ψ_v, respectively.

Irrespectively of the choice of the basis, the conduction and valence components are determined by solving a Schrödinger-like equation of the form

$$i\hbar\frac{d}{dt}\begin{pmatrix}\psi_c\\ \psi_v\end{pmatrix} = H\begin{pmatrix}\psi_c\\ \psi_v\end{pmatrix}, \tag{1}$$

where H is an approximation of a full-band Hamiltonian. Its diagonal components correspond to uncoupled bands, and the off-diagonal terms account for interband effects. This type of approximation can be performed in different ways [4], and the method of approximating the Bloch basis affects the specific form of H deeply, not only from a formal point of view, but also from a physical one.

Using the classical Kane basis, Ψ can be approximated by

$$\Psi(x,t) \simeq \psi_c^K(x,t)u_c^K(x) + \psi_v^K(x,t)u_v^K(x). \tag{2}$$

Here, u_a^K is the periodic part of the Bloch function $b_a(k,x)$, $a = c, v$, evaluated at $k = 0$. Instead, the MEF model uses an expansion in the Wannier basis, approximating the conduction and valence components up to the first order in $|k|$ [10]. Then, Ψ can be approximated by

$$\Psi(x,t) \simeq \psi_c^M(x,t)u_c^M(x) + \psi_v^M(x,t)u_v^M(x). \tag{3}$$

It is well known that each Wannier basis element arises from applying the Fourier transform to the Bloch functions related to the same band index n. The envelope functions ψ_c^M and ψ_v^M are the projections of Ψ on the Wannier basis, and therefore the corresponding multi-band densities represent the (cell-averaged) probability amplitude of finding an electron on the conduction or valence bands, respectively.

This simple picture does not apply to the Kane model. In fact, in the Kane approach, the periodic part $u_n(k,x)$ of each element of the Bloch basis, $b_n(k,x) = e^{ikx}u_n(k,x)$, is projected on the same basis but calculated for $k = 0$. Thus, the generic element of the Kane basis, defined by $b_n^K(k,x) = e^{ikx}u_n^K(0,x)$, is no more linked to the Bloch basis by a diagonal transformation. This fact can be simply verified by introducing a unitary operator $\Theta_{n,n'}$ such that $u_n(k,x) = \sum_{n'}\Theta_{n,n'}(k)u_{n'}^K(0,x)$. Then, we have

$$b_n(k,x) = e^{ikx}u_n(k,x) = \sum_{n'} e^{ikx}\Theta_{n,n'}(k)u_{n'}^k(0,x) = \sum_{n'}\Theta_{n,n'}(k)b_{n'}^K(k,x).$$

$\Theta_{n,n'}$ written for two bands and approximated up to the first order in $|k|$ is

$$\Theta_{n,n'}(k) = \delta_{n,n'} + \frac{\hbar^2}{m_0}\sum_{n\neq n'}\frac{P_{n,n'}}{E_n - E_{n'}}k. \tag{4}$$

At the envelope function level, (4) implies that Kane envelope functions and MEF envelope functions are connected by the relation $\psi_a^K = \sum_b \Theta_{b,a}\psi_b^M$. Using transformation (4) at the first order in $|k|$, we can write explicitly [10]

$$\psi_a^K = \psi_a^M + i\frac{\hbar^2}{m_0}\sum_{b\neq a}\frac{P_{a,b}}{E_a - E_b}\nabla\psi_b^M. \tag{5}$$

Going back to (1), for the Kane model the Hamiltonian takes the form

$$H = H^K := \begin{pmatrix} -\dfrac{\hbar^2}{2m_0}\Delta + E_c + V & -\dfrac{\hbar^2}{m_0}P\cdot\nabla \\ \dfrac{\hbar^2}{m_0}P\cdot\nabla & -\dfrac{\hbar^2}{2m_0}\Delta + E_v + V \end{pmatrix}, \tag{6}$$

where E_c is the minimum of the conduction band energy, E_v is the maximum of the valence band energy, m_0 is the bare electron mass and $P := P_{c,v}$ is the coupling coefficient between the two bands given by the matrix element of the gradient operator between u_c^K and u_v^K [7, 3].

For the MEF model in a semiconductor with isotropic effective mass tensor, the Hamiltonian is

$$H = H^M := \begin{pmatrix} -\dfrac{\hbar^2}{2m_c^*}\Delta + E_c + V & -\dfrac{\hbar^2 P\cdot\nabla V}{m_0 E_g} \\ -\dfrac{\hbar^2 P\cdot\nabla V}{m_0 E_g} & \dfrac{\hbar^2}{2m_v^*}\Delta + E_v + V \end{pmatrix}, \tag{7}$$

where $E_g = E_c - E_v$ and m_c^*, m_v^* are the effective masses for the conduction and valence bands, respectively. In the following we will assume $m_c^* = m_v^* = m^*$ and, for simplicity, we will focus on one-dimensional systems.

3 Hydrodynamic models

In this section, following the technical procedure proposed in [1], we compare the hydrodynamic formulations of the Kane and MEF models.

In general, we expect a straightforward extension of the hydrodynamical formalism for a single-band semiconductor to multi-band framework, provided that each component of the wave function behaves like the electron wave function in a single-band whenever no interband effects are present.

In this work we apply the WKB method, which is a standard way to write the Schrödinger equation in hydrodynamic form [6]. Extending this approach to two-band models, we look for solutions to the system (1), written with $H = H^A$, $A = K, M$ (see (6) and (7)), of the form

$$\psi_a^A(x,t) := \sqrt{n_a^A(x,t)} \exp\left(\frac{im^A}{\hbar} S_a^A(x,t)\right), \quad a = c, v, \tag{8}$$

with $m^K = m_0$, $m^M = m^*$. In the following, we will not specify the attributions of the indices a and A. The squared amplitude n_a^A can be immediately regarded as a probability density of the electron in the band a, and the gradient of the phase corresponds to the velocity of the electron in the same band. We remark that both $n_c^M + n_v^M = |\psi_c^M|^2 + |\psi_v^M|^2$ and $n_c^K + n_v^K = |\psi_c^K|^2 + |\psi_v^K|^2$ can be interpreted as approximations of the true total density number of electrons, which in principle are different, due to the different type of expansion used in the Kane and MEF approaches. Using (8), we can transform system (1), written with $H = H^A$, for the variables ψ_a^A, to a formally equivalent system for the variables n_a^A, S_a^A. To derive a hydrodynamical formulation of (1), we introduce the complex velocities $u_a^A := \frac{\hbar}{m^A}\nabla \log \psi_a^A$, and write

$$u_a^A = u_{\mathrm{os},a}^A + i u_{\mathrm{el},a}^A := \frac{\hbar}{m^A}\frac{\nabla\sqrt{n_a^A}}{\sqrt{n_a^A}} + i\nabla S_a^A. \tag{9}$$

The real and imaginary parts of u_a^A are named osmotic velocities and current velocities, respectively. Also, we introduce the electron current densities $J_a^A := \frac{\hbar}{m^A}\,\mathrm{Im}\,(\overline{\psi_a^A}\nabla\psi_a^A) = n_a^A u_{\mathrm{el},a}^A$ and the interband particle densities $n_{cv}^A = \overline{n_{vc}^A} = \overline{\psi_c^A}\psi_v^A = \sqrt{n_c^A}\sqrt{n_v^A}\,e^{i\sigma^A}$, with $\sigma^A := \frac{m^A}{\hbar}(S_v^A - S_c^A)$.

Using the above definitions in (1), we can derive equations for the particle densities n_c^A, n_v^A, and the currents J_c^A, J_v^A,

$$\frac{\partial n_c^A}{\partial t} + \nabla\cdot J_c^A = S_{cv}^A, \qquad \frac{\partial n_v^A}{\partial t} + \alpha\nabla\cdot J_v^A = S_{vc}^A, \tag{10}$$

$$\frac{\partial J_c^A}{\partial t} + \mathrm{div}\left(\frac{J_c^A\otimes J_c^A}{n_c^A}\right) + \frac{n_c^A}{m^A}\nabla\left(E_c + V + V_c^A + V_{cv}^A\right) = S_{cv}^A\frac{J_c^A}{n_c^A}, \tag{11}$$

$$\frac{\partial J_v^A}{\partial t} + \alpha\,\mathrm{div}\left(\frac{J_v^A\otimes J_v^A}{n_v^A}\right) + \frac{n_v^A}{m^A}\nabla\left(E_v + V + \alpha V_v^A + V_{vc}^A\right) - S_{vc}^A\frac{J_v^A}{n_v^A}, \tag{12}$$

with $\alpha = 1$ for the Kane model and -1 for the MEF model. Here, $V_a^A = -\frac{\hbar^2\Delta\sqrt{n_a^A}}{2m^A\sqrt{n_a^A}}$ are the Bohm potentials for each band, the interband potentials V_{ab}^A are given by

$$V_{cv}^K = -\hbar\,\mathrm{Re}\left(\frac{n_{cv}^K P\cdot u_v^K}{n_c^K}\right), \qquad V_{vc}^K = \hbar\,\mathrm{Re}\left(\frac{n_{vc}^K P\cdot u_c^K}{n_v^K}\right),$$

$$V_{cv}^M = V_{vc}^M = -\frac{\hbar^2 P\cdot\nabla V}{m_0 E_g n_c^M}\,\mathrm{Re}\,n_{cv}^M,$$

and we have introduced

$$S_{cv}^K = -2\,\mathrm{Im}\left(n_{cv}^K P\cdot u_v^K\right), \qquad S_{vc}^K = 2\,\mathrm{Im}\left(n_{vc}^K P\cdot u_c^K\right),$$

$$S_{cv}^M = -S_{vc}^M = -\frac{2\hbar P\cdot\nabla V}{m_0 E_g}\,\mathrm{Im}\,n_{cv}^M.$$

In order to close the system and obtain an extension of the classical Madelung fluid equations, we need to add an equation for σ^A,

$$\frac{\hbar}{m^A}\nabla\sigma^A = \frac{J_v^A}{n_v^A} - \frac{J_c^A}{n_c^A}. \tag{13}$$

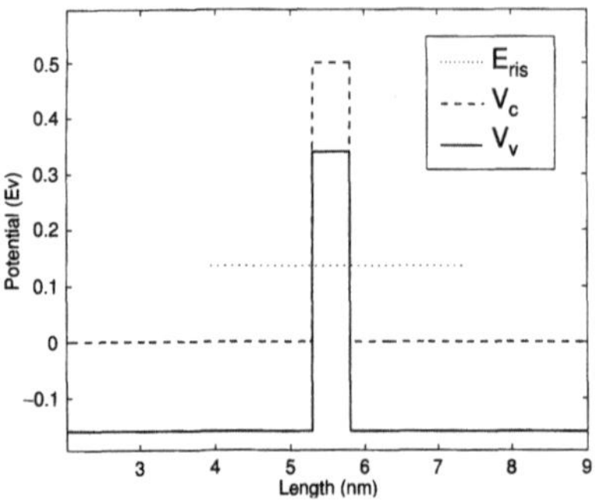

Fig. 1. Band diagram of the simulated heterostructure. The dotted line denotes the energy of the resonant state in the valence quantum well

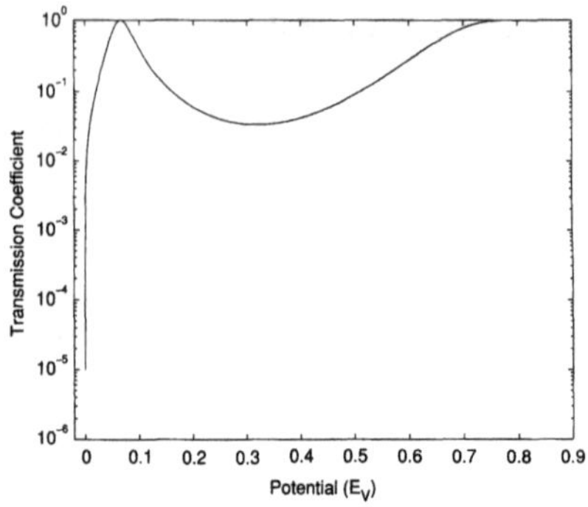

Fig. 2. Plot of the transmission coefficient of the heterostructure as a function of the E_{inc}

Summing the equations for the densities, we obtain the balance law for the total density (continuity equation). We see that for the MEF model, the total current is the sum of the currents for valence and conduction band. In the Kane model, the continuity equation reads

$$\frac{\partial}{\partial t}(n_c^K + n_v^K) + \nabla\cdot(J_c^K + J_v^K + 2\frac{\hbar}{m_0} P \,\mathrm{Im} n_{cv}^K) = 0.$$

The appearance of an additional interband term for the current is an indication of the inadequacy of the Kane-based hydrodynamical model. This topic will be discussed in details in the following section.

4 Numerical results

In this section we show some numerical results arising from the two proposed approaches. Our aim is to show that a more direct physical meaning can be ascribed to the hydrodynamical variables derived from the MEF approach.

We consider a heterostructure which consists of two homogeneous regions separated by a potential barrier and which realizes a single quantum well in valence band. In Fig. 1 we have marked the energy of resonant state $E_{ris} = 0.066\ eV$, which is given by the solution of an eigenvalue problem for the Hamiltonian operator. In our simulation, we have used the following parameters: $E_g = E_c - E_v = 0.16\ eV$, $m^* = 0.023\ m_0$, $P = 5 \cdot 10^9 m^{-1}$.

A conduction electron beam (i.e. a plane wave envelope function with positive momentum k and energy $\hbar^2 k^2/2m^* + E_c$) is injected in the heterostructure from the left. Then, the analytical solution for eq. (1) in the regions $x < 0$ and $x > L$ is explicitly given by $\psi = e^{-iE_{inc}t/\hbar}\psi^A$ were

$$\psi^A = \begin{cases} \mathbf{e}_c^A \left\{e^{ikx} + r_c\, e^{-ikx}\right\} + \mathbf{e}_v^A\, r_v\, e^{ik_{rv}x}, & x \in (-\infty, 0] \\ \mathbf{e}_c^A\, t_c\, e^{ikx} + \mathbf{e}_v^A\, t_v\, e^{-ik_{rv}x}, & x \in [L, \infty) \end{cases} \qquad A = M, K$$

where $\psi^A = \begin{pmatrix} \psi_c^A \\ \psi_v^A \end{pmatrix}$, $k_{rv} = -\frac{i}{\hbar}\sqrt{2m^*(E_{inc} - E_v)}$, and $\mathbf{e}_c^A$, $\mathbf{e}_v^A$ are unitary vectors defined as follows: $\mathbf{e}_c^M = \begin{pmatrix} 1 \\ 0 \end{pmatrix}$, $\mathbf{e}_v^M = \begin{pmatrix} 0 \\ 1 \end{pmatrix}$ for the MEF model, and $\mathbf{e}_c^K = \begin{pmatrix} \sqrt{\frac{\sqrt{\eta}+E_g}{2\sqrt{\eta}}} \\ i\sqrt{\frac{\sqrt{\eta}-E_g}{2\sqrt{\eta}}} \end{pmatrix}$, $\mathbf{e}_v^K = \begin{pmatrix} \sqrt{\frac{\sqrt{\eta}-E_g}{2\sqrt{\eta}}} \\ -i\sqrt{\frac{\sqrt{\eta}+E_g}{2\sqrt{\eta}}} \end{pmatrix}$, with $\eta = E_g^2 + 4\frac{\hbar^2 k^2 P^2}{m_0^2}$, for the Kane model. Furthermore, $r_c(t_c)$ and $r_v(t_v)$ are the reflection (transmission) coefficients in the conduction and valence bands, respectively. They depend on the detailed shape of the heterostructure, and are numerically evaluated by a Runge-Kutta scheme which solves directly the eigenvalue problem related to eq. (1), obtained, as usual, by formally replacing $i\hbar\frac{d}{dt}$ with E [8] .

We calculate the envelope function solution in the region $0 < x < L$ for incremental values of the electron beam energy. The results are plotted in Fig. 3-8 (left-hand side for the MEF model, and right-hand side for the Kane model).

When the electron energy is well below of the resonant energy E_{ris} (Fig. 3-4) the incident conduction plane wave is reflected: r_c approaches 1 and, consequently, the transmission coefficient t_c tends to 0. In this case the valence states are almost unexcited and a small amount of charge cumulates in the valence quantum well. Instead, when the electron energy approaches the resonant level, the transmission coefficient enhances and the electron can travel from the left to the right of the heterostructure by using the bounded valence resonant state as a "bridge" state. Identifying ψ_c^M and ψ_v^M with the components of the electron wave function in conduction and valence band, it is immediate to verify how their behaviour reflects the previous considerations.

Further, since in the MEF model the coupling coefficient of conduction and valence band is proportional to ∇V, interband current flow arises only in proximity of the interfaces, when both ψ_c^M and ψ_v^M are not vanishing.

On the other hand, even in absence of an external potential, when no interband transition can occur, the Kane model exhibits a coupling of all the envelope functions. Then, the naive interpretation of the envelope functions which we have ascertained for the MEF model, cannot be directly extended to the Kane model.

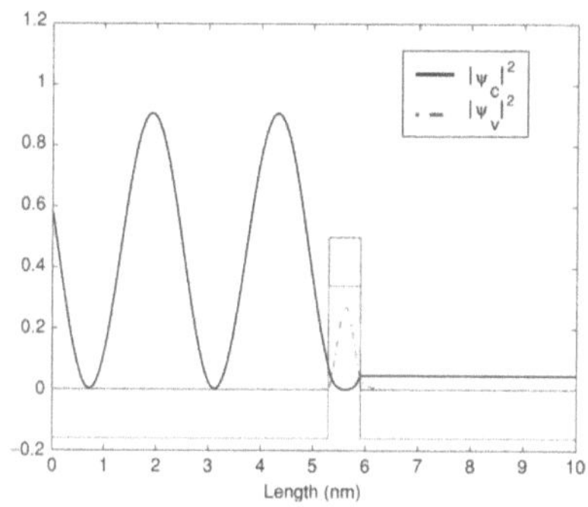

Fig. 3. MEF model: $E_{inc} = 0.028\ eV$

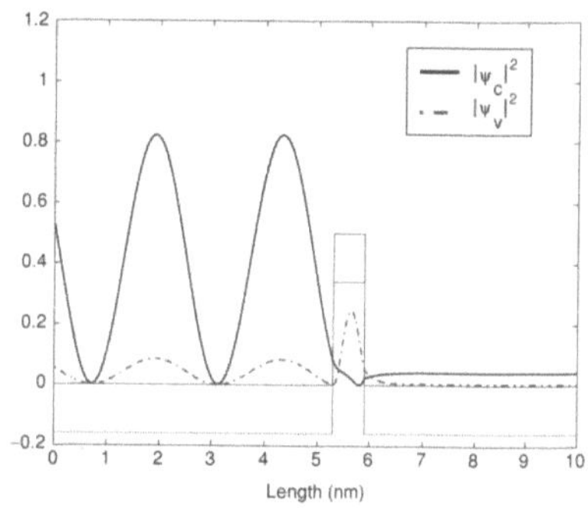

Fig. 4. Kane model: $E_{inc} = 0.028\ eV$

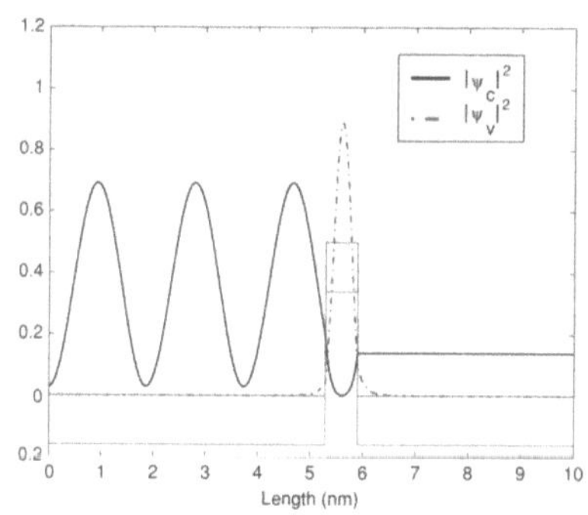

Fig. 5. MEF model: $E_{inc} - 0.046\ eV$

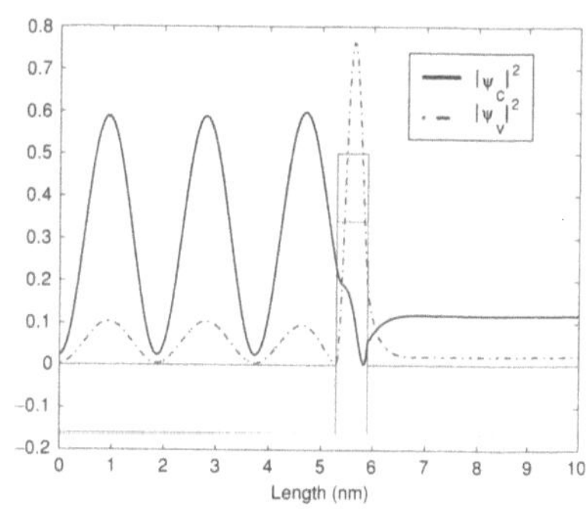

Fig. 6. Kane model: $E_{inc} = 0.046cV$

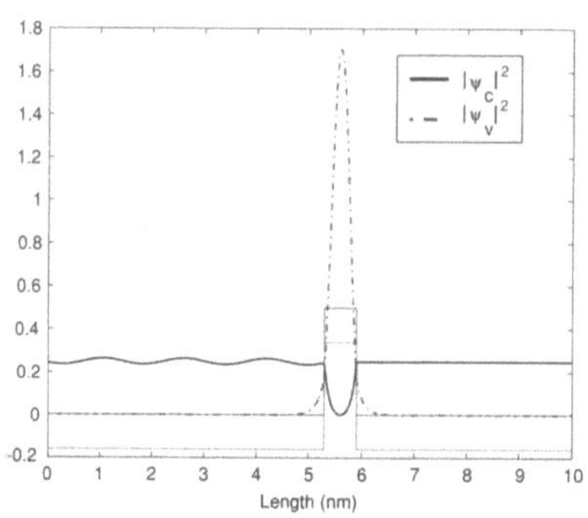

Fig. 7. MEF model: $E_{inc} = 0.066\ eV$

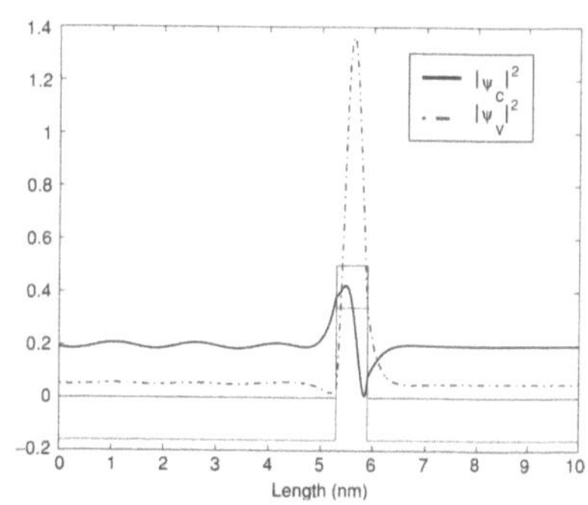

Fig. 8. Kane model: $E_{inc} = 0.066\ eV$

References

1. G. Alì, G. Frosali, On quantum hydrodynamic models for the two-band Kane system, 2004 (preprint)
2. G. Alì, G. Frosali and C. Manzini, On the drift-diffusion model for a two-band quantum fluid at zero-temperature, *Ukrainian Math. J.* **57(6)**, 723-730 (2005)
3. G. Borgioli, G. Frosali, P.F. Zweifel, Wigner approach to the two-band Kane model for a tunneling diode, *Transport Theory Stat. Phys.* **32(3&4)**, 347-366 (2003)
4. G. Borgioli, O. Morandi, G. Frosali, and M. Modugno, Different approaches for multi-band transport in semiconductors, *Ukrainian Math. J.* **57(6)**, 742-748 (2005)
5. L. Demeio, L. Barletti, A. Bertoni, P. Bordone and C. Jacoboni, Wigner function approach to multi-band transport in semiconductors, *Physica B,* **314** 104-107 (2002)
6. A. Jüngel, *Quasi-hydrodynamic Semiconductor Equations*, Birkhäuser, Basel, 2001
7. E. O. Kane, Energy band structure in p-type Germanium and Silicon, *J. Phys. Chem. Solids* **1**, 82–89 (1956)
8. J. Kefi, *Analyse mathématique et numérique de modèles quantiques pour les semiconducteurs* PhD Thesis, Toulouse University (2003)
9. J. M. Luttinger and W. Kohn, Motion of electrons and holes in perturbed periodic fields, *Phys. Rev.* **97**, 869-883 (1955)
10. M. Modugno, O. Morandi, A multi-band envelope function model for quantum transport in a tunneling diode, *Phys. Rev. B* (2005) (to appear)
11. R.Q. Yang, M. Sweeny, D. Day and J.M. Xu, Interband tunneling in heterostructure tunnel diodes, *IEEE Transactions on Electron Devices,* **38**(3) 442-446 (1991)
12. W.T. Wenckebach, *Essential of Semiconductor Physics*, J.Wiley & Sons, Chichester, 1999

Mixed Finite Element Numerical Simulation of a 2D Silicon MOSFET with the Non-Parabolic MEP Energy-Transport Model

A. M. Anile[1], A. Marrocco[2], V. Romano[1], and J. M. Sellier[1]

[1] Dipartimento di Matematica e Informatica, Università di Catania, Viale A.Doria 6, I-95125 Catania, Italy, {anile,romano,sellier}@dmi.unict.it
[2] INRIA, Domaine de Voluceau, Rocquencourt BP 105, 78153, Le Chesnay, France, americo.marrocco@inria.fr

Abstract The Mixed Finite Element scheme presented in [Raviart et al. (1997), Marrocco et al. (1996)] is used to simulate a 2D silicon MOSFET with a consistent energy-transport model for electron in semiconductors, free of any fitting parameters, formulated on the basis of the maximum entropy principle (MEP) in [Anile et al. (1999), Romano (2000), Romano (2001), Anile et al. (2003)]. Comparison with MC data shows the superiority of the model with respect to the standard models known in literature.

Key words: semiconductors, energy-transport model, mixed finite elements, MOSFET

1 The MEP energy-transport model in the Kane dipersion relation case

In this section we give a cursory presentation of the Energy-Transport model based on MEP. For more details the interested reader is referred to [Anile et al. (1999), Romano (2000), Romano (2001)].

One assumes that the conduction band is described around each minimum (valley) by the Kane dispersion relation approximation

$$\mathcal{E}(\mathbf{k})[1+\alpha\mathcal{E}(\mathbf{k})] = \frac{\hbar^2 k^2}{2m^*}, \qquad \mathbf{k} \in \mathbb{R}^3 \tag{1}$$

where $\mathcal{E}$ is the electron energy, m^* is the effective electron mass (which is 0.32 m_e in Silicon, with m_e the electron mass in the vacuum), $\hbar\mathbf{k}$ is the crystal momentum, $\hbar$ is the Planck constant divided by 2π and α is the non-parabolicity factor (α=0.5 eV^{-1} for Silicon).

The energy-transport model, obtained for silicon semiconductor in [Romano (2001)] starting from the hydrodynamical model based on the maximum entropy principle [Anile et al. (1999), Romano (2000)], is given by the following set of balance equations for the electron density n and energy W, coupled to the Poisson equation for the electric potential ϕ

$$\frac{\partial n}{\partial t} + \mathrm{div}(n\mathbf{V}) = 0, \quad \frac{\partial (nW)}{\partial t} + \mathrm{div}(n\mathbf{S}) - ne\mathbf{V}\cdot\nabla\phi = nC_W, \tag{2}$$

$$\epsilon\Delta\phi = -e(N_D - N_A - n). \tag{3}$$

where N_D and N_A are the donor and acceptor densities respectively, e is the elementary charge, ϵ is the dielectric constant while div, ∇ and Δ are the divergence, gradient and laplacian operators. The generation-recombination terms are neglected because they are of the order of 10^{-9} sec while we will consider devices with characteristic time of about few picoseconds. The evolution equations are closed with the constitutive relations for the velocity $\mathbf{V}$ and the energy-flux $\mathbf{S}$

$$\mathbf{V} = D_{11}(W)\nabla\log n + D_{12}(W)\nabla W + D_{13}(W)\nabla\phi, \tag{4}$$

$$\mathbf{S} = D_{21}(W)\nabla\log n + D_{22}(W)\nabla W + D_{23}(W)\nabla\phi. \tag{5}$$

The elements of the diffusion matrix $D = (D_{ij})$ read

$$D_{11} = \frac{c_{22}U - c_{12}F}{c_{11}c_{22} - c_{12}c_{21}}, \quad D_{12} = \frac{c_{22}U' - c_{12}F'}{c_{11}c_{22} - c_{12}c_{21}}, \quad D_{13} = -e\frac{c_{22} - c_{12}G}{c_{11}c_{22} - c_{12}c_{21}},$$
$$D_{21} = \frac{c_{11}F - c_{21}U}{c_{11}c_{22} - c_{12}c_{21}}, \quad D_{22} = \frac{c_{11}F' - c_{21}U'}{c_{11}c_{22} - c_{12}c_{21}}, \quad D_{23} = e\frac{c_{21} - c_{11}G^{(0)}}{c_{11}c_{22} - c_{12}c_{21}}.$$

All the coefficients c_{ij} and the functions U, F, G depend on the energy W. The prime denotes derivative with respect to W.

The energy production term has a relaxation form $C_W = -\frac{W-W_0}{\tau_W}$ where τ_W is the energy relaxation time, which depends also on W, and $W_0 = 3/2k_BT_L$ is the energy at equilibrium, with T_L the lattice temperature, here assumed to be constant.

The expressions of U, F, G, C_W, c_{ij}, D_{ij} have been obtained in [Anile et al. (1999), Romano (2000)] both for parabolic band and Kane's dispersion relation. In the case that the conduction energy bands of electrons are described by the Kane dispersion relation, the expressions of U, F, G, C_W, c_{ij}, D_{ij} require a numerical evaluation of some integrals and for them an analytical expression is not available. These computations have been done in [Anile et al. (1999), Romano (2000)] and, in order to improve the efficiency of the simulation code, discrete data have been approached by splines. For the details one can see [Anile et al. (2004)].

In our simulations the holes will be considered at equilibrium.

In order to use the numerical method we shown in section 2, the MEP Energy-Transport model must be formulated in an equivalent form in the framework of linear irreversible thermodynamics in terms of the so-called entropic variables. We skip all the details. The interested reader is referred to [Anile et al. (2004)].

First we introduce the quantities

$$N_c = 2(\frac{2\pi k_B m^* T_n}{\hbar^2})^{\frac{3}{2}}, \quad n = N_c(T_n)\exp{(e\frac{\phi + \varphi_n}{k_B T_n})}$$

where φ_n is the so-called electron Fermi quasi-level and T_n is the absolute electron temperature, which outside of equilibrium is assumed to be related to the energy lagrangian multiplier λ^W through the relation $\frac{1}{T_n} = \lambda^W k_B$.

We want to transform, in the stationary case, the system (2)-(3) in the following form (the generation-recombination effects have been neglected)

$$-\text{div } \mathbf{J_n} = 0, \quad -\text{div } \mathbf{J_n^T} + n\frac{W - W_0}{\tau_W} = 0, \tag{6}$$

$$\text{div } \mathbf{D} = e(N_D - N_A - n + p), \quad \mathbf{J_n} = A_{11}\nabla\left(\frac{\varphi_n}{T_n}\right) + A_{12}\nabla\left(\frac{-1}{T_n}\right), \tag{7}$$

$$\mathbf{J_n^T} = A_{21}\nabla\left(\frac{\varphi_n}{T_n}\right) + A_{22}\nabla\left(\frac{-1}{T_n}\right) \tag{8}$$

where $\mathbf{J_n}$ is the electron current $-e\mathcal{J}_\mathbf{n}$, $\mathbf{D}$ is the electric displacement vector and $\mathbf{J_n^T} = -\mathcal{J}_\mathbf{n}^\mathbf{u} - \phi\mathbf{J_n}$, with $\mathcal{J}_\mathbf{n}^\mathbf{u}$ energy-flux density.

By noting that

$$\mathbf{J_n} = -en\mathbf{V} = -en\left[D_{11}\nabla \log n + D_{12}\nabla W + D_{13}\nabla\phi\right] \tag{9}$$

$$\nabla \log n = \frac{1}{n}\nabla n = \frac{1}{n}\nabla\left[N_c(T_n)\exp\left(e\frac{\varphi_n + \phi}{k_B T_n}\right)\right] \tag{10}$$

and comparing equations (9)-(10) with equations (7), we get

$$A_{11} = e^2 L_{11}, \quad A_{12} = -e^2 L_{11}\phi - \frac{en}{k_B}\left\{D_{11}W - \frac{D_{12}}{\frac{d\lambda^W}{dW}}\right\} \tag{11}$$

$$A_{21} = e^2 L_{11}\varphi_n + eL_{21}, \quad A_{22} = e^2 L_{11}\varphi_n^2 + e(L_{21} + L_{12})\varphi_n + L_{22}, \tag{12}$$

where

$$L_{11} = -\frac{nD_{11}}{k_B}, \; L_{12} = +\frac{nD_{12}}{k_B\frac{d\lambda^W}{dW}} + \frac{nD_{11}}{k_B}(\nu_n - W), \; L_{21} = -\frac{nD_{21}}{k_B} + \frac{nD_{11}}{k_B}\nu_n,$$

$$L_{22} = +\frac{nD_{22}}{k_B\frac{d\lambda^W}{dW}} + \frac{nD_{11}}{k_B}\nu_n(\nu_n - W) - L_{12}\left[(\nu_n - W) + \nu_n\right].$$

Here ν_n is a sort of chemical potential and it is given by $\nu_n = k_BT_n \log n + k_BT_nF(W)$, after introducing the primitive (defined up to a constant) $F(W) = \int W\frac{d\lambda^W}{dW}dW$. Note that the matrix A is not symmetric, unless the parabolic band approximation is adopted. However it is possible to prove that A is positive definite [Romano (2001)].

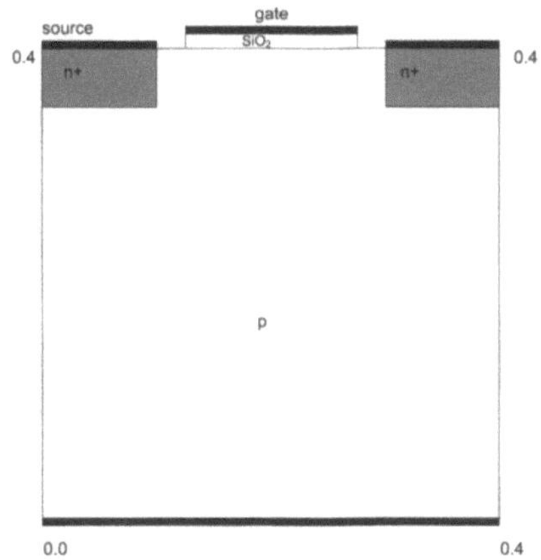

Fig. 1. Schematic representation of a bidimensional MOSFET

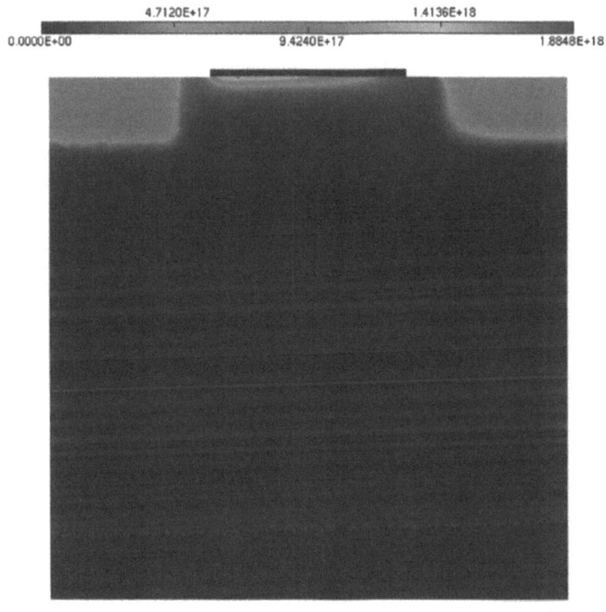

Fig. 2. Stationary solution for the electron density in cm^{-3}

2 Sketch of the numerical scheme

Let us recall some key features of such a numerical scheme. For more details see [Anile et al. (2004)].

- As mixed finite element approximation the classical Raviart-Thomas RT_0 is used for space discretization (see [Montarnal (1997), Marrocco et al. (1996), Brezzi et al. (1991)] for more details).
- Operator-splitting techniques for solving saddle point problems arising from mixed finite elements formulation [Glowinski et al. (1989)].
- Implicit scheme (backward Euler) for time discretization of the artificial transient problems generated by operator splitting techniques.
- A block-relaxation technique, at each time step, is implemented in order to reduce as much as possible the size of the successive problems we have to solve, by keeping at the same time a large amount of the implicit character of the scheme.
- Each non-linear problem coming from relaxation technique is solved via the Newton-Raphson method.

Concerning the block relaxation technique, three main steps have to be considered

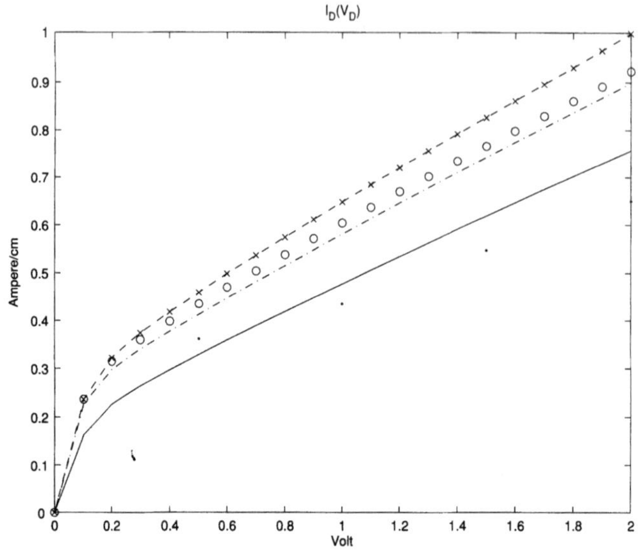

Fig. 3. Characteristic curve for the MOSFET. The drain current I_D versus the drain-source applied voltage V_D at $V_G = 0$. The dots are the MC solution, the continuous line is the MEP model, the xxx and dashed line are the Stratton and Chen model respectively, the ooo line is the SHE model, the dotted dashed line is the reduced hydrodynamical model

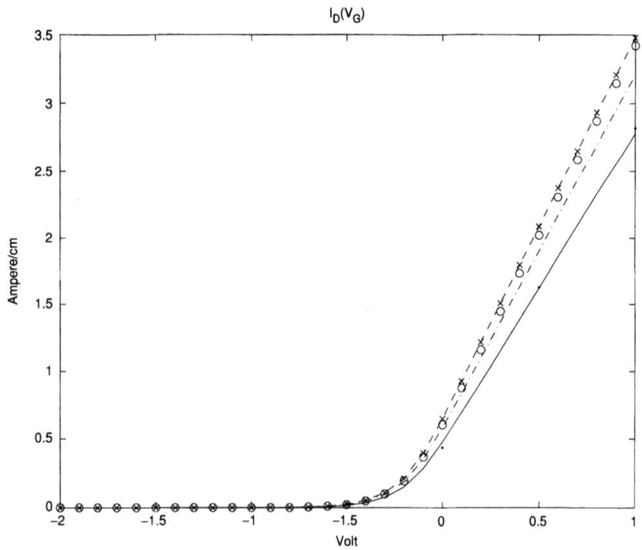

Fig. 4. The drain current I_D versus the gate applied voltage V_G) at $V_S = 0$. The notation is as figure 3

- A step related to the Poisson equation for the computation of φ^{k+1} and $\mathbf{D}^{k+1}$ with the other unknowns, frozen at the last known values, i.e. φ_p^k,$\mathbf{J}_p^k$,φ_n^k,$\mathbf{J}_n^k$,T_n^k,$\mathbf{J}_{T_n}^k$).
- A second step related to the hole continuity equation (if needed) for the computation of φ_p^{k+1} and $\mathbf{J}_p^{k+1}$.
- A third step in which the variables φ_n^{k+1},$\mathbf{J}_n^{k+1}$,T_n^{k+1}, $\mathbf{J}_{T_n}^{k+1}$ are computed simultaneously.

The reasons of such a procedure are explained in [Montarnal (1997)] and are essentially the strong coupling between the equations.

3 Simulation of a 2D silicon MOSFET

In this section we check the validity of our energy-transport model and the efficiency of the numerical method by simulating a bidimensional Metal Oxide Semiconductor Field Effect Transistor (MOSFET). The shape of the device is pictured in Fig. 1. The axes of reference frame are chosen parallel to the edges of the device. The silicon part of the MOSFET is represented by the numerical domain $[0, 0.4] \times [0, 0.4]$ and at the top of the silicon part the silicon oxide domain is $[0.125, 0.275] \times [0.4, 0.406]$ where the unit is the micron. The regions of high-doping n^+ are the subset $[0, 0.1] \times [0.35, 0.4] \cup [0.3, 0.4] \times [0.35, 0.4]$. The contacts at the source and drain are $0.1\mu m$ wide and the contact at the gate is $0.15\mu m$ wide. The distance between the gate and the other two contacts is $0.025\mu m$. A grid of 4644 elements has been used: 3344 in the bulk zone, 343 in the n^+ source zone, 357 in the n^+ drain zone and 600 in the oxide zone. The doping concentration, with abrupt junctions, is

$$n_D(x) - n_A(x) = \begin{cases} 10^{18} cm^{-3} \text{in the } n^+ \text{ regions} \\ -10^{14} cm^{-3} \text{in the } p \text{ region} \end{cases}$$

We assume ohmic contacts on the source, drain, gate and base while Neumann conditions are imposed on the remaining part of the boundary. In order to reach the desired bias, $V_d = 1.0$ $V_s = 0$ and $V_g = 0.5$, we first compute the equilibrium state and then use a continuation method on the applied potential. First, we go to $V_d = 1.0$ by steps of 0.1 Volt and after we go to $V_g = 0.5$ within two steps of 0.25 Volt. The total amount of computational time to reach the desired bias for the simulation reported in the figures was about 18 minutes on a laptop computer IBM ThinkPad A31.

In Fig. (2) the stationary electron density is plotted. In Figs. (3)-(4) the characteristic curves obtained with the MEP model are compared with those given by a MC simulation [Archimedes code] and by the standard energy transport models known in literature: that derived by the spherical harmonic expansion [Ben Abdallah et al. (1996a), Ben Abdallah et al. (1996b)], the Stratton one [Stratton et el. (1962)], that proposed by Chen et al [Chen et el. (1992)] and by Lyumkis et al [Lyumkis et al. (1992)], the energy-transport limit of the hydrodynamical model of Blotekjaer-Baccarani-Wordeman [Blotekjaer (1970), Baccarani et al. (1982)]. For the details the interested reader is referred to [Anile et al. (2004)]. It is evident that the MEP model is the most accurate since it gives the results closest to the MC data.

References

[Glowinski et al. (1989)] Glowinski, R., and Le Tallec, P.: Augmented Lagrangian and Operator Splitting Methods in Nonlinear Mechanics. SIAM, Studies in Applied Mathematics, Philadelphia (1989)

[Brezzi et al. (1991)] Brezzi,F., and Fortin, M.: Mixed and Hybrid Finite Element Methods. Springer Series in Computational Mathematics 15. Springer-Verlag, Berlin Heidelberg New York (1991)

[Marrocco et al. (1996)] Marrocco, A., and Montarnal, Ph.: Simulation des modèles energy-transport à l'aide des éléments finis mixtes. C.R. Acad. Sci. Paris, **323**, Serie I, 535–541 (1996)

[Raviart et al. (1997)] Raviart, P. A., and Thomas, J. M.: A mixed finite element method for 2nd order elliptic problems. In: Galligani, I., and Magenes, E. (eds) Mathematical Aspects of Finite Element methods, Lecture Notes 606. Springer-Verlag, Berlin Heidelberg New York (1977)

[Montarnal (1997)] Montarnal, Ph.: Modèles de transport d'énergie des semi-conducteurs, études asymptotiques et résolution par éléments finis mixtes. Thèse de doctorat, Université Paris VI, Paris (1997)

[Anile et al. (1999)] Anile, A.M., and Romano, V.: Non parabolic band transport in semiconductors: closure of the moment equations. Continuum Mech. Thermodyn., **11**, 307–325 (1999)

[Romano (2000)] Romano, V.: Non parabolic band transport in semiconductors: closure of the production terms in the moment equations. Continuum Mech. Thermodyn., **12**, 31–51 (2000)

[Romano (2001)] Romano, V.: Non-parabolic band hydrodynamical model for silicon semiconductors and simulation of electron devices. Math. Meth. Appl. Sci., **24**, 439–471 (2001)

[Anile et al. (2003)] Anile, A.M., Mascali, G., and Romano, V.: Recent developments in hydrodynamical modeling of semiconductors. In: Anile (ed) Mathematical Problems in Semiconductor Physics. Lecture Notes in Mathematics 1832. Springer, Berlin Heidelberg New York (2003)

[Anile et al. (2004)] Anile, A. M., Marrocco, A., Romano, V., Sellier, J. M.:Numerical Simulation of the 2D Non-Parabolic MEP Energy-Transport Model with a Mixed Finite Elements Scheme. Preprint (2004) available at the web site: www.dmi.unict.it/~romano

[Archimedes code] GNU/Archimedes - web-site: www.gnu.org (Free Directory). It is possible to get the package by e-mail to sellier@dmi.unict.it

[Ben Abdallah et al. (1996a)] Ben Abdallah, N., Degond, P., and Genieys, S.: An energy-transport model for semiconductors derived from the Boltzmann equation. J. Stat. Phys., **84**, 205–231 (1996)

[Ben Abdallah et al. (1996b)] Ben Abdallah, N., and Degond, P.: On a hierarchy of macroscopic models for semiconductors. J. Math. Phys., **37**, 3306–3333 (1996)

[Stratton et el. (1962)] Stratton, R.: Diffusion of hot and cold electrons in semiconductor barriers. Phys. Rev., **126**, 2002–2014 (1962)

[Chen et el. (1992)] Chen, D., Kan, E. C., Ravaioli, U., Shu, C-W., Dutton, R.: An improved energy-transport model including nonparabolicity and non-maxwellian distribution effects. IEEE on Electron Device Letters, **13**, 26–28 (1992)

[Lyumkis et al. (1992)] Lyumkis, E., Polsky, B., Shir, A., Visocky, P.: Transient semiconductor device simulation including energy balance equation. Compel **11**, 311–325 (1992)

[Blotekjaer (1970)] Blotekjaer,K.: Transport equations for electron in two-valley semiconductors. IEEE Trans. on Electron Devices **17**, 38–47 (1970)

[Baccarani et al. (1982)] Baccarani, G., Wordeman, M. R.: An investigation on steady-state velocity overshoot in silicon. Solid-state Electronics **29**, 970–977 (1982)

Comparison of Different Methodologies for Parameter Extraction in Circuit Design

A. M. Anile[1], S. Rinaudo[2], A. Ciccazzo[2], V. Cinnera Martino[2], C. Milazzo[1], and S. Spinella[3]

[1] Università di Catania, Dipartimento di Matematica e Informatica, viale A. Doria 6, Catania, Italy
[2] ST Microelectronics,Stradale Primosole 50, 95121 Catania, Italy
[3] Università della Calabria, Dipartimento di Linguistica, Ponte P. Bucci 17B, 87036 Arcavacata di Rende, Italy

Abstract This preliminary work concerns parameter extraction for electronic device circuit models. The reliability of electronic circuit design simulators depends crucially on the validity of the parameters which appear in the circuit models. These parameters must fit the measurements of a real device and measured data not be to sensitive to small data perturbation (*robustness*). We compare standard fitting techniques and possible alternatives in order to investigate the connection between fitting and robustness in parameter extraction.

1 Introduction

The reliability of electronic circuit design simulators depends crucially on the validity of the mathematical models which are implemented in the simulators and on the accurate knowledge of the parameters which appear in the mathematical models. The best values for the model parameters are found by fitting the measured data as closely as possible to the simulated data in the sense of a suitable weighted l^2 metric and this process is usually performed as a sequence of optimizations, usually based on the Levenberg-Marquard algorithm, which require a good initial guess and yield only local minima (corresponding to different set of parameters).
In this context two problems arise:

- When there several measurement curves (e.g. different components of the Y matrix for small signal analysis) whether more accurate results could be obtained by applying the concepts of multi-objective optimization to the different measurement curves instead of lumping them together in a single objective function.This seems to be the case when fitting compact models to MOS devices DC measurements
- When several optima are obtained how to choose the most convenient set of parameter values.

We surmise that the criterion to use in order to choose among the several local minima must be related to the robustness of the extracted parameters. In fact since the experimental data are subjected to measurement uncertainties, the parameters of choice **must be the least sensitive to the data variations**, a concept related to the Taguchi **quality concept** [6]. In this paper we investigate both problems in the case of parameter extraction for a 3-cell inductor model. For two algorithms (the third and the fourth), instead of the weighted l^2 metric we use the multiobjective goal attainment metric [3], which reflects better the multi-objective nature of the problem in this case (comparison of the elements of the Y or S matrix of the circuit with the corresponding experimental data) For all algorithms we compute the variance of the obtained minima in order to assess the robustness of the results.

2 The circuit

Inductor devices hold a fundamental role in the radiofrequency field and it is important to develop models which represent the intrinsic characteristics correctly at every working frequencies.
In STMicroelectronics a new method was implemented which models a distributed integrated inductor on a buried layer as a Spice model with its parameters determined by a suitable optimization algorithm [1].

The inductor has the following structure:

Shape	Number of turns	Outer dim. (μm)
Octag.	2.5	200.0
	Width (μm)	**Spacing** (μm)
	16.0	8.0
	SiO_2 Thick. (μm)	**Al Thick.** (μm)
	1.8	3.0

Table 1 - Test inductor dimensions

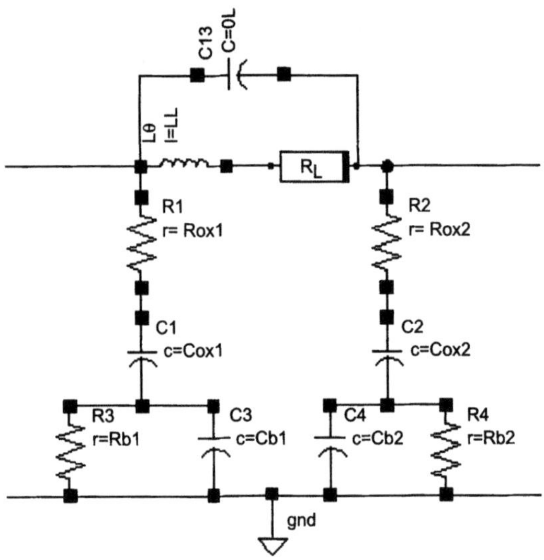

Fig. 1. This figure shows the inductor cell circuit schema

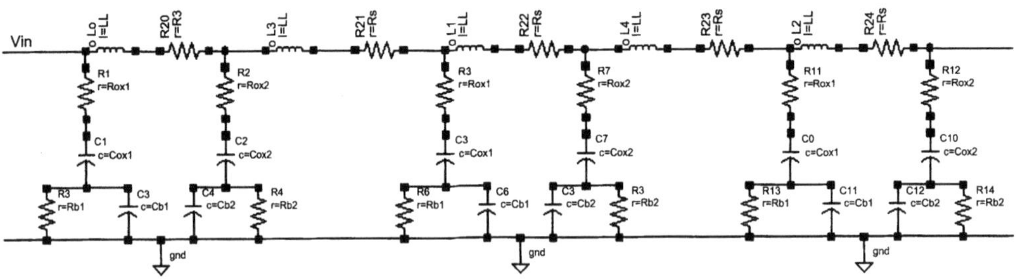

Fig. 2. This figure shows the distributed circuit schema of the inductor

Figure 1 shows the base cell circuit of the inductor. The base cell considers the buried layer contribution through the parallel between the resistor R_b and the capacity C_b. This split contribution is justified for the effects of the field oxide and the epitaxy resistance. Likewise the sublayer contribution is split as showed in Fig. 1. It is also possible take into account the skin effect with the equation

$$R_L(f) = R_0(1 + K_1 f^{K_2}) \tag{1}$$

where the coefficients K_1 and K_2 are referred to the physical characteristics of the material.

The cell is the base module of the distributed model showed in Fig. 2, where the inductor is represented through a series of cells in order to characterize the high frequency behavior. The linking of these elementary cells is made by

	TMG initial estimation	ADS	Opsim
RL (Ω)	0.1548	1.47	0.21
LL ($10^{-9}H$)	0.21	1.79	0.25
Rox1 (ohm)	20.41	10.27	4.60
Cox1 ($10^{-12}F$)	5.19	0.78	0.05
Rox2 (ohm)	17.67	8.87	0.0
Cox2 ($10^{-12}F$)	5.37	0.27	0.04
Rb1 (ohm)	16.49	0.56	6.57
Cb1 ($10^{-12}F$)	0.002	2.33	1.77
Rb2 (ohm)	0.6	0.58	0.03
Cb2 ($10^{-12}F$)	0.51	0.25	0.07
K1	0.6	0.3	0.3
K2	0.87	1.44	1.42

Table 2. Preliminary results of parameter extraction

a coupled net resulting from the contribution of the R_L and L series. Finally R_{in} and R_{out} represent the contribution due to the contact resistance.

Preliminary investigations have been carried out with commercial tools and in-house (STMicroelectronics) optimization software and they have yielded different sets of parameters [2]. These results are summarized in Table 2. The first column shows the initial estimation computed by the TMG fitting [5]. This fitting is based on a initial estimation of the parameter. For instance we consider the inductor of the Fig. 1. This device is constructed on a silicon sublayer without buried layer, thus the contribution of the parallel between R_b and C_b can be removed. This choice makes the computation easier.

The parameter extraction for this device concerns the fitting estimation of 12 parameters that are shown in the circuits Fig. 1. A set of Y-parameters are given in the range of frequency between 100 Mhz and 20 Ghz. These data can be real measurements or given by an automatic simulation flow.

Notice that the capacitance effects represented by the C_L parallel are negligible with respect to the device length and to the frequencies taken into account.

Comparison of previous data (table 2) shows large variations of the extracted parameter. Causes of these behaviours could be due to the non-homogeneous kind of variables (variables can converge with different speed rates) and to the different merit functions of optimization in l^2 which can find different balancing among errors.

Previous remarks compel us to consider the quality of results in the sense of robustness.

3 The algorithms

In order to test the robustness of the sets of parameters and also to investigate the performance of multi-objective approaches the following four methods have been considered

- The function **lsqnonlin** of MATLAB (LSQ) [7] with the default option of large scale optimization, which uses the subspace trust method based on the interior-reflective Newton method. The structure of the nonlinear least-squares problem $f(x) = \frac{1}{2}\|F(x)\|_2^2$ is exploited to enhance efficiency. In particular, an approximate Gauss-Newton direction, i.e., a solution s to min $\|Js + F\|$ (where J is the Jacobian of F) is used to help to define the subspace S. Second derivatives of the component function are not used (see Matlab documentation). This is a sophisticated routine and similar algorithms are implemented in the commercial simulators used in the microelectronics industry.
- The **DIRECT** method (DIR), which is a global search method described in [4] and applies to Lipschitz continuous functions and, after an initial implicit estimate of the Lipschitz constant chooses the potentially optimal rectangles and resamples them along their axis. Afterward it divides these rectangles and proceed by sampling and dividing until a stop criterion is met. This method exploits the estimation of Lipschitz constant to balance global and local search and reaches a quasi-global solution in large domain.
- the **fgoalattain** function of MATLAB (MUL) [7] which is a multiobjective goal optimization algorithm and uses sequential quadratic programming (SQP). The goal programming problem claims a set $F^* = \{F_1^*, F_2^*, \ldots, F_n^*\}$ of targets for the vector function

$$F(x) = \{F_1(x), F_2(x), \ldots, F_n(x)\}.$$

The relative degree of under- or overachievement of the goals is controlled by a vector of weighting coefficients, $w = \{w_1, w_2, \ldots, w_n\}$, and is expressed as a standard optimization problem using the following formulation

$$\min_{\gamma \in R, x \in \Omega} \gamma$$

$$\text{such that } F_i(x) - w_i\gamma \leq F_i^*, \quad i = 1, \ldots, n$$

- The heuristic (HYB) combining few initial steps of **DIRECT** in order to obtain a reasonable initial guess and subsequently **fgoalattain**. The first step detects a suitable region to start the fgoalattain method, which can use this initial information to set up good constraint systems.

These algorithms were set up with a maximum number of 6500 function evaluations allowed and a termination tolerance on the function value of 10^{-8}. Generally the optimization process finished when reaching a small gradient for the target function in advance. The mean error reached on each of these points is of the order of 10^-3.

4 Results

A Montecarlo simulation tests are performed for each algorithm in order to compare the reliability of the methods. Through the Montecarlo simulation it is possible to repeat virtually an experiment and to get a quality measure of fitting robustness. The simulation starts with a initial fitting in order to identify a possible set of parameters $\widetilde{\mathbf{x}}$. This set of parameter is used to synthesize a new surrogated set of data $\mathcal{D}_{\widetilde{\mathbf{x}}}$ which are perturbed by a white noise. In this study the noise is a gaussian error with $\mu = 0$ and $\sigma = \frac{1}{10}$ of data magnitude. This process mimics artificially the statistical properties of real data. Then the fitting is processed on the surrogated data to get a new set of parameters. This kind of artificial process is repeated many times to get a large class of parameters. Finally, classical statistics are performed on this class of parameter set and confidence limits on parameters are calculated from these simulations.

Table 3 shows the results for each variable of 100 montecarlo simulations. Mean and standard deviation are shown for each methods: LSQ stands for nonlinear least squares methods, DIR stands for global search DIRECT, MUL stands for multiobjective goal programming and HYB stands for the combination of DIR and MUL. The column of standard deviation shows the best results for DIRECT methods, which seems insensitive to the data variations.

It is apparent that the DIRECT algorithm selects the most robust set of parameters. It is well known that genetic algorithms for global optimization are robust but the are much more computationally intensive than DIRECT.

The reasons of this result can be found in the interpretation of this kind of global search. DIRECT leads the optimization towards a basin of convergence for the objective function. In this basin DIRECT can find good solutions rather than optimal solutions but in this subregion there will be small variations of the objective function because the Lipschitz constant will be small.

Table 3. Montecarlo simulation of the extraction parameter process. Synthetic sets has an additive Normal Error with mean 0 and standard deviation of $\frac{1}{10}$ of the range for each variables

	Mean				STD			
	LSQ	DIR	MUL	HYB	LSQ	DIR	MUL	HYB
RL	4.88	1.51	8.86	1.03	2.50	0.15	2.33	0.16
LL (10^{-9})	2.41	1.53	3.34	1.72	2.04	0.09	3.29	0.73
Rox1	8.24	8.54	7.02	8.67	2.22	0.76	3.33	1.08
Cox1 (10^{-12})	0.38	0.25	0.82	0.96	0.23	0.05	0.35	0.15
Rox2	8.92	7.54	6.92	6.23	1.78	0.38	3.33	3.63
Cox2 (10^{-12})	0.42	0.06	0.92	0.003	0.25	0.004	0.2	0.02
Rb1	2	0.66	3.39	2.39	2.61	0.54	3.55	1.63
Cb1 (10^{-12})	0.76	0.92	0.38	0.26	0.24	0.04	0.37	0.26
Rb2	1.15	2.73	3.54	2.85	2.01	0.42	3.67	2.15
Cb2 (10^{-12})	5.3	0.93	0.34	0.68	0.27	0.42	0.36	0.23
K1	4.96	0.35	8.89	0.21	2.52	0.14	2.26	0.99
K2	4.93	1.44	6.98	1	2.47	0.13	2.14	0.0002

These results bring forth the hypothesis that good solution must be found with a tradeoff between best fitting and robustness of the solution. In this sense optimization methods must take into account these two aspects in order to find solutions which are stable.

5 Conclusion

Previous tests show that:

- In this case the multi-objective approach in the sense of considering separately the various components of the Y matrix, does not lead to improved performance over the case when the components are lumped together in a l^2 metric.
- The DIRECT global optimization algorithm seems to be the most robust among those we have considered. It is well known that genetic a algorithms are robust optimizers but they are also very demanding on the computational resources. The DIRECT optimization algorithm seems to combine the required robustness with a limited demand on the computational resources, at least for the type of problems we have considered here.

Since robustness is a major issue, a different multi-objective approach could be considerd, one in which the two main objectives are the average and the variance of the l^2 metric. These concepts are under current investigation and the results will be reported elsewhere.

References

1. A. Ciccazzo, G. Greco, S. Rinaudo, A New Scalable Spice Model for Spiral Inductors in Substrate with Buried Layer, Radio and Wireless Conference, 2003, Proceedings, pp. 345-348
2. K.Doganis & D.L.Scharfetter, General Optimization and Extraction of IC Device Model Parameters, IEEE Transaction on Electron Device, **Volume ED-30**, n. 9, (1983)
3. Gembicki, F.W., Vector Optimization for Control with Performance and Parameter Sensitivity Indices, Ph.D. Thesis, Case Western Reserve Univ., Cleveland, Ohio, 1974
4. D. R. Jones, C. D. Perttunen, B. E. Stuckman, Lispchitz Optimization Without The Lipschitz Constant, Journal of Optimization Theory and Application **Volume, 79** (1993) pp. 157-181
5. S. Rinaudo, G. Privitera, G. Ferla and A. Galluzzo, "*Small-Signal and Noise Two-Dimensional Modeling of Submicrometer High Speed Bipolar Transistor*" Proceedings of GaAS 97 - 5th European Gallium Arsenide and related III-V compounds Applications Symposium. Bologna September 3-5, 1997
6. G. Taguchi, *Introduction to quality engineering* (Krauss International Publications, White Plains NY 1986)
7. The Mathworks, Inc., Matlab(R), The Language of technical computing, release 13

Sound Synthesis and Chaotic Behaviour in Chua's Oscillator

E. Bilotta[1], R. Campolo[2], P. Pantano[2], and F. Stranges[1]

[1] Dipartimento di Linguistica, Università della Calabria, Via P. Bucci, Cubo 17/B, Arcavacata di Rende (CS), {bilotta,f.stranges}@unical.it
[2] Dipartimento di Matematica, Università della Calabria, Via P. Bucci, Cubo 30/B, Arcavacata di Rende (CS), {r.campolo,piepa}@unical.it

Abstract Chua's oscillator is a dynamical system by which it is possible to investigate chaos both from the theoretical and the experimental point of view. Studying this system, many strange attractors have been observed and many routes to chaos have been discovered. Furthermore generalizations of Chua's oscillator have been found which present n-scroll attractors. In this paper we propose a methodology for reading the complexity of such systems. We have analyzed the bifurcation map of a system with a 4-scroll attractor and we have been able to perform sound analysis and synthesis of its solutions and to construct 3D images and musical pieces which follow the relevant changes in the behavior of this dynamical system.

1 Chua's oscillator

Chua's Oscillator has been widely investigated at experimental [ZA85] and numerical [Mat85] levels and, since in this circuit the presence of chaos, strange attractors and bifurcations has been proved, it has become a paradigm for the study of chaos [Mad93]. The dimensionless equations for Chua's oscillator can be written as follows [CWHZ93]

$$\begin{cases} \frac{dx}{dt} = k\alpha(y - x - f(x)) \\ \\ \frac{dy}{dt} = k(x - y + z) \\ \\ \frac{dz}{dt} = k(-\beta y - \gamma z) \end{cases} \tag{1}$$

where

$$f(x) = bx + \frac{1}{2}(a - b)\{|x + 1| - |x - 1|\}. \tag{2}$$

This dynamical system has three degrees of freedom, and six control parameters: a, b, α, β, γ and $k \in \{-1, 1\}$. Chua's oscillator has been recently generalized in many directions. For example:

- Introducing additional break points in the piecewise linear function of Chua's oscillator, dynamical systems have been obtained which present many strange attractors called *n-double scroll* attractors [SV91, SV93, SHC97]. An experimental confirmation of n-double scroll attractors has been given in [ABFM96, YSV00]. In Aziz-Alaoui's paper [Azi00] a method for generating a 10-scroll *multispiral attractor* has been proposed.
- Chua's circuits with smooth nonlinearities (e.g. cubic nonlinearity) have been studied [Alt93, KRC93].
- Systems with hyperchaotic attractors have been obtained by means of three coupled Chua's circuits with sine-type functions as nonlinearities [CG03].
- In Yang's et *al.* paper [YC00], a new class of piecewise-linear three-dimensional autonomous systems has been studied. These systems present a three-segment piecewise-linear function and a single equilibrium point.

In a recent paper [BGP05], some of the authors of the present work realized an extensive tutorial on Chua's attractors, where a new methodology is presented, which makes use of sound and music in order to understand some of the main features of chaos. The authors have developed many software packages for creating 3D images of Chua's attractors (Figure 1), for listening to the sound, by using sound synthesis (Figure 2) and producing musical pieces (Figure 3) on the basis of the x, y and z curves of these dynamical systems.

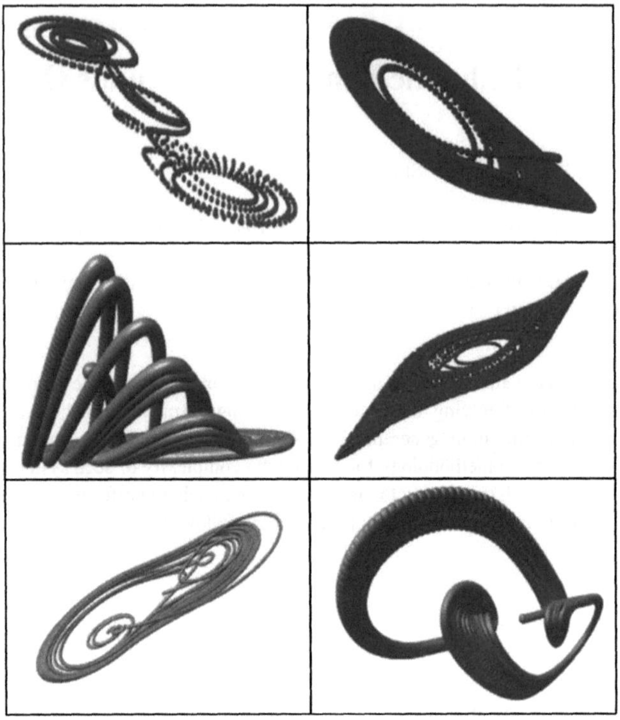

Fig. 1. These images show some Chua's attractors

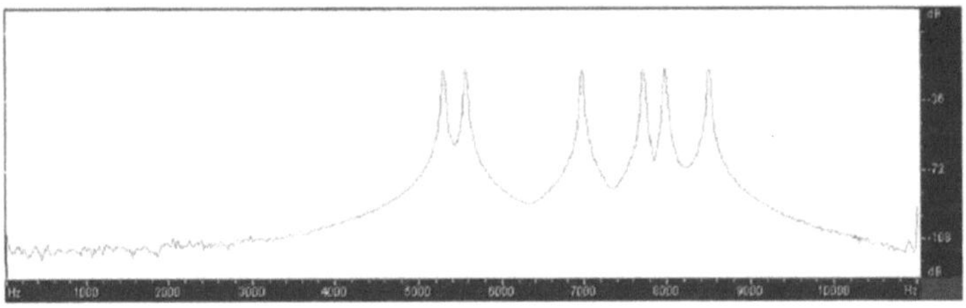

Fig. 2. Sound synthesis of one Chua's attractor

Fig. 3. Staff of the generated piece of music

The main aims of this paper are:

- to visualize, by means of 3D software tools, Chua's n-scroll attractors and to study the qualitative changes of their behavior, realizing bifurcation maps;
- to read the complexity of these systems by using sound and music.

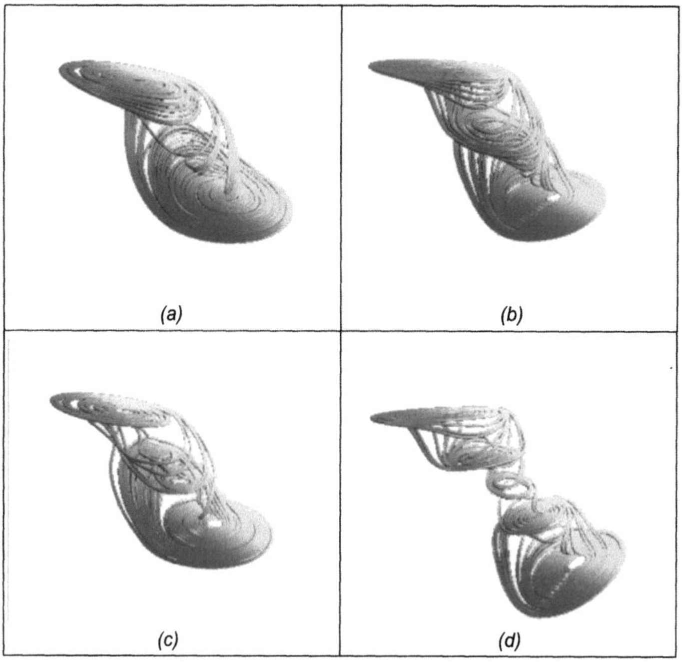

Fig. 4. n-scroll attractors: (a) 2-scroll; (b) 3-scroll; (c) 4-scroll; (d) 5-scroll

2 n-scroll attractors

Chua's oscillators which exhibit even or odd numbers of scroll attractors are described by the following evolution equations:

$$\begin{cases} \dfrac{dx}{dt} = \alpha(y - h(x)) \\ \\ \dfrac{dy}{dt} = (x - y + z) \\ \\ \dfrac{dz}{dt} = -\beta y - \gamma z \end{cases} \tag{3}$$

where $h(x)$ is given by:

$$h(x) = m_{2q-1}x + \frac{1}{2}\sum_{i=1}^{2q-1}(m_{i-1} - m_i)\{|x + c_i| - |x - ci|\}, \tag{4}$$

q being a natural number and $\mathbf{m}$ and $\mathbf{c}$ $2q$ and $(2q - 1)-$dimensional vectors, respectively. As well known [SHC97, YSV00], using the following values of the parameters: $\alpha = 9$, $\beta = 14.286$, $\gamma = 0$, one obtains systems with a different number of scroll attractors according to the choice of q, $\mathbf{m} = [m_0; m_1; \ldots; m_{2q-1}]$ and $\mathbf{c} = [c_1; c_2; \ldots; m_{2q-1}]$. In particular

- 2-scroll attractor: $q = 1$, $\mathbf{m} = [-1/7; 2/7]$, $c = 1$ (Figure 4.a);
- 3-scroll attractor: $q = 2$, $\mathbf{m} = [0.9/7; -3/7; 3.5/7; -2.4/7]$, $\mathbf{c} = [1; 2.15; 4]$ (Figure 4.b);
- 4-scroll attractor: $q = 2$, $\mathbf{m} = [-1/7; 2/7; -4/7; 2/7]$, $\mathbf{c} = [1; 2.15; 3.6]$ (Figure 4.c);
- 5-scroll attractor: $q = 3$, $\mathbf{m} = [0.9/7; -3/7; 3.5/7; -2.7/7; 4/7; -2.4/7]$, $\mathbf{c} = [1; 2.15; 3.6; 6.2; 9]$ (Figure 4.d);

We have introduced a new parameter ρ in order to study the qualitative changes of (3), by considering a vector $\mathbf{m}' = \rho\mathbf{m}$. In this way the slope of the function h is changed, thus creating bifurcation maps at the varying of the parameter ρ.

In this work, we have considered the case of a system with a 4-scroll attractor, varying the parameter ρ in the interval $[0.5, 3.5]$. Figure 5 shows the bifurcation diagram. We have analyzed this diagram in order to detect qualitative changes in the system under consideration. In particular it presents: a period 1-limit cycle for $\rho = 0.51$; a period

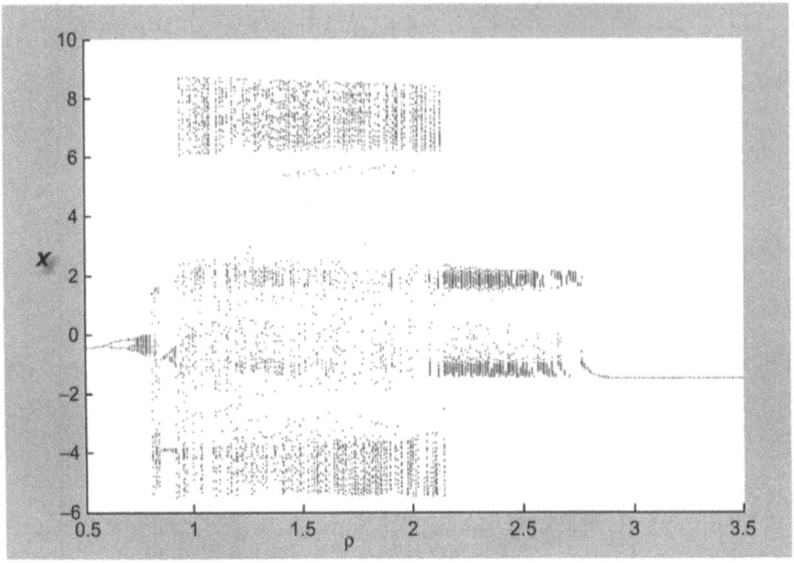

Fig. 5. Bifurcation map for a system with a 4-scroll attractor

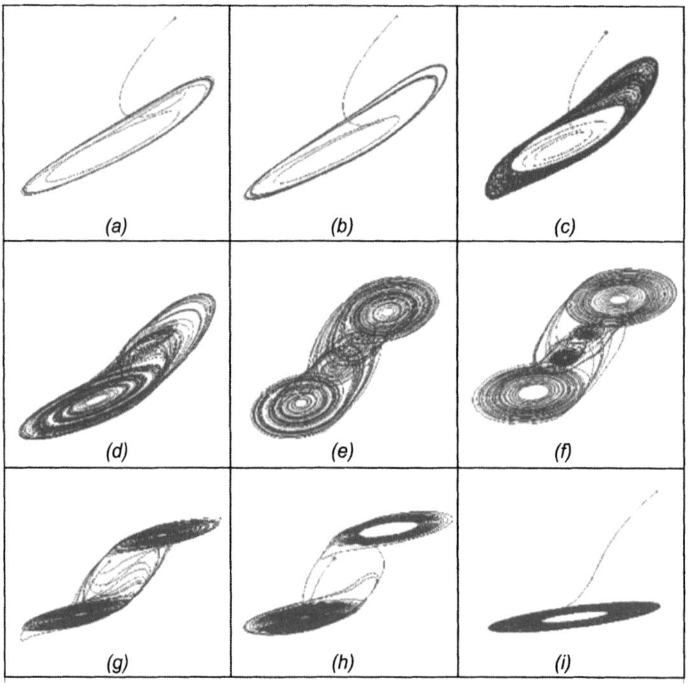

Fig. 6. Phases portraits in the x-y-z space of a system with a 4-scroll attractor: (a) $\rho = 0.51$; (b) $\rho = 0.61$; (c) $\rho = 0.75$; (d) $\rho = 0.808$; (e) $\rho = 1$; (f) $\rho = 2.16$; (g) $\rho = 2.17$; (h) $\rho = 2.71$; (i) $\rho = 2.84$

2-limit cycle for $\rho = 0.61$; a "*spiral Chua's attractor*" for $0.72 \leq \rho \leq 0.8$; a 4-scroll attractor for $0.808 \leq \rho \leq 2.16$. The 4-scroll attractor gives its way to a 2-scroll attractor for $2.17 \leq \rho \leq 2.80$. The 2-scroll attractor disappears and a periodic orbit continues to exist for $\rho \geq 2.81$. The changes at the varying of the parameter ρ are reported in Figure 6.

Fig. 7. Staff of the piece of music generated by a system with a 4-scroll attractor, $\rho = 1$

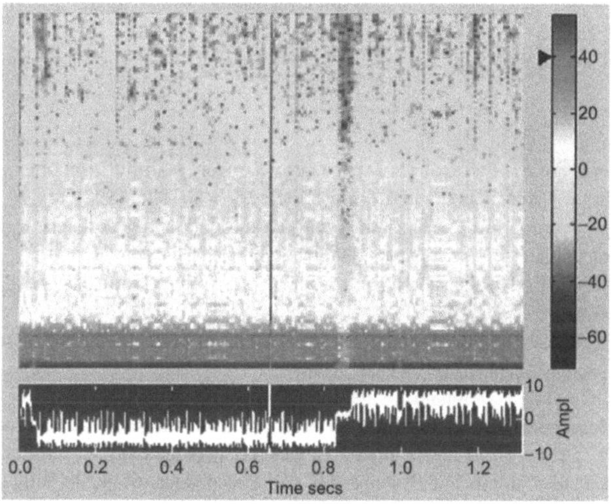

Fig. 8. Spectrogram for $x(t)$, $\rho = 1$

3 Sounds and music

We have translated the behavior of the above-considered system into music and sounds. Figure 7 shows a musical piece realized in correspondence of $\rho = 1$. Figure 8 and Fig. 9 present the spectrograms for $\rho = 1$ (4-scroll) and $\rho = 2.71$ (2-scroll). Sounds, music and images are available at the following web site:
http://galileo.cincom.unical.it/Esg/PlayChaos/index.htm.

The methodology, which we have used in this paper, can also be applied in cultural industry [BFP04]. In conclusion, we have introduced bifurcation maps for Chua's systems with n-scroll attractors, this gives us the possibility of studying the morphogenesis of these dynamical systems at a deeper level.

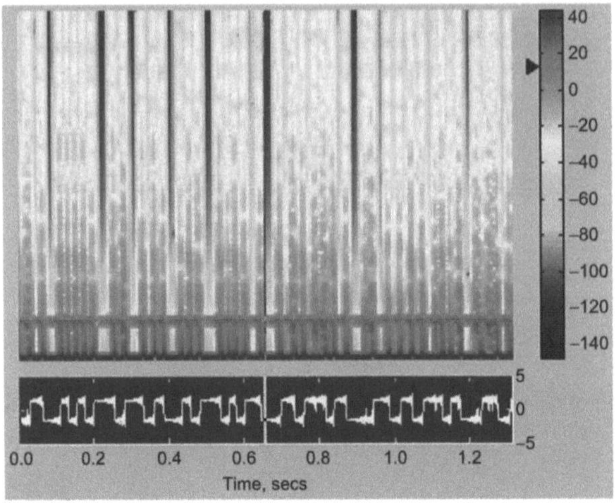

Fig. 9. Spectrogram for $x(t)$, $\rho = 2.71$

References

[ABFM96] Arena, P., Baglio, S., Fortuna, L., Manganaro, G.: Generation of n-double scrolls via cellular neural networks. Int. J. Circuit Theory and Applications, **24**, 241–252, (1996)

[Alt93] Altman, E.J.: Normal form analysis of Chua's circuit with applications for trajectory recognition. IEEE Transactions on Circuits and Systems, **40**, 675–682, (1993)

[Azi00] Aziz-Alaoui, M.A.: Multispiral chaos. Proc. 2nd Int. Conf. Control of Oscillations and Chaos, vol. 1, 88–91, (2000)

[BFP04] Bilotta, E., Francaviglia, M., Pantano, P.: Applications of Mathematics to Cultural Industry. Proceedings Mini-Symposium SIMAI 2004, Venice, (2004)

[BGP05] Bilotta, E., Gervasi, S., Pantano, P.: Reading Complexity in Chuas Oscillator through Music. Part I: A new way of understanding chaos. Int. J. of Bifurcation and Chaos, 15, **2**, in press, (2005)

[CG03] Cafagna, D., Grassi, G.: A new approach to generate hyperchaotic 3D-scroll attractors in a closed chain of Chua's circuits Circuits and Systems. ISCAS . Proceedings of the 2003 International Symposium on, **3**, 60–63 (2003)

[CWHZ93] Chua, L.O., Wu, C.W., Huang, A., Zhong, G.: A Universal Circuit for Studying and Generating Chaos-Part I: Routes to Chaos. IEEE Transctions on Circuits and Systems, **40**, (1993)

[KRC93] Khibnik, A.I., Roose, D., Chua, L.O.: On periodic orbits and homoclinic bifurcations in Chuas circuit with a smooth nonlinearity. Int. J. of Bifurcation and Chaos, **3**, 363–384, (1993)

[Mad93] Madan, R.N.: Chua's Circuit: A Paradigm for Chaos. World Scientific, Singapore (1993)

[Mat85] Matsumoto, T.: A Chaotic Attractor from Chua's Circuit. IEEE Transctions on Circuits and Systems, **12**, 1055–1059 (1985)

[SHC97] Suykens, J.A.K., Huang, A., Chua, L.O.: A family of n-scroll attractors from a generalized Chua's circuit. Archiv für Elektronik und Ubertragungstechnik, **51**, 131–138, (1997)

[SV91] Suykens, J.A.K., Vandewalle, J.: Quasilinear approach to nonlinear systems and the design of n-double scroll (n=1,2,3,4,...). IEEE Proceedings-G, **138**, 595–603, (1991)

[SV93] Suykens, J.A.K., Vandewalle, J.: Generation of n-double scrolls (n=1,2,3,4,...). IEEE Trans. Circuits and Systems, **40**, 861–867, (1993)

[YC00] Yang, T., Chua L.O.: Piecewise-linear chaotic systems with a single equilibrium point. International Journal of Bifurcation and Chaos, **10**, 2015–2060 (2000)

[YSV00] Yalçin, M.E., Suykens, J.A.K., Vandewalle, J.: Experimental confirmation of 3- and 5-scroll attractors from a generalized Chuas circuit. IEEE Trans. Circuits and Systems, **47**, 425–429, (2000)

[ZA85] Zhong, G.Q, Ayrom, F.: Experimental confermation of chaos from Chua's circuit. Int. J. Circuits Theory App., **13**, 93–98 (1985)

A Kinetic Type Extended Model for Polarizable and Magnetizable Fluids.

M. C. Carrisi, F. Demontis, S. Pennisi, and A. Scanu

Dipartimento di Matematica, Università di Cagliari, Via Ospedale 72, spennisi@unica.it

Abstract An elegant formulation of thermodynamics in electromagnetic fields has been provided by Liu and Müller and is based upon the conservation laws of mass, momentum and energy as well as on Maxwell's equations. However, in other physical context it has been shown the opportunity of considering an extended set of independent variables. Therefore, it is fitting to follow an extended approach also for charged fluids in electromagnetic fields; in literature this methodology has already been used, but only for the case of negligible effects of polarization and magnetization; here this restriction is removed and the general case treated. The entropy principle and the principle of material frame indifference are imposed; by using the methods of Extended Thermodynamics, we can see that they give very strong restrictions on the constitutive functions appearing in these balance laws.

1 Introduction.

In ordinary Thermodynamics, the conservation laws of mass (with density F), momentum (with density F_i) and energy (with density $\frac{1}{2}F_{ll}$) are used as field equations; in these equations, also the momentum flux density F_{ij} and the energy flux density $\frac{1}{2}F_{ill}$ occur, and they are linked to the independent variables F, F_i, $\frac{1}{2}F_{ll}$ and to their gradients through the state equations and the Navier-Stokes and Fourier laws. But in this way parabolic equations are obtained which yield infinite speeds of shocks propagation. In extended thermodynamics (see [1] and subsequent papers summarized in [2]) the aim has been realized to obtain an iperbolic set of field equations (and symmetric too) in the following way

- Consider as independent variables F, F_i, F_{ij}, F_{ill} (in other words, also the above fluxes have been included); for this increased number of independent variables, consider also a corresponding increased number of field equations.
- Link the new fluxes, which appear in these equations, only to the independent variables and not to their gradients. Restrict the generality of these links, or constitutive equations, by imposing the principle of entropy and that of Galilean invariance.

In this way a symmetric hyperbolic set of field equations are obtained, consequently yielding finite speeds of shocks propagation and continuous dependence on the initial conditions; therefore, they are more physically significant than those of ordinary Thermodynamics. This last one can also be recovered from those of Extended Thermodynamics as first approximation of a particular iterative procedure.
However, in [1], the flux appearing in a field equation is the independent variable of the subsequent equation; it follows that the original model describes only mono-atomic gases. We have verified that, also from the mathematical point of view, this structure leads to results which are too much restrictive for polarizable and magnetizable fluids; for example, also at equilibrium we obtain polarization effects without magnetization, which fact is physically unacceptable. This shows that the theory knows how polarization and magnetization cannot occur in mono-atomic gases! The reason is that in this case the model doesn't take into account the interactions between atoms and molecules.
In [3] it is shown how, also in Extended Thermodynamics, field equations can be considered which overcome this problem, and the fluxes are called F_k, G_{ik}, G_{ijk}, G_{ikll} (the first of these is still the momentum density, obviously); however, in [3] such constitutive functions have not been found by imposing the principles of entropy and that of Galilean invariance. This result has been recently achieved by some of us in [4] with a method akin to that of the kinetic theory, so that it has been called "A kinetic type extended model ...".
Here we want furtherly improve the model so that it may well describe also polarizable and magnetizable fluids. To this end we have in literature only models in the framework of ordinary Thermodynamics, such us [5]; here we want to obtain a model in the framework of Extended Thermodynamics, because it leads to more physically significant results

as seen above. We consider the following extended set of field equations. The first four of these have been found by applying the general guidelines of ref. [3]; note the contribution of the Lorentz force in the right-hand sides, and that of a term (in third and fourth equation) which takes into account external supplies other than body forces, according to the note on page 129 of ref. [3]. The eqs. $(1)_{1,2}$, (2), (3) and the trace of eq. $(1)_3$ are those studied by Liu and Müller [5] in the non extended approach.
The subsequent four equations are the Maxwell equations with electric field E_i and density of magnetic flux B_i, while the last two are definitions of the current J_i and of the charge density q in terms of the Polarization P_k, Magnetization M_j, the free current j_i^F and the free charge density q^F.

$$\begin{aligned}
&\partial_t F + \partial_k F_k = 0 \quad , \quad \partial_t F_i + \partial_k G_{ik} = qE_i + \epsilon_{iqp} J_q B_p \, , \\
&\partial_t F_{ij} + \partial_k G_{ijk} = \frac{2}{F} F_{(i}(qE_{j)} + \epsilon_{j)qp} J_q B_p) + \\
&\quad + \frac{2}{3}(E_r + \epsilon_{rqp} v_q B_p)(J_r - q v_r)\delta_{ij} + P_{<ij>} \, , \\
&\partial_t F_{ill} + \partial_k G_{ikll} = \frac{3}{F} F_{(il}(qE_{l)} + \epsilon_{l)qp} J_q B_p) + \\
&\quad + \frac{10}{3F}(E_r + \epsilon_{rqp} v_q B_p)(J_r - q v_r) F_i + P_{ill} \, ,
\end{aligned} \tag{1}$$

$$\begin{aligned}
&\partial_t B_i + \epsilon_{ijk}\partial_j E_k = 0 \quad , \quad -\mu_0\epsilon_0 \partial_t E_i + \epsilon_{ijk}\partial_j B_k = \mu_0 J_i \, , \\
&\partial_k B_k = 0 \; , \; \epsilon_0 \partial_k E_k = q \, ,
\end{aligned} \tag{2}$$

$$\partial_t P_i + \partial_k(\epsilon_{ijk} M_j + 2P_{[i} v_{k]}) = J_i - J_i^F \quad , \quad \partial_k P_k = q^F - q \, , \tag{3}$$

where $v_k = \frac{F_k}{F}$ is the velocity, ϵ_{ijk} is the Levi-Civita symbol, μ_0 the vacuum permeability, ϵ_0 the dielectric constant. The conservation of charge $\partial_t(q - q^F) + \partial_k(J_k - J_k^F) = 0$ is a consequence of (3). We note that another possible approach is to consider (3) not as field equations, but as definitions of J_i and q; the remaining eqs. (1), (2) are still a system of first order partial differential equations, even if the time and space derivatives occur also in the right-hand sides, through J_i and q. But in this way the divergence form is lost; for this reason we have chosen a different approach. We stress, once again, that in this set of equations the independent variables are F, F_i, F_{ij}, F_{ill}, B_i, E_i and P_i; but also the quantities G_{ik}, G_{ijk}, G_{ikll}, $P_{<ij>}$, P_{ill}, M_i, $J_i^* = J_i^F - q^F v_i$ occur in this system, so that they are unknown quantities for which closure relations are needed. The main result of this work are the expressions of these constitutive functions. They can be found by eliminating the parameters λ, λ_i, λ_{ij}, λ_{ill}, β_i, ϵ_i, π_i between the subsequent eqs. (6), $(7)_{2-4}$, $(8)_3$ which are expressed in terms of the functions h' and ϕ'_k, whose expressions are reported in the subsequent eqs. (16) and (17).
The arguments which allows us to find them are usual in Extended Thermodynamics, i.e., to impose that every solution of our system (1), (2), (3) satisfies a supplementary conservation law $\partial_t h + \partial_k \phi_k = \sigma \geq 0$. This amounts in assuming the existence of Lagrange multipliers λ, λ_i, λ_{ij}, λ_{ill}, β_i, ϵ_i, π_i, b, ϵ, π such that

$$\begin{aligned}
dh &= \lambda dF + \lambda_i dF_i + \lambda_{ij} dF^{ij} + \lambda_{ill} dF^{ill} + \beta_i dB_i + \epsilon_i dE_i(-\mu_0\epsilon_0) + \pi_i dP_i \, , \\
d\phi_k &= \lambda dF_k + \lambda_i dG_{ik} + \lambda_{ij} dG_{ijk} + \lambda_{ill} dG_{illk} + \beta_i \epsilon_{ikj} dE_j + \\
&\quad + \epsilon_i \epsilon_{ikj} dB_j + \pi_i d(\epsilon_{ikj} M_j + 2P_{[i} v_{k]}) + b dB_k + \epsilon\epsilon_0 dE_k + \pi dP_k \, ,
\end{aligned} \tag{4}$$

besides a residual inequality which we leave out for the sake of brevity.
By taking λ, λ_i, λ_{ij}, λ_{ill}, β_i, ϵ_i, π_i as independent variables, and by defining

$$\begin{aligned}
h' &= \lambda F + \lambda_i F^i + \lambda_{ij} F^{ij} + \lambda_{ill} F^{ill} + \beta_i B_i - \mu_0\epsilon_0 \epsilon_i E_i + \pi_i P_i - h \, , \\
\phi'_k &= \lambda F_k + \lambda_i G_{ik} + \lambda_{ij} G_{ijk} + \lambda_{ill} G_{illk} \\
&\quad + \beta_i \epsilon_{ikj} E_j + \epsilon_i \epsilon_{ikj} B_j + \pi_i(\epsilon_{ikj} M_j + 2P_{[i} v_{k]}) - \phi_k \, , \text{they become}
\end{aligned} \tag{5}$$

$$\begin{aligned}
&F = \frac{\partial h'}{\partial \lambda}, \quad F^i = \frac{\partial h'}{\partial \lambda_i}, \quad F^{ij} = \frac{\partial h'}{\partial \lambda_{ij}}, \quad F^{ill} = \frac{\partial h'}{\partial \lambda_{ill}} \, , \\
&B_i = \frac{\partial h'}{\partial \beta_i}, \quad -\mu_0\epsilon_0 E_i = \frac{\partial h'}{\partial \epsilon_i}, \quad P_i = \frac{\partial h'}{\partial \pi_i} ;
\end{aligned} \tag{6}$$

$$\begin{aligned}
\frac{\partial \phi'_k}{\partial \lambda} &= \frac{\partial h'}{\partial \lambda^k} - b\frac{\partial^2 h'}{\partial \lambda \partial \beta_k} + \frac{\epsilon}{\mu_0}\frac{\partial^2 h'}{\partial \lambda \partial \epsilon_k} - \pi\frac{\partial^2 h'}{\partial \lambda \partial \pi_k} \, , \\
G_{ik} &= \frac{\partial \phi'_k}{\partial \lambda_i} + b\frac{\partial^2 h'}{\partial \lambda_i \partial \beta_k} - \frac{\epsilon}{\mu_0}\frac{\partial^2 h'}{\partial \lambda_i \partial \epsilon_k} + \pi\frac{\partial^2 h'}{\partial \lambda_i \partial \pi_k} \, , \\
G_{ijk} &= \frac{\partial \phi'_k}{\partial \lambda_{ij}} + b\frac{\partial^2 h'}{\partial \lambda_{ij} \partial \beta_k} - \frac{\epsilon}{\mu_0}\frac{\partial^2 h'}{\partial \lambda_{ij} \partial \epsilon_k} + \pi\frac{\partial^2 h'}{\partial \lambda_{ij} \partial \pi_k} \, , \\
G_{illk} &= \frac{\partial \phi'_k}{\partial \lambda_{ill}} + b\frac{\partial^2 h'}{\partial \lambda_{ill} \partial \beta_k} - \frac{\epsilon}{\mu_0}\frac{\partial^2 h'}{\partial \lambda_{ill} \partial \epsilon_k} + \pi\frac{\partial^2 h'}{\partial \lambda_{ill} \partial \pi_k} \, ,
\end{aligned} \tag{7}$$

$$\begin{aligned}
\frac{\epsilon_{ikj}}{-\mu_0\epsilon_0}\frac{\partial h'}{\partial \epsilon_j} &= \frac{\partial \phi'_k}{\partial \beta_i} + b\frac{\partial^2 h'}{\partial\beta_i\partial\beta_k} - \frac{\epsilon}{\mu_0}\frac{\partial^2 h'}{\partial\beta_i\partial\epsilon_k} + \pi\frac{\partial^2 h'}{\partial\beta_i\partial\pi_k}, \\
\epsilon_{ikj}\frac{\partial h'}{\partial \beta_j} &= \frac{\partial \phi'_k}{\partial \epsilon_i} + b\frac{\partial^2 h'}{\partial\epsilon_i\partial\beta_k} - \frac{\epsilon}{\mu_0}\frac{\partial^2 h'}{\partial\epsilon_i\partial\epsilon_k} + \pi\frac{\partial^2 h'}{\partial\epsilon_i\partial\pi_k}, \\
\epsilon_{ikj}M_j + 2P_{[i}v_{k]} &= \frac{\partial \phi'_k}{\partial \pi_i} + b\frac{\partial^2 h'}{\partial\pi_i\partial\beta_k} - \frac{\epsilon}{\mu_0}\frac{\partial^2 h'}{\partial\pi_i\partial\epsilon_k} + \pi\frac{\partial^2 h'}{\partial\pi_i\partial\pi_k}, \\
0 &= \frac{\partial \phi'_{(k}}{\partial \pi_{i)}} + b\frac{\partial^2 h'}{\partial\pi_{(i}\partial\beta_{k)}} - \frac{\epsilon}{\mu_0}\frac{\partial^2 h'}{\partial\pi_{(i}\partial\epsilon_{k)}} + \pi\frac{\partial^2 h'}{\partial\pi_i\partial\pi_k}.
\end{aligned} \tag{8}$$

The eq. $(8)_4$ is the symmetric part of $(8)_3$, after that $(8)_3$ remains simply the definition of magnetization M_i. We note that, by dropping eqs. $(6)_{5,6,7}$, (8) and calculating the remaining ones in $\beta_i = 0$, $\epsilon_i = 0$, $\pi_i = 0$, $b = 0$, $\epsilon = 0$, $\pi = 0$, we obtain an important subsystem, i.e., the equation of the extended approach to dense gases and macromolecular fluids. These have been studied in [4] and we can use here the results. Similarly, by dropping eqs, $(6)_{1-4,7}$, (7), $(8)_{3,4}$ and calculating the remaining ones in $\lambda = 0$, $\lambda_i = 0$, $\lambda_{ij} = 0$, $\lambda_{ill} = 0$, $\pi = 0$, $\pi_i = 0$, we obtain the Maxwell equations. In the next section we will exploit their implications to eqs. (6), (7) and (8). At last, in section 3, we will consider the general case.

2 A supplementary conservation law for Maxwell equations

We have to consider the eq. $(6)_{5,6}$ and $(8)_{1,2}$ with $\pi = 0$ i.e.

$$\begin{aligned}
B_i &= \frac{\partial h'}{\partial \beta_i}; \quad -\mu_0\epsilon_0 E_i = \frac{\partial h'}{\partial \epsilon_i}, \\
\frac{1}{-\mu_0\epsilon_0}\epsilon_{ikj}\frac{\partial h'}{\partial \epsilon_i} &= \frac{\partial \phi'_k}{\partial \beta_i} + b\frac{\partial^2 h'}{\partial\beta_i\partial\beta_k} - \frac{\epsilon_0}{\mu_0}\frac{\partial^2 h'}{\partial\beta_i\partial\epsilon_k}, \\
\epsilon_{ikj}\frac{\partial h'}{\partial \beta_j} &= \frac{\partial \phi'_k}{\partial \epsilon_i} + b\frac{\partial^2 h'}{\partial\epsilon_i\partial\beta_k} - \frac{\epsilon_0}{\mu_0}\frac{\partial^2 h'}{\partial\epsilon_i\partial\epsilon_k};
\end{aligned} \tag{9}$$

clearly, here we haven' t to impose the Galilean invariance principle, with decomposition in velocity dependent and independent parts; in fact, the velocity doesn't occur in this equations. For this reason we have assumed a supplementary conservation law and not an entropy principle. From the representation theorems [6], [7] and [8] we know that $\phi'_k = \varphi_1\epsilon_k + \varphi_2\beta_k + \varphi_3\epsilon_{krs}\beta_r\epsilon_s$ with φ_1, φ_2, φ_3, h', b and ϵ functions of $G_{11} = \epsilon_i\epsilon_i$, $G_{12} = \epsilon_i\beta_i$, $G_{22} = \beta_i\beta_i$. After that the symmetric parts with respect to i and k of $(9)_{3,4}$ give 2 linear combinations of $\epsilon_i\epsilon_k$, $\epsilon_{(i}\beta_{k)}$, $\beta_i\beta_k$, δ_{ik}, $\epsilon_{(i}\epsilon_{k)rs}\epsilon_r\beta_s$ and $\beta_{(i}\epsilon_{k)rs}\epsilon_r\beta_s$ which must be zero; by setting equal to zero the coefficients of the last 2 of the above tensors, we find that φ_3 is a constant. The skew-symmetric parts, with respect to i and k, of eqs. $(9)_{3,4}$ are linear combinations of $\epsilon_{[i}\beta_{k]}$, $\epsilon_{ikj}\epsilon_j$, $\epsilon_{ikj}\beta_j$; putting equal to zero the coefficients of these last 2 tensors, we find

$$\begin{aligned}
\frac{\partial h'}{\partial G_{11}} &= \frac{\mu_0\epsilon_0}{2}\varphi_3; \quad \frac{\partial h'}{\partial G_{12}} = 0; \quad \frac{\partial h'}{G_{22}} = \frac{1}{2}\varphi_3 \quad \text{i.e.,} \\
h' &= \frac{1}{2}\varphi_3(G_{22} + \mu_0\epsilon_0 G_{11}) + \text{const} = \frac{1}{2}\varphi_3(\beta_i\beta_i + \mu_0\epsilon_0\epsilon_i\epsilon_i) + \text{const}
\end{aligned}$$

After that, what remains of eq. $(9)_3$ shows that

$$\frac{\partial \varphi_1}{\partial G_{12}} = 0, \quad \frac{\partial \varphi_1}{\partial G_{22}} = 0, \quad \frac{\partial \varphi_2}{\partial G_{12}} = 0, \quad \frac{\partial \varphi_2}{\partial G_{22}} = 0, \quad \varphi_2 = -b\varphi_3$$

and what remains of $(11)_4$ gives $\quad \dfrac{\partial \varphi_1}{\partial G_{11}} = 0, \quad \dfrac{\partial \varphi_2}{\partial G_{11}} = 0; \quad \varphi_1 = \epsilon\epsilon_0\varphi_3;$

in other words ϵ, b, φ_1, φ_2 and φ_3 are constant and $\varphi_1 = \epsilon\epsilon_0\varphi_3$, $\varphi_2 = -b\varphi_3$.

3 The case with polarization and magnetization

Consider now the general case, the problem of finding the functions

$$h'(\lambda, \lambda_i, \lambda_{ij}, \lambda_{ill}, \beta_i, \epsilon_i, \pi_i) \quad \text{and} \quad \phi'_k(\lambda, \lambda_i, \lambda_{ij}, \lambda_{ill}, \beta_i, \epsilon_i, \pi_i) \quad \text{satisfying}$$

eqs. (6), (7) and (8). We have already determined, in ref. [4] , their expressions in $\beta_i = 0$, $\epsilon_i = 0$, $\pi_i = 0$. Let us define now the functions $\Delta h'$ and $\Delta\phi'_k$ from

$$h'(\lambda, \lambda_i, \lambda_{ij}, \lambda_{ill}, \beta_i, \epsilon_i, \pi_i) = h'(\lambda, \lambda_i, \lambda_{ij}, \lambda_{ill}, 0, 0, 0) + \Delta h' \tag{10}$$
$$\phi'_k(\lambda, \lambda_i, \lambda_{ij}, \lambda_{ill}, \beta_i, \epsilon_i, \pi_i) = \phi'_k(\lambda, \lambda_i, \lambda_{ij}, \lambda_{ill}, 0, 0, 0) + \Delta\phi'_k ,$$

and note that they become zero when calculated in $\beta_i = 0$, $\epsilon_i = 0$, $\pi_i = 0$. Substitute eqs. (10) in the conditions emerging from (6), (7) and (8), i.e., $(7)_1$, $(8)_{1,2,4}$ thus obtaining

$$\frac{\partial \Delta\phi'_k}{\partial \lambda} = \frac{\partial \Delta h'}{\partial \lambda^k} - b\frac{\partial^2 \Delta h'}{\partial \lambda \partial \beta_k} + \frac{\epsilon}{\mu_0}\frac{\partial^2 \Delta h'}{\partial \lambda \partial \epsilon_k} - \pi\frac{\partial^2 \Delta h'}{\partial \lambda \partial \pi_k}, \tag{11}$$
$$\frac{\epsilon_{ikj}}{-\mu_0\epsilon_0}\frac{\partial \Delta h'}{\partial \epsilon_j} = \frac{\partial \Delta\phi'_k}{\partial \beta_i} + b\frac{\partial^2 \Delta h'}{\partial \beta_i \partial \beta_k} - \frac{\epsilon}{\mu_0}\frac{\partial^2 \Delta h'}{\partial \beta_i \partial \epsilon_k} + \pi\frac{\partial^2 \Delta h'}{\partial \beta_i \partial \pi_k},$$
$$\epsilon_{ikj}\frac{\partial \Delta h'}{\partial \beta_j} = \frac{\partial \Delta\phi'_k}{\partial \epsilon_i} + b\frac{\partial^2 \Delta h'}{\partial \epsilon_i \partial \beta_k} - \frac{\epsilon}{\mu_0}\frac{\partial^2 \Delta h'}{\partial \epsilon_i \partial \epsilon_k} + \pi\frac{\partial^2 \Delta h'}{\partial \epsilon_i \partial \pi_k},$$
$$0 = \frac{\partial \Delta\phi'_{(k}}{\partial \pi_{i)}} + b\frac{\partial^2 \Delta h'}{\partial \pi_{(i} \partial \beta_{k)}} - \frac{\epsilon}{\mu_0}\frac{\partial^2 \Delta h'}{\partial \pi_{(i} \partial \epsilon_{k)}} + \pi\frac{\partial^2 \Delta h'}{\partial \pi_i \partial \pi_k}.$$

When considering only the Maxwell equations, we have obtained that b and ϵ are constants; this suggests to restrict ourselves, also in this general case, to the solutions with b, ϵ and π not depending on λ, β_i, ϵ_i, π_i, for the sake of simplicity. By defining

$$\phi''_k = \Delta\phi'_k + b\frac{\partial \Delta h'}{\partial \beta_k} - \frac{\epsilon}{\mu_0}\frac{\partial \Delta h'}{\partial \epsilon_k} + \pi\frac{\partial \Delta h'}{\partial \pi_k} \tag{12}$$

the eqs. (11) become

$$\frac{\partial \Delta h'}{\partial \lambda^k} = \frac{\partial \phi''_k}{\partial \lambda}; \quad \frac{-\epsilon_{ikj}}{\mu_0\epsilon_0}\frac{\partial \Delta h'}{\partial \epsilon_j} = \frac{\partial \phi''_k}{\partial \beta_i}; \tag{13}$$
$$\epsilon_{ikj}\frac{\partial \Delta h'}{\partial \beta_j} = \frac{\partial \phi''_k}{\partial \epsilon_i}; \quad 0 = \frac{\partial \phi''_{(k}}{\partial \pi_{i)}}.$$

The symmetric parts with respect to i and k of $(13)_{2-4}$ show (with the same proof which deduces a rigid motion if the deformation tensor is zero) that ϕ''_k is linear both in ϵ_i that in β_i and in π_i ,i.e.,

$$\phi''_k = \phi_{kabc}\beta_a\epsilon_b\pi_c + \overset{3}{\phi}_{kab}\beta_a\epsilon_b + \overset{2}{\phi}_{kab}\pi_a\epsilon_b + \overset{1}{\phi}_{kab}\pi_a\beta_b +, \tag{14}$$
$$+ \overset{1}{\phi}_{ka}\epsilon_a + \overset{2}{\phi}_{ka}\beta_a + \overset{3}{\phi}_{ka}\pi_a + \overset{''0}{\phi_k}$$

where ϕ_{kabc}, $\overset{i}{\phi}_{kab}$, $\overset{i}{\phi}_{ka}$, $\overset{''0}{\phi_k}$ doesn't depend on β_i, ϵ_i and π_i; moreover, still the symmetric parts of $(13)_{2-4}$ show that ϕ_{kabc}, $\overset{i}{\phi}_{kab}$, $\overset{i}{\phi}_{ka}$ change sign when we exchange the index k with whatever of the other indices. But we can exchange whatever couple of indices trough 3 changes of indices involving the first one; it follows that ϕ_{kabc}, $\overset{i}{\phi}_{kab}$, $\overset{i}{\phi}_{ka}$ are skew-symmetric tensors for every couple of indices. But in ϕ_{kabc} at least one of the indices 1 2 3 occurs 2 times; therefore, we have $\phi_{kabc} = 0$. Moreover, $\overset{i}{\phi}_{kab}$ is not zero only when $k\ a\ b$ is 1 2 3 or anyone of its permutations; therefore, it is proportional to ϵ_{kab}. In other words, the scalars $\overset{i}{\varphi}(\lambda, \lambda_r, \lambda_{rs}, \lambda_{rll})$ and the vectors $\overset{i}{v_b}(\lambda, \lambda_r, \lambda_{rs}, \lambda_{rll})$ exist, such that

$$\overset{i}{\phi}_{kab} = \overset{i}{\varphi}\epsilon_{kab}; \quad \overset{i}{\phi}_{ka} = \epsilon_{kab}\overset{i}{v_b}. \tag{15}$$

These partial results simplify very much the exploitation of conditions (13), although the passages remain long and tedious, so that we report simply the final results, i.e., the expressions for the functions $\Delta h'$ and $\Delta\phi'_k$; they are the first three rows of the following eq. (16) and the first five rows of the eq. (17), respectively. The remaining rows are the expressions of $h'(\lambda, \lambda_i, \lambda_{ij}, \lambda_{ill}, 0, 0, 0)$ and $\phi'_k(\lambda, \lambda_i, \lambda_{ij}, \lambda_{ill}, 0, 0, 0)$ found in ref. [4] (up to second order in the variables λ_i, $\lambda_{<ij>}$, λ_{ill}); their sum, according to eq. (10), gives the functions h' and ϕ'_k, i.e.,

$$h' = \frac{1}{2}\overset{3}{\varphi}(\mu_0\epsilon_0\epsilon_j\epsilon_j + \beta_j\beta_j) - \mu_0\epsilon_0\overset{1}{\varphi}\epsilon_j\pi_j + \overset{2}{\varphi}\beta_j\pi_j + \epsilon_{rjb}\lambda_r\epsilon_j\overset{11}{v_b} + \tag{16}$$
$$+\epsilon_{rjb}\lambda_r\beta_j\overset{21}{v_b} + \mu_0\epsilon_0\epsilon_j(\overset{20}{v_j} + \overset{21}{v_j}\lambda) - \beta_j(\overset{10}{v_j} + \overset{11}{v_j}\lambda) +$$
$$+\frac{\partial \overset{3}{v_{b_0}}}{\partial \lambda}\epsilon_{kab}\lambda_k\pi_a + \pi_r H_r(\lambda, \lambda_{ia}, \lambda_{ill}, \pi_k) +$$
$$-\frac{8}{27\cdot 35}G'(\lambda)\lambda_{ll}^{-3/2} - \frac{2}{21}G'(\lambda)\lambda_{ll}^{-7/2}\lambda^i\lambda_{ill} +$$
$$\frac{2}{9\cdot 35}G''(\lambda)\lambda_{ll}^{-5/2}\lambda_i\lambda^i - \frac{2}{105}G'(\lambda)\lambda_{ll}^{-7/2}\lambda_{<ij>}\lambda_{<ij>} +$$
$$\frac{1}{2}G(\lambda)\lambda_{ll}^{-9/2}\lambda_{ill}\lambda_{ill},$$

$$\begin{aligned}
\phi'_k = {} & \varphi^3(\epsilon_{kab}\beta_a\epsilon_b - b\beta_k + \epsilon\epsilon_0\epsilon_k) + \varphi^2(\epsilon_{kab}\pi_a\epsilon_b - b\pi_k - \pi\beta_k) + \\
& + \varphi^1(\epsilon_{kab}\pi_a\beta_b - \epsilon\epsilon_0\pi_k + \epsilon_0\mu_0\pi\epsilon_k) + v_b^{10}\epsilon_{kab}\epsilon_a + v_b^{20}\epsilon_{kab}\beta_a + \\
& + v_b^{21}(\epsilon_{kab}\beta_a\lambda + \epsilon_b\lambda_k - \delta_{kb}\epsilon_r\lambda_r) + \\
& + v_b^{11}\left(\epsilon_{kab}\epsilon_a\lambda - \beta_b\lambda_k\frac{1}{\epsilon_0\mu_0} + \delta_{kb}\beta_r\lambda_r\frac{1}{\epsilon_0\mu_0}\right) + \\
& - \pi\left[\pi_r\frac{\partial H_r}{\partial \pi_k} + H_k - (H_k)_{\pi_r=0}\right] + \epsilon_{kab}\pi_a[v_{b1}^3 + v_{b0}^3] + \\
& \frac{4}{9\cdot 35}G'(\lambda)\lambda_{ll}^{-5/2}\lambda_k + \left[-\frac{2}{21}G(\lambda)\lambda_{ll}^{-7/2} + f_1(\lambda_{ll})\right]\lambda_{kll} + \\
& \left[\frac{2}{5}G(\lambda)\lambda_{ll}^{-3/2} + f_2(\lambda_{ll})\right]\lambda_{<kr>}\lambda_{rll} - \frac{4}{105}G'(\lambda)\lambda_{ll}^{-7/2}\lambda_{<kr>}\lambda_r\,.
\end{aligned} \tag{17}$$

Here, φ^i, v_b^{11}, v_b^{21}, v_b^{10}, v_b^{20} are functions of λ_{rs}, λ_{rll},
b, ϵ, π, v_{b1}^3 are functions of λ_r, λ_{rs}, λ_{rll},
v_{b0}^3 is function of λ, λ_{rs}, λ_{rll},
H_r is function of λ, λ_{rs}, λ_{rll}, π_r ;
$G(\lambda)$ is function of λ, $f_1(\lambda_{ll})$ and $f_2(\lambda_{ll})$ are functions of λ_{ll};
they are arbitrary functions restricted only by

$$\pi\frac{\partial^2 v_{b_0}^3}{\partial\lambda^2} = 0 \text{ and } \pi\left(H_k\right)_{\pi_r=0} = (bv_k^{11} + \epsilon\epsilon_0 v_k^{21})\lambda + \pi H_k^*(\lambda_{ia}, \lambda_{ill}), \tag{18}$$

with H_k^* another arbitrary function of its variables.
The first terms of eq. (16) and the first one of eq. (17) are the same of the corresponding ones in sect 2, for the Maxwell equations. The only difference is that here φ^3 may depend on λ_{ij}, λ_{ill} , while in sect 2 it was a constant. It is easy to verify that, in this way, eqs. (11) are satisfied.
The expressions (16) and (17) can now be inserted in eqs. (6), $(7)_{2-4}$, $(8)_3$ and give F, F_i, F_{ij}, F_{ill}, B_i, E_i, P_i, G_{ik}, G_{ijk}, G_{ikll}, M_i as functions of the parameters λ, λ_i, λ_{ij}, λ_{ill}, β_i, ϵ_i, π_i. The first ones of these functions can be used to obtain the parameters as functions of F, F_i, F_{ij}, F_{ill}, B_i, E_i, P_i; by inserting these in the remaining ones, we obtain the constitutive functions G_{ik}, G_{ijk}, G_{ikll}, M_i as functions of the independent variables F, F_i, F_{ij}, F_{ill}, B_i, E_i, P_i. In this way the requested closure has been obtained. We apologize because we cannot report these passages in the only 8 pages allowed for these proceedings; the interested reader may do them by himself, because they are straightforward, or may ask us to send them privately. The same thing we have to say for the other constitutive functions $P_{<ij>}$, P_{ill}, $J_i^* = J_i^F - q^F v_i$.

Conclusions
We retain the results of the present paper very satisfactory, because they allow to study also polarizable and magnetizable fluids in the framework of the well established theory of Extended Thermodynamics. The field equations to be solved are (1), (2) and (3) closed in the above mentioned way; although apparently complicate they can be put in the symmetric hyperbolic form by simply changing the independent variables, so predicting finite speeds of wave propagations. There remains to understand the physical meaning of the arbitrary functions still remaining in our closure. Some of them depend upon the particular fluid treated, and are related to the state functions; and the others? are zero, perhaps? This will be argument of further investigation.

Thanks: We thank anonymous referees; they helped us in improving the presentation of this article.

References

1. Liu, I-S., Muller I.: Extended thermodynamics of classical and degenerate gases, Arch. Rational Mech. Anal 83 (1983)
2. Müller, I., Ruggeri, T.: Rational Extended Thermodynamics. Spinger-Verlag, New York, Berling Heidelberg. (1998)
3. Liu, I-S.: On the Structure of Balance Equations and Extended Field Theories of Mechanics. IL NUOVO CIMENTO, **92B**, 121 (1986)
4. Carrisi, M.C., Demontis, F., Scanu, A.: A kinetic type extended model for dense gases and macromolecular solution, accepted for publication in LE MATEMATICHE
5. Liu, I-S.,Müller,I: On the Thermodymamics and Thermostatics of Fluids in Electromagnetic Fields. Arch. Rational Mech. Anal. **46**,149. (1972)

6. Smith, G.F.: On Isotropic Functions of Symmetric Tensors, Skew-symmetric Tensors and Vectors. Int J. Engng. Sci., **9**, 899 (1971)
7. Pennisi, S.,Trovato, M.: On the Irreducibility of Professor G.F. Smith's Representation of Isotropic Functions. Int J. Engng. Sci., **25**, 1059 (1987)
8. Pennisi, S.: Representation Theorems for Isotropic Functions: Estension to the Case of Pseudo-tensors. Ricerche di Matematica. **47**, 181 (1998)

Quantum Corrected Drift–Diffusion Modeling and Simulation of Tunneling Effects in Nanoscale Semiconductor Devices

G. Cassano[1], C. de Falco[2], C. Giulianetti[1], and R. Sacco[1]

[1] Dipartimento di Matematica "F. Brioschi", Politecnico di Milano, Via Bonardi 9, 20133 Milano, Italy
[2] Dipartimento di Matematica "F. Enriques", Università degli Studi di Milano, via Saldini 50, 20133 Milano, Italy

Abstract In this communication, we deal with the numerical approximation of a Quantum Drift–Diffusion model capable of describing tunneling effects through the thin oxide barrier in nanoscale semiconductor devices. We propose a novel formulation of the mathematical model, based on a spatially heterogeneous approach, and a generalization of the Gummel decoupled algorithm, widely adopted in the case of the Drift-Diffusion system. Then, we address the finite element discretization of the linearized problems obtained after decoupling, proving well-posedness and a discrete maximum principle for each of them. Finally, we validate the physical accuracy and numerical stability of the proposed algorithms on the numerical simulation of a real-life nanoscale device.

1 Introduction and Motivation

In this work, we propose a novel mathematical formulation and numerical approximation of the Quantum Drift–Diffusion model with Tunneling (QDDT). This model was introduced in [1, 2], and is a suitable generalization of the Quantum Drift–Diffusion system (QDD) including a macroscopic description of tunneling through a thin oxide barrier. A possible approach to this latter problem is shown in [3], where a fitting parameter is introduced in the constitutive relation for the quantum correction to the electric potential. Another modeling approach was recently proposed in [2] and applied in [1] to Gate–Oxide tunneling in a MOS structure. The model in [1] handles the different phenomena which govern transport in the different regions of the device, distinguishing between the semiconductor and polysilicon regions, where the scattering mechanisms are more relevant (*viscous flow*), and the oxide region where transport is essentially inertia dominated (*ballistic flow*). In the present article, we focus on a novel mathematical reformulation of the model of [1] and on its proper numerical discretization. With this aim, we devise an efficient and stable simulation procedure based on a suitable generalization of Gummel's decoupled algorithm, which has several advantages:

- the solution of the full nonlinear QDDT system (that could be faced, for instance, by resorting to Newton's method, as done in [1]) is reduced into the successive solution of linearized differential subproblems of smaller size;
- each subproblem can be properly treated by adopting a suitable mathematical and numerical formulation, in order to easily enforce some prescribed constraint on the solution (in particular the strict positivity of the computed carrier densities).

The physical accuracy of the novel formulation is successfully validated in the simulation of a realistic nanoscale device, for which experimental measurements are available for comparison.

2 Quantum Drift–Diffusion Model Including Tunneling

In this section, we present the QDDT model as proposed in [1] for computing the tunneling current through the ultra-thin gate oxide of a MOSFET device. Considering the device represented in Fig. 1 (left), we are interested in simulating the 1–d cross–section B–B. The computational domain Ω is composed of three parts: $\Omega_{poly} = (0, \alpha)$ representing the polysilicon gate contact, $\Omega_{ox} = (\alpha, \beta)$, representing the silicon–dioxide gate insulator and $\Omega_{Si} = (\beta, L)$, representing the silicon bulk of the device.
The QDDT model consists of the following set of equations to be solved in Ω_{Si} and Ω_{poly}

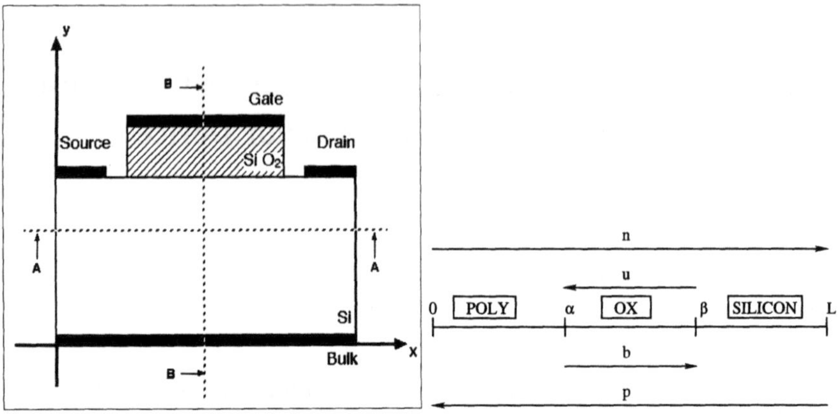

Fig. 1. Left: two–dimensional cross–section of a MOS transistor. The B–B section indicates the heterostructure. Right: one–dimensional scheme of the heterostructure and convention chosen to describe charge transport

$$
\begin{cases}
-\left(\varepsilon\varphi_x\right)_x = q\left(r^2 - s^2 + D\right) \\
-\left(s^2\mu_n\left(\varphi_n\right)_x\right)_x = U \\
\left(r^2\mu_p\left(\varphi_p\right)_x\right)_x = U \\
-\alpha_n s_{,xx} + s\left(-\varphi + 2V_{th}\ln\left(\frac{s^2}{\sqrt{n_{int}}}\right) + \varphi_s\right) = 0 \\
-\alpha_p r_{,xx} + r\left(\varphi + 2V_{th}\ln\left(\frac{r^2}{\sqrt{n_{int}}}\right) - \varphi_r\right) = 0
\end{cases} \tag{1}
$$

and the following set of equations to be solved in Ω_{ox}

$$
\begin{cases}
-\left(\varepsilon\varphi_x\right)_x = q\left(r^2 - s^2 + v^2 - z^2\right) \\
\varphi_{\nu,x} = 0 \qquad \nu = s, r, v, z \\
-\alpha_n s_{,xx} + s\left(-\varphi + 2V_{th}\ln\left(\frac{s^2}{\sqrt{n_{int}}}\right) + \varphi_s\right) = 0 \\
-\alpha_p r_{,xx} + r\left(\varphi + 2V_{th}\ln\left(\frac{r^2}{\sqrt{n_{int}}}\right) - \varphi_r\right) = 0 \\
-\alpha_n z_{,xx} + z\left(-\varphi + 2V_{th}\ln\left(\frac{z^2}{\sqrt{n_{int}}}\right) + \varphi_s\right) = 0 \\
-\alpha_p v_{,xx} + v\left(\varphi + 2V_{th}\ln\left(\frac{v^2}{\sqrt{n_{int}}}\right) - \varphi_r\right) = 0,
\end{cases} \tag{2}
$$

where the notation $f_{,x}$, $f_{,xx}$, etc., has been used to indicate the derivative(s) of a function $f = f(x)$ with respect to the spatial coordinate x.

We notice that $(1)_1$ and $(2)_1$ are the Poisson equations relating the electric potential φ to the charge density in the device. We also set

$$
s = \sqrt{n},\ r = \sqrt{p},\ v = \sqrt{b},\ z = \sqrt{u}, \tag{3}
$$

where n, p, b and u are the charge densities shown in Fig. 1 (right). As a matter of fact, the physical description of charge transport in the oxide requires to introduce two kinds of carriers flowing throughout the device, electrons and holes, as divided into two distinct populations, each associated with the device contact from which it is emitted and mathematically described by an individual statistics. As a consequence, we can define: a) *backward tunneling electrons* u and *backward tunneling holes* b; b) *forward tunneling electrons* n and *forward tunneling holes* p.

Moreover, φ_n, φ_p, φ_s, φ_r, φ_z and φ_v are the quasi–Fermi potentials associated with n, p, s, r, z and v. The function U is the net recombination rate (see [1] for its physical modeling)

$$
U = \frac{np - n_{eq}p_{eq}}{\tau_p(n + \sqrt{n_{eq}p_{eq}}) + \tau_n(p + \sqrt{n_{eq}p_{eq}})}, \tag{4}
$$

where the (non spatially constant) equilibrium electron and hole concentrations n_{eq} and p_{eq} are such that U vanishes at thermal equilibrium, as physically expected. The given function D is the net doping profile of the device. The constants q and V_{th}, are the electron charge and the thermal voltage, $\alpha_n = \hbar^2/(6\,q\,m_n^*)$ and $\alpha_p = \hbar^2/(6\,q\,m_p^*)$, where $\hbar$, m_n^* and m_p^* represent the modified Planck constant and the effective masses for electrons and holes, while ε is the permittivity of each material in the heterostructure and is a piecewise constant function over Ω. Finally μ_n

and μ_p are the electron and hole mobilities. A Dirichlet boundary condition is imposed for the electric potential φ at the device boundaries $x = 0$ and $x = L$, while at the material interfaces $x = \alpha$ and $x = \beta$ continuity of the electric displacement $\varepsilon\varphi_{,x}$ is enforced. Analogously, both carrier densities and quasi-Fermi levels satisfy a Dirichlet condition at the device boundaries.

To properly describe tunneling effects, in [1] it was assumed that particles coming through the oxide interface do not interact with the *upstream* barrier (at $x = \alpha$ for s electrons and v holes, at $x = \beta$ for z electrons and r holes). This amounts to stating that carriers do not experience electric forces able to modify their energy, during their travel across the potential barrier arising at the silicon and polysilicon dioxide interfaces. Therefore, a reasonable approach to account for the presence of an energy discontinuity (the barrier) is to enforce suitable conditions at the *downstream* interface (at $x = \beta$ for s electrons and v holes, at $x = \alpha$ for z electrons and r holes), where the particles enter the semiconductor and the polysilicon regions, respectively. These *upstream* conditions for densities and quasi-Fermi potentials read

$$\begin{aligned} s(\alpha^-) &= s(\alpha^+), & \alpha^n_{poly} s_{,x}(\alpha^-) &= \alpha^n_{ox} s_{,x}(\alpha^+), \\ r(\beta^-) &= r(\beta^+), & \alpha^p_{ox} r_{,x}(\beta^-) &= \alpha^p_{Si} r_{,x}(\beta^+), \\ r(\alpha^-) &= v(\alpha^+), & \alpha^p_{poly} r_{,x}(\alpha^-) &= \alpha^p_{ox} v_{,x}(\alpha^+), \\ z(\beta^-) &= s(\beta^+), & \alpha^n_{ox} z_{,x}(\beta^-) &= \alpha^n_{Si} s_{,x}(\beta^+), \end{aligned} \tag{5}$$

and

$$\begin{aligned} \varphi_s(\alpha^-) &= \varphi_s(\alpha^+), & \varphi_s(\beta^+) &= \varphi_v(\beta^-), \\ \varphi_r(\beta^-) &= \varphi_r(\beta^+), & \varphi_r(\alpha^+) &= \varphi_v(\alpha^-). \end{aligned} \tag{6}$$

In a similar way, the *downstream* conditions for the carrier densities read

$$\begin{aligned} s_{,x}(\beta^-) &= 0, & z_{,x}(\alpha^+) &= 0 \\ r_{,x}(\alpha^+) &= 0, & v_{,x}(\beta^-) &= 0, \end{aligned} \tag{7}$$

while the corresponding expressions for the current densities are

$$\begin{aligned} -\mu_n s^2 \varphi_{s,x}(\alpha^-) &= \gamma_z z^2(\alpha^+) - \gamma_s s^2(\beta^-), \\ -\mu_p r^2 \varphi_{r,x}(\alpha^-) &= \gamma_r r^2(\alpha^+) - \gamma_v v^2(\beta^-), \\ -\mu_n s^2 \varphi_{s,x}(\beta^+) &= \gamma_z z^2(\alpha^+) - \gamma_s s^2(\beta^-), \\ -\mu_p r^2 \varphi_{r,x}(\beta^+) &= \gamma_r r^2(\alpha^+) - \gamma_v v^2(\beta^-), \end{aligned} \tag{8}$$

γ_s, γ_z, γ_r and γ_v being the tunneling velocities associated with each carrier type that are in general used as fitting parameters in the numerical simulation. According to the model described in this section, transport is *scattering dominated* in polysilicon and silicon, while it is *inertia dominated* in the gate–oxide. One remarkable feature of this model is that the *inertia dominated* transport in the oxide region is accounted for by imposing the non-local interface conditions (7) and (8).

3 Mathematical Reformulation and Functional Iteration Technique

In this section we provide a suitable mathematical reformulation of the QDDT model presented in Sect. 2. The main novelty compared to [1] is that the present approach lends itself in a very natural way towards a *fully decoupled* iterative solution of the whole system. A flow–chart of the modified Gummel iteration scheme (which is described in more detail in [4]) is presented in Fig. 2.

In what follows we briefly describe the subproblems that constitute each block of the iteration procedure. For sake of simplicity, we consider only the equations for the electrons, upon a suitable scaling is performed (see [4]), as a completely similar treatment holds for the holes.

The first step in the decoupled scheme consists of the solution of the nonlinear Poisson equation in the whole device domain Ω, supplied with Dirichlet boundary conditions

$$\begin{cases} -(\lambda^2(x)\varphi_{,x})_{,x} = \rho(\varphi) & \text{in } \Omega = (0, L) \\ \varphi(0) = \varphi_0, \qquad \varphi(L) = \varphi_L & \end{cases} \tag{9}$$

where ρ is the net charge density including the electron and hole concentrations (which nonlinearly depend on φ) and the doping concentration. For the solution of (9) we adopt a standard Newton linerization and a piecewise linear finite element discretization.

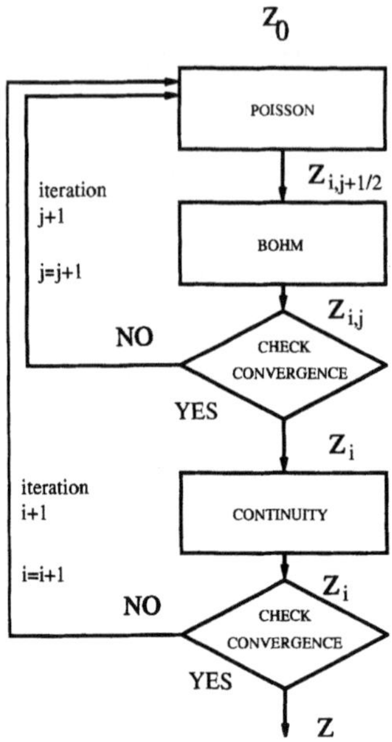

Fig. 2. Flow–chart of the modified Gummel iteration scheme

For the solution of the Bohm equation, which constitutes the second step in the iteration procedure, we define the new variables w and φ_w as follows

$$w = \begin{cases} z \text{ in } \Omega_{Ox} \\ s \text{ in } \Omega_{Si} \end{cases} \qquad \varphi_w = \begin{cases} \varphi_z \text{ in } \Omega_{Ox} \\ \varphi_s \text{ in } \Omega_{Si} \end{cases} \tag{10}$$

and we denote by s and φ_s the restrictions of s and φ_s (as previously defined) to $\Omega_{Poly} \cup \Omega_{Ox}$. After the above change of variables, the Bohm problem for electrons reduces to the successive solution of the following nonlinear boundary value problems supplied with Dirichlet–Neumann boundary conditions

$$\begin{cases} -\delta_s^2\, s_{,xx} + s\,(-\varphi + 2\ln(s) + \varphi_s) = 0 \text{ in } \Omega_{Poly} \cup \Omega_{Ox} \equiv (0,\beta) \\ s(0) = s_0, \qquad s_{,x}(\beta) = 0 \end{cases}$$
$$\begin{cases} -\delta_w^2\, w_{,xx} + w\,(-\varphi + 2\ln(w) + \varphi_w) = 0 \text{ in } \Omega_{Ox} \cup \Omega_{Si} \equiv (\alpha, L) \\ w_{,x}(\alpha) = 0, \qquad w(L) = w_L. \end{cases} \tag{11}$$

For the solution of (11) we adopted the modified Newton iteration described in [4] to preserve the positivity of s and w at each step, and a piecewise linear finite element discretization.

To describe the last step of the iteration scheme, which consists of the solution of the continuity equations, let us now introduce the following new variables

$$n = \begin{cases} s^2 \text{ in } \Omega_{Poly} \\ w^2 \text{ in } \Omega_{Si}, \end{cases} \qquad G_n = \begin{cases} G_s = -\varphi + \varphi_s + 2\ln s \quad \text{in } \Omega_{Poly} \\ G_w = -\varphi + \varphi_w + 2\ln w \text{ in } \Omega_{Si}, \end{cases}$$

where G_s and G_w are the quantum corrections to the electric potential (Bohm potentials). Then, the electron continuity equations reduce into the following one

$$\begin{cases} -(J_n)_{,x} = -U \qquad \text{in } \Omega_{Poly} \cup \Omega_{Si} \\ J_n = \mu_n\,(n_{,x} - (\varphi + G_n)_{,x}) \\ n(0) = s_0^2, \qquad n(L) = w_L^2, \\ J_n(\alpha) = J_n(\beta) = \left(\gamma_z w^2(\alpha) - \gamma_s s^2(\beta)\right). \end{cases} \tag{12}$$

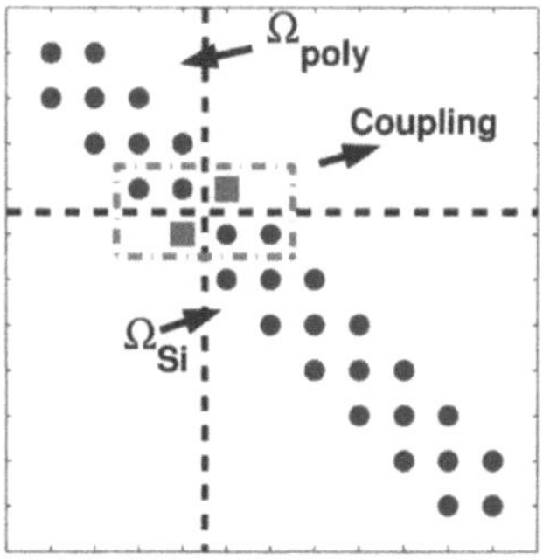

Fig. 3. Structure of the Matrix deriving from the FEM discretization of continuity equation

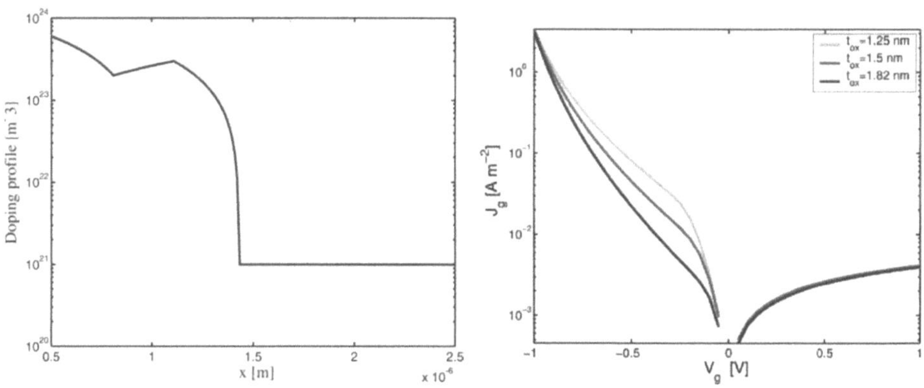

Fig. 4. Left: gate doping. Right: I-V characteristics

Let $V \equiv \varphi + G_n$. Then, since the quasi–Fermi levels are constant in Ω_{Ox}, we have

$$w^2(\alpha) = w^2(\beta)e^{V(\alpha)-V(\beta)} = n(\beta)e^{V(\alpha)-V(\beta)}$$

$$s^2(\beta) = s^2(\alpha)e^{V(\beta)-V(\alpha)} = n(\alpha)e^{V(\beta)-V(\alpha)},$$

from which

$$J_n(\alpha) = J_n(\beta) = \left(\gamma_z n(\beta)e^{V(\alpha)-V(\beta)} - \gamma_s n(\alpha)e^{V(\beta)-V(\alpha)}\right).$$

Note that the above reformulation of the interface condition yields a maximum principle in the discrete version of the continuity equation provided that a Scharfetter–Gummel finite element scheme is adopted [5, 6]. As a matter of fact, denoting by M the number of internal nodes in the polysilicon mesh and by N the number of internal nodes in the silicon mesh, then the matrix stemming from the discretization of (12) is of the form

$$A = \begin{bmatrix} A_1 & A_2 \\ A_3 & A_4 \end{bmatrix} \tag{13}$$

where the diagonal bloks A_1 (of size $M+1$) and A_4 (of size $N+1$) are the same as produced by the discretization of $(12)_1$ in Ω_{poly} and Ω_{Si} respectively, and the coupling between the two subdomains is expressed by rows $M+1$ and $M+2$ which read

$$\begin{array}{c|c|c||c|c|c}
0 & -\frac{\mu_M}{h_M}B_M^- & \frac{\mu_M}{h_M}B_M^+ + \gamma_{M+1} & -\gamma_{M+2} & 0 & \ldots \\
\hline
\ldots & 0 & -\gamma_{M+1} & \frac{\mu_{M+1}}{h_{M+1}}B_{M+1}^+ + \gamma_{M+2} & -\frac{\mu_{M+1}}{h_{M+1}}B_{M+1}^+ & 0
\end{array}$$

where $B_M^{\pm}$ denotes the inverse of the Bernoulli function evaluated at $\pm(\varphi_{M+1} - \varphi_M)$. It is easy to check that the complete matrix A is strictly diagonally dominant by columns, that the diagonal entries are positive and that the off–diagonal ones are negative. This implies that A is an M–matrix, and that a discrete maximum principle holds for the computed electron density, provided that the standard splitting of the generation/recombination term is adopted [5].

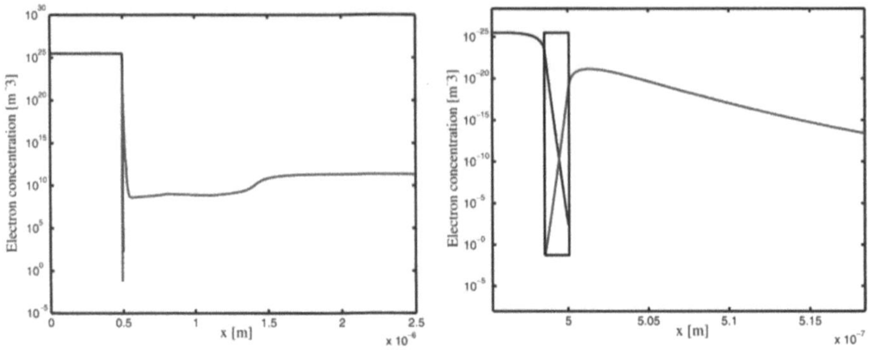

Fig. 5. Electron concentration at thermal equilibrium

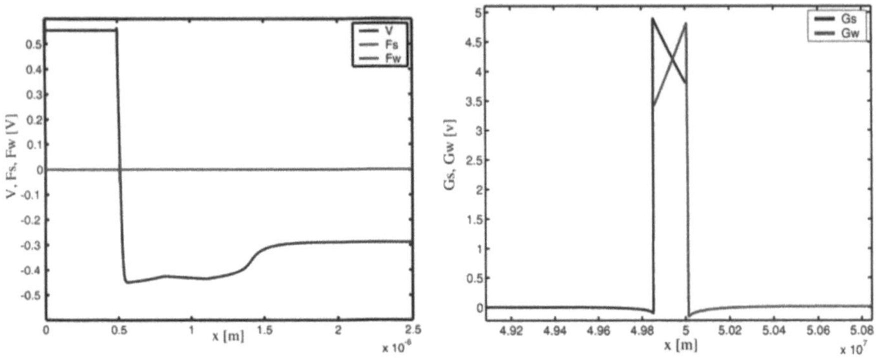

Fig. 6. Electric potential, Bohm potentials and quasi–Fermi potentials

4 Numerical Results

As a benchmark for the physical assessment of the model described above we performed a simulation of a 1–d MOS structure similar to that discussed in [1] for which experimental data are available. This structure represents a cross section in the Bulk–Gate direction of a MOS transistor with Source and Drain contacts floating. The sizes of the subregions of the simulated device are as follows:

- $t_{Poly} = 500nm$
- $t_{ox} = 1.25,\ 1.5,\ 1.82\, nm$
- $t_{Si} = 2\,\mu\, m,$

and the doping profile (of n^+-type, with $N_D^+ = 3 \cdot 10^{25}\, m^{-3}$) is shown in Fig. 4 (left). Notice the ability of the formulation in capturing the extremely steep layers arising in the electron density at material interfaces (Fig. 5). Moreover, the very high value attained by the quantum correction inside the oxide region is comparable to the height of the oxide barrier (Fig. 6, right). Finally, it is remarkable to point out that the computed I-V curves shown in Fig. 4 (right) are in very good agreement with the measurement results reported in [1].

5 Acknowledgments

This research was partially supported by the INDAM 2003 Grant "Modellistica Numerica per il Calcolo Scientifico e Applicazioni Avanzate".

References

1. M. G. Ancona, Z. Yu, R. W. Dutton, P. J. V. Voorde, M. Cao, D. Vook, Density-Gradient Analysis of MOS Tunneling, IEEE Trans. Electron Devices 47 (12) (2000) 2310–2319

2. M. Ancona, Macroscopic description of quantum–mechanical tunneling, Physical Review B 42 (2) (1990) 1222–1233
3. T. Höhr, A. Schenk, A. Wittestein, W. Fichtner, On density–gradient modeling of tunneling through insulator, IEICE Trans. Electron. E86–C (3)
4. C. de Falco, E. Gatti, A. L. Lacaita, R. Sacco, Quantum-corrected drift-diffusion models for transport in semiconductor devices, Mox Technical Report n.40, May 2004. To appear in Journal of Computational Physics
5. J. W. Jerome, Analysis of Charge Transport, Springer Verlag, New York, 1996
6. E. Gatti, S. Micheletti, R. Sacco, A New Galerkin Framework for the Drift-Diffusion Equation in Semiconductors, East West J. Numer. Math. 6 (2) (1998) 101–135

Reverse Statistical Modeling for Analog Integrated Circuits

A. Ciccazzo, V. Cinnera Martino, A. Marotta, and S. Rinaudo

ST Microelectronics, Stradale Primosole 50, 95121, Catania, Italy,
{angelo.ciccazzo,valeria.cinnera-martino,angelo.marotta,salvatore.rinaudo}@st.com

Abstract As the IC manufacturing process becomes more complex, circuit performance becomes more sensitive to statistical process variations. Therefore, it is essential to be able to statistically characterize IC manufacturing process fluctuations and to reliably predict circuit performance spreads at the design stage. A full statistical modeling flow for integrated circuits, which uses the information related to the measurements of device performance and which the aim is to extract a Spice like statistical model, is presented. The technique shown, innovative compared to the existent ones, is based on several Monte Carlo simulation steps, in order to estimate the second order moments for every statistical model parameter; afterwards, an optimization phase follows, with the aim to identify the cross-correlation among the Spice parameters. The operations flow has been validated on a diode and IGBT device.

Key words: Statistical modeling, integrated circuits, parameters extraction, optimization.

1 Introduction

Analog integrated circuits are characterized by a series of performances that are measured at the end of their production in order to test whether their values satisfy the design constraints. A device, usually, is replied on several dies of a wafer and on several wafers but every retort does not result in compliance with the others, in terms of electrical performance, because the fabrication steps are affected by various factors that make aleatory the outcome. Some of these factors are: imperfections that characterize the masks and tolerances in their positioning, various effects of the ionic implantation, variation of the temperature during the production, tolerances in the dimensions, etc.

Generally the process fluctuations produce fluctuations in the electrical performances; consequently, during circuit design the device statistical modeling is fundamental in order to estimate and to take into account the fluctuations that would characterize the electrical behavior.

In order to check the device fabrication process, each wafer contains few sites with special test structures, which enable the measurement of device performances and constitute a statistical database for the electrical behavior of the device in issue ([CC01]). The statistical database of electrical measurements is named, in this paper, T84, or experimental statistics T84, from the testing machine name used in STMicroelectronics.

The aim of the proposed flow is the following: on the base of this information, that constitutes the experimental statistics, we want to map the performances space $T = \{T_1, \ldots, T_t\}$ such as gain and bandwidth, to circuit parameters space $P = \{P_1, \ldots, P_n\}$ e.g. Spice parameters or circuit components values (see Fig. 1). Variations in the fabrications process cause random fluctuations in T space, which in turn cause P to fluctuate ([KC93],[MD02]).

In other words, we want to extract a Spice model whose parameters are aleatory variables; each variable is characterized by a probability distribution function supposed Gaussian, in agreement with Central Limit Theorem; therefore, for the model parameters which have to be statistically described, it is necessary to identify the medium value, the standard deviation and the correlation coefficients. In order to carry out the statistical modeling, we have thought a flow of operations more innovative than those currently well-known in literature. The classical approach can be summarized as following:

- It has N measures which constitute the experimental statistics;
- It makes for N times the M model parameters extraction;
- Following the N parameters extraction steps, it has, for each parameter $P_i, i = 1, \ldots, n$ of the M model, N values which allow to estimate the statistical distribution;

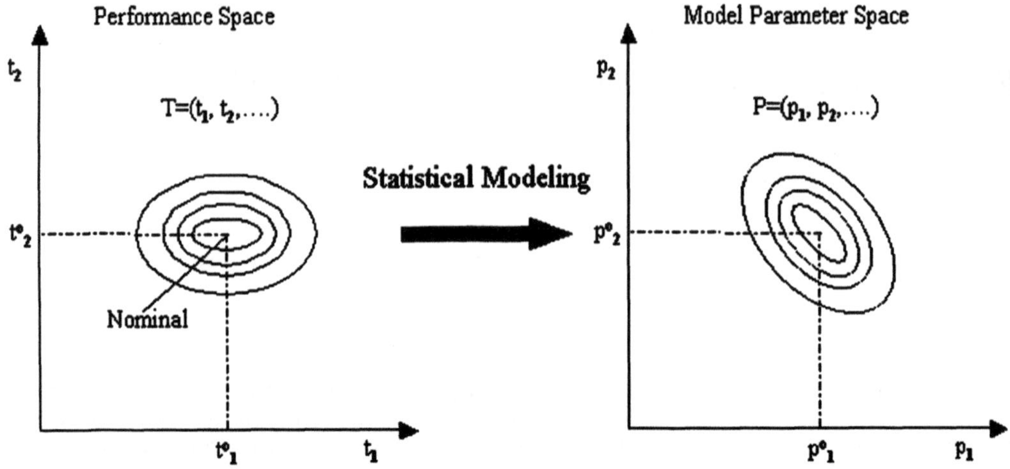

Fig. 1. Proposed flow: from the experimental statistics we determine a statistical Spice model for the device

The previous approach is theoretically valid but it is unproposable from a practical point of view; in fact, this technique repeats the parameters extraction phase for N times, already very heavy from the computational point of view.

2 Statistical Modeling Flow

The flow described below allows to obtain accurate results in less time and with a clearly inferior computational cost compared to the classical approach. The proposed method uses the experimental statistics as a target to be satisfied and, above all, as a selectivity factor for device model: a device model will be accepted only if it is characterized by parameters values that allow to obtain, through electrical simulations, some performances which are included in the experimental statistics. The flow of statistical modeling is based on several Monte Carlo simulation steps, in order to estimate the second order moments for every statistical model parameter; afterwards an optimization phase follows with the aim to identify the cross-correlation coefficients among the parameters. The statistical modeling flow is described in detail as follows:

- **Step 1: Start up**
  We have a typical (nominal) model of the device, M_0, where $\{v_{0,1}, \ldots, v_{0,n}\}$ are the values of the parameters $\{P_1, \ldots, P_n\}$; we consider the device statistical model M, that is a model whose parameters are random variables; during this start up step, each parameter $P_i \in P$ is modelled with a normal distribution with the medium value equal to the nominal value, that is $\mu_i = v_{0,i}$, and a very big standard deviation σ_i (for example, equal to the double of the medium value); a null cross-correlation among the parameters $\{P_1, \ldots, P_n\}$ is bond.
- **Step 2: Instances Generation**
  We generate m instances of the statistical model: so, we have m models $\{M_i\}_{i=1}^m$, with the parameters values extracted according to the distribution imposed and to the grade of cross-correlation established.
- **Step 3: Performances Target Calculation**
  Through circuit simulations, for each m model $\{M_i\}_{i=1}^m$ we calculate the performances considered in the experimental statistics T84.
- **Step 4: Selection (Filtering)**
  We will accept only the models which among $\{M_i\}_{i=1}^m$, have such parameters values to reproduce statistically accceptable performances; a model M_i satisfies the experimental statistics, that is, it generates performance targets acceptable according to this statistics, only if it reproduces some performances that have values included in the range $[\mu \pm 3\sigma]$ (probability of 99%) considered by the experimental statistics. On the base of this criterion we will select S, $S < m$, models among the m $\{M_i\}_{i=1}^m$.
- **Step 5: Standard Deviations Calculation**
  For each $P_j, j = 1, \ldots, n$, parameter we consider the S values obtained in the S selected models ($\{v_{i,j}\}_{i=1}^S$) and on the base of them we estimate the standard deviation from its normal distribution:

$$\sigma_j = \sqrt{\frac{1}{(S-1)} \sum_{i=1}^{S} (v_{i,j} - \mu_j)^2} \quad j = 1, \ldots, n \tag{1}$$

So, we will generate the statistical model $\widetilde{M}$ with the statistical distribution of the parameters calculated on the base of the S selected values, that is, the $\widetilde{M}$ model parameters statistically described have medium value equal to the nominal M_0 model values, standard deviation updated with the (1) and null cross-correlation.

- **Step 6: Stop or Reiteration**
  If the S number of selected instances, which have generated the $\widetilde{M}$ model calculated in the step 5, were sufficiently high, that is, for example, if the 99% of the m instances would have been accepted, the flow would stop and the statistical model in output would be the $\widetilde{M}$ calculated in the step 5, if S does not reach the acceptance threshold we will have to repeat the flow starting from step 2.
- **Step 7: Correlation Coefficients Determination**
  After obtaining the model $\widetilde{M}$, that is having estimated the parameters standard deviations, we will make an optimization process to determine the cross-correlation coefficients which were null until now; taking as target the cross-correlation coefficients among the T84 electrical performances, we will optimize the value of the Spice model parameters cross-correlation coefficients. The optimization method used is the Direct in the Jones et al. version described in [JPS93].

3 Validation of the proposed statistical modeling method

The aim of this section is to validate the operative flow, described in the section 2, which extracts the statistical model of a device of which we have performances that derive from the T84 experimental statistics.
As this flow output is a circuit model whose parameters are described by a probability distribution estimated starting from a certain experimental statistics, to test the flow convergence, we have generated a fictitious experimental statistics, using a well-known Spice M_{target} statistical model; we call it fictitious statistics because it is obtained through Monte Carlo simulations and not through a device measures. If the proposed statistical modeling method was valid, in that case the statistical distribution of the Spice parameters, obtained through our flow, it would converge on the statistical distribution of the starting M_{target} model parameters. The operations flow used to validate the statistical modeling method is described in detail as follows:

(i) **Choice of a M_{target} Statistical Model**
We consider a M_{target} statistical model with the parameters $\{P_1, \ldots, P_n\}$ statistically described by a normal distribution with $\mu_{i_{target}}$ and $\sigma_{i_{target}}$ mean value and standard deviations respectively; we impose a determinate cross-correlation among these parameters.

(ii) **Generation of the Fictitious Experimental Statistics**
Through N Monte Carlo simulations, using the M_{target} model, we generate N value of the performances target $T_j, j = 1, \ldots, t$. For each t performance target, on the base of its simulated N values, we estimate the mean value μ_{T_j} and the standard deviation σ_{T_j} of the normal relative distribution; moreover, we collect the correlation coefficients among these electrical performances. After that, for each considered electrical performance we will have a fictitious experimental statistics (fictitious as it has been collected from a simulation data and not from the real device measures trough the T84 machine).

(iii) **Statistical Model Extraction using the Proposed Flow**
The fictitious experimental statistics will be given in input, as a target of T84, to the proposed statistical modeling flow, which will extract the statistical model $\widetilde{M}$ having the parameters $P_i, i = 1, \ldots, n$ with normal distribution, that is with mean value $\widetilde{\mu_i}$ (which represent the nominal value of M_{target} model parameters), $\widetilde{\sigma_i}$ standard deviation and a determinate cross-correlation.

(iv) **Comparison**
If the extraction flow of the statistical model is convergent, so for each $P_i, i = 1, \ldots, n$ of the model, the probability distribution estimated with $\widetilde{\mu_i}$ and $\widetilde{\sigma_i}$, will have to converge on the target distribution with $\mu_{i_{target}}$ and $\sigma_{i_{target}}$; furthermore, the estimated cross-correlation grade among the parameters has to converge on that imposed for the M_{target} model.

4 Validation Tests

Firstly, we have done several tests on a simple device, such as a diode, and, secondly, we have tested the flow on an IGBT.

4.1 Diode test

In this section, we consider a validation test related to a diode to be statistically characterized. To generate the fictitious experimental statistics, we have done a Monte Carlo simulation through a Spice like circuit simulator using a well known statistical diode model. The Spice parameters of this model, described with a normal statistical distribution, are shown in table 1.

We have fixed the cross-correlation coefficient $\rho(Is, Rs)_{target} = 0.5$. We speak about target values because, through the proposed flow, we want to obtain them. As electrical measures, we have considered V_{th} and G_m: in Table 2 we show the experimental statistics related to them, obtained through simulation.

The cross-correlation coefficient calculted among the electrical performances is $\rho(V_{th}, G_m) = -0.0253$. After 11 iterations of the proposed flow we have reached the estimation of the standard deviations of the Spice parameters, sufficiently accurate, as we reached the 98% of the selected instances (take in mind that $\widetilde{\mu_i} = \mu_{i_{target}}$). The result has been reported in the 3-th column of Table 1.

After we have done the final optimization phase, with the aim to optimize the value of the correlation coefficient among the Spice parameters Is and Rs, (until now null), comparing the current value of $\rho(V_{th}, G_m)$ with its target value equal to $\rho(V_{th}, G_m) = -0.0253$: for each iteration of the optimization process we have done a Monte Carlo simulation of N steps.

We have performed the optimization process many times with different Monte Carlo steps and different model parameter standard deviations in order to understand the result accuracy.

The optimized $\rho(Is, Rs)$ is shown in table 3, related to different optimization process setting: taking into account that $\rho(Is, Rs)_{target}$ is equal to 0.5, we can notice that a small Monte Carlo steps such as N=150 does not lead to an accurate $\rho(Is, Rs)$ estimation, even if we have performed the optimization process take into account the ideal $\sigma_{i_{target}}$ of Spice model parameters, i.e null error on them standard deviations. The optimization process performed with a greater Mote Carlo steps, such as N=500, even if we use the extracted model parameter standard deviations, leads to a good $\rho(Is, Rs)$ estimation, as we have obtained optimized $\rho(Is, Rs) = 0.4223$ vs 0.5 target value; considering N=500, we have repeated the optimization with a null error on the model parameter standard deviations, i.e. taking into account the ideal $\sigma_{i_{target}}$, and the correlation coefficient estimation has been 0.4935, very close to the target value 0.5. Looking at table 3, it is clear that the estimate accuracy of the correlation coefficient among the Spice parameters, through the optimization process, depends on the accuracy with which we have estimated the standard deviations, through the filtering step, and the N steps number which characterizes the Monte Carlo simulations in the optimization phase.

Table 1. Diode Spice parameters statistically described

Parameters	$\mu_{i_{target}}$	$\sigma_{i_{target}}$	$\widetilde{\sigma_i}$
Is[A]	1E-13	2E-14	2.1582E-14
Rs[Ω]	2	0.3	0.2853

Table 2. Fictitious experimental statistics calculated for the diode

Performances target	Measure conditions	μ	σ
V_{th}[V]	I_D=0.01A	0.675587	0.488877E-02
$G_m[\Omega^{-1}]$	I_D=1A	0.505821	0.812661E-01

Table 3. Summary of the Optimization process performed many times with different Monte Carlo steps and different model parameter standard deviations

MC steps N	Consider Extracted $\widetilde{\sigma_i}$	Consider Target σ_i	Optimized $\rho(Is, Rs)$
N=150	×		0.3807
N=150		×	0.4572
N=500	×		0.4223
N=500		×	0.4935

4.2 IGBT test

In this section we consider a validation test related to an IGBT to be statistically characterized. To generate a fictitious experimental statistics we have done a Monte Carlo simulation through a Spice like circuit simulator. The Spice parameters of the IGBT model, to which we have assigned a normal statistical distribution, are shown in Table 4. We established the target cross-correlation among the parameters showed in Table 5.

We speak about target values because, through the proposed flow, we want to obtain them. The circuit simulation, in the statistical model instances generation, will generate the parameters values, statistically described, according to the imposed distribution and to the cross-correlation grade.

In table 6, we show the fictitious experimental statistics calculated for the electrical performances $Bvdss$, $Vdson$, Gmp and Vth. The cross-correlation coefficients among these electrical performances are shown in Table 7. We have applied the flow described in section 2 and, after 6 iterations, by executing the filtering with 3σ threshold, we have selected the 98% of the instances. On the base of these instances we have estimated the standard deviations of the Spice parameters showed in the last column of Table 4. Using the statistical model with the updated standard deviations, we have done the final optimization phase to optimize the correlation coefficients values among the Spice parameters (until now null), taking as a target the correlation coefficients of the experimental statistics related to the electrical performances (see Table 7). For each iteration of the optimization process we have done a Monte Carlo simulation of N=500 steps. The best values found for the cross-correlation coefficients among the Spice parameters are shown in last column of Table 5.

Table 4. IGBT Spice parameters statistically described

Parameters	$\mu_{i_{target}}$	$\sigma_{i_{target}}$	$\widetilde{\sigma_i}$
Vz[V]	7.75	0.8	0.8800
Ron[Ω]	1e-3	1e-4	1.8121E-04
K0[AV^{-2}]	7.925	0.8	0.7023
VT0[V]	1.81	0.2	0.2146

Table 5. Cross-correlation coefficients among the Spice parameters of the IGBT

Cross-correlation	Target values	Extracted values
$\rho(Ron, Vz)$	0.6	0.5974
$\rho(K0, Vz)$	0.4	0.3711
$\rho(K0, Ron)$	0.5	0.5562
$\rho(VT0, Vz)$	0.6	0.6674
$\rho(VT0, Ron)$	0.8	0.5123
$\rho(VT0, K0)$	0.3	0.2737

Table 6. Fictitious experimental statistics calculated for the IGBT

Target Performances	Measure conditions	μ	σ
Bvdss[V]	I_D=250μA	381.883	38.6796
Vdson[V]	I_D=5A , V_{GE}=5V	1.04981	0.7401E-02
Gmp[AV^{-1}]	V_{DE}=15V	16.9108	0.920343
Vth[V]	I_D=250μA , V_{DG}=0V	1.61821	0.200153

Table 7. Cross-correlation coefficients for electrical performances considered for IGBT device

Cross-correlation	Target values
$\rho(Vdson, Bvdss)$	-0.1013
$\rho(Gmp, Bvdss)$	0.2554
$\rho(Gmp, Vdson)$	-0.9634
$\rho(Vth, Bvdss)$	0.5936
$\rho(Vth, Vdson)$	0.1995
$\rho(Vth, Gmp)$	0.0419

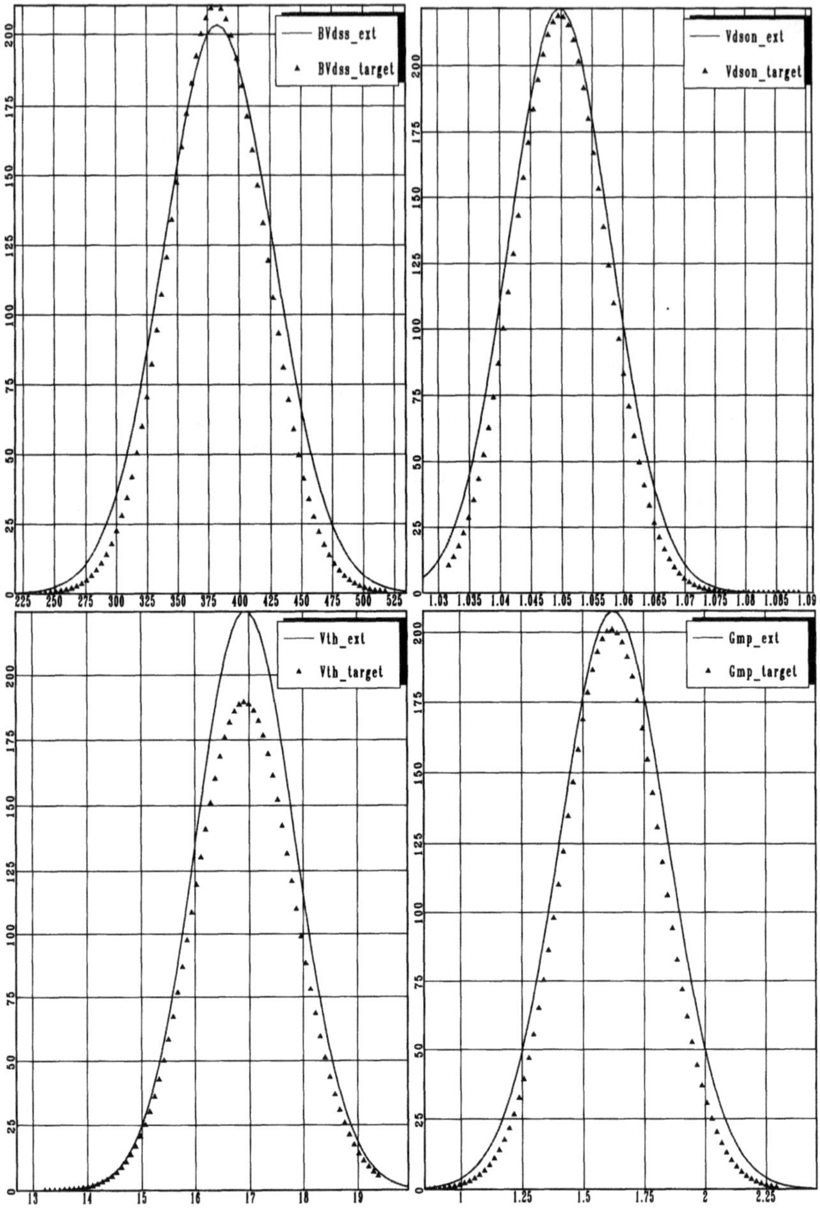

Fig. 2. Distribution of electrical performances of IGBT device: Bvdss, Vdson, Vth, Gmp. The extracted distributions (red line), result of our flow, are compared to their target distributions (blue triangles)

As example of the statistical modeling flow convergence, in Fig. 2 are showed the electrical performances (Bvdss, Vdson, Vth, Gmp) distributions considered for the IGBT test, very close to their target distributions: our technique allows to obtain a statistical Spice model which reproduces the electrical experimental performances distributions.

5 Conclusion

We have shown a statistical modeling technique which extracts a statistical model for a given device, by using the information included in the experimental statistics on electrical performances, called T84, corresponding to the device in issue. The technique shown is innovative compared to the existent ones; the latter, valid from a theoretical point of view, are practically unacceptable, as they need to repeat the parameters extraction phase for a thousand times, which is already very heavy from a computational point of view.

The new method is based on the use of the experimental statistics T84, as a target to be satisfied, and, above all, as a selectivity factor of the device models; the latter will be accepted if the parameters values are such to characterize a model which can supply simulated performances which are included in the experimental statistics. In the end, we do an optimization phase to estimate the correlation coefficients among the Spice model parameters, taking as a target those among the electrical performances. The operations flow has been validated on a diode and IGBT device.

References

[KC93] Koskinen, T., Cheung, P.Y.K.: Statistical and behavioural modelling of analogue integrated circuits, IEE Proceedings-G, **140**, 171–176 (1993)

[JPS93] Jones, D.R., Perttunen, C.D., Stuckman, B.E.: Lipschitzian Optimization without the Lipschitz Constant, Journal Optmization Theory and Application, **79**, 157–181 (1993)

[CC01] Cherubal, S., Chatterjee,. A.: Test generation based diagnosis of device parameters for analog circuits. DATE 2001, München, pp. 596-602

[MD02] McAndrew, C., Drennan, P.G.: Unified Statistical Modeling for Circuit Simulation. Nanotech 2002, Vol.1, pp. 715 - 718

Coupled EM & Circuit Simulation Flow for Integrated Spiral Inductor

A. Ciccazzo, G. Greco and S. Rinaudo

STMicroelectronics, Stradale Primosole,50 - 95100 Catania, Italy,
{angelo.ciccazzo,giuseppe-cad.greco,salvatore.rinaudo}@st.com

Abstract At present, a practicable way to design IC custom inductors involves EM simulators that are able, for frequencies below 10GHz, to reproduce quite faithfully the behaviour of RF IC structures. The extracted model is based on lumped elements in a SPICE subcircuit format or S-parameter representation and requires no adjustment after fabrication and measurement. To help designers developing their projects, an automatic simulation flow has been implemented for the modelling of planar and multi-layer polygonal integrated inductors on silicon substrates based on the Cadence (Virtuoso) - Agilent (Momentum) environment.

A computer program which extracts a physical-based model of inductor components that is suitable for circuit (ELDO) simulation has been used to evaluate the effect of variations in metallization, layout geometry, and substrate parameters upon monolithic inductor performance. Planar (2.5-D) numerical simulations (MOMENTUM) have been used to extract the S-Parameter based model. Square, octagonal, hexagonal and circular inductors could be designed and simulated. Experimental results confirm the accuracy of the flow. This flow is based on HSB3 technology developed by ST Microelectronics.

Key words: Integrated inductor, EM simulation, Simulation flow

1 Introduction

Smaller and smaller integrated RF circuits are going to replace discrete and hybrid components in wireless portable communication applications where high levels of integration are, nowadays, more requested . The silicon technologies, till now have provided only integrated transistors, resistors, and capacitors for RF IC designers and adding planar inductors to the list would allow designers to implement fully integrated solutions for all that RF circuits which include inductance, but the lack of an accurate and generic model of a monolithic inductor on silicon substrate has often prevented designers from employing them.

Using large pre-characterized inductor libraries, in RF IC design flows, often could represent a very useful solution because the designer can choose the more appropriate inductors amongst those available in the library but often, for reasons tied to very restricted design requirements, the designer can not find the useful one amongst them. For this reason, the necessity to make custom inductors, whose characteristics can be determined in synthesis phase, often arises.

A practicable way to design IC custom inductors, at present, involves using EM simulators that, for frequency below 10GHz are able to reproduce the behaviour of RF IC structures quite faithfully. The extracted model comprises lumped elements in a SPICE subcircuit format or a S-parameter representation and requires no further adjustment after fabrication and measurements. An automatic simulation flow has been implanted in order to help our designers to build their inductors easily. The flow is based both on commercial and custom tool developed in STMicroelectronics and it is available for the RF technology HSB3. The tools involved in the flow are the commercial drawing software Virtuso, developed by Cadence and the EM 2.5D simulator Momentum, developed by Agilent Technologies while the custom tools developed in STMicroelectronics are the estimation tool Pcell Parametric Inductor, the optimisation program ToolsMG and the graphic processor tool Imago. In the figure below the proposed simulation flow is shown.

In the next chapters we will see some details about the flow steps.

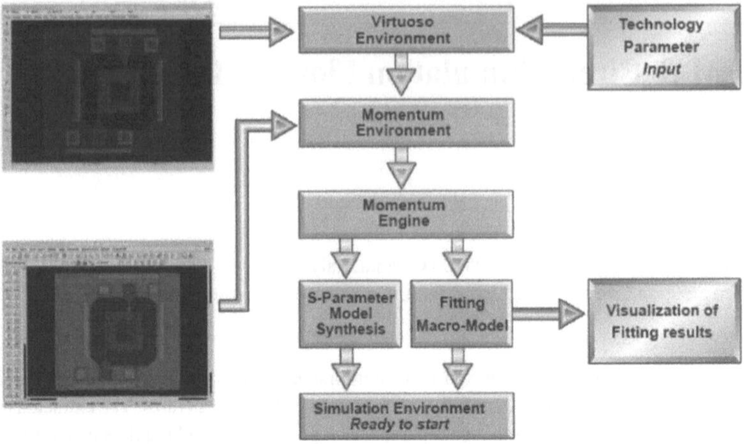

Fig. 1. Simulation Flow

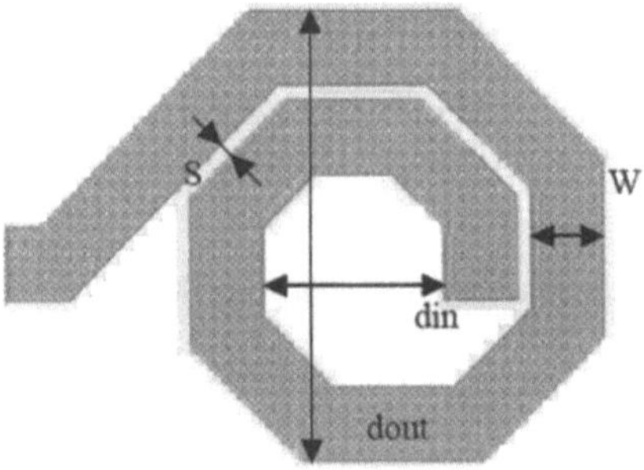

Fig. 2. Spiral inductor and main parameters used in the estimation formulas

2 Drawing the inductor and having the initial inductance estimation

Using a parametric cell developed in STMicroelectronics and implemented in Cadence Skill Languages, the designer can easily draw the desired inductor obtaining a preliminary estimation of the inductance. This estimation come out from an implementation of the Modified Wheeler Formula and Geometric Mean Distance [1] equation.

Wheeler presented several formulas for planar spiral inductors, which were intended for discrete inductors but simple modification of the original formulas allows obtaining an expression that is valid for planar spiral integrated inductors. The user can choose the inductor topology and possible shape choices are: square, octagonal, hexagonal and circular.

$$L_0 = K_1 \cdot \mu_0 \frac{n^2 \cdot d_{avg}}{1 + K_2 \cdot \rho} \tag{1}$$

where d_{in} is the inner diameter of coil, d_{out} is the outer diameter of the coil

$$d_{avg} = \frac{d_{in} + dout}{2} \tag{2}$$

is the average diameter,

$$\rho = \frac{d_{out} - din}{d_{out} + d_{in}} \tag{3}$$

Table 1. Modified Wheeler expression K_1 and K_2

Layout	K_1	K_2
Square	2.34	2.75
Hexagonal	2.33	3.82
Octagonal	2.25	3.55

Table 2. Coefficients for current sheet expression

Layout	C_1	C_2	C_3	C_4
Square	1.27	2.07	0.18	0.13
Hexagonal	1.09	2.23	0.00	0.17
Octagonal	1.07	2.29	0.00	0.19
Circle	1.00	2.46	0.00	0.20

is the fill ratio of coil. Coefficients for Modified Wheeler expression shape dependent K_1 and K_2 are shown in Table 1.

Besides this first formula iis possible to use another simple expression to estimate the inductance of a planar spiral inductor. This can be obtained by approximating the sides of the spirals by symmetrical current sheets of equivalent current densities.

$$L_{gmd} = \frac{\mu \cdot n^2 \cdot d_{avg} \cdot c_1}{2} + \left(\ln\left(\frac{c_2}{\rho}\right) + c_3 \cdot \rho + c_4 \cdot \rho^2 \right) \tag{4}$$

where the coefficients c_i are shape dependent and are shown in Table 2.

3 Exporting Layout and s-parameters simulation

In order to calculate S-Parameters from the drawn inductor structure, the user must export the layout from the drawing environment to the electromagnetic simulation environment. All the necessary environment setting, the technology information of the substrate and the characteristics of the inductor metals necessary to execute the simulation are tied to the technology and supplied by the program flow in a transparent way for the designer.

The only important action the user has to do is to define the number and position of signal ports where s-parameters must be calculated through.

The scattering parameters simulation step follows the project exportation and port definition phases. When the simulation is completed, the simulator will automatically show a series of Smith diagram pertinent to all S parameters extracted. After viewing the results the user can leave the electromagnetic environment and come back to the designing environment in order to begin the model synthesis phase.

4 Synthesis and simulation of the S-parameters based model

From the Virtuoso environment, after selecting an item placed on a custom menu, the user can generate an S-parameter model for Spectre and Ads simulator automatically. These models are written with a syntax useful to the simulators and are directly based on s-parameters. During the synthesis phase, the flow uses a data file in Touchstone format generated by EM simulation and containing the S-parameters representation of simulated inductors.

When this phase is completed, the models are created under the "Predefined" models directory of simulator and, at the same time all necessary views for simulations are also generated. Once the previous step is completed, the user can perform the simulation of implemented devices.

5 Macromodel Fitting

By selecting the item "Fit Macro Model" from a customized menu a lumped component inductor is created. The inductor S-parameter representation will be synthesized in a two-port network consisting of lumped elements.

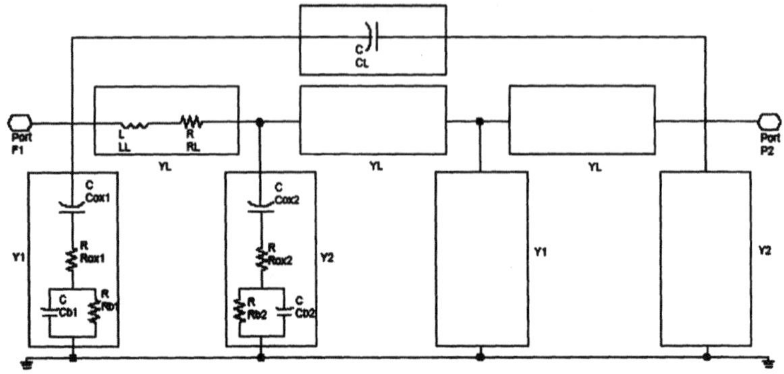

Fig. 3. Two-cell Macro Model

The electrical equivalent model of a two cell subdivision of the inductor is shown in Fig. 3 The inductor can be represented as an equivalent distributed inductor model with a variable number of cells to describe better the inductor behaviour at high frequencies [2] [3]. The main elements of the two ports are the series inductance L_L, the resistance R_L of the segment, the capacitance C_{ox1} , C_{ox2} and resistance R_{ox1}, R_{ox2} formed by the insulating S_i0_2 between the inductor and the silicon substrate. The two elements modelling the substrate layers under the insulator named, C_{b1}, C_{b2}, R_{b1} and R_{b2}. This step could be performed to know the capacitance and resistance effects from the inductor versus substrate.

The user could select a topology according to the technology needs. This subcircuit is now described in analytical form to represent the small-signal characterization to be used to fit the AC representation coming from electromagnetic simulation. The fitting method is based on a multi-objective optimisation algorithm developed on a controlled random search method [4] [5] [6]. At the end the model is stored in a file with spice like syntax, to be used in the next step of the flow. The synthesis phase is based on the program ToolsMG and its operations are transparent to the user. Once the fitting phase is completed, through an option available on a customized menu it is possible to ascertain visually the fitting quality by comparing results obtained by simulation and results produced through fitting.

6 Simulation versus Measurement

The proposed inductor simulation has been tested on an example application where s-parameters measured and simulated have been compared. The inductor being tested has the specifications shown below.

- Shape: Octagonal
- Number of turns: 2.5
- Outer dimension (μm): 200.0
- W (μm): 16.0
- S (μm): 8.0
- S_iO_2 Thick. (μm): 1.8
- A_l Thick. (μm): 3.0

The inductors were measured using a network analyser and high frequency probes. The two port S-parameter measurements and simulation were per-formed over frequency range of $100MHz$ to $20GHz$. A set of unconnected probe pads was also measured to determine the parasitic of the pads. The pad parasitics were de-embedded from the measured data by subtracting the Y-parameters of the pads from the Y-parameters of the inductor and converting the results back to S-parameters. The Q value of the inductor [7] [8] is calculated as:

$$Q = \frac{Im\{\frac{1}{Y_{11}}\}}{Re\{\frac{1}{Y_{11}}\}} \tag{5}$$

The inductance value of the inductor is calculated as:

$$L = \frac{Im\{\frac{1}{Y_{11}}\}}{2 \cdot \pi \cdot f} \tag{6}$$

The resistance value of the inductor is calculated as:

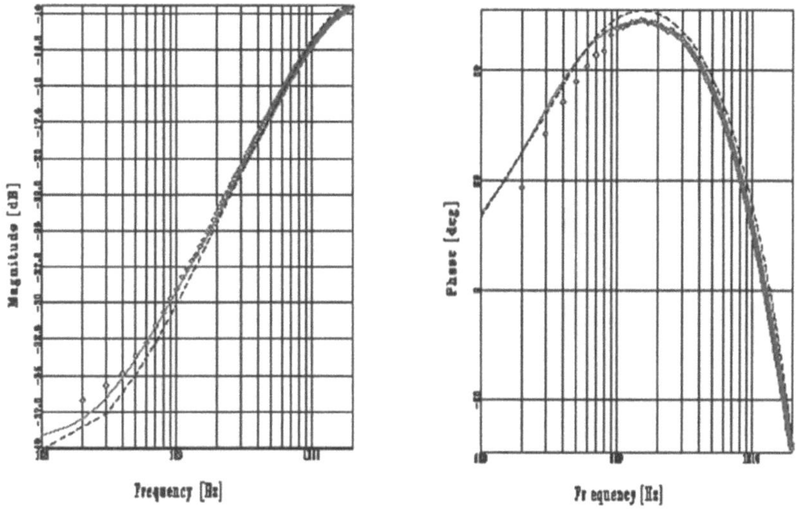

Fig. 4. S_{11} Parameters: ∘ Measure, - EM Simulation, ⋯ Macromodel

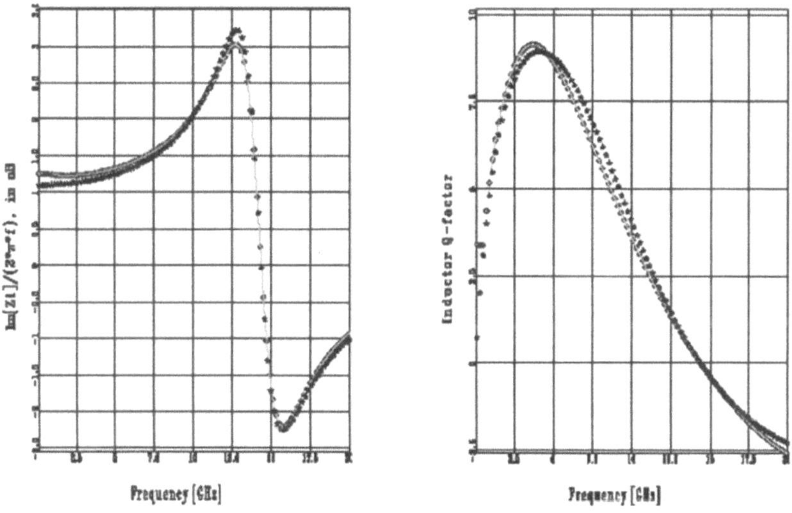

Fig. 5. Q-Factor and Inductance value: ∘ Measure, - EM Simulation, ⋆ Macromodel

$$R = Re\{\frac{1}{Y_{11}}\} \tag{7}$$

The planar spiral inductor was synthesized using a semi empirical model of inductor based on 3 cell division, that means that the cell is composed by 5 Y_L, 3 Y_1 and 3 Y_2 (Fig. 3). The measured value of inductance was 1.18 nH, simulation results and synthesized model give value about 1.1 nH. Fig. 4 shows the measured and simulated S-parameters for the inductor, which was fabricated using the HSB3 process. For S-parameters based model we intend the S-parameters extracted with EM Momentum simulations. The simulation results have been obtained using the S-parameter representation (Momentum) and the semi empirical inductor model. The overall agreement between simulation results and measured data is very good. Fig. 5 shows the Q-factor and inductance value versus frequency where at low frequency the value is about 1 nH in agreement with the predicted value given by equation (1 and (2). Finally Fig. 6 shows the resistance effect of the inductor, the self-resonant effect is present.

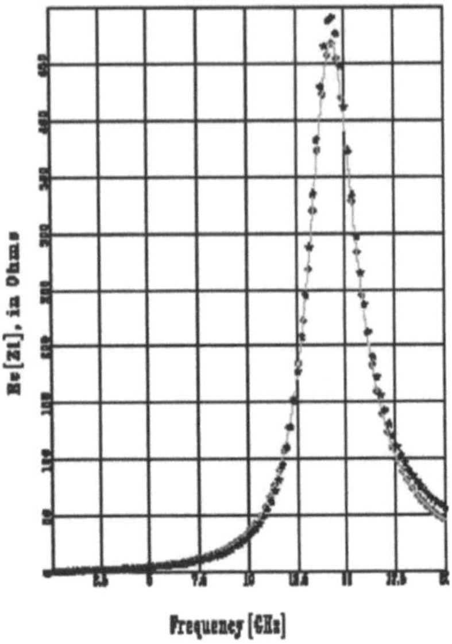

Fig. 6. Resistance value: ∘ Measure, - EM Simulation, ⋆ Macromodel

7 Conclusion

The tool enables the designer to safely analyse the geometry, the type and the positioning of all inductors in a single RF IC prior to fabrication. The developed flow allows automation of a drawing, synthesis and simulation procedure that would, otherwise, be very complicated and difficult to manage. A complete interface structure guides the users through each step making the whole flow very easy to use. The product is available as a STMicroelectronics Unicad Tool and can be, at the moment, used with the HSB3 family design kits.

References

1. S. Mohan, M. Hershenson, S. P. Boyd, and T. H. Lee Simple Accurate Expressions for Planar Spiral Inductances IEEE Journal of solid-state circuits, vol. 34, N. 10, October 1999
2. H.Ronkainen, H. Kattelus, E. Tarvainen, T. Riihisaari, M. Andersson, P. Kuivalainen IC compatible planar inductors on silicon IEE Proc. Circuits Devices Svst., vol. 144, N. 1, February 1997
3. R.D. Lutz, Y. Hahm, A. Weisshaar, V.K. Tripathi Modeling of spiral inductors on lossy substrates for RFIC applications IEEE Radio Frequency Integrated Circuits Symposium, 1998
4. W.L. Price, Global Optimization by Controlled Random Search Journal of Optimization Theory and Application, vol. 40, N. 3, July 1983
5. W.L. Price, Global Optimization Algorithms for a CAD Workstation Journal of Optimization Theory and Application, vol. 55, N. 1, October 1987
6. J.A. Nelder and R. Mead, A simplex method for function minimization Computer Journal, vol. 7 pp. 308 (1965)
7. P. Arcioni, R. Castello, L. Perregrini, E. Sacchi, F. Svelto, An Improved Lumped-Element Equivalent Circuit for on Silicon Integrated Inductors RAWCON 98 Proceedings $0-7803-4439-1/98/c1998$ IEEE
8. John R. Long, Miles A. Copeland, The Modeling, Characterization, and Design of Monolithic Inductors for Silicon RF IC's, IEEE Journal of solid-state circuits, vol. 32, N. 3, March 1997

An Optimal Control approach for an Energy Transport Model in Semiconductor Design

C. R. Drago and A. M. Anile

University of Catania, Department of Mathematics and Computer Sciences, Viale A. Doria 6, I-95125 - Catania, {drago,anile}@dmi.unict.it

Abstract. In this paper an optimal control approach for the Energy Transport model in semiconductor device design is presented. After proving an existence result for the minimization problem, the first-order optimality system is derived and an existence result of Lagrange-multipliers is established.

1 Introduction

Recently there has been an increasing interest in optimal design of semiconductor devices. A major objective in the optimal design is to improve the current flow over some contacts by modifying the device doping profile, which enters as a source term in the mathematical model for semiconductor devices.

In this framework, Pinnau et al. have presented an optimal control approach for the standard Drift Diffusion model (see [1]). In this paper the same optimal control approach is investigated by considering the Energy-Transport Model The dimensionless stationary Energy-Transport (E.T.) model for charge carriers in a semiconductor enclosed in a bounded domain $\Omega \subset \mathbb{R}^d$, $d = 1, 2, 3$, is given, by the following equations for electron density n, electron temperature T, coupled to the Poisson equation for the electric potential V [4]:

$$\begin{aligned} \mathrm{div} J_1 &= 0 \\ \mathrm{div} J_2 &= J_1 \cdot \nabla V + W(\mu, T) \\ \lambda^2 \Delta V &= n(\mu, T) - C(x) \end{aligned} \tag{1}$$

here J_1 is the carrier flux density, J_2 the energy flux density, W the energy production, μ the chemical potential, λ the Debye length and $C(x)$ the doping concentration. Assuming the parabolic state equation one has $n(\mu, T) = T^{3/2} e^{\mu/T}$. The general form of the constitutive equations is given by:

$$\begin{aligned} J_1 &= -L_{11}\left(\nabla \frac{\mu}{T} - \frac{\nabla V}{T}\right) - L_{12} \nabla\left(-\frac{1}{T}\right) \\ J_2 &= -L_{21}\left(\nabla \frac{\mu}{T} - \frac{\nabla V}{T}\right) - L_{22} \nabla\left(-\frac{1}{T}\right). \end{aligned}$$

The coefficients L_{ij} depend on μ and T and the diffusion matrix $L = (L_{ij})$ is symmetric and positive definite. Moreover we assume $W(\mu, T) = -\frac{3}{2}\frac{n(\mu,T)(T-T_L)}{\tau_w(T)}$, where $\tau_w(T)$ is the scaled energy relaxation time, which depends also on T, and T_L is the lattice temperature. By introducing the dual entropy variables:

$$w_1 = \mu/T - V/T \quad \text{and} \quad w_2 = -1/T,$$

one obtains the following symmetric equations (see [4] for a review):

$$\begin{aligned} \mathrm{div} I_1 &= 0 \\ \mathrm{div} I_2 &= Q(w, V) \\ \lambda^2 \Delta V &= N(w, V) - C(x) \\ I_1 &= -\sum_{k=1}^{2} D_{1k}(w, V) \nabla w_k \\ I_2 &= -\sum_{k=1}^{2} D_{2k}(w, V) \nabla w_k \end{aligned} \tag{2}$$

where $w = (w_1, w_2)$, the matrix $D = (D_{ij})$ is still symmetric and positive definite, $Q(w,V) = W(\mu,T)$ and $N(w,V) = n(\mu,T)$.

The system (2) is supplemented by homogeneous Neumann boundary conditions on a part $\Gamma_N \subset \partial\Omega$, modelling the insulating parts of the boundary, and Dirichlet conditions on the remaining part, which models the Ohmic contacts of the device:

$$\begin{aligned} w_1 = w_{1D},\ w_2 = w_{2D},\ V = V_D \quad & \text{on } \Gamma_D \\ I_i \cdot \nu = \nabla V \cdot \nu = 0 \quad i = 1,2 \quad & \text{on } \Gamma_N, \end{aligned} \tag{3}$$

here ν denotes the unit outward normal vector along the boundary and w_{1D}, w_{2D} and V_D are the $H^1(\Omega)$-extensions of fixed functions defined on Γ_D.

Let $\overline{C}$ be a given reference doping profile and let Γ_O be a portion of the Ohmic contacts Γ_D, at which we can measure the current I_1. At the contact Γ_O we prescribe a gained current density I_g and allow deviations, in some suitable norm, of the doping profile from $\overline{C}$ in order to gain this current flow. In other words we intend to minimize cost functionals of the form

$$F(w,V,C) = \frac{1}{2}\|(I_1 - I_g)\cdot\nu\|^2_{H^{-1/2}(\Gamma_0)} + \frac{\gamma}{2}\int_\Omega |\nabla(C - \bar{C})|^2 dx \tag{4}$$

where $\gamma > 0$. C enters as a source term in the E.T. model, which can be interpreted as a constraint, to the minimization problem, determining the current I_1, by the state variables (w_1, w_2, V).

After discussing, in sec. 2 some analytical questions of the optimal control problem, in sec. 3 we prove an existence result and in sec. 4 we derive the first-order optimality necessary conditions. Finally, in sec. 5, we establish an existence and uniqueness result for the Lagrange multipliers.

2 Problem Formulation and Analytic Setting

We make the following assumptions.

(H.1) $\Omega \subset \mathbb{R}^d, d = 1, 2$ or 3 is a bounded domain with lipschitzian boundary $\partial\Omega = \Gamma_N \cup \Gamma_D$, $\Gamma_N \cap \Gamma_D = \emptyset$, Γ_N closed, $\text{meas}_{d-1}(\Gamma_D) > 0$

(H.2) $D_{ij} \in L^\infty(\mathbb{R}^2 \times \mathbb{R}, \mathbb{R}^{2\times 2})$ is a symmetric uniformly positive definite 2×2 matrix.

(H.3) For all w, V, $\hat{\mathrm{w}}$, $\hat{\mathrm{V}}$ the function $Q : \mathbb{R}^2 \times \mathbb{R} \to \mathbb{R}$ satisfies:

$$\begin{aligned} \sum_{k=1}^{2}(Q(w,V) - Q(\hat{\mathrm{w}},\hat{\mathrm{V}}))(w_k - \hat{\mathrm{w}}_k) &\le 0 \\ Q(w,V)(w_2 - \bar{w}) &\le 0 \\ |Q(w,V)| &\le c(1 + |w| + |V|) \end{aligned}$$

with $\bar{w} < 0$ and $c > 0$ a real constant.

(H.4) $N \in L^\infty(\mathbb{R}^2 \times \mathbb{R})$; $w_{iD}, V_D \in H^1(\Omega) \cap L^\infty(\Omega)$ and $w_{2D} = \bar{w}$.

(H.5) We assume that the functions $D_{ij}, Q, N \in C^1(\mathbb{R}^2 \times \mathbb{R})$.

(H.6) For the gained current we require $I_g \cdot \nu \in H^{-1/2}(\partial\Omega)$ and $\overline{C} \in H^1(\Omega)$ for the reference doping profile.

Remarks: Assumption (H.2) follows from basic physical principles. Assumption (H.4) on N is too restrictive, but the existence result for the solution of (2), given by the following theorem 3.1, is still assured, relaxing it by assuming the following condition for the source term Q (as shown in [4]):

$$\sum_{k=1}^{2}(Q(w,V) - Q(\hat{\mathrm{w}},\ \hat{\mathrm{V}}))(w_k - \hat{\mathrm{w}}_k) \le -c|w - \hat{\mathrm{w}}|^2.$$

In order to introduce a functional analytic framework, we consider the space of states

$$X = y_D + X_0$$

where $y_D \stackrel{\text{def}}{=} (w_D, V_D)$ denotes the boundary data, introduced above, and $X_0 = \left[H_0^1(\Omega \cup \Gamma_N)\right]^2 \times \left(H_0^1(\Omega \cup \Gamma_N)\right.$ $\left.\cap L^\infty(\Omega)\right)$ is equipped with the norm $\|y\|_{X_0} \stackrel{\text{def}}{=} \|w\|_{[H^1(\Omega)]^2} + \|V\|_{H^1(\Omega)} + \|V\|_{L^\infty(\Omega)}$.

The space $H_0^1(\Omega \cup \Gamma_N) = \{u \in H^1(\Omega) : u = 0 \text{ on } \Gamma_D\}$ can be considered as the closure of $C_0^\infty(\Omega \cup \Gamma_N)$ with respect to the $H^1(\Omega)$-norm (see [8]). The set of co-states will be $Z \stackrel{\text{def}}{=} [H^1(\Omega)]^3$ and the set of admissible controls is given by

$$\mathcal{C} = \{C \in H^1(\Omega) : C = \bar{C} \text{ on } \Gamma_D\}. \tag{5}$$

Let us define $y \stackrel{\text{def}}{=} (w_1, w_2, V)$. We can rewrite the state equations (2) as $f(y, C) = 0$, where the nonlinear mapping $f : X \times \mathcal{C} \to Z^*$ is defined by

$$f(y, C) \stackrel{\text{def}}{=} \begin{pmatrix} div\left(\sum_{k=1}^2 D_{1k}(w, V)\nabla w_k\right) \\ div\left(\sum_{k=1}^2 D_{2k}(w, V)\nabla w_k\right) + Q(w, V) \\ \lambda^2 \Delta V - N(w, V) + C(x) \end{pmatrix}. \tag{6}$$

Theorem 2.1 The mapping f defined by (6) is Fréchet differentiable. The action of the first derivative at a point $(y, C) \in X \times \mathcal{C}$ in a direction $\hat{y} = (\hat{w}_1, \hat{w}_2, \hat{V}) \in X_0$ is given by:

$$< f_y(y, C)\hat{y}, z >= \tag{7}$$

$$=< div\left[\sum_{k=1}^2 \left(\left(\frac{\partial D_{1k}}{\partial w_1}\hat{w}_1 + \frac{\partial D_{1k}}{\partial w_2}\hat{w}_2 + \frac{\partial D_{1k}}{\partial V}\hat{V}\right)\nabla w_k\right)\right], z^{w_1} > +$$

$$+ < div\left[\sum_{k=1}^2 \left(\left(\frac{\partial D_{2k}}{\partial w_1}\hat{w}_1 + \frac{\partial D_{2k}}{\partial w_2}\hat{w}_2 + \frac{\partial D_{2k}}{\partial V}\hat{V}\right)\nabla w_k\right)\right], z^{w_2} > +$$

$$< div\left[\sum_{k=1}^2 (D_{1k}(w, V)\nabla \hat{w}_k)\right], z^{w_1} > +$$

$$< div\left[\sum_{k=1}^2 (D_{2k}(w, V)\nabla \hat{w}_k)\right], z^{w_2} > + < \frac{\partial Q}{\partial w_1}\hat{w}_1 + \frac{\partial Q}{\partial w_2}\hat{w}_2 + \frac{\partial Q}{\partial V}\hat{V}, z^{w_2} > +$$

$$+ \lambda^2 < \Delta\hat{V}, z^V > - < \frac{\partial N}{\partial w_1}\hat{w}_1 + \frac{\partial N}{\partial w_2}\hat{w}_2 + \frac{\partial N}{\partial V}\hat{V}, z^V >,$$

for all $z = (z^{w_1}, z^{w_2}, z^V) \in Z$ and

$$< f_C(y, C)\hat{C}, z >=< \hat{C}, z^V > \tag{8}$$

for all $\hat{C} \in H_0^1(\Omega \cup \Gamma_N)$ and $z \in Z$. The symbol $< \cdot, \cdot >$ denotes the dual pairing of Z^* and Z.

Proof. First of all let us prove the continuity of the nonlinear mapping $(y, C) \to f(y, C)$. Let $(y, C), (\tilde{y}, \tilde{C}) \in X \times \mathcal{C}$ and let $z \in Z$. By the definition of the mapping f one gets

$$< f(y, C) - f(\tilde{y}, \tilde{C}), z >=$$

$$=< div\left(\sum_{k=1}^2 D_{1k}(w, V)\nabla w_k - \sum_{k=1}^2 D_{1k}(\tilde{w}, \tilde{V})\nabla \tilde{w}_k\right), z^{w_1} > +$$

$$+ < div\left(\sum_{k=1}^2 D_{2k}(w, V)\nabla w_k - \sum_{k=1}^2 D_{2k}(\tilde{w}, \tilde{V})\nabla \tilde{w}_k\right), z^{w_2} > +$$

$$+ < Q(w, V) - Q(\tilde{w}, \tilde{V}), z^{w_2} > +\lambda^2 < \Delta(V - \tilde{V}), z^V > +$$
$$- < N(w, V) - N(\tilde{w}, \tilde{V}), z^V > + < C - \tilde{C}, z^V > .$$

Let us observe that

$$div\left(\sum_{k=1}^2 D_{ik}(w, V)\nabla w_k - \sum_{k=1}^2 D_{ik}(\tilde{w}, \tilde{V})\nabla \tilde{w}_k\right) =$$

$$div\left(\sum_{k=1}^2 \left(D_{ik}(w, V) - D_{ik}(\tilde{w}, \tilde{V})\right)\nabla w_k\right) + div\left(\sum_{k=1}^2 D_{ik}(\tilde{w}, \tilde{V})\nabla(w_k - \tilde{w}_k)\right).$$

After several integration by parts, by considering the continuity of the trace operators and by the assumption (H.5) we can estimate

$$\|f(y,C) - f(\tilde{y},\tilde{C})\|_{Z^*} \le K\left(\|y-\tilde{y}\|_{X_0} + \|C-\tilde{C}\|_{\mathcal{C}}\right),$$

where $K = K(\|D_{ik}\|_{L^\infty}, \Omega)$ denotes a strictly positive constant.

Secondly, (7) is easily obtained, after observing that for each functions D_{ik}, Q and N it holds

$$\Im(w_1 + t\hat{w}_1, w_2 + t\hat{w}_2, V + t\hat{V}) = \\ = \Im(w_1, w_2, V) + \frac{\partial \Im}{\partial w_1} t\hat{w}_1 + \frac{\partial \Im}{\partial w_2} t\hat{w}_2 + \frac{\partial \Im}{\partial V} t\hat{V} + o(t^2),$$

where $\Im : \mathbb{R}^2 \times \mathbb{R} \to \mathbb{R}$, $\Im \in C^1(\mathbb{R}^2 \times \mathbb{R})$. Moreover one has

$$< f(y,C) - f(\tilde{y},C) - f_y(y,C)(y-\tilde{y}), z >= \\ =< div\left[\sum_{k=1}^{2}(D_{1k}(w,V) - D_{1k}(\tilde{w},\tilde{V}))\nabla(w_k - \tilde{w}_k)\right], z^{w_1} > + \\ + < div\left[\sum_{k=1}^{2}(D_{2k}(w,V) - D_{2k}(\tilde{w},\tilde{V}))\nabla(w_k - \tilde{w}_k)\right], z^{w_2} > + \\ + o(\sum_{k=1}^{3} < (y_k - \tilde{y}_k)^2, z_k >)$$

thus after several integration by parts and observing hypothesis (H.5) one gets

$$\|f(y,C) - f(\tilde{y},C) - f_y(y,C)(y-\tilde{y})\|_{Z^*} \le K \cdot o(\|y-\tilde{y}\|^2_{X_0})$$

and thus the existence and continuity of the mapping $(y,C) \to f_y(y,C)$.

Finally, (8) is a direct consequence of the linearity of the mapping $(y,C) \to f(y,C)$ w.r.t C.

3 Existence of Solutions

In order to establish the existence of a solution to the optimal control problem, we require standard regularity properties for the cost functional F.

(H.7) Let $F : X \times \mathcal{C} \to \mathbb{R}$ denote a cost functional which is assumed to be twice continuously Fréchet differentiable with Lipschitz continuous second derivatives. Further let F be of separated type i.e $F(y,C) = F_1(y) + F_2(C)$ and radially unbounded w.r.t C (i.e. $|F| \to +\infty$ as $\|C\| \to +\infty$) for every $y \in X$. Moreover let us assume that F is bounded from below and weakly lower semicontinuous.

Remark: Clearly, the cost functional (4) fits into this setting.

We now consider the minimization problem

$$\min_{X \times \mathcal{C}} F(w,V,C) \qquad \text{s.t.} \qquad f(w,V,C) = 0 \tag{9}$$

The solvability of the state equations (2) for every $C \in L^\infty(\Omega)$ has been established in [3] by the following result:

Theorem 3.1 Let $C \in L^\infty(\Omega)$. Under the assumptions (H.1)-(H.5), there exists a weak solution (w, V) of (2)-(3), in the sense that $w_i - w_{iD} \in H_0^1(\Omega \cup \Gamma_N)$ $i = 1, 2$, $V - V_D \in H_0^1(\Omega \cup \Gamma_N) \cap L^\infty(\Omega)$, and equations (2) are satisfied in the usual weak sense.

For the proof see [4]. We observe that the assumption $C \in L^\infty(\Omega)$ is too strict for our purposes as we seek a minimizer in $H^1(\Omega)$, nevertheless the previous result still holds by assuming $C \in H^1(\Omega)$. Clearly, for $d = 1$ from Sobolev's immersion theorem, one can easily obtain that $H^1(\Omega) \subset L^\infty(\Omega)$ and the proof is the same as in [4]. For $d = 2$ or 3, the existence of a weak solution $V - V_D \in H_0^1(\Omega \cup \Gamma_N)$ is still guaranteed by standard existence result, since $C \in L^2(\Omega)$. On the other hand, in order to get an L^∞-bound for V, we follow the same idea in [4] (pag. 176), where an L^2-bound for the entropy variables $u = (\frac{\mu}{T}, -\frac{1}{T})$ is obtained. This guarantees, (see also pag. 148, [4]), the desired L^∞- bound for V, by Stampacchia elliptic estimates, since $H^1(\Omega) \hookrightarrow L^p(\Omega)$ $(p \in [1, 6[)$ holds.

Theorem 3.2 Assume (H.1)-(H.6) and (H.7) then the constrained minimization problem (9) admits a solution $(w^*, V^*, C^*) \in X \times \mathcal{C}$.

Proof. Let $\{(w_n, V_n, C_n)\} \subseteq X \times \mathcal{C}$ be a minimizing sequence, i.e

$$F(w_n, V_n, C_n) \to inf_{X \times \mathcal{C}} F(w, V, C).$$

where the infimum is finite, as F is bounded frow below. From the radial unboundedness of F we infer that $\{C_n\}$ is bounded in $\mathcal{C}$. Hence there exists a weakly-convergent subsequence, again denoted by $\{C_n\}$, such that

$$C_n \rightharpoonup C^* \quad \text{weakly in } \mathcal{C}.$$

Since $\mathcal{C}$ is weakly closed with respect to the $H^1(\Omega)$ norm, we have that $C^* \in \mathcal{C}$.

Moreover, by the continuous embedding $H^1(\Omega) \hookrightarrow L^p(\Omega)$ $(p \in [1,6[)$, the sequence $\{C_n\}$ is also bounded in $L^p(\Omega)$.

Let (w_n, V_n) denote a solution of (2)-(3). Because of the boundness of N, we have, by employing the Stampacchia's method (see [7]), the following estimates

$$\|V_n\|_{H^1(\Omega)} + \|V_n\|_{L^\infty(\Omega)} \leq K(\sqrt[p]{\text{mis}\Omega}\|N\|_{L^\infty} + \|C_n\|_{L^p(\Omega)} + \|V_D\|_{L^\infty(\Gamma_D)})$$

where $K = K(\Omega) > 0$.

Moreover in [4], Jüngel proves that $\|w_n\|_{H^1(\Omega)} \leq c_1$, where $c_1 > 0$ depends on the $L^\infty(\Omega)$-norm of V_n and the $H^1(\Omega)$-norm of w_D.

Hence there exists a subsequence, again denoted by $\{(w_n, V_n)\}$, such that

$$(w_n, V_n) \rightarrow (w^*, V^*) \text{ weakly in } [H^1(\Omega)]^3$$

which, by Rellich theorem, implies strong convergence in $[L^2(\Omega)]^3$. We also have due to the uniform L^∞-bounds

$$V_n \rightarrow V^* \text{weakly-* in } L^\infty(\Omega)$$

This is sufficient to pass to the limit in the state equation (2)-(3)

$$div\left(-\sum_{k=1}^{2} D_{ik}(w^*, V^*)\nabla w_k^*\right) = Q_i(w^*, V^*) \quad i = 1, 2$$
$$\lambda^2 \triangle V^* = N(w^*, V^*) - C(x).$$

On the other hand $F(w^*, V^*, C^*) \leq \lim \inf F(w, V, C)$ Hence $F(w^*, V^*, C^*) \leq \inf F(w, V, C)$ and so necessarily $F(w^*, V^*, C^*) = \inf F(w, V, C)$.

Remark: Here we have used the assumption on the boundedness of N. We plan to drop this hypothesis is future developments.

4 First-order optimality system

The Lagragian $\mathcal{L} : X \times \mathcal{C} \times Z \rightarrow \mathbb{R}$ associated to the minimization problem (9) is given by

$$\mathcal{L}(y, C, \xi) \stackrel{\text{def}}{-} F(y, C) + < f(y, C), \xi >,$$

where $\xi \stackrel{\text{def}}{=} (\xi^{w_1}, \xi^{w_2}, \xi^V)$. By Theorem 2.1 and (H.7), the Lagrangian $\mathcal{L}$ is continuously Fréchet differentiable. The first order optimality system related to problem (9) is given by

$$\nabla_{(y,C,\xi)}\mathcal{L}(y, C, \xi) = 0. \tag{10}$$

It is easy to show that variations of $\mathcal{L}$ with respect to ξ yield the state equations $f(y, C) = 0$. Moreover taking variation of $\mathcal{L}$ w.r.t the control $C \in \mathcal{C}$ leads to the optimality condition given by: (see [1])

$$\xi^V = \gamma\triangle(C - \bar{C}), \tag{11}$$

with the following boundary conditions,

$$C = \bar{C} \quad \text{on} \quad \Gamma_D, \qquad \nabla C \cdot \nu = \nabla\bar{C} \cdot \nu \quad \text{on} \quad \Gamma_N. \tag{12}$$

Finally, the variations of the lagragian w.r.t the state y yield the co-state equations

$$f_y^*(y, C)\xi = -F_y(y, C) \qquad \text{in } X^* \tag{13}$$

where $f_y^*(y, C) \in L(Z, X^*)$ denote the adjoint operators associated to $f_y(y, C)$.

For the derivation of the co-state equations, let us begin by considering the l.h.s of equation (13). Let $\tilde{y} \in X_0 \cap C_0^\infty(\Omega)$ denote a test function. After several integration by parts, in order to remove all derivatives from the test function $\tilde{y}$, we obtain the adjoint system

$$div[D_{11}(w,V)\nabla\xi^{w_1}] + div[D_{21}(w,V)\nabla\xi^{w_2}] + \tag{14}$$
$$-\sum_{k=1}^{2}\left(\frac{\partial D_{1k}}{\partial w_1}\nabla w_k\right)\cdot\nabla\xi^{w_1} - \sum_{k=1}^{2}\left(\frac{\partial D_{2k}}{\partial w_1}\nabla w_k\right)\cdot\nabla\xi^{w_2} + \frac{\partial Q}{\partial w_1}\xi^{w_2} = \frac{\partial N}{\partial w_1}\xi^V$$

$$div[D_{12}(w,V)\nabla\xi^{w_1}] + div[D_{22}(w,V)\nabla\xi^{w_2}] + \tag{15}$$
$$-\sum_{k=1}^{2}\left(\frac{\partial D_{1k}}{\partial w_2}\nabla w_k\right)\cdot\nabla\xi^{w_1} - \sum_{k=1}^{2}\left(\frac{\partial D_{2k}}{\partial w_2}\nabla w_k\right)\cdot\nabla\xi^{w_2} + \frac{\partial Q}{\partial w_2}\xi^{w_2} = \frac{\partial N}{\partial w_2}\xi^V$$

$$\lambda^2\Delta\xi^V - \frac{\partial N}{\partial V}\xi^V = \tag{16}$$
$$= \sum_{k=1}^{2}\left(\frac{\partial D_{1k}}{\partial V}\nabla w_k\right)\cdot\nabla\xi^{w_1} + \sum_{k=1}^{2}\left(\frac{\partial D_{2k}}{\partial V}\nabla w_k\right)\cdot\nabla\xi^{w_2} - \frac{\partial Q}{\partial V}\xi^{w_2}$$

Since $\tilde{y} \in X_0$, we can choose $\tilde{y}$ arbitrarily on Γ_N and $\nabla\tilde{y}\cdot\nu$ can be chosen arbitrarily on Γ_D. First, if we assume $\tilde{y}$ arbitrary on Γ_N and $\nabla\tilde{y}\cdot\nu = 0$ on $\partial\Omega$ and by observing that $\lambda^2 > 0$, we get the following boundary conditions

$$\nabla\xi^{w_1}\cdot\nu = \nabla\xi^{w_2}\cdot\nu = \nabla\xi^V\cdot\nu = 0 \qquad \text{on } \Gamma_N. \tag{17}$$

Therefore, assuming $\nabla\tilde{y}\cdot\nu$ arbitrary on Γ_D we obtain

$$\xi^V = \xi^{w_2} = 0 \quad \text{on } \Gamma_D \quad \text{and} \quad \xi^{w_1} = \begin{cases} 0 & \text{on } \Gamma_D\backslash\Gamma_O, \\ -\varphi & \text{on } \Gamma_O \end{cases} \tag{18}$$

where, (see [1]) φ is the $H^1(\Omega)$-solution of

$$\begin{aligned} -\Delta\varphi + \varphi &= 0 \text{ in } \Omega \\ \varphi &= 0 \text{ on } \Gamma_N \\ \nabla\varphi\cdot\nu &= \begin{cases} 0 & \text{on } \Gamma_D\backslash\Gamma_O \\ < (I_1(y) - I_g)\cdot\nu, \cdot >_{H^{-1/2}(\Gamma_0),H^{1/2}(\Gamma_0)} & \text{on } \Gamma_O. \end{cases} \end{aligned} \tag{19}$$

Thus the first-order necessary optimality condition (10) consists of the state equations (2)-(3) and the adjoint system (14)-(18). The adjoint and the control are coupled via the optimality condition (11).

5 Existence of Lagrange-multipliers

In this section we are going to establish an existence and uniqueness result for the Lagrange multipliers $(\xi^{w_1}, \xi^{w_2}, \xi^V)$.

The first two equations of system (14)-(16) can be written in the simplified form

$$\text{div}\left(-\sum_{k=1}^{2} D_{ki}(w,V)\nabla\xi^{w_k}\right) + \sum_{k=1}^{2}\mathbf{b}_{ki}\cdot\nabla\xi^{w_k} - \mathbf{c}_i\cdot\xi^w = -s_i\xi^V, \quad i = 1,2 \tag{20}$$

where $\mathbf{b}_{ki} = \sum_{j=1}^{2}\frac{\partial D_{kj}}{\partial w_i}\nabla w_j$, $\mathbf{c}_i = \left(0, \frac{\partial Q}{\partial w_i}\right)$, $s_i = \frac{\partial N}{\partial w_i}$ and $\xi^w = (\xi^{w_1}, \xi^{w_2})$.

From (H.2) it follows that (D_{ki}) is symmetric positive definite and there exists $\delta(V) > 0$ such that

$$\sum_{i,k=1}^{n+1} D_{ki}\xi_k\xi_i \geq \delta(V)|\xi|^2 \quad \forall\xi \in \mathbb{R}^2.$$

Taking into account the L^∞-bound of V, there exists $\delta_0 > 0$ such that $\delta(V) \geq \delta_0$ (see [4]).

Moreover if we define $\mathbf{h} = \left(\sum_{k=1}^{2}(\frac{\partial D_{1k}}{\partial V}\nabla w_k), \sum_{k=1}^{2}(\frac{\partial D_{2k}}{\partial V}\nabla w_k)\right)$ and $\mathbf{g} = (0, \frac{\partial Q}{\partial V})$, equation (16) can be written as

$$-\lambda^2 \triangle \xi^V + \frac{\partial N}{\partial V}\xi^V = -\mathbf{h}\cdot\nabla\xi^w + \mathbf{g}\cdot\xi^w, \tag{21}$$

where $\nabla\xi^w = (\nabla\xi^{w_1}, \nabla\xi^{w_2})$.

Theorem 5.1 Assume (H.1)-(H.7). Then there exist two constants $l = l(\Omega, C, \|\mathbf{b}_{ik}\|_{L^\infty(\Omega)}, \delta_0) > 0$ and $\sigma = \sigma(\Omega, \lambda, C, \left\|\frac{\partial N}{\partial V}\right\|_{L^\infty(\Omega)}, \|\mathbf{b}_{ik}\|_{L^\infty(\Omega)}, \|s_i\|_{L^\infty(\Omega)}, \delta_0, l) > 0$, such that for each state $(w_1, w_2, V) \in X$ with

$$\sum_{i=1}^{2} \|\mathbf{c}_i\|_{L^\infty(\Omega)} \le l \qquad \text{and} \qquad \sum_{j=1}^{2}\left(\|h_j\|_{L^\infty(\Omega)} + \|g_j\|_{L^\infty(\Omega)}\right) \le \sigma$$

and

$$\left\|\frac{\partial N}{\partial V}\right\|_{L^\infty(\Omega)} < \frac{\lambda^2}{C}, \qquad \sum_{i,k=1}^{2} \|\mathbf{b}_{ik}\|_{L^\infty(\Omega)} < \frac{\delta_0}{C}$$

(where $C = C(\Omega) > 0$ is the Poincaré constant), system (14)-(16) supplemented with boundary conditions (17)-(18) admits a unique solution $(\xi^{w_1}, \xi^{w_2}, \xi^V) \in Z$.

Proof. In order to reduce the linear system (20), (21) to a single elliptic equation, let us define $\xi^{\bar{w}} \in [H^1(\Omega)]^2$ as the unique solution of

$$\mathrm{div}\left(-\sum_{k=1}^{2} D_{ki}(w,V)\nabla\xi^{\bar{w}_k}\right) + \sum_{k=1}^{2}\mathbf{b}_{ki}\cdot\nabla\xi^{\bar{w}_k} - \mathbf{c}_i\cdot\xi^{\bar{w}} = 0, \quad i = 1,2$$

$$\xi^{\bar{w}_1} = \begin{cases} 0 \text{ on } \Gamma_D\backslash\Gamma_O \\ -\varphi \quad \text{on } \Gamma_O \end{cases}, \quad \xi^{\bar{w}_2} = 0 \quad \text{on } \Gamma_D \text{ and } \nabla\xi^{\bar{w}_j}\cdot\nu = 0 \quad \text{on } \Gamma_N,\ j = 1,2$$

where φ is the solution of (19). Then, let us introduce the solution operator $T_{\xi^w}(\xi^V) : L^2(\Omega) \to [H_0^1(\Omega\cup\Gamma_N)]^2$ by

$$\mathrm{div}\left(-\sum_{k=1}^{2} D_{ki}(w,V)\nabla T_{\xi^{w_k}}(\xi^V)\right) + \sum_{k=1}^{2}\mathbf{b}_{ki}\cdot\nabla T_{\xi^{w_k}}(\xi^V) - \mathbf{c}_i\cdot T_{\xi^w}(\xi^V) = -s_i\xi^V$$

$$(i = 1,2), \qquad T_{\xi^w}(\xi^V) = (T_{\xi^{w_1}}(\xi^V), T_{\xi^{w_2}}(\xi^V))^T \in [H_0^1(\Omega\cup\Gamma_N)]^2.$$

Hence the system (20), (21) assumes the equivalent form

$$-\lambda^2\triangle\xi^V + \frac{\partial N}{\partial V}\xi^V + \mathbf{h}\cdot\nabla T_{\xi^w}(\xi^V) - \mathbf{g}\cdot T_{\xi^w}(\xi^V) = -\mathbf{h}\cdot\nabla\xi^{\bar{w}} + \mathbf{g}\cdot\xi^{\bar{w}} \tag{22}$$

subject to $\xi^V \in H_0^1(\Omega\cup\Gamma_N)$, and ξ^{w_j} given by

$$\xi^{w_j} = \xi^{\bar{w}_j} + T_{\xi^{w_j}}(\xi^V) \qquad j = 1,2. \tag{23}$$

From elliptic estimates, we have

$$\|T_{\xi^{w_1}}(\xi^V)\|_{H^1(\Omega)} + \|T_{\xi^{w_2}}(\xi^V)\|_{H^1(\Omega)} \le \varepsilon\sum_{i=1}^{2}\|s_i\|_{L^\infty(\Omega)}\|\xi^V\|_{L^2(\Omega)},$$

for some constant $\varepsilon = \varepsilon(\Omega, C, \|\mathbf{b}_{ik}\|_{L^\infty(\Omega)}, \delta_0, l) > 0$, provided we have

$$\sum_{i=1}^{2}\|\mathbf{c}_i\|_{L^\infty(\Omega)} \le l,$$

for $l = l(\Omega, C, \|\mathbf{b}_{ik}\|_{L^\infty(\Omega)}, \delta_0) > 0$ small enough and $\sum_{i,k=1}^{2}\|\mathbf{b}_{ik}\|_{L^\infty(\Omega)} < \frac{\delta_0}{C}$.

In order to apply the Lax-Milgram theorem to equation (22), let us define the bilinear form $a : H_0^1(\Omega\cup\Gamma_N)\times H_0^1(\Omega\cup\Gamma_N) \to \mathbb{R}$ and the linear functional $G : H_0^1(\Omega\cup\Gamma_N) \to \mathbb{R}$ by:

$$a(\xi^V,\phi) = \lambda^2\int_\Omega \nabla\xi^V\cdot\nabla\phi\,dx + \int_\Omega \frac{\partial N}{\partial V}\xi^V\phi\,dx + $$
$$+\int_\Omega \mathbf{h}\cdot\nabla T_{\xi^w}(\xi^V)\phi\,dx - \int_\Omega \mathbf{g}\cdot T_{\xi^w}(\xi^V)\phi\,dx$$

$$G(\phi) = -\int_\Omega \mathbf{h}\cdot\nabla\xi^{\bar{w}}\phi\,dx + \int_\Omega \mathbf{g}\cdot\xi^{\bar{w}}\phi\,dx$$

It is a simpler matter to prove that a and G are continuous. Let us prove the coercivity of a. By using the Schwarz and Poincaré inequality we get

$$\begin{aligned}
a(\xi^V,\xi^V) \geq \lambda^2 \int_\Omega |\nabla\xi^V|^2\,dx - \left\|\frac{\partial N}{\partial V}\right\|_{L^\infty(\Omega)} \|\xi^V\|^2_{L^2(\Omega)} + \\
-\sum_{j=1}^{2} \|h_j\|_{L^\infty(\Omega)} \|\nabla T_{\xi^{w_j}}(\xi^V)\|_{L^2(\Omega)} \|\xi^V\|_{L^2(\Omega)} + \\
-\sum_{j=1}^{2} \|g_j\|_{L^\infty(\Omega)} \|T_{\xi^{w_j}}(\xi^V)\|_{L^2(\Omega)} \|\xi^V\|_{L^2(\Omega)} \geq \\
\geq \lambda^2 \|\nabla\xi^V\|^2_{L^2(\Omega)} - C\left\|\frac{\partial N}{\partial V}\right\|_{L^\infty(\Omega)} \|\nabla\xi^V\|^2_{L^2(\Omega)} - \\
-\varepsilon \sum_{j=1}^{2} \left(\|h_j\|_{L^\infty(\Omega)} + \|g_j\|_{L^\infty(\Omega)}\right) \sum_{i=1}^{2} \|s_i\|_{L^\infty(\Omega)} \|\xi^V\|^2_{L^2(\Omega)} \geq \kappa \|\xi^V\|^2_{H^1(\Omega)},
\end{aligned}$$

for some constant $\kappa = \kappa(\Omega, \lambda, C, \|\frac{\partial N}{\partial V}\|_{L^\infty(\Omega)}, \|\mathbf{b}_{ik}\|_{L^\infty(\Omega)}, \|s_i\|_{L^\infty(\Omega)}, \delta_0, l, \sigma)$ > 0 provided we have

$$\sum_{j=1}^{2} \left(\|h_j\|_{L^\infty(\Omega)} + \|g_j\|_{L^\infty(\Omega)}\right) \leq \sigma$$

for $\sigma = \sigma(\Omega, \lambda, C, \|\frac{\partial N}{\partial V}\|_{L^\infty(\Omega)}, \|\mathbf{b}_{ik}\|_{L^\infty(\Omega)}, \|s_i\|_{L^\infty(\Omega)}, \delta_0, l) > 0$ small enough and $\left\|\frac{\partial N}{\partial V}\right\|_{L^\infty(\Omega)} < \frac{\lambda^2}{C}$. Thus the existence of ξ^V is a consequence of the Lax-Milgram theorem. Further, the existence of ξ^w is uniquely determined by (23).

References

[1] M. Hinze, R. Pinnau, An optimal control approach to semiconductor design. Math. Mod. Meth. Appl. Sc., 12 (1) pp.. 89-107, 2002

[2] M. Burger, R. Pinnau, Fast optimal design for semiconductor devices. SIAM J. Appl. Math., 64 (1) pp. 108-126, 2003

[3] P. Degond, S. Génieys, A. Jüngel, A system of parabolic equations in nonequilibrium thermodynamics including thermal and electrical effects, J. Math Pure Appl. 76 (1997) pp. 991-1015

[4] A. Jüngel, Quasi-hydrodynamic Semiconductor Equations, Progress in Nonlinear Differential Equation and Their Application (2001) Birkhäuser

[5] W. R. Lee, S. Wang, K. L. Teo, An optimization approach to a finite dimensional parameter estimation problem in semiconductor device design. Journal of Computational Physics, 156 pp. 241-256, 1999

[6] M. Stockinger, R. Strasser, R. Plasun, A. Wild, S. Selberherr. A qualitative study on optimized MOSFET doping profiles. In Proocedings SISPAD 98 Conf., pp. 77-80, 1998

[7] G. Stampacchia, Contributi alla regolarizzazione delle soluzioni dei problemi al contorno del secondo ordine ellittiche. Ann. Scuola Norm. Sup. Pisa 12 (1958) pp. 223-245

[8] G. M. Troianello, Elliptic Differential Equations and Obstacle Problems, Plenum Press, New York, first edition, 1987

A Multigroup-WENO Solver for the Non-Stationary Boltzmann-Poisson System for Semiconductor Devices

M. Galler[1], A. Majorana[2], and F. Schürrer[1]

[1] Institute of Theoretical and Computational Physics, Graz University of Technology, Graz, Austria,
{galler,schuerrer}@itp.tu-graz.ac.at
[2] Dipartimento di Matematica e Informatica, Università di Catania, Catania, Italy, majorana@dmi.unict.it

Abstract We present a multigroup-WENO solver for the non-stationary Boltzmann-Poisson system for semiconductor device simulation. The proposed numerical technique is applied for investigating the carrier transport in bulk silicon, in a silicon $n^+ - n - n^+$ diode and in a silicon MESFET. Additionally, the obtained results are compared to those of a full WENO solver.

1 Introduction

In modern highly integrated devices, a consistent description of the dynamics of carriers is essential for a deeper understanding of the observed transport properties. In this paper we propose a deterministic multigroup-WENO solver for the coupled Boltzmann-Poisson system, which describes semiconductor devices on a mesoscopic level. Our numerical scheme is based on the combination of the multigroup method [1] for treating the dependence of the electron distribution function on the three-dimensional wave vector and a fifth-order WENO solver [2], [3] for dealing with the two-dimensional physical space. The resulting transport equations are used for simulating the charge transport in bulk silicon, in a silicon $n^+ - n - n^+$ diode and in a silicon MESFET. Moreover, the relation of these results to those of a full WENO solver are discussed.

2 The Boltzmann-Poisson System

The evolution of the electron distribution function $f(t, \mathbf{x}, \mathbf{k})$ in semiconductors in dependence of time t, the position $\mathbf{x}$ and the electron wave vector $\mathbf{k}$ is governed by the Boltzmann transport equation (BTE) [4]

$$\frac{\partial f}{\partial t} + \frac{1}{\hbar}\nabla_{\mathbf{k}}\varepsilon \cdot \nabla_{\mathbf{x}} f - \frac{q}{\hbar}\mathbf{E} \cdot \nabla_{\mathbf{k}} f = Q(f), \tag{1}$$

where q denotes the positive electric charge. The function $\varepsilon(\mathbf{k})$ is the energy of the considered crystal conduction band, measured from the band minimum; according to the Kane dispersion relation, ε is the positive root of

$$\varepsilon(1 + \alpha\varepsilon) = \frac{\hbar^2 k^2}{2m^*}, \tag{2}$$

where α is the non-parabolic factor and m^* the effective electron mass. The electric field $\mathbf{E}$ is related to the donor density N_D and the electron density n, as the zero-order moment of the electron distribution function f, by the Poisson's equation

$$\epsilon\Delta V = q\left[n(t, \mathbf{x}) - N_D(\mathbf{x})\right], \quad \mathbf{E} = -\nabla_{\mathbf{x}} V, \tag{3}$$

where ϵ is the dielectric constant and V the electric potential. The collision operator $Q(f)$ takes into account acoustic deformation potential and optical intervalley scattering [5]. For low electron density, it reads

$$Q(f)(t, \mathbf{x}, \mathbf{k}) = \int_{\mathbb{R}^3} \left[S(\mathbf{k}', \mathbf{k}) f(t, \mathbf{x}, \mathbf{k}') - S(\mathbf{k}, \mathbf{k}') f(t, \mathbf{x}, \mathbf{k})\right] d\mathbf{k}',$$

where

$$S(\mathbf{k},\mathbf{k}') = K\left[(n_q+1)\delta(\varepsilon(\mathbf{k}')-\varepsilon(\mathbf{k})+\hbar\omega_p) + n_q\delta(\varepsilon(\mathbf{k}')-\varepsilon(\mathbf{k})-\hbar\omega_p)\right] + K_0\delta(\varepsilon(\mathbf{k}')-\varepsilon(\mathbf{k}))$$

and K and K_0 being constant for silicon semiconductors. The symbol δ indicates the usual Dirac distribution and ω_p is the constant phonon frequency. Moreover,

$$n_q = \left[\exp\left(\frac{\hbar\omega_p}{k_B T_L}\right) - 1\right]^{-1}$$

is the occupation number of phonons, k_B the Boltzmann constant and T_L the lattice temperature. As in [6] and [2], we introduce dimensionless quantities and perform a coordinate transformation for $\mathbf{k}$ according to

$$\mathbf{k} = \frac{\sqrt{2m^* k_B T_L}}{\hbar}\sqrt{w(1+\alpha_K w)}\left(\mu, \sqrt{1-\mu^2}\cos\varphi, \sqrt{1-\mu^2}\sin\varphi\right), \tag{4}$$

where the new independent variables are the dimensionless energy $w = \varepsilon/(k_B T_L)$, the cosine of the polar angle μ and the azimuth angle φ with $\alpha_K = k_B T_L \alpha$. It is useful to consider the new unknown function Φ related to the electron distribution function via

$$\Phi(t,\boldsymbol{x},w,\mu,\varphi) = s(w) f(t,\boldsymbol{x},\mathbf{k})|_{\boldsymbol{k}=\ldots\sqrt{1-\mu^2}\sin\varphi)}$$

where $s(w) = \sqrt{w(1+\alpha_K w)}(1+2\alpha_K w)$ is, apart from a dimensional constant factor, the density of states. The use of this unknown gives a new dimensionless Boltzmann equation, where the free streaming operator can be written in a conservative form. The explicit expression of this equation, details on the used material parameters, the dimensionless quantities and the coordinate transformation are found in [3].

3 The Multigroup-WENO Model Equations

For deducing our model equation to the coupled Boltzmann-Poisson system, we proceed as follows. The first step is to fix a maximum value w_{max} for the dimensionless energy. Of course, w_{max} must be related to the physically studied process, and we must check that $\Phi(t,\boldsymbol{x},w_{max},\mu,\varphi)$ is negligible for all t, $\boldsymbol{x}$, μ and φ.

If, for instance, the distribution function depends only on two spatial coordinates (x,y), we must choose three suitable integer N, M and R. Hence, the independent variables w, μ and φ are discretized via

$$\begin{aligned} w_{i+1/2} &= i\Delta w, & i &= 0,1,\ldots,N,\ \Delta w = w_{max}/N,\\ \mu_{j+1/2} &= -1 + j\Delta\mu, & j &= 0,1,\ldots,M,\ \Delta\mu = 2/M,\\ \varphi_{k+1/2} &= k\Delta\varphi, & k &= 0,1,\ldots,R,\ \Delta\varphi = \pi/R, \end{aligned}$$

where we take into account that, due to the 2D spatial geometry and the symmetry of the collision operator, $\varphi \in [0,\pi]$. It is important to remark that N must be chosen in such way that $\zeta = \hbar\omega_p/(k_B T_L \Delta w) \in \mathbb{N}$ in order to treat the Dirac distribution in the collision operator correctly.

The unknown function Φ is approximated by the finite sum

$$\Phi(t,x,y,w,\mu,\varphi) \approx \sum_{i=1}^{N}\sum_{j=1}^{M}\sum_{k=1}^{R} n_{ijk}(t,x,y)\lambda_{w_i}(w)\,\lambda_{\mu_j}(\mu)\,\lambda_{\varphi_k}(\varphi) \tag{5}$$

containing $N \times M \times R$ coefficients n_{ijk} and the characteristic functions $\lambda_{w_i}(w)$, $\lambda_{\mu_j}(\mu)$ and $\lambda_{\varphi_k}(\varphi)$. The first one is defined by

$$\lambda_{w_i}(w) = \begin{cases} \dfrac{1}{\Delta w}, & \text{if } w \in [w_{i-1/2}, w_{i+1/2}],\\ 0, & \text{otherwise,} \end{cases} \tag{6}$$

and the other functions analogously. The evolution equations for the coefficients n_{ijk} are constructed as suggested by the method of weighted residuals [7]. The ansatz (5) is inserted into the dimensionless Boltzmann equation and the result is integrated over the cells

$$\mathcal{Z}_{ijk} = [w_{i-1/2}, w_{i+1/2}] \times [\mu_{j-1/2}, \mu_{j+1/2}] \times [\varphi_{k-1/2}, \varphi_{k+1/2}].$$

This procedure yields a set of $N \times M \times R$ partial differential equations for the n_{ijk} [1]. The physical interpretation of the unknowns reveals that the n_{ijk} equal, except for a constant factor, the density of electrons with wave vectors $\mathbf{k}(w, \mu, \varphi) \in \mathcal{Z}_{ijk}$. Consequently, macroscopic quantities are simply given as weighted sums of the n_{ijk}.

Details of the full general procedure are given in a forthcoming paper [9]. Here, the main ideas of the numerical scheme are shown in the Appendix, where, for sake of clearness, we consider a simple model equation. The extension to the Boltzmann equation is straightforward.

4 Numerical Results

In this section we present numerical results obtained by the help of our multigroup-WENO solver. All the calculations are carried out for silicon at the temperature $T_L = 300$ K. The initial data for the coefficients n_{ijk} at time $t_0 = 0$ ps are chosen as integrated Maxwellians normalized to the donor density at the considered positions.

4.1 Electron Transport in Bulk Silicon

In the Fig. 1, we illustrate the dependence of the stationary-state drift velocity and the mean energy on the applied electric field strength. Moreover, the inserts show the temporal evolution of these quantities in response to the onset of an electric field pulse for the field strengths $E_x = 10$ kV cm^{-1} and $E_x = 50$ kV cm^{-1}. The parameters used in these calculations are set to $N = 100$, $M = 22$, $\zeta = 4$. Our results are compared to those of a full WENO solver proposed in [2]. Here, we observe very good agreement between the results for both the steady state values and the transients.

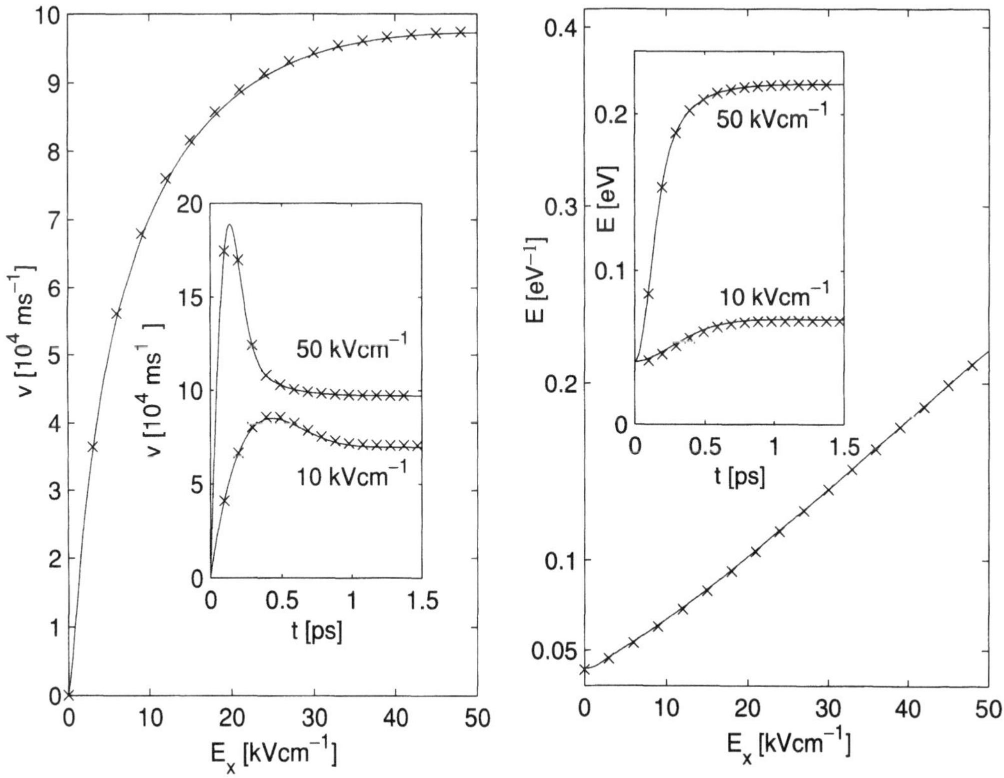

Fig. 1. Stationary-state drift velocity v and stationary-state mean energy E versus the electric field E_x in silicon at T_L=300 K. The inserts illustrate v and E as functions of time t in response to the onset of an electric field pulse. (—): multigroup-WENO model; ($\times$): WENO solver [2]

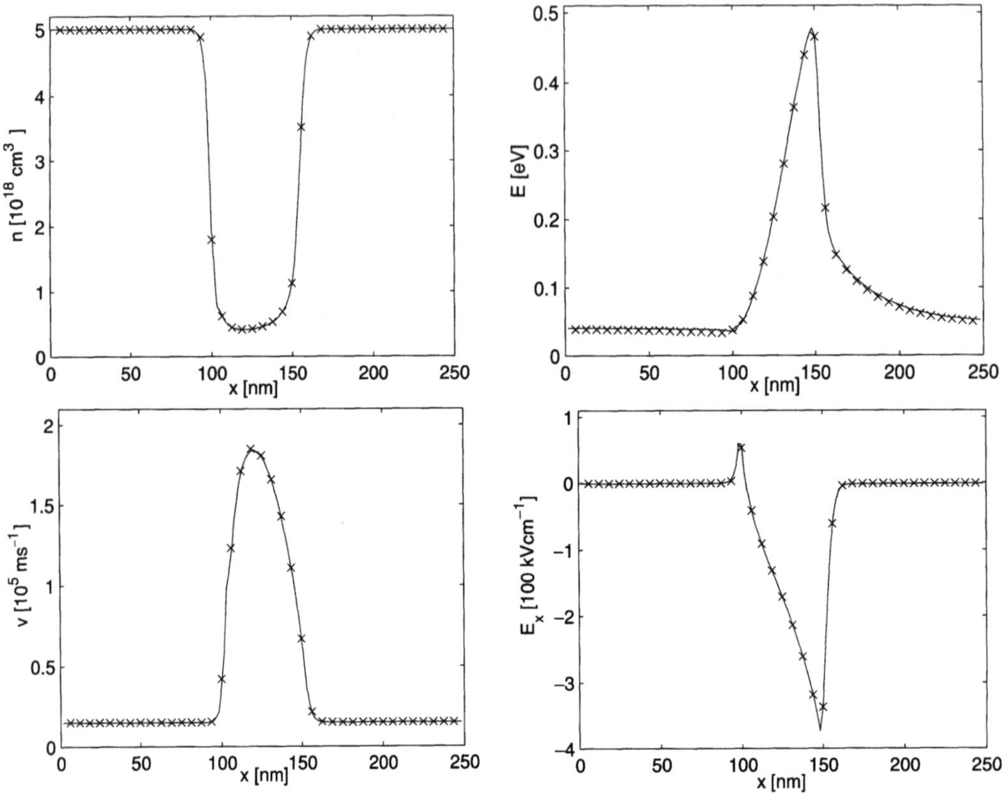

Fig. 2. Steady state electron density n, drift velocity v, mean energy E and electric field strength E_x as a function of position x in the n^+-n-n^+ diode.(—): multigroup-WENO model; ($\times$): WENO solver [2]

4.2 Electron Transport in a Silicon $n^+ - n - n^+$ Diode

The considered $n^+ - n - n^+$ diode has a total length of 250 nm with a 50 nm active channel located at the middle of the device. The doping concentrations are set to $N_D = 5 \times 10^{18}$ cm^{-3} in the n^+ region and $N_D = 10^{15}$ cm^{-3} in the n region. The applied voltage is $V_{bias} = 1$ V and the parameters of the grid are chosen as $N = 100$, $M = 22$, $\zeta = 4$ together with 150 grid points in real space. Figure 2 displays the stationary-state values of the electron density, the drift velocity, the mean energy and the electric field strength as a function of position in the n^+-n-n^+ diode. Moreover, we compare our results with those of the full WENO solver [2] and find that they coincide in the whole x range.

4.3 Electron Transport in a Silicon MESFET

For the simulation of the Si-MESFET, we use the geometry shown in Fig. 3 with the potentials at source $V_s = 0$ V, gate $V_g = -0.8$ V and drain $V_d = 1$ V. The donor densities are chosen as $n = 10^{17}$ cm^{-3} and $n^+ = 3 \times 10^{17}$ cm^{-3}. The ohmic contacts at source and drain act as particle reservoirs. The Schottky contact at the gate is assumed to be an absorbing boundary, whereas perfectly reflecting boundary conditions are imposed at the non-contact surfaces. Concerning the boundary conditions for the Poisson equations, we apply the Neumann condition (vanishing electric field in the direction normal to the surface) on those boundary regions, where there are no contacts. These regions act as insulating boundaries, while the source, gate and drain contacts are treated as Dirichlet boundaries, where the bias voltages are applied. The parameters of the grid are chosen as $N = 75$, $M = 8$, $R = 8$, $\zeta = 3$ together with 48×32 grid points in real space. Figure 4 illustrates the steady state electron density and the electrostatic potential versus position. We observe highly accurate non-oscillatory behavior near the junctions. In Fig. 5, we compare the cuts of the stationary-state electron density, the energy density and the x-components of the momentum and the electric field obtained with the multigroup-WENO solver with those from the full WENO solver [3] for several y-positions. Again, we observe good agreement between the results with the CPU time about a factor 2 lower when applying of the multigroup-WENO procedure instead of the full WENO technique for the same grid. Hence, we believe that the treatment of the momentum dependence of the electron distribution function, which does not show steep gradients,

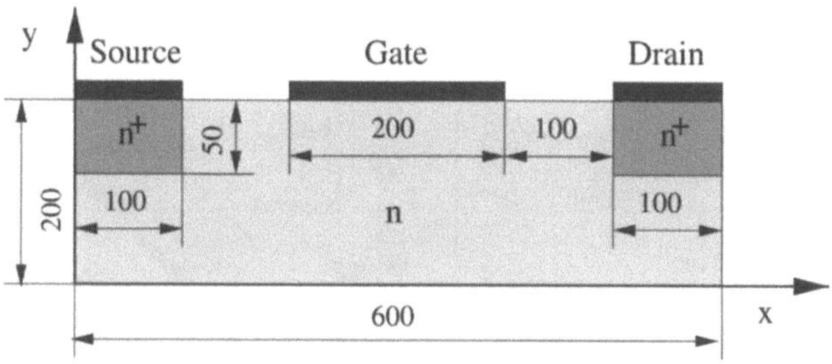

Fig. 3. Schematic illustration of a 2D-MESFET. Lengths are given in nm

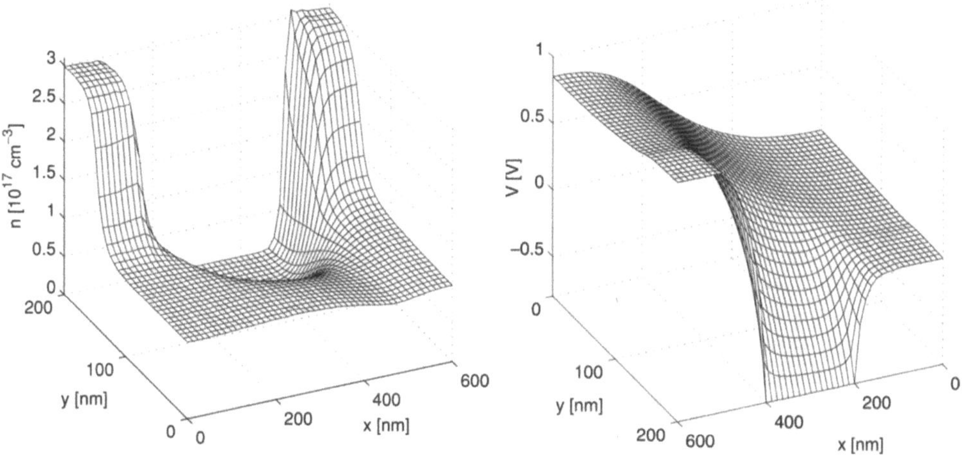

Fig. 4. Steady state electron density n and electrostatic potential V versus position in the Si-MESFET

with the multigroup method and the spatial dependence with a high-order WENO scheme to cope with sharp doping profiles is an appropriate approach for the deterministic simulation of semiconductor devices.

5 Conclusion

A multigroup-WENO solver for the non-stationary Boltzmann-Poisson system is applied for simulating the electron transport in bulk silicon, in the spatially one-dimensional $n^+ - n - n^+$ diode and in the spatially two-dimensional MESFET. The comparison of these results with those obtained by full WENO schemes [2], [3] clarifies that the proposed multigroup-WENO solver is certainly a powerful tool for the accurate simulation of the carrier transport in semiconductor devices. The use of our numerical scheme for approximating the partial derivatives with respect to w, μ and φ requires less CPU time amount than the full WENO scheme. Although this new scheme is of lower order then the previous one, we do not observe a lost of accuracy in the moments of the distribution function.

6 Appendix

Consider the equation

$$u_t + \partial_x[a(w)u] + \partial_w[b(t,x,w)u] = 0 \tag{7}$$

where $u(t,x,w)$ is the unknown and the function a and b are given. If we integrate (7) over the interval $[w_{i-\frac{1}{2}}, w_{i+\frac{1}{2}}]$, then we have

$$\int_{w_{i-\frac{1}{2}}}^{w_{i+\frac{1}{2}}} \left[\frac{\partial}{\partial t}u(t,x,w) + \frac{\partial}{\partial x}a(w)\,u(t,x,w)\right] dw = b(t,x,w)u(t,x,w)\Big|_{w_{i+\frac{1}{2}}}^{w_{i-\frac{1}{2}}}.$$

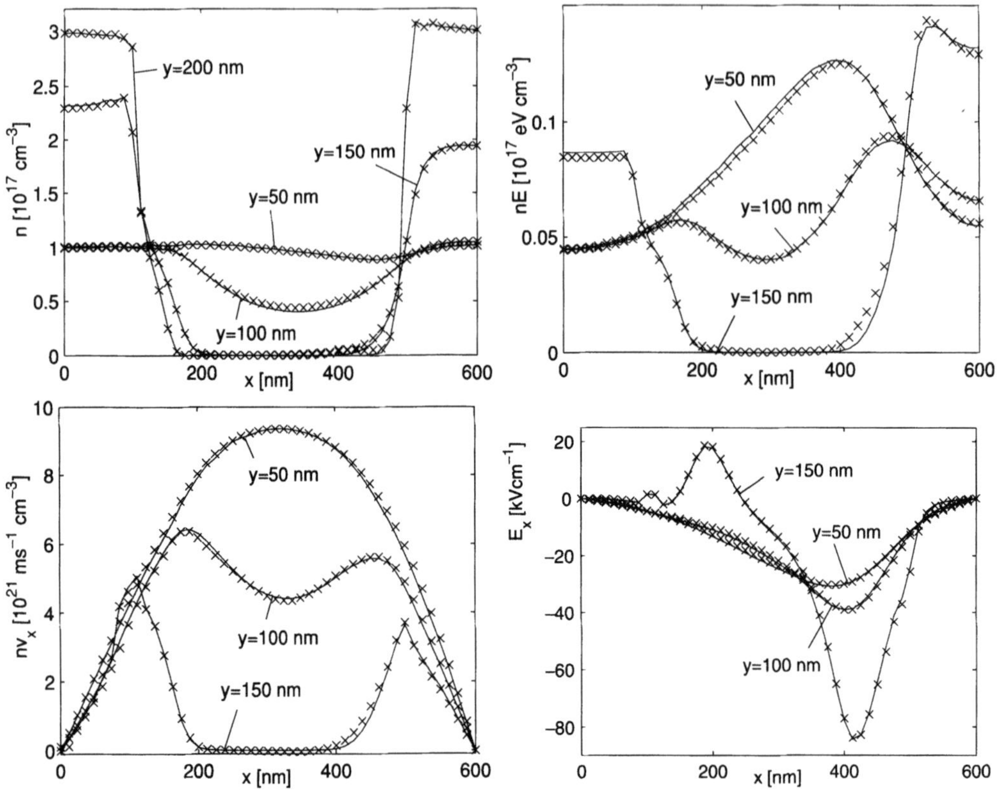

Fig. 5. The stationary-state electron density n, the energy density nE and the x-components of the momentum nv_x and of the electric field E_x versus position x in the Si-MESFET. (—): multigroup-WENO model; ($\times$): WENO solver [3]

Now, assuming that

$$u(t,x,w) \approx \sum_{i=1}^{N} u_i(t,x)\lambda_{w_i}(w)\,,$$

where the characteristic function λ_{w_i} is defined by Eq. (6), we obtain under reasonable assumpions

$$\frac{\partial u_i}{\partial t} + \left(\int_{w_{i-\frac{1}{2}}}^{w_{i+\frac{1}{2}}} a(w)\, dw \right) \frac{\partial u_i}{\partial x} = b(t,x,w)u(t,x,w)\Big|_{w_{i+\frac{1}{2}}}^{w_{i-\frac{1}{2}}}\,.$$

For fixed (t,x) and i, if $b(t,x,w_{i+\frac{1}{2}}) \neq 0$, then the term $u(t,x,w_{i+\frac{1}{2}})$ is treated with the help of an upwind scheme with a linear approximation using a MinMod slope limiter [8]. In fact, a simple Taylor expansion results in

$$u(t,x,w_{i+\frac{1}{2}}) \approx \begin{cases} u_i(t,x) + \dfrac{\Delta w}{2}\dfrac{\partial u}{\partial w}(t,x,w_i) & \text{if } b(t,x,w_{i+\frac{1}{2}}) > 0 \\[2ex] u_{i+1}(t,x) - \dfrac{\Delta w}{2}\dfrac{\partial u}{\partial w}(t,x,w_{i+1}) & \text{if } b(t,x,w_{i+\frac{1}{2}}) < 0 \end{cases}.$$

For $b(t,x,w_{i+1/2}) > 0$ (otherwise, a similar formula holds), we approximate the partial derivative according to

$$\frac{\partial u}{\partial w}(t,x,w_i) \approx \begin{cases} \min\{|d_-|,|d_+|\}\,\mathrm{sgn}(d_-) & \text{if } d_- d_+ > 0 \\ 0 & \text{otherwise} \end{cases}.$$

where $d_- = \dfrac{u_i(t,x) - u_{i-1}(t,x)}{\Delta w}$ and $d_+ = \dfrac{u_{i+1}(t,x) - u_i(t,x)}{\Delta w}$.

Then, a set on N partial differential equations for the unknowns u_i is derived. The spatial dependence is considered with the help of the fifth-order WENO method [2] and the time integration is performed by applying a third-order TVD Runge Kutta scheme [10].

Acknowledgments

This work was supported by the Fond zur Förderung der wissenschaftlichen Forschung, Vienna, under contract number P14669-TPH, by the European community programme IHP, contract number HPRN-CT-2002-00282 on behalf of the CNR and by M.U.R.S.T. (Cofin 2002) *Mathematical problem of Kinetic Theories*.

References

1. Galler, M., Schürrer, F.: A deterministic solution method for the coupled system of transport equations for the electrons and phonons in polar semiconductors. J. Phys. A: Math. Gen., **37**, 1479–1497 (2004)
2. Carrillo, J.A., Gamba, I.M., Majorana, A., Shu, C.-W.: A WENO-solver for the transients of Boltzmann-Poisson system for semiconductor devices. Performance and comparisons with Monte Carlo methods. J. Comp. Phys., **184**, 498–525 (2003)
3. Carrillo, J.A., Gamba, I.M., Majorana, A., Shu, C.-W.: A direct solver for 2D non-stationary Boltzmann-Poisson systems for semiconductor devices: a MESFET simulation by WENO-Boltzmann schemes, J. Comput. Electronics, **2**, 375–380 (2003)
4. Lundstrom, M.: Fundamentals of Carrier Transport. Cambridge University Press, Cambridge (2000)
5. Ziman, J.M.: Electrons and Phonons. The Theory of Transport Phenomena in Solids. Oxford University Press, Oxford (2000)
6. Majorana, A. and Pidatella, R.M.: A finite difference scheme solving the Boltzmann-Poisson system for semiconductor devices, Journal of Computational Physics, **174**, 649–668 (2001)
7. Lapidus, L., Pinder, G.F.: Numerical Solution of Partial Differential Equations in Science and Engineering. Wiley, New York (1982)
8. LeVeque, R.J.: Numerical Methods for Conservation Laws. Birkhäuser, Basel (1992)
9. Galler, M., Majorana, A.: Deterministic and Stochastic Simulations of Electron Transport in Semiconductors, to appear in Transp. Theory and Stat. Phys
10. Shu, C.-W., Osher, S.: Efficient implementation of essentially non-oscillatory shock capturing schemes. J. Comp. Phys., **77**, 439–471 (1988)

Deterministic Numerical Simulation of 1d Kinetic Descriptions of Bipolar Electron Devices

P. González[1], J. A. Carrillo[2] and F. Gámiz[3]

[1] Dpt. of Applied Mathematics - UGR, Granada, Spain, prodelas@ugr.es
[2] ICREA - Dpt. Matemàtiques - UAB, Barcelona, Spain, carrillo@mat.uab.es
[3] Dpt. of Electronics - UGR, Granada, Spain, fgamiz@ugr.es

1 Introduction

In this work we deal with the deterministic simulation of some electronic bipolar devices; in particular, p-n junctions and bipolar junction transistors (BJT's), among others. Essentially, a BJT transistor consists of the inverse union of two diodes of type p-n, being able to form devices of type p^+-n-p or n^+-p-n, where superscript $^+$ indicates a strongly doped region. The intermediate region between the highly doped emitter and the lower doped collector serves as a base whose applied voltage controls the carriers flux between emitter and collector. These bipolar transistors constitute a basic element in manufacturing modern electronic devices as tiny rectifiers, luminance photocells and many others.

It is well known that in order to explain some of the phenomena observed in these devices, it is necessary to consider both the flow of electrons in the conduction band and holes in the valence band. In each band the particles may undergo collisions due to dispersion mechanisms (scattering). Also, we are able to include electron-hole pairs generation and recombination, whose effects in certain devices are not negligible.

We take into account acoustic phonons in the elastic approximation and optical non-polar phonons in the inelastic approximation with a unique frequency ω both for electrons and for holes. These scattering phenomena are the most relevant for silicon (Si) (see [Tom93] for its derivation and to find the physical parameters of the material) and were used in the case of electron transport in [MP01, CGMS03]. In addition, we may include generation-recombination processes: band to band, Auger recombination, ... [Sze85, MRS90, SB00].

The simulation of this type of devices have been undertaken by means of the numerical resolution of the corresponding system of partial differential equations of Boltzmann-Poisson type using deterministic methods.

2 Boltzmann-Poisson system for bipolar devices

As we have already indicated, from a mathematical point of view, we have to deal with a system of two transport equations of Boltzmann type: one for the electrons f_e (with negative charge) and another one for the holes f_h (with positive charge), coupled with a Poisson's equation for the potential, from which the corresponding electric field is calculated $\widetilde{\mathbf{E}} = -\nabla_{\widetilde{\mathbf{x}}} V$

$$\begin{cases} \partial_{\widetilde{t}} f_l + \dfrac{1}{\hbar}\nabla_{\mathbf{k}}\varepsilon_l(\mathbf{k})\cdot\nabla_{\widetilde{\mathbf{x}}} f_l + sig(l)\,\dfrac{\mathfrak{e}}{\hbar}\widetilde{\mathbf{E}}\cdot\nabla_{\mathbf{k}} f_l = Q(f_l) + R_l(f_e, f_h) \\ (l = e, l = h;\ \text{where } sig(l) := -1 \text{ when } l = e \text{ and } 1 \text{ when } l = h) \\ \\ f_l(0, \widetilde{\mathbf{x}}, \mathbf{k}_l) = \mathcal{M}_l(\mathbf{k}_l)\ \widetilde{\mathcal{N}}_l(\widetilde{\mathbf{x}});\ \mathbf{k}_l \in \mathbb{R}^3,\ \widetilde{\mathbf{x}} \in [0, \widetilde{L}]\subset\mathbb{R} \\ \\ \Delta V \equiv \nabla^2 V(\widetilde{t}, \widetilde{\mathbf{x}}) = \dfrac{\mathfrak{e}}{\epsilon}\left(\rho_e - \rho_h - \widetilde{\mathcal{N}}_e + \widetilde{\mathcal{N}}_h\right);\ \widetilde{t} \in \mathbb{R}_0^+,\ \widetilde{\mathbf{x}} \in [0, \widetilde{L}]\subset\mathbb{R} \end{cases} \tag{1}$$

where, $f_l \equiv f_l(\widetilde{t}, \widetilde{\mathbf{x}}, \mathbf{k}_l)$, $\widetilde{t} \in \mathbb{R}_0^+$, $\widetilde{\mathbf{x}} \in [0, \widetilde{L}]\subset\mathbb{R}$, $\mathbf{k}_l\in\mathbb{R}^3$ represent the probability density functions of finding an electron ($l = e$) or a hole ($l = h$) with wave vector $\mathbf{k}_l$, located in the spatial point $\widetilde{\mathbf{x}}$ at time $\widetilde{t}$; $\hbar$ is the constant of Planck divided by 2π, $\mathfrak{e}$ is the electron charge, ϵ is the permittivity's constant of the crystal and $\widetilde{L}$ is the length of the device. The tilde everywhere emphasizes the fact that we are considering dimensional variables.

The initial condition for each transport equation consists of "maxwellians" (distributions in the kernel of the collision operators), so that the initial value of the density (at $\widetilde{t}=0$) in $\rho_l\left(\widetilde{t},\widetilde{\mathbf{x}}\right) \equiv \int_{\mathbb{R}^3} f_l\left(\widetilde{t},\widetilde{\mathbf{x}},\mathbf{k}_l\right) d\mathbf{k}_l$ agrees with the corresponding dopant functions $\widetilde{\mathcal{N}}_l\left(\widetilde{\mathbf{x}}\right)$. We consider non-parabolic (Kane model) bands for both electrons and holes:

$$\varepsilon_l\left(\mathbf{k}_l\right) = E_l - sig(l)\frac{\frac{\hbar^2}{m_l^*}\left|\mathbf{k}_l\right|^2}{1+\sqrt{1+2\frac{\widetilde{\alpha}}{m_l^*}\hbar^2\left|\mathbf{k}_l\right|^2}} \tag{2}$$

where $E_l \equiv E_C$ (for $l=e$) or E_V (for $l=h$) are the minimum and the maximum of the energy in the conduction and in the valence band respectively; m_e^* and m_h^* are the effective masses for electrons and holes respectively and $\widetilde{\alpha}$ is the non-parabolicity factor. Therefore, the maxwellian distributions for electrons $\mathcal{M}_e$ and holes $\mathcal{M}_h$ are

$$\mathcal{M}_l\left(\mathbf{k}_l\right) = \mathcal{M}\exp\left(-sig\left(l\right)\frac{E_l}{k_B T_L}\right)\left(\frac{\sqrt{2k_B T_L m_l^*}}{\hbar}\right)^{-3}\exp\left(sig\left(l\right)\frac{\varepsilon_l\left(\mathbf{k}_l\right)}{k_B T_L}\right) \tag{3}$$

with $\mathcal{M}$ the corresponding factor to have unit mass (see next section), k_B the Boltzmann factor and T_L the lattice temperature.

The operator $Q\left(f_l\right)$ includes the scattering phenomena and we refer to [MP01, CGMS03] for its exact form. $R_l\left(f_e,f_h\right)$ represents the generation-recombination (GR) mechanisms in this type of devices. This GR terms are defined, for $l=e$ and $l=h$, as $R_l \equiv R_l^{RF}+R_l^{AU}+R_l^{RL}+R_l^{SRH}$; i.e., the sum of different GR mechanisms:

$$R_l^{RF}\left(f_e,f_h\right) = -\frac{1}{\tau_{RF}\,\rho_i}\left(\rho_{\bar{l}}\, f_l - \rho_i^2\,\mathcal{M}_l\right)$$

is the band to band mechanism by photons (with $\tau_{RF}=0.1\,\mu\mathrm{s}=10^5$ ps, noting $\bar{l}=e$ when $l=h$ and $\bar{l}=h$ when $l=e$);

$$R_l^{AU}\left(f_e,f_h\right) = -\Gamma_e\left(\rho_e\,\rho_h\, f_e - \rho_e\,\rho_i^2\,\mathcal{M}_e\right) - \Gamma_h\left(\rho_e\,\rho_h\, f_h - \rho_h\,\rho_i^2\,\mathcal{M}_h\right)$$

is the Auger GR mechanism (with typical values for Si: $\Gamma_e = 2.8\times10^{-31}\,\mathrm{cm}^6\,\mathrm{s}^{-1}$ and $\Gamma_h = 9.9\times10^{-32}\,\mathrm{cm}^6\,\mathrm{s}^{-1}$); R_l^{SRH} is the so-called Schockley-Read-Hall GR mechanism (see [MRS90, Sze85, SB00])

$$R_e^{SRH}\left(f_e,f_h\right) = n_{tr}C_c\mathcal{M}_e - \left(N_{tr}-n_{tr}\right)C_a\,f_e$$
$$R_h^{SRH}\left(f_e,f_h\right) = \left(N_{tr}-n_{tr}\right)C_d\mathcal{M}_h - n_{tr}\,C_b\,f_h$$

with $n_{tr} = N_{tr}\frac{C_a\,\rho_e + C_d}{C_a\,\rho_e + C_c + C_b\rho_h + C_d}$ and the following values for the constants: $N_{tr} = 5\times10^{16}\,\mathrm{cm}^{-3}$, $\tau_{n,p} = 10^5$ ps; $C_a \equiv \frac{1}{\tau_n\,N_{tr}} = \frac{1}{\tau_p\,N_{tr}} \equiv C_b = 2.0\times10^{-22}\,\frac{\mathrm{cm}^3}{\mathrm{ps}}$; $C_c \equiv C_a\,n_i = C_b\,n_i \equiv C_d \simeq 2.894\,5\times10^{-12}\,\mathrm{ps}^{-1}$. And finally the simpler linear GR terms (with $\tau_{RL} = 1.\,\mu\mathrm{s} = 10^6$ ps)

$$R_l^{RL}\left(f_e,f_h\right) = -\frac{1}{\tau_{RL}}\left(f_l - \mathcal{M}_l\left(\mathbf{k}_l\right)\widetilde{\mathcal{N}}_l\left(\widetilde{\mathbf{x}}\right)\right)$$

also valid for small deviations with respect to the equilibrium. Here, ρ_i is the intrinsic concentration given by

$$\rho_i \equiv n_i\left(T_L\right) = 2\left(\frac{k_B T_L\,\sqrt{m_e^*\,m_h^*}}{2\pi\,\hbar^2}\right)^{\frac{3}{2}} e^{-\frac{E_g}{2k_B\,T_L}}$$

depending on the jump energy $E_g = E_C - E_V$ (typical values for Si at room temperature are $E_g \simeq 1.08$ eV and $n_i\left(300\ \mathrm{K}\right) \simeq 1.44725\times10^{10}\ \mathrm{cm}^{-3}$).

Transformation of the problem

In the one-dimensional case we perform the same change of variables to pseudo-spherical coordinates introduced in [MP01] in each equation, noting that the effective mass for the conduction and valence band carriers are different: $m_e^* \equiv 0.31\,m_0$ for the electrons, $m_h^* \equiv 0.5\,m_0$ for the holes where m_0 is the mass of the electron at rest:

$$\mathbf{k}_l \equiv \mathbf{k}_l\left(w_l,\phi,\mu\right) \equiv \sqrt{2}\frac{\sqrt{m_l^* k_B T_L}}{\hbar}\sqrt{w_l\left(1+\alpha_K w_l\right)}\begin{pmatrix}\sqrt{1-\mu^2}\cos\phi\\ \sqrt{1-\mu^2}\sin\phi\\ \mu\end{pmatrix}$$

where the jacobian of this transformation is given by $\sqrt{2}\,(m_l^* T_L k_B)^{\frac{3}{2}}\,\hbar^{-3}\, s\,(w_l)$ with

$$s\,(w) := \sqrt{w\,(w\alpha_K + 1)}\,(1 + 2w\alpha_K)\,.$$

This change of variables allows a simple expression for the band energies $\varepsilon_l(\mathbf{k}_l) = E_l - sig(l) k_B T_L w_l$ that implies also a simple form of the maxwellians (3) (with $\mathcal{M} := (2\pi \int_0^{+\infty} s(w)\, e^{-w}\, dw)^{-1}$) as

$$\mathcal{M}_l \equiv \mathcal{M}_l\,(w_l) = \left(\sqrt{2k_B T_L m_l^*}/\hbar\right)^{-3} \mathcal{M}\, e^{-w_l}.$$

Therefore, in the case of axial symmetry in space with respect to the $\widetilde{z}$ axis, we reduce ourselves to one variable in space: $\widetilde{z} \in \left[0, \widetilde{L}\right] \subset \mathbb{R}$. The new unknowns will be the following functions (for $l = e$ and $l = h$)

$$\Phi_l \equiv \Phi_l\left(\widetilde{t}, \widetilde{z}, w_l, \mu\right) \equiv s\,(w_l)\;F_l\left(\widetilde{t}, \widetilde{z}, w_l, \mu\right)$$

where $F_l\left(\widetilde{t}, \widetilde{z}, w_l, \mu\right)$ denote the functions $f_l\left(\widetilde{t}, x_3, \mathbf{k}_l\right)$ once the corresponding change of variables (2) is carried out. Now the convective terms of each transport equation is written as follows

$$s\,(w_l)\,\frac{1}{\hbar}\frac{\partial \varepsilon_l}{\partial k_3}\frac{\partial f_l}{\partial x_3} = \frac{\partial}{\partial \widetilde{z}}\left(a_1^{(l)}(w_l,\mu)\,\Phi_l\right)$$

with

$$a_1^{(l)} \equiv \sqrt{2\frac{k_B T_L}{m_l^*}}\,\frac{\mu\, s\,(w_l)}{(1 + 2\alpha_K w_l)^2}.$$

The force term is written as

$$\left(\frac{\mathfrak{e}}{\hbar}\widetilde{\mathbf{E}} \cdot \nabla_{\mathbf{k}} f_l\right)\, s\,(w_l) \equiv \left(\frac{\mathfrak{e}}{\hbar}\widetilde{E}_3 \frac{\partial f_l}{\partial k_3}\right)\, s\,(w_l)$$
$$= \frac{\partial}{\partial w_l}\left(a_2\left(\widetilde{t}, \widetilde{z}, w_l, \mu\right)\,\Phi_l\right) + \frac{\partial}{\partial \mu}\left(a_3\left(\widetilde{t}, \widetilde{z}, w_l, \mu\right)\,\Phi_l\right)$$

with

$$a_2^{(l)} \equiv \frac{\mathfrak{e}\,\widetilde{E}_3\left(\widetilde{t}, \widetilde{z}\right)}{\sqrt{2k_B T_L m_l^*}}\,\frac{2\mu\, s\,(w_l)}{(1 + 2\alpha_K w_l)^2}\;;\; a_3^{(l)} \equiv \frac{\mathfrak{e}\,\widetilde{E}_3\left(\widetilde{t}, \widetilde{z}\right)}{\sqrt{2k_B T_L m_l^*}}\,\frac{\left(1 - \mu^2\right)(1 + 2\alpha_K w_l)}{s\,(w_l)}.$$

In this way, the two Boltzmann's equations can be written in a totally conservative form, using these new variables:

$$\frac{\partial \Phi_l}{\partial \widetilde{t}} + \frac{\partial}{\partial \widetilde{z}}\left(a_1^{(l)}\,\Phi_l\right) + sig\,(l)\left(\frac{\partial}{\partial w_l}\left(a_2^{(l)}\,\Phi_l\right) + \frac{\partial}{\partial \mu}\left(a_3^{(l)}\,\Phi_l\right)\right) = \widehat{Q}\,(\Phi_l) + \widehat{R}_l\,(\Phi_e, \Phi_h)$$

where the generation-recombination operators are (for $l = e$ and $l = h$)

$$\widehat{R}_l\,(\Phi_e, \Phi_h) \equiv s\,(w_l)\;R_l\,(F_e, F_h)$$

and the collision operator takes the same form as in [CGMS03]. The energy intervals $w_l \in \left[0, w_l^{(\max)}\right]$ must be adjusted in the numerical experiments so that $F_l\left(\widetilde{t}, \widetilde{z}, w_l, \mu\right) \simeq 0$ for any $w_l \geq w_l^{(\max)}$ and $\forall \widetilde{t}, \widetilde{z}, \mu$, both for $l = e$ and $l = h$.

Coupling

The coupling between both transport equations is done by means of the appropriate Poisson's equation from which the electric field $\widetilde{E}_3\left(\widetilde{t}, \widetilde{z}\right) = -\partial_{\widetilde{z}} V\left(\widetilde{t}, \widetilde{z}\right)$ is obtained. Using dimensional coordinates in space $\widetilde{z} \in [0, \widetilde{L}]$ and time $\widetilde{t} \in \mathbb{R}_0^+$

$$\begin{cases} \dfrac{\partial^2 V}{\partial \widetilde{z}^2}\left(\widetilde{t}, \widetilde{z}\right) = \dfrac{\mathfrak{e}}{\epsilon}\left(\mathcal{F}_e\left(\widetilde{t}, \widetilde{z}\right) - \mathcal{F}_h\left(\widetilde{t}, \widetilde{z}\right)\right),\; \widetilde{z} \in [0, \widetilde{L}] \\ V\left(\widetilde{t}, 0\right) = V_{left}, \qquad V\left(\widetilde{t}, \widetilde{L}\right) = V_{right} \end{cases}$$

where $\mathcal{F}_l\left(\widetilde{t}, \widetilde{z}\right) \equiv \rho_l\left(\widetilde{t}, \widetilde{z}\right) - \widetilde{\mathcal{N}}_l\left(\widetilde{z}\right)$ in which the density of carriers for electrons ($l = e$) and holes ($l = h$) is given by

$$\rho_l\left(\widetilde{t}, \widetilde{z}\right) = \left(\frac{\sqrt{2m_l^* k_B T_L}}{\hbar}\right)^3 \pi \int_0^{w_l^{(\max)}} \int_{-1}^{1} \Phi_l\left(\widetilde{t}, \widetilde{z}, w_l, \mu\right)\, d\mu\, dw_l.$$

Here, the boundary conditions V_{left}, V_{right} have to be settled according with the applied V_{bias} and the corresponding built-in potentials due to the p-n or n-p junctions, generically given by $V_{bi} = \frac{k_B T_L}{e} \ln\left(\frac{N_a N_d}{n_i^2}\right)$ where N_a and N_d denote the constant values of the dopants concentrations (donors and receivers) in both sides of the junction. In the case of a BJT transistor, we are forced to solve separately two Poisson's equations to include the potential at the base; then also V_{BE} and V_{CE} have to be considered.

3 Numerical resolution

The simulations have been undertaken using deterministic methods: weigthed essentially non oscillatory finite differences (FD-WENO [Shu98]) of high order for the advection terms together Runge-Kutta third order in time. As it is well known, the basic idea of the WENO methods, consists of using a convex nonlinear combination of all the possible candidates obtained for the so called Essentially Non Oscillatory (ENO) reconstruction technique instead of taking only one, that would provide the more regular. In this way, when k possible reconstruction candidates (obtained from $2k-1$ cells) are taking into account, but only one of them is used for the reconstruction, finally we will be able to reach an order of precision at most k, with the ENO technique; whereas with the WENO procedure, it is possible to reach a precision order of $2k-1$ (at least in the regular regions of the function), because all the possible candidates are used. One of the main drawbacks of FD-WENO methods is the need of a uniform mesh.

However, dealing with devices of a certain length, the consideration of a uniform mesh forces to take an undesirable high number of points to obtain an appropriate resolution in the depletion zone, where the more intense fluxes take place. Therefore, we have also implemented the possibility of using nonuniform meshes in the spatial variable $\widetilde{z}$ by

$$\widetilde{z}_i \equiv \xi^{-1}(\xi_i) \equiv \mathcal{X}(\xi_i), \qquad i = 1, \ldots, M \tag{4}$$

originating from some regular transformation (with $\xi \equiv \xi(\widetilde{z})$ and $\mathcal{X} \equiv \mathcal{X}(\xi)$ smooth enough).

For this purpose, we can take a transformation $\varphi : [-1,1] \longrightarrow [-1,1]$ that is an uneven function (symmetrical with respect to the origin) sufficiently regular and with inverse φ^{-1} also regular, so that $|\varphi(\xi)| \ll 1$ for $|\xi| \ll 1$; that is to say, with the objective of which the images of a uniformly distributed set of points near the origin are more accumulated near the origin. Therefore, if we take $\widetilde{z} = \mathcal{X}(\xi) \equiv \frac{\widetilde{L}}{2^s}(\varphi(\xi)+1)^s \in [0, \widetilde{L}]$ with $s = \log_2\left(\frac{\widetilde{L}}{\widetilde{z}_o}\right)$, then $\xi \equiv \xi(\widetilde{z}) \equiv \mathcal{X}^{-1}(\widetilde{z}) = \varphi^{-1}\left(2\left(\frac{\widetilde{z}}{\widetilde{L}}\right)^{\frac{1}{s}} - 1\right) \in [-1,1]$ and now the accumulation of points will take place around $\widetilde{z}_o \in [0, \widetilde{L}]$ (the case $s = 1$ corresponds with $\widetilde{z}_o = \frac{\widetilde{L}}{2}$).

In this way, any conservation law of the form $\partial_t u + \partial_{\widetilde{z}} f(u) = 0$ is transformed into $\partial_t u + \xi'(\widetilde{z})\, \partial_\xi f(u) = 0$ and it would suffice to apply the ENO or WENO technique to it in order to approach the first derivative $\partial_\xi f(u)$, keeping the conservativity of the scheme and the convergence, assured by a theorem of Lax-Wendroff type (consult ([OC83])).

Numerical Results

We have verified how the numerical results obtained correspond qualitatively and quantitatively well to the behavior expected for p-n junctions. Extensive comparisons with results of the drift-diffusion equations have been performed using the software PISCES. The only problem in some specific cases (depending on the length and doping differences) is the need of fine grids for obtaining sufficiently decreasing residues in order to have a good convergence to the final steady state. To try to overcome this problem we have used the indicated space transformation of coordinates (4) to have the adapted resolution in the depletion or transition zone, where almost all the particle interchange really takes place and where it appears a non null electric field.

Although this method is well adapted to p-n junctions it is difficult to be generalized to other geometries as BJT transistors. In this case, we are exploring the use of some sort of multidomain/multigrid technique inspired in a FD-WENO method with interpolation at subdomain interfaces ([SS03]).

The numerical results we show correspond to a BJT (n^+-p-n) device of $2\mu m$ with abrupt dopants jumps at 0.49 and 0.59 μm of $N_{DE} = 5 \times 10^{18}$, $N_{AB} = 5 \times 10^{17}$ and $N_{DC} = 1 \times 10^{16}$ cm^{-3} in the emitter, base and collector, respectively. In this case, a uniform grid of 1000 points in space, 40 in energy and 8 in the angle variable have been used for the computations. The potential drop is of 3 V between emitter and collector and 0.62 V between emitter and base. Built-in potentials of the corresponding p-n junctions have been taken also into account.

We have modeled the base contact at $\widetilde{z}_b = 0.54$ μm by subdividing the device in two regions $A = [0, \widetilde{z}_b]$ and $B = [\widetilde{z}_b, \widetilde{L}]$. In region A we impose the base potential drop and we impose maxwellian distribution for holes with incoming velocities at $\widetilde{z}_b$. In region B we take into account base and collector potential drops in the Poisson equation and we impose incoming boundary conditions in $\widetilde{z}_b$ from the results of region A. This coupling procedure of the computation in regions A and B is performed at each stage of the RK solver and at each time we need to compute the electric potential in the whole device.

Figures 1a–c shows the evolution of the electron and hole densities and the electric potential. The final results are compared (using diamonds) to the drift diffusion results from PISCES software and they are in good agreement as one can expect in this particular device. Probability density functions are shown in next figures for electrons and holes in different points of the device from which one can observe the response of electrons and holes to the electric field.

Acknowledgements

The first author acknowledges support from the Ministry of Science and Technology of Spain (BFM2002-02649) and the second, from the European IHP network HYKE (HPRN-CT-2002-00282) and DGI-MCYT/FEDER project BFM2002-01710. We also thank the referees for useful comments and suggestions.

Figures

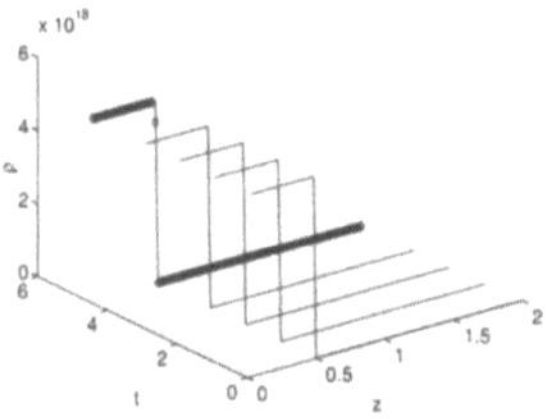

Fig. 1a. Electron Density

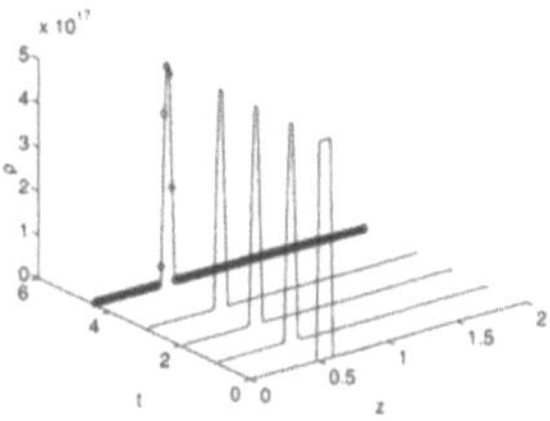

Fig. 1b. Hole Density

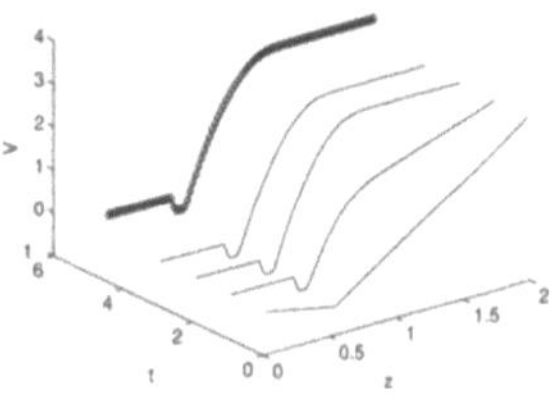

Fig. 1c. Electric Potential

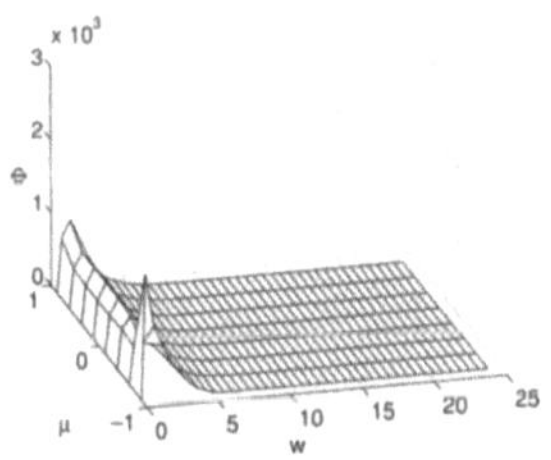

Electron pdf's at 0.49 μm

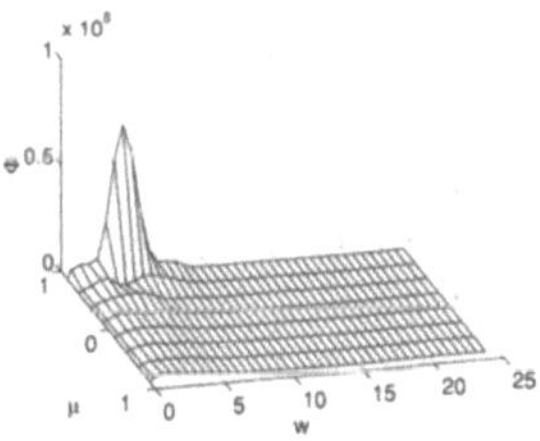

... at 0.59 μm

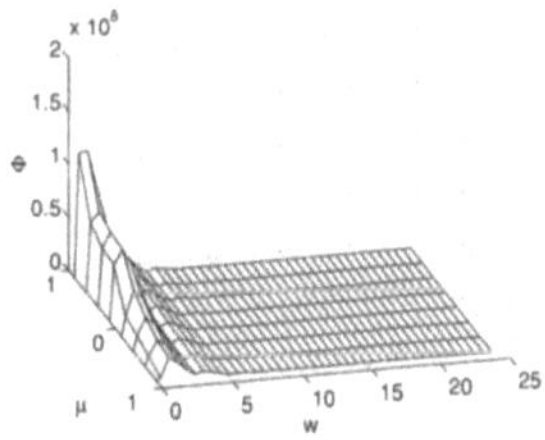

... at 1.0 μm

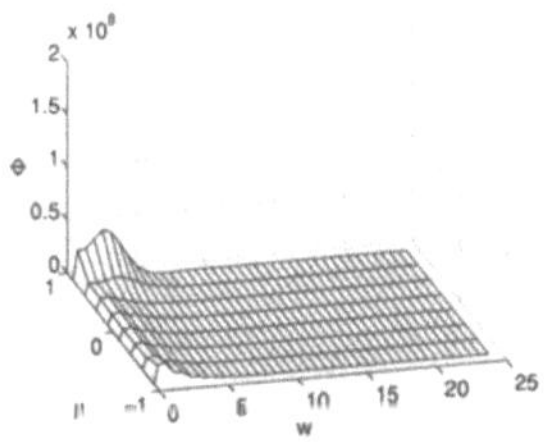

Hole pdf's at 0.49 μm

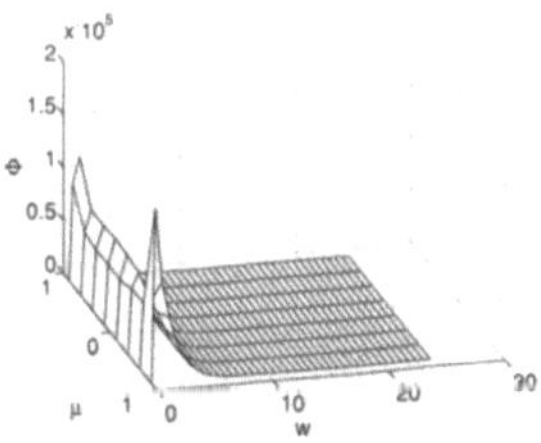

... at 0.59 μm

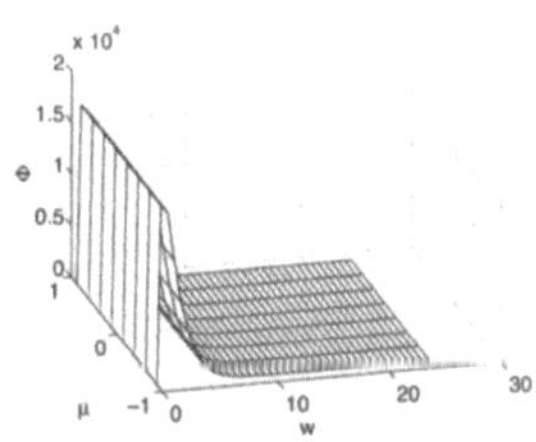

... at 1.0 μm

References

[CGMS03] Carrillo, J.A., Gamba, I.M., Majorana, A., Shu, C.W.: A WENO-solver for the transients of Boltzmann-Poisson system for semiconductor devices. Performance and comparisons with Monte Carlo methods. J. C. P., **184**, 498–525 (2003)

[MP01] Majorana, A., Pidatella, R.M.: A Finite Difference Scheme Solving the Boltzmann-Poisson System for Semiconductor Devices. J. C. P., **174**, 649–668 (2001)

[OC83] Osher, S., Chakravarthy, S.: Upwind schemes and boundary conditions with applications to Euler equations in general geometries, J.C.P., **50**, 12–49 (1983)

[Shu98] Shu, C.-W.: Essentially Non-oscillatory and Weighted Essentially Non-oscillatory Schemes for Hyperbolic Conservation Laws. In: Cockburn, B., Johnson, C., Shu, C.-W. and Tadmor, E. (eds) Advanced Numerical Approximation of Nonlinear Hyperbolic Equations. Lecture Notes in Mathematics. Springer-Verlag, Berlin/New York (1998)

[SS03] Sebastien, K., Shu, C.W.: Multidomain WENO Finite Difference Method with Interpolation at Subdomain Interfaces. Journ. Scient. Comput. **19**, 405–439 (2003)

[MRS90] Markowich, P.A., Ringhofer, C.A., Schmeiser, C.: Semiconductor Equations. Springer-Verlag, Wien (Austria) (1990)

[SB00] Streetman, B. G., Banerjee, S.: Solid State Electronic Devices (Fifth Edition). Prentice Hall International, Inc. New Jersey (2000)

[Sze85] Sze, S.M: Semiconductor Devices. John Wiley & Sons, AT&T Bell Laboratories. Murray Hill-New Jersey (1985)

[Tom93] Tomizawa, K.: Numerical simulation of submicron semiconductor devices. Artech House, Boston (1993)

A Hybrid Intelligent Computational Methodology for Semiconductor Device Equivalent Circuit Model Parameter Extraction

Y. Li

Department of Communication Engineering &
Microelectronics and Information Systems Research Center,
National Chiao Tung University, Hsinchu 300, Taiwan,
E-mail: ymli@faculty.nctu.edu.tw

Abstract In this paper, a hybrid intelligent computational methodology is presented for the parameter extraction of compact models. This solution technique integrates the genetic algorithm (GA), the neural network (NN), and the Levenberg-Marquardt (LM) method for current-voltage (I-V) curves characterization, optimization, and parameter extraction of deep-submicron metal-oxide-semiconductor field effect transistors (MOSFETs). For a specified compact model, this unified optimization technique extracts a set of corresponding parameters with respect to measured data. The GA is performed to search solutions according to the feedback of the NN, where the LM solves a local optimization problem with the input of the GA. The well-known BSIM and EKV compact models of MOSFETs have been studied and implemented for automatic parameters extraction. In terms of accuracy and convergence of score, the proposed optimization technique is computationally verified to show its advantages for parameter extraction of MOSFETs. Comparisons among pure GA approach, solution with GA and NN, solution with GA and LM, and the proposed method are also discussed.

1 Introduction

Understanding electrical characteristics of various semiconductor devices is one of important issues in modern electronic industry. In semiconductor device modelling, setting on each construction parameter is always a complicated problem which significantly affects the results of designed and fabricated very large scale integration (VLSI) device and circuit. Computer simulator together with a set of optimal parameters is the right issue for VLSI circuit design, in particular for the nanoelectronics era. For a specified compact model, the process that fits the simulation data as closely as possible to the measured data is the so-called parameter extraction in electrical engineering. It is not only a time-consuming task but also requires engineering expertise to find a proper configuration of parameters with reasonable physical meanings. Model parameter extraction has been of great interest in both the design and fabrication communities in the last decade; nevertheless, it still has room to improve the performance of extraction methods for searching model solutions in semiconductor industry.

An equivalent circuit model together with a set of proper parameters intrinsically characterizes the electrical characteristics of designed and fabricated VLSI devices. Various compact models have been studied for deep-submicron and sub-100 nanometer devices [1][2][3][4][5]. Problem of model parameter extractions can be regarded as a multidimensional optimization problem which minimizes the error between the measured data and simulated result. Numerical and evolutional methods, such as Newton-liked method and genetic algorithm (GA), have been considered in the characteristic optimization of VLSI devices, but numerical methods in general require an accurate initial guess to perform a local optimization. Solution with a pure GA [6][7][8] suffers a long time evolution process. It may take days even weeks to find suitable parameters for several devices within a single model, for example. Any accurate and robust solution methodologies will benefit parameter extraction of equivalent circuit models of VLSI device [9][10].

We in this paper develop a flexible hybrid intelligent computational methodology. Based on the GA, the neural network (NN), and the Levenberg-Marquardt (LM) method, this solution technique for parameter extraction of deep-submicron MOSFETs shows its superiority over the other approaches. Starting from GA for a rough estimation on the solution, the LM method will enable a local optimization, where the NN investigates the quality of solutions and suggests searching directions for the GA. This unified optimization technique extracts optimal parameters for a given compact model from measured current-voltage (I-V) curves in a computationally cost-effective manner. The well-known BSIM and EKV compact models of metal-oxide-semiconductor field effect transistors (MOSFETs) have been implemented and verified for the automatic parameters extraction using the proposed optimization technique. Accuracy and convergence of score are verified to show the advantages of the developed method for parameter extraction of MOSFETs with respect to the BSIM and EKV compact models.

This paper is organized as follow. Section 2 briefly states the concept of GA, NN, LM, and the proposed hybrid intelligent computational methodology. Section 3 shows computer experimental results for MOSFETs' parameter extraction with different compact models. Finally, we draw conclusions.

2 The Methodology of Hybrid Optimization

In this section, we state the hybrid intelligent computational methodology for semiconductor device equivalent circuit model parameter extraction. The solution methodology integrates the GA, NN, and LM methods [6][7][8]. First of all, we briefly state the basic concept of the GA, NN, and LM methods. The configuration of the the proposed hybrid intelligent computational methodology is then described.

It is known that GA is a globally searching optimization method which is based on the mechanism of natural selection and natural genetics. It works with a code of parameter strings called chromosome instead of the solutions themselves. Each chromosome represents a solution set, and the fitness functions is adopted to measure the survival scores of all chromosomes in the population. Then the GA will accord its selection scheme to select several chromosomes for copulation, discard unwanted chromosomes, and select the crossover scheme to produce the new generation. Then the GA will apply fitness function for the new population again and loop this cycle until certain stop criteria is achieved [9][10].

NN is an adaptive learn network which has the remarkable ability to derive meaning from complicated or imprecise data. It has been wildly used in various rages especially in pattern reorganizations and the image processing. In this work, we adopt Hamming net to guide GAs to search the better solutions. The Hamming net is a supervised feedback NN which contains two sub-networks, the matching score net and the maximum net. When the training patterns stores into matching score net, it measures the differences between input patterns and training patterns. After grabbing the output of each node in the matching score net, the maximum net is functioned to determine which training pattern is the most similar to the input pattern. Once there is a unique restrained output above the threshold, the Hamming net terminated, and considers the training pattern represented by the node which provided the outstanding output is most similar to the input pattern, thus the input pattern can be clustered into this training pattern.

In contrast with the GA and NN above, the LM method is a quasi-Newton method to accelerate the Gauss-Newton method. It belongs to one of numerical optimization methods. The Gauss-Newton method is the basic algorithm for solving the nonlinear optimization problem. Due to the nonlinear property of the problem, a gradient for each variable can be obtained. It starts from an initial guess, and follows the direction of the normal of the gradient to find the optimal solution. Therefore, the initial guess must be chosen carefully, or the solution may fell into a local optima. Unlike the Gauss-Newton method has the fixed steps toward the solution, LM optimization method detects that some regions with monotonic variation property can be speed up by increasing the step size. On the other hand, when the optimization process encounters a sensitive region, the step should be shorten to avoid skipping the optimum.

An execution flowchart of the hybrid intelligent computational technique for the parameter extraction task is shown in Fig. 1. As shown in this figure, the GA firstly searches the entire problem space to get a set of roughly estimated solutions. After a roughly computed solution is obtained, the LM method performs a local optima search and sets the local optima as the suggested values for the GA to perform further optimizations. Meanwhile, the NN is applied to investigate the influence of parameters on the optimized functions, and guides the GA to focus on those significant parameters to obtain the better solutions instead of performing blind search. The NN compares the difference of the physical characteristics of the measured data and the simulated I-V curves. According to the examined results of the original and the computed first derivatives of I-V curves, the NN will suggests that the GA should focus on the evolution of those corresponding parameters. Conventional GA-based methods are plagued by problems such as rapid decreases in the population diversity and disproportionate exploitation and exploration of the solution space with multiple dimensions. The results are frequent premature convergence and inefficient search. Compared with the pure GA-based global optimization techniques, the LM method finds a solution rapidly with an accurate initial guesses. We have to note that the LM method, a modified Gauss-Newton method, is still a local method and is easily trapped into local optima. With a proper integration of the LM method in the optimization process, the GA saves much unnecessary efforts to search optima. Furthermore, the most significant parameters that influence physical quantities of VLSI devices have also be detected and monitored. If physical quantities are intolerant, other electrical characteristics will also lose their accuracy. Therefore, the parameters which affect those major quantities should be extracted firstly and the priority of optimization sequence of the model parameters should be considered. Besides, each physical quantity affects some specified I-V curves characteristics such that we can be conscious of the intolerance of physical quantities through investigating the characteristics of I-V curves. The information described above is built in our NN. Under the guidance of the NN, the GA emphasizes the most important parameters and corrects physical quantities one by one. A hybrid optimization algorithm is shown below.

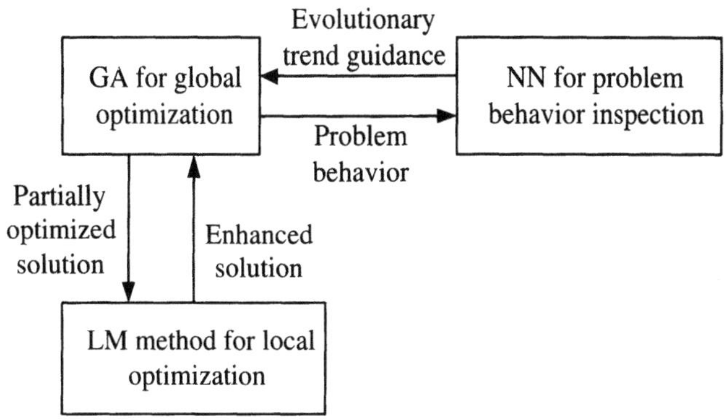

Fig. 1. An illustration of the proposed hybrid intelligent computational methodology. In the beginning, the GA performs search of parameters in large. According to the electrical characteristics of the problem and the obtained rough results from the GA, the NN plays a role in identifying the physical meaning of computed data and guiding the direction of search of the GA. The LM simultaneously solves the corresponding optimization problem with the input data from the GA. It returns the optimized results to the GA for the next process of evolution

```
Begin Hybrid Optimization Algorithm
    Initialize parameters extraction environment
    Begin GA optimization
      Initialize GA
      While EstimatedError(BestSolution) > ToleranceError
        GA performs ParametersExtractionOptimization
        GA obtain BestSolution
        LM ParametersExtractionOptimization(BestSolution)
        NN ModelInspection(BestSolution)
      End While
    End GA optimization
End Hybrid Optimization Algorithm
```

This unified framework exhibits effective optimization in automatic parameter extraction of VLSI device. It shows robust automatic optimization performance for different parameter extraction of compact models. The BSIM and EKV models are the well-known compact models in VLSI industry; for a DC base band characterization, there are more than one hundred parameters in the BSIM model and 30 parameters in the EKV model have to be extracted. We investigate the accuracy and efficiency of the proposed hybrid intelligent computational technique by considering both the BSIM-4 and EKV compact models. Shown in the next section, numerical results confirm that the proposed methodology is superior to the other approaches, compared with the pure numerical and evolutional methods.

3 Results and Discussion

With the hybrid intelligent computational methodology, the accuracy of the extracted parameters, shown in Figs. 2 and 3, are obtained for both the BSIM-4 and EKV compact models, respectively. We note that the BSIM-4 compact

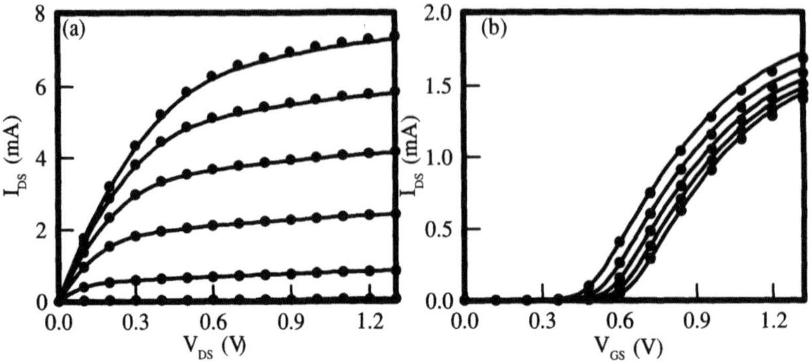

Fig. 2. The BSIM-4 extracted (solid-line) and measured (dot-lines) (a) $I_{DS} - V_{DS}$ and (b) $I_{DS} - V_{GS}$ curves of the 90 nm MOSFET (width = 10.0 μm), where V_{BS} = 0 V and V_{GS} varies from 0.4 V to 1.4 V in the $I_{DS} - V_{DS}$ curves, and V_{DS} = 0.1 V and V_{BS} varies is 0.0 V to -1.5 V in the $I_{DS} - V_{GS}$ curves

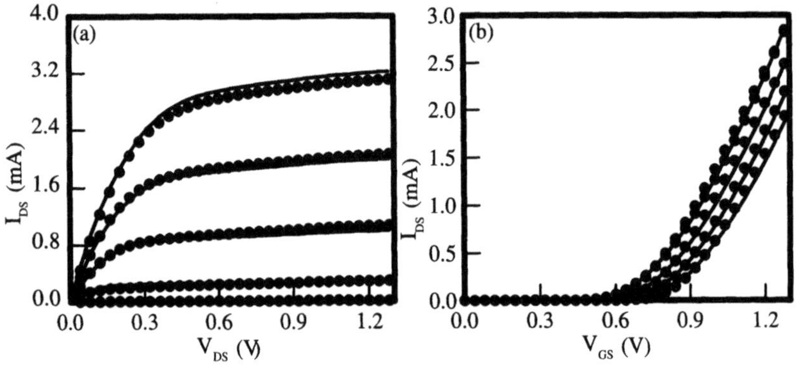

Fig. 3. The EKV extracted (solid-line) and measured (dot-lines) (a) $I_{DS} - V_{DS}$ and (b) $I_{DS} - V_{GS}$ curves are of the 0.18 μm MOSFET, where V_{BS} = -0.6V and V_{GS} migrates from 0.4 V to 1.4 V in the $I_{DS} - V_{DS}$ curves, and V_{DS} = 1.3 V and V_{BS} is from 0.0 V to -0.9 V in the $I_{DS} - V_{GS}$ curves

model has a capability of characterizing sub-100 nm MOSFETs. The EKV compact model features the modelling of deep-submicron MOSFETs. Therefore, 90 nm MOSFETs are adopted for the BSIM-4 compact model and 180 nm MOSFETs are considered for the EKV compact model in our following investigations. As shown in Fig. 2a, it represents the $I_{DS} - V_{DS}$ curves and Fig. 2b stands for $I_{DS} - V_{GS}$ curves; the width and length of the N-MOSFET is equal to 10 μm and 90 nm. For another testing example, the width and length of the target device is equal to 10 μm and 180 nm in Fig. 3. Similarly, Fig. 3a shows the $I_{DS} - V_{DS}$ curves and Fig. 3b stands for $I_{DS} - V_{GS}$ curves. The errors between the measured data and the extracted I-V curves are less than 3% for both the BSIM-4 and EKV compact models. Accuracy of the proposed method is confirmed through the two examples above. Compared with the BSIM-4 compact model, under the same stopping criterion, the EKV compact model has a faster convergence property, but the former has a higher accuracy of extraction.

With the same setting on the device dimension, Tab. 1 summarizes a comparison of the time cost of the hybrid intelligent computational methodology and standard GA for parameter extraction of single and multiple devices with respect to the different compact models. The hybrid technique shows no advantage than others in extracting fewer devices on both compact models as we expected before. However, the more target devices optimized, the superiority of the hybrid technique becomes more efficient apparently. The table 1 suggests the hybrid technique reveals its excellent performance. Figure 4 is a comparison of the fitness score versus the number of generation with respect to different extraction methods for the BSIM-4 compact model. As shown in Fig. 4, without the guidance of the NN, the methods of the GA only and the GA+LM spend a lot of time to reduce the fitness score down to 0.5, while the ones with the NN can fast shot this problem. Comparing the GA+LM and GA+NN+LM methods, shown in Fig. 4, the NN detects the difference between the measured data and the extracted I-V curves, and suggest a better extraction direction to the GA to perform the fine tune task among the I-V curves and its corresponding relative parameters. The results indicate that the evolutionary process with the guidance of the NN shows the better convergence behavior, and they confirm the

Table 1. Comparison of the time cost among different methods for the BSIM-4 and EKV compact models with different number of targets to be extracted

Number of devices	BSIM-4		EKV	
to be extracted	this work (sec.)	pure GA (sec.)	this work (sec.)	pure GA (sec.)
1	354	348	108	125
2	986	998	684	798
4	8451	9841	4587	5981
8	90984	11845	15240	17549
16	260772	290187	41251	46587

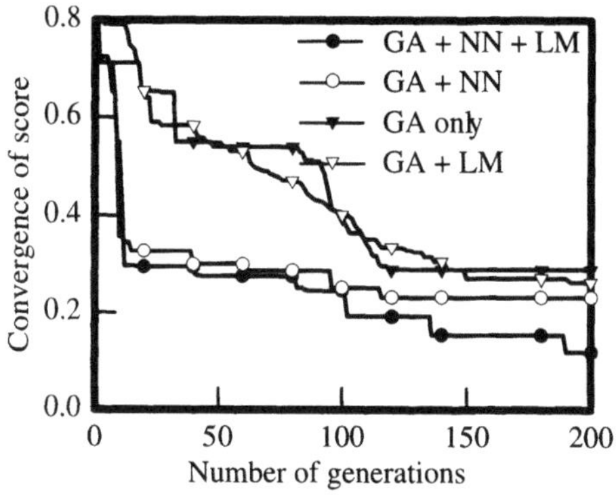

Fig. 4. The convergence of score versus the number of generations with respect to different extraction methods, where the testing is with the BSIM-4 compact model applying to 4 N-MOSFETs with different length and width, where the smallest dimension of device is 130 nm

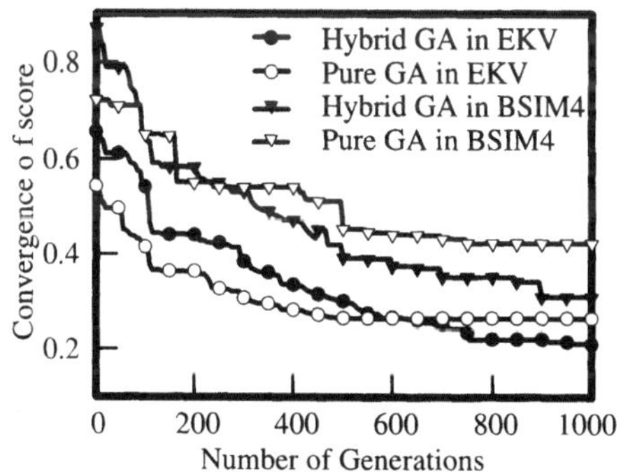

Fig. 5. Comparisons of the convergence of score versus the number of generations between the proposed method and pure GA with the BSIM-4 and EKV compact models. In this experiment, 16 N-MOSFETs with different length and width are optimized in a global sense, where the smallest dimension of device is 130 nm

great efficiency of the method. By extracting a set of global parameters for 16 N-MOSFETs with different width and length, Fig. 5 shows a comparison of the convergence of score of the proposed method with pure GA for different two compact models. The smallest channel length is 130 nm among the extracting 16 devices. The EKV compact model can more quickly achieve a lower score than the BSIM-4 model due to less parameters, but it's final results are not better than the results of the BSIM-4 model after lots of generations. It is due to the intrinsic limitation of EKV compact model. We note that the proposed method can continuously improve the fitness score when the others saturate, shown in Figs. 4 and 5.

4 Conclusions

In this paper, based on the GA, the NN, and the LM methods, we have developed a hybrid intelligent computational technique for model parameter extraction of modern VLSI devices. This automatic optimization technique has been successfully developed and implemented for the BSIM-4 and EKV compact models of VLSI MOSFETs. Preliminary experiments have confirmed that the proposed method can solve complicated multidimensional optimization problem. It may provide a cost-effective way to parameter extraction of deep-submicron and nanoscale VLSI devices. This optimization technique can also be applied to extract parameters of other semiconductor devices. We are currently extend this approach to VLSI circuit design optimization.

Acknowledgment

This work is supported in part by the National Science Council (NSC) of TAIWAN under Contract NSC-93-2215-E-492-008, Contract NSC-94-2215-E-492-005, and Contract NSC 94-2752-E-009-003-PAE, by the grant of the Ministry of Economic Affairs, Taiwan under Contract No. 93-EC-17-A-07-S1-0011, and by the Taiwan Semiconductor Manufacturing Company, under a 2004-2005 grant.

References

1. Hu, C.: BSIM model for circuit design using advanced technologies. Digest of Technical Papers 2001 Symposium on VLSI Circuits, 5–10 (2001)
2. Bucher, M., Lallement, C., Enz, C., Theodoloz, F., and Krummenacher, F.: The EPFL-EKV MOSFET Model Equations for Simulation. Swiss Federal Institute of Technology (EPFL) (1999)
3. BSIM 4.2.1 MOSFET Model User's Manual, U.C. Berkeley, CA (2001)
4. Bendix P.: Tech. Proc. Int. Conf. Modeling and Simulation of Microsystems, 649 (2002)
5. Xi, X., Cao, K., He J., Wan, H., Chan, M., Hu, C.: Symmetry realization of BSIM model with dynamic reference method for circuit simulation. The 60th Annual Device Research Conf. Dig., 65–66 (2002)
6. Holland, J.: Adaptation in Natural and Artificial Systems, University of Michigan Press, Ann Arbor, Mich. (1975)
7. Goldberg, D.E.: Genetic Algorithm Search, Optimization and Machine Learning, Reading, MA: Addison-Wesley (1989)
8. Hung, C.-A., Lin, S.-F.: Adaptive Hamming Net: A Fast-Learning ART 1 Model Without Searching. Neural Networks, **8**, no. 4, 605–618 (1995)
9. Li, Y., Sun, C.-T., Chen, C.K.: A Floating-Point Based Evolutionary Algorithm for Model Parameters Extraction and Optimization in HBT Device Simulation. Neural Networks and Soft Computing, Physica-Verlag, Germany, 364–369 (2003)
10. Li, Y., Cho, Y.-Y., Wang, C.S., and Huang, K. Y.: A Genetic Algorithm Approach to InGaP/GaAs HBT Parameters Extraction and RF Characterization. Japan Journal of Applied Physics, **42**, 2371–2374 (2003)

A SPICE-Compatible Mobility Function for Excimer Laser Annealed LTPS TFT Analog Circuit Simulation

Y. Li[1,2] and C.-S. Wang[2,3]

[1] Department of Communication Engineering, National Chiao Tung University, Hsinchu 300, Taiwan, ymli@faculty.nctu.edu.tw
[2] Microelectronics and Information Systems Research Center, National Chiao Tung University, Hsinchu 300, Taiwan
[3] Department of Mathematics, National Tsing Hua University, Hsinchu 300, Taiwan

Abstract In this paper, we present a calibrated SPICE-compatible mobility function for modeling and simulation of the excimer leaser annealed lower temperature polycrystalline silicon (LTPS) thin film transistors (TFTs). Compared with the conventional mobility function in the well-known RPI TFT equivalent circuit model, the proposed mobility function exhibits accurate results in terms of several DC characteristics. A physical-based model parameter extraction procedure is also proposed for studying the RPI TFT model with the proposed mobility. The model implemented in a circuit simulator for LTPS TFT analog circuit simulation shows reasonable outputs and encounters no any convergence problems.

1 Introduction

Excimer laser annealing technique has recently been proposed in the fabrication of LTPS TFTs, in particular for it application to active-matrix liquid crystal display (AMLCD) [1] and system on panel (SOP). The laser annealed polycrystalline silicon has relatively larger grain size and relatively exhibits higher electron and hole mobility functions than the conventional ones. Therefore, an embedded driving circuit can be achieved by replacing additional driving integration circuits in LCDs. It is known that an equivalent circuit model plays an important role in designing embedded driving circuits using laser annealed LTPS TFTs. Unfortunately, currently reported mobility functions are valid for some conventional TFTs and can not have reasonable prediction in the circuit simulation of the laser annealed LTPS TFTs [2].

In this work, we propose a calibrated SPICE-compatible mobility function which is suitable for the simulation of the n- and p-type laser annealed LTPS TFTs. This mobility is mainly considering the channel mobility degradation from the vertical electric field. With the well-known RPI TFT model, numerical results using the conventional TFT mobility and ours are performed and compared with the measured data for different dimension of LTPS TFTs. It is found that the RPI TFT model with our mobility shows several improved characteristics. To extract model parameters, a physical-based extraction procedure for the explored model is also discussed. To further verify the validity of the mobility function in circuit design, we perform a two-stage common source amplifier of LTPS TFTs using the RPI TFT model with the conventional and our mobility functions. Numerical results show that the proposed mobility predicts reasonable output than that of the conventional one. We note that it can be directly incorporated into circuit simulator without numerical convergence problems.

In Sec. 2, we state the mobility function in the RPI TFT model for modeling and simulation of LTPS TFTs. A model parameter extraction procedure is also presented. In Sec. 3, we report and discuss the simulation results obtained with the conventional and our mobility functions. Comparison between simulation and measurement is performed to show the accuracy of the model. Finally, we draw the conclusions.

2 Modeling and Simulation

The RPI TFT model is a compact model developed on the single crystalline MOSFET model [3, 4]. It has following properties

(i) field effect mobility is a function of gate bias;
(ii) effective mobility accounts for trap states;

(iii) reverse bias drain current function of electric field near drain and temperature;
(iv) a design independent on channel length;
(v) a unified DC model consists of four parts (leakage, subthreshold, above threshold, and kink effect parts) for channel length down to 4 μm;
(vi) an AC model accurately reproduces C_{gc} frequency dispersion; and
(vii) an automatic scaling of model parameters that accurately models a wide range of device geometries.

The conventional mobility function used in the RPI TFT model is represented as

$$\frac{1}{\mu_{FET}} = \frac{1}{MU0} + \frac{1}{Tmu1 \cdot (\frac{2V_{GTE}}{V_{sth}})^{MMU}}. \quad (1)$$

Though the vertical electrical field induced mobility degradation effect has been introduced in the RPI TFT model, this expression can not successfully model the mobility of LTPS TFTs. According to our observation that a higher order effect of electric field on the electron and hole transport is significantly; therefore, a unified carrier's mobility function is phenomenologically proposed.

$$\mu_{FET} = \frac{U_0}{1 + U_a(\frac{V_{GTE}+2V_{sth}}{T_{OX}}) + U_b(\frac{V_{GTE}+2V_{sth}}{T_{OX}})^2}. \quad (2)$$

We note that the appearance of Eq. (2) is similar to the mobility of BSIM-4 model [7]. However, the physical and mathematical meaning between ours and BSIM-4 are different. By introducing the higher order term with parameter U_b in Eq. (2), the mobility successfully considers the effects of the TFT in high electric field, the substrate bias, and the temperature. It improves the correctness of the circuit simulation, without increasing any complexity and convergence of the circuit simulation. The proposed mobility is useful in precisely simulating the circuit characteristics of the complementary SOP circuit. The V_{GTE} appearing in Eqs. (1) and (2) is given by [4]

$$V_{GTE} = V_{sth}[1 + \frac{V_{GT}}{2V_{sth}} + \sqrt{DELTA^2 + (\frac{V_{GT}}{2Vsth} - 1)^2}]. \quad (3)$$

Notations used in Eqs. (1)-(3) have their conventional meanings in [4].

To extract the parameters of the RPI TFT model, the developed extraction strategy consists of the following five steps sequentially, and each step optimizes different parameters among different regions. Firstly, a set of characterized mobility parameters, U_0, U_a, U_b, can be extracted precisely in the linear region of $I_{DS} - V_{GS}$ curves. Second, we extract the sub-threshold regions of the $I_{DS} - V_{GS}$ curves by selecting the threshold and flat band parameters: VTO, and ETA. Third, by choosing the saturation and the kink effect parameters, AT, BT, we extract the saturation region of the $I_{DS} - V_{DS}$ curves. Furthermore, we fit the saturation region on both $I_{DS} - V_{DS}$ and $I_{DS} - V_{GS}$ curves by tuning ASAT, LASAT, and VST. Finally, we extract the leakage region of the $I_{DS} - V_{GS}$ in the log scale with respect to parameters: IO, IOO, DD, DG. The $I_{DS} - V_{GS}, I_{DS} - V_{DS}$, output conductance ($G_{ds}$), and transconductance (G_m). In Sec. 3, characteristics calculated with the conventional [3, 4] and our mobility functions are examined, where a two-stage common source amplifier is also explored.

3 Results and Discussion

In this section, based on the proposed extraction procedure we first verify the accuracy of the mobility function used in RPI TFT model for two different LTPS TFT devices. The verification is performed with comparison between simulation and measurement. The first sample is with the channel width (W) and length (L) of 20 μm and 4.5 μm, respectively. For the second sample, the device's W/L = 4/12 $[\mu m/\mu m]$. The extracted results of $I_{DS} - V_{DS}$ and $I_{DS} - V_{GS}$ curves are shown in Figs. 1a and 1b. Their correspondingly calculated first derivatives are shown in Figs. 1c and 1d. Similarly, the second example presents its optimization results in Fig. 2. Both examples have shown good accuracy of the RPI TFT model with our mobility for different dimension of LTPS TFT sample. Maximum average errors are computed for all calculations. Errors of all simulated I-V curves are within 1% and errors of the calculated first derivatives are about 5%. The detail maximum average errors of the examined devices are summarized in Table 1. The piecewise curves of G_{ds}, shown in Figs. 1c and 2c, are due to the nature of the RPI TFT model [3, 4]. We note that the calculated first derivatives of $I_{DS} - V_{GS}$ curves, shown in Figs. 1d and 2d, exhibit nonphysical variations. It is because we do not simultaneously take the variation of the first derivatives of I-V curves into consideration in our extraction process. However, to further eliminate these phenomena and improve the extraction accuracy of the calculated first derivatives of $I_{DS} - V_{GS}$ curves, optimization process should be done by simultaneously considering errors of the original curves and their first derivatives. Tables 2 and 3 list some extracted parameters for the first and the second samples, respectively.

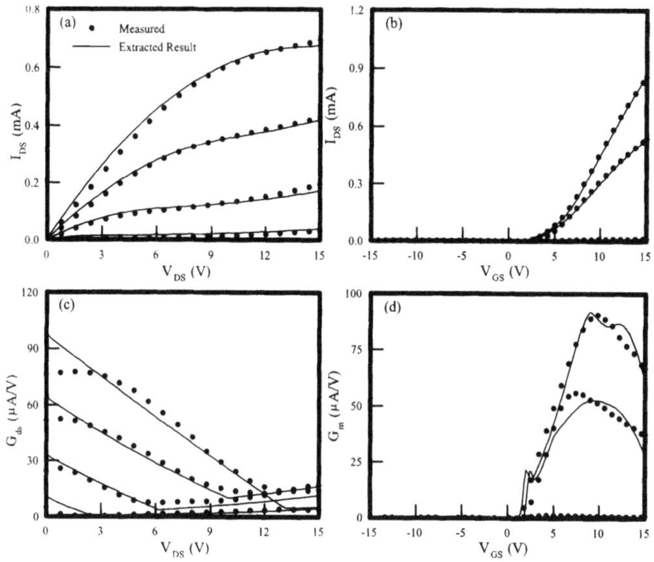

Fig. 1. Simulated original (a) $I_{DS} - V_{DS}$ of the LTPS TFT with W/L = 20/4.5 [$\mu m/\mu m$], where V_{GS} = 3.0 V, 6.0 V, 9.0 V, and 12.0 V, from bottom to top, respectively. (b) Plot of $I_{DS} - V_{GS}$, where V_{DS} = 0.1 V, 5.1 V, and 10.1 V, from bottom to top, respectively. The corresponding first derivatives are shown in (c) $I_{DS} - V_{DS}$ and (d) $I_{DS} - V_{GS}$

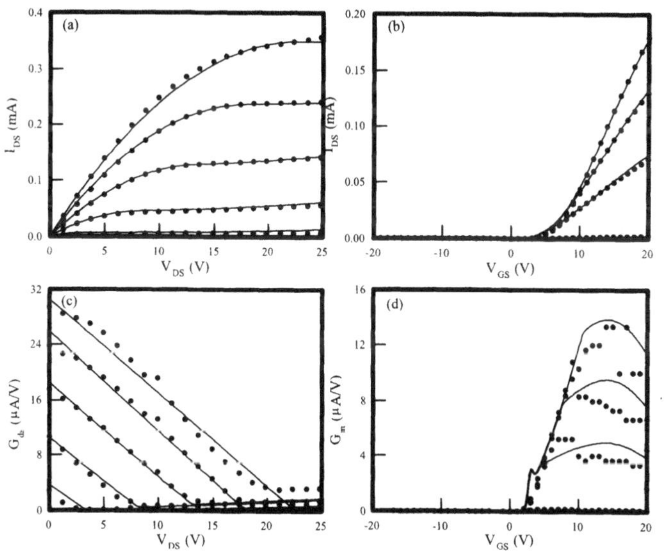

Fig. 2. Simulated original (a) $I_{DS} - V_{DS}$ of the device with W/L = 4/12 [$\mu m/\mu m$], where V_{GS} is from 4.0 V (bottom) to 20.0 V (top) with step 4.0 V. (b) Plot of $I_{DS} - V_{GS}$, where V_{DS} are 0.1 V, 3.1 V, 6.1 V, and 9.1 V, from bottom to top, respectively. The corresponding first derivatives are shown in (c) $I_{DS} - V_{DS}$ and (d) $I_{DS} - V_{GS}$

Table 1. The maximum average errors of the extracted devices

	Sample-1	Sample-2
$I_D - V_D$	1.68 %	0.91 %
$I_D - V_G$	0.60 %	0.67 %
G_{ds}	5.54 %	3.07 %
G_m	3.12 %	5.08 %
$Log(I_D - V_G)$	2.82 %	1.96 %

Table 2. A list of the extracted parameters of the LTPS TFT with W/L = 20/4.5 and 4/12 $[\mu m/\mu m]$, respectively

Name	W/L=20/4.5	W/L=4/12	Name	W/L=20/4.5	W/L=4/12
ASAT	1.286	0.412	AT	$3.1415e^{-8}$	$3.0281e^{-8}$
LASAT	$8.6953e^{-7}$	$-5.4417e^{-6}$	LKINK	$1.9e^{-5}$	$1.9e^{-5}$
MMU	2.9062	2.79015	MU0	76.63075	124.69635
ASAT	0.412	0.412	AT	$3.02817e^{-8}$	$3.02817e^{-8}$
MUS	1	1	VKINK	100	100
VSI	2	2	VST	11.51466	21.046579
BLK	0.001	0.001	DD	$1.4e^{-7}$	$4.1945e^{-6}$
EB	0.68	0.68	IO	6	613.297
BT	$1.05188e^{-6}$	$-3.2749e^{-6}$	MK	1.3	1.3
MU1	0.0022	0.0022	BT	$-3.27459e^{-6}$	$-3.2748e^{-6}$
VON	0	0	VTO	1.8932	1.14956256
DG	$2.0e^{-7}$	$3.70455e^{-7}$	IOO	150	1776.80049

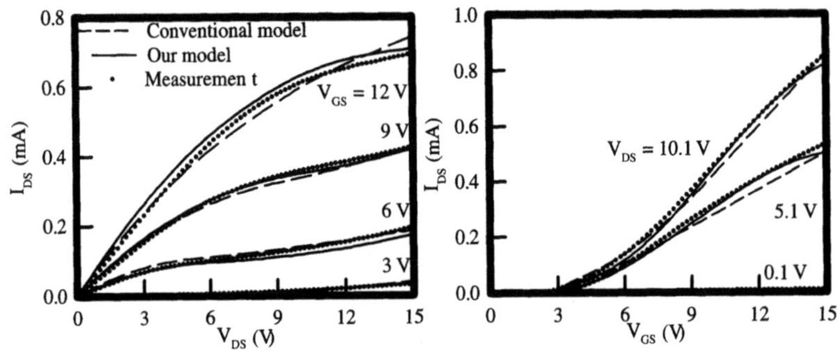

Fig. 3. Comparison of $I_{DS} - V_{DS}$ (the left figure) and $I_{DS} - V_{GS}$ (the right one) among the quasi-static measurement and the simulations of the RPI model with the conventional and our mobility functions

To validate the proposed mobility function used in the RPI TFT model, we further compare the accuracy of the RPI TFT model using the conventional and our mobility function in the next example.

Comparison of the simulated and calculated first derivatives of two sets of I-V curves for the RPI TFT model with the conventional and our mobility functions is shown in Figs. 3 and 4, where the measurement is performed for the LTPS TFT with W/L = 20/4.5 $[\mu m/\mu m]$. The dotted lines are measured data, the dashed lines are the result of the RPI TFT model with the conventional mobility, and the solid lines are the outcome of our mobility. As shown in Fig. 3, the proposed mobility has a good agreement with the measured data; moreover, the accuracy of the calculated physical quantity is further achieved. The conventional RPI mobility is poor in modeling the effect of high gate bias, shown in the right figure of Fig. 3. Without a correct formulation for the device operated under high gate bias regime, shown in the right figure of Fig. 4, output characteristics calculated by the RPI TFT model with the conventional mobility is inaccurate at high gate biases. According to the right figure of Fig. 3, the RPI TFT model with the conventional mobility overestimates the current level at the high gate biases and underestimates drain current at the low gate biases. The deviated output characteristics is further observed from Fig. 4 that the modified mobility does improve the accuracy of the calculated first derivatives of $I_D - V_D$ curves.

The proposed mobility function in the RPI TFT model shows its accuracy for the quasi-static characteristic simulations; in particular, for the high field properties of Gm and Gds. The improved accuracy of the calculated Gm implies a reliable TFT circuit simulation. It is one of important issues for the analog circuit simulation. By implementing the mobility function in SPICE3f5, we have performed a circuit simulation with LTPS TFTs. As shown in Fig. 5, the input signal $Vin = 0.05\sin(2\pi ft)$, where f = 1 KHz and the DC bias is at 6.0 V. The simulation results are shown in Fig. 6, the output gain calculated from the proposed mobility is equal to 8; however, the simulation of output gain with the conventional one is 4. It is found that the proposed mobility function predicts a reasonable higher output gain than that of the conventional one. This result is mainly from the different estimation of the transconductance, which plays a

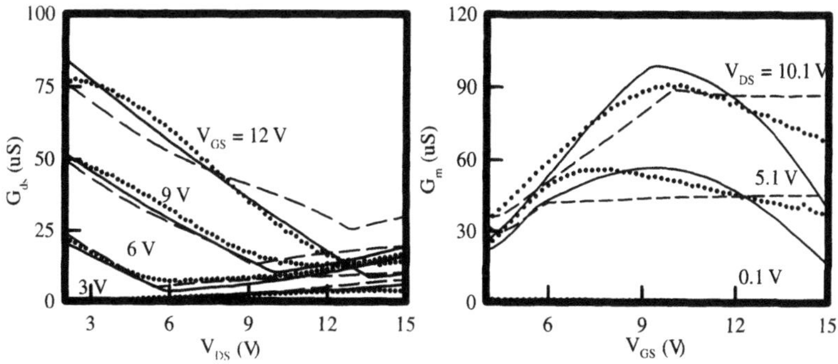

Fig. 4. Comparison of the output conductance (the left figure) and the transconductance (the right one) among the quasi-static measurement and the simulations of the RPI model with the conventional and our mobility functions

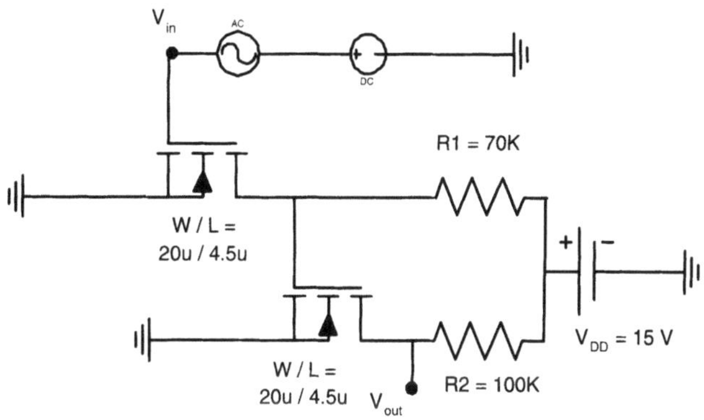

Fig. 5. A two-stage common source amplifier with LTPS TFTs

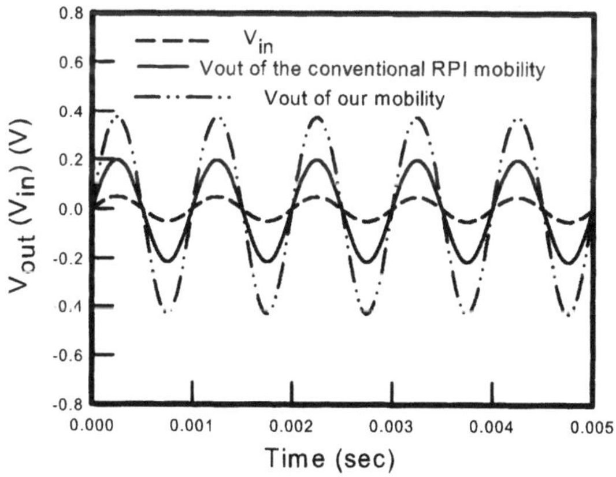

Fig. 6. A plot of the input and output signals of the simulated circuit

crucial factor in analog circuit simulation. A 50% discrepancy observed from Fig. 6 should be clarified when designing SOP circuits.

4 Conclusions

We have presented a mobility function together with a parameter extraction procedure for the excimer laser annealed LTPS TFTs simulation. It has been successfully implemented into the well-known RPI TFT model and performed device and circuit simulation without numerical difficulties. Comparison among the results of the conventional mobility function, the measured data, and our results has confirmed that the proposed mobility function can model LTPS TFT's characteristics and have better simulation accuracy. This SPICE-compatible mobility function has also applied to LTPS TFTs analog circuit simulation and demonstrated reasonable output results. We are currently fabricating and measuring the electrical characteristics of LTPS TFT circuit and compare the obtained data with the simulated results.

Acknowledgment

This work is supported in part by the National Science Council of Taiwan under Contract NSC-93-2215-E-429-008 and Contract NSC 94-2752-E-009-003-PAE, by the Ministry of Economic Affairs, Taiwan under Contract 93-EC-17-A-07-S1-0011, and by the Toppoly Optoelectronics Corp under a 2003-2005 grant.

References

1. Jagar, S., Cheng, C.F., Zhang, S., Wang, H., Poon, M.C., Kok, C.W., Chan, M.: A SPICE model for thin-film transistors fabricated on grain-enhanced polysilicon film. IEEE Trans. Electron Devices, 50, 1103 (2003)
2. Wang, A. and Saraswat, K. C.: A strategy for modeling of variations due to grain size in polycrystalline thin-film transistors. IEEE Trans. Electron Devices, 47, 1035 (2000)
3. Farmakis, F.V., Brini, J., Kamarinos, G., Angelis, C.T., Dimitriadis, C.A., Miyasaka, M.: On-current modeling of large-grain polycrystalline silicon thin-film transistors. IEEE Trans. Electron Devices, 48, 701 (2001)
4. Shur, M. S., Slade, H. C., Ytterdal, T. *etal*: Modeling and scaling of a-Si:H and Poly-Si Thin Film Transistors. in: Material Research Society Proceeding, Amorphous and Microcrystalline Silicon Technology, 467 (1997)
5. Armstrong, G.A., Uppal, S., Brotherton, S.D., Ayres, J.R.: Modeling of Laser-Annealed Polysilicon TFT Characteristics. IEEE Electron Device Letters, 18, 315 (1997)
6. Li, Y., Cho, Y.-Y., Wang, C.-S., Huang, K.-Y.: A Genetic Algorithm Approach to InGaP/GaAs HBT Parameters Extraction and RF Characterization. Jpn. J. Appl. Phys., 42, 2371 (2003)
7. Li, Y. and Cho, Y.-Y.: Intelligent BSIM4 Model Parameter Extraction for Sub-100 nm MOSFETs era. Jpn. J. Appl. Phys., 43, 1717 (2004)

Parallelization of WENO-Boltzmann Schemes for Kinetic Descriptions of 2D Semiconductor Devices

J. M. Mantas[1], J. A. Carrillo[2], and A. Majorana[3]

[1] Software Engineering Department - UGR, Granada, Spain, jmmantas@ugr.es
[2] ICREA - Dpt. Matemàtiques - UAB, Barcelona, Spain, carrillo@mat.uab.es
[3] Dip. Matematica e Informatica - UCT, Catania, Italy, majorana@dmi.unict.it

Abstract We present the results obtained with a data parallelization of the WENO-Boltzmann numerical scheme for semiconductors introduced in [CGMS03A, CGMS03B]. We show that an efficient and parallel implementation of this deterministic method allows the computation of macroscopic quantities and IV curves of a realistic 2D device: a Si MESFET. The execution time and speedup results demonstrate the good scalability of the proposed paralell implementation.

1 Introduction

The parallelization of a direct WENO (Weighted Essentially Non-Oscillatory) solver for the 2D-spatial Boltzmann-Poisson system describing electron transport in Si-based semiconductor devices has been addressed. A non-parabolic Kane energy-band and elastic acoustic and inelastic non-polar optical phonon operators have been used [CGMS03A] in the physical description of the electron transport in the device. This choice is by no means restrictive and more complicated band structures, including several valleys, and different scattering mechanisms, both intervalley and intravalley ones, can be included in a flexible way both in the numerical method and its parallelization [CCM04].

The numerical scheme which has been parallelized [CGMS03A, CGMS03B] uses a formulation of the Boltzmann-Poisson system in spherical coordinates for the wave vector space. After adimensionalization one is reduced to simulate the evolution in time t of the distribution function Φ in the five-dimensional space $(x, y, \omega, \mu, \phi)$, where x and y are the spatial coordinates, $\omega \geq 0$ is a dimensionless energy, $\mu \in [-1, 1]$ is the cosine of the angle with respect to the x-axis and $\phi \in [0, \pi]$ the azimuthal angle. The resulting Boltzmann equation reads

$$\frac{\partial \Phi}{\partial t} + \frac{\partial}{\partial x}(a_1 \Phi) + \frac{\partial}{\partial y}(a_2 \Phi) + \frac{\partial}{\partial \omega}(a_3 \Phi) + \frac{\partial}{\partial \mu}(a_4 \Phi) + \frac{\partial}{\partial \phi}(a_5 \Phi) = s(\omega) C(\Phi) \tag{1}$$

where the flux functions a_i and the Jacobian factor $s(w)$ can be seen in [CGMS03B]. The dimensionless collision operator $C(\Phi)$ is given by:

$$\begin{aligned} C(\Phi)(t, x, y, \omega, \mu, \phi) = {} & \frac{1}{2\pi t^*} \int_0^{\pi} \int_{-1}^{1} [\beta \Phi(t, x, y, \omega, \mu', \phi') \\ & + a \Phi(t, x, y, \omega + \alpha, \mu', \phi') + \Phi(t, x, y, \omega - \alpha, \mu', \phi')] d\phi' d\mu' \\ & - \frac{1}{s(\omega)\, t^*} [\beta s(\omega) + a s(\omega - \alpha) + s(\omega + \alpha)] \Phi(t, x, y, \omega, \mu, \phi) \,. \end{aligned} \tag{2}$$

where the constant parameters t^*, α, a and β depend on scattering mechanisms (see [CGMS03A] for a more detailed description).

Flux functions a_3, a_4 and a_5 depend on the electric field vector $\mathbf{E}$ which is computed self-consistently by solving the dimensionless Poisson equation in the 2D spatial domain

$$\Delta V = \epsilon \left[n(t, x, y) - N_D(x, y) \right], \tag{3}$$

$$\mathbf{E} = -\nabla V,$$

where ϵ is a dimensionless parameter, $N_D(x, y)$ the dimensionless doping profile, V the electric potential and $n(t, x, y)$ the electron density computed from Φ by

$$\int_0^{\pi} \int_0^{\infty} \int_{-1}^{1} \Phi \, d\mu \, d\omega \, d\phi. \tag{4}$$

In the Boltzmann equation (1) the advection part is treated with a 5th order non-oscillatory finite difference WENO scheme [SHU98], the collision operator (2) is approximated by means of a quadrature formula and the time-dependent part is solved by an explicit Runge-Kutta method. Poisson's equation (3) is solved by an iterative standard SOR method computed at each Runge-Kutta step and a midpoint quadrature formula is used in (4) resulting in a charge-conservative method.

We refer to [CGMS03A, CGMS03B, CGMS04] for the complete description of the numerical method, a discussion of this particular choice of the numerical scheme and the comparison of the results with respect to Monte Carlo simulations. Let us briefly mention that high-gradient regions in macroscopic quantities, in particular: density, energy and mean velocity (see Fig. 2-3) clearly imply the existence of high-gradient regions both in physical space (x, y) and in velocity space (ω, μ, ϕ) for the unknown distribution function Φ (see also Fig. 5). In order to accurately approximate the derivatives in these regions, we use WENO reconstruction methods which are well fitted for this purpose [SHU98] and, moreover, they produce a high order approximation in smooth regions.

Results of the 1D case [CGMS03A] and of the 2D case [CGMS04] have been validated by comparing them to Monte Carlo methods, and they give excellent validation results even with coarse grids. Moreover, in the 2D case Monte Carlo methods are not well apt for resolving almost empty regions in the device (close to the gate in a MESFET) while deterministic methods do. Therefore, these deterministic results, although not competitive with Monte Carlo methods at the level of the execution time in 2D, should be used as *benchmarks* for Monte Carlo, hydrodynamic or drift-diffusion results. We refer to [CGMS04] for a more detailed discussion about WENO-Boltzmann versus Monte Carlo methods in 2D.

Other advantages of this method are the transient computation, the knowledge of the distribution function itself and not only of their moments as well as the absence of oscillations or numerical noise even close to regions between different boundary conditions. A drawback of the use of WENO schemes is that we are reduced to uniform grid sizes and almost rectangular type geometries. Nevertheless, these drawbacks can be overcome by using an interpolation between different uniform grid sizes as in [GCG04, SS03].

2 Parallelization and numerical results

The aim of this work has been to obtain efficient parallel implementations of this 2D-space solver for a PC cluster because this scheme demands a great deal of computing power. The parallelization is based on domain decomposition techniques. An analysis of the scheme reveals that the best choice to decompose the data structure of the solution Φ is by splitting only the physical space dimensions among the processors by following a 2D block-decomposition. For this purpose, a logical 2D grid of processors is automatically defined according to the actual spatial grid size and the available number of processors. In this way, only the flux terms for the physical space require communication among the processors. Moreover, an overlapping of communication and computation has been enabled to improve the performance in the computations of these fluxes. The rest of fluxes can be parallelized in a straightforward manner because remote communication is not required. This choice leads to a relatively low communication cost and involves an important reuse of the existing sequential code.

To solve the Poisson equation in order to compute the electric potential, a parallel red-black Successive Over-Relaxation (SOR) scheme is implemented following a data distribution which matches with the distribution followed by the WENO-Boltzmann solver. The time discretization is carried out by using a parallel implementation of a third-order low-Storage Runge-Kutta method [GS98] in order to save memory resources. An scalable message-passing implementation of the solver has been obtained for any number of processors by making use of the Message Passing Interface (MPI).

The parallel scheme has been applied to the simulation of a MESFET device (see Fig. 1) used in several works as a benchmark for testing electron transport solvers [JS94, CGMS03B, AMRS04]. The doping profile is given by $3 \times 10^{17} cm^{-3}$ in the n^+ regions and $10^{17} cm^{-3}$ in the n region.

Results for the density, the potential and the electric current field can be seen in Fig. 2-3 with a potential of $-0.8V$ at the gate and $1V$ at the drain with respect to the source. Several numerical experiments have been made on a cluster of 8 dual 2.5 Ghz AMD processors connected via a Gigabit ethernet switch. Numerical results for macroscopic quantities are shown at 5ps, where we use $48 \times 32 \times 102 \times 12 \times 12$ grid points for the $(x, y, \omega, \mu, \phi)$ domain.

Insulating boundaries are treated mimicking reflecting boundaries for the Boltzmann equation (1) and Neumann boundary conditions for the electric potential (3). As a result zero normal component to the insulating boundary of the

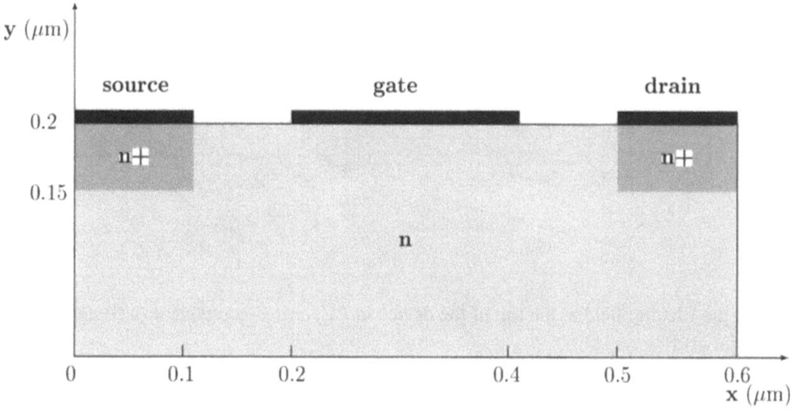

Fig. 1. Schematic representation of a 2D silicon MESFET device

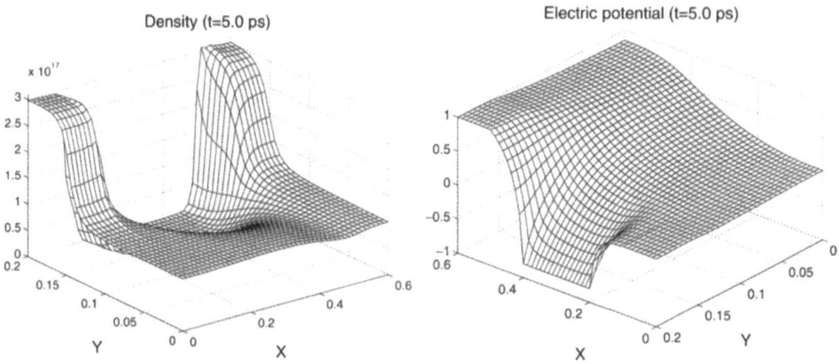

Fig. 2. Density (cm^{-3}) and electric potential (V) at 5 ps with a 48×32 spatial grid

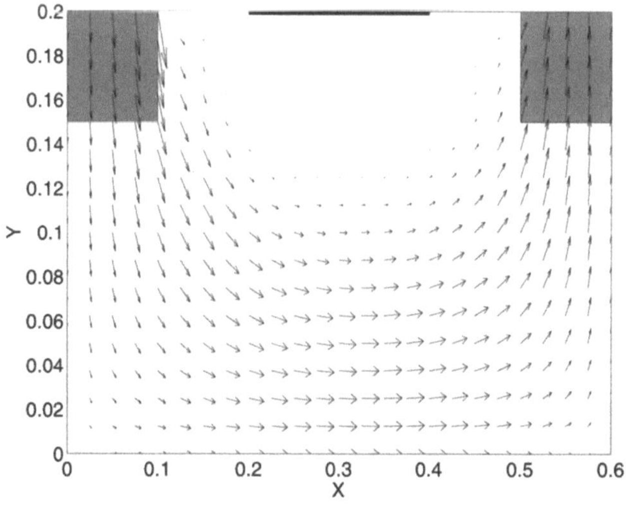

Fig. 3. Electric current field at 5 ps with a 48×32 spatial grid

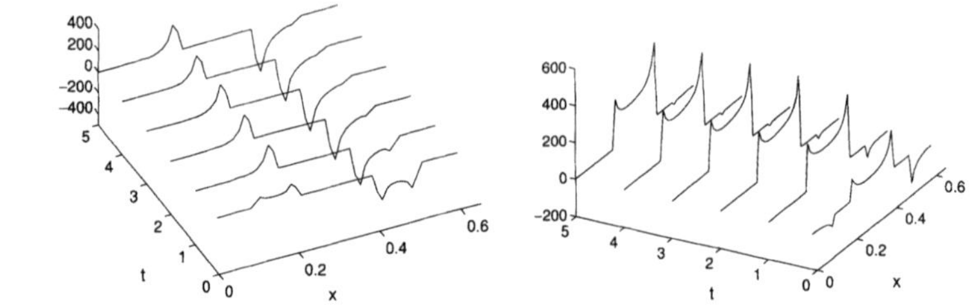

Fig. 4. Evolution of the Electric field at the top of the device in kV/cm: x-component (left) and y-component (right) with a 48×32 spatial grid

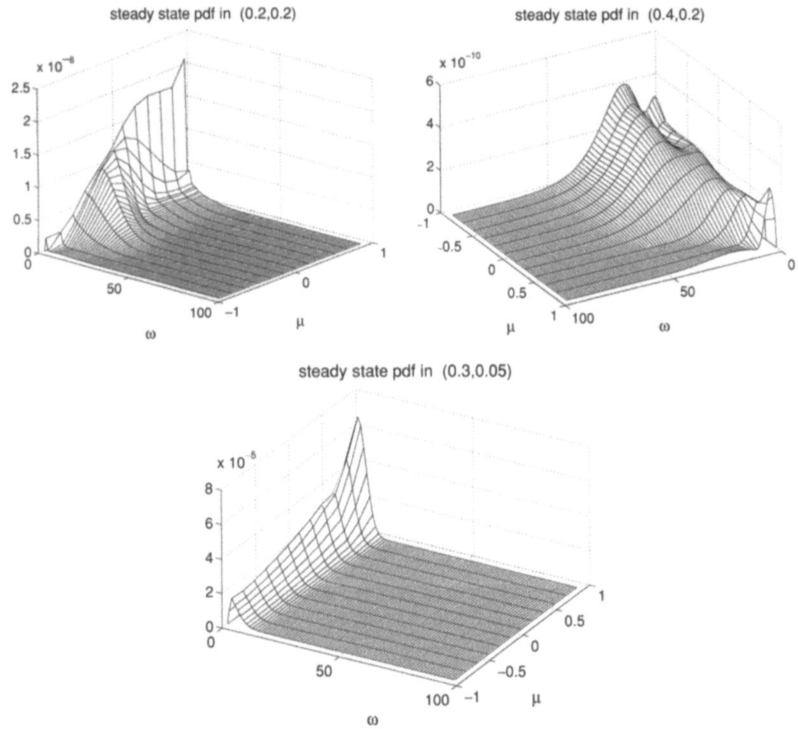

Fig. 5. Normalized PDF at 3 points of the device: top left figure at $(0.2, 0.2)$ (left corner of the gate), top right at $(0.4, 0.2)$ (right gate corner), bottom at $(0.3, 0.05)$

electric current (see Fig. 3) is well resolved. Source and drain are Ohmic contacts and therefore we impose Dirichlet boundary conditions for the electric potential and inflow boundary conditions for the Boltzmann equation implying local neutrality. The gate is simulated as a Schottky contact and thus it should repel electrons. The gate region should have a very low density; and the electrons may only leave the gate region to enter the device. As a consequence, the density at the gate is extremely small: more than 8 orders of magnitude lower than the doping profile. Therefore, we consider Dirichlet boundary conditions for the electric potential and zero charge density for the Boltzmann equation. We again refer to [CGMS04] for a detailed explanation of the implementation of the boundary conditions.

We clearly observe the appearance of singularities for the electric field at the points between insulating and contact boundary conditions at the top of the device. These singularities can be seen in Fig. 4 where the evolution of the electric field on the top of the device is plotted up to 5ps. They are not numerical artifacts and they do not disappear by grid refinement but on the contrary, they become larger in value resulting in a small CFL number for the time solver and therefore slowing the pace of the code. This is by no means a failure of the approach but a real success since it captures these singularities that appear even in drift-diffusion approximations [G93]. Other numerical methods to approximate

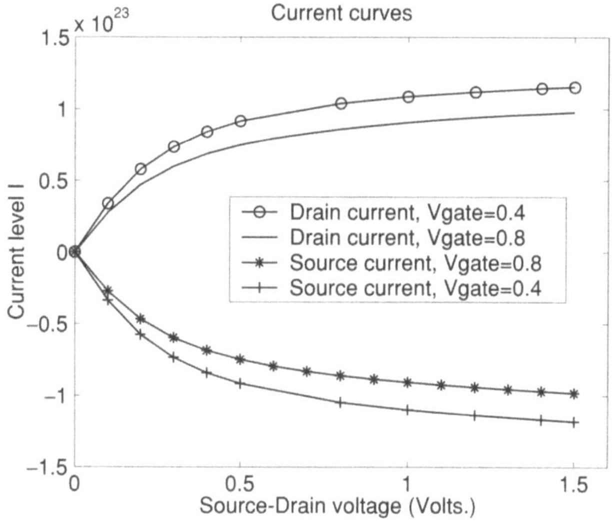

Fig. 6. Voltage-current curves at 5 ps with a 42×24 spatial grid

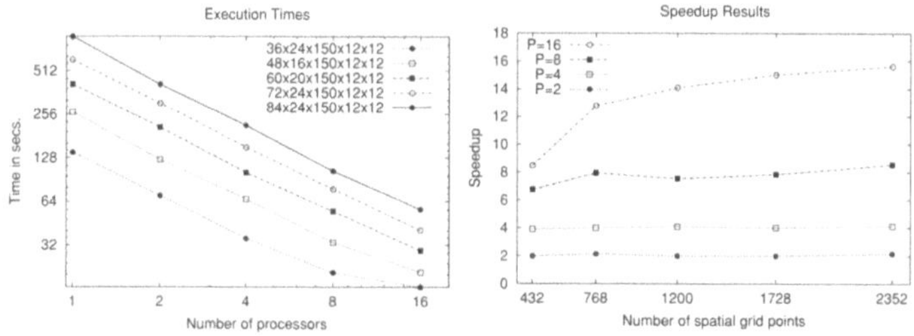

Fig. 7. Execution time results and speedup results for several spatial grid sizes

the transport phase based on characteristics and dimensional splitting (see for instance [FSB01]) may improve the number of necessary time-steps since they avoid a restrictive CFL. This direction will be further investigated in forthcoming works.

In Fig. 5 we observe the probability density function (PDF) at three points of the device as a function of (w, μ) averaged over ϕ. We observe how the flow of electrons takes place since $\mu = 1$ is the direction of increasing x and $\mu = -1$ is the opposite.

The efficient implementation of the scheme allows us to compute current-voltage characteristics for this device. The current-voltage characteristic curves when a voltage of -0.8 V and -0.4 V is applied at the gate are shown in Fig. 6. These results have been obtained by simulating the MESFET device at each of the drain voltage values corresponding to stars in Fig. 6 up to 5ps. Here, we use $42 \times 24 \times 80 \times 12 \times 12$ grid points for discretizing the $(x, y, \omega, \mu, \phi)$ domain. In this figure we show the computed current both at the drain and at the source since they should be equal up to a sign for the stationary solution and this is so up to 5 digits in all experiments performed.

Figure 7 shows execution times and speedup results obtained with several grid sizes and number of processors for the time integration of 0.01 picoseconds. The results show a good scalability in the range of processors [2,16] and a parallel efficiency close to 100 %.

3 Conclusions

A flexible parallelization of WENO-Boltzmann schemes for the kinetic description of realistic semiconductor devices has been performed. This method is flexible in band structure, scattering mechanisms and boundary conditions for the kinetic description and fairly flexible regarding the device geometry. Shown numerical results, in the particular case of a MESFET, reveal that these simulations although still computationally expensive and not competitive with Monte

Carlo methods, provide useful benchmarks for all known solvers for charge particle transport in semiconductors and they are the most accurate simulations to date up to our knowledge.

Acknowledgements

The authors acknowledge support from the European IHP network HYKE "Hyperbolic and Kinetic Equations: Asymptotics, Numerics, Applications" HPRN-CT-2002-00282. JM and JAC acknowledge partial support from DGI-MCYT/FEDER project BFM2002-01710. A part of the present work was carried out during a visit of the first author at the Dipartimento di Matematica e Informatica della Università di Catania. He is grateful to A. Majorana and G. Russo for their hospitality.

References

[AMRS04] Anile, A. M., Marrocco, A., Romano, V., Sellier, J.-M., Numerical simulation of 2D Silicon MESFET and MOSFET described by the MEP based energy-transport model with a mixed finite element scheme, preprint (2003)

[CGMS03A] Carrillo, J. A., Gamba, I., Majorana, A., Shu, C.-W. A WENO-solver for the transients of Boltzmann-Poisson system for semiconductor devices: performance and comparisons with Monte Carlo methods, Journal of Comp. Physics, **184**, 498–525 (2003)

[CGMS03B] Carrillo, J. A., Gamba, I., Majorana, A., Shu, C.-W. A direct solver for 2D non-stationary Boltzmann-Poisson systems for semiconductor devices: a MESFET simulation by WENO-Boltzmann schemes, Journal of Comp. Electronics, **2**, 375–380 (2003)

[CCM04] Cáceres, M. J., Carrillo, J. A., Majorana, A., Deterministic simulation of the Boltzmann-Poisson system in GaAs-based semiconductors, submitted to SIAM Journal on Scientific Computing, (2004)

[CGMS04] Carrillo, J. A., Gamba, I., Majorana, A., Shu, C.-W., 2D Device simulations by WENO finite differences approximations of the Boltzmann-Poisson system for semiconductors, work in preparation, (2004)

[FSB01] Filbet, F.; Sonnendrücker, E.; Bertrand, P., Conservative numerical schemes for the Vlasov equation, J. Comput. Phys. **172**, 166–187 (2001)

[G93] Gamba, I., Asymptotic behavior at the boundary of a semiconductor device in two space dimensions, Ann. di Mat. Pura App. (IV) Vol. CLXIII, 43-91, (1993)

[GCG04] González-Rodelas, P., Carrillo, J. A., Gámiz, F. Deterministic Numerical Simulation of 1d kinetic descriptions of Bipolar Electron Devices, in proceedings of the Int. Workshop on Scientific Computing in Electrical Engineering (2004)

[GS98] Gottlieb, S., Shu, C.-W., Total variation diminishing Runge-Kutta schemes, Math. Comp., **67**, 73-85 (1998)

[JS94] Jerome, J.W. and and Shu, C.-W., "Energy models for one-carrier transport in semiconductor devices", in IMA Volumes in Mathematics and Its Applications, v59, W. Coughran, J. Cole, P. Lloyd and J. White, editors, Springer-Verlag, 1994, pp.185-207

[SHU98] Shu, C.-W. Essentially non-oscillatory and weighted essentially non-oscillatory schemes for hyperbolic conservation laws, Lecture Notes in Mathematics **1697**, 325-432 (1998)

[SS03] Sebastian, K., Shu, C.-W., Multidomain WENO finite difference method with interpolation at subdomain interfaces, J. Scientific Computing, **19**, 405-438 (2003)

Hole Mobility in Silicon Semiconductors.

G. Mascali[1], V. Romano[2], and J. M. Sellier[2]

[1] Università della Calabria, Via Bucci cubo 30B, I-87036 Arcavacata di Rende (Cs) & INFN-Gruppo c. Cosenza, Italia, g.mascali@unical.it
[2] Università di Catania, Viale A. Doria 6, I-95125 Catania, Italia

Abstract In several bipolar electronic devices holes give a relevant contribution to the total current. Therefore it is important to take into account also hole transport besides that of electrons. In this work we present a hydrodynamical model of hole transport in silicon semiconductors based on the maximum entropy principle following the approach used in [3] for electrons. We employ this model for studying the hole mobility and a 1-D n^+-p-n^+ structure

1 Introduction and physical setting

In industrial applications the simulation of hole transport, in bipolar devices, is usually obtained by numerically integrating the drift-diffusion model, which is based on the assumption of isothermal charge flow. This is well justified in devices such as MOSFET (Metal Oxide Field Effect Transistors) since the contribution of holes to the total current is marginal. However in bipolar heterojunctions the role of holes in charge transport is of the same order or even greater than that of electrons. In such situations more sophisticated models, which include at least the average energy as fundamental variable, are needed. These models are usually known as hydrodynamical models.

In Si three valence bands are present [3]. The first two are the heavy and light valence bands which are degenerate in correspondence of their maximum at $\mathbf{k} = 0$, $\mathbf{k}$ being the hole wave vector. The third band is the so-called *split-off band* which is separated from the first two by the spin-orbit energy $\Delta = 0.0443$ eV at $\mathbf{k} = 0$. Because of its low density of states and its energy separation the split-off valence band is usually neglected.

The dispersion relations for the two degenerate energy bands of light and heavy holes have quite difficult analytical expressions. For this reason, here, we consider a simplified energy band model usually employed in order to get a macroscopic description of hole transport. It consists of a single spherical parabolic band with an effective mass related to some plausible average in the $\mathbf{k}$ space. Therefore the hole energy is approximated by the expression $\mathcal{E} = \frac{\hbar^2 k^2}{2m_H^*}$, where m_H^* is the heavy hole effective mass and $\mathbf{k}$ can vary on all $\mathbb{R}^3$, $\hbar$ is the reduced Planck constant.

In the semiclassical approximation holes are considered as particles of mass m_H^* and charge e having the same magnitude as that of electrons but positive sign. Their behavior inside the crystal is described by a distribution function $f_H(\mathbf{x}, \mathbf{k}, t)$ which satisfies the hole Boltzmann transport equation

$$\frac{\partial f_H}{\partial t} + \mathbf{v}_H \cdot \nabla_{\mathbf{x}} f_H + \frac{e\mathbf{E}}{\hbar} \cdot \nabla_{\mathbf{k}} f_H = \mathcal{C}[f_H]. \tag{1}$$

$\mathbf{v}_H$ is the hole velocity which is related to the energy by the relation $\mathbf{v}_H = \frac{1}{\hbar}\nabla_{\mathbf{k}}\mathcal{E}(\mathbf{k})$, [3], which in the parabolic band approximation reads $\mathbf{v}_H = \frac{\hbar\mathbf{k}}{m_H^*}$. $\mathbf{E}$ represents the electric field and $\mathcal{C}[f_H]$ the collision term. The electric field $\mathbf{E}$ satisfies the Poisson equation

$$-\nabla\left(\epsilon(\mathbf{x})\nabla\phi\right) = e\left(N_D(\mathbf{x}) - N_A(\mathbf{x}) - n(\mathbf{x}) + p(\mathbf{x})\right), \tag{2}$$

where ϵ is the dielectric constant of the material, ϕ the electric potential and n, p N_D, N_A the electron, hole, donor and acceptor densities.

As regards the collision term, in silicon holes interact with two types of phonons: the non-polar optical and the acoustic phonons. Moreover one has to take into account also the scattering with the impurities of the crystal. See [1] for details.

2 The hydrodynamical model

Starting from the Boltzmann equation (1), it is possible to obtain macroscopic equations describing hole transport. To this end, it is sufficient to multiply equation (1) by suitable weight functions $\psi = \psi(\mathbf{k})$ and integrate with respect to $\mathbf{k}$ on $\mathbb{R}^3$. If we set

$$M_\psi = \int_{\mathbb{R}^3} \psi(\mathbf{k}) f_H(\mathbf{x}, \mathbf{k}, t) d\mathbf{k},$$

which is the moment of f_H relative to the weight function $\psi(\mathbf{k})$, we get, after some manipulation, the following equation ‡

$$\frac{\partial M_\psi}{\partial t} + \frac{\partial}{\partial x^i} \int_{\mathbb{R}^3} \psi(\mathbf{k}) v^i f_H d\mathbf{k} - \frac{eE^i}{\hbar} \int_{\mathbb{R}^3} \frac{\partial \psi}{\partial k^i} f_H d\mathbf{k} = \int_{\mathbb{R}^3} \psi(\mathbf{k}) \mathcal{C}[f_H] d\mathbf{k}. \tag{3}$$

In particular we have used the weight functions $\psi_A = (1, \mathbf{v}, \mathcal{E}, \mathcal{E}\mathbf{v})$, obtaining a set of equations for the hole density p, the average velocity $\mathbf{V}_H$, the energy W_H and the energy flux $\mathbf{S}_H$

$$\frac{\partial p}{\partial t} + \frac{\partial (p\, V_H^i)}{\partial x^i} = 0, \tag{4}$$

$$\frac{\partial (p\, P_H^j)}{\partial t} + \frac{\partial (p\, U_H^{ij})}{\partial x^i} - p\, e\, E^j = p\, C_{P_H}^j, \qquad j = 1, 2, 3, \tag{5}$$

$$\frac{\partial p\, W_H}{\partial t} + \frac{\partial (p\, S_H^i)}{\partial x^i} - p\, e\, E_i V_H^i = p\, C_{W_H}, \tag{6}$$

$$\frac{\partial (p\, S_H^j)}{\partial t} + \frac{\partial (p\, F_H^{ij})}{\partial x^i} - p\, e\, E_i G_H^{ji} = p\, C_{S_H}^j, \qquad j = 1, 2, 3. \tag{7}$$

$$p = \int_{\mathbb{R}^3} f_H d\mathbf{k}, \qquad V_H^i = \frac{1}{m_H} P_H^i = \frac{1}{p} \int_{\mathbb{R}^3} v^i f_H d\mathbf{k},$$

$$W_H = \frac{1}{p} \int_{\mathbb{R}^3} \mathcal{E} f_H d\mathbf{k}, \qquad S_H^i = \frac{1}{p} \int_{\mathbb{R}^3} v^i \mathcal{E} f_H d\mathbf{k}.$$

This set of equations is not closed because of the presence of the fluxes

$$\begin{aligned} U_H^{ij} &= \frac{1}{p} \int_{\mathbb{R}^3} f_H v^i \hbar k^j d\mathbf{k} \qquad \text{crystal momentum flux,} \\ F_H^{ij} &= \frac{1}{p} \int_{\mathbb{R}^3} \mathcal{E}(\mathbf{k}) v^i v^j f d\mathbf{k} \qquad \text{flux of energy flux,} \\ G_H^{ij} &= \frac{1}{p} \int_{\mathbb{R}^3} \frac{1}{\hbar} f_H \frac{\partial (\mathcal{E} v^i)}{\partial k^j} d\mathbf{k}, \end{aligned} \tag{8}$$

and the production terms

$$\begin{aligned} C_{P_H}^j &= \frac{1}{p} \int_{\mathbb{R}^3} \hbar k^j \mathcal{C}[f_H](\mathbf{x}, \mathbf{k}, t) d\mathbf{k} \qquad \text{average crystal momentum production,} \\ C_{W_H} &= \frac{1}{p} \int_{\mathbb{R}^3} \mathcal{E}(\mathbf{k}) \mathcal{C}[f_H](\mathbf{x}, \mathbf{k}, t) d\mathbf{k} \qquad \text{the energy production,} \\ C_{S_H}^j &= \frac{1}{p} \int_{\mathbb{R}^3} \mathcal{E}(\mathbf{k}) v^j \mathcal{C}[f_H] d\mathbf{k} \qquad \text{flux energy production.} \end{aligned} \tag{9}$$

The closure has been obtained by employing the maximum entropy principle (MEP) and expanding with respect to a formal anisotropy parameter. Here we skip all the details and refer the reader to [3]. The resulting approximated MEP distribution function depends linearly on the fundamental variables $\mathbf{V}_h$, and $\mathbf{S}_h$,

$$\begin{aligned} f_H^{ME} = \frac{\exp(-\frac{3}{2W_H}\mathcal{E})}{\left(\frac{4}{3}\pi m^* W_H\right)^{3/2}} p \Bigg[1 - \left(-\frac{21 m_H^*}{4W_H} \mathbf{V}_H + \frac{9 m_H^*}{4W_H^2} \mathbf{S}_H \right) \cdot \mathbf{v} \\ - \mathcal{E} \left(\frac{9 m_H^*}{4W_H^2} \mathbf{V}_H - \frac{27 m_H^*}{20 W_H^3} \mathbf{S}_H \right) \cdot \mathbf{v} \Bigg]. \end{aligned} \tag{10}$$

‡Einstein summation over repeated letters is understood

Substituting the distribution function (10) in (8) and (9), one gets the following closure relations

$$U_H^{ij} = \frac{2}{3} W_H \delta^{ij}, \quad m_H^* F_H^{ij} = \frac{10}{9\, m_H^*} W_H^2 \delta^{ij}, \quad G_H^{ij} = \frac{5}{3\, m_H^*} W_H \delta^{ij},$$

and

$$\begin{aligned}
C_{P_H}^i &= C_{P_H}^{i(op)} + C_{P_H}^{i(ac)} + C_{P_H}^{i(imp)}, \\
C_{W_H} &= C_{W_H}^{(op)} + C_{W_H}^{(ac)} + C_{W_H}^{(imp)}, \\
C_{S_H}^i &= C_{S_H}^{i(op)} + C_{S_H}^{i(ac)} + C_{S_H}^{i(imp)},
\end{aligned}$$

where

$$\begin{aligned}
C_{P_H}^{i(op)} &= c_{11}^{(op)} V^i + c_{12}^{(op)} S^i, \quad C_{S_H}^{i(op)} = c_{21}^{(op)} V^i + c_{22}^{(op)} S^i, \\
C_{W_H}^{(op)} &= \frac{3}{2} \hbar \omega_{op} \tilde{\mathcal{K}}_{op}\, W_H^{-3/2} \left[N_{op}\, B_1 - (N_{op}+1) B_2 \right], \\
C_{P_H}^{i(ac)} &= c_{11}^{(ac)} V^i + c_{12}^{(ac)} S^i, \quad C_{S_H}^{i(ac)} = c_{21}^{(ac)} V^i + c_{22}^{(ac)} S^i, \\
C_{W_H}^{(ac)} &= -\frac{4096}{27} \pi^2 m_H^* v_s^2 \mathcal{K}'_{ac}\, W_H^{-1/2} \left(W_H - \frac{3}{2} k_B T_L \right), \\
C_{P_H}^{i(imp)} &= c_{11}^{(imp)} V^i + c_{12}^{(imp)} S^i, \quad C_{S_H}^{i(imp)} = c_{21}^{(imp)} V^i + c_{22}^{(imp)} S^i, \\
C_{W_H}^{(imp)} &= 0.
\end{aligned}$$

The c_{ij}'s, B_1, and B_2 are functions of the average energy and they can be found in [3], while

$$\tilde{\mathcal{K}}_{op} = \frac{8 (m_H^*)^{3/2} \sqrt{3\pi}}{\hbar^3} \mathcal{K}_{op}, \quad \mathcal{K}'_{ac} = \frac{3\sqrt{3} m_H^{*3/2}}{16 \pi^{3/2} \hbar^4 v_s} \mathcal{K}_{ac}.$$

$\mathcal{K}_{op} = \frac{(D_t K)^2}{8\pi^2 \rho \omega_{op}}$, where $D_t K$ is the optical deformation potential, ω_{op} is the optical phonon frequency, ρ is the silicon density. $\mathcal{K}_{ac} = \frac{\Xi_d^2}{8\pi^2 \rho v_s}$, Ξ_d being the acoustic deformation potential, v_s the longitudinal component of the sound velocity.

3 Drift-Diffusion model and Mobility

Introducing an energy relaxation time by means of the formula $C_{W_H} = \frac{W_H - W_0}{\tau_{W_H}}$, where $W_0 = 3/2\, k_B T_L$, and applying the drift-diffusion scaling

$$t = \mathcal{O}(\frac{1}{\delta^2}), \quad x^i = \mathcal{O}(\frac{1}{\delta}), \quad V^i = \mathcal{O}(\delta), S^i = \mathcal{O}(\delta), \quad \tau_W = \mathcal{O}(\frac{1}{\delta}).$$

to the system (4)-(7), it is possible to obtain the limiting drift-diffusion model [3] valid in the low field regime

$$\frac{\partial p}{\partial t} + \nabla \cdot \mathbf{J}_H = 0, \quad \mathbf{J}_H = p \mathbf{V}_H = D_{11}(W_0) \nabla p + p D_{13}(W_0) \nabla \phi, \tag{11}$$

where $\mathbf{J}_H$ is the hole current, while

$$D_{11} = \frac{\frac{2}{3} c_{22} W_H - \frac{10}{9} c_{12} \frac{W_H^2}{m_H^*}}{c_{11} c_{22} - c_{12} c_{21}} \quad \text{and} \quad D_{13} = e \frac{c_{22} - \frac{5}{3} c_{12} \frac{W_H}{m_H^*}}{c_{11} c_{22} - c_{12} c_{21}},$$

with

$$c_{ij} = c_{ij}^{(op)} + c_{ij}^{(ac)} + c_{ij}^{(imp)},$$

$D_{11}(W_0)$ and $D_{13}(W_0)$ are therefore two explicit functions of W_0.

By comparing $(11)_2$ with the expression of the particle current $\mathbf{J}$ in the form

$$\mathbf{J} = -D_p \nabla p - \mu_{p0} p \nabla \phi,$$

one can identify the diffusivity coefficient D_p and the low field mobility μ_{p0} as

$$D_p = -D_{11}(W_0), \quad \mu_{p0} = -D_{13}(W_0). \tag{12}$$

If we neglect the quadratic terms in the velocity then $W_0 = \frac{3\ k_B T_L}{2}$, consequently

$$D_p = \mu_{p0}\frac{2W_0}{3e} = \mu_{p0}\frac{k_B T_L}{e} \tag{13}$$

and the Einstein relation is verified for each W_0.

As regards the high field mobility, it is obtained by numerically solving the full system (4)-(7). The results are shown in fig. 1, where a comparison with the Caughey-Thomas formula of the mobility

$$\mu_p = \mu_{p0}\left[1 + \left(\frac{\mu_{p0} \mid \mathbf{E} \mid}{v_s}\right)^2\right]^{-1/2}$$

is reported. The influence of impurities is also taken into account, see Fig. 2.

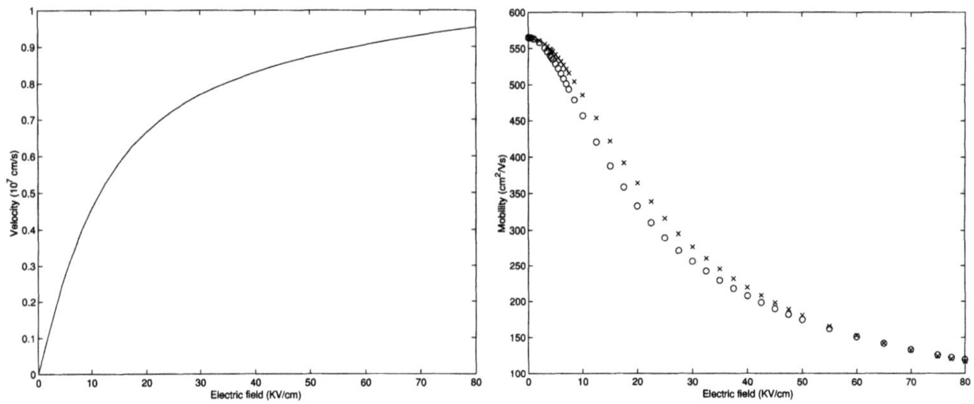

Fig. 1. Left: Hole velocity as function of the electric field. Right: Hole mobility as function of the electric field, crosses: Caughey-Thomas formula, circles: Hydrodynamical model

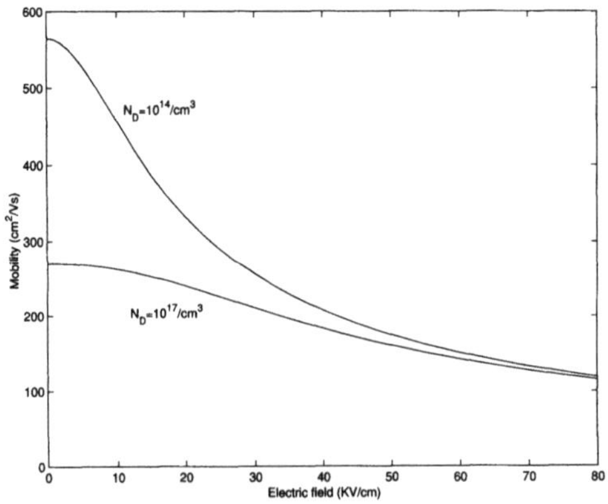

Fig. 2. Mobility as function of the electric field for different doping concentrations

4 n^+-p-n^+ device

Now we consider a one dimensional n^+-p-n^+ Si structure of length 250 nm with a 50 nm channel, which is intended to represent the channel of a MOSFET with an active region of 50 nm. The donor doping profile is

$$\begin{cases} N_D(x) = 5 \times 10^{18}/cm^3 \text{ if } \quad 0 \le x \le 100\text{nm} \quad \text{or} \quad 150\text{nm} \le x \le 250\text{nm}, \\ N_D(x) = 10^{15}/cm^3 \qquad \text{ if } \quad 100 \le x \le 150\text{nm}, \end{cases}$$

while the acceptor concentration N_A is identically equal to $10^{16}/\text{cm}^3$. Previous simulations of this device have been performed by keeping the holes at equilibrium [5], here we take into account the dynamics both of the electrons and the holes. The electron transport is modeled by means of the hydrodynamical model developed in [1]. The physical paremeters used for the simulation can be found in [4] as regards the electrons and in [3] as regards the holes. Since the time scale of the device is of the order of 1 ps, we neglect the generation-recombination terms. In Figs. 3-6a we report the electron and hole densities, the mean velocities, energies and the electric field for an applied bias of 0.6V. The contribution of the holes to the total current is negligible but their distribution at the steady state is not trivial. There is an accumulation of holes in the region of the channel close to the first junction, followed by a depletion zone close to the second junction. The depletion region is wider and therefore less deep than the accumulation one as determined by the electric field. Major differences with respect to the electrons are present in the stationary velocity profile, which has a maximum at about x=0.11 μm and a minimum at about x=0.15 μm as a consequence of the hole distribution and the conservation of the partial currents.

In Fig. 6b the IV characteristic-curve is shown. The results are in good agreement with the Monte Carlo ones, see Fig. 2_6 of ref. [5].

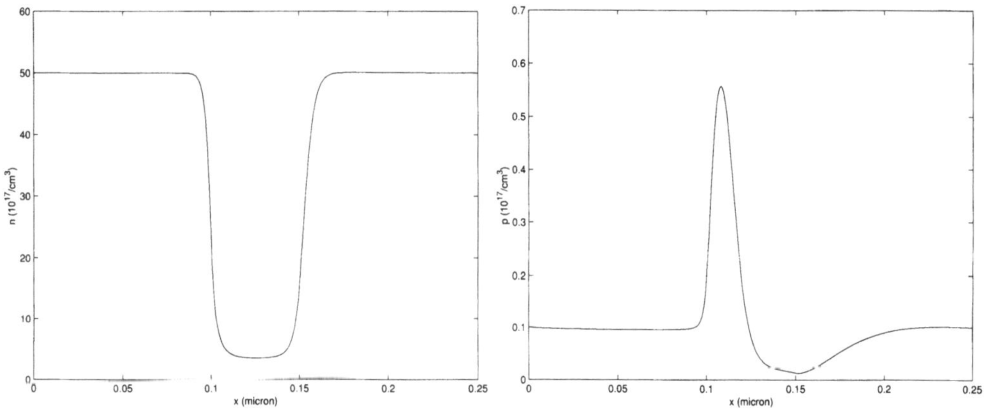

Fig. 3. Electron and hole concentrations vs x

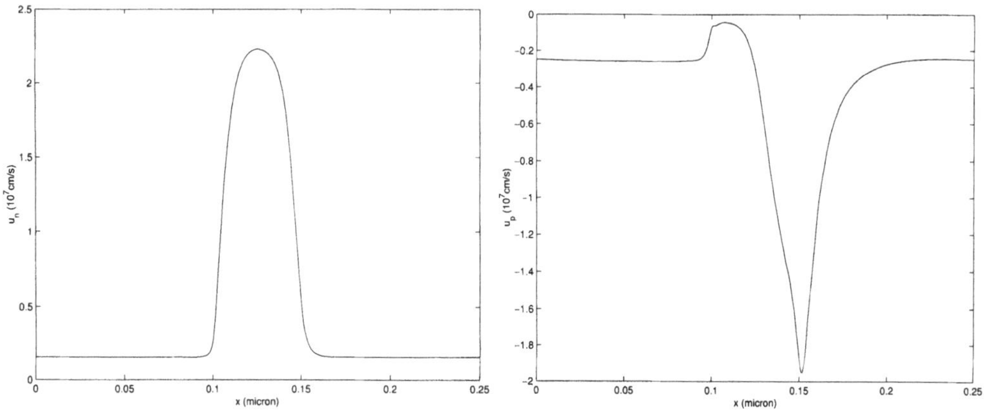

Fig. 4. Electron and hole velocities vs x

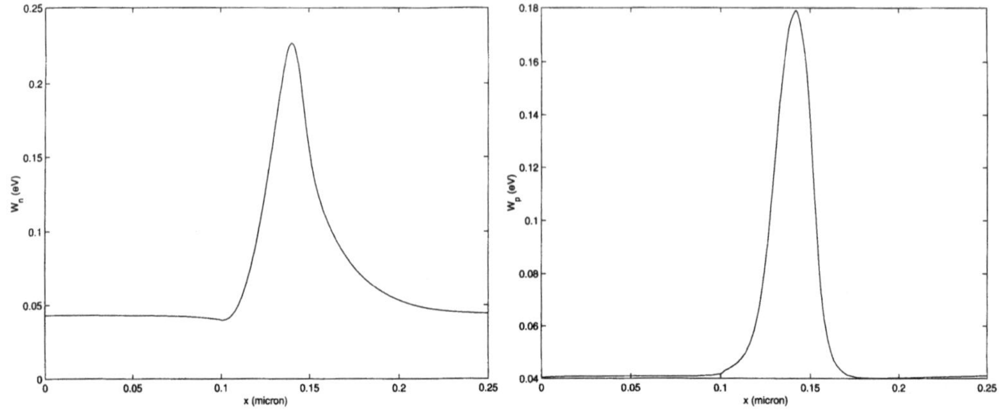

Fig. 5. Electron and hole energies vs x

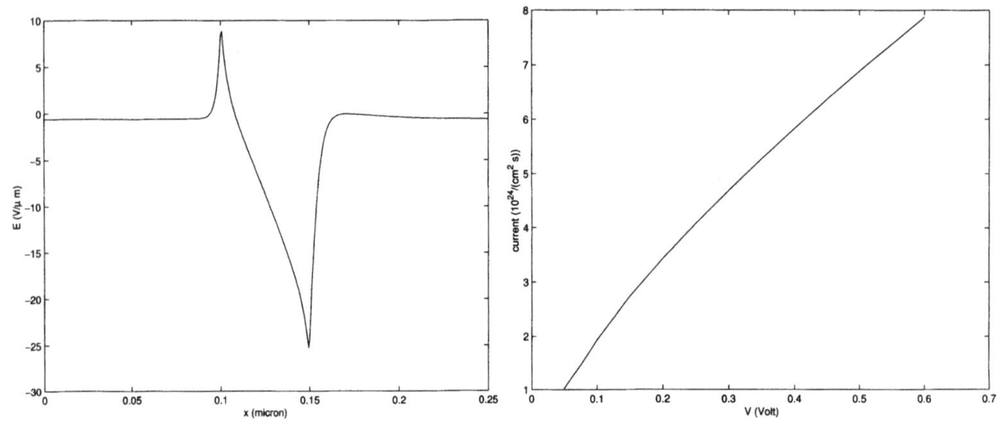

Fig. 6. Electric field vs x and current vs applied bias

References

1. G. Mascali and V. Romano: Hydrodynamical model of charge transport in GaAs based on the maximum entropy principle, Cont. Meh. Thermodyn., (2002) 14: 405-423
2. I. Müller and T. Ruggeri, *Rational Extended Thermodynamics*, Berlin, Springer-Verlag 1998
3. G. Mascali, V. Romano, J. M. Sellier, *MEP Parabolic Hydrodynamical Model for Holes in Silicon Semiconductors*, Il Nuovo Cimento B (2005) 120 (2): 197-215
4. A. M. Anile, G. Mascali, V. Romano, *Recent developments in hydrodynamical modeling of semiconductors*, in Lecture Notes in Mathematics, vol. 1823, Springer-Verlag 2003
5. A. M. Anile, J. A. Carrillo, I. M. Gamba, C. W. Shu *Approximation of the BTE by a relaxation time operator: simulations for a 50 nm channel silicon diode*, VLSI Design Journal (2001) 13: 349-354

Anisotropic Mesh Adaptivity Via a Dual-Based A Posteriori Error Estimation for Semiconductors *

S. Micheletti and S. Perotto

MOX - Modeling and Scientific Computing, Dipartimento di Matematica "F. Brioschi", Politecnico di Milano, via Bonardi 9, 20133 Milano, Italy, {stefano.micheletti,simona.perotto}@mate.polimi.it

1 Introduction

The accurate computation of a physically meaningful quantity (the goal quantity) associated with the solution of a given problem is of paramount importance in engineering applications. For example, in micro- or nano-electronics the output current is a fundamental quantity to assess the performance of the device at hand. In particular, we consider the Drift-Diffusion (DD) model for semiconductors describing the charge-transport in a device in terms of the electric potential (ψ), electron (n) and hole (p) concentrations [Sel84]. Thus, the goal quantity can be described by a suitable functional J, either linear or nonlinear, of the variables ψ, n and p. The accurate approximation of J can be dealt with in the framework of the optimal control theory. In particular, we consider an anisotropic a posteriori error estimation relying on the dual-based approach of [BR01, GS02]. We solve an adjoint (dual) linearized problem while employing anisotropic interpolation estimates [FP01, FP03] to bound the approximation error associated with the solution of the dual problem with respect to a suitable finite dimensional space. Thus, the parameters describing the distribution and shaping of the elements of the computational mesh act as control parameters, through which it is possible to approximate the goal quantity as accurate as needed.

The outline of the paper is as follows. In Sect. 2 the DD model for semiconductors is introduced. In Sect. 3 we sketch the abstract framework on which the anisotropic analysis is based. In particular, in Sect. 3.1 we recall some anisotropic interpolation error estimates which are the basic tool linking the dual-based a posteriori analysis of Sect. 4 to the anisotropic mesh adaptivity procedure. In Sect. 5 we derive the desired anisotropic a posteriori error estimator, while in Sect. 6 we address the iterative adaptive procedure used to construct the anisotropic meshes. A numerical validation is carried out on some test cases dealing with a pn-junction diode in Sect. 7. Finally, some conclusions and open issues are drawn in Sect. 8.

2 The Drift-Diffusion model

In this section we recall the stationary Drift-Diffusion charge transport model (see e.g. [Sel84]), consisting of the conservation laws for charge and for electron and hole concentrations (1)(left)

$$\left\{\begin{array}{l} \mathrm{div}(\epsilon \boldsymbol{E}) - \rho = 0, \\ -\,\mathrm{div}\boldsymbol{J}_n + qR = 0, \\ \mathrm{div}\boldsymbol{J}_p + qR = 0, \end{array}\right. \qquad \left\{\begin{array}{ll} \boldsymbol{E} & = -\nabla\psi, \\ \boldsymbol{J}_n & = q(D_n \nabla n - \mu_n\, n\, \nabla\psi), \\ \boldsymbol{J}_p & = -q(D_p \nabla p + \mu_p\, p\, \nabla\psi), \\ \rho & = q(p - n + D), \\ D_n & = \mu_n V_{\mathrm{th}}, \\ D_p & = \mu_p V_{\mathrm{th}}. \end{array}\right. \tag{1}$$

completed by the constitutive relations (1)(right). In (1), ψ, n and p are the unknowns, i.e. the electric potential, and the electron and hole concentrations, while $\boldsymbol{J}_n, \boldsymbol{J}_p$ are the electron and hole current densities, $\boldsymbol{E}$ is the electric field,

*This work was partially supported by the INDAM 2003 Project "Modellistica Numerica per il Calcolo Scientifico e Applicazioni Avanzate".

ρ is the net charge density, D is the given doping profile, R is the recombination/generation rate, D_n, μ_n (D_p, μ_p) are the electron (hole) diffusion coefficient and mobility, $V_{\rm th}$ is the thermal voltage, ϵ is the semiconductor dielectric permittivity, and q is the positive electron charge.

As typical expression for R, we consider henceforth the so-called Shockley-Read-Hall form, given by $R = (pn - n_i^2)/[(p+n_i)\tau_n + (n+n_i)\tau_p]$, where n_i is the electron/hole intrinsic concentration, and τ_n and τ_p are suitable relaxation times (see, e.g., [Sel84]). The whole system is completed by suitable boundary conditions, usually of mixed type. For simplicity, we consider the case where the device is made up of a homogeneous semiconductor material, occupying the computational domain Ω, that is we do not deal with metal-semiconductor or metal-oxide-semiconductor structures. Thus, the boundary $\partial\Omega$ of Ω is split into two non-overlapping parts, $\Gamma_{\rm D}$ and $\Gamma_{\rm N}$, where Dirichlet and Neumann boundary conditions are imposed, respectively. For instance, in the case of the pn junction diodes of Fig. 1, we have $\Gamma_{\rm D} = \overline{\rm AG} \cup \overline{\rm CD}$ while $\Gamma_{\rm N} = \partial\Omega \backslash \Gamma_{\rm D}$. The boundary conditions characterizing the devices in Fig. 1 are thus given by $\psi = \psi_{\rm D}$, $n = n_{\rm D}$ and $p = p_{\rm D}$ on $\Gamma_{\rm D}$, while $\nabla\psi \cdot \boldsymbol{n} = \nabla n \cdot \boldsymbol{n} = \nabla p \cdot \boldsymbol{n} = 0$, on $\Gamma_{\rm N}$, where $\boldsymbol{n}$ is the unit outward normal vector to $\partial\Omega$, $\psi_{\rm D} = V_{\rm app} + V_{\rm bi}$, with $V_{\rm app}$ the external applied voltage and $V_{\rm bi} = V_{\rm th} \sinh^{-1}(D/(2n_i))|_{\Gamma_{\rm D}}$, the so-called built-in voltage, while $n_{\rm D} = [(D + \sqrt{D^2 + 4n_i^2})/2]|_{\Gamma_{\rm D}}$ and $p_{\rm D} = -[(D + \sqrt{D^2 + 4n_i^2})/2]|_{\Gamma_{\rm D}}$.

3 The anisotropic "tool box"

Let us introduce a conformal partition $\mathcal{T}_h$ of Ω, in the usual sense [Cia78], consisting of triangular elements and let K denote the general triangle. Let $T_K : \widehat{K} \to K$ be the standard affine mapping between the reference triangle $\widehat{K}$ (e.g. the unit equilateral one) and the general one K, with $\boldsymbol{x} = (x_1, x_2)^{\rm T} = T_K(\widehat{\boldsymbol{x}}) = M_K \widehat{\boldsymbol{x}} + \boldsymbol{t}_K$. Then let us introduce the polar decomposition of M_K, i.e. $M_K = B_K Z_K$, with B_K symmetric positive definite and Z_K orthogonal matrices, respectively. Decomposing B_K in terms of its eigenvectors $\boldsymbol{r}_{i,K}$ and eigenvalues $\lambda_{i,K}$, with $i = 1, 2$, yields $B_K = R_K^{\rm T} \Lambda_K R_K$, where $R_K^{\rm T} = [\mathbf{r}_{1,K}, \mathbf{r}_{2,K}]$ and $\Lambda_K = {\rm diag}[\lambda_{1,K}, \lambda_{2,K}]$. Throughout we assume $\lambda_{1,K} \geq \lambda_{2,K}$, that is $s_K = \lambda_{1,K}/\lambda_{2,K} \geq 1$, s_K being the so called stretching factor (see Fig. 2 for the geometrical meaning of the quantities $\lambda_{i,K}, \boldsymbol{r}_{i,K}$).

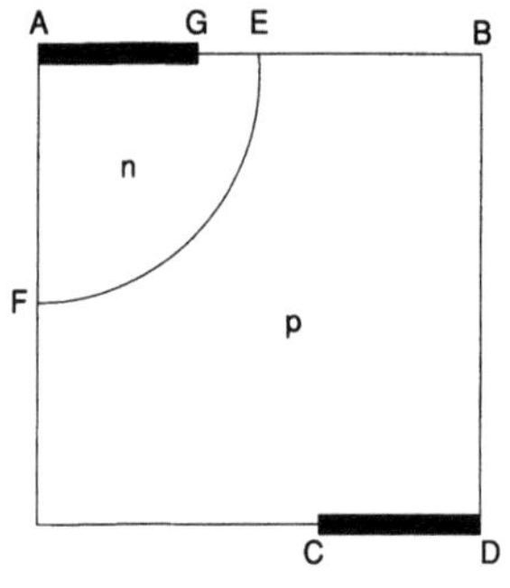

Fig. 1. Geometry of a *pn* junction diode

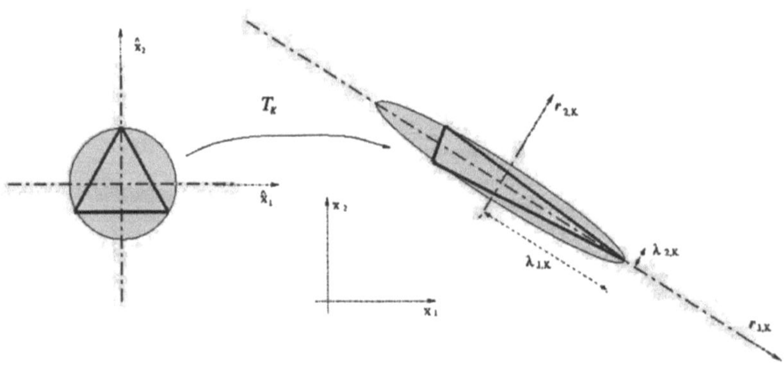

Fig. 2. Geometrical quantities related to the affine mapping T_K

3.1 Anisotropic interpolation error estimates

Moving from the above abstract framework, we now recall some anisotropic interpolation error estimates, introduced in [FP01, FP03]. We assume a standard notation for the Lebesgue and Sobolev spaces, see, e.g., [Cia78]. For any function $v \in H^1(\Omega)$, let $G_K(v)$ be the symmetric positive semi-definite matrix with entries $(G_K(v))_{i,j} = \int_{\Delta_K} \partial_{x_i} v \, \partial_{x_j} v \, d\boldsymbol{x}$, and let $I_K(v) \in \mathbb{P}^1(K)$ be a Clément-like interpolant of v on K, where $\mathbb{P}^1(K)$ is the space of polynomials of degree less than or equal to one on K, Δ_K being a suitable patch of elements surrounding K. Then the following estimates can be proved:

$$\|v - I_K(v)\|^2_{L^2(K)} \le C_1 \sum_{i=1}^{2} \lambda^2_{i,K} \, (\boldsymbol{r}^{\mathrm{T}}_{i,K} \, G_K(v) \boldsymbol{r}_{i,K}), \tag{2}$$

$$\|v - I_K(v)\|^2_{L^2(e)} \le C_2 \, \frac{1}{\lambda_{2,K}} \sum_{i=1}^{2} \lambda^2_{i,K} \, (\boldsymbol{r}^{\mathrm{T}}_{i,K} \, G_K(v) \boldsymbol{r}_{i,K}), \tag{3}$$

where the edge $e \in \partial K$ and C_1, C_2 suitable constants (see [MPP03] for more details).

4 Dual-based a posteriori analysis

Suppose that we are interested in approximating the goal quantity $J(u)$ by $J(u_h)$ such that $|J(u) - J(u_h)| \le \tau$, with J a continuous functional, possibly nonlinear, u and u_h the exact and approximate solutions to the problem at hand, and τ a given tolerance. In electronics, J can be, for example, the total current in a device, or in Computational Fluid Dynamics, it can represent the kinetic energy or the vorticity of a fluid, the lift or drag in a flow past a body, while in structural mechanics, it can be the torsion moment, the stress values or the total surface tension.

With this aim we can follow the so-called dual approach in [BR01, GS02]. In a general setting, let $a(u; v)$ and $J(u)$ be semilinear forms, where it is understood that, when more than one argument is present, the forms are linear with respect to all the arguments on the right of the semicolon. The problem at hand can be formulated as the following control problem: find $u \in V$ such that

$$J(u) = \min_{v \in M} J(v) \quad \text{with} \quad M = \{w \in V : a(w; v) = F(v), \quad \forall v \in V\},$$

where F is a linear form and V is a suitable Hilbert space. Let $\mathcal{L}(u; z) = J(u) + F(z) - a(u; z)$, for any $u, z \in V$, be the corresponding Lagrangian. The condition for finding the critical points of $\mathcal{L}$, that is

$$\mathcal{L}'(u, z; \varphi, v) = J'(u; \varphi) + F(v) - a(u; v) - a'(u; \varphi, z) = 0, \qquad \forall \varphi, v \in V,$$

yields the primal problem (P.P.): find $u \in V$ such that

$$a(u; v) = F(v), \qquad \forall v \in V,$$

and the adjoint problem (A.P.): find $z \in V$ such that

$$a'(u; \varphi, z) = J'(u; \varphi), \qquad \forall \varphi \in V,$$

where $a'(u; \varphi, z) = \lim_{\epsilon \to 0} \frac{1}{\epsilon} [a(u + \epsilon\varphi; z) - a(u; z)]$ is the Gâteaux derivative of $a(u; z)$ with respect to its first argument, and likewise for $J'(u; \varphi)$. Then let us consider the Galerkin approximation (G.A.) of (P.P.): find $u_h \in V_h$ such that

$$a(u_h; v_h) = F(v_h), \qquad \forall v_h \in V_h,$$

where V_h is a suitable finite dimensional subspace of V. The problems (P.P.) and (G.A.) are linked by the Galerkin orthogonality (G.O.) condition:

$$a(u; v_h) - a(u_h; v_h) = 0, \qquad \forall v_h \in V_h.$$

In the case when both a and J are linear, from (P.P.), (A.P.) and (G.O.) the following error representation holds: $J(u) - J(u_h) = F(z - \varphi_h) - a(u_h; z - \varphi_h)$, for any $\varphi_h \in V_h$. Otherwise in the more general case when either a or J are nonlinear, it can be proved that

$$J(u) - J(u_h) = F(z - \varphi_h) - a(u_h; z - \varphi_h) + \mathcal{R} \qquad \forall \varphi_h \in V_h, \tag{4}$$

where

$$\mathcal{R} = \int_0^1 [a''(u_h + se; e, e, z) - J''(u_h + se; e, e)] s\, ds,$$

is a remainder term quadratic with respect to $e = u - u_h$ (see Propositions 2.2 and 2.3 in [BR01]). In practice, neglecting $\mathcal{R}$, choosing φ_h as a suitable interpolant of z and integrating by parts over the elements of the mesh, we obtain an estimate of the form

$$|J(u) - J(u_h)| \leq C \sum_{K \in \mathcal{T}_h} \rho_K(u_h)\, \omega_K(z),$$

where $\rho_K(u_h)$ is a residual term depending only on the approximate solution u_h and $\omega_K(z)$ is a weighting term taking into account the dual solution z.

5 Goal-oriented a posteriori analysis for the DD model

In this section we apply the general framework of the previous section to the Drift-Diffusion model (1) (more details can be found in [MP04]). For this purpose, let $U = (\psi, n, p)$ and $Z = (z_1, z_2, z_3)$ be the primal and dual solution triplets, respectively, and let $(u, v) = \int_\Omega u\, v\, d\Omega$ denote the standard $L^2(\Omega)$-scalar product. Then problem (1) can be cast in the abstract framework above by defining

$$\begin{aligned} a(U; Z) = &(\epsilon \nabla \psi, \nabla z_1) - q(p - n + D, z_1) + q(D_n \nabla n - \mu_n n \nabla \psi, \nabla z_2) + q(R, z_2) \\ &+ q(D_p \nabla p + \mu_p p \nabla \psi, \nabla z_3) + q(R, z_3), \end{aligned}$$

while $F(U) = 0$. Thus $\mathcal{L}(U; Z) = J(U) - a(U; Z)$, for any $(U, Z) \in W \times \widetilde{W}$, where W is the affine space of functions in $[H^1(\Omega)]^3$ taking into account the nonhomogeneous Dirichlet boundary conditions, while $\widetilde{W} = [H^1_{\Gamma_D}(\Omega)]^3$. Letting $V = (v_1, v_2, v_3)$, we have

$$\begin{aligned} a'(U; V, Z) = &(\epsilon \nabla v_1, \nabla z_1) - q(v_3 - v_2, z_1) \\ &+ q(D_n \nabla v_2 - \mu_n n \nabla v_1 - \mu_n v_2 \nabla \psi, \nabla z_2) + q(R'_n(U) v_2, z_2) + q(R'_p(U) v_3, z_2) \\ &+ q(D_p \nabla v_3 + \mu_p p \nabla v_1 + \mu_p v_3 \nabla \psi, \nabla z_3) + q(R'_n(U) v_2, z_3) + q(R'_p(U) v_3, z_3), \end{aligned}$$

$R'_n(U), R'_p(U)$ being the derivatives of the recombination/generation term with respect to n and p, respectively. Let $J(U)$ be the quantity we are interested in and let us introduce the Galerkin approximation $U_h = (\psi_h, n_h, p_h) \in W_h$ of the primal solution, such that $a(U_h, V_h) = 0$, for any $V_h \in \widetilde{W}_h$, where W_h and $\widetilde{W}_h$ are finite dimensional subspaces of W and $\widetilde{W}$, respectively. Moving from equality (4) and neglecting the remainder term $\mathcal{R}$, it holds

$$J(U) - J(U_h) \simeq -a(U_h; Z - V_h), \qquad \forall V_h \in \widetilde{W}_h.$$

In more detail, by splitting the integrals over Ω as sums over the elements K of the mesh $\mathcal{T}_h$, we get

$$J(U) - J(U_h) \simeq \sum_{i=1}^{3} \sum_{K \in \mathcal{T}_h} \{(\rho_K^i, z_i - v_{h,i})_K + \frac{1}{2}(j_e^i, z_i - v_{h,i})_{\partial K}\},$$

where $\rho_K^i = \rho_K^i(\psi_h, n_h, p_h)$ and $j_e^i = j_e^i(\psi_h, n_h, p_h)$, with $i = 1, 2, 3$, are the internal and edge residuals, respectively, defined by

$$\begin{cases} \rho_K^1 = [\,\mathrm{div}(\epsilon \boldsymbol{E}_h) - q(p_h - n_h + D)\,]|_K, \\ \rho_K^2 = [\,-\mathrm{div} \boldsymbol{J}_{n,h} + qR(U_h)\,]|_K, \\ \rho_K^3 = [\,\mathrm{div} \boldsymbol{J}_{p,h} + qR(U_h)\,]|_K, \end{cases} \quad \text{and } j_e^i = \begin{cases} [\boldsymbol{j}^i \cdot \boldsymbol{n}]_e, & \forall e \in \mathcal{E}_h, \\ 2\boldsymbol{j}^i \cdot \boldsymbol{n}, & \forall e \in \Gamma_N, \\ 0, & \forall e \in \Gamma_D, \end{cases}$$

where $\boldsymbol{j}^1 = -\epsilon \boldsymbol{E}_h = \epsilon \nabla \psi_h$, $\boldsymbol{j}_h^2 = \boldsymbol{J}_{n,h} = q(D_n \nabla n_h - \mu_n n_h \nabla \psi_h)$ and $\boldsymbol{j}_h^3 = -\boldsymbol{J}_{p,h} = q(D_p \nabla p_h + \mu_p p_h \nabla \psi_h)$ are the discrete displacement, electron and hole current densities, respectively, $R(U_h)$ is the recombination/generation term evaluated at U_h, $\mathcal{E}_h$ is the set of the internal edges of $\mathcal{T}_h$ and $[\,v\,]_e$ denotes the jump of the function v across the edge e. Now choosing $V_h|_K = I_K(Z)$, i.e. by identifying the test function V_h with the Clément-like interpolant of the dual solution Z, and thanks to the anisotropic interpolation error estimates (2)-(3), we obtain

$$|J(U) - J(U_h)| \leq C \sum_{i=1}^{3} \sum_{K \in \mathcal{T}_h} \alpha_K \, R_K^i(U_h) \, \omega_K^i(z_i), \tag{5}$$

where $C = C(C_1, C_2)$, $\alpha_K = (\lambda_{1,K}\lambda_{2,K})^{3/2}$, and for $i = 1, 2, 3$,

$$R_K^i(U_h) = \frac{1}{(\lambda_{1,K}\lambda_{2,K})^{1/2}} \left(\|\rho_K^i\|_{L^2(K)} + \frac{1}{2\lambda_{2,K}^{1/2}} \|j_e^i\|_{L^2(\partial K)}\right),$$

$$\omega_K^i(z_i) = \frac{1}{(\lambda_{1,K}\lambda_{2,K})^{1/2}} \left[s_K (\boldsymbol{r}_{1,K}^{\mathrm{T}} G_K(z_i) \boldsymbol{r}_{1,K}) + \frac{1}{s_K} (\boldsymbol{r}_{2,K}^{\mathrm{T}} G_K(z_i) \boldsymbol{r}_{2,K})\right]^{1/2},$$

with $R_K^i(U_h)$ and $\omega_K^i(z_i)$ independent of $|K|$, at least asymptotically, i.e. when the mesh is sufficiently fine.

6 Generation of the mesh

The technique used to compute the adapted meshes is a metric-based, adaptive iterative procedure that, starting from a given mesh, $\mathcal{T}_h^{(k)}$, consisting of $N_h^{(k)}$ elements, finds the new mesh $\mathcal{T}_h^{(k+1)}$ by exploiting the error estimator (5). In practice, the anisotropic quantities describing the new mesh $\mathcal{T}_h^{(k+1)}$ are approximated by functions piecewise constant on $\mathcal{T}_h^{(k)}$. Since we are dealing with a vector problem, each of the three terms in (5) yields a contribution to the adaptive procedure, i.e. a corresponding mesh. As the procedure to obtain each mesh is the same for all the three contributions, we detail in the following the general procedure for a given $i \in \{1, 2, 3\}$. For this purpose, let $\eta_K^i = \alpha_K \, R_K^i(U_h) \, \omega_K^i(z_i)$ be the local error estimator. We impose that:

i) $\eta_K^i = \tau / N_h^{(k)}$, for any $K \in \mathcal{T}_h^k$, where τ is the given tolerance (equidistribution criterion);
ii) $|K|$ be as large as possible (mesh elements minimization criterion).

Requirement *ii)* amounts to solving the minimization problem for the quantities at step $k + 1$:

find the optimal values $\widetilde{s}_K$ *of* s_K *and* $\widetilde{\boldsymbol{r}}_{1,K}$ *of* $\boldsymbol{r}_{1,K}$ *such that* $\omega_K^i(z_i)$ *be minimized, subject to the constraints* $s_K \geq 1$, $\boldsymbol{r}_{1,K}, \boldsymbol{r}_{2,K} \in \mathbb{R}^2$, $\boldsymbol{r}_{i,K} \cdot \boldsymbol{r}_{j,K} = \delta_{ij}$, *for* $i, j = 1, 2$, *with* δ_{ij} *the Kronecker symbol,*

where the dependence on $k + 1$ is understood. Let μ_m and μ_M be the (positive) minimum and maximum eigenvalues of $G_K(z_i)/(\lambda_{1,K}^{(k)}\lambda_{2,K}^{(k)})$, respectively. Then the solution of this minimization problem yields $\widetilde{\boldsymbol{r}}_{1,K}$ parallel to the eigenvector associated with μ_m and $\widetilde{s}_K = \widetilde{\lambda}_{1,K}/\widetilde{\lambda}_{2,K} = (\mu_M/\mu_m)^{1/2}$. Finally, requirement *i)* allows us to obtain the specific values for $\widetilde{\lambda}_{1,K}$ and $\widetilde{\lambda}_{2,K}$, as

$$\widetilde{\lambda}_{1,K}\widetilde{\lambda}_{2,K} \simeq \left(\frac{\tau}{N_h^{(k)}}\right)^{2/3} \left(R_K^i(U_h) \left(\widetilde{s}_K \, \mu_m + \frac{1}{\widetilde{s}_K} \, \mu_M\right)^{1/2}\right)^{-2/3}.$$

The above quantities define in a unique way the size and shape of the elements of the new mesh $\mathcal{T}_h^{i,(k+1)}$. Following the same above procedure, once the three metrics for $\mathcal{T}_h^{1,(k+1)}, \mathcal{T}_h^{2,(k+1)}, \mathcal{T}_h^{3,(k+1)}$ have been obtained, the final mesh $\mathcal{T}_h^{(k+1)}$ may be obtained by computing the intersection of the three metrics, as described in [GB98].

7 Numerical results

We asses the procedure outlined in the previous sections on some test cases. Firstly, let us provide some computational details:

- the Scharfetter-Gummel node-centred box method is used as numerical approximation scheme, thus, only the current densities along the edges are meaningful [BBFS90]. This scheme guarantees a discrete maximum principle for the unknowns;
- the reconstruction of the current densities inside each triangle is carried out by the lowest order edge elements of Nédélec's first family [Ned80];
- the Newton method is used to solve the whole system;
- the stiffness matrix for the dual problem is just the transpose of the Jacobian associated with the non-linear system of the primal problem, so that the overhead of solving the dual problem is approximately the same as that of one further iteration of the Newton method;
- the software BAMG [Hec98] is used to compute all the meshes.

We consider the step-junction diode of Fig. 1, with the following choice for the data: $\Omega = (0, 10) \times (0, 10)\,\mu\text{m}$, symmetric doping, i.e. $D = 10^{17}\,\text{cm}^{-3}$ in the curved polygonal n-region A-G-E-F-A, and $D = -10^{17}\,\text{cm}^{-3}$ in the remaining part, contact length $|\overline{\text{AG}}| = |\overline{\text{CD}}| = 4\,\mu\text{m}$, and junction radius $|\overline{\text{AE}}| = |\overline{\text{AF}}| = 5\,\mu\text{m}$ centred at A, $\tau_n = \tau_p = 10^{-9}\,\text{s}$, $\mu_n = 1300\,\text{cm}^2\text{V}^{-1}\text{s}^{-1}$, and $\mu_p = 400\,\text{cm}^2\text{V}^{-1}\text{s}^{-1}$.

7.1 Control of total current

As first choice, we identify the goal functional J with the total current, i.e. with the flux of the total current density $\boldsymbol{J} = \boldsymbol{J}_n + \boldsymbol{J}_p$, either at the n-side contact $\overline{\mathrm{AG}}$, or at the p-side edge $\overline{\mathrm{CD}}$. In the case of the n-side contact, for example, the functional J is computed as $J(U) = \int_{\overline{AG}} \boldsymbol{J} \cdot \boldsymbol{n}\, ds = \int_{\partial\Omega} \boldsymbol{J} \cdot \boldsymbol{n}\, \omega\, ds = \int_\Omega \omega\, \mathrm{div}\boldsymbol{J}\, d\Omega + \int_\Omega \boldsymbol{J} \cdot \nabla\omega\, d\Omega = \int_\Omega \boldsymbol{J} \cdot \nabla\omega\, d\Omega$, for any function ω smooth enough, such that $\omega|_{\overline{\mathrm{AG}}} = 1$ and $\omega|_{\overline{\mathrm{CD}}} = 0$. Notice that, we have used the divergence theorem and, from (1), the property that $\mathrm{div}\, \boldsymbol{J} = 0$, so that, thanks also to the boundary conditions, the flux of $\boldsymbol{J}$ at the two contacts is equal and of opposite sign. This escamotage holds for the weak formulation but it generally fails in the discrete case. It is shown however, to provide rather accurate results in the FEM context (see [BR01, GS02]). In Fig. 3, both rows show the evolution of the meshes at the first three iterates at $V_{\mathrm{app}} = 0.9\,\mathrm{V}$. The top row refers to the control on the n-side and the meshes are those corresponding to the dual variable z_1 only, while the bottom row deals with the control on the p-side contact, respectively, and the meshes are associated with z_3.

7.2 Control of pointwise electron concentration

As second test case, we consider the control of the electron concentration at the point $(4.167, 8.638)\,\mu\mathrm{m}$ at the two biases corresponding to a forward $V_{\mathrm{app}} = 0.7V$ and a reverse $V_{\mathrm{app}} = -5V$. Figure 4 shows the meshes corresponding to the dual variable z_2 at the first iteration (top row) and the corresponding plot of z_2 (bottom row). The left column refers to the forward-bias case while the right column is associated with the reverse-bias polarization.

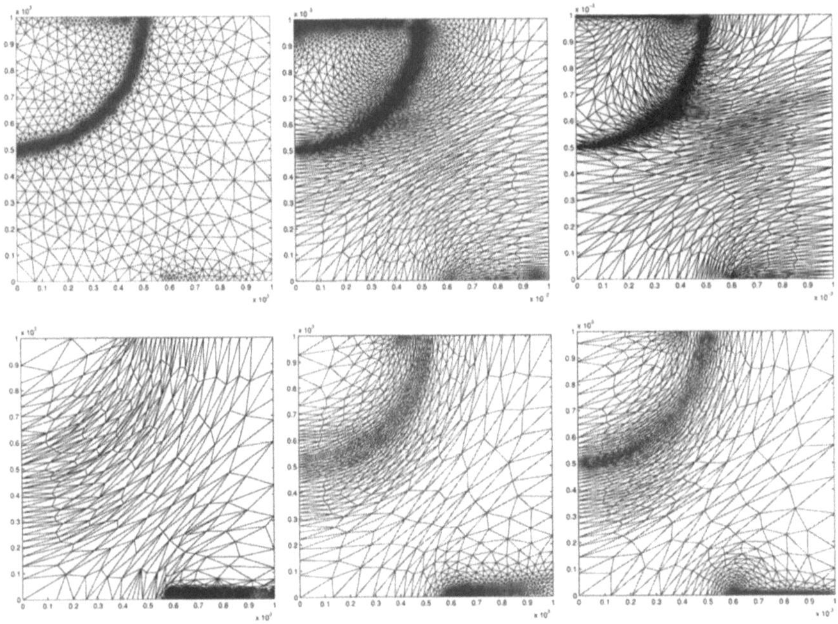

Fig. 3. Control of total current: sequence of adapted meshes

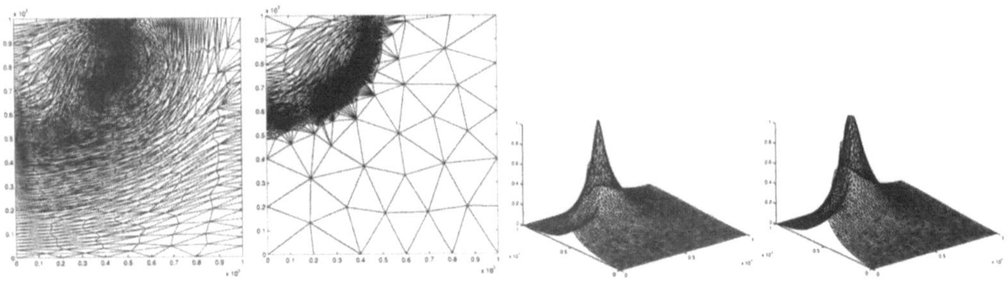

Fig. 4. Control of pointwise electron concentration: sequence of meshes for $V_{\mathrm{app}} = 0.7V$ (left) and $V_{\mathrm{app}} = -5V$ (right) at the first iteration

8 Conclusions

We have dealt with a dual-based anisotropic a posteriori error estimation for the Drift-Diffusion model in semiconductors. This allows us to control suitable goal quantities via the optimal control theory where the controls are essentially the geometrical quantities describing the mesh. By an appropriate distribution and shaping of the elements we can guarantee that the error in the desired output functional is below a given tolerance. Several open issues are in oder: the validation on other functionals and on other devices; the extension to the time-dependent problem.

References

[BBFS90] Bank, R.E., Bürgler, J.F., Fichtner, W., Smith, R.K.: Some upwinding techniques for finite element approximations of convection-diffusion equations. Numer. Math., **58**, 185–202 (1990)

[BR01] Becker, R., Rannacher, R.: An optimal control approach to a posteriori error estimation in finite element methods. Acta Numer., **10**, 1–102 (2001)

[Cia78] Ciarlet, Ph.: The Finite Element Method for Elliptic Problems. North-Holland Publishing Company, Amsterdam (1978)

[FP01] Formaggia, L., Perotto, S.: New anisotropic a priori error estimates. Numer. Math., **89**, 641–667 (2001)

[FP03] Formaggia, L., Perotto, S.: Anisotropic error estimates for elliptic problems. Numer. Math., **94**, 67–92 (2003)

[GB98] George, P.L., Borouchaki, H.: Delaunay Triangulation and Meshing - Application to Finite Elements. Editions Hermes, Paris (1998)

[GS02] Giles, M.B., Süli, E.: Adjoint methods for PDEs: a posteriori error analysis and postprocessing by duality. Acta Numer., **11**, 145–236 (2002)

[Hec98] Hecht, F.: http://www-rocq.inria.fr/gamma/cdrom/www/bamg/eng.htm (1998)

[MP04] Micheletti, S., Perotto, S.: Goal-oriented a posteriori error estimation and anisotropic mesh adaptivity for semiconductors. In preparation.

[MPP03] Micheletti, S., Perotto, S., Picasso, M.: Stabilized finite elements on anisotropic meshes: a priori error estimates for the advection-diffusion and Stokes problems. SIAM J. Numer. Anal., **41** (3), 1131–1162 (2003)

[Ned80] Nédélec, J.C.: Mixed finite elements in $\mathbb{R}^3$. Numer. Math., **35** (3), 315–341 (1980)

[Sel84] Selberherr, S.: Analysis and Simulation of Semiconductor Devices. Springer Verlag, Wien (1984)

Kinetic Relaxation Models for the Boltzmann Transport Equation for Silicon Semiconductors

O. Muscato

Dipartimento di Matematica e Informatica, Viale Andrea Doria 6, 95125 Catania, Italy, muscato@dmi.unict.it

Abstract An overview on the relaxation-time approximations of the collisional operator of the Boltzmann transport equation for semiconductors is given. Solutions of these kinetic models are obtained through the use of exact-integral representations in the stationary and homogeneous regime. Some properties of these solutions have been discussed and their validity have been assessed by Monte Carlo simulations in bulk silicon.

1 Introduction

The semiclassical Boltzmann transport equation (hereafter BTE) coupled with Poisson equation, provides the natural environment for modeling submicron semiconductor devices. Solving it numerically is not an easy task, because the BTE is an integro-differential equation with six dimensions in the phase space and one in time. Recently, finite difference scheme [1, 2], discontinuous spline approximations of the distribution function [3] have been introduced, but with a heavy computational cost. Following the experience of the kinetic theory of gases [4], simpler expressions have been proposed for the BTE collisional operator. The most widely known collision model is usually called Bhatnagar, Gross and Krook (hereafter BGK) model, where the fine structure of the collisional operator is replaced by a blurred image, based upon a simple operator, which retains only the qualitative and average properties of the true collisional operator.

Usually it is assumed that the distribution function relaxes to its equilibrium value determined by the local density and the lattice temperature, and that this process can be characterized by a relaxation time. This approach is certainly not exact for scattering in semiconductors, because neglects the energy and angular dependence of the scattering rate as well as the discrete amounts of energy lost in the scattering from optical phonons.

This model is deeply influenced by the choice of the relaxation time: it can be taken constant [5, 6], or function of the electric field [7, 8] or of the electron momentum [9, 10]. In the stationary homogeneous regime, by using the BGK approximation, an analytic solution of the distribution function can be obtained, and consequently its moments can be evaluated numerically. In this paper, we want to asses the validity of these models, by comparing the moments of the distribution function with those obtained by MC simulations for bulk silicon, in the stationary regime.

2 Basic equations

The BTE for electrons and one conduction band writes [11]:

$$\frac{\partial f}{\partial t} + \mathsf{v}(\mathbf{k}) \cdot \nabla_{\mathsf{x}} f - \frac{q}{\hbar} \mathsf{E} \cdot \nabla_{\mathbf{k}} f = Q[f] \tag{1}$$

where the unknown function $f(t, \mathsf{x}, \mathbf{k})$ represents the probability density of finding an electron at time t in the position $\mathsf{x} \equiv (x_1, x_2, x_3)$ with the wave-vector $\mathbf{k} \equiv (k_1, k_2, k_3)$, $\hbar$ is the Planck constant divided by 2π, and q is the absolute value of the electron charge. The domain of $\mathbf{k}$ can be the three-dimensional space or the first Brillouin zone. We denote by Ω the $\mathbf{k}$-domain. The energy of the considered crystal conduction band structure $\varepsilon(\mathbf{k})$ is measured from the band minimum. In the neighborhood of the band minimum a good dispersion relation is given by the *quasi-parabolic* approximation:

$$\varepsilon(\mathbf{k})\left[1+\alpha\varepsilon(\mathbf{k})\right]=\frac{\hbar^2\mathbf{k}^2}{2m^*},\quad \mathbf{k}\in\Omega \tag{2}$$

where α is the non-parabolicity parameter (for silicon $\alpha = 0.5\ eV^{-1}$), m^* denotes the effective electron mass, which is 0.32 m_e (free electron mass) in silicon. The electron group velocity $\mathsf{v} \equiv (v_1, v_2, v_3)$ is given by

$$\mathsf{v}(\mathbf{k})=\frac{1}{\hbar}\nabla_{\mathbf{k}}\varepsilon=\frac{\hbar\mathbf{k}}{m^*\sqrt{1+4\alpha\frac{\hbar^2\mathbf{k}^2}{2m^*}}}\ .$$

The electric field $\mathsf{E}(t,\mathsf{x}) \equiv (E_1, E_2, E_3)$ is related to electronic distribution f by the usual Poisson equation.
Concerning the collision term, the main scattering mechanisms in a silicon semiconductor are the electron-phonon interactions, the interaction with impurities, the electron-electron scatterings and the interaction with stationary imperfections of the crystal, as vacancies. In general, for low electron density, the collision operator can be schematically written as

$$Q[f]=\int_{\mathbb{R}^3}\left[w(\mathbf{k}',\mathbf{k})f(\mathbf{k}')-w(\mathbf{k},\mathbf{k}')f(\mathbf{k})\right]\,d\mathbf{k}', \tag{3}$$

where $w(\mathbf{k},\mathbf{k}')$ is the transition probability per unit time from a state $\mathbf{k}$ to a state $\mathbf{k}'$. The first term in (3) represents the gain and the second one the loss. In silicon, the main scattering phenomena are due to the electron-phonon interactions which can be modeled as [12]

$$\begin{aligned} w(\mathbf{k},\mathbf{k}') = {} & K_0(\mathbf{k},\mathbf{k}')\,\delta(\varepsilon(\mathbf{k}')-\varepsilon(\mathbf{k}))+\sum_{i=1}^{6}K_i(\mathbf{k},\mathbf{k}')\times \\ & \left[\delta(\varepsilon(\mathbf{k}')-\varepsilon(\mathbf{k})+\hbar\omega_i)(\mathsf{n}_{q_i}+1)+\delta(\varepsilon(\mathbf{k}')-\varepsilon(\mathbf{k})-\hbar\omega_i)\mathsf{n}_{q_i}\right] \end{aligned} \tag{4}$$

where $\hbar\omega_i$ is a optical phonon energy and n_{q_i} the phonon equilibrium distribution which, according to the Bose-Einstein statistics, is given by

$$\mathsf{n}_{q_i}=\frac{1}{\exp(\hbar\omega_i/k_BT_0)-1}\ .$$

where T_0 is the lattice temperature, K_0 and K_i represent respectively the elastic and inelastic scattering probabilities. The electron-electron interaction is taken into account in the framework of the mean field approximation through the Poisson equation. This is reasonable since we consider the case of low electron density and, therefore, we can neglect the short range collisions between electrons.
An analysis of the collisional operator Q can be found in [13]. In this paper the H-theorem is established, and the null space of Q[f] * is determined by the functions

$$f(\mathbf{k})=\Gamma(\varepsilon)exp\left[-\frac{\varepsilon}{k_BT_0}\right] \tag{5}$$

where Γ is a constant function (which gives the well-known lattice maxwellian distribution function) or is periodic, supposing that all phonon energies $\hbar\omega_i$ are commensurable. We point out that in silicon the six optical phonon energies are non commensurable.

3 BGK models

The first BGK model for semiconductors was introduced by Trugman and Taylor [5] in *the parabolic band* approximation (i.e. α=0 in eq.(2))

$$Q[f]=-\frac{f-nf_0}{\tau} \tag{6}$$

where n is the electron density, and f_0 is the lattice maxwellian

$$f_0=\sqrt{\frac{m^*}{2\pi k_BT_0}}\quad exp\left[-\frac{m^*\mathsf{v}^2}{2k_BT_0}\right]\ .$$

*i.e. the functions such that Q[f]=0.

In the following we shall consider the so called *bulk silicon*, which is an homogeneous piece of silicon where an external homogeneous electric field is frozen in the material,i.e. E=(E,0,0) where E is a parameter. In this case the BTE, with the BGK collisional operator (6), reduces to a linear first order PDE ,whose solution is [5]

$$f_B(\mathbf{k}) = \sqrt{\frac{\pi}{2}} exp\left(\frac{1}{2}\eta^2 - \eta k_x\right) exp\left(-\frac{k_y^2 + k_z^2}{2}\right) erfc\left[-\frac{1}{\sqrt{2}}\left(k_x - \eta\right)\right]$$

where

$$\eta = -\frac{\sqrt{m^\star k_B T_0}}{qE\tau} \quad .$$

Let us introduce the moments of the distribution function:

$$V_i = \frac{1}{n}\int_\Omega f v_i d^3\mathbf{k} \tag{7}$$

$$W = \frac{1}{n}\int_\Omega \varepsilon(\mathbf{k}) f d^3\mathbf{k}, \tag{8}$$

$$S_i = \frac{1}{n}\int_\Omega f v_i \varepsilon(\mathbf{k}) d^3\mathbf{k}. \tag{9}$$

which represent respectively the average velocity, the average energy and the energy-flux. By using the distribution function f_B after some calculations one obtains

$$V_x = \mu E \quad , \mu = \frac{q}{m^\star}\tau \tag{10}$$

$$W = \frac{3}{2} k_B T_0 + m^\star V_x^2, \tag{11}$$

$$S_x = V_x\left(\frac{5}{2} k_B T_0 + 3m^\star V_x^2\right) \quad . \tag{12}$$

If τ is constant, the average velocity (10) is inconsistent with the velocity saturation phenomena, observed experimentally for high electric fields.
In order to overcome to this difficulty, an electric field dependent relaxation time has been considered [7] :

$$\tau(E) = \frac{m^\star}{q}\mu(E) = \frac{m^\star}{q}\frac{2\mu_0}{1 + \sqrt{1 + 4\left(\frac{\mu_0 E}{v_0}\right)^2}}$$

where the parameters μ_0 and v_0 are obtained as **fitting parameters** with MC data in bulk silicon. Now V_x coincides with the MC data, but the average energy (11) and the energy-flux (12) differ completely respect to the MC data, as shown in Figs. 1 and 2. For the quasi-parabolic case, a similar procedure could be applied, supposing that a new fitting function τ(E) could be determined.
An alternative is to consider the relaxation time as function of the momentum. In fact the collisional operator (3) can be easily written as :

$$\begin{aligned} Q[f] &= \int_{\mathbb{R}^3} w(\mathbf{k}', \mathbf{k}) f(\mathbf{k}')\, d\mathbf{k}' - f(\mathbf{k}) \int_\Omega w(\mathbf{k}, \mathbf{k}')\, d\mathbf{k}' \\ &= \int_{\mathbb{R}^3} w(\mathbf{k}', \mathbf{k}) f(\mathbf{k}')\, d\mathbf{k}' - \frac{1}{\tau(\mathbf{k})} f(\mathbf{k}) \end{aligned} \tag{13}$$

where

$$\frac{1}{\tau(\mathbf{k})} = \int_\Omega w(\mathbf{k}, \mathbf{k}')\, d\mathbf{k}' \tag{14}$$

is the total scattering rate, which is not a fitting parameter but it is explicitly determined by the physical transition probability eq.(3). The equation (13) suggests us the following approximation :

$$Q[f] \sim Q_\mathbf{k}[f] = -\frac{f(\mathbf{k}) - F_0(\mathbf{k})}{\tau(\mathbf{k})} \tag{15}$$

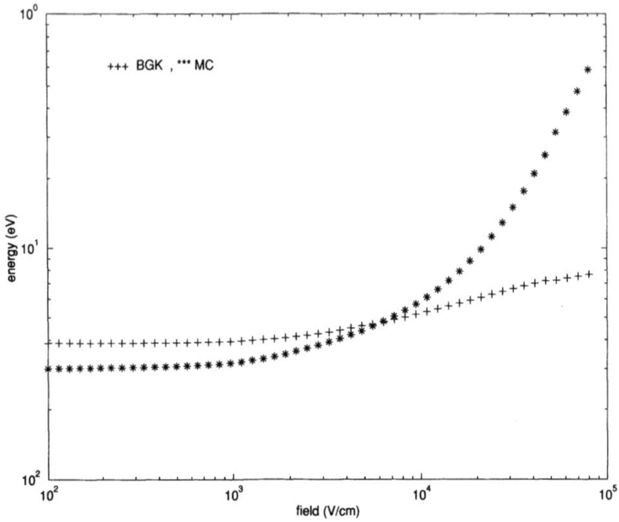

Fig. 1. The BGK energy (11) (with +++) versus the electric field and MC data (with ***), in the parabolic band approximation

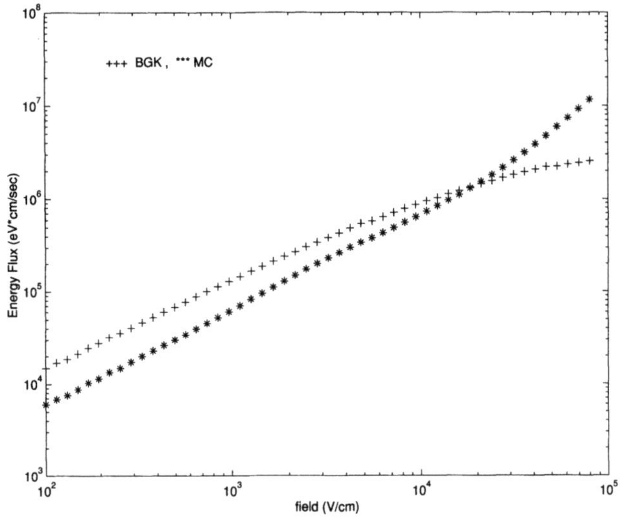

Fig. 2. The BGK energy-flux (12) (with +++) versus the electric field and MC data (with ***), in the parabolic band approximation

where $F_0(\mathbf{k})$ is again the lattice maxwellian, i.e.

$$F_0(\mathbf{k}) = \exp\left(\alpha_0 - \frac{\varepsilon}{k_B T_0}\right) \tag{16}$$

with another normalization constant α_0, chosen in such a way to maintain the mass conservation, i.e.

$$\int_\Omega Q_\mathbf{k}[f] d\mathbf{k} = \int_\Omega \frac{F_0(\mathbf{k}) - f(\mathbf{k})}{\tau(\mathbf{k})} d\mathbf{k} = 0 \,. \tag{17}$$

In [10] we proved that this operator fulfills an H-theorem. The BTE with the collisional operator (15) reduces, for bulk silicon, to a linear first order partial differential equation, whose solution, for an electric $\mathsf{E} = (E,0,0)$, is:

$$f(\mathbf{k}) = \frac{1}{E} \int_{k_x}^{+\infty} \frac{F_0(\eta, k_y, k_z)}{\tau(\eta, k_y, k_z)} \quad \exp\left[-\frac{1}{E} \int_{k_x}^{\eta} \frac{d\beta}{\tau(\beta, k_y, k_z)}\right] d\eta \quad . \tag{18}$$

Since F_0 depends on f via eq.(17), the previous equation is an highly nonlinear integral equation to be solved numerically by quadratures.

Simulation results and conclusions

We have tested the BGK solution eq.(18), in the *quasi-parabolic* band approximation, with our MC code [14]. Silicon was at room temperature (T_0=300^0 K) and a homogeneous electric field was frozen in the material along the x direction (E=100 div 120,000 V/cm): we gathered statistics after the transient regime, i.e. our simulation results are valid in the stationary regime. We have evaluated numerically the moments (7)-(9) by using eq.(18), and the results are shown in Figs. 3,4,5. The energy-flux MC data are very well fitted by the corresponding data obtained by using the BGK solution, as shown in Fig. 5. From Figs. (3) and (4) we notice that the BGK model is able to capture the drift velocity and the

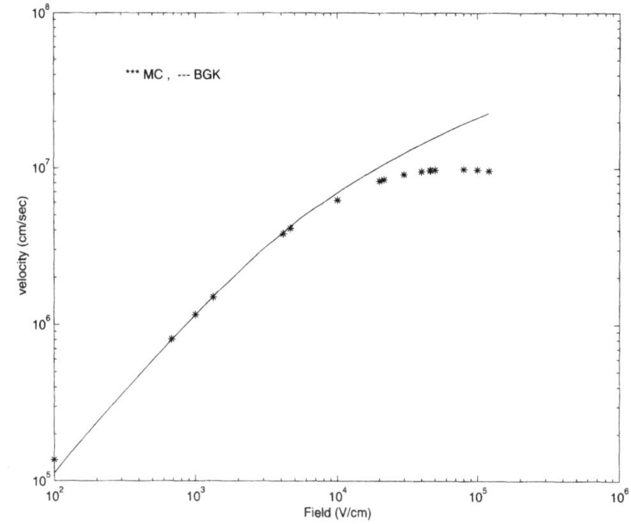

Fig. 3. The mean velocity versus electric field obtained with the BGK model eq.(18) (solid line) and MC data (with ***), in the quasi-parabolic band approximation

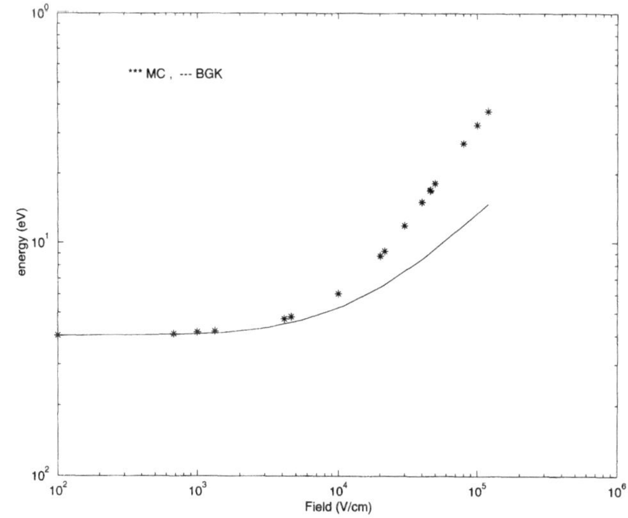

Fig. 4. The mean energy versus electric field obtained with the BGK model eq.(18) (solid line) and MC data (with ***), in the quasi-parabolic band approximation

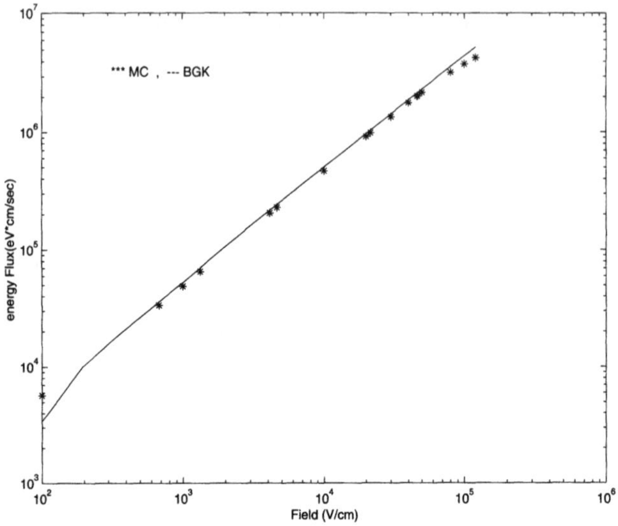

Fig. 5. The energy-flux versus electric field obtained with the BGK model eq.(18) (solid line) and MC data (with ***), in the quasi-parabolic band approximation

average energy with a good precision, only for moderate electric fields ($\leq$ 30,000 V/cm), but fails for high-fields regimes. This behaviour can be explained as follows: the BGK scheme in rarified gas dynamics is based on the fact that the distribution function relaxes to a well known distribution function (i.e. the local maxwellian). In the semiconductor case, the lattice maxwellian is not the equilibrium distribution function (in the stationary and homogenous regime) due to the presence on the electric field in the BTE. Numerical experiments confirm that, for moderate electric field, the equilibrium distribution function is maxwellian, but not for high fields where anisotropies appear [15].
We guess that, for moderate electric fields, the BGK model (15) can be used for simulating real devices and efficiency comparisons can be performed with solvers of the full BTE. These will be the topics of future researches.

Acknowledgement. This work has been supported by COFIN 2002, MURST 60%, European Community's Human Potential Programme under contract HPRN-CT-2002-00282 (HYKE)

References

1. A. Majorana and R.M. Pidatella, *J. Comp. Phys.* **174**, 649 (2001)
2. J.A. Carrillo, I.M. Gamba,A. Majorana, C.-W Shu,*J.Comp. Electr.*, **2**,375 (2003)
3. C. Ertler and F. Schürrer, *J. Phys. A* **36**, 8759 (2003)
4. C. Cercignani, *The Boltzmann equation and its applications*, Springer-Verlag, Berlin, 1988.
5. S.A. Trugman and A.J. Taylor,*Phys. Rev. B* **33**, 5575, (1986)
6. H.U. Baranger and J. W. Wilkins, *Phys. Rev. B* **36**, 1436, 1987
7. A. Schenk,*Advanced physical models for silicon device simulation*, Springer-Verlag Wien, 1998
8. J.A. Carrillo, I.M. Gamba, O. Muscato and C.-W. Shu,*The IMA volumes in Mathematics and its applications*, eds. N. Ben Abdallah et al., Vol. 135, pp. 75-84, Springer (2003)
9. L.R. Logan, H.H.K. Tang and G.R. Srinivasan, *Phys. Rev. B*, 6581, (1991)
10. O. Muscato, *Physica A*,**317**, 113, (2003)
11. P.A. Markowich , C. Ringhofer and C. Schmeiser , *Semiconductor equations,* Springer-Verlag, New York (NY), 1991
12. C. Jacoboni and L. Reggiani, *Rev. Modern Phys.* **55**, 654, (1983)
13. A. Majorana, Il Nuovo Cimento **108**, 871-877, (1993)
14. O. Muscato, COMPEL, **19**, 812-828, (2000)
15. O. Muscato, Rendiconti Circ. Matematico Palermo, Serie II, Suppl. 57, 357, (1998)

Exact Solutions for the Drift-Diffusion Model of Semiconductors via Lie Symmetry Analysis

V. Romano, J. M. Sellier, M. Torrisi

Dipartimento di Matematica e Informatica, Universita' di Catania, Italy,
{romano,sellier,torrisi}@dmi.unict.it

Abstract The symmetry analysis of the drift-diffusion models for semiconductors is performed and examples of exact invariant solutions are obtained. These latter ones can be used as useful benchmarks for testing numerical codes.

1 Introduction

Simple macroscopic models widely used in engineering applications for the description of charge carrier transport in semiconductors are the *drift-diffusion* ones [1, 3, 2]. They are based on the assumption of isothermal motion and are constituted by the balance equation for electron density and the Poisson equation for the electric potential. In these models there is the presence of some arbitrary functions as the mobilities, whose expression is based on fitting of experimental data or Monte Carlo simulations.

A first symmetry analysis [4, 5, 6, 7, 8] of the drift-diffusion models has been performed in [9] for a simplified version with the use of weak equivalence classification. The most general unipolar model is has been investigated in [10] in the one dimensional case. Here we recall the symmetry classification and give some examples of invariant exact solutions for suitable doping profiles and mobilities (for a similar study of the energy-transport model see [11, 12]).

Apart from the mathematical interest, the obtained results are of a certain relevance for the applications because they furnish example of benchmark solutions useful for testing the numerical code simulating electron devices by the drift-diffusion model.

2 The mathematical model

The unipolar drift-diffusion model for electrons in semiconductors is given by the balance equation for electron (or hole) density coupled to the Poisson equation for the electric potential [1, 2, 3]

$$\frac{\partial n}{\partial t} + \nabla \cdot \mathbf{J} = 0, \quad \lambda^2 \Delta \Phi = n - c(x) \tag{1}$$

where n is the electron density, $\mathbf{J}$ the electron momentum density, λ^2 the dielectric constant divided by the elementary charge, Φ the electric potential and $c(x)$ the doping concentration that is a function of the position x. Δ is the Laplacian operator and ∇ the divergence operator. The drift-diffusion models can be considered as a simplified macroscopic description of charge transport in semiconductors. Their theoretical foundation is based on the moment method applied to the Boltzmann transport equation for charge carriers.

The system (1) is supplemented by a constitutive relation for the momentum density $\mathbf{J}$, which is expressed as the sum of a diffusion and a drift term as $\mathbf{J} = K\nabla n + \mu n \nabla \Phi$ where K is the diffusion coefficient and μ the mobility.

It is usually assumed that K and μ are related by the Einstein relation $K = -U_0\mu$ where $U_0 = \frac{k_B T_L}{e}$ is the (constant) thermal potential with k_B, T_L, e Boltzmann constant, lattice temperature kept at equilibrium, absolute value of the elementary charge, respectively. The mobility μ is considered to be a function of the modulus $|E|$ of the electric field $\mathbf{E} = -\nabla \Phi$, that is $\mu = \mu(|E|)$. In more sophisticated approaches also a dependence on the donor and acceptor concentration is taken into account. The explicit expressions of $\mu(|E|)$ are obtained in the existing drift-diffusion models by a fitting of Monte Carlo simulations or experimental data.

Case	Forms of $c(x)$ and $\mu(E^2)$	Extensions of $\mathcal{L}_P$
I	$c(x) = c_0,\ \mu = \mu(E^2)$	$X_1 = \frac{\partial}{\partial x}$
II	$c(x) = \frac{c_0}{(k_0 x + k_1)^2},\ \mu = \mu_0$	$X_1 = 2k_0 t \frac{\partial}{\partial t} + (k_0 x + k_1)\frac{\partial}{\partial x} - 2nk_0 \frac{\partial}{\partial n} - k_0 E \frac{\partial}{\partial E}$
III	$c(x) = c_0,\ \mu = \mu_0$	$X_1 = B(t)\frac{\partial}{\partial x} - \frac{B'(t)}{\mu_0}\frac{\partial}{\partial E}$

Table 1. Lie group classification for the drift-diffusion model

3 The symmetry classification in the one-dimensional case

In the one dimensional case the Poisson equation can be rewritten in terms of the relevant component of the electric field. The complete system becomes

$$\frac{\partial n}{\partial t} + \frac{\partial J}{\partial x} = 0, \quad -\lambda^2 \frac{\partial E}{\partial x} = n - c(x), \tag{2}$$

$$J = -\mu(|E|)\, U_0 \,\frac{\partial n}{\partial x} - \mu(|E|)\, n\, E. \tag{3}$$

We will get the symmetry classification of the systems (2)-(3) by the infinitesimal Lie method [4, 5, 6, 7, 8]. Our goal is to determine the functional forms of mobility and doping profile for which the system (2)-(3) does admit symmetries.

Let us consider the one-parameter Lie group of infinitesimal transformations in (x, t, n, E)-space given by

$$\hat{t} = t + \epsilon\, \xi^1(x,t,n,E) + \mathcal{O}(\epsilon^2), \quad \hat{x} = x + \epsilon\, \xi^2(x,t,n,E) + \mathcal{O}(\epsilon^2), \tag{4}$$

$$\hat{n} = n + \epsilon\, \eta^1(x,t,n,E) + \mathcal{O}(\epsilon^2), \quad \hat{E} = E + \epsilon\, \eta^2(x,t,n,E) + \mathcal{O}(\epsilon^2), \tag{5}$$

where ϵ is the group parameter and the associated Lie algebra $\mathcal{L}$ is the set of vector fields of the form $X = \xi^1 \frac{\partial}{\partial t} + \xi^2 \frac{\partial}{\partial x} + \eta^1 \frac{\partial}{\partial n} + \eta^2 \frac{\partial}{\partial E}$.

One requires that the transformations (4)-(5) leave invariant the set of solutions of the system (2)-(3). In other words, one requires that the transformed system has the same form as the original one.

The analysis of the invariance conditions leads to the following classification (for the details see [10]). The *principal Lie algebra* $\mathcal{L}_P$ is spanned by

$$X = \frac{\partial}{\partial t}. \tag{6}$$

In the cases summarized in Table 1, one has also the following extensions whose generator is indicated by X_1.

4 Reduction to ODE systems

One of the advantages of the symmetry analysis is the possibility of finding solutions of the original system of PDEs by solving a system of ODEs. These systems of ODEs, called *reduced systems*, are obtained by introducing suitable similarity variables, determined as invariant functions with respect to the infinitesimal generator of the symmetry transformation.

On the basis of the infinitesimal generators reported in the previous section, we have the following reduced systems.

4.1 Case 1

The generator is $c_1 \frac{\partial}{\partial t} + c_2 \frac{\partial}{\partial x}$ with c_1 and c_2 arbitrary real constant. The invariance conditions lead to $\frac{dt}{c_1} = \frac{dx}{c_2}$ and gives the similarity variable $\sigma = c_2 t - c_1 x$. Since for $c_1 = 0$ one has the homogeneous case, that is with solutions depending only on t, putting $\alpha = \frac{c_2}{c_1}$ $(c_1 \neq 0)$, one gets $\sigma = x - \alpha t$ and the similarity solutions

$$n = \tilde{n}(\sigma), \quad E = \tilde{E}(\sigma) \tag{7}$$

where $\tilde{n}$ and $\tilde{E}$, after suppressing the symbol "tilde" for simplicity, are solutions of the reduced system

$$\alpha n + \mu(E^2) U_0 n' + \mu(E^2) n E = \text{const} = k, \quad \lambda^2 E' = c_0 - n \tag{8}$$

4.2 Case 2

In this case the infinitesimal generator is

$$X = (2k_0 t + c_1)\frac{\partial}{\partial t} + (k_0 x + k_1)\frac{\partial}{\partial x} - 2nk_0\frac{\partial}{\partial n} - k_0 E\frac{\partial}{\partial E} \tag{9}$$

The invariance conditions reads

$$\frac{dt}{2k_0 t + c_1} = \frac{dx}{k_0 x + k_1} = -\frac{dn}{2k_0 n} = -\frac{dE}{k_0 E} \tag{10}$$

from which one obtains the following similarity variable $\sigma(x,t) = \frac{2k_0 t + c_1}{(k_0 x + k_1)^2}$ and the similarity solution

$$n = \frac{N(\sigma)}{(k_0 x + k_1)^2}, \quad E = \frac{F(\sigma)}{k_0 x + k_1} \tag{11}$$

where $N(\sigma)$ and $F(\sigma)$ are solution of the equations

$$\lambda^2[F(\sigma) + 2k_0 F'(\sigma)\sigma] + c_0 - N(\sigma) = 0 \tag{12}$$
$$2N'(\sigma) - 2\,\mu_0 k_0 U_0\,(3 + 2\sigma)\,\frac{d}{d\sigma}(N\sigma) - 2\mu_0\sigma\frac{d}{d\sigma}(NF) + 3\mu_0 NF = 0 \tag{13}$$

4.3 Case 3

In this case the generator is

$$X = c_1\frac{\partial}{\partial t} + B(t)\frac{\partial}{\partial x} - \frac{B'(t)}{\mu_0}\frac{\partial}{\partial E} \tag{14}$$

with c_1 arbitrary real constant. The invariance conditions read

$$\frac{dt}{c_1} = \frac{dx}{B(t)} = -\frac{\mu_0 dE}{B'(t)} \tag{15}$$

from which we get the following similarity variable $\sigma = G(t) - c_1 x$ where $G(t) = \int B(t)$. In this case, the variables n and E are given by

$$n(x,t) = N(\sigma), E(x,t) = -\frac{G'(t)}{\mu_0 c_1} + F(\sigma) \tag{16}$$

After some computation, the Poisson equation and the density balance equation read

$$-\lambda^2 c_1 F'(\sigma) = c_0 - N(\sigma) \tag{17}$$
$$\frac{d}{d\sigma}\left[-c_1 U_0 N'(\sigma) + N(\sigma)F(\sigma)\right] = 0 \tag{18}$$

5 Examples of exact solutions

Here we give some example of exact solutions, obtained by solving the reduced system showed in the previous section (for more details see [10]). The cases refer to those reported in the table.

5.1 Case 2

For bulk silicon $c_0 = 0$, if the relation $5/2\mu_0 = \lambda^2$ is valid, by setting $k_0 = 5/2$ we have the exact solution of the reduced system (12)-(13)

$$N(\sigma) = \frac{1}{\sigma}, \quad F(\sigma) = \frac{2}{7\,\mu_0\sigma} + C\,\sigma^{5/2} \tag{19}$$

where C is an arbitrary constant and $\sigma(x,t) = 4\frac{5\,t + c_1}{(5\,x + 2\,k_1)^2}$.

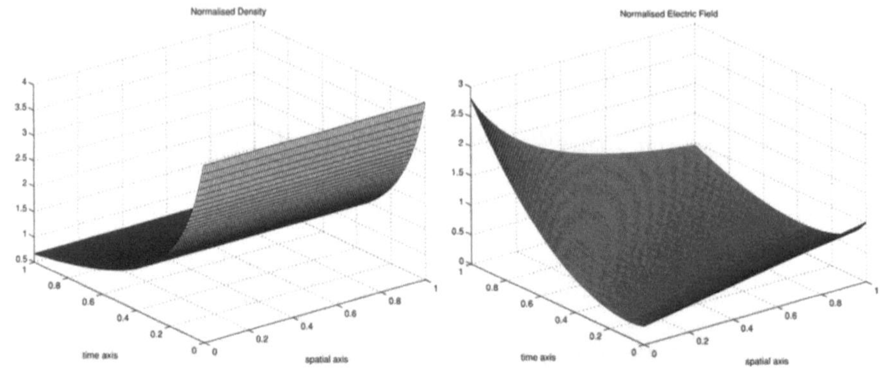

Fig. 1. Plot of the solution 20

In terms of the variable n and E one has

$$n = \frac{4}{5t + c_1}, \quad E = \frac{2}{5x + 2k_1}\left[\frac{(5x + 2k_1)^2}{14\mu_0(5t + c_1)} + 32C\left(\frac{5t + c_1}{4}\right)^{5/2}\right] \tag{20}$$

In fig. 1 the previous solution is plotted for $c_1 = 1$, $k_1 = 1$ and $C = 1/32$.

Another class of solutions has been obtained also for $c_0 \neq 0$ as follows. First from (12)-(13) we get a single second order ODE for F

$$\begin{aligned} &4\lambda^2\, k_0\, \sigma\, F''\left[-\mu_0\sigma F + 1 - 3\mu_0 k_0 U_0\sigma - 2\mu_0 k_0 U_0\sigma^2\right] \\ &+\lambda^2 F'\left[2 + 4k_0 - 2\mu_0\sigma c_0/\lambda^2 - 24\mu_0 k_0^2 U_0\lambda^2 - 6\mu_0 k_0 U_0\lambda^2\sigma - 16\mu_0 k_0^2 U_0\right. \\ &\left.-4\mu_0 k_0 U_0\sigma^2 - 4\mu_0\sigma F + 2\mu_0\sigma k_0 F - 4\mu_0\sigma^2 k_0 F'\right] \\ &+\mu_0 F\left[3\lambda^2 - 2\mu_0 k_0 U_0\lambda^2(3 + 2\sigma) + 3c_0\right] - 2\mu_0 k_0 U_0 c_0(3 + 2\sigma) = 0. \end{aligned} \tag{21}$$

If without loss of generality we set $k_0 = 2$ and look for solutions of (21) linear in σ, one finds

$$n(x,t) = \frac{1}{(2x + k_1)^2}\left\{\lambda^2\left[-80\frac{U_0(4t + c_1)}{(2x + k_1)^2} + \frac{3\,(80U_0\lambda^2 - c_0)}{13\lambda^2} + c_0\right]\right\}, \tag{22}$$

$$E(x,t) = \frac{1}{(2x + k_1)}\left[-16\frac{U_0(4t + c_1)}{(2x + k_1)^2} + \frac{3\,(80U_0\lambda^2 - c_0)}{13\lambda^2}\right], \tag{23}$$

provided that

$$\mu_0 = \frac{2704\lambda^2 U_0}{3\,(4512\lambda^4 U_0^2 + 116\lambda^2 U_0 c_0 - 3c_0^2)}.$$

The latter requires, in order to have a positive low field mobility μ_0,

$$c_1^* < c_0 < c_2^*$$

c_1^*, c_2^* being the two (real) roots of $4512\lambda^4 U_0^2 + 116\lambda^2 U_0 c_0 - 3c_0^2 = 0$.

In fig. 2 the solution (22)-(23) is plotted for $c_0 = 1$, $c_1 = 1$ and $k_1 = 5$.

5.2 Case 3

Let us combine the equations (17),(18) into the single relation

$$c_1^2 U_0\lambda^2 F'' - (c_1\lambda^2 F' + c_0)F + J_0 = 0, \tag{24}$$

where the constant J_0 arises from a first integration of (18). For bulk silicon, $c_0 = 0$, if $J_0 = 0$ the previous equation becomes $F' = \frac{F^2}{2c_1 U_0} + F_0$ with F_0 an integration constant. The general solution of this latter equation is $F(\sigma) = -\gamma\tanh\left[\frac{\gamma(\sigma + c_2)}{2\,c_1\,U_0}\right]$ where $\gamma = \sqrt{2c_1 U_0 F_0}$ and c_2 a further integration constant.

For $N(\sigma)$ one has $N(\sigma) = \lambda^2 c_1\left(\frac{F^2}{2c_1 U_0} + F_0\right)$. For periodic $G(t)$ the solution is periodic in time.

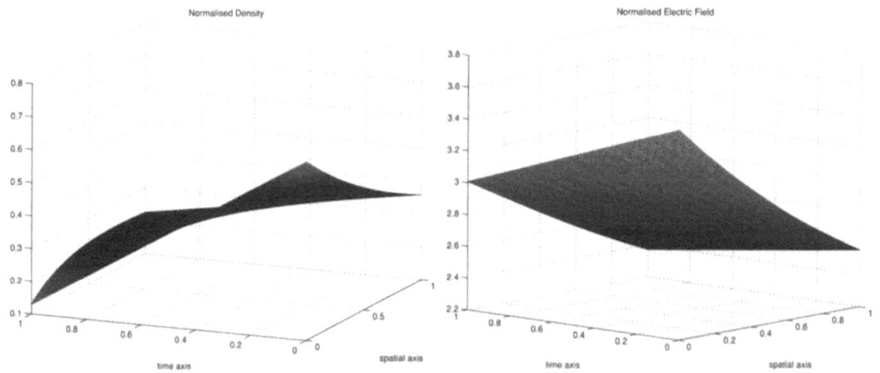

Fig. 2. Plot of the solution (22)-(23)

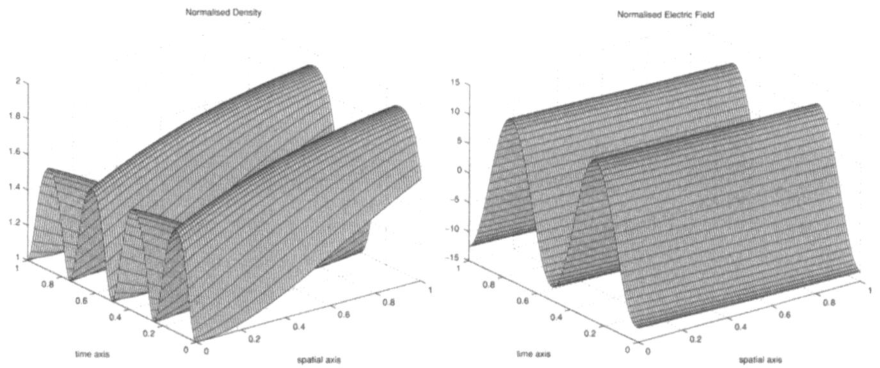

Fig. 3. Plot of the solution (25)-(26) for $c_2 = 0$, $\omega = 4\pi$, $l = 1\mu$m, $\omega l/\mu_0 = \gamma = 2$ Volt/μm. In the top figure the normalized density $n/\lambda^2 F_0$ is shown while in the bottom one the normalized electric field E/γ is reported

If, for example, l is the length of the device, with the choice $G(t) = l \sin \omega t$, $c_1 = 1$ and $F_0 > 0$, the similarity variable reads $\sigma = l \sin \omega t - x$ and the solution becomes

$$n(x,t) = \lambda^2 F_0 \left\{ 1 + \tanh^2 \left[\frac{\gamma (l \sin \omega t + x + c_2)}{2U_0} \right] \right\}, \tag{25}$$

$$E(x,t) = -\frac{\omega l \cos \omega t}{\mu_0} - \gamma \tanh \left[\frac{\gamma (L \sin \omega t + x + c_2)}{2U_0} \right]. \tag{26}$$

In fig. 3 the previous solution is plotted for $c_2 = 0$ and $\omega = 4\pi$.

Acknowledgements

The author V.R. acknowledges the financial support by M.I.U.R. (COFIN 2002 *Problemi Matematici delle teorie cinetiche*) and P.R.A. (ex 60 %). The author J.M.S acknowledges the financial support by M.I.U.R. (COFIN 2002 *Problemi Matematici delle teorie cinetiche*). The author M. T. acknowledges the support by P.R.A. (ex 60 %), by CNR through G.N.F.M. and by M.I.U.R. (Project: *Non Linear Mathematical Problems of Wave Propagation and Stability in Models of Continuous Media*).

References

1. S. Selberherr, Analysis and simulation of semiconductor devices, Springer-Verlag, Wien - New York, 1984
2. W. Hänsch, The drift-diffusion equation and its applications in MOSFET modeling, Springer-Verlag, Wien, 1991

3. P. Markowich, C. A. Ringhofer and C. Schmeiser, Semiconductor equations, Springer-Verlag, Wien, 1990
4. L. V. Ovsiannikov, Group Analysis of Differential Equations, Academic Press, New York, 1982
5. P. J. Olver, Applications of Lie Groups to Differential Equations, Springer-Verlag, New York, 1986
6. G. W. Bluman and S. Kumei, Symmetries and Differential Equations, Springer-Verlag, New-York, 1989
7. N. H. Ibragimov, CRC Hanbook of Lie Group Analysis of Differential Equations, CRC Press, Boca Raton FL, 1994
8. W. I. Fushchych and W. M. Shtelen, Symmetry Analysis and Exact Solutions of Nonlinear Equations of Mathematical Physics, Kluwer, Dordrecht , 1993
9. V. Romano and M. Torrisi, Application of weak equivalence transformations to a group analysis of a drift-diffusion model, J. Phys. A (1999) 32 7953-7963
10. V. Romano, J. M. Sellier and M. Torrisi, Symmetry analysis and exact invariant solutions for the drift-diffusion models of semiconductors, J. Physics A (2004) 341 62–72
11. V. Romano and A. Valenti, Symmetry analysis and exact solutions for a class of energy-transport models of semiconductors, J. Phys. A (2002) 35 1751-1762
12. V. Romano and A. Valenti, Exact invariant solutions for a class of energy-transport models of semiconductors in the two dimensional case, to appear in Communications in Nonlinear Science and Numerical Simulations (2003)

Different Extrapolation Strategies in Implicit Newmark-Beta Schemes for the Solution of Electromagnetic High-Frequency Problems*

A. Skarlatos[1], M. Clemens[2], and T. Weiland[1]

[1] Technische Universität, Schlossgartenstr. 8, Darmstadt, Germany,
{skarlatos,thomas.weiland}@temf.tu-darmstadt.de
[2] Helmut Schimdt Universität, Holstenhofweg 85, 22043 Hamburg, Germany, m.clemens@hsu-hh.de

Abstract The construction of good starting values for iterative solvers applied an implicit Newmark-Beta time stepping scheme for the simulation of transient electromagnetic problems, can significantly improve the convergence of those solvers, and consequently decrease the total solution time. Especially in high frequency problems, this is an important issue since simple extrapolation schemes, like the zero-order Taylor approximation, do not seem to bring significant improvements on the convergence of the solver. Different extrapolation strategies applied to high frequency problems simulated using perfectly matched boundary conditions are compared in terms of their effect in the speed-up of the solver.

1 Introduction

For the simulation of high frequency problems in Time Domain (TD) systems of ordinary differential equations are derived by volume discretization techniques like the Finite Elements Method (FEM) [1], or the Finite Integration Technique (FIT) [2]. Implicit time stepping schemes as the popular conservative Newmark-Beta scheme applied to these systems express the field in each time step in the form of an algebraic system of linear equations. These linear systems are symmetric and positive definite and they can be efficiently solved using iterative solvers like the Conjugate Gradient method (CG). The choice of a good starting value for such solvers is important for the overall efficiency of the scheme since it can reduce the time needed for the solution of the equation system in each time step. An initial approximation for the solution can be constructed using information obtained from the solution of the system in previous time steps.

Extrapolation strategies based on a Taylor expansion have been already successfully applied in quasi-static electric and magnetic transient simulations [3]. The Subspace Projection Extrapolation Scheme (SPE) presented in [4] achieves an optimal combination of different approximations for the initial value by exactly solving the original problem projected onto a subspace. These two schemes together with some other techniques based on the splitting of the system matrix in diagonal and non-diagonal terms, and the projection of the excitation vector on a subspace constructed from the previous solutions are examined in this paper for the high frequency formulation. The considered techniques are tested in two important cases, the resonator and the scattering problem. For the last one, the introduction of an Absorbing Boundary Condition (ABC) is necessary for the truncation of the computational domain. This is achieved by the utilization of Perfectly Matched Layers (PML), which has been proven a very efficient technique in the case of leapfrog formulation [6]. Its application however in implicit Newmark-Beta formulations becomes cumbersome due to the dispersive nature of the PML materials which introduces convolution terms into the wave equation. To model the PML material correctly, modifications to the usual matrix formulation of the closed problem have to be carried out.

2 The FIT-TD Implicit Formulation

The spatial discretization of the Maxwell equations using the FIT scheme leads to the following discrete form of the wave equation

*A. Skarlatos is supported by the graduate student program "Modelierung, Simulation und Optimierung in Ingenieurswissenschaften" of the Deutsche Forschungsgemeinschaft (DFG) under grant GK-GRK 853.

$$\mathbf{M}_{\varepsilon}\frac{d^2}{dt^2}\widehat{\mathbf{e}} + \mathbf{M}_{\kappa}\frac{d}{dt}\widehat{\mathbf{e}} + \widetilde{\mathbf{C}}\mathbf{M}_{\mu}^{-1}\mathbf{C}\widehat{\mathbf{e}} = -\frac{d}{dt}\widehat{\widehat{\mathbf{j}}}_{e}, \tag{1}$$

where $\mathbf{C}, \widetilde{\mathbf{C}}$ are topological matrices which resemble the continuous topological operator curl and $\mathbf{M}_{\varepsilon}, \mathbf{M}_{\mu}^{-1}$, and $\mathbf{M}_{\kappa}$ are material matrices [2]. For the sake of simplicity we shall call $\mathbf{M} = \mathbf{M}_{\varepsilon}$, $\mathbf{D} = \mathbf{M}_{\kappa}$, $\mathbf{K} = \widetilde{\mathbf{C}}\mathbf{M}_{\mu}^{-1}\mathbf{C}$ and $\mathbf{f} = -\frac{d}{dt}\widehat{\widehat{\mathbf{j}}}_{e}$. The wave equation is then simplified into

$$\mathbf{M}\frac{d^2}{dt^2}\widehat{\mathbf{e}} + \mathbf{D}\frac{d}{dt}\widehat{\mathbf{e}} + \mathbf{K}\widehat{\mathbf{e}} = \mathbf{f}. \tag{2}$$

Discretization of the above equation in time using the Newmark-Beta scheme [5] yields

$$\begin{aligned}\left(\mathbf{M} + \gamma\Delta t\mathbf{D} + \beta\Delta t^2\mathbf{K}\right)\widehat{\mathbf{e}}^{(n+1)} &= \left(2\mathbf{M} + (1-2\gamma)\Delta t\mathbf{D} - (0.5 - 2\beta + \gamma)\Delta t^2\mathbf{K}\right)\widehat{\mathbf{e}}^{(n)} \\ &+ \left(\mathbf{M} + (1-\gamma)\Delta t\mathbf{D} - (0.5 + \beta - \gamma)\Delta t^2\mathbf{K}\right)\widehat{\mathbf{e}}^{(n-1)} \\ &+ \Delta t^2\mathbf{f}^{(n)}.\end{aligned} \tag{3}$$

An unconditionally stable, second-order accurate scheme is obtained if $\beta \geq 0.25$. The choice of parameters $\beta = 0.25, \gamma = 0.5$ guarantees that the time discrete electromagnetic energy is conserved in each new time step [2]. For this reason the Newmark-Beta scheme is the most widely used implicit time integrator for electromagnetic high frequency problems [1]. The system matrix $\mathbf{M} + \gamma\Delta t\mathbf{D} + \beta\Delta t^2\mathbf{K}$ is symmetric and positive definite and it can be efficiently solved by the Conjugate Gradient method (CG).

Extrapolation Strategies

The construction of a start value for the solver in order to achieve a faster convergence can be done using different extrapolation strategies.

Taylor Expansion

The simplest way to extrapolate the field $\widehat{\mathbf{e}}$ at the $n+1$ timestep is to express the field in the form of a Taylor series [3]

$$\widehat{\mathbf{e}}^{(n+1)} = \widehat{\mathbf{e}}^{(n)} + \frac{d}{dt}\widehat{\mathbf{e}}^{(n)}\Delta t + \frac{d^2}{dt^2}\widehat{\mathbf{e}}^{(n)}\frac{\Delta t^2}{2!} + \dots. \tag{4}$$

Keeping the first terms of the series only we obtain a zero-th order extrapolation

$$\widehat{\mathbf{e}}_0^{(n+1)} \approx \widehat{\mathbf{e}}^{(n)}, \tag{5}$$

or the first order approximation

$$\widehat{\mathbf{e}}_0^{(n+1)} \approx \widehat{\mathbf{e}}^{(n)} + \frac{d}{dt}\widehat{\mathbf{e}}^{(n)}\Delta t, \tag{6}$$

and approximating the first derivative with a higher order finite differences scheme [3] we get

$$\frac{d}{dt}\widehat{\mathbf{e}}^{(n)} \approx \frac{1}{\Delta t}\left(\frac{3}{2}\widehat{\mathbf{e}}^{(n)} - 2\widehat{\mathbf{e}}^{(n-1)} + \frac{1}{2}\widehat{\mathbf{e}}^{(n-2)}\right). \tag{7}$$

We would expect that the more terms we take into account, the more accurate approximation of the field at the $n+1$ time step we achieve. However this is not the case here since we are not dealing with exact arithmetics. Given that the field values in the previous time steps are just numerical approximations of the real ones and since we are working with finite precision arithmetic, we are not able to know a-priory which order will provide the best approximation. A quite efficient way to overcome this problem is to combine the different extrapolated values obtained by applying different orders of Taylor expansion, in order to get an optimal approximation for the solution. This is the main idea behind the Subspace Projection Extrapolation (SPE) technique [4].

The Subspace Projection Extrapolation (SPE) Scheme

In the SPE scheme the extrapolated start vectors $\widehat{\mathbf{e}}_i^{(n+1)}, i = 1, ..., m$ obtained by different orders of Taylor expansion are linearly combined to yield an optimal start vector [4]. To do this an orthonormalization process is applied to the extrapolation vectors (e.g. by applying a Modified Gram-Schmidt (MGS) algorithm) to get a set of orthonormal vectors $\mathbf{v}_1, ...\mathbf{v}_{\tilde{m}}$. These vectors define a basis of a subspace, on which the original problem is projected and solved. Consider the projection operator

$$\mathbf{V} := \{\mathbf{v}_1|...|\mathbf{v}_{\tilde{m}}\} \in \mathbb{R}^{N\times\tilde{m}}, \tag{8}$$

where N is the size of the original problem. The problem is projected onto the above constructed subspace and is solved according to

$$\mathbf{V}^T\mathbf{A}\mathbf{V}\mathbf{z} = \mathbf{V}^T\mathbf{q}, \tag{9}$$

where $\mathbf{A} = \mathbf{M} + \gamma\Delta t\mathbf{D} + \beta\Delta t^2\mathbf{K}$, $\mathbf{q}$ the right hand side (excitation) vector of (3), and $\mathbf{z} \in \mathbb{R}^{\tilde{m}}$. The time step index was omitted. After solving the reduced problem, the vector $\mathbf{z}$ is extracted back into the original space and set as start value for the original problem

$$\widehat{\mathbf{e}}_{0,SPE} = \mathbf{V}\mathbf{z}. \tag{10}$$

Diagonal Extraction Approximation

Let us return to the Newmark formulation for the wave equation

$$\left(\mathbf{M} + 0.5\Delta t\mathbf{D} + \beta\Delta t^2\mathbf{K}\right)\widehat{\mathbf{e}}^{(n+1)} = \mathbf{q}^{(n)}, \tag{11}$$

where γ was set to 0.5. The matrices $\mathbf{M}$ and $\mathbf{D}$ are diagonal. Bringing the non-diagonal matrix $\mathbf{K}$ on the right hand side (rhs) we obtain

$$(\mathbf{M} + 0.5\Delta t\mathbf{D})\,\widehat{\mathbf{e}}^{(n+1)} = \mathbf{q}^{(n)} - \beta\Delta t^2\mathbf{K}\widehat{\mathbf{e}}^{(n+1)}. \tag{12}$$

Since the $\widehat{\mathbf{e}}^{(n+1)}$ is not known, we can approximate it on the rhs with the field value in the previous timestep, namely $\widehat{\mathbf{e}}^{(n)}$

$$(\mathbf{M} + 0.5\Delta t\mathbf{D})\,\widehat{\mathbf{e}}^{(n+1)} \approx \mathbf{q}^{(n)} - \beta\Delta t^2\mathbf{K}\widehat{\mathbf{e}}^{(n)}. \tag{13}$$

The matrix on the left hand side is diagonal, and it can be inverted very easily and we get

$$\widehat{\mathbf{e}}_0^{(n+1)} := (\mathbf{M} + 0.5\Delta t\mathbf{D})^{-1}\left(\mathbf{q}^{(n)} - \beta\Delta t^2\mathbf{K}\widehat{\mathbf{e}}^{(n)}\right)$$

as initial guess for (3).

Subspace Projection of the Right Hand Side

The solution of the system in each time step is actually equivalent to applying the inverse matrix on the excitation vector. If we could thus express the excitation vector as a linear combination of vectors on which the operation of the inverse matrix was known, we could construct a good approximation to the solution by combining those "inverted" vectors. Let us consider once again the equation system arising from the Newmark formulation

$$\mathbf{A}\widehat{\mathbf{e}}^{(n+1)} = \mathbf{q}^{(n+1)}, \tag{14}$$

where both system matrix and excitation vectors are abbreviated by the terms $\mathbf{A}$ and $\mathbf{q}^{(n+1)}$. The solution of that equation can be written as

$$\widehat{\mathbf{e}}^{(n+1)} = \mathbf{A}^{-1}\mathbf{q}^{(n+1)}. \tag{15}$$

Normalizing the rhs vector to $\mathbf{v}_1 := \mathbf{q}^{(n+1)}/\parallel\mathbf{q}^{(n+1)}\parallel$, we also have

$$\mathbf{A}^{-1}\mathbf{v}_1 = \frac{\widehat{\mathbf{e}}^{(n+1)}}{\parallel\mathbf{q}^{(n+1)}\parallel}. \tag{16}$$

The excitation vector for the next time step now can be written as the sum of a vector parallel to $\mathbf{v}_1$, and a remaining "residual" term $\mathbf{r}^{(n+2)}$ with

$$\mathbf{q}^{(n+2)} = \alpha_1 \mathbf{v}_1 + \mathbf{r}^{(n+2)}, \tag{17}$$

where $\alpha_1 := \mathbf{v}_1^T \mathbf{q}^{(n+2)}$ arises from the projection of the $\mathbf{q}^{(n+2)}$ to the $\mathbf{v}_1$ vector. The thus defined "residual" vector $\mathbf{r}^{(n+2)}$ should not be confused with the CG residual vector. The solution for the next time step can be thus written as

$$\widehat{\mathbf{e}}^{(n+2)} = \alpha_1 \mathbf{A}^{-1} \mathbf{v}_1 + \mathbf{A}^{-1} \mathbf{r}^{(n+2)}, \tag{18}$$

and since the vector $\mathbf{A}^{-1}\mathbf{v}_1$ is known from the previous time step, we can approximate $\widehat{\mathbf{e}}^{(n+2)}$ to get an initial guess in the CG solver according to

$$\widehat{\mathbf{e}}_0^{(n+2)} := \alpha_1 \mathbf{A}^{-1} \mathbf{v}_1 = (\alpha_1 / \parallel \mathbf{q}^{(n+1)} \parallel) \widehat{\mathbf{e}}^{(n+1)}, \tag{19}$$

With this start vector we get the new time step solution $\widehat{\mathbf{e}}^{(n+2)}$ by solving the Newmark system.
Normalizing the residual vector as the new basis vector

$$\mathbf{v}_2 := \mathbf{r}^{(n+2)} / \parallel \mathbf{r}^{(n+2)} \parallel = (\mathbf{q}^{(n+2)} - \alpha_1 \mathbf{v}_1) / \parallel \mathbf{r}^{(n+2)} \parallel, \tag{20}$$

the vector $\mathbf{A}^{-1}\mathbf{v}_2$ can be cheaply evaluated by

$$\mathbf{A}^{-1}\mathbf{v}_2 = (\widehat{\mathbf{e}}^{(n+2)} - \alpha_1 \mathbf{A}^{-1}\mathbf{v}_1) / \parallel \mathbf{r}^{(n+2)} \parallel . \tag{21}$$

The next excitation vector $\mathbf{q}^{(n+3)}$ can now be decomposed into components in the vector subspace spanned by the normalized vectors $\mathbf{v}_1$ and $\mathbf{v}_2$ plus a new residual vector $\mathbf{r}^{(n+3)}$ with

$$\mathbf{q}^{(n+3)} = \beta_1 \mathbf{v}_1 + \beta_2 \mathbf{v}_2 + \mathbf{r}^{(n+3)}, \quad \beta_i := \mathbf{v}_i^T \mathbf{q}^{(n+3)}, \; i = 1, 2. \tag{22}$$

Since the vectors $\mathbf{A}^{-1}\mathbf{v}_1, \mathbf{A}^{-1}\mathbf{v}_2$ are already known, they can be used to yield an initial approximation for the solution

$$\widehat{\mathbf{e}}_0^{(n+3)} := \beta_1 \mathbf{A}^{-1} \mathbf{v}_1 + \beta_2 \mathbf{A}^{-1} \mathbf{v}_2. \tag{23}$$

Solving now the system for the $n+3$ time step allows us to determine the residual term $\mathbf{r}^{(n+3)}$ and so on. Continuing this procedure in the next time steps, we can increase the dimension of the constructed subspace by one in each time step. From numerical experiments it turns out that each basis vector can provide a good description of the excitation term (and consequently of the solution vector) in the proximity of the point where it was constructed. After some time steps it becomes nearly uncorrelated with them, which is expected due to the fast variation of the signals. In practice, the best performance for the proposed algorithm is obtained if the described procedure is restarted after a number of time steps.

3 The Open Problem: PML Boundary Conditions

For the sake of simplicity we shall restrict ourselves in the 2D-TM problem. The wave equation for the TM case can be written in Frequency Domain (FD)

$$\varepsilon L_2(\omega) E_z \boldsymbol{e}_z + \nabla \times \left[\overline{\overline{L}}_1(\omega)^{-1} \nabla \times (E_z \boldsymbol{e}_z) \right] = -J_z \boldsymbol{e}_z. \tag{24}$$

$\overline{\overline{L}}_1, L_2$ are the PML parameters for the 2D case given by

$$\overline{\overline{L}}_1(\omega) = \frac{s_x}{s_y} \boldsymbol{e}_x \boldsymbol{e}_x + \frac{s_y}{s_x} \boldsymbol{e}_y \boldsymbol{e}_y \tag{25}$$

$$L_2(\omega) = s_x s_y, \tag{26}$$

with

$$s_x = 1 + \frac{\sigma_x}{j\omega\varepsilon_0}, \qquad s_y = 1 + \frac{\sigma_y}{j\omega\varepsilon_0}. \tag{27}$$

The notation $\sigma_{x,y}$ summarizes the PML conductivities σ_x, σ_y in the layers at the x and y boundaries, respectively. In the rest domain (computational domain) they have zero value, so the parameters at these cells become 1 and (24) reduces to the normal wave equation. Transforming (24) into the TD we get

$$\varepsilon\left[\frac{\partial^2 E_z(t)}{\partial t^2}+\frac{\sigma_x+\sigma_y}{\varepsilon_0}\frac{\partial E_z(t)}{\partial t}+\frac{\sigma_x\sigma_y}{\varepsilon_0^2}E_z(t)\right]\mathbf{e}_z+$$
$$\nabla\times\left[\mu^{-1}\overline{\overline{L}}_1(t)*\nabla\times\left[E_z(t)\mathbf{e}_z\right]\right]=-\frac{\partial J_z(t)}{\partial t}\mathbf{e}_z, \quad (28)$$

where $\overline{\overline{L}}_1(t)$ is the transformed tensor $\overline{\overline{L}}_1(\omega)$, and $*$ denotes the convolution operator. Discretizing the above modified wave equation using the FIT discretization scheme, and after some manipulations we obtain

$$\mathbf{M}_\varepsilon\frac{d^2}{dt^2}\widehat{\mathbf{e}}+\mathbf{D}_{\sigma_1}\mathbf{M}_\varepsilon\frac{d}{dt}\widehat{\mathbf{e}}+\left(\widetilde{\mathbf{C}}\mathbf{M}_\mu^{-1}\mathbf{C}+\mathbf{D}_{\sigma_2}\mathbf{M}_\varepsilon\right)\widehat{\mathbf{e}}+\widetilde{\mathbf{C}}\mathbf{M}_\mu^{-1}\widetilde{\mathbf{L}}_1(t)\mathbf{C}*\widehat{\mathbf{e}}=-\frac{d}{dt}\widehat{\widehat{\mathbf{j}}}_e, \quad (29)$$

where $\mathbf{D}_{\sigma_1}$, $\mathbf{D}_{\sigma_2}$ are diagonal matrices which contain the discretized PML conductivities in (28) along the grid edges and

$$\widetilde{\mathbf{L}}_{1_{i,j}}(t)=\frac{\sigma_{x,y_i}-\sigma_{y,x_i}}{\varepsilon_0}e^{-\sigma_{y,x_i}t/\varepsilon_0}u(t)\delta_{i,j}, \quad (30)$$

where $u(t)$ is the Heaviside (step) function, and $\delta_{i,j}$ is the Kronecker delta. The first subscript of the σ parameter is used for the x-directed edges (i.e. σ_x for the $\sigma_{x,y}$ and σ_y for the $\sigma_{y,x}$), and the second for the y-directed ones. Discretization of (29) using the Newmark-Beta scheme and for $\gamma=0.5$ gives

$$\begin{aligned}\left(\mathbf{M}+0.5\Delta t\mathbf{D}+\beta\Delta t^2\mathbf{K}\right)\widehat{\mathbf{e}}^{(n+1)}=&\left(2\mathbf{M}+(1-2\beta)\,\Delta t^2\mathbf{K}\right)\widehat{\mathbf{e}}^{(n)}\\&+\left(\mathbf{M}+0.5\Delta t\mathbf{D}+\beta\Delta t^2\mathbf{K}\right)\widehat{\mathbf{e}}^{(n-1)}\\&+\Delta t^2\left(\beta\mathbf{w}^{(n+1)}+(1-2\beta)\,\mathbf{w}^{(n)}+\beta\mathbf{w}^{(n-1)}\right)\\&+\Delta t^2\mathbf{f}^{(n)},\end{aligned} \quad (31)$$

with $\mathbf{M}=\mathbf{M}_\varepsilon$, $\mathbf{D}=\mathbf{M}_\kappa+\mathbf{D}_{\sigma_1}\mathbf{M}_\varepsilon$, $\mathbf{K}=\widetilde{\mathbf{C}}\mathbf{M}_\mu^{-1}\mathbf{C}+\mathbf{D}_{\sigma_2}\mathbf{M}_\varepsilon$ and $\mathbf{f}=-\frac{d}{dt}\widehat{\widehat{\mathbf{j}}}_e$, and where $\mathbf{w}^{(n)}$ denotes the convolution term at n-th time step, namely

$$\mathbf{w}^{(n)}=\left[\widetilde{\mathbf{C}}\mathbf{M}_\mu^{-1}\widetilde{\mathbf{L}}_1(t)\mathbf{C}*\widehat{\mathbf{e}}(t)\right]^{(n)}. \quad (32)$$

Based on the properties of the exponential function we can show that the convolution term can be evaluated recursively according to the following update scheme

$$\mathbf{w}^{(n)}=\widetilde{\mathbf{C}}\mathbf{M}_\mu^{-1}\mathbf{v}^{(n)}, \quad (33)$$
$$\mathbf{v}^{(n)}=\mathbf{D}_1\mathbf{v}^{(n-1)}+\mathbf{D}_2\mathbf{C}\widehat{\mathbf{e}}^{(n)}, \quad (34)$$

with

$$\mathbf{D}_{1_{i,j}}=e^{-\sigma_{y,x_i}\Delta t/\varepsilon_0}\delta_{i,j}, \quad (35)$$

$$\mathbf{D}_{2_{i,j}}=\begin{cases}\frac{\sigma_{x,y_i}-\sigma_{y,x_i}}{\sigma_{y,x_i}}\left(1-e^{-\sigma_{y,x_i}\Delta t/\varepsilon_0}\right)\delta_{i,j} & \text{, if } \sigma_{y,x_i}\neq 0\\ \frac{\sigma_{x,y_i}-\sigma_{y,x_i}}{\varepsilon_0}\Delta t\delta_{i,j} & \text{, if } \sigma_{y,x_i}=0\end{cases}. \quad (36)$$

4 Examples

As first example we shall consider the resonator structure shown in Fig. 1a. The resonator contains two dielectric cylinders, one with $\varepsilon_r=15$ and radius 1 cm, and one with $\varepsilon_r=3$ and radius 10 cm. The second cylinder has a conductivity of $\kappa=0.08$ S/m. The choice of the material parameters was made in such way, to get a system of relatively bad condition. Each side of the resonator has a length of 1 m, which brings the first resonance at 210 MHz. The structure is excited by a small dipole located at the point (0.233, 0.233, 0.489). The excitation signal is a modulated Gaussian pulse with bandwidth 150 MHz. As second example we shall examine the scattering from a dielectric cylinder with $\varepsilon_r=10$ and radius 15 cm. The configuration of the problem is shown in Fig. 1. The field is produced by an infinite electric line (TM case). The bandwidth of the excitation pulse is 400 MHz. A PML is applied for the truncation of the computational domain.

Both problems were solved with CG using the above presented extrapolation techniques for the construction of start values. The total number of iterations as well as the solution time needed for each case are compared in Table 1.

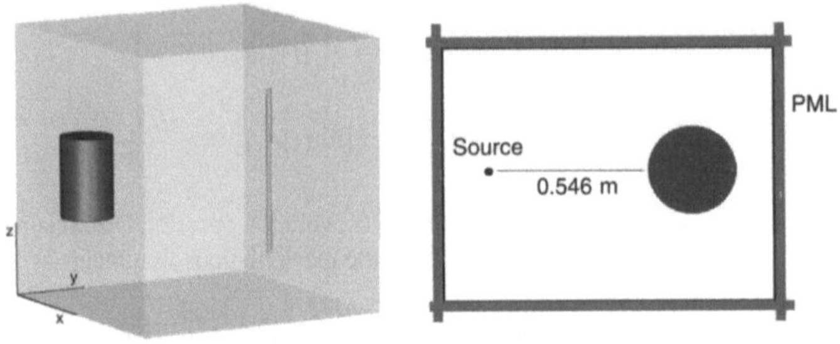

(a) (b)

Fig. 1. a) Lossy Resonator b) Scattering from dielectric cylinder (TM case)

	Resonator Problem		Scattering Problem	
Extrapolation Scheme	Num. of Iterations	Time (s)	Num. of Iterations	Time (s)
Taylor 0	50,109	13,349	21,930	6,077
SPE	31,689	8,703	17,611	5,184
Matrix Splitting	39,000	10,640	19,798	5,861
Rhs projection	27,936	8,272	12,740	4,054

Table 1. Comparison of the different extrapolation schemes

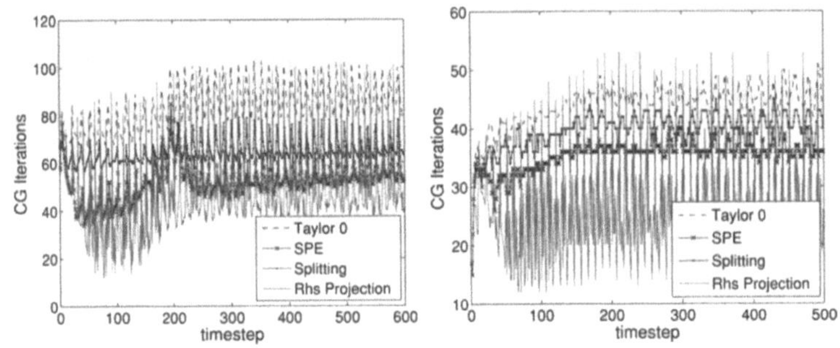

Fig. 2. Solver convergence with the different extrapolation schemes: a) Resonator Problem b) Scattering Problem

We notice that the SPE technique and the subspace projection of the rhs vector provide the best results in terms of the solver speed-up for both cases. In Fig. 2 the CG iterations for each timestep and for the different schemes are illustrated. Again we notice that the number of iterations applying the SPE and rhs projection schemes is clearly below those of the rest extrapolation schemes for most of the timesteps. The rather rapid variation of the number of iterations for the rhs projection algorithm is due to the restart procedure which was described above. For these examples the use of extrapolation techniques with implicit Newmark-Beta schemes offers a significant improvement in terms of computational simulation time over the standard approach.

5 Conclusions

In this paper a number of different extrapolation techniques for the construction of start values for implicit Newmark-Beta time-stepping schemes applied to electromagnetic high frequency problems were examined. Applied to electromagnetic wave propagation problems, were a perfectly matched layer (PML) absorbing boundary condition was used, the solution time could be significantly reduced by constructing good approximations of the solution vectors by using the information obtained in the previous time steps. This improvement in the solution time can be very useful in cases were no explicit formulation is available (e.g. in the FEM-TD formulation).

References

1. Volakis, J.L., Chatterjee, A., Kempel, L.C.: Finite Elements in Electromagnetics. Wiley, New York (1993)
2. Clemens, M., Weiland, T.: Discrete Electromagnetism with the Finite Integration Technique, PIER Monograph Series **32**, 63–85, (2001)
3. Clemens, M., Wilke M., and Weiland T.: Extrapolation strategies in numerical schemes for transient magnetic field simulations. IEEE Trans. Magn. **39**, 1171–1174, (2003)
4. Clemens, M., Wilke M., and Weiland T.: Subspace projection extrapolation scheme for transient field simulations, IEEE Trans. Magn. **40**, 934–937, (2004)
5. Edelvik, F., Ledfelt, G., Lötstedt, P., and Riley, D. J.: An unconditionally stable subcell model for arbitrarily oriented thin wires in the FETD method, IEEE Trans. on Antennas and Propagat., **51**, 1797–1805, (2003)
6. Gedney, S. D.: An anisotropic perfectly matched layer-absorbing medium for the truncation of FDTD lattices, IEEE Tr.A.P. **44**, 1630–1639, (1996)

Part V

Basic Research for Software Tools and Work in Progress

Electromagnetic Characterization Flow of Leadless Packages for RF Applications

G. Alessi

ST Microelectronics, Stradale Primosole 50, 95121, Catania, Italy, gesualdo.alessi@st.com

Abstract The aim of this paper is to show a characterization flow which integrates the 3D electromagnetic simulator Ansoft High Frequency Structure Simulator (HFSS) in Cadence Virtuoso layout editor for leadless packages modeling. A set of Cadence Skill language and Ansoft Macro Language procedures makes easier the leadless package simulations with HFSS inside Cadence Design Framework II. The tool lets HFSS draw the 3D model of a package with the bond wires from a 2D view in the layout editor Virtuoso. The correct settings for boundaries and ports are fixed as well. At the end of the electromagnetic simulation, HFSS produces a scattering matrix that can be associated to a new cell package defined in the DFII Library Manager. For this cell four cell views are created: a symbol view and the views for the three circuit simulators Mentor Graphics Eldo, Cadence Spectre and Agilent ADS.

Key words: electromagnetic simulation, package, hfss, virtuoso.

1 Introduction

Monolithic microwave integrated circuits (MMIC) performance is greatly influenced by the package, so it is necessary to account for its parasitic effects. An electrical model, generated using electromagnetic simulators, can be incorporated into a circuit simulator, such as Eldo, Spectre and ADS, which can predict the overall behavior of the circuit. The 3D electromagnetic simulator Ansoft HFSS [1] calculates the scattering matrix that can be exported into a file, in the Touchstone standard format.

The developed tool is a set of Cadence Skill Language [2] procedures and macros written in Ansoft Macro Language which allows the user to extract an electrical model of a package with its bond wires, from the Cadence environment. These procedures realize a graphic user interface for the following input data: package database (packages geometry and electromagnetic characteristics), bond wire models database (containing the bond wires geometric models), project data (die and board dimensions and electromagnetic characteristics, bond wires diameter), bond wires data (coordinates of the wires start-end points, bond wire models, leads and ports connected to the wires).

A geometric model is drawn in the HFSS 3D Modeler; materials and boundaries are assigned as well. The user specifies the connections between bond wires and ports, whereas the other necessary settings (port type and position) are automatically fixed. The block simulated in HFSS includes board, package and bond wires. At the end of the simulation a cell *package*, that is a *black box* element linked to the scattering matrix of the model, is introduced in the Cadence Library Manager with four cell views: a symbol view (an n-port element) that can be instantiated in a circuit schematic and three views for the above-mentioned simulators.

2 A simulation flow overview

The simulation flow is divided in three main steps:

(i) data input from Cadence Virtuoso;
(ii) electromagnetic simulation with Ansoft HFSS;
(iii) creation of a cell *package* in Cadence Library Manager.

Data on geometry and electromagnetic characteristics of package, die and bond wires is introduced by means of forms created in Cadence Skill language and saved in files readable by the HFSS macros. It is also necessary to introduce the coordinates of the starting and ending points of the wires and the numbers of the leads they are connected to ("0" is assigned to the die paddle, for downbonding wires). This operation can be tedious for large number of wires, so it has been automatized by using a fictitious *bondwire* layer: the user draw the bond wires selecting the shape *path* and the layer *bondwire* inside Virtuoso layout editor and a Skill procedure grabs the coordinates and the lead number.

The files created by the skill interface (package, bond wire models, project and bond wires data) are imported by the HFSS macro; the 3D model is drawn and the user has only to specify the EM simulation settings (e.g. frequency range, number of simulation steps).

3 The simulated model

The tool models QFN (Quad Flat No-Lead) packages: ASAT LPCC (Leadless Plastic Chip Carrier) [3] and Carsem MLPQ (Micro Leadframe Package Quad) [4]. These packages have exposed die paddle for mechanical and thermal integrity. The die paddle is soldered to the board and is connected to the metal plate on the opposite side of the board (the ground reference) through vias. The whole vias are considered as an only metal contact in the 3D model (*die paddle & gnd contacts* in Fig. 1).

The ports (Lumped Gap Sources) are connected to the leads across the board and to the wires across the die. Not used leads are connected to ground through perfect conductor surfaces. The bond wires connected to the die-paddle (downbonding) have only one port.

The package is surrounded by an air volume which has, on its external surface, a radiation boundary condition; this condition allows the electromagnetic field to radiate freely in the space. Surfaces defined as radiation boundaries absorb the electromagnetic field, moving the boundary to an infinite distance from the structure. On the lower surface of the board a perfectly conducting boundary (perfect E boundary) is chosen. This plane is a common reference both for the ports on the leads, directly connected to it, and for the ports on the bond wires, connected to it through the die paddle and the ground contacts.

The package database file contains the dimensions and the electromagnetic characteristics of leadless packages, whereas the board data is contained in the project data file.

The bond wire model implemented in the macros is the Philips/T.U. Delft [5] (Fig. 2). The following parameters are defined:

- **Stop height**: the height of the bond wire end points;
- **Start height**: the height of the bond wire start point; it corresponds to the die thickness if $Stop\ \ height = 0$; if this value is "0" a connection between the die paddle and the leads is modeled;
- **Max height**: the maximum height of the bond wire;

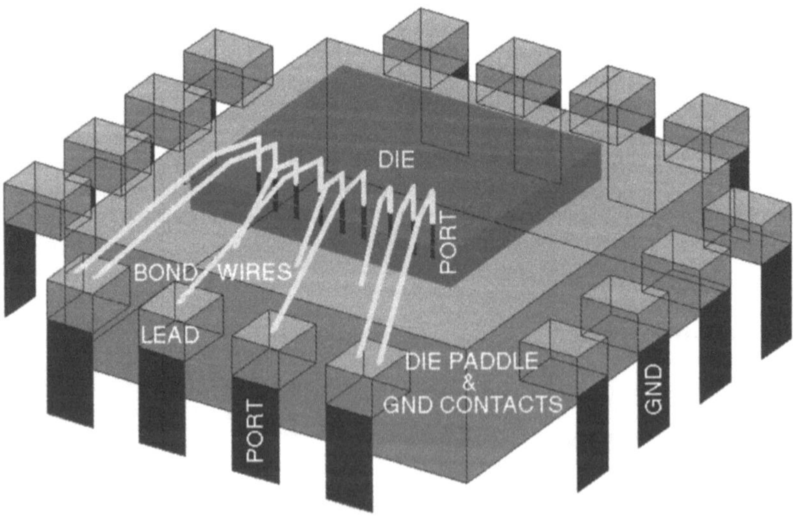

Fig. 1. Package model

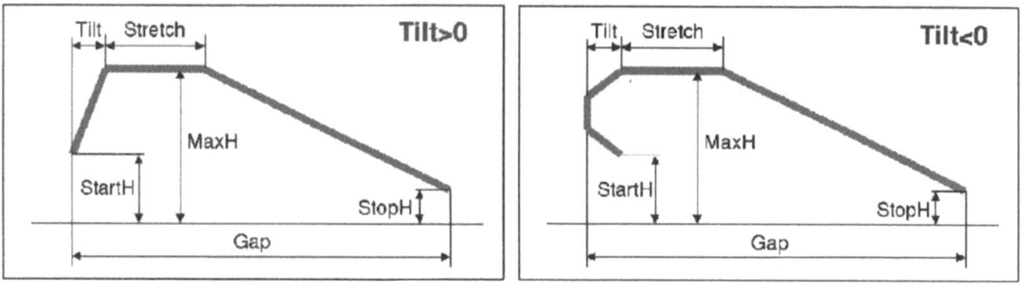

Fig. 2. Philips/T.U. Delft model

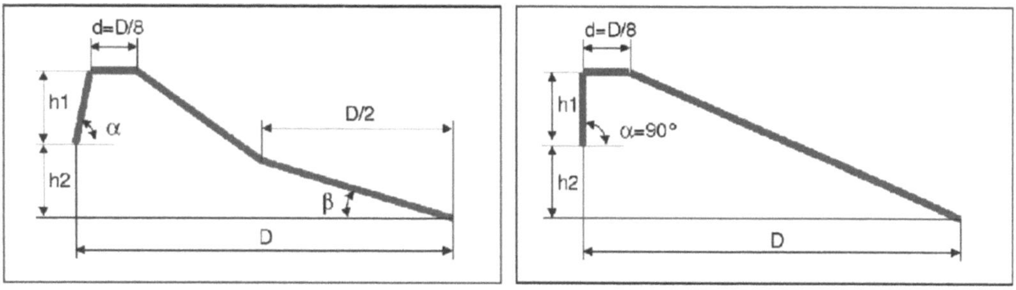

(a) (b)

Fig. 3. EIA/JEDEC bond wire model: (a) complete model, (b) simplified model

- **Gap**: the total distance covered by the wire;
- **Stretch**: the length of the horizontal section that models the wire loop;
- **Tilt**: allows to model a wire bending after its start point. By choosing $Tilt = 0$ the simplified model described by the EIA/JEDEC Standard n.59 [6] (Fig. 3) is obtained.

The tool takes only some of the above mentioned quantities as wire model parameters:

- **Gap/Stretch**: the *Gap* is different from wire to wire, so it cannot be considered a wire model parameter;
- **Loop height**: corresponds to the difference between *Max height* and *Stop height*;
- **Tilt**: has the meaning previously explained;
- **Start height**: is a boolean whose only two possible values are "0" for a lead-die paddle connection whereas "1" corresponds to the die thickness ($Stop\ height = 0$).

It is possible to define different models which are saved in a database file; the user can choose for each bond wire a model among the ones included in the database or define new models.

The implementation of the bond wire geometry in HFSS is accomplished by the translation of a polygon, representing the cross-section of the bond wire, along the one-dimensional path specified by the model; the polygon has the same cross-sectional area as the wire itself. The number of sides the polygon contains (at least four, according to the EIA/JEDEC standard) and the diameter are project parameters: the model parameters only establish the shape of the wires but not the size. Choosing a different model for each bond wire allows the user to define different shapes for different types of wires; for example if there are more than one row of connections to the die paddle, the wires in the more inner downbonding rows are higher than the others (Fig. 4).

4 The graphic user interface

The user selects four layers among the technology layers, which comprehend the design-kit layers and the Virtuoso system layers. These layers are associated to bond wires, package layout, die and text labels.

After the definition of a layout Cell View a menu is introduced in the Virtuoso window.

It is possible to choose a leadless package among the ones contained in the package database. The correspondent layout is drawn, using the layer specified in a layers configuration form.

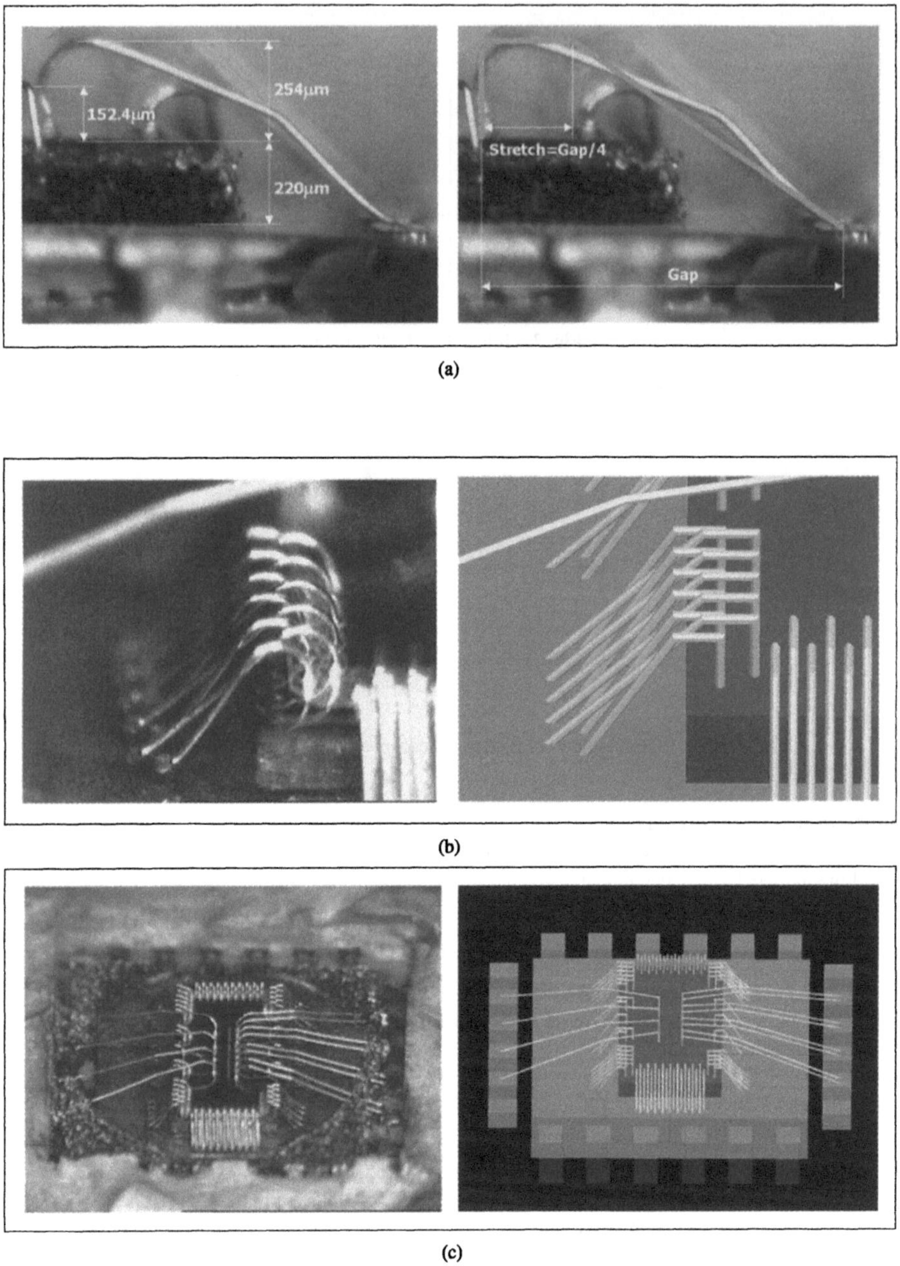

Fig. 4. Bond wires and their geometric models: (a) bond wire on a lead, (b) bond wires on the die paddle (downbonding), (c) wire bonded die

After choosing the layer associated to bond wires, the user draws the bond wires and grabs their position from the Virtuoso window; these coordinates are saved on a file.

In Fig. 5 the forms regarding data about bond wires (a), die, bond wires section and board (b), packages (c) and wire models (d), are shown.

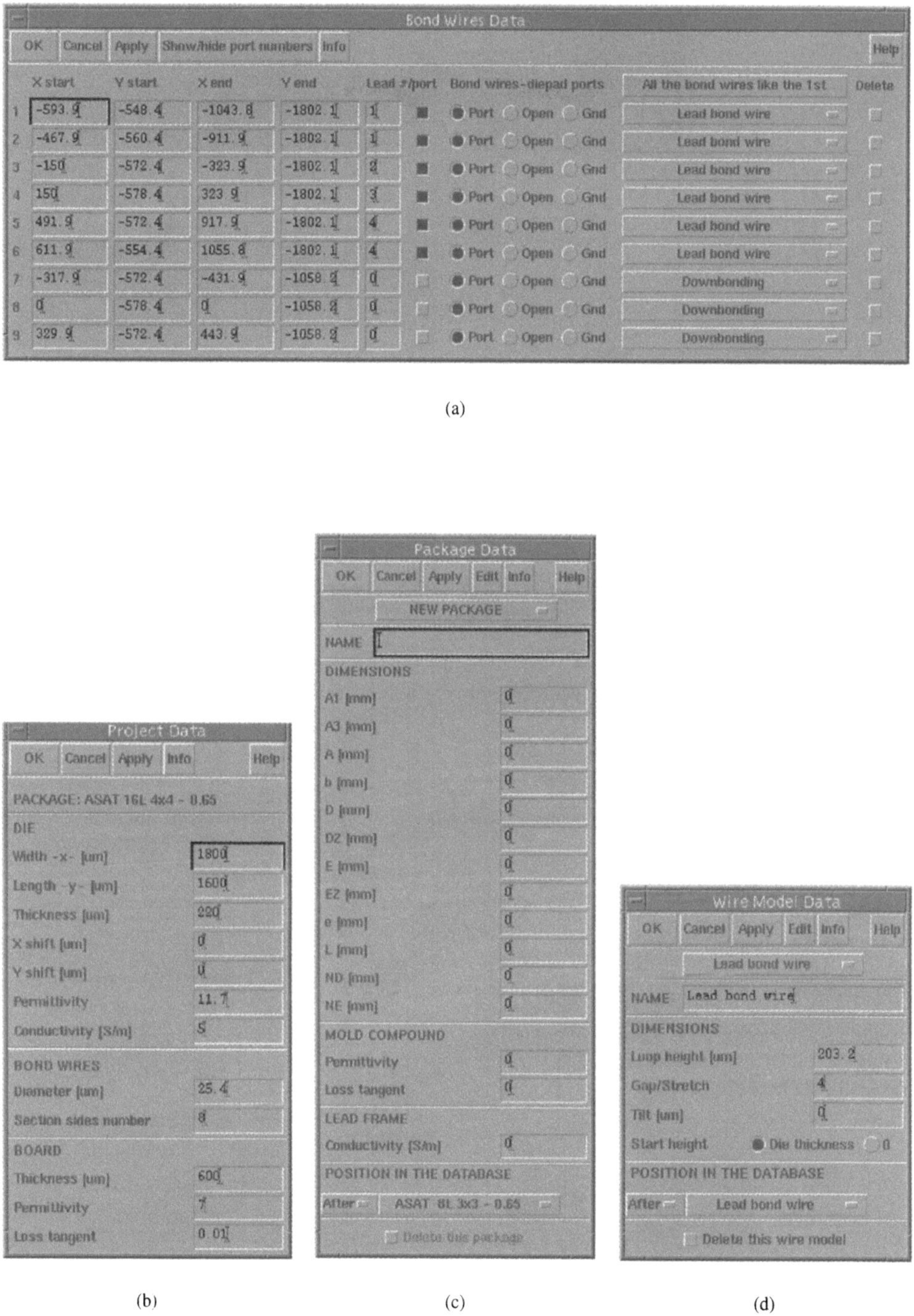

Fig. 5. Data forms: (a) bond wires, (b) project, (c) package database, (d) bond wire model database

5 Package symbol CDF

The CDF (Component Description Format [7]) describes the parameters and the attributes of parameters of individual components and libraries of components. The CDF contains also the simulation information for the simulators the component can work with.

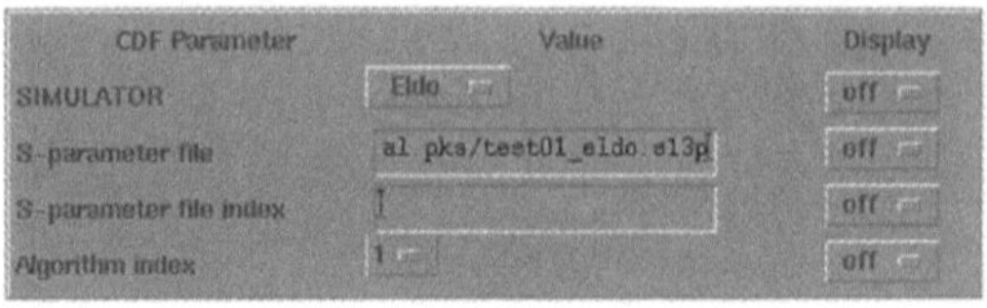

(a)

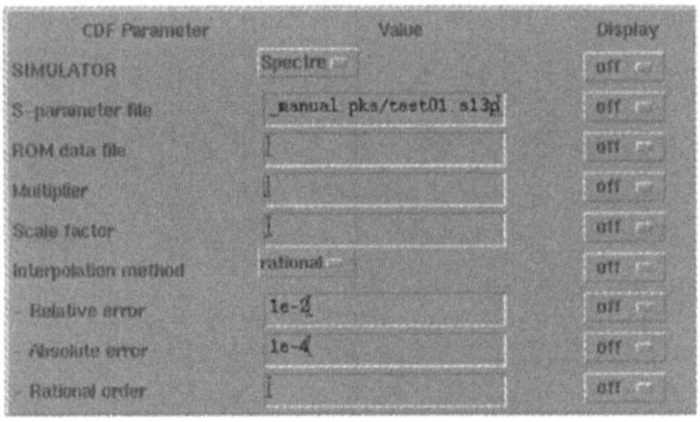

(b)

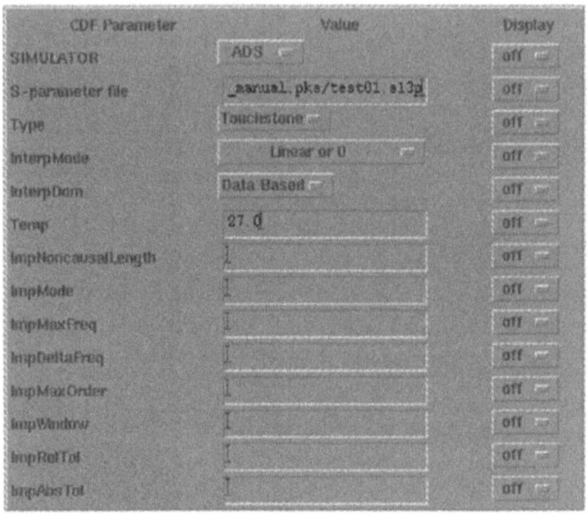

(c)

Fig. 6. Package symbol properties for (a) Eldo, (b) Spectre, (c) ADS

The symbol associated to the package, contains the simulation information concerning the following circuit simulators:

- Mentor Graphics Eldo,
- Cadence Spectre,
- Agilent ADS (Advanced Design System).

The simulation information inserted in the package symbol CDF were extracted from:

- the Eldo Special Component S-Model (Fig. 6 (a));
- the Spectre NPORT component (Fig. 6 (b));
- the ADS Data Item (Fig. 6 (c)).

6 Conclusions

In this paper a system that integrates Ansoft HFSS in the DFII environment has been shown. The tool makes easier the electromagnetic simulation of leadless packages.

The construction of a 3D model, including the package and the bond wires, can be very difficult if no automatic means is used. The tool presented allows the designer to modify the bond wires configuration by simply introducing, moving or deleting wires. Besides the correct simulation settings are chosen for the ports and the boundaries.

Finally a symbol for the package is created; it is linked to the S-parameter file obtained from HFSS and its CDF contains the simulation information for the circuit simulators Eldo, Spectre and ADS.

References

[1] Ansoft Corporation: Introduction to the Ansoft Macro Language, HFSS v8.5 (2002)
[2] Cadence: Skill Language Reference, Skill Language User Guide, Opus 4.4.6 (2001)
[3] http://www.asat.com/products/leadless/MO-220I.pdf
[4] http://www.carsem.com/Capability/Drwgs&Data_old/mlpq_PKGML00001-T.pdf
[5] Agilent: Online Documentation, ADS2002 (2002)
[6] EIA/JEDEC Standard: Bond Wire Modeling Standard, EIA/JESD59, (1997)
[7] Cadence: Component Description Format User Guide, Opus 4.4.6 (2001)

Domain Decomposition Techniques and Coupled PDE/ODE Simulation of Semiconductor Devices

G. Alì[1] and S. Micheletti[2]

[1] Istituto per le Applicazioni del Calcolo "M. Picone", sez. di Napoli, Via P. Castellino 111, 80131 Napoli, Italy, & INFN-Gruppo c. Cosenza, g.ali@iac.cnr.it

[2] MOX - Modeling and Scientific Computing, Dipartimento di Matematica "F. Brioschi", Politecnico di Milano, via Bonardi 9, 20133 Milano, Italy, stefano.micheletti@mate.polimi.it

1 Introduction

In the limit of an infinitesimal semiconductor region, a rigorous treatment due to Sah [Sah70] provided a circuit representation both for the Poisson and the transport equations. In [PML00] this model was extended to regions of arbitrary size, opening the way for automatic generation of circuits from the device simulation. In this paper we apply the above simplifying procedure to a selected region of a semiconductor device, keeping the full PDE treatment in the other regions.

Similar ideas were applied successfully to other areas, such as haemodynamics [QV03] and fluiddynamics, with application to the study of river bifurcation [MPS04]. From a strictly mathematical viewpoint, analytical results on coupled PDE/ODE systems (as arising in integrated circuit simulation) can be found in [ABGT04].

2 Equivalent circuit formulation

In this section we outline a general procedure to derive an equivalent circuit formulation of the drift-diffusion equations. For simplicity, we consider a one-dimensional device, modelled by a space interval $I = (x_A, x_B)$, and characterized by a doping profile $D(x)$, $x \in I$. The behavior of the device is described by the transient drift-diffusion system [Sel84],

$$\begin{cases} q\dfrac{\partial n}{\partial t} - \partial_x J_n = -qR, & q\dfrac{\partial p}{\partial t} + \partial_x J_p = -qR, \\ J_n = -q\mu_n n \partial_x \phi_n, & J_p = -q\mu_p p \partial_x \phi_p, \\ -\partial_x(\epsilon \partial_x \phi) = q(D + p - n), \end{cases} \tag{1}$$

where n, p are the number densities of electrons and holes, with charge $-q$ and q, respectively, J_n, J_p and ϕ_n, ϕ_p are the corresponding current densities and quasi-Fermi potentials, and ϕ is the electrostatic potential. The number densities are expressed in terms of the quasi-Fermi and electrostatic potentials by means of the Maxwell-Boltzmann relations,

$$n = n_i \exp\left(\frac{\phi - \phi_n}{U_{\rm th}}\right), \quad p = n_i \exp\left(\frac{\phi_p - \phi}{U_{\rm th}}\right), \tag{2}$$

where the constant $U_{\rm th}$ is the thermal potential. In (1), μ_n, μ_p are the mobilities for electrons and holes, respectively. They are bounded, strictly positive functions, which depend on x, and $E = -\partial_x \phi$. The dependency on the particle densities n, p is usually neglected [Sel84, MRS00]. The generation-recombination term can be modeled as a given function $R = R(x, n, p, J_n, J_p)$.

System (1) is considered in a given time interval $(0, T)$, and supplemented with appropriate initial conditions, and with Dirichlet conditions on the Ohmic contacts of the device,

$$\phi_n(x,t) = \phi_p(x,t) = \phi(x,t) - V_{\rm bi}(x) = U_{\rm ap}(x,t), \text{ in } \{x_A, x_B\} \times (0,T), \tag{3}$$

where $V_{\rm bi}$ is the built-in potential and $U_{\rm ap}$ the applied potential.

We assume that a certain region (x_a, x_b) of the device has a behavior "almost linear", in a sense that will be clarified later. Then, we consider a decomposition $\{x_a \equiv x_0, x_1, \ldots, x_m \equiv x_b\}$, together with the intermediate nodes

$\{x_{\frac{1}{2}}, x_{\frac{3}{2}}, \ldots, x_{m-\frac{1}{2}}\}$. Each inner node x_k, with $k = 1, \ldots, m-1$ determines a cell $\Delta x_k = (x_{k-\frac{1}{2}}, x_{k+\frac{1}{2}})$. We also introduce the region boundary cells $\Delta x_0 = (x_0, x_{\frac{1}{2}})$, $\Delta x_m = (x_{m-\frac{1}{2}}, x_m)$.
we find

$$\begin{cases} \frac{d}{dt}Q_n^k = J_n^{k+\frac{1}{2}} - J_n^{k-\frac{1}{2}} - U^k, \\ \frac{d}{dt}Q_p^k = J_p^{k-\frac{1}{2}} - J_p^{k+\frac{1}{2}} - U^k, \\ \frac{d}{dt}Q_d^{k+\frac{1}{2}} - \frac{d}{dt}Q_d^{k-\frac{1}{2}} = \frac{d}{dt}Q_p^k - \frac{d}{dt}Q_n^k, \end{cases} \tag{4}$$

where $Q_n^k := \int_{x_{k-\frac{1}{2}}}^{x_{k+\frac{1}{2}}} qn\,dx$, $Q_p^k := \int_{x_{k-\frac{1}{2}}}^{x_{k+\frac{1}{2}}} qp\,dx$ represent the total charge (in absolute value) carried in the cell Δx_k by electron and holes, respectively, $Q_d^{k+\frac{1}{2}}(t) := \epsilon E(x_{k+\frac{1}{2}}, t)$, $J_d^{k+\frac{1}{2}} := \frac{d}{dt}Q_d^{k+\frac{1}{2}}$ are the displacement charge and current, respectively, $J_d^{k+\frac{1}{2}} = J_d(x_{k+\frac{1}{2}}, t)$, $J_n^{k+\frac{1}{2}} = J_n(x_{k+\frac{1}{2}}, t)$, $J_p^{k+\frac{1}{2}} = J_p(x_{k+\frac{1}{2}}, t)$, are the displacement, electron and hole currents, respectively, through the branch connecting the node x_k to the node x_{k+1}, and $U^k := \int_{x_{k-\frac{1}{2}}}^{x_{k+\frac{1}{2}}} qR\,dx$.

3 Closure relations and calibration

Up to this point, (4) is an exact equation, but it is not closed and cannot be used in this form for numerical simulation. To proceed with, we need to make some constitutive assumptions for the quantities appearing in (4).

The main ansatz is that all quantities depend on the three potentials ϕ, ϕ_n and ϕ_p evaluated at the nodes. More precisely, the quantities with integer index, say k, depend on the potentials

$$V^k(t) := \phi(x_k, t), \quad V_n^k(t) := \phi_n(x_k, t), \quad V_p^k(t) := \phi_p(x_k, t), \tag{5}$$

evaluated at the node with the same index. Instead, the quantities with fractional index, say $k - \frac{1}{2}$, depend on the potentials at the neighboring nodes, V^{k-1}, V_n^{k-1}, V_p^{k-1} and V^k, V_n^k, V_p^k.

Explicitly, for the electron, hole and displacement charge, we postulate closure relations of the following type:

$$\begin{aligned} Q_n^k &= f_n^k(V^k - V_n^k), & Q_p^k &= f_p^k(V_p^k - V^k), \\ Q_d^{k-\frac{1}{2}} &= f_d^{k-\frac{1}{2}}(V^{k-1} - V^k),\ Q_d^{k+\frac{1}{2}} = f_d^{k+\frac{1}{2}}(V^k - V^{k+1}). \end{aligned} \tag{6}$$

The first two closure relations in (6) come from an extension of the formulas (2), an extension of the formulas (2), which express the $\phi - \phi_n$ and $\phi_p - \phi$, respectively. The other two closure relations come from the expression of the electric field as obtained by Poisson equation.

For the electron and hole currents, recalling $(1)_2$, we assume closure relations which are the product of three terms, the first one accounting for the mobility, the second for the carrier density, and the third for the gradient of the quasi-Fermi potential (or "voltage"). Then, the resulting closure relations take the general form:

$$\begin{aligned} J_n^{k-\frac{1}{2}} &= g_{mn}^{k-\frac{1}{2}}(V^{k-1} - V^k) g_{dn}^{k-\frac{1}{2}}(V^{k-1} - V_n^{k-1}, V^k - V_n^k) g_{vn}^{k-\frac{1}{2}}(V_n^{k-1} - V_n^k), \\ J_p^{k-\frac{1}{2}} &= g_{mp}^{k-\frac{1}{2}}(V^{k-1} - V^k) g_{dp}^{k-\frac{1}{2}}(V^{k-1} - V_p^{k-1}, V^k - V_p^k) g_{vp}^{k-\frac{1}{2}}(V_p^{k-1} - V_p^k). \end{aligned} \tag{7}$$

Finally, we consider the generation-recombination term U^k. Neglecting for simplicity impact ionization effects, this term depends on the carrier densities n, p. Then, it is simple to generalize this dependency by assuming

$$U^k = h^k(V^k - V_n^k, V_p^k - V^k). \tag{8}$$

So far, we have expressed (at least formally) all quantities in (4) in terms of voltage differences, once the functions f's, g's and h's are explicitly given. Anyway, the voltage differences are not independent. In particular, we have

$$\begin{aligned} (V^{k-1} - V^k) - (V^{k-1} - V_n^{k-1}) + (V^k - V_n^k) - (V_n^{k-1} - V_n^k) = 0, \\ (V^{k-1} - V^k) + (V_p^{k-1} - V^{k-1}) - (V_p^k - V^k) - (V_p^{k-1} - V_p^k) = 0. \end{aligned}$$

Thus, we can express the voltage differences $V_n^{k-1} - V_n^k$ and $V_p^{k-1} - V_p^k$ in terms of the other voltage differences.

Using the closure relations (6)–(8), the integrated system (4) becomes

$$\begin{cases} \dfrac{d}{dt} f_n^k(V^k - V_n^k) = -h^k(V^k - V_n^k, V_p^k - V^k) \\ \qquad + g_n^{k+\frac{1}{2}}(V^k - V^{k+1}, V^k - V_n^k, V^{k+1} - V_n^{k+1}) \\ \qquad - g_n^{k-\frac{1}{2}}(V^{k-1} - V^k, V^{k-1} - V_n^{k-1}, V^k - V_n^k), \\ \dfrac{d}{dt} f_p^k(V_p^k - V^k) = -h^k(V^k - V_n^k, V_p^k - V^k) \\ \qquad - g_p^{k+\frac{1}{2}}(V^k - V^{k+1}, V_p^k - V^k, V_p^{k+1} - V^{k+1}) \\ \qquad + g_p^{k-\frac{1}{2}}(V^{k-1} - V^k, V_p^{k-1} - V^{k-1}, V_p^k - V^k), \\ \dfrac{d}{dt} f_d^{k+\frac{1}{2}}(V^k - V^{k+1}) - \dfrac{d}{dt} f_d^{k-\frac{1}{2}}(V^{k-1} - V^k) \\ \quad = \dfrac{d}{dt} f_p^k(V_p^k - V^k) - \dfrac{d}{dt} f_n^k(V^k - V_n^k). \end{cases} \tag{9}$$

Here, we have introduced the shorthand notation $g_n^{k-\frac{1}{2}} := g_{mn}^{k-\frac{1}{2}} g_{dn}^{k-\frac{1}{2}} g_{vn}^{k-\frac{1}{2}}$, $g_p^{k-\frac{1}{2}} := g_{mp}^{k-\frac{1}{2}} g_{dp}^{k-\frac{1}{2}} g_{vp}^{k-\frac{1}{2}}$.

Clearly, equation (9) is not sufficient to determine the unknown potentials at the nodes, unless the closure relations (6)–(8) are explicitly given. relations (6)–(8) are explicitly given. The determination of the closure relations seems to be an to be an impossible task. Anyway, in some cases, it is possible to assume these relations to be linear, and to calibrate the

$$U_{\mathrm{ap}}(x,t) = V_{\mathrm{ap}}(x) + v_{\mathrm{ap}}(t), \qquad |v_{\mathrm{ap}}| \ll |V_{\mathrm{ap}}|(\ll U_{\mathrm{th}}). \tag{10}$$

At first approximation, the solution corresponding to the applied potential $V_{\mathrm{ap}} + v_{\mathrm{ap}}$ can be thought as the sum of the steady-state solution corresponding to V_{ap} and the time-dependent solution corresponding to v_{ap}. Moreover, the time-dependent solution can be evaluated by using the linearized equations around the steady-state solution. This approximation can be justified, at least formally, by invoking the smallness of the potential v_{ap}.

Bearing this in mind, we propose a procedure to calibrate the closure relations in the small signal regime. First, we evaluate the steady-state solution $(\phi_s, \phi_{ns}, \phi_{ps})$ corresponding to the applied potential V_{ap}, in the whole domain. Then, we can determine the value of this solution at the k-th node, $(V_s^k, V_{ns}^k, V_{ps}^k) := (\phi_s(x_k), \phi_{ns}(x_k), \phi_{ps}(x_k))$, and evaluate also the total electron, hole and displacement charge inside the k-th cell, Q_{ns}^k, Q_{ns}^k, Q_{ns}^k, the currents at the intermediate nodes, $J_{ns}^{k-\frac{1}{2}}$, $J_{ps}^{k-\frac{1}{2}}$, and the generation-recombination term, U_s^k. It is pretty natural to identify

$$Q_{ns}^k \equiv f_n^k(V_s^k - V_{ns}^k), \quad Q_{ps}^k \equiv f_p^k(V_{ps}^k - V_s^k), \quad Q_{ds}^{k-\frac{1}{2}} \equiv f_d^{k-\frac{1}{2}}(V_s^{k-1} - V_s^k).$$

Thus, we can expand the closure relations (6) around the steady-state solution, introducing the potential differences $(v^k, v_n^k, v_p^k) := (V^k - V_s^k, V_n^k - V_{ns}^k, V_p^k - V_{ps}^k)$, and rewriting them as

$$\begin{aligned} &Q_n^k = Q_{ns}^k + C_n^k\left(v^k - v_n^k\right), \quad Q_p^k = Q_{ps}^k + C_p^k\left(v_p^k - v^k\right), \\ &Q_d^{k-\frac{1}{2}} = Q_{ds}^{k-\frac{1}{2}} + C_d^{k-\frac{1}{2}}\left(v^{k-1} - v^k\right). \end{aligned} \tag{11}$$

Here, we have introduced the capacitances

$$\begin{aligned} C_n^k &:= \frac{Q_n^k - Q_{ns}^k}{(V^k - V_n^k) - (V_s^k - V_{ns}^k)}, C_p^k := \frac{Q_p^k - Q_{ps}^k}{(V_p^k - V^k) - (V_{ps}^k - V_s^k)}, \\ C_d^{k-\frac{1}{2}} &:= \frac{Q_d^{k-\frac{1}{2}} - Q_{ds}^{k-\frac{1}{2}}}{(V^{k-1} - V^k) - (V_s^{k-1} - V_s^k)}. \end{aligned} \tag{12}$$

If the functional dependencencies in (11) are approximately linear, we can evaluate the constant capacitances by comparing two steady-states solutions and using relations (12), which become exact.
In a similar way, we can write

$$\begin{aligned} J_n^{k-\frac{1}{2}} &= J_{ns}^{k-\frac{1}{2}} + G_{mn}^{k-\frac{1}{2}}\left(v^{k-1} - v^k\right) \\ &\quad + G_{fn}^{k-\frac{1}{2}}\left(v^{k-1} - v_n^{k-1}\right) + G_{rn}^{k-\frac{1}{2}}\left(v^k - v_n^k\right) + G_{vn}^{k-\frac{1}{2}}\left(v_n^{k-1} - v_n^k\right), \\ J_p^{k-\frac{1}{2}} &= J_{ps}^{k-\frac{1}{2}} + G_{mp}^{k-\frac{1}{2}}\left(v^{k-1} - v^k\right) \\ &\quad + G_{fp}^{k-\frac{1}{2}}\left(v_p^{k-1} - v^{k-1}\right) + G_{rp}^{k-\frac{1}{2}}\left(v_p^k - v^k\right) + G_{vp}^{k-\frac{1}{2}}\left(v_p^{k-1} - v_p^k\right), \\ U^k &= h_s^k + h^k(v^k - v_n^k, v_p^k - v^k) \end{aligned}$$

and, again assuming linear closure relations, we can determine the coefficients $G_{mn}^{k-\frac{1}{2}}$, $G_{fn}^{k-\frac{1}{2}}$, $G_{rn}^{k-\frac{1}{2}}$, $G_{vn}^{k-\frac{1}{2}}$, $G_{mp}^{k-\frac{1}{2}}$, $G_{fp}^{k-\frac{1}{2}}$, $G_{rp}^{k-\frac{1}{2}}$, $G_{vp}^{k-\frac{1}{2}}$ and h^k, by using steady-state numerical simulations at different bias points. Once the constants are determined, we can use the linearized form of the reduced system (9) and solve for the potentials (v^k, v_n^k, v_p^k). We also notice that this general technique, based on the calibration of a physically-based equivalent circuit by a numerical solution of the "relevant" equations, can be extended to higher space dimensions.

4 A model problem: the one-dimensional P-N diode

In this section we propose a mixed equivalent circuit-PDE formulation of a P-N diode, by applying the ideas expounded in the previous section on a subdomain of the diode.

A schematic sketch of a P-N diode is shown in Fig. 1.

The device is composed of two regions, Ω_n and Ω_p, which are separated by a regular, connected hypersurface Γ. The two regions are positively and negatively doped, respectively.

We model the P-N junction by a segment $I = (x_A, x_B)$ on which the drift-diffusion equations (1) hold. The hypersurface Γ reduces to a point $x_\Gamma \in I$. Inside the diode (x_A, x_B), we can distinguish three regions: the P-doped quasi-neutral region (x_A, x_a), the depletion region (x_a, x_b), and the N-doped quasi-neutral region (x_b, x_B). Although the locations of the separation points x_a and x_b, shown in Fig. 2, can be estimated by a simplified approximation analysis, it is more efficient to determine them by appropriate steady-state numerical simulations. We would like to apply the equivalent circuit theory to the depletion region, for a forward biased diode, that is, with positive external voltage U_{ap}.

The simplest decomposition of this region is based on two nodes, $x_0 \equiv x_a$, $x_1 \equiv x_b$. The separation point x_Γ between the differently doped regions will be identified with the intermediate node $x_{\frac{1}{2}}$. There are two possible ways of considering the discretization blocks associated to the to two nodes we have fixed. They are shown in Fig. 2, denoted by $\{\Delta x_0, \Delta x_1\}$ and $\{\Delta x_a, \Delta x_b\}$. The first two blocks use the additional nodes $x_{-\frac{1}{2}}$ and $x_{\frac{3}{2}}$, which are inside the P-doped quasi-neutral region and the N-doped quasi-neutral region, respectively.

Following the theory expounded in the previous section, we propose a coupled model, which uses the drift-diffusion equations in the quasi-neutral regions and the equivalent circuit formulation in the depletion region.

To start with, we sort out the unknowns of the problem. In the P-doped region (x_A, x_a), the unknowns are the three potentials $(\phi^P, \phi_n^P, \phi_p^P)$, which satisfy system (1) with $(x, t) \in (x_A, x_a) \times (0, T)$, with boundary data

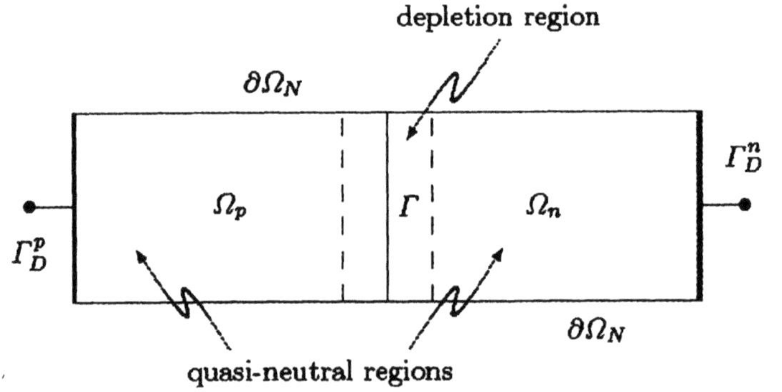

Fig. 1. P-N diode (two-dimensional cross section)

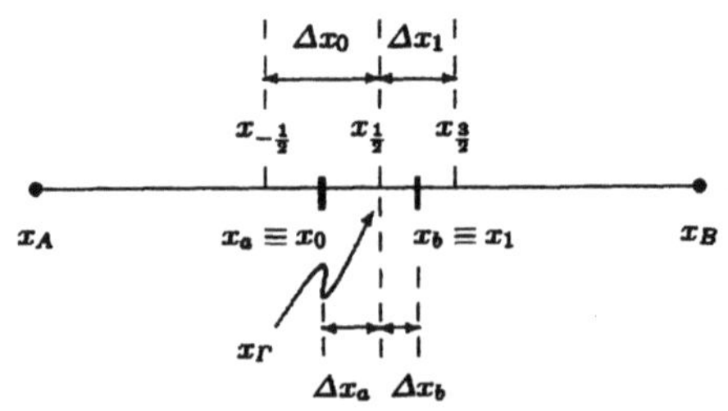

Fig. 2. P-N diode (one-dimensional approximation and domain decomposition)

$$\begin{cases} \phi_n^P(x_A,t) = \phi_p^P(x_A,t) = \phi^P(x_A,t) - V_{\text{bi}}(x_A) = U_{\text{ap}}(x,t), \\ \phi_n^P(x_a,t) = \phi_n^a(t), \quad \phi_p^P(x_a,t) = \phi_p^a(t), \quad \phi^P(x_a,t) = \phi^a(t). \end{cases} \tag{13}$$

Here, the potentials ϕ_n^a, ϕ_p^a, and ϕ^a, are also unknowns.

In a similar way, in the negatively doped region (x_b, x_B), the unknowns are the three potentials $(\phi^N, \phi_n^N, \phi_p^N)$, which satisfy system (1) with $(x,t) \in (x_b, x_B) \times (0,T)$, with boundary data

$$\begin{cases} \phi_n^N(x_b,t) = \phi_n^b(t), \phi_p^N(x_b,t) = \phi_p^b(t), \phi(x_b,t) = \phi^b(t), \\ \phi_n^N(x_B,t) = \phi_p^N(x_B,t) = \phi^N(x_B,t) - V_{\text{bi}}(x_B) = 0, \end{cases} \tag{14}$$

where, the potentials ϕ_n^b, ϕ_p^b, and ϕ^b are unknowns.

In the depletion region (x_a, x_b), we consider the unknowns (V^0, V_n^0, V_p^0) on the node x_0, and (V^1, V_n^1, V_p^1) on the node x_1, defined by (5). For the node x_0, we consider equation (9), written for $k = 0$ (9), written for $k = 0$ modified to take care of the boundary current sources at $x_{-\frac{1}{2}}$, now denoted by For the node x_1, we use equation (9), written for $k = 1$ (9), written for $k = 1$ and we let

To close the above coupled model, we need to assign coupling conditions at the boundary. Clearly, the unknown potentials in the boundary conditions (14) and (13) can be identified with the potentials at the

$$(\phi^a, \phi_n^a, \phi_p^a) = (V^0, V_n^0, V_p^0), \quad (\phi^b, \phi_n^b, \phi_p^b) = (V^1, V_n^1, V_p^1). \tag{15}$$

As for the current sources at the boundary, we can identify them with the currents coming through the boundary from the quasi neutral regions, that is

$$\begin{aligned} (J_d^a(t), J_n^a(t), J_p^a(t)) &= (J_d^P(x_a,t), J_n^P(x_a,t), J_p^P(x_a,t)), \\ (J_d^b(t), J_n^b(t), J_p^b(t)) &= (J_d^N(x_b,t), J_n^N(x_b,t), J_p^N(x_b,t)), \end{aligned} \tag{16}$$

or, depending on the choice of the discretization boxes,

$$\begin{aligned} (J_d^a(t), J_n^a(t), J_p^a(t)) &= (J_d^P(x_{-\frac{1}{2}},t), J_n^P(x_{-\frac{1}{2}},t), J_p^P(x_{-\frac{1}{2}},t)), \\ (J_d^b(t), J_n^b(t), J_p^b(t)) &= (J_d^N(x_{\frac{3}{2}},t), J_n^N(x_{\frac{3}{2}},t), J_p^N(x_{\frac{3}{2}},t)). \end{aligned} \tag{17}$$

5 Summary and perspectives

We have presented a general model reduction technique for device simulation based on the box integration method. By combining this technique with a domain decomposition approach, we have derived a coupled PDAE model for the whole device. Finally, we have proposed a model problem (one-dimensional P-N diode) to validate our approach. Numerical results for the reduced, coupled equations for the model problem are on the way.

The main open problems which we would like to face is the automatic subdivision, by domain decomposition, of the device, and the automatic calibration of the reduced equations (not necessarily under small signal hypothesis). This work is meant to be a preliminary step towards this goal.

References

[ABGT04] Alì, G., Bartel, A., Günther, M., Tischendorf, C.: Elliptic partial differential algebraic multiphysics models in electrical network design, M^3AS, **13** 9, 1261–1278 (2003)

[MRS00] Markowich, P. A., Ringhofer C. A., Schmeiser, C.: Semiconductor Equations. Springer (1990)

[MPS04] Miglio, E., Perotto, S., Saleri, F.: A multiphysics strategy for free surface flows. To appear in Lecture Notes in Computational Sciences and Engineering, Proceedings of the 15th International Conference on Domain Decomposition Methods. R. Kornhuber, R.H.W. Hoppe, D.E. Keyes, J. Periaux, O. Pironneau, J. Xu, Eds., Springer-Verlag (2004)

[PML00] Pacelli, A., Mastropasqua M., Luryi, S.: Generation of equivalent circuits from physics-based device simulation. IEEE Trans. Computer Aided Design of Integrated Circuits, **19**, 1241–1250 (2000)

[QV03] Quarteroni, A., Veneziani, A.: Analysis of a geometrical multiscale model based on the coupling of ODEs and PDEs for blood flow simulations. Multiscale Model. Simul., **1** 2, 173–195 (2003)

[Sah70] Sah, C. T.: The equivalent circuit model in solid-state electronics–III. Conduction and displacement currents. Solid State Electronics, **13**, 1547–1575 (1970)

[Sel84] Selberherr, S.: Analysis and Simulation of Semiconductor Devices. Springer (1984)

Interconnection Modeling Challenges in System-in-Package (SiP) Design

S. Castorina and R. A. Ene

Synapto s.r.l., stradale Vincenzo Lancia 57, 95100 Catania - Italy, {scastorina, rene}@synapto.com

1 Introduction

The possibility to combine multiple functionalities in one electronic device is directly related to the possibility of integrating more electronic components into a unique system. From the technological point of view there are two alternatives: integration at the level of the semiconductor in order to build a system-on-chip (SoC), or integration at the package level in order to obtain the system-in-package (SiP). The SiP alternative, which consists in integrating various components (semiconductor devices, resistors, inductors, capacitors, sensors, antennas, etc.) using advanced printed circuit technologies, requires less development time and resources compared with the ones needed for SoC implementation. This has not only the advantage of reduced costs, but gives the possibility to implement in a very short time various functionalities in different models, facilitating the customization of the product, making possible to satisfy specific requests, thus widening the market opportunities of a given product. Many of the issues related to interconnect substrates for today's high-frequency/high-speed mixed signal applications, are also of concern in the case of SiPs. For example, performance of SiPs are affected by the electromagnetic properties of the interconnect circuit, which are of increasing importance as the frequency rises. These are, in particular, distributed electromagnetic effects, which manifest as interconnect-induced delay, reflection, radiation, and long-range nonlocal coupling. Such a complexity is further complicated by embedded passive devices, sensors and exotic materials, which introduce more discontinuities, inhomogeneities, anisotropy and nonlinearities in the electromagnetic behavior of the system. Moreover, the trend in integrated circuits design is to take advantage of miniaturization to pack more functionalities in a chip; this results in increased chip and package size and complexity, and such complexity is ultimately reflected at the interconnect level[1]. The SiP is a viable alternative to the SoC [2], but it requires the development of a correct methodology of design [3] and characterization of the interconnection in order to enter the mainstream of the electronic industry technologies. In this paper, the main issues related to SiP modeling and design will be addressed, with particular attention devoted to the requirements that integrated packages for SiPs simulation and design must satisfy.

2 Embedded passives in SiPs

A very important issue related with SiPs is the development of technologies and processes for the integration of passive components in the PCB substrate (embedded passives), i.e. resistors, capacitors and inductors.
The choice of using embedded passives technologies in SiP is dictated by two necessities. The first one regards the impossibility to obtain good passives, especially inductors, on semiconductors substrates, while many materials with very good electrical and high frequency properties are available for printed circuit technologies. Furthermore, embedded passive components exhibit enhanced electrical characteristics compared to their discrete counterpart. This is especially true for the parasitic inductance and resistance of discrete components. In fact, due to the presence of the connecting leads, and thus to their parasitic inductance, a discrete capacitor behaves like a resonant circuit, and exhibits a capacitive behavior only if the operating frequency is well below the resonance frequency of the equivalent circuit. An embedded capacitor, instead, has no connecting leads, thus has a pure capacitive impedance for a wider range of frequencies. Important benefits in terms of noise filtering can, therefore, be obtained by using embedded by-pass capacitors for the power and ground lines on the board. Embedded inductors can be implemented with many different shapes, exploiting either single or multiple metal layers, in coil or transmission lines fashion. Such inductors may be used in RF applications, since they show better characteristics than those obtained on semiconductor substrates.
The second necessity arises from the continuous increment in number of passives required by new semiconductor devices (for example a Pentium IV requires approximately 550 passives per chip, and a PC motherboard can arrive at

more than 2300 passives) combined with the need to shrink the dimensions of the final product. The solution to these problems lays in the massive use of passive embedded technologies, integrating resistors, capacitors and inductors, into the interconnection on the printed circuit board. Although the advantages in terms of component densities and high frequency properties of the embedded passives are known, the problem of closely integrating these devices in a three-dimensional structure have not been yet systematically studied [3].
The use of new or exotic materials could be exploited to implement sensory functions in the board. For example, a resistive layer used to produce embedded resistors may exhibit a sufficiently high temperature coefficient which would make it suitable as an embedded temperature sensor. Similarly materials used as dielectric in embedded capacitors or magnetic cores in embedded inductors may have potential sensory properties which could be suitably used. Embedded inductors can also be used as antennas. The use of special materials embedded in the SiP structure enhances the characteristics of the inductors, but also increases the parasitic couplings with the surrounding environment.

3 Practical issues in modeling SiPs

SiP substrate is usually a multilayer PCB where interconnections between elements are realized by means of conductive tracks. Such interconnections are photo-defined and etched in the conductive layers of the PCB, and each conductive layer is separated from the other by one or more dielectric layers. Conductive tracks are used to carry energy and information, i.e. as power and signal lines, and one or more conductive planes are realized on the board to provide a reference potential for power and signal lines. The structure composed by conductive strip parallel to a reference (or ground) plane, separated by a dielectric layer, acts as a transmission line (a microstrip, in particular) and therefore can be described by applying transmission lines' theory. However such a structure has a dispersive behavior and many parasitic effects which also depend on frequency. As an example, consider the signal return path on the ground plane can be taken into account: at low frequencies the return current follows the lower resistance path on the ground plane and is uniformly distributed in the track cross section, while at higher frequencies current tends to accumulate on the bottom part of the track due to skin effect and the return signal path tends to align with the track due to proximity effect, leading to substantial resistance variation which need to be taken into account in a correct model. Other issues are represented by the modeling of two or more such parallel lines, taking into account the couplings between conductors. Equivalent circuit models usually depend on the physical configuration of conductors, i.e. if they are broad-side or edge coupled, parallel or not, etc., but also on the propagation modes configuration. Thus, for the same structure the circuit models may differ for different configurations of modes and for the different simulation engines (for example the even and odd modes in coupled transmission lines versus a complete multi-modal equivalent). In structures with further complex geometries, as in the case of planar inductors, which can be seen as a set of coupled conductors, more complicated coupling schemes and phenomena may arise, as shown in Fig. 1, thus the '2D' or '2.5D' models used for the previous structures will no longer be suitable for the reliable simulation of such complex structures, which will therefore require '3D' field modeling, at least for the circuit parameters estimation. Among the parasitic couplings in planar embedded inductors are those with the conductive planes which will be present in the board and/or package. The electromagnetic field generated by the inductors will couple with such conductive planes inducing parasitic current paths through them.

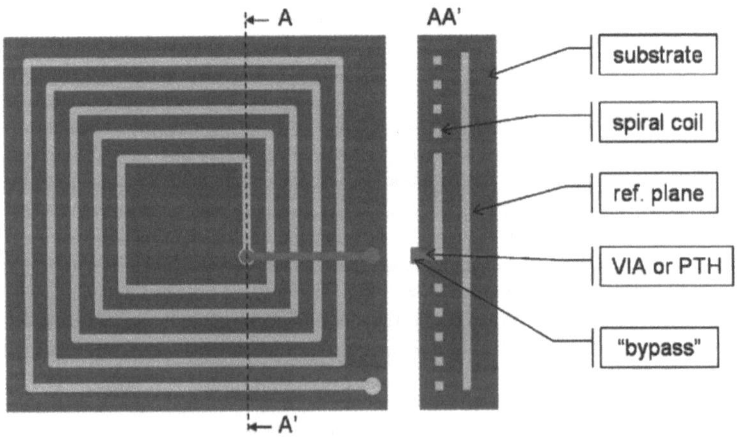

Fig. 1. Schematic structure of a planar embedded inductor

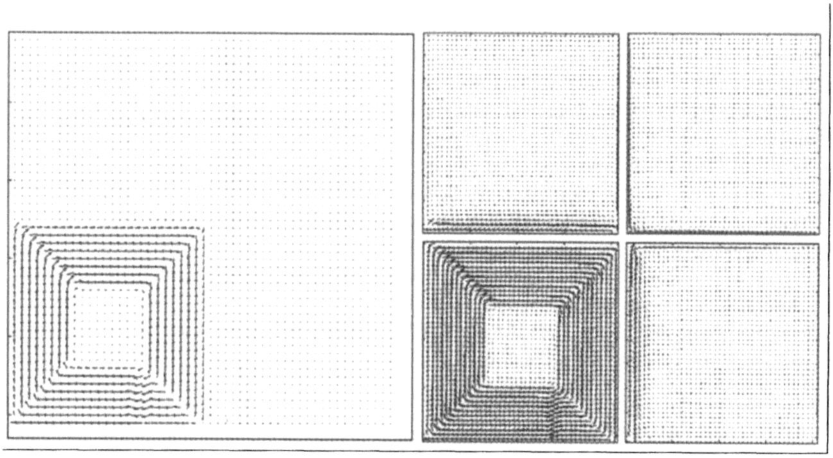

Fig. 2. Simulation results showing the induced current distribution on both continuous (left) and splitted (right) ground planes for an inductor of the kind shown in Fig. 1. Note the opposite currents on domain borders in the splitted ground plane

The inductance value, the auto-resonance frequency and the quality factor of the inductor are affected by the presence and the distance of these conductive planes. Moreover, neighbor circuits must be protected and shielded against the parasitic currents induced on conductive planes to avoid undesired couplings and/or malfunctioning of the electronics. Preliminary simulations have shown that such parasitic currents are well confined in the portion of the conductive plane superimposed to the coil, and therefore a splitted reference plane should provide a sufficient degree of electrical isolation between the coil and the rest of the electronics. Nonetheless, a careless use of partitioning in simulation tools gives the result shown in Fig. 2. The use of the "distributed" results in a "breakdown" system solving approach can lead to errors, although the aggregate result of the intermediary simulation steps is correct.

By generalizing the considerations made for the aforementioned specific cases, the tasks that simulation tools dedicated to SiP design should be able to perform are the following:

- element value calculation
- coupling and parasitic effects evaluation;
- analysis of signal propagation;
- power system design;
- full circuital simulation;
- EM simulation (partial and/or full wave);
- thermal simulation;
- other application specific tasks.

An important feature that a complete simulation environment should have is the possibility to work with different manufacturing technologies, and be open to future upgrades and enhancements of these technologies.

4 Simulation strategies for SiP

The high integration achieved in SiP technologies creates a highly dense three-dimensional circuit, in which active semiconductor devices are interconnected between them and with the external world using a web of thin conductors and embedded passives. The close integration of active and passive devices requires a careful study of the couplings inside SiP structures at the development stage. The correct estimation of electromagnetic effects not only enhances the overall quality of the SiP, but also provides mandatory information to the designer who needs to integrate the SiP in an even more complex system. Electromagnetic phenomena can be described very accurately by the solution of Maxwell's equations. Unfortunately, the equations cannot be solved exactly, excluding some special cases that have little importance in practice. The very strong predictive power of the equations can be unleashed only with the development of efficient numerical solutions. Main issues are relative to the numerical analysis of the electromagnetic behavior of complex three-dimensional interconnect structures, in particular in the area of interconnect analysis, signal integrity issues and behavioral modeling (needed for extraction techniques and model order reduction).

Beyond the numerical method employed, electromagnetic solvers may also be classified on the basis of the domain in which solution is calculated: frequency or time domain.

The most successfully commercial electromagnetic simulation tools are frequency domain solvers. In particular, those based on the Finite Element Method (FEM), which provide good accuracy in reasonable simulation times. For a first level simulation, or for relatively simple structures, they are a good option for SiP design, due to their stability and availability of interfaces to/from other design tools. However, they have problems for complex structures.
Among the numerical methods for the solution of electromagnetic problems, time-domain solution methods are receiving increasing interest, mainly thanks to their more relaxed CPU and memory requirements, with respect to frequency domain methods, in the analysis of high-frequency/wideband applications. Moreover, time domain methods can fully capture the physics behind electromagnetic radiation phenomena, thus providing detailed indications to the designer on the entity, locations and countermeasures about the effects of discontinuities, radiation/interference phenomena, shielding structures, etc.[1]. A further advantage of time domain solution methods is that nonlinear devices can only be accurately described in the time-domain, thus board level or system level simulations can be easily performed in transient mode[4] The Finite Difference Time Domain (FDTD) method is a widely used numerical algorithm to solve Maxwell's equations. However it has some drawbacks, in particular:

- the entire computational domain must be meshed, the step size must be small compared to the smallest wavelength and the smallest feature in the model;
- if the field values are required at some distance the computational domain is excessively large;
- the method requires strictly stability conditions.

Because of these drawbacks it is difficult to take into account the very local effects like the skin effect or current crowding. To deal with the very fine structures needed to model this effect, a very small time-step must be taken, and the matrices grow to unmanageable sizes. To tackle these problems local refinement of the mesh in the existing FDTD method and accelerated time integration processes might be applied [5]. To deal with these computational disadvantages of FDTD a fundamental research on the compact representation of FDTD simulation results is needed. Also in other areas of application Reduced Order Modeling can solve the problem of prohibitively large computations. Consider, for instance, the influence of interconnect structures on chips and printed circuit boards. Although we are interested in the coupling effect of these structures, a detailed simulation of the electromagnetic effects is not needed. Of real interest are mainly the inputs and outputs of the interconnect. Therefore, one is able to capture the main electromagnetic effects into a compact model. The model behaves like a black-box and the inputs and outputs can be coupled to the circuit simulation program. In this process numerical mathematics plays an important role, but also the theory of electromagnetism is used. Several of these techniques are very promising[1]. They, for instance, are very accurate and preserve the stability and the passivity of the underlying model. Although promising, not one of the methods is implemented in a stand-alone application which simultaneously satisfies all the requirements of a commercial integrated software package.

For the purpose of comparison, the simulated frequency response of a relatively simple, multilayer, resonating structure, obtained using several commercially available electromagnetic solvers that implement different solving methods, are presented in Fig. 3, together with experimental results. The electromagnetic solvers used to achieve this result make use of time domain (TD) methods (Transmission Line Matrix Method, or TLMM) and frequency domain (FD) methods (Method of Moments, or MoM; Fast Multipole expansion Method, or FMM; and Finite Element Method, or FEM). As it can be seen from such a comparison, for not very intricate structures, the accuracy of all tools is equivalent. The computational differences become evident only when simulating complex structures.

Beside the functional blocks composing the whole system, accurate modeling is also needed for all the elementary structures realized at the board or package level and which represent discontinuities from the straight interconnection (i.e. a straight segment of transmission line). Such structures include corners and bends, vias and holes, embedded elements, test points, connectors, etc. The detailed characterization of such discontinuities is needed to predict and ensure proper devices and systems operation.
A possible design strategy for SiP analysis and design can consists in the following steps:

- Divide the complex problem into simpler 'basic blocks'.
- Solve each block by applying the suitable model and procedures.
- 'Merge' the individual block results to reconstruct the original system's model and simulate it.

To solve each basic block several simulation engines and modeling techniques are used:

- Circuit simulators (Spice-like).
- Transmission Lines simulators.
- 2D, 2.5D and 3D field solvers generating lumped, distributed or mixed models.
- Models (exact, simplified, empiric, etc.).
- Electro-Magnetic simulators.
- Thermal Analysis and simulation tools.
- etc.

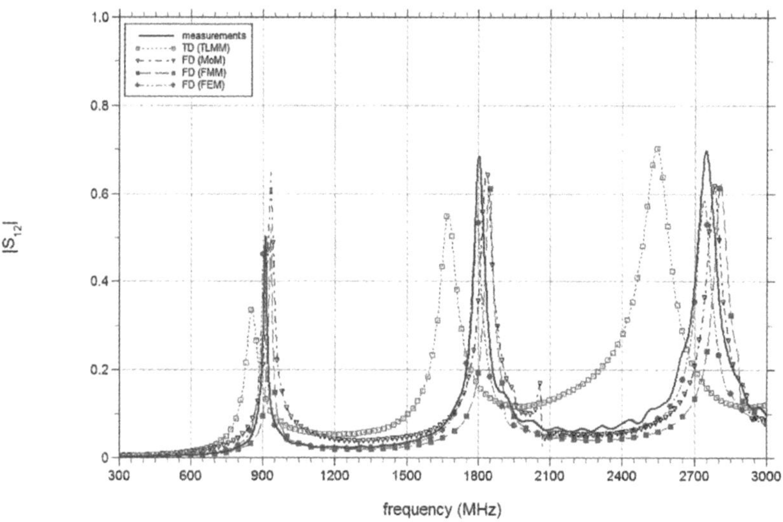

Fig. 3. Comparison of the measured frequency response of a resonating structure with the simulated responses calculated by means of different commercial electromagnetic solvers

A hypothetical, future, integrated analysis and design software package for SiP applications would have to satisfy also the following requirements:

- Perform in automatic the partitioning of the complex system.
- Automatically selects the appropriate simulation tool for each basic block.
- Provide robust simulation algorithms.
- Good integration with all other CAD tools used in the design process without operator intervention.
- Last, but not least, short simulation time is a stringent requirement.

5 Conclusions

This paper has addressed the issues related to the electromagnetic modeling and simulation of interconnects in System-in-Package (SiP) applications. SiP shows potentials to be a valuable approach to electronic systems integration. It has many advantages over other approaches, like System-on-Chip: higher flexibility and modularity, shorter development time and lower cost, ability to integrate several different technologies on the same board/package. However, SiP performance are strongly affected by the electromagnetic properties of interconnect structure, embedded passive devices, exotic materials, etc. Therefore a suitable modeling, simulation and computer-aided design framework is required to allow exploiting the full potentials of these technologies. The requirements and characteristics of such a framework have been listed here. A number of software package dedicated to electromagnetic numerical analysis, circuit simulation, layout design and so on are commercially available, each with peculiar characteristics which make them particularly suitable for a given set of functions, frequency range, application. However, a development framework specifically dedicated to SiP design, fully integrated in the design and manufacturing process flow, and which can satisfy all the requirements listed above, has yet to come, even if significant research efforts are devoted to this problem.

References

1. Ruehli, A. E., Cangellaris, A. C.: Progress in the methodologies for the electrical interconnects and electronic packages. Proceedings of the IEEE, vol. 89, No. 5, May 2001, pp. 740-771
2. R. Compañó (Editor): Technology Roadmap for Nanoelectronics, 2nd Ed., November 2000, European Communities

3. International Technology Roadmap for Semiconductors, Modeling and Simulation, 2001 Edition, ITRS
4. Ramachandra, A., Nakhla, M. S.: Simulation of High-Speed Interconnect. Proceedings of the IEEE, vol. 89, No. 5, May 2001, pp. 693-728
5. Horváth, R.: A review and comment of the recent FDTD literature from the point of view of the numerical solution fastness. Reports on Applied and Numerical Analysis (RANA), Eindhoven University of technology, report 01-23, 2001

General Linear Methods for Nonlinear DAEs in Circuit Simulation

S. Voigtmann

Humboldt University Berlin, Institute of Mathematics, Unter den Linden 6, D-10099 Berlin, Germany,
steffen@math.hu-berlin.de

Abstract The modified nodal analysis (MNA) leads to differential algebraic equations with properly stated leading terms. In this article a special structure of the DAEs modelling electrical circuits is exploited in order to derive a new decoupling for nonlinear index-2 DAEs. This decoupling procedure leads to a solvability result and is also used to study general linear methods, a class of numerical schemes that covers both Runge-Kutta and linear multistep methods. Convergence for index-2 DAEs is proved.

1 Introduction

When simulating electrical circuits, one is confronted with solving differential algebraic equations (DAEs) of the form

$$A(t)\frac{d}{dt}d(x(t),t) + b(x(t),t) = 0, \qquad t \in \mathcal{I}. \tag{1}$$

In case of the charge oriented modified nodal analysis the vector d contains charges and fluxes while x represents all node potentials and currents of voltage defining elements like voltage sources and inductors. Typically the index of (1) does not exceed 2 [Est00, Theorem 3.1.3].

Common circuit simulators like SPICE or TITAN use the so-called charge oriented approach $A(t)(R(t)y(t))' + b(x(t),t) = 0$, $y(t) - d(x(t),t) = 0$, where charges and fluxes are introduced as a new variable y. Notice that this enlarged system is of the form

$$A(t)(D(t)x(t))' + b(x(t),t) = 0, \qquad t \in \mathcal{I}. \tag{2}$$

Solutions lie in the linear space $C_D^1(\mathcal{I},\mathbb{R}^m) := \{ z \in C(\mathcal{I},\mathbb{R}^m) \mid Dz \in C^1(\mathcal{I},\mathbb{R}^n) \}$.

Using the concept of the tractability index [Mär03] we study DAEs (2) having index $\mu \in \{1,2\}$. In Sect. 2 we will exploit the specific structure of the MNA equations to derive a decoupling procedure for nonlinear index-2 DAEs. This will enable us to prove existence and uniqueness of solutions. In Sect. 3 we study general linear methods for (2) and prove convergence.

We assume that $\mathcal{I}$ is a compact interval, $\mathcal{D} \subset \mathbb{R}^m$ a domain and that $A : \mathcal{I} \to L(\mathbb{R}^n,\mathbb{R}^m)$, $D : \mathcal{I} \to L(\mathbb{R}^m,\mathbb{R}^n)$ and $b : \mathcal{D} \times \mathcal{I} \to \mathbb{R}^m$ are continuous. Let b_x' exist and be continuous. Finally, the leading term of (2) is supposed to be properly stated, i.e. $\ker A(t) \oplus \operatorname{im} D(t) = \mathbb{R}^n$ for $t \in \mathcal{I}$ and there is a smooth projector function $R \in C^1(\mathcal{I}, L(\mathbb{R}^n,\mathbb{R}^n))$ such that $\ker R(t) = \ker A(t)$, $\operatorname{im} R(t) = \operatorname{im} D(t)$ (see [HM04]).

For analysing (2) we introduce the following sequence of matrix functions and subspaces defined pointwise for $t \in \mathcal{I}$ and $x \in \mathcal{D}$.

$$\left.\begin{aligned}
G_0(t) &= A(t)D(t), \qquad B_0(x,t) = b_x'(x,t)\\
N_0(t) &= \ker G_0(t), \qquad S_0(x,t) = \{ z \in \mathbb{R}^m \mid B_0(x,t)z \in \operatorname{im} G_0(t) \},\\
Q_0(t) &\text{ is a projector onto } N_0(t),\ P_0(t) = I - Q_0(t),\\
G_1(x,t) &= G_0(t) + B_0(x,t)Q_0(t),\\
N_1(x,t) &= \ker G_1(x,t),\\
S_1(x,t) &= \{ z \in \mathbb{R}^m \mid B_0(x,t)z \in \operatorname{im} G_1(x,t) \}.
\end{aligned}\right\} \tag{3}$$

Let $Q_1(x,t)$ be a projector function onto N_1 and $P_1(x,t) = I - Q_1(x,t)$. Finally denote with $D^-(t)$ the generalised reflexive inverse of $D(t)$ defined by

$$DD^-D = D, \quad D^-DD^- = D^-, \quad D^-D = P_0, \quad DD^- = R.$$

Definition 1. (see [Mär03]) *The DAE (2) with a properly stated leading term is regular with tractability index $\mu \in \{1,2\}$ on $\mathcal{D}\times\mathcal{I}$ if there is a sequence (3) such that for $(x,t) \in \mathcal{D}\times\mathcal{I}$*

(i) G_i has constant rank $r_i < m$ for $0 \le i < \mu$,
(ii) Q_i is continuous for $i = 0,\dots,\mu-1$, $Q_1(x,t)Q_0(t) = 0$ and DN_1, DS_1 are spanned by continously differentiable basis functions,
(iii) $N_{\mu-1} \cap S_{\mu-1} = \{0\}$.

Observe that for index-1 equations, $Q_1 = 0$ and $Q_1Q_0(t) = 0$ trivially holds. For index-2 DAEs $G_2(x,t) = G_1(x,t) + B_0(x,t)P_0(t)Q_1(x,t)$ remains nonsingular on $\mathcal{D}\times\mathcal{I}$ and we have $N_1(x,t) \oplus S_1(x,t) = \mathbb{R}^m$. In the following we will adopt the convention to choose Q_1 to be the canonical projector onto N_1 along S_1. Due to $N_0 \subset S_1$ the property $Q_1Q_0 = 0$ is then always given.

The space $N_0(t) \cap S_0(x,t) = \operatorname{im} Q_0(t)Q_1(x,t)$ (see [Est00]) is of vital importance as it describes the so-called index-2 components, i.e. the particular part of the solution that can be calculated only by performing an inherent differentiation process. In [Est00] it is shown that the circuit's layout determines this subspace. Thus it is independent of x. We choose a projector function $T(t)$ onto $N_0(t) \cap S_0(x,t)$ that depends on t only. Note that $U = I - T$ satisfies $\ker U(t) = \operatorname{im} Q_0(t)Q_1(x,t)$. T can be chosen such that $TP_0 = 0$ and $P_0T = 0$ for $t \in \mathcal{I}$. Then the following properties are valid: $Q_0T = T = TQ_0$, $Q_1T = 0$, $P_0U = P_0 = UP_0$, $Q_1UQ_0 = 0$.

2 Decoupling nonlinear index-2 equations

From [Est00, Corollary 3.2.8] it is well known that for the charge oriented modified nodal analysis the index-2 components Tx enter the equations only in a linear way, i.e. (2) has the structure

$$A(t)(D(t)x(t))' + b(U(t)x(t),t) + \mathfrak{B}(t)T(t)x(t) = 0. \tag{4}$$

This particular form of the DAEs arising in circuit simulation makes it possible to develop a new decoupling procedure for index-2 DAEs. For a given solution $x(\cdot)$ of (4) denote $x_0 = x(t_0)$ and introduce the new variable

$$w = \bar{P}_1D^-(Dx)' + (Q_0 + \bar{Q}_1)x \tag{5}$$

where $\bar{P}_1(t) = P_1(x_0,t)$ and $\bar{Q}_1(t) = Q_1(x_0,t)$. Here and in the sequel t arguments are generally omitted for better readability. Notice that

$$\bar{Q}_1w = \bar{Q}_1x, \qquad Q_0w = -Q_0\bar{Q}_1D^-(Dx)' + Q_0x + Q_0\bar{Q}_1x,$$
$$D\bar{P}_1w = D\bar{P}_1D^-(Dx)'.$$

From $G_1(x_0,\cdot)\bar{Q}_1 = 0$ we infer $A(Dx)' + \mathfrak{B}Tx = (AD + \mathfrak{B}T)w$ and, denoting $u = D\bar{P}_1x$, we find

$$x = P_0\bar{P}_1x + P_0\bar{Q}_1x + Q_0x = D^-u + (P_0\bar{Q}_1 + Q_0\bar{P}_1)w + Q_0\bar{Q}_1D^-(Dx)'.$$

The component $Ux = D^-u + (P_0\bar{Q}_1 + UQ_0)w$ can be written in terms of u and w such that (4) is equivalent to

$$F(u,w,t) := (AD + \mathfrak{B}T)w + b(D^-u + (P_0\bar{Q}_1 + UQ_0)w,\cdot) = 0. \tag{6}$$

Lemma 1. *Let (2) be a regular DAE with index $\mu \in \{1,2\}$. Let $y_0 \in \operatorname{im} D(t_0)$, $(x_0,t_0) \in \mathcal{D}\times\mathcal{I}$ be given such that $A(t_0)y_0 + b(U(t_0)x_0,t_0) + \mathfrak{B}(t_0)T(t_0)x_0 = 0$. Denote*

$$u_0 = D(t_0)\bar{P}_1(t_0)x_0, \qquad w_0 = \bar{P}_1(t_0)D^-(t_0)y_0 + (Q_0 + \bar{Q}_1)(t_0)x_0$$

and consider F to be defined on a neighbourhood $\mathcal{N}_0 \subset \mathbb{R}^n \times \mathbb{R}^m \times \mathbb{R}$ of (u_0,w_0,t_0). Then there is a neighbourhood $\mathcal{N}_1 \subset \mathbb{R}^n \times \mathbb{R}$ of (u_0,t_0) and a continuous mapping $\mathfrak{z} : \mathcal{N}_1 \to \mathbb{R}^m$ such that $F(u,\mathfrak{z}(u,t),t) = 0 \quad \forall\ (u,t) \in \mathcal{N}_1$.

Proof. Due to (6) we have $F(u_0,w_0,t_0) = 0$ and

$$F'_w(u,w,\cdot) = AD + \mathfrak{B}T + b'_x(D^-u + (P_0\bar{Q}_1 + UQ_0)w,\cdot)\,(P_0\bar{Q}_1 + UQ_0)$$

implies that $F'_w(u_0,w_0,t_0) = G_2(x_0,t_0)$ is nonsingular. Thus the assertion follows from the implicit function theorem. □

Notice that the mapping $\precapprox$ from the previous lemma is defined only locally around (u_0, t_0). For simplicity we assume that the interval $\mathcal{I}$ is sufficiently small such that $\precapprox$ is defined for all $t \in \mathcal{I}$.

We arrive at the following representation of the solution:

$$x = D^- u + (Q_0 \bar{P}_1 + P_0 \bar{Q}_1) \precapprox(u, \cdot) + Q_0 \bar{Q}_1 D^- (u + D\bar{Q}_1 \precapprox(u, \cdot))'. \tag{7}$$

The component $u = D\bar{P}_1 x$ satisfies the ordinary differential equation

$$D\bar{P}_1 D^- u' = D\bar{P}_1 \precapprox(u, \cdot) - D\bar{P}_1 D^- (D\bar{Q}_1 \precapprox(u, \cdot))'. \tag{8}$$

As for linear DAEs this equation will be called the inherent regular ODE. Since, by the index-2 condition, $D\bar{P}_1 D^- \in C^1(\mathcal{I}, \mathbb{R}^n)$, we may rewrite (8) as

$$u' = (D\bar{P}_1 D^-)' u + D\bar{P}_1 \precapprox(u, \cdot) + (D\bar{P}_1 D^-)' D\bar{Q}_1 \precapprox(u, \cdot). \tag{9}$$

Similar to [HM04] we will now study (9) without assuming the existence of a solution.

Theorem 1. *Let the assumptions of Lemma 1 be satisfied. Then*

(i) $\mathrm{im}\, D\bar{P}_1 D^-$ *is a (time-varying) invariant subspace of the inherent ODE* (9)*, i.e.* $u(t_0) \in \mathrm{im}\,(D\bar{P}_1 D^-)(t_0)$ *implies* $u(t) \in \mathrm{im}\,(D\bar{P}_1 D^-)(t)\ \forall\ t \in \mathcal{I}$.

(ii) If the subspaces $\mathrm{im}\, D\bar{P}_1 D^-$ *and* $\mathrm{im}\, D\bar{Q}_1 D^-$ *are constant, then* (9) *simplifies to* $u' = D\bar{P}_1 \precapprox(u, \cdot)$, $u(t_0) \in \mathrm{im}\,(D\bar{P}_1 D^-)(t_0)$.

Proof. Similar to [HM04, Theorem 2.2]. Replace R by $D\bar{P}_1 D^-$.

Theorem 2. *Let the assumptions of Lemma 1 be satisfied. Assume that the mapping* $t \mapsto D(t) Q_1(t)(x_0, t) \precapprox(u(t), t)$ *is* C^1*, where* u *is the solution of the inherent regular ODE* (9) *with initial value* $u(t_0) = DP_1(x_0, t_0) x_0$*. Then there is a unique solution* $x \in C^1_D(\mathcal{I}, \mathbb{R}^m)$ *of the initial value problem*

$$A(t)(Dx)'(t) + b((Ux)(t), t) + (\mathfrak{B}T)(t) x(t) = 0, \qquad DP_1(x_0, t_0)(x_0 - x(t_0)) = 0.$$

Proof. From Lemma 1 we get the mapping $\precapprox(u, t)$ and thus the solution u of the inherent regular ODE (9). Due to Theorem 1 $u(t) \in \mathrm{im}\, D(t) P_1(x_0, t) D^-(t)$ holds for all t where u is defined. Then the mapping x as defined in (7) is a solution since

$$\begin{aligned} A(Dx)' + b(Ux, \cdot) + \mathfrak{B}Tx &= A(Dx)' + b(Ux, \cdot) + \mathfrak{B}Tx - F(u, \precapprox(u, \cdot), \cdot) \\ &= (AD + \mathfrak{B}T)\bar{Q}_1 D^- (Dx)' + AD\bar{P}_1 D^- (Dx)' - AD\bar{P}_1 \precapprox(u, \cdot) = 0. \end{aligned}$$

□

Remark 1. If (2) was an index-1 DAE, then $\bar{P}_1 = I$, $\bar{Q}_1 = 0$ and all results can be reinterpreted also for index-1 equations. In particular, (5) reduces to $w = D^-(Dx)' + Q_0 x$. This is exactly the mapping studied in [HM04] and the decoupling procedure presented here generalises [HM04].

3 Numerical Integration by General Linear Methods

For the numerical solution of index-2 equations (4) we investigate general linear methods (GLM). This class of methods seems to be very attractive for circuit simulation since there are methods with diagonally implicit structure that have high stage order and a stability behaviour similar to fully implicit Runge-Kutta methods. Examples of such methods are given in [But03, Wri03]. The diagonally implicit structure yields a very efficient implementation. We will demonstrate this by studying an example at the end of this section. Also, recall that both Runge-Kutta and linear multistep methods can be cast into general linear form.

A GLM is given by a partitioned matrix $M = \left[\begin{array}{c|c} \mathcal{A} & \mathcal{U} \\ \hline \mathcal{B} & \mathcal{V} \end{array}\right] \in L(\mathbb{R}^{s+r}, \mathbb{R}^{s+r})$. We will always assume that $\mathcal{A}$ is nonsingular. The discretisation of the DAE (4) using the general linear method M reads

$$A_{ni}[DX]'_{ni} + b(U_{ni} X_{ni}, t_{ni}) + \mathfrak{B}_{ni} T_{ni} X_{ni} = 0, \quad i = 1, \ldots, s, \tag{10a}$$

$$[DX]_n = h(\mathcal{A} \otimes I_m)[DX]'_n + (\mathcal{U} \otimes I_m)[Dx]^{[n-1]}, \tag{10b}$$

$$[Dx]^{[n]} = h(\mathcal{B} \otimes I_m)[DX]'_n + (\mathcal{V} \otimes I_m)[Dx]^{[n-1]}. \tag{10c}$$

For better readability we will drop the Kronecker products in the sequel. As in the case of linear multistep methods r pieces of information $[Dx]_k^{[n-1]} \in \mathbb{R}^m$, $k = 1, \dots, r$, are passed on from step to step. These quantities represent some approximations to $D(t)x(t)$ or it's derivative. See [But03] for more details. Observe that only information about the exact solution's D component is carried on. Thus errors in the null-space of D are not propagated.

Similar to Runge-Kutta methods internal stages $X_{ni} \in \mathbb{R}^m$, $i = 1, \dots, s$, are calculated at intermediate time points $t_{ni} = t_{n-1} + c_i h$ within every step. In (10) we wrote $A_{ni} = A(t_{ni})$ and used similar notations for $\mathfrak{B}$, T and U. For compactness of notation we introduced $X_n = \left(X_{n1}^T \cdots X_{ns}^T\right)^T \in \mathbb{R}^{ms}$ and similarly $[DX]_{ni} = D_{ni}X_{ni}$. The initial input vector $[Dx]^{[0]}$ can be calculated by generalised Runge-Kutta methods [But03].

From [HMT03] it is well known that one should investigate numerically qualified DAEs in order to get good numerical results. We will therefore restrict attention to DAEs where the subspaces $\operatorname{im} D\bar{P}_1 D^-$ and $\operatorname{im} D\bar{Q}_1 D^-$ are constant. Recall from Theorem 1 that the inherent regular ODE (9) now reads

$$u' = D\bar{P}_1 \precapprox(u, \cdot), \qquad u(t_0) \in \operatorname{im}(D\bar{P}_1 D^-)(t_0). \tag{11}$$

We want to apply the decoupling procedure to the discretised problem (10). Therefore we need to split the vector $[Dx]^{[n-1]}$ into it's $D\bar{P}_1$ and $D\bar{Q}_1$ parts. If $[Dx]^{[0]}$ was calculated by a generalised Runge-Kutta method, then

$$[Dx]_k^{[0]} = \mathbf{u}_k^{[0]} + \mathbf{v}_k^{[0]} \in \operatorname{im}(D\bar{P}_1)(t_0) \oplus \operatorname{im}(D\bar{Q}_1)(t_0), \qquad k = 1, \dots, r. \tag{12}$$

Splitting the stages $\mathbf{U}_{ni} = D_{ni}\bar{P}_{1,ni}X_{ni}$, $\mathbf{V}_{ni} = D_{ni}\bar{Q}_{1,ni}X_{ni}$ and defining $\mathbf{U}_n'$, $\mathbf{V}_n'$ by $\mathbf{U}_n = h\mathcal{A}\mathbf{U}_n' + \mathcal{U}\mathbf{u}^{[n-1]}$ and $\mathbf{V}_n = h\mathcal{A}\mathbf{V}_n' + \mathcal{U}\mathbf{v}^{[n-1]}$, respectively, we find that (12) holds not only for the first but for every step, since

$$\mathbf{u}^{[n]} = \mathcal{B}\mathcal{A}^{-1}\mathbf{U}_n + \mathcal{M}_\infty \mathbf{u}^{[n-1]}, \qquad \mathbf{v}^{[n]} = \mathcal{B}\mathcal{A}^{-1}\mathbf{V}_n + \mathcal{M}_\infty \mathbf{v}^{[n-1]}.$$

Notice that $\mathcal{M}_\infty = \mathcal{V} - \mathcal{B}\mathcal{A}^{-1}\mathcal{U}$ is the methods stability matrix $\mathcal{M}(z)$ evaluated at infinity. This matrix plays a role similar to $R(\infty) = 1 - b^T \mathcal{A}^{-1}\mathbb{1}$ for Runge-Kutta methods.

As in (5) we define $\mathbf{W}_{ni} = P_{1,ni}D_{ni}^-[DX]_{ni}' + (Q_{0,ni} + \bar{Q}_{1,ni})X_{ni}$ such that

$$X_{ni} = D_{ni}^-\mathbf{U}_{ni} + (Q_{0,ni}\bar{P}_{1,ni} + P_{0,ni}\bar{Q}_{1,ni})\mathbf{W}_{ni} + Q_{0,ni}\bar{Q}_{1,ni}D_{ni}^-[DX]_{ni}'. \tag{13}$$

From (10a) it follows that $F(\mathbf{U}_{ni}, \mathbf{W}_{ni}, t_{ni}) = 0$. Thus $\mathbf{W}_{ni} = \precapprox(\mathbf{U}_{ni}, t)$ is given by the mapping $\precapprox$ from Lemma 1. Here we have to assume that the stepsize h is small enough to guarantee that $(\mathbf{U}_{ni}, t_{ni})$ remains in the neighbourhood $\mathcal{N}_1$ of (u_0, t_0) where $\precapprox$ is defined.

Theorem 3. *Let M be a stiffly accurate general linear method with nonsingular $\mathcal{A}$. Assume that $\mathcal{V}$ is power bounded and that the spectral radius of $\mathcal{M}_\infty = \mathcal{V} - \mathcal{B}\mathcal{A}^{-1}\mathcal{U}$ is less than 1. Then M is convergent for numerically qualified DAEs (2) with index $\mu \in \{1, 2\}$.*

If M has order p and stage order q for ordinary differential equations, then the order of convergence is (at least) q.

Proof. Since $\mathbf{U}_{ni}' = D_{ni}\bar{P}_{1,ni}D_{ni}^-[DX]_{ni}' = D_{ni}P_{1,ni}\precapprox(\mathbf{U}_{ni}, t_{ni})$ holds for numerically qualified DAEs, the decoupling procedure shows that (10) is equivalent to the split system

$$\left.\begin{aligned} \mathbf{U}_{ni}' &= D_{ni}\bar{P}_{1,ni}\precapprox(\mathbf{U}_{ni}, t_{ni}), & \mathbf{V}_{ni} &= D_{ni}\bar{Q}_{1,ni}\precapprox(\mathbf{U}_{ni}, t_{ni}), \\ \mathbf{U}_n &= h\mathcal{A}\,\mathbf{U}_n' + \mathcal{U}\,\mathbf{u}^{[n-1]}, & \mathbf{V}_n &= h\mathcal{A}\,\mathbf{V}_n' + \mathcal{U}\,\mathbf{v}^{[n-1]}, \\ \mathbf{u}^{[n]} &= h\mathcal{B}\,\mathbf{U}_n' + \mathcal{V}\,\mathbf{u}^{[n-1]}, & \mathbf{v}^{[n]} &= h\mathcal{B}\,\mathbf{V}_n' + \mathcal{V}\,\mathbf{v}^{[n-1]}. \end{aligned}\right\} \tag{14}$$

The left hand block of equations is exactly the numerical scheme resulting from applying M directly to the inherent regular ODE (11). Thus ODE theory for general linear methods [But03] yields

$$\mathbf{U}_{ni} = u(t_{n-1} + c_i h) + \mathcal{O}(h^{\tilde{q}+1}), \qquad \mathbf{U}_{ni}' = u'(t_{n-1} + c_i h) + \mathcal{O}(h^{\tilde{q}+1}),$$

where we denoted $\tilde{q} = \min(p-1, q)$. Let u be the inherent ODE's exact solution and introduce

$$v(t) = D(t)\bar{Q}_1(t)\precapprox(u(t), t).$$

Then

$$\|\mathbf{V}_{ni} - v(t_{ni})\| \le \int_0^1 \|\precapprox_u'(\tau\mathbf{U}_{ni} + (1-\tau)u(t_{ni}), t_{ni})\| \mathrm{d}\tau \, \|\mathbf{U}_{ni} - u(t_{ni})\|$$

and thus $\mathbf{V}_{ni} = v(t_{n-1} + c_i h) + \mathcal{O}(h^{\tilde{q}+1})$. Denoting exact input quantities by $\hat{\mathbf{v}}^{[n]}$ and using techniques from [HLR89, Theorem 3.1] we obtain the recursion

$$\Delta \mathbf{v}^{[n]} = \mathcal{M}_\infty^n \Delta \mathbf{v}^{[0]} + \sum_{i=1}^{n} \mathcal{M}_\infty^{n-i} \delta_i$$

where $\Delta \mathbf{v}^{[n]} = \mathbf{v}^{[n]} - \hat{\mathbf{v}}^{[n]}$ and $\delta_{ij} = \mathcal{B}\mathcal{A}^{-1}(\mathbf{V}_{ij} - v(t_{i-1} + c_j h)) = \mathcal{O}(h^{\tilde{q}+1})$. Given that the spectral radius of $\mathcal{M}_\infty$ is less than 1 and $\Delta \mathbf{v}^{[0]} = \mathcal{O}(h^{\tilde{q}+1})$ we find $\Delta \mathbf{v}^{[n]} = \mathcal{O}(h^{\tilde{q}+1})$ and, consequently, $\mathbf{V}'_{ni} = v'(t_{n-1} + c_i h) + \mathcal{O}(h^{\tilde{q}})$.

By assumption the general linear method has stiff accuracy, i.e. the numerical result $x_n = X_{ns}$ coincides with the last stage. Thus we can use (7), (13) to find a bound for the global error

$$\begin{aligned} \|x_n - x(t_n)\| \leq \; & C_1 \|\mathbf{U}_{ns} - u(t_n)\| + C_2 \|\mathfrak{z}(\mathbf{U}_{ns}, t_n) - \mathfrak{z}(u(t_n), t_n)\| \\ & + C_3 (\|\mathbf{U}'_{ns} - u'(t_n)\| + \|\mathbf{V}'_{ns} - v'(t_n)\|) = \mathcal{O}(h^{\min(p-1, q)}) \end{aligned}$$ □

If $p > q$, Theorem 3 predicts order q behaviour for the global error. This agrees with the results in [HLR89]. However, since the proof above is given for general linear methods it not only covers Runge-Kutta methods but also linear multistep methods and even more general methods such as those studied in [Wri03]. In particular, for general linear methods $p = q$ is possible even for diagonally implicit methods. Also the BDF methods have the same property. With a global error of order $\mathcal{O}(h^p)$ they indeed have a higher order than predicted by Theorem 3. From [BCP96] it is known that the k step BDF methods exhibit the true order of convergence for index-2 DAEs only after $k + 1$ steps.

For general linear methods a similar statement holds. For completeness we formulate this result in the following remark.

Remark 2. Let the assumptions of theorem 3 hold. Assume that, in addition, $p = q \geq 2$ and $\mathcal{M}_\infty^{k_0} = 0$. Then M is convergent for (2) with order p after $k_0 + 1$ steps.

Notice that (14) is the general linear method's discretisation of a Hessenberg index-1 DAE

$$u'(t) = f(u(t), t), \qquad\qquad v(t) = g(u(t), t) \tag{15}$$

using the direct approach as in [HLR89]. In order to prove the statement of remark 2 one needs to show that $\mathbf{V}'$ is calculated with order $p = q$ when M is applied to (15).

To reach this goal a careful analysis of numerical methods for fully implicit index-1 DAEs can be performed using the language of B-series for differential algebraic equations. In [Kvæ90] Kværnø studied the case of Runge-Kutta methods, but her approach has to be generalised for the much larger class of general linear methods.

The technical effort introducing elementary differentials and B-series for fully implicit index-1 DAEs is far too much to be presented here. We will therefore skip the proof which will be given in [Voi05]. We conclude this article by studying a benchmark circuit.

Example 1. Consider the ring modulator depicted in Fig. 1 that mixes a low-frequency signal $U_{in\,1}$ with a high-frequency input signal $U_{in\,2}$. The circuit is modelled by a system of 11 ODEs and 4 algebraic equations [DR89, Pul02]:

$$\begin{aligned} C\dot{U}_1 &= I_1 - I_3/2 + I_4/2 + I_7 - U_1/R, & L_H \dot{I}_1 &= -U_1, \\ C\dot{U}_2 &= I_2 - I_5/2 + I_6/2 + I_8 - U_2/R, & L_H \dot{I}_2 &= -U_2, \\ 0 &= I_3 - d(U_{D_1}) + d(U_{D_4}), & L_{S_2} \dot{I}_3 &= U_1/2 - U_3 - R_{G_2} I_3, \end{aligned}$$

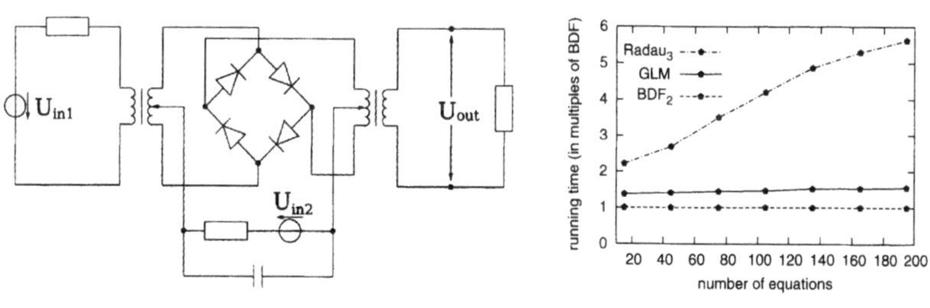

circuit diagram efficiency of numerical schemes

Fig. 1. The ring modulator circuit

$$\begin{aligned}
0 &= -I_4 + d(U_{D_2}) - d(U_{D_3}), & L_{S_3}\dot{I}_4 &= -U_1/2 + U_4 - R_{G_3}I_4,\\
0 &= I_5 + d(U_{D_1}) - d(U_{D_3}), & L_{S_2}\dot{I}_5 &= U_2/2 - U_5 - R_{G_2}I_5,\\
0 &= -I_6 - d(U_{D_2}) + d(U_{D_4}), & L_{S_3}\dot{I}_6 &= -U_2/2 + U_6 - R_{G_3}I_6,\\
C_p\dot{U}_7 &= -U_7/R_p + d(U_{D_1}) + d(U_{D_2}) & L_{S_1}\dot{I}_7 &= -U_1, \ + U_{in\,1}\\
&\quad - d(U_{D_3}) - d(U_{D_4}), & &\quad - (R_J + R_{G_1})I_7,\\
& & L_{S_1}\dot{I}_8 &= -U_2 - (R_C + R_{G_1})I_8.
\end{aligned}$$

As usual we used the abbreviations $U_{D_1} = U_3 - U_5 - U_7 - U_{in\,2}$, $U_{D_2} = -U_4 + U_6 - U_7 - U_{in\,2}$, $U_{D_3} = U_4 + U_5 + U_7 + U_{in\,2}$, $U_{D_4} = -U_3 - U_6 + U_7 + U_{in\,2}$ and $d(U) = \gamma(\exp(\delta U) - 1)$ with $\gamma = 40.67286402 \cdot 10^{-9} A$, $\delta = 17.7493332\, V^{-1}$. Notice that the index-2 model of the ring modulator is used here.

We solved the circuit equations using the general linear method

$$M = \left[\begin{array}{cc|cc} 1-\sqrt{2}/2 & 0 & 1 & 1-\sqrt{2}/2 \\ \sqrt{2}/4 & 1-\sqrt{2}/2 & 1 & \sqrt{2}/4 \\ \hline \sqrt{2}/4 & 1-\sqrt{2}/2 & 1 & \sqrt{2}/4 \\ 0 & 1 & 0 & 0 \end{array}\right]$$

in Nordsieck form having stiff accuracy and order and stage order $p = q = 2$. The diagonally implicit structure of M makes an efficient implementation possible. In order to prove this claim we solved several instances of the ring modulator simultaneously in order to produce problems of arbitrary size.

The solution was calculated on the interval $[0, 10^{-3}]$ using $rtol = atol = 10^{-4}$ and constant stepsize $h = 10^{-6}$. Fig. 1 contains a plot of the running time against the problem size. For comparison we chose the BDF method with order 2 and the RadauIIA scheme with two stages (order 3). Since we are solving an index-2 problem the order of convergence for the Radau method is 2 as well (the stage order) [HLR89]. Notice that we normalised the running time of the BDF method to 1.

It is clearly visible that the general linear method requires only a multiple of about 1.4 of the BDF's running time. This factor grows very slowly with the problem size. In contrast to this, the running time of the fully-implicit Runge-Kutta scheme grows much more rapidly.

References

[BCP96] K. E. Brenan, S. L. Campbell, and L. R. Petzold. *Numerical solution of initial-value problems in differential-algebraic equations*, volume 14 of *Classics in Applied Mathematics*. Society for Industrial and Applied Mathematics (SIAM), 1996

[But03] J. C. Butcher. *Numerical methods for ordinary differential equations*. John Wiley & Sons Ltd., Chichester, 2003

[DR89] G. Denk and P. Rentrop. Mathematical models in electrical circuit simulation and their numerical treatment. In Strehmel, editor, *Numerical Treatment of differential equations*, Fifth Seminar 'Numdiff-5', Halle, 1989

[Est00] D. Estévez Schwarz. *Consistent initialization for index-2 differential algebraic equations and it's application to circuit simulation*. PhD thesis, Humboldt Universität zu Berlin, 2000

[HLR89] E. Hairer, C. Lubich, and M. Roche. *The numerical solution of differential-algebraic systems by Runge-Kutta methods*, volume 1409 of *Lecture Notes in Mathematics*. Springer-Verlag, Berlin, 1989

[HM04] I. Higueras and R. März. Differential algebraic equations with properly stated leading terms. *Comp. & Math. with Appl.*, 48(1-2):215–235, 2004

[HMT03] I. Higueras, R. März, and C. Tischendorf. Stability preserving integration of index-2 DAEs. *Appl. Numer. Math.*, 45(2-3):201–229, 2003

[Kvæ90] A. Kværnø. Runge-Kutta methods applied to fully implicit differential-algebraic equations of index 1. *Math. Comp.*, 54(190):583–625, 1990

[Mär03] R. März. Differential algebraic systems with properly stated leading term and MNA equations. In *Modeling, simulation, and optimization of integrated circuits (Oberwolfach, 2001)*, volume 146 of *Internat. Ser. Numer. Math.*, pages 135–151. Birkhäuser, Basel, 2003

[Pul02] R. Pulch. A finite difference method for solving multirate partial differential algebraic equations. *ZAMM*, 2002

[Voi05] S. Voigtmann. General linear methods for nonlinear DAEs with properly stated leading terms. in preparation, 2005

[Wri03] W. Wright. *General linear methods with inherent Runge-Kutta stability*. PhD thesis, The University of Auckland, New Zealand, 2003

Colour Figures

Modeling and Simulation for Thermal-Electric Coupling in an SOI-Circuit

A. Bartel, U. Feldmann

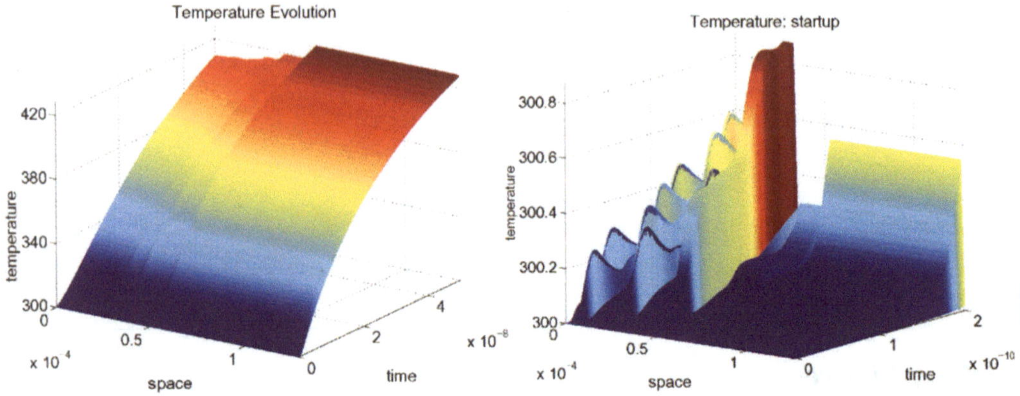

Fig. 3. (p. 31) Temperature distribution [0. s, 50.ns] (left), startup [0. s, 0.2ns] (right)

A Staggered ALE Approach for Coupled Electromechanical Systems

M. Greiff , U. Binit Bala, W. Mathis

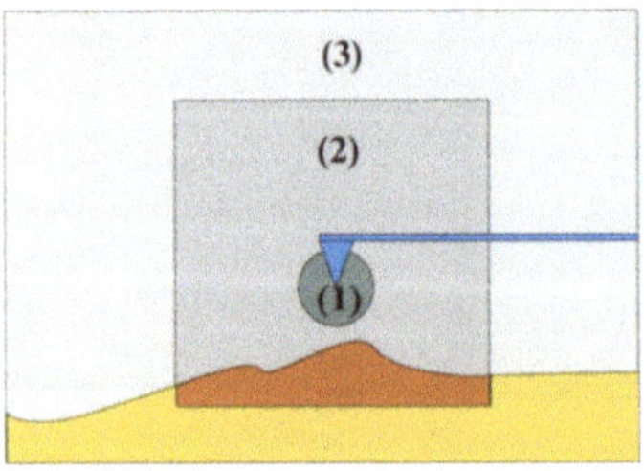

Fig. 3. (p. 36) 2D model including various numerical methods

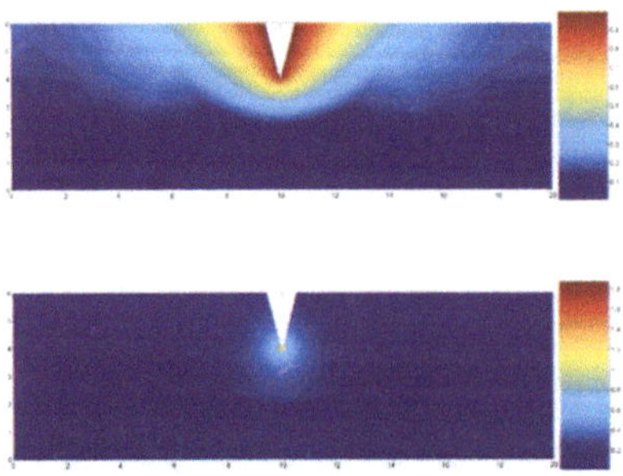

Fig. 4. (p. 38) Simulated potential and electric field by coupled FEM-RGM

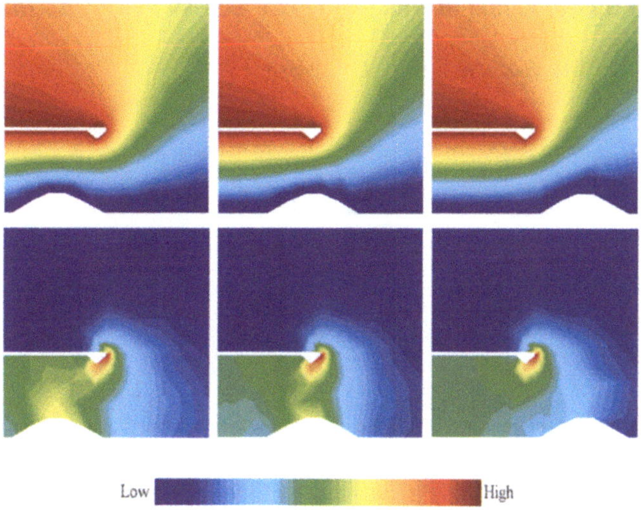

Fig. 5. (p. 38) Simulated potential and electric field by BEM

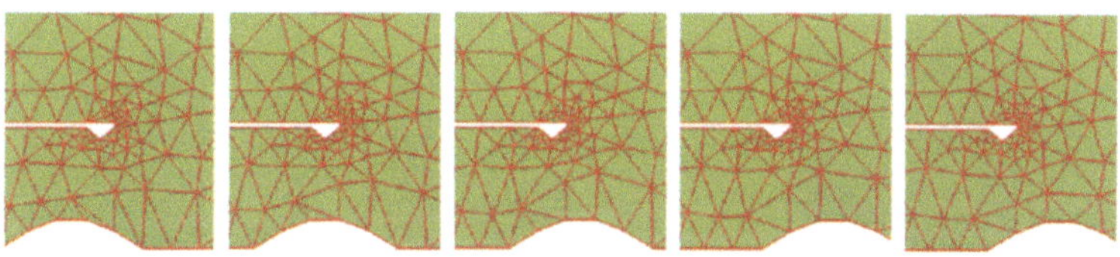

Fig. 6. (p. 39) ALE mesh update

Algebraic Sparsefied Partial Equivalent Electric Circuit (ASPEEC)

D. Ioan, G. Ciuprina, M. Rădulescu

Fig. 5. (p. 51) Codestar meander resistor benchmark - RPOLY2_ME

3-D FE Particle Based Model of Ion Transport Across Ionic Channels

M. E. Oliveri, S. Coco, D. S. M. Gazzo, A. Laudani, G. Pollicino

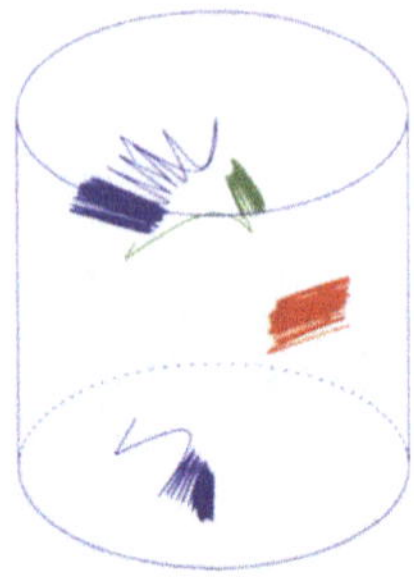

Fig. 3. (p. 62) Ion trajectories all confined inside the channel for a membrane voltage of 0mV (simulation interval 10ps)

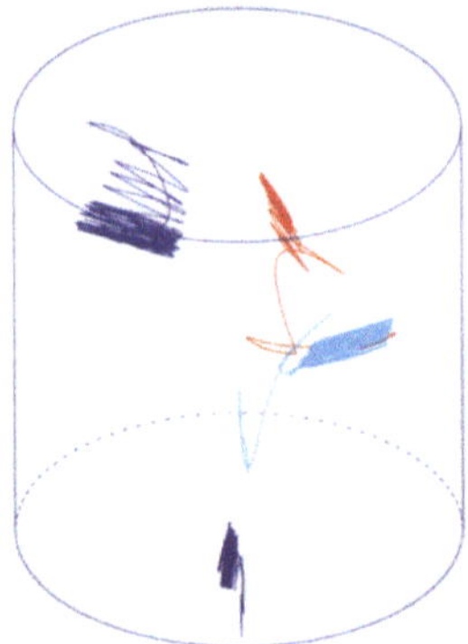

Fig. 4. (p. 63) Ion trajectories inside the channel for a membrane voltage of 100mV in the event of an ion exiting the channel (simulation interval 10ps)

Coupled Calculation of Electromagnetic Fields and Mechanical Deformation

U. Schreiber, U. van Rienen

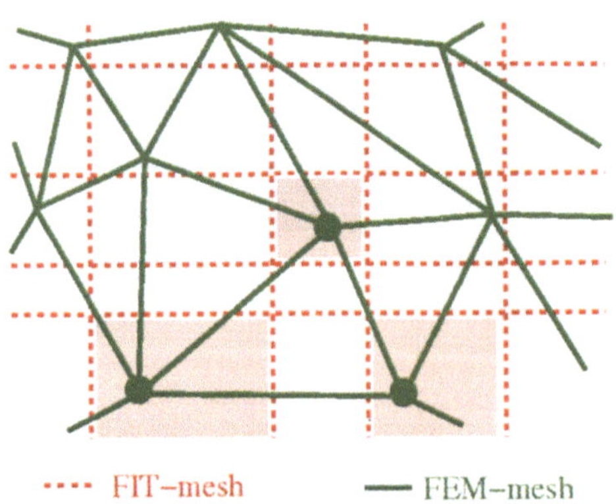

Fig. 1. (p. 66) Coupled Simulation Meshes

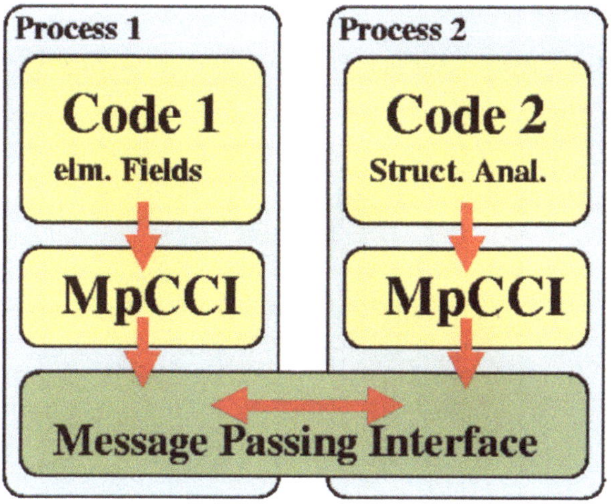

Fig. 2. **(p. 67)** MpCCI Software-Layers

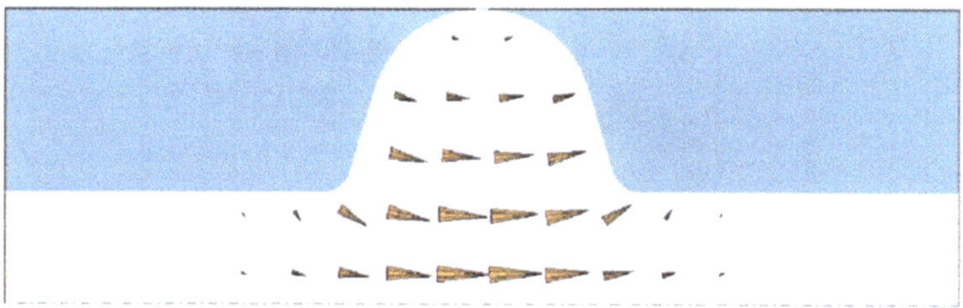

Fig. 3. **(p. 67)** Electric field distribution of the fundamental mode at f = 1.3 GHz in a one-cell-cavitiy type TESLA. The maximal field strength on axis is 25 MV/m

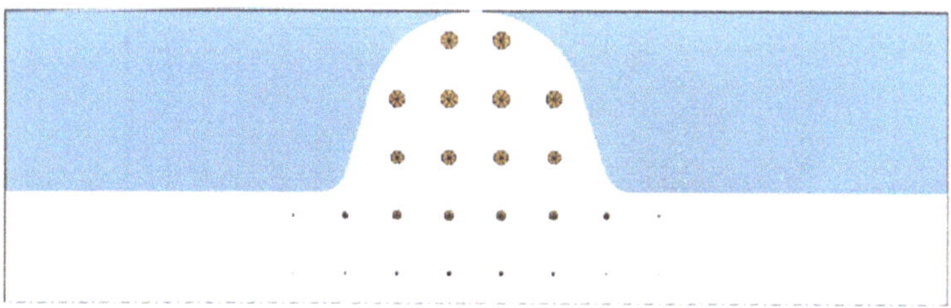

Fig. 4. **(p. 67)** Magnetic field distribution of the fundamental mode at f = 1.3 GHz in a one-cell-cavitiy type TESLA. The maximal field strength on axis is 25 MV/m

Challenging Coupled Problems in TCAD

A. Benvenuti, L. Bortesi, G. Carnevale, A. Ghetti, A. Pirovano, L. Vendrame, L. Zullino

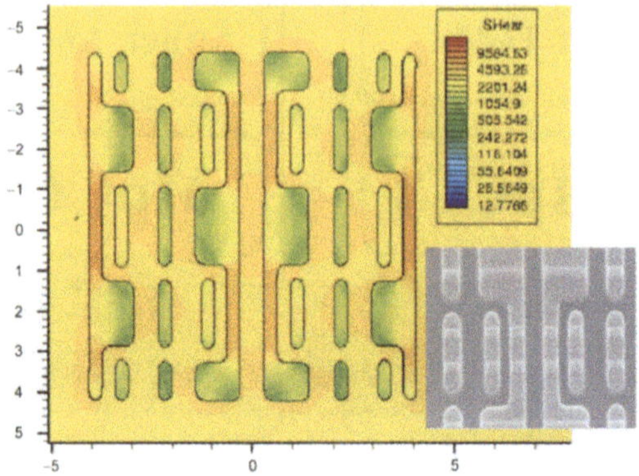

Fig. 1. (p. 74) Simulated final shear stress distribution on SRAM layout (planar stress approximation). Inset: SEM planar view of active area layout after delayering

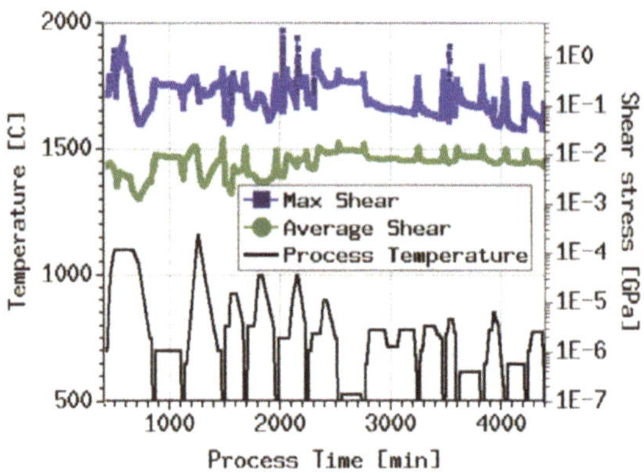

Fig. 2. (p. 75) Simulated maximum and average shear stress (right Y axis) during full process flow (left Y axis: thermal budget profile)

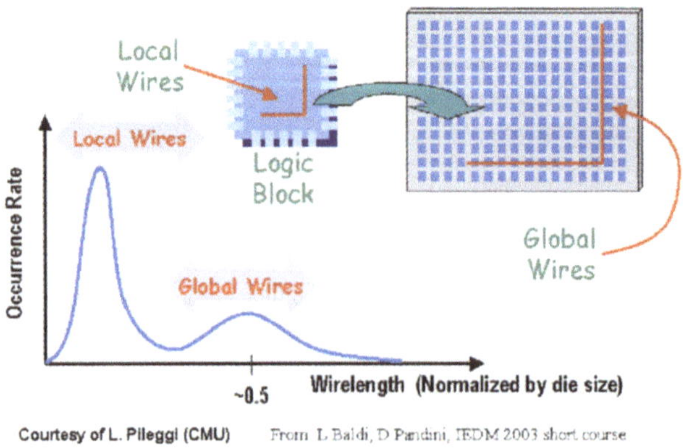

Fig. 3. (p. 75) Typical distribution of wire lengths (normalized to chip size) for a block-based design

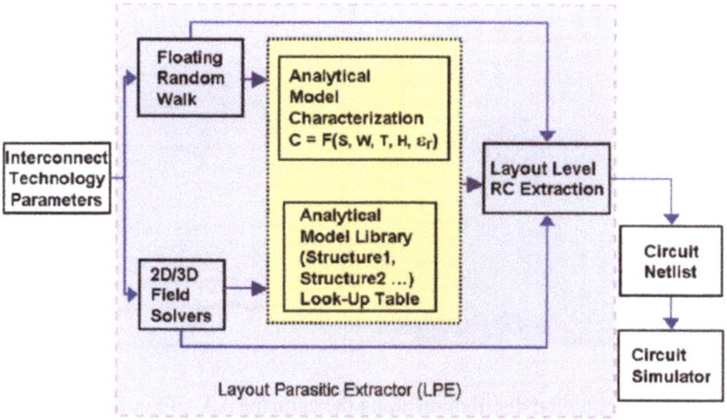

Fig. 4. (p. 76) Schematic of RC parasitics extraction methodology

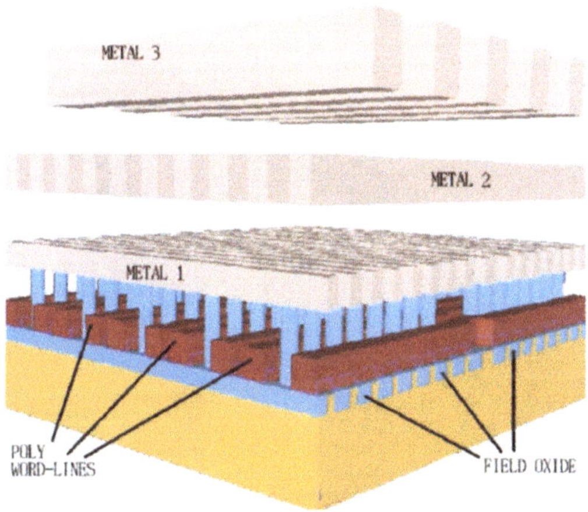

Fig. 5. (p. 77) 3D view of interconnect lines on a portion of Flash array (10 x 16 cells). The parasitic capacitances have been extracted both with a conventional field solver and with an efficient Floating Random Walk code [Bra03]

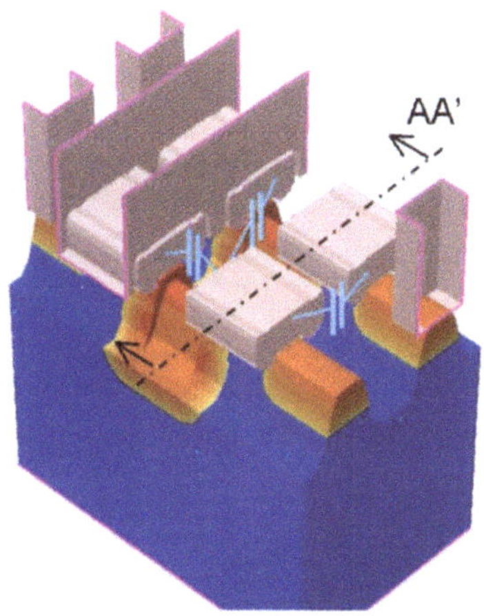

Fig. 7. (p. 77) Pictorial view of four adjacent Flash cells illustrating parasitic capacitive coupling between floating gates. Dielectrics, Word Line along cutplane AA' and one drain contact are not shown to allow better visibility of the floating gates

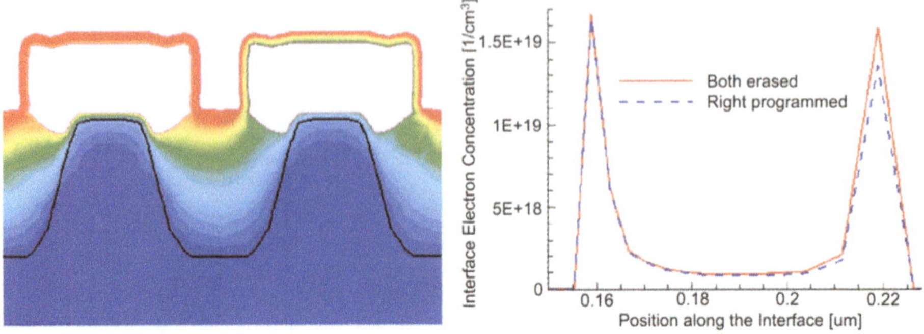

Fig. 8. (p. 78) Left: electrostatic potential distribution along the cutplane AA' of Fig. 7, when the disturbed cell (left) is erased and the disturbing cell (right) is programmed. An asymmetry in the channel potential of the disturbed cell can be qualitatively seen. Right: corresponding electron concentration along the channel of the disturbed cell as a function of the charge stored in the floating gate of the disturbing cell

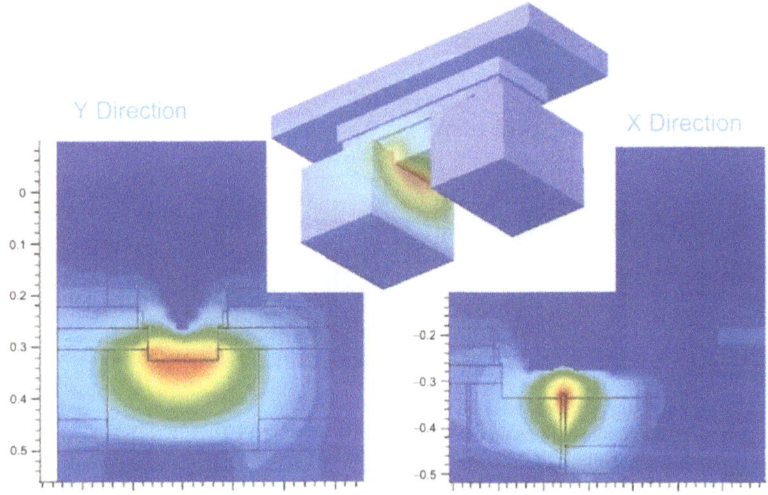

Fig. 9. (p. 79) 3D temperature distribution (top) and 2D cross sections on two adjacent Phase Change Memory (PCM) cells during "reset" operation on the left cell

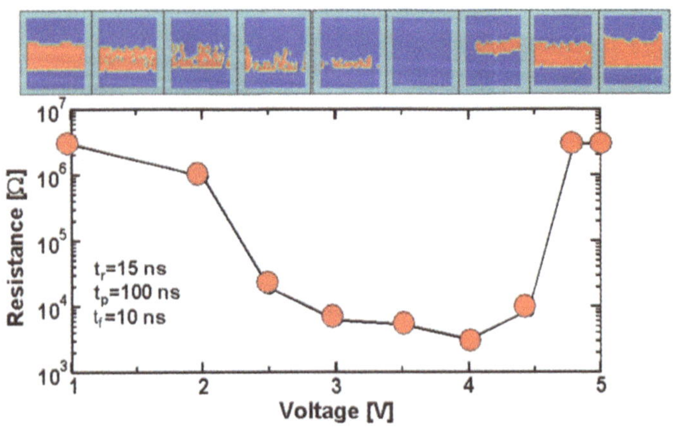

Fig. 10. (p. 79) Simulated "programming curve" for a PCM cell, showing the programmed resistance as a function of the programming pulse voltage. On the top the corresponding self-consistently simulated amorphous region is shown in red for each of the nine bias points

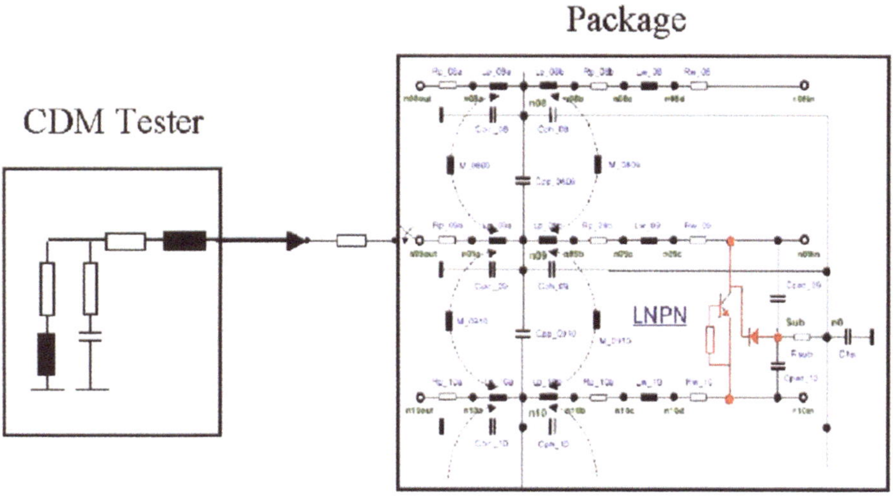

Fig. 11. (p. 80) Simplified schematic of equivalent circuit used in Charged Device Model ESD event simulation. The lateral npn BJT (in red) is described numerically with a Drift-Diffusion model

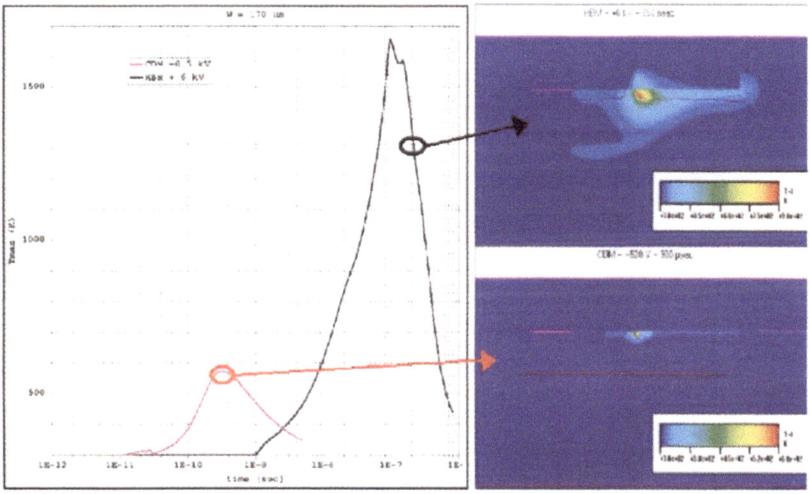

Fig. 12. (p. 80) Maximum device temperature during ESD events for Human-Body Model (black, top inset) and Charged Device Model (red, bottom inset) discharge event simulation

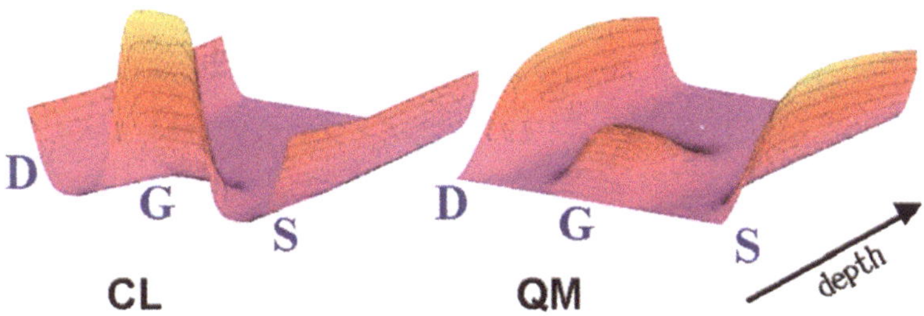

Fig. 13. (p. 80) Comparison between electron distribution in a MOS vertical cross section under low V_{ds} bias condition. Left: classic result; right: full-2D Quantum-Mechanical solution. The position of gate (G), source (S) and drain (D) electrodes is schematically marked on both figures; the depth direction has been stretched to highlight the different distance from the gate oxide of the peak channel concentration

Symbolic Methods in Industrial Analog Circuit Design

T. Halfmann, T. Wichmann

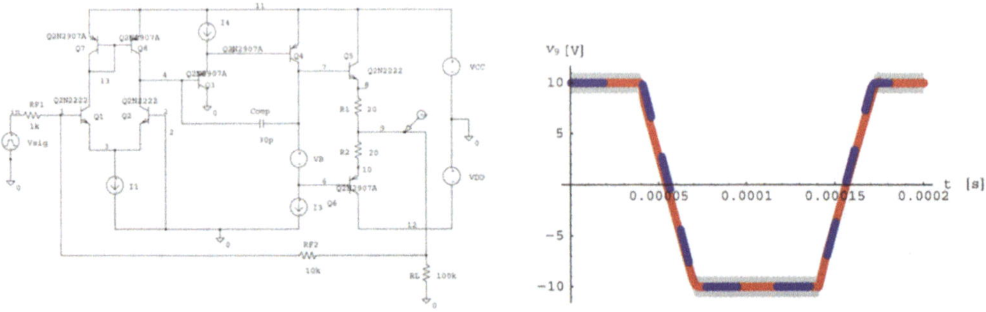

Fig. 3. (p. 93) Bipolar operational amplifier

Stochastic Differential Algebraic Equations in Transient Noise Analysis

R. Winkler

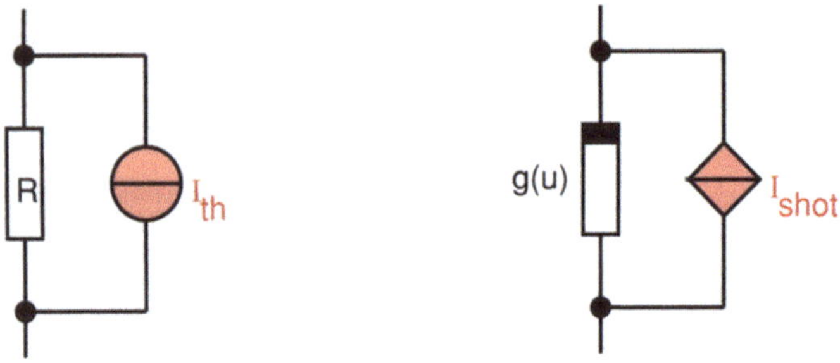

Fig. 1. (p. 153) Thermal noise of a resistor and shot noise of a pn-junction

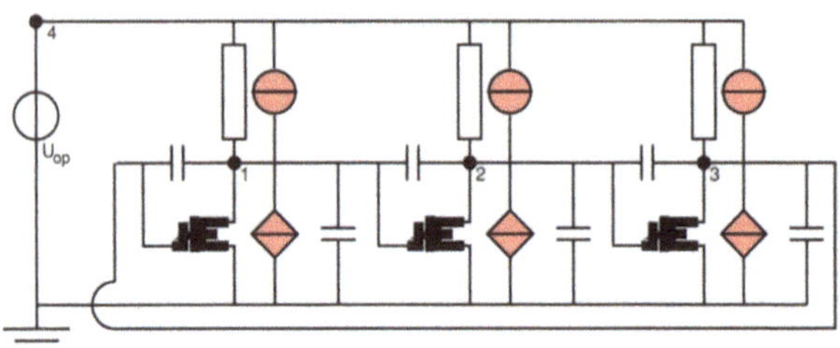

Fig. 2. (p. 155) Thermal noise sources in a MOSFET ring-oscillator model

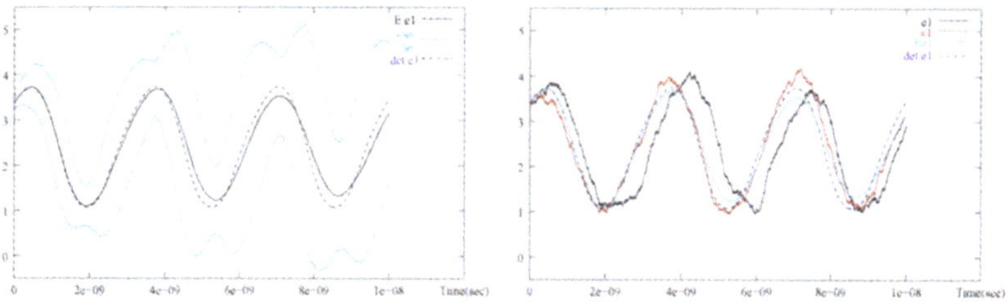

Fig. 3. (p. 155) Statistical parameters and solution paths for the nodal potential at node 1

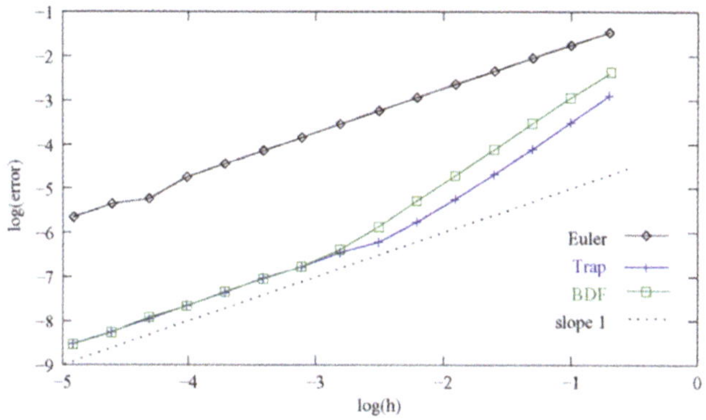

Fig. 4. (p. 157) Global error vs. step-size in logarithmic scale

COLLGUN: a 3D FE Simulator for the Design of TWTs Electron Guns and Multistage Collectors

S. Coco, S. Corsaro, A. Laudani, G. Pollicino, R. Dionisio, R. Martorana

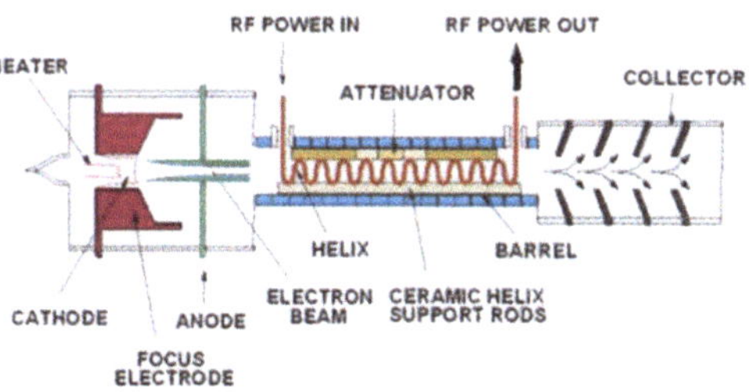

Fig. 1. (p. 177) Helix TWT schematic

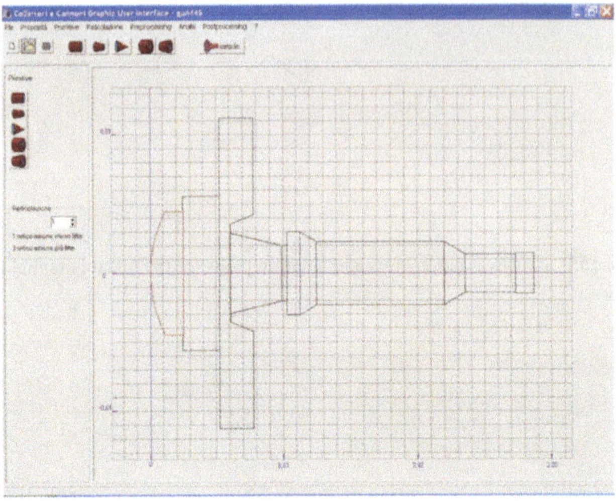

Fig. 2. (p. 178) GUI main window

Fig. 4. (p. 181) 3-D plot of emitted electron trajectories

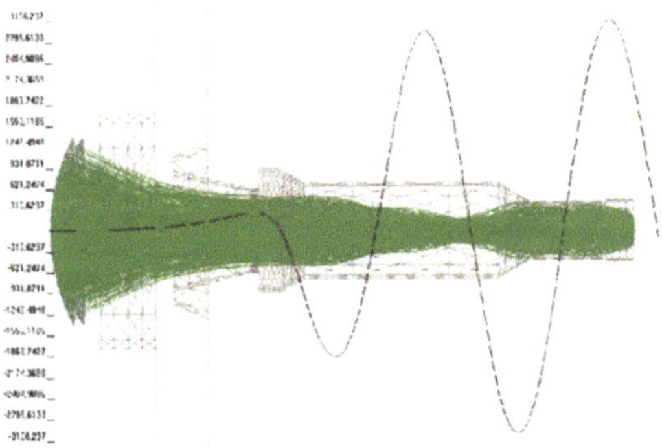

Fig. 5. (p. 181) 2-D projection of trajectories and on-axis profile of the focusing magnetic field

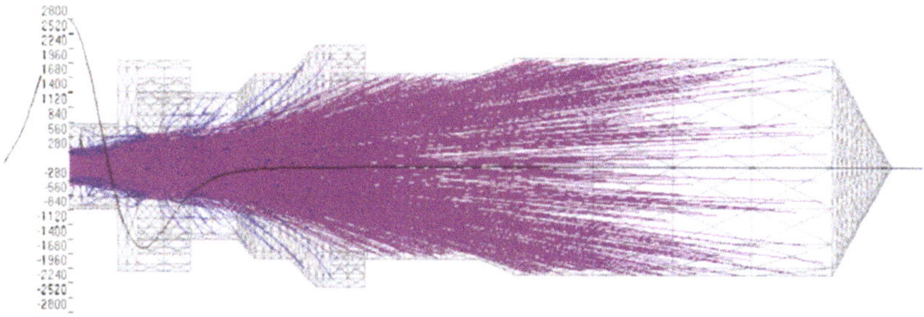

Fig. 6. (p. 181) 2-D projection of trajectories and on-axis profile of the focusing magnetic field

Fig. 7. (p. 182) 3-D plot of electron trajectories

A New Thin-Solenoid Model for Accurate 3-D Representation of Focusing Axisymmetric Magnetic Fields in TWTs

S. Coco, A. Laudani, G. Pollicino

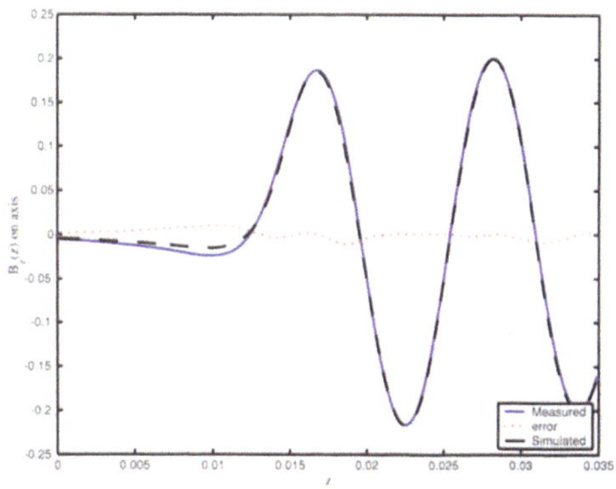

Fig. 3. (p. 187) On-axis profile of the magnetic field for the second example

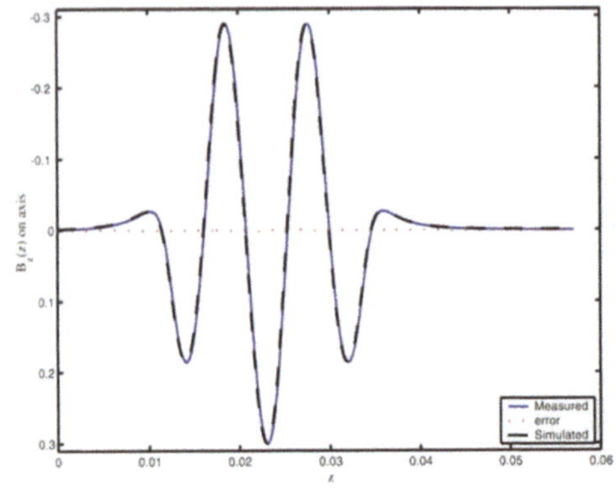

Fig. 4. (p. 187) On-axis profile of the magnetic field for the 3rd example

Numerical Computation of Magnetic Field and Inductivity of Power Reactor with Respect of Real Magnetic Properties of Iron Core

M. Marek

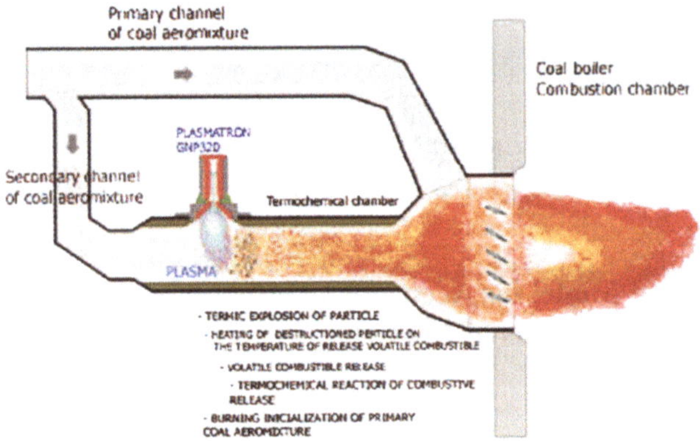

Fig. 1. (p. 236) Initialization principle of coal mixture burning by the help of plasma

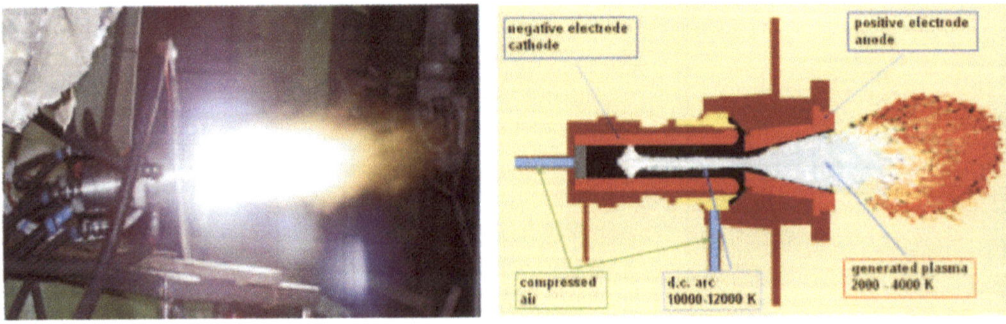

Fig. 2. (p. 236) Generator of low-temperature plasma plasmatron GNP320

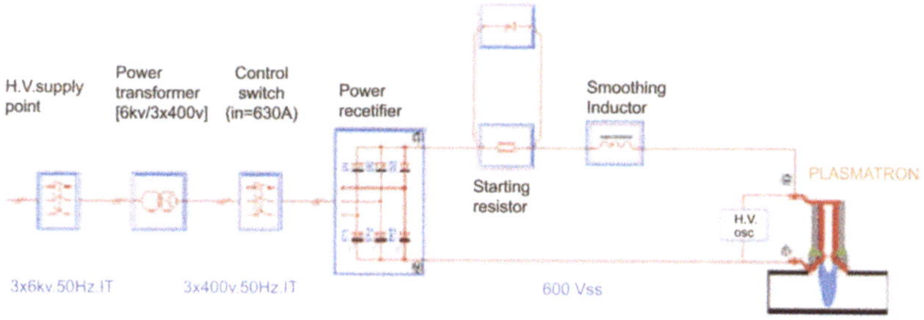

Fig. 3. (p. 236) Block diagram of plasmatron GNP320 power supply system

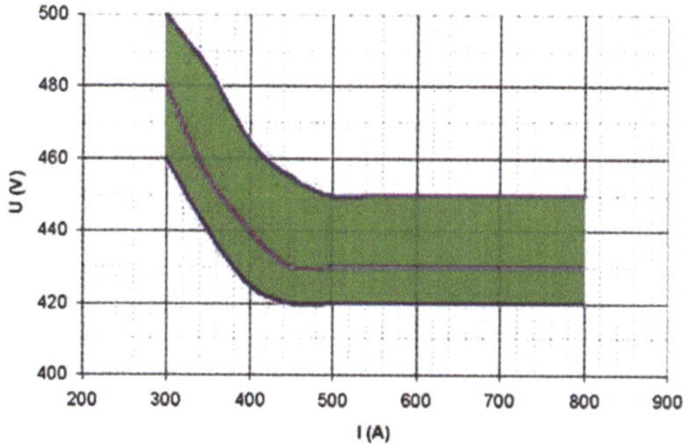

Fig. 4. (p. 237) VA characteristics of plasmatron GNP320 (various pressure relations)

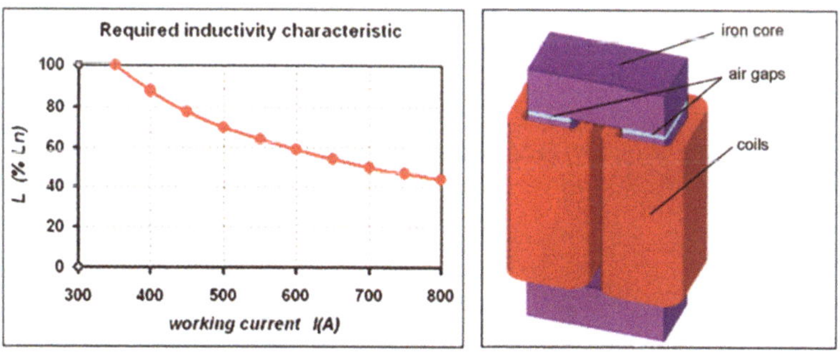

Fig. 5. (p. 237) (left) Required percent inductivity size in dependence on working current, (right) Designed inductor construction

Fig. 6. (p. 238) REMACOMP Gauging system for magnetic characteristics measuring of constructional materials in dynamic fields 1Hz 10 kHz. Epstein frame (in the middle). SST yoke (right)

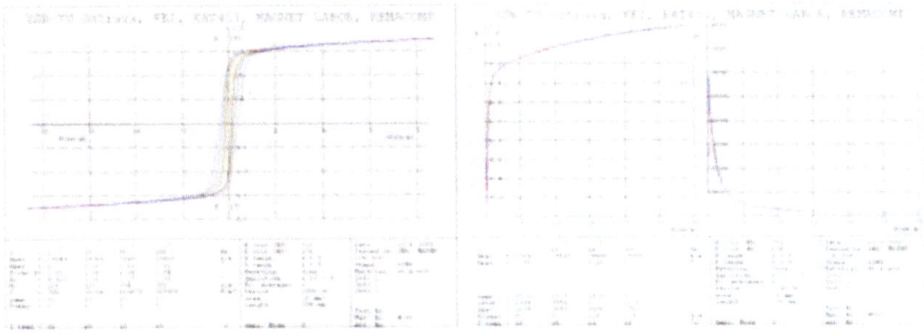

Fig. 7. (p. 238) Chosen BH and magnetization characteristics measured onto electrotechnical sheets metal samples of core material for the frequency of 1,10,50,100 Hz

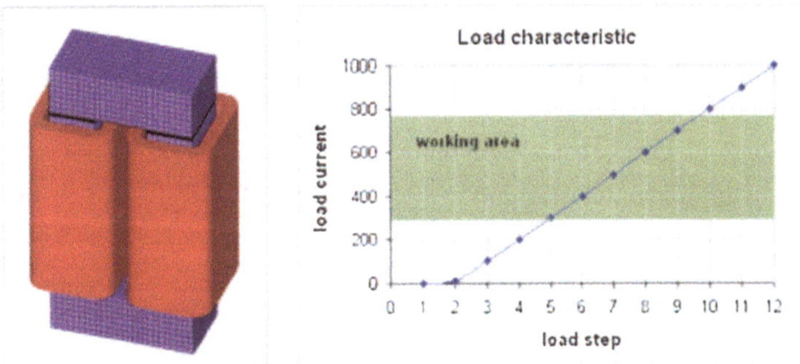

Fig. 8. (p. 239) FEM inductor model and Load characteristic

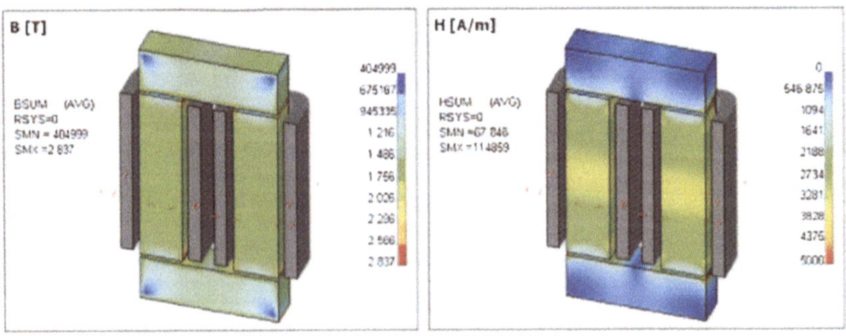

Fig. 9. (p. 240) Computed magnetic field lay-out in the core (I = 800A)

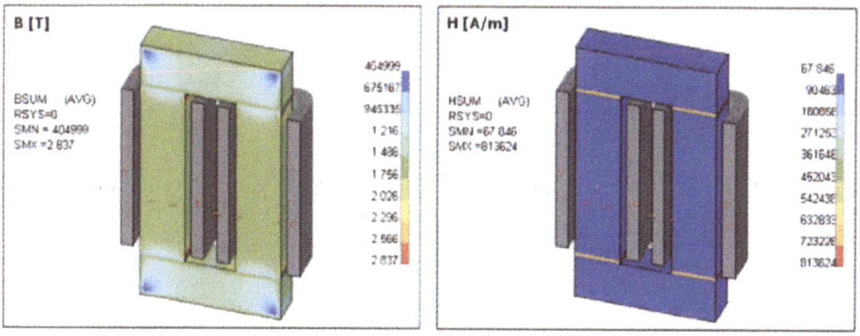

Fig. 10. (p. 240) Computed magnetic field lay-out in the core and in the air gap (I = 800A)

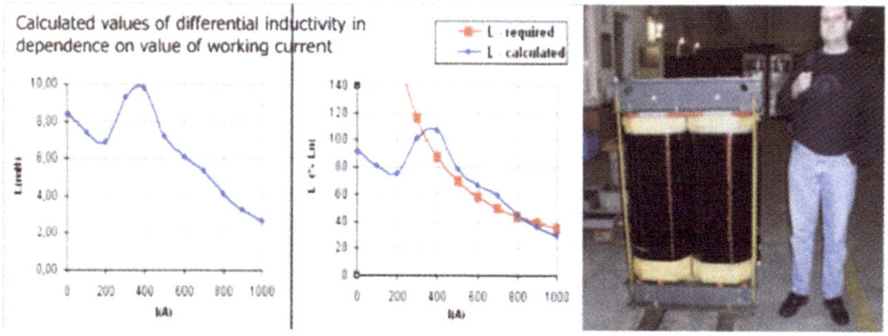

Fig. 11. (p. 241) (left) Computed inductance, (middle) Comparing of inductance computed values with required percent values (right) The real inductor design

Two-Band Quantum Models for Semiconductors Arising from the Bloch Envelope Theory

G. Alì, G. Frosali, O.Morandi

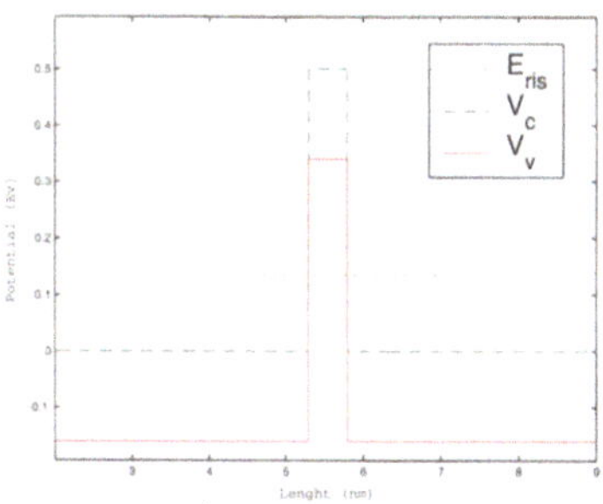

Fig. 1. (p. 276) Band diagram of the simulated heterostructure. The dotted line denotes the energy of the resonant state in the valence quantum well

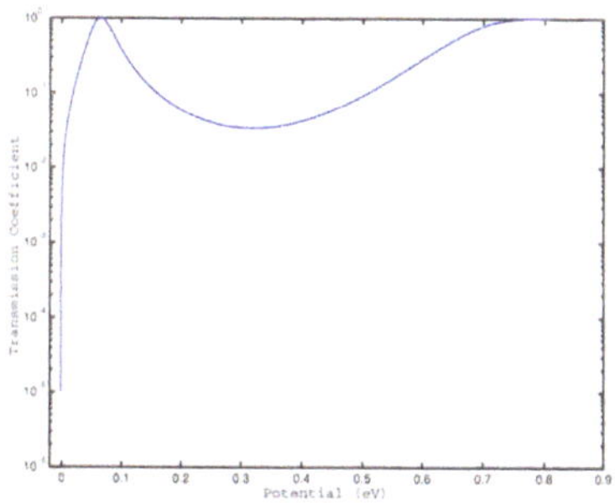

Fig. 2. (p. 276) Plot of the transmission coefficient of the heterostructure as a function of the E_{inc}

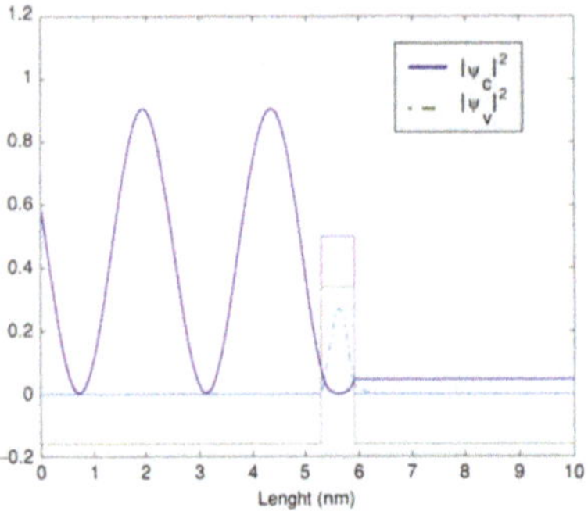

Fig. 3. (p. 277) MEF model: $E_{inc} = 0.028\ eV$

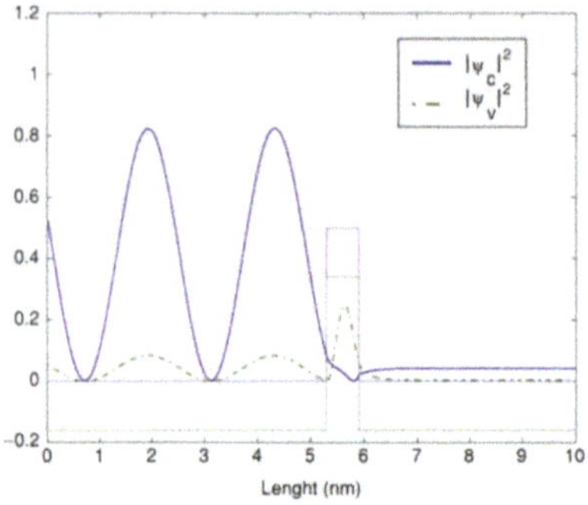

Fig. 4. (p. 277) Kane model: $E_{inc} = 0.028\ eV$

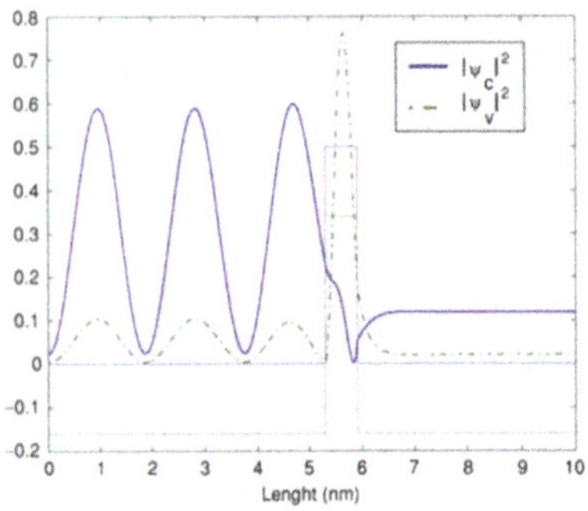

Fig. 5. (p. 277) MEF model: $E_{inc} = 0.046\ eV$

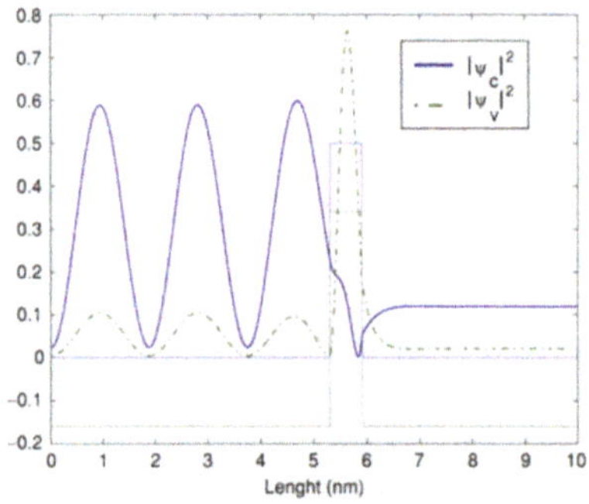

Fig. 6. (p. 277) Kane model: $E_{inc} = 0.046 eV$

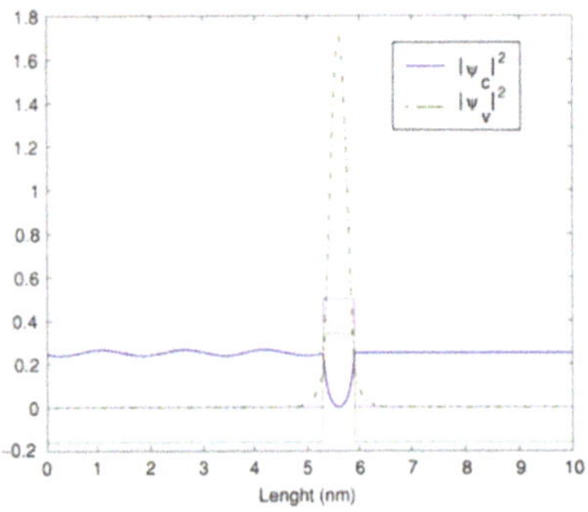

Fig. 7. (p. 277) MEF model: $E_{inc} = 0.066\ eV$

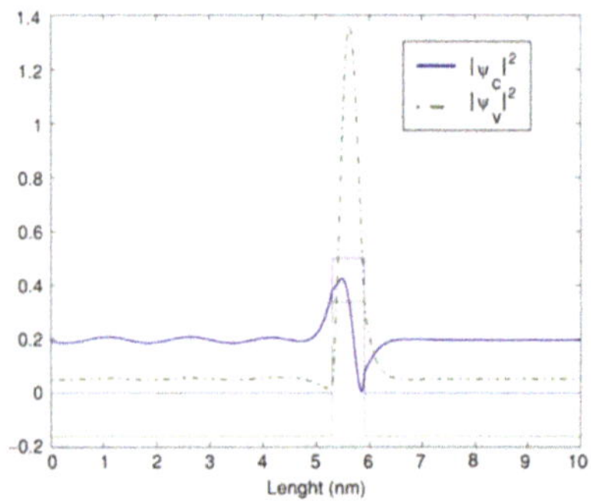

Fig. 8. (p. 277) Kane model: $E_{inc} = 0.066\ eV$

Mixed Finite Element Numerical Simulation of a 2D Silicon MOSFET with the Non-Parabolic MEP Energy-Transport Model

A. M. Anile, A. Marrocco, V. Romano, J. M. Sellier

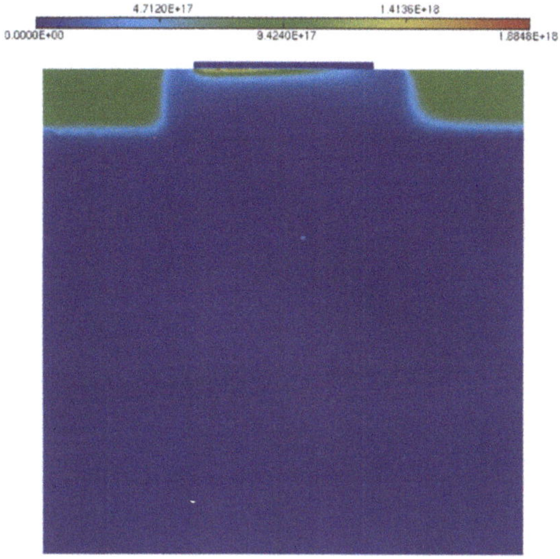

Fig. 2. (p. 282) Stationary solution for the electron density in cm^{-3}

Sound Synthesis and Chaotic Behaviour in Chua's Oscillator.

E. Bilotta, R. Campolo, P. Pantano, F. Stranges

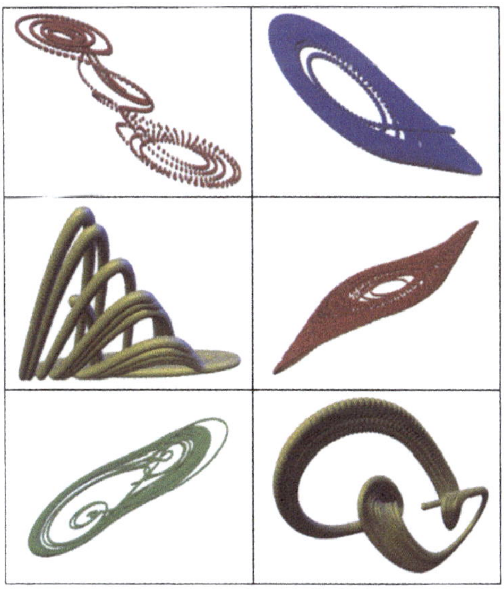

Fig. 1. (p. 292) These images show some Chua's attractors

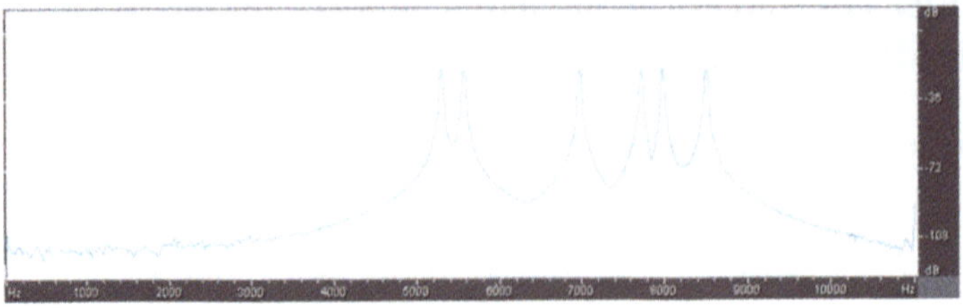

Fig. 2. (p. 292) Sound synthesis of one Chua's attractor

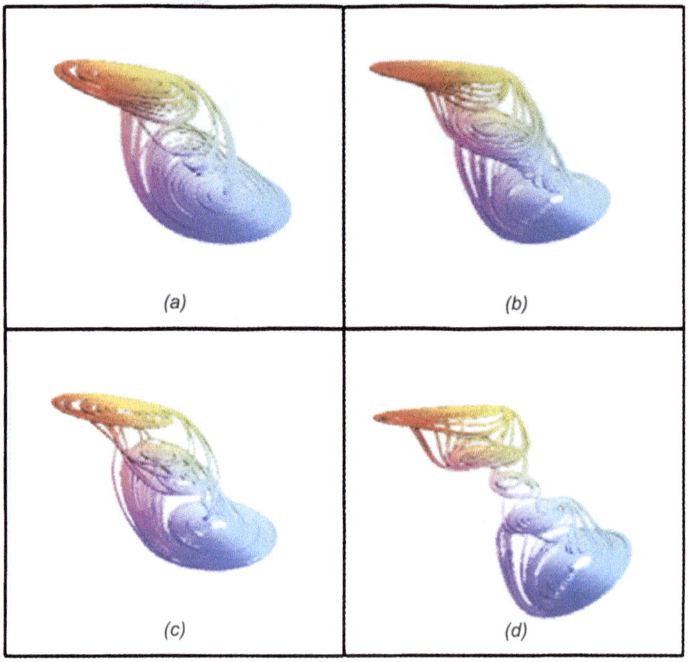

Fig. 4. (p. 293) n-scroll attractors: (a) 2-scroll; (b) 3-scroll; (c) 4-scroll; (d) 5-scroll

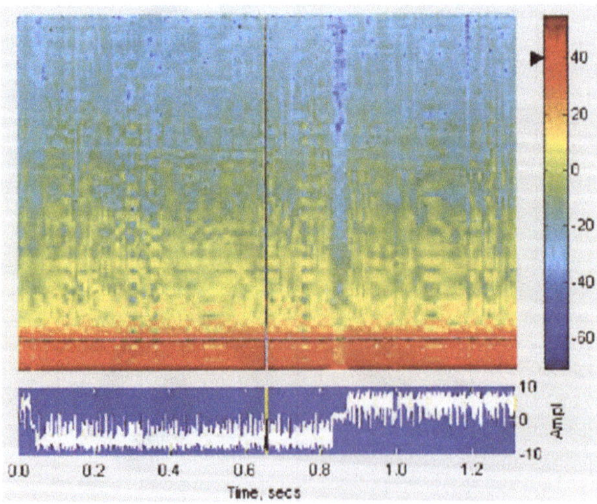

Fig. 8. (p. 295) Spectrogram for $x(t)$, $\rho = 1$

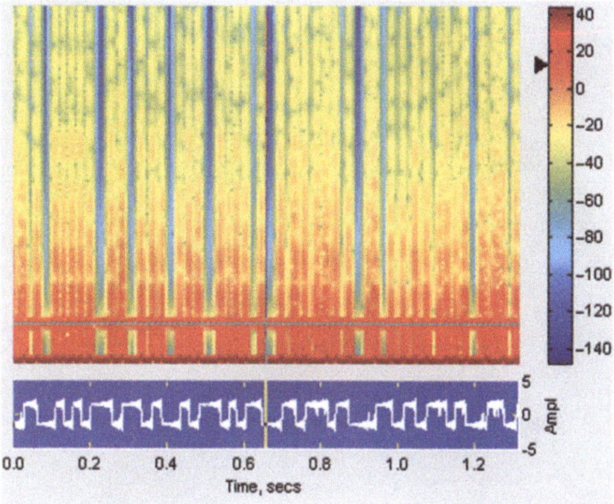

Fig. 9. (p. 296) Spectrogram for $x(t)$, $\rho = 2.71$

Quantum Corrected Drift–Diffusion Modeling and Simulation of Tunneling Effects in Nanoscale Semiconductor Devices

G. Cassano, C. de Falco, C. Giulianetti, R. Sacco

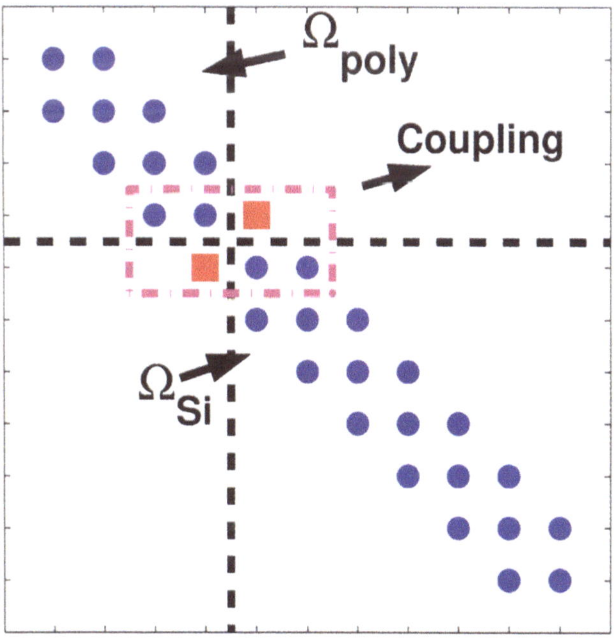

Fig. 3. (p. 307) Structure of the Matrix deriving from the FEM discretization of continuity equation

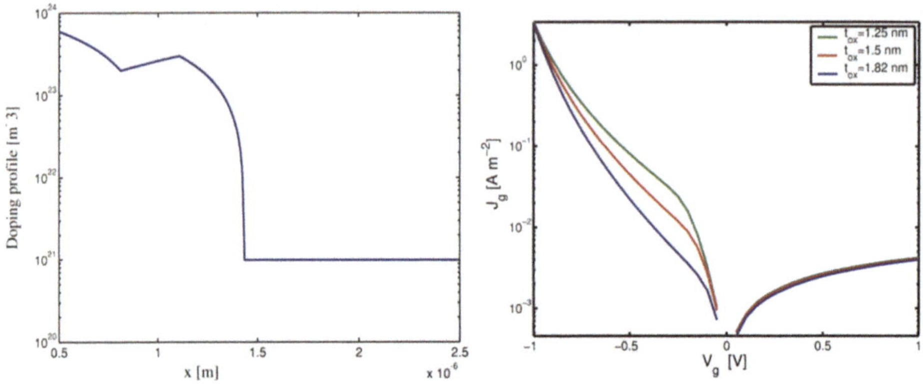

Fig. 4. (p. 308) Left: gate doping. Right: I-V characteristics

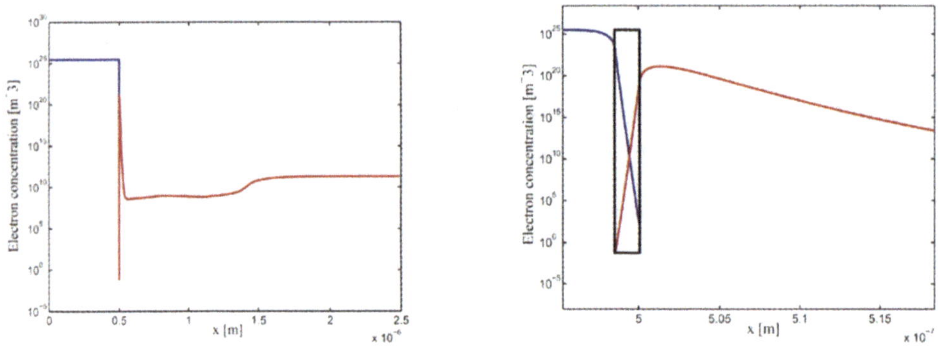

Fig. 5. (p. 308) Electron concentration at thermal equilibrium

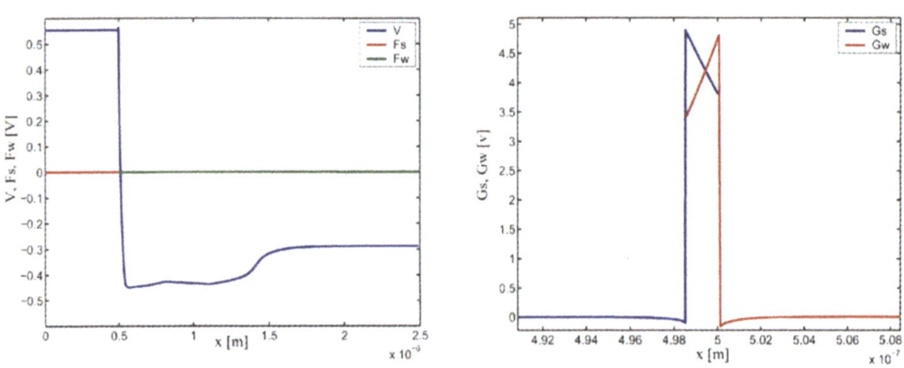

Fig. 6. (p. 308) Electric potential, Bohm potentials and quasi–Fermi potentials

Reverse Statistical Modeling for Analog Integrated Circuits

A. Ciccazzo, V. Cinnera Martino, A. Marotta, S. Rinaudo

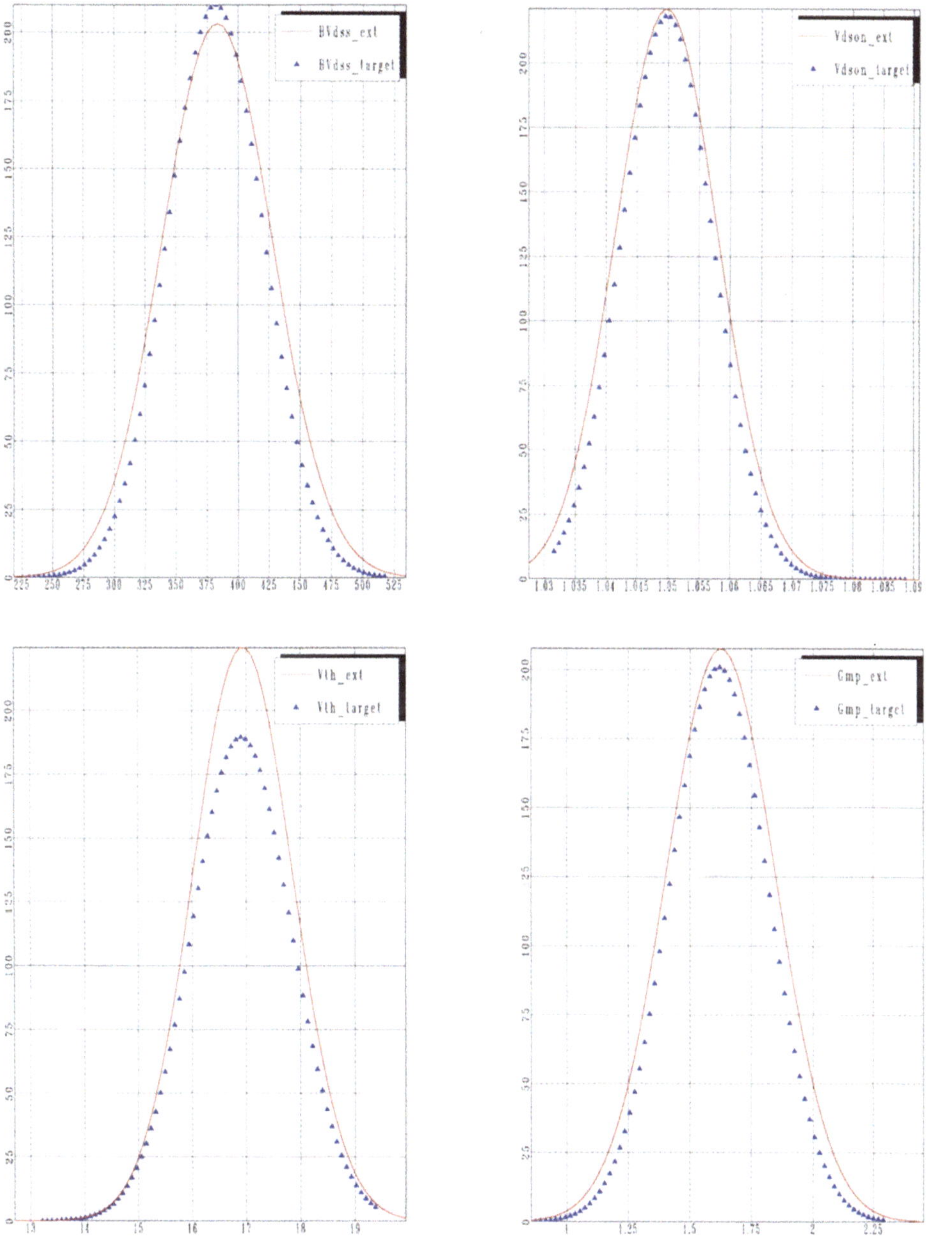

Fig. 2. (p. 316) Distribution of electrical performances of IGBT device: Bvdss, Vdson, Vth, Gmp. The extracted distributions (red line), result of our flow, are compared to their target distributions (blue triangles)

Coupled EM & Circuit Simulation Flow for Integrated Spiral Inductor

A. Ciccazzo, G. Greco, S. Rinaudo

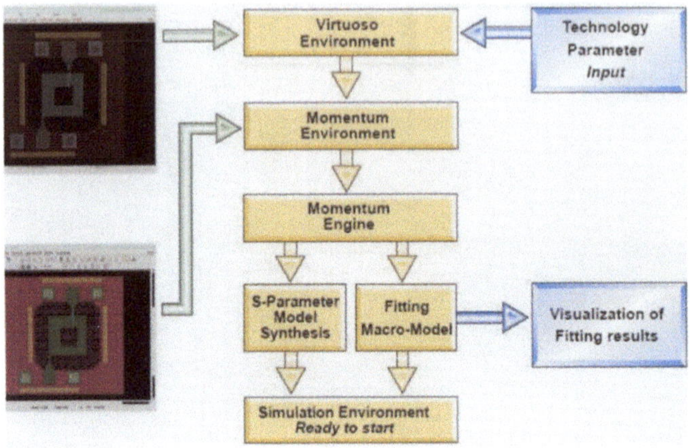

Fig. 1. (p. 320) Simulation Flow

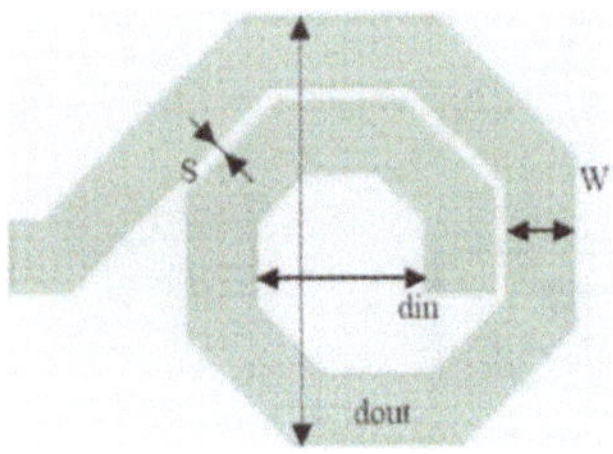

Fig. 2. (p. 320) Spiral inductor and main parameters used in the estimation formulas

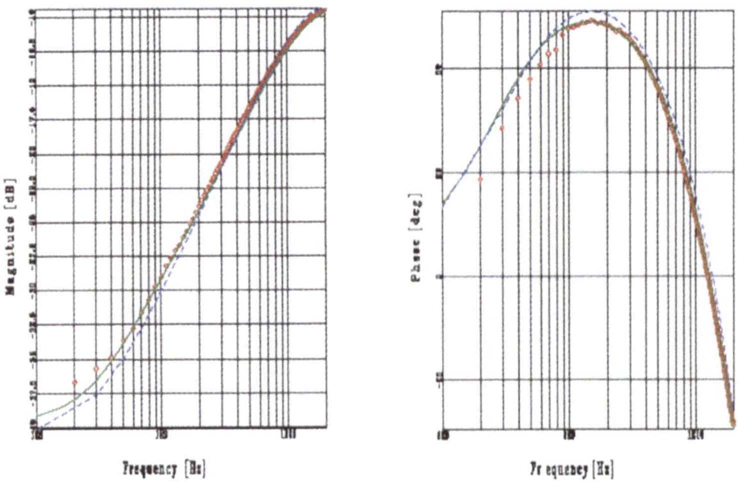

Fig. 4. (p. 323) S_{11} Parameters: ∘ Measure, - EM Simulation, $\cdots$ Macromodel

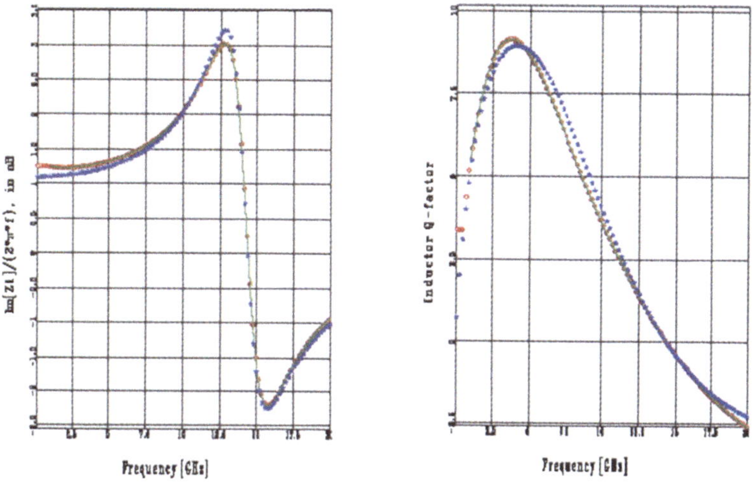

Fig. 5. (p. 324) Q-Factor and Inductance value: ∘ Measure, - EM Simulation, ⋆ Macromodel

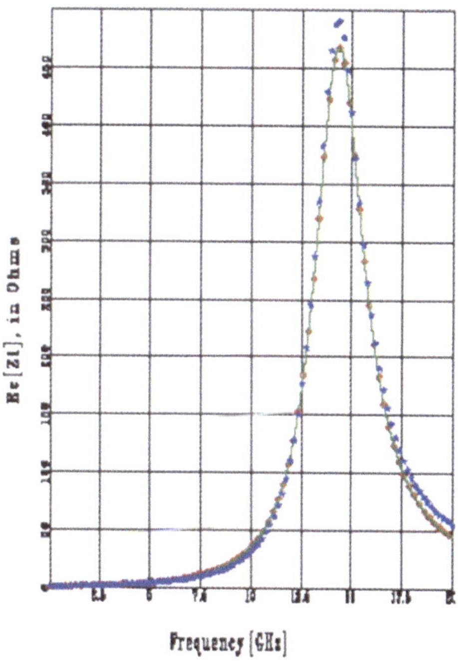

Fig. 6. (p. 324) Resistance value: ∘ Measure, - EM Simulation, ⋆ Macromodel

A Multigroup-WENO Solver for the Non-Stationary Boltzmann-Poisson System for Semiconductor Devices

M. Galler, A. Majorana, F. Schürrer

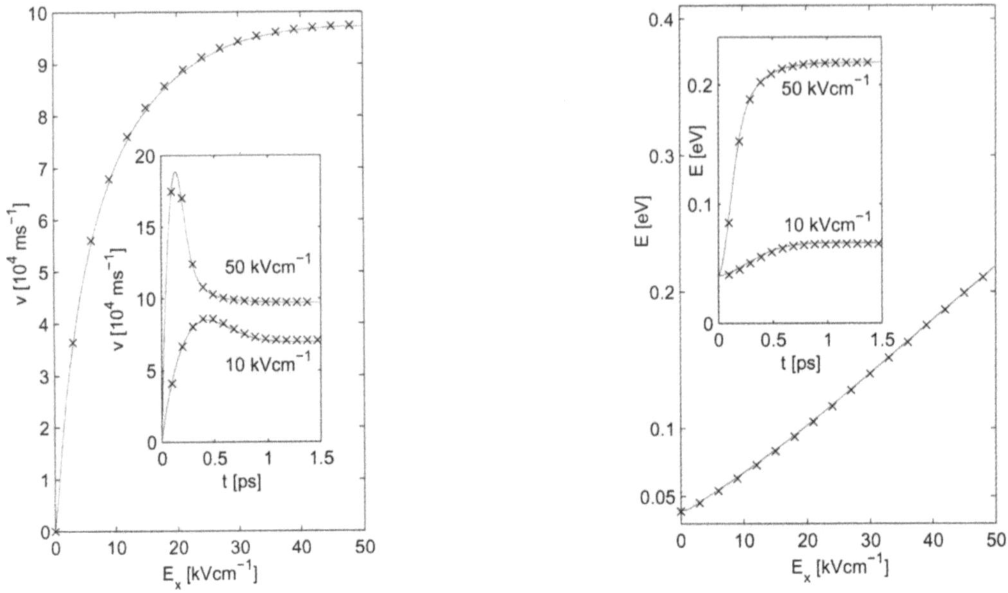

Fig. 1. (p. 337) Stationary-state drift velocity v and stationary-state mean energy E versus the electric field E_x in silicon at T_L=300 K. The inserts illustrate v and E as functions of time t in response to the onset of an electric field pulse. (—): multigroup-WENO model; (×): WENO solver [2]

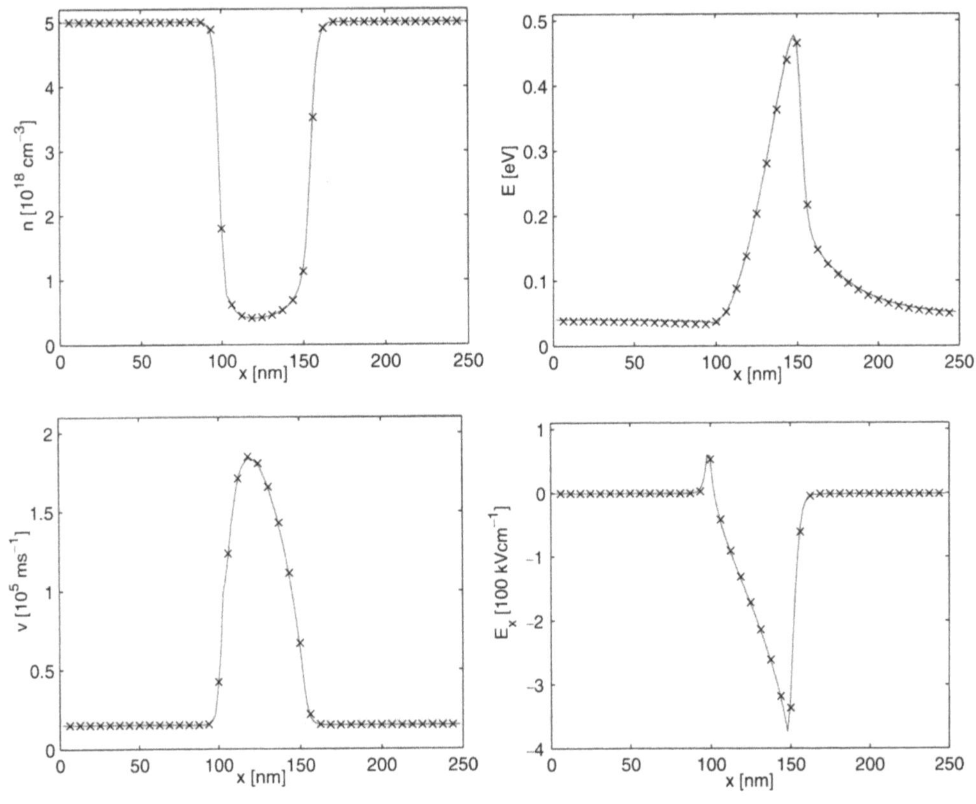

Fig. 2. (p. 338) Steady state electron density n, drift velocity v, mean energy E and electric field strength E_x as a function of position x in the n^+-n-n^+ diode.(—): multigroup-WENO model; (×): WENO solver [2]

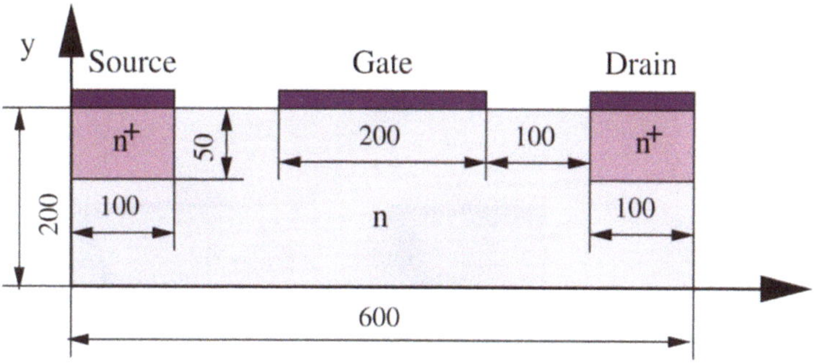

Fig. 3. (p. 339) Schematic illustration of a 2D-MESFET. Lengths are given in nm

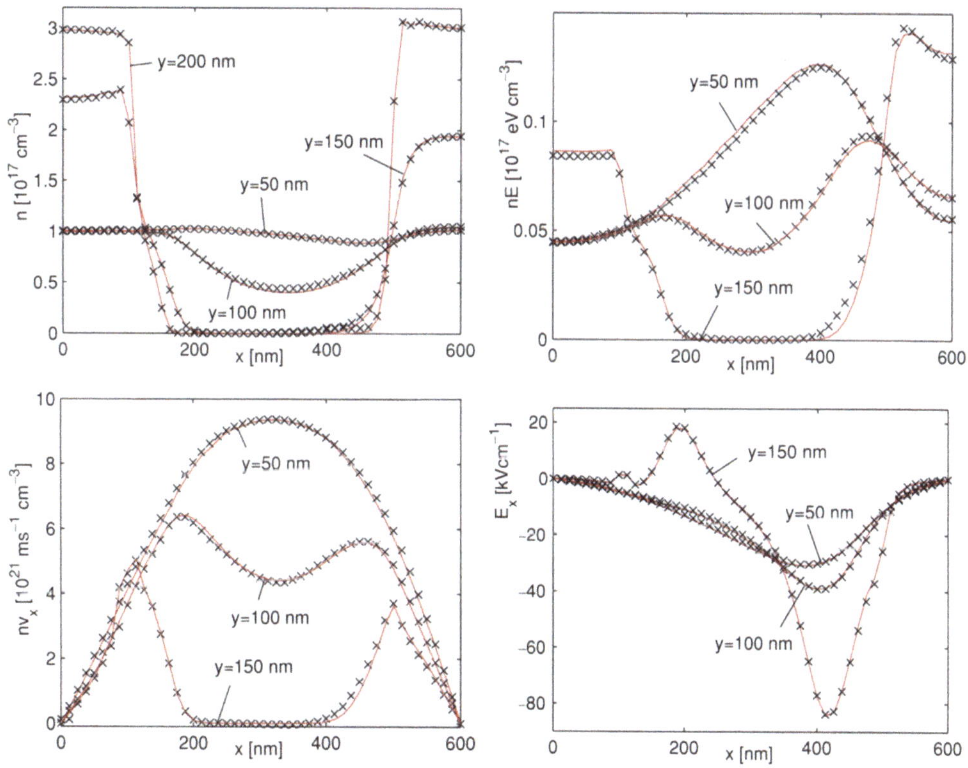

Fig. 4. (p. 340) The stationary-state electron density n, the energy density nE and the x-components of the momentum nv_x and of the electric field E_x versus position x in the Si-MESFET. (—): multigroup-WENO model; ($\times$): WENO solver [3]

Anisotropic Mesh Adaptivity Via a Dual-Based A Posteriori Error Estimation for Semiconductors

S. Micheletti, S. Perotto

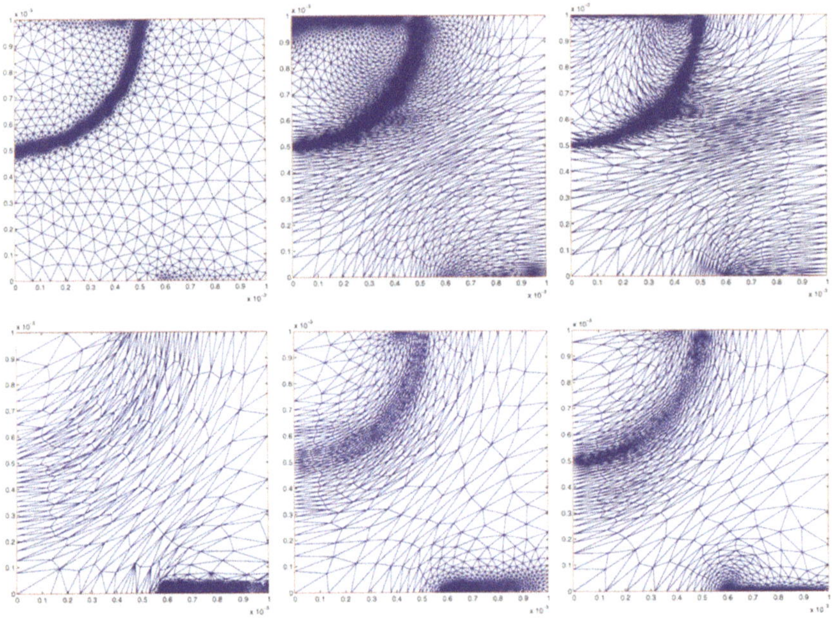

Fig. 3. (p. 380) Control of total current: sequence of adapted meshes

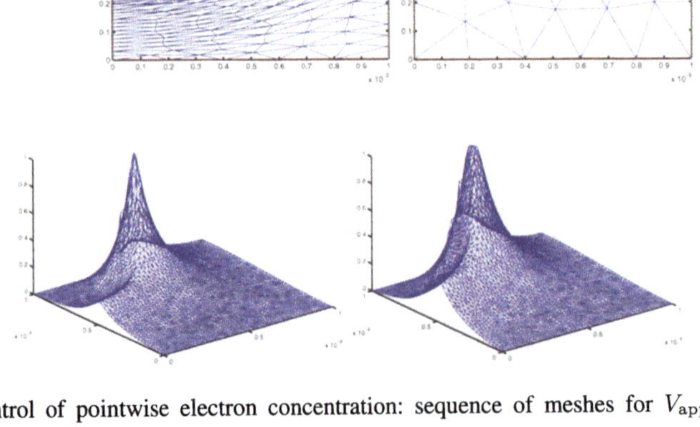

Fig. 4. (p. 381) Control of pointwise electron concentration: sequence of meshes for $V_{\mathrm{app}} = 0.7V$ (left) and $V_{\mathrm{app}} = -5V$ (right) at the first iteration

Electromagnetic Characterization Flow of Leadless Packages for RF Applications

G. Alessi

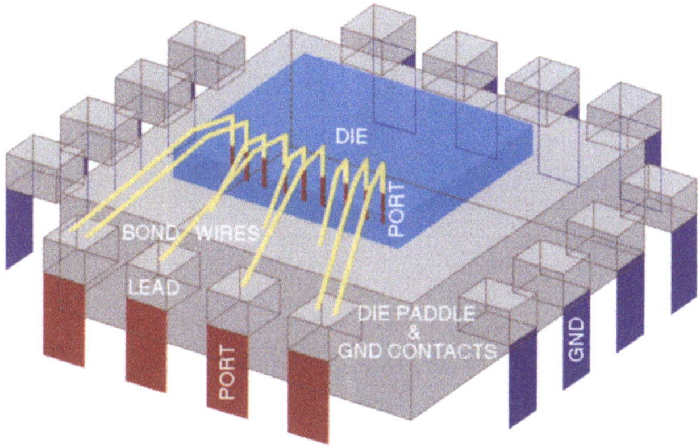

Fig. 1. (p. 406) Package model

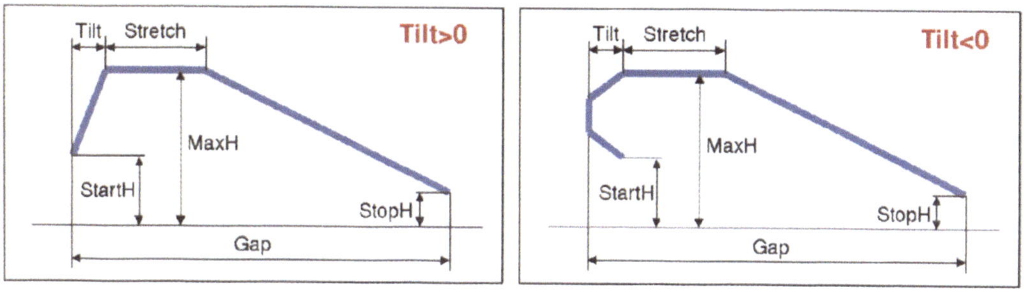

Fig. 2. (p. 407) Philips/T.U. Delft model

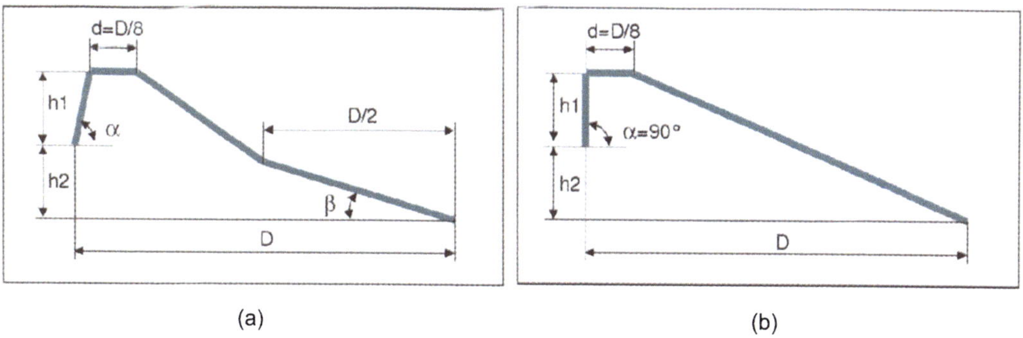

Fig. 3. (p. 407) EIA/JEDEC bond wire model: (a) complete model, (b) simplified model

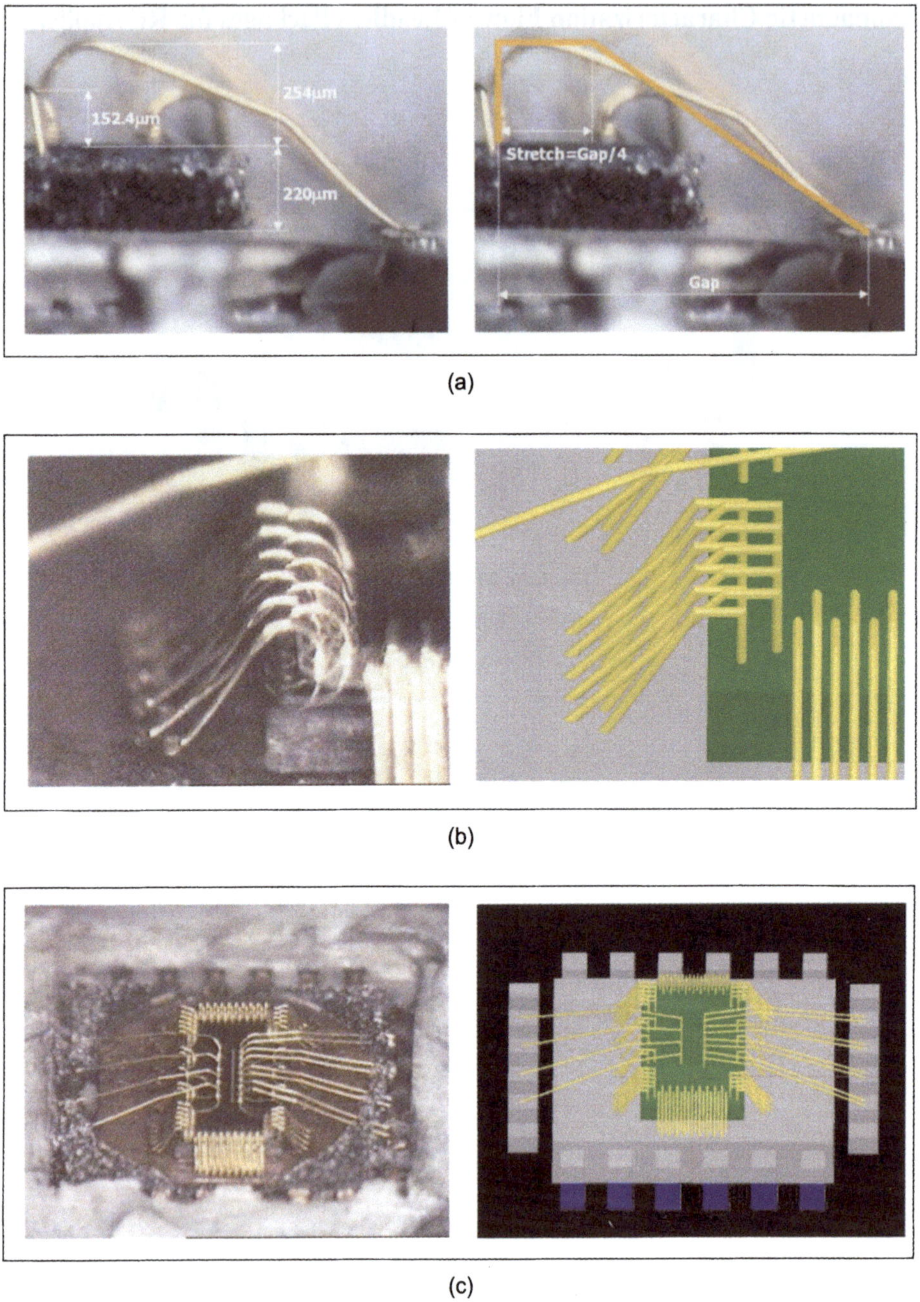

Fig. 4. (p. 408) Bond wires and their geometric models: (a) bond wire on a lead, (b) bond wires on the die paddle (downbonding), (c) wire bonded die

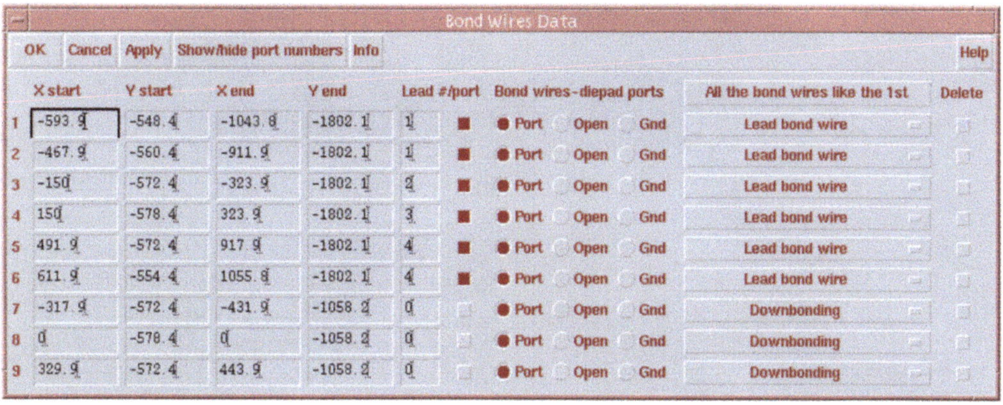

(a)

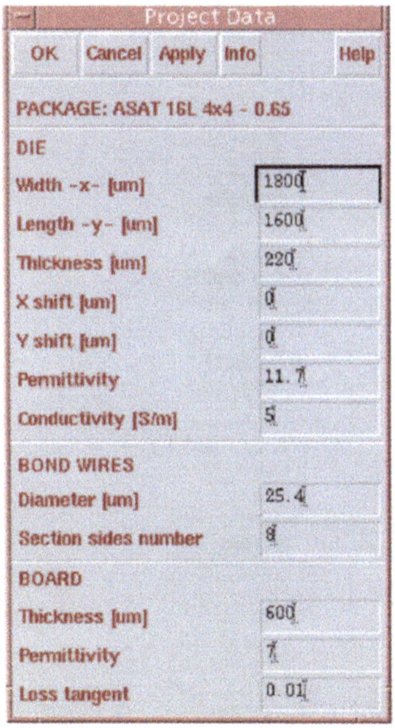

(b)

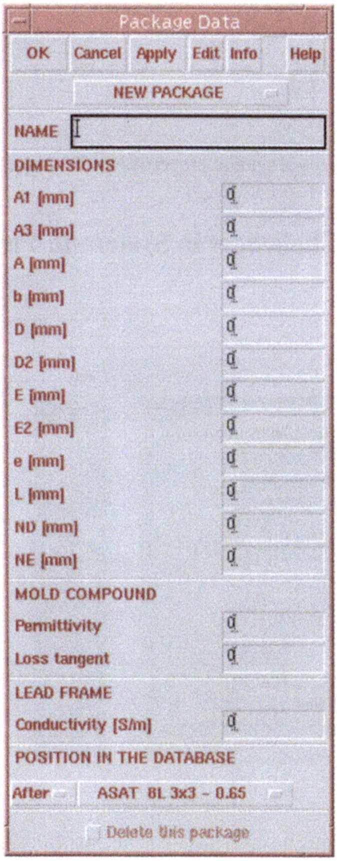

(c)

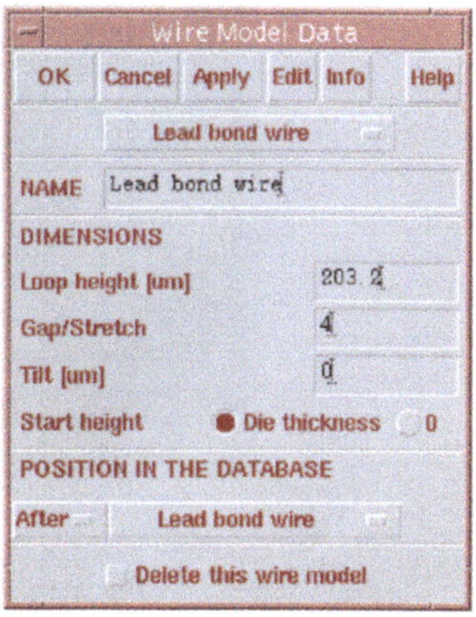

(d)

Fig. 5. (p. 409) Data forms: (a) bond wires, (b) project, (c) package database, (d) bond wire model database

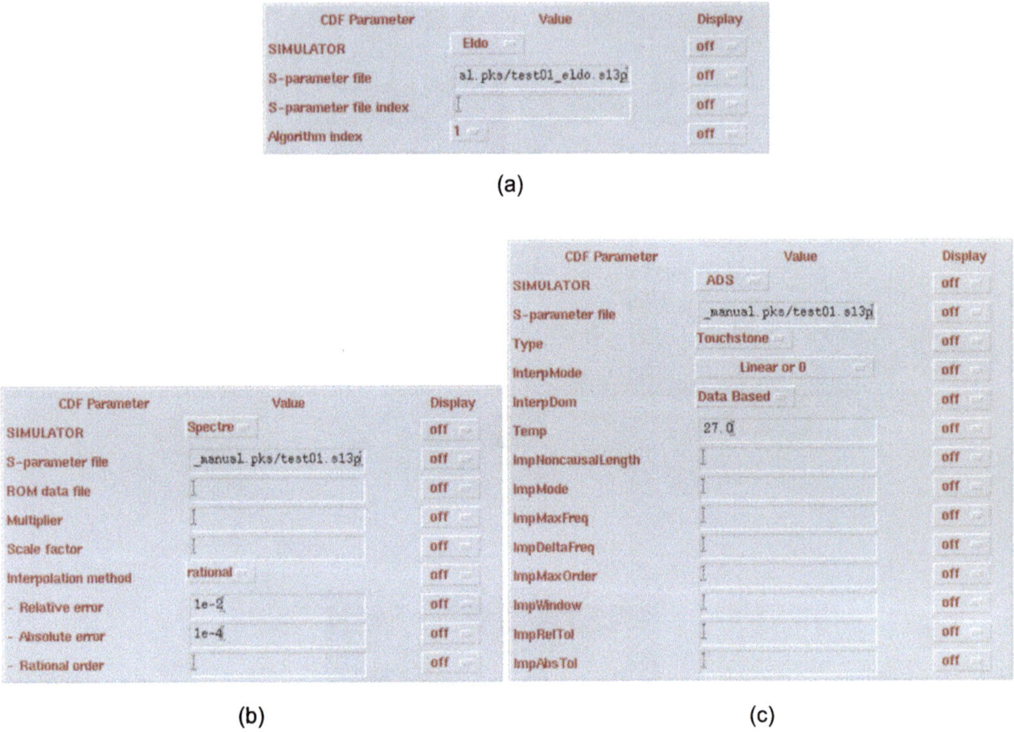

Fig. 6. (p. 410) Package symbol properties for (a) Eldo, (b) Spectre, (c) ADS

Interconnection Modeling Challenges in System-in-Package (SiP) Design

S. Castorina, R. A. Ene

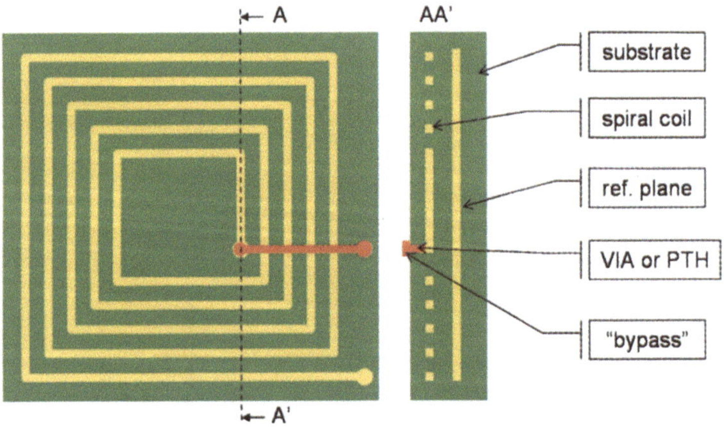

Fig. 1. (p. 420) Schematic structure of a planar embedded inductor

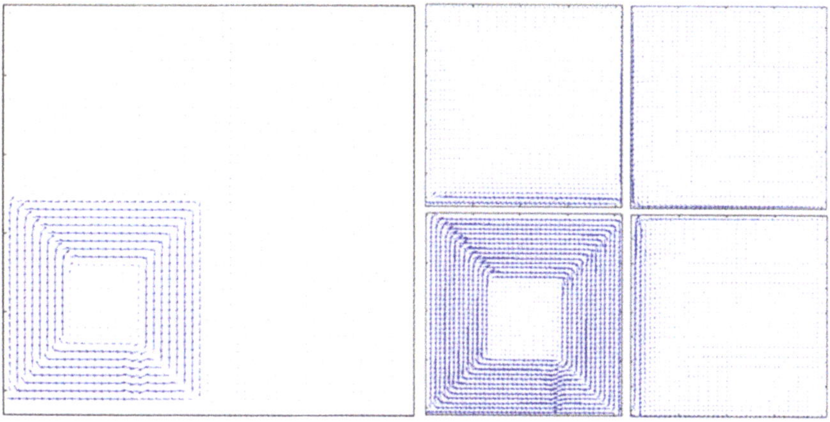

Fig. 2. (p. 421) Simulation results showing the induced current distribution on both continuous (left) and splitted (right) ground planes for an inductor of the kind shown in Fig. 1. Note the opposite currents on domain borders in the splitted ground plane

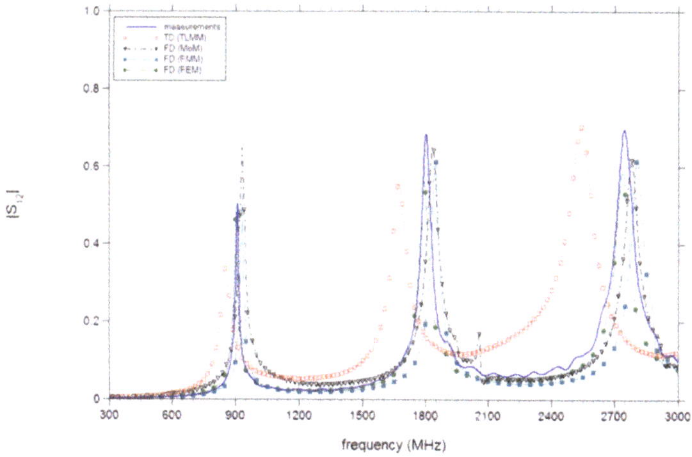

Fig. 3. (p. 423) Comparison of the measured frequency response of a resonating structure with the simulated responses calculated by means of different commercial electromagnetic solvers

Author Index

MIX
Papier aus verantwortungsvollen Quellen
Paper from responsible sources
FSC® C105338

If you have any concerns about our products,
you can contact us on
ProductSafety@springernature.com

In case Publisher is established outside the EU,
the EU authorized representative is:
Springer Nature Customer Service Center GmbH
Europaplatz 3, 69115 Heidelberg, Germany

Printed by Libri Plureos GmbH
in Hamburg, Germany